Explorations
in
Numerical Analysis

Explorations
in
Numerical Analysis

James V Lambers
Amber C Sumner

The University of Southern Mississippi, USA

World Scientific

NEW JERSEY · LONDON · SINGAPORE · BEIJING · SHANGHAI · HONG KONG · TAIPEI · CHENNAI · TOKYO

Published by

World Scientific Publishing Co. Pte. Ltd.

5 Toh Tuck Link, Singapore 596224

USA office: 27 Warren Street, Suite 401-402, Hackensack, NJ 07601

UK office: 57 Shelton Street, Covent Garden, London WC2H 9HE

Library of Congress Cataloging-in-Publication Data

Names: Lambers, James V., author. | Sumner, Amber C., author.

Title: Explorations in numerical analysis / by James V. Lambers and Amber C. Sumner.

Description: New Jersey : World Scientific, [2018] | Includes bibliographical references and index.

Identifiers: LCCN 2018011145| ISBN 9789813209961 (hardcover : alk. paper) |

    ISBN 9789813209978 (pbk. : alk. paper) | ISBN 9789813220027 (e-book)

Subjects: LCSH: Numerical analysis--Textbooks. | LCGFT: Textbooks.

Classification: LCC QA297 .L335 2018 | DDC 518.0285/536--dc23

LC record available at https://lccn.loc.gov/2018011145

British Library Cataloguing-in-Publication Data

A catalogue record for this book is available from the British Library.

For any available supplementary material, please visit
https://www.worldscientific.com/worldscibooks/10.1142/10446#t=suppl

Printed in Singapore

We dedicate this book
to Dianna Lambers
and Race Robertson

Preface

This book evolved over a period of several years from lecture notes written by the first author. The original notes for a course at Stanford University in 2001 were based on [Golub and Ortega (1993)], then significantly expanded at the University of California at Irvine in 2004, influenced by [Burden and Faires (2004)]. At the University of Southern Mississippi, additional notes for undergraduate and graduate courses taught by both authors, and guided by textbooks in numerical analysis [Ascher and Greif (2011); Burden and Faires (2004); Sauer (2012); Süli and Mayers (2003)], as well as numerical linear algebra [Demmel (1997); Golub and van Loan (2012)], and numerical methods for partial differential equations [Leveque (2007); Trefethen (2000)], completed the foundation for this book. It is written for a year-long sequence of numerical analysis courses for either advanced undergraduate or beginning graduate students. Part II is suitable for a semester-long first course on numerical linear algebra.

One of the aspects of numerical analysis that appeals to the authors, and we hope appeals to students as well, is its variety. Problems can range from entirely theoretical to utterly devoid of theory, rather focused on aspects from computer science. One goal of this book is to show students how the theoretical and practical aspects of numerical analysis not only complement, but also reinforce each other. What can be proven about algorithms drives their enhancement, and what is observed about algorithms guides the development of the theoretical foundation.

In our view, the best way to get students engaged with the various aspects of numerical analysis is to invite them to "get their hands dirty" with all of it. As such, this book is not organized in the traditional "theorem-proof" format that is often seen in mathematics textbooks. In fact, "*Proof:*" is not seen once in this book following any theorem. Instead, many proofs are left to the reader as exercises, but with guidance that serves as "training wheels" to help students slowly but surely develop their proof-writing skills.

On the programming side, prior experience with programming is not assumed, but students are guided, by tutorials, demonstrations, and sequences of exercises, through programming tasks ranging from the most basic to very elaborate. By the time they have reached the end of the book, they will have at their disposal an extensive library of numerical routines, that they have written themselves, and can enhance and customize for further study or research.

Coding examples throughout the book are written in MATLAB. MATLAB has been a vital tool throughout the numerical analysis community since its creation over thirty years ago, and its syntax that is oriented around vectors and matrices greatly accelerates the prototyping of algorithms compared to other programming environments.

How to Use This Book

Simply reading about the design, analysis, and implementation of the many algorithms covered in this book does not suffice the reader to fully understand these aspects or be able to efficiently and effectively work with these algorithms in code, especially as part of a larger application. A "hands-on" approach is needed to achieve this level of proficiency, and this book is written with this necessity in mind. To encourage taking such an approach, there are frequent "interruptions" in the text at which some coding task is required. The purposes of these coding tasks are to get you accustomed to programming, particularly in MATLAB, but also to reinforce the concepts of numerical analysis that are presented throughout the book. This is not the kind of book with which one should get comfortable in their favorite chair to read–it is one that they should have open along with their laptop to, as the title suggests, explore.

While the text is primarily intended for a one-year sequence in numerical analysis, naturally instructors will have their own priorities in terms of the depth and breadth with which the various topics are covered. With this in mind, sections or subsections that are considered optional are marked with a * in the heading and in the Table of Contents. Furthermore, there is some intentional overlap; for example, an instructor who wishes to include Chapter 8, which includes least squares problems for approximating functions, can do so without also having to include Chapter 4 that covers least squares in a manner that is more appropriate for a course on numerical linear algebra. The appendices, which provide a review of calculus and linear algebra, can help ensure students understand the basics of eigenvalues for the purpose of studying numerical methods for PDEs in Chapter 14, even if Chapter 6 on eigenvalue problems, again more fitting for a numerical linear algebra course, is omitted. The dependence of sections within the main text on sections in the appendices will also be indicated explicitly.

To foster exploring rather than just reading, the book includes three different types of exercises, designed with a progression through cognitive levels in mind. At the end of each section, except in Chapter 1, "Concept Check" questions aid students in ensuring that their reading has been sufficiently thorough by quizzing them on the main ideas of the section. "Explorations" are problems that are interspersed throughout the text, to encourage students to work with concepts and algorithms in a hands-on manner immediately, to reinforce comprehension and appreciation. Finally, exercises at the end of each chapter build on the knowledge acquired while reading the chapter, to either introduce students to applications of interest or illustrate how the various concepts from different sections fit together.

The first chapter includes a tutorial on MATLAB. Depending on how much time the instructor has, they can elect to have their students work through the entire tutorial at the beginning of the first course, or defer coverage of certain sections until the relevant functions are needed–for example, covering polynomial functions before polynomial interpolation in Chapter 7. It is recommended that students review essential material from calculus and linear algebra that is presented in Appendices A and B, respectively.

Acknowledgments

The authors would express their gratitude to Rochelle Kronzek, Benny Lim Jing Quan, and the rest of the very helpful and accommodating staff at World Scientific for their assistance with and faith in this project. We are also very thankful for the feedback provided by anonymous reviewers on a draft. Finally, we are indebted to the students in the authors' MAT 460/560 and 461/561 courses, taught in 2015-16 and 2017-2018, who were subjected to drafts of this book. In particular, we would like to thank Kwesi Acheampong, Gokul Bhusal, Khadijah Bordley, Laken Camp, Carina Chen, Kaitlin Cooksey, Theresa Coumbe, Jade Dedeaux, Kelsey Fairley, Ross Grisham, Erica Keys, Aaditya Kharel, Jonathan Kolb, Christen Leggett, Timber Lott, Shiron Manandhar, Jeremy Norris, Megan Payne, Caridae Pittman, Bailey Rester, Chloe Richards, Jesse Robinson, Hamas Tahir, Carley Walker, and Orlala Wentink for calling our attention to mistakes in the text and helping to improve the clarity and quality of the explorations. Special thanks are owed to David Patterson for taking the time to read through the text and examine the MATLAB examples.

J. V. Lambers
A. C. Sumner

Contents

Part IV   Nonlinear Equations and Optimization   389

10. Zeros of Nonlinear Functions   391

11. Optimization   435

Part V   Differential Equations   467

12. Initial Value Problems   469

PART I
Preliminaries

Chapter 1

What is Numerical Analysis?

This book provides a comprehensive introduction to the subject of **numerical analysis**, which is the study of the design, analysis, and implementation of numerical methods for solving mathematical problems that arise in science and engineering. These numerical methods differ from the *analytical* methods that are presented in other mathematics courses, in that they rely exclusively on the four basic arithmetic operations–addition, subtraction, multiplication and division–so that they can be implemented on a computer. As such, numerical analysis, a branch of mathematics, is an essential component of **scientific computing**, also known as **computational science**, which is a multidisciplinary field devoted to the solution of problems in science and engineering through computing technology.

Numerical analysis is employed to develop and analyze numerical methods for solving problems that arise in other areas of mathematics, such as calculus, linear algebra, or differential equations. Of course, these areas already include methods for solving such problems, but these are analytical in nature. Examples of analytical methods are:

- applying the Fundamental Theorem of Calculus to evaluate a definite integral,
- using Gaussian elimination, with exact arithmetic, to solve a system of linear equations, and
- using the Method of Undetermined Coefficients to solve an inhomogeneous ordinary differential equation.

Such analytical methods have the benefit that they yield exact solutions, but the drawback is that they can only be applied to a limited range of problems. Numerical methods, on the other hand, can be applied to a much wider range of problems, but only yield approximate solutions. Fortunately, in many applications, one does not necessarily need very high accuracy, and even when such accuracy is required, it can still be obtained, if one is willing to expend the extra computational effort (or, really, have a computer do so).

The goal in numerical analysis is to develop numerical methods that are effective, in terms of the following criteria:

3

- *A numerical method must be accurate.* While this seems like common sense, careful consideration must be given to the notion of accuracy. For a given problem, what level of accuracy is considered sufficient? As will be discussed in Section 2.1, there are many sources of error, and reducing error from one source might increase error from another. As such, it is important to question whether it is prudent to expend resources to reduce one type of error, when another type of error is already more significant. This will be illustrated in, for example, Section 9.1.

- *A numerical method must be efficient.* Although computing power has been rapidly increasing in recent decades, this has resulted in expectations of solving larger-scale problems. Therefore, it is essential that numerical methods produce approximate solutions with as few arithmetic operations or data movements as possible. Efficiency is not only important in terms of time; memory is still a finite resource and therefore methods must also aim to minimize data storage needs.

- *A numerical method must be robust.* A method that is highly accurate and efficient for some problems, but performs poorly on others, is unreliable and therefore not likely to be used in applications, even if any alternative is not as accurate and efficient. The user of a numerical method needs to know that the result produced can be trusted.

These competing criteria should be balanced according to the requirements of the application. For example, if lower accuracy is acceptable, then greater efficiency can be achieved. This can be the case, for example, if there is so much uncertainty in the underlying mathematical model that there is no point in obtaining high accuracy.

1.1   Overview

In this section, we provide an overview of the topics that will be covered in this book. While this selection of topics is not intended to be an exhaustive list of topics that could be covered in an undergraduate numerical analysis sequence, it does provide the reader with a sufficiently broad and deep background to pursue further study through more advanced coursework or research.

1.1.1   Error Analysis

Because solutions produced by numerical methods are not exact, we will begin our exploration of numerical analysis with one of its most fundamental aspects, which is *error analysis*. The primary goal of error analysis is to quantify, as much as possible, the deviation of a computed solution from the exact solution, which is the error. However, out of necessity, error *analysis* goes far beyond mere error *measurement*. As explained above, effective numerical methods must not only be *accurate*, but they must also be *efficient* and *robust*. In view of these criteria, error analysis relates error to parameters of numerical methods that affect accuracy and efficiency, and often in opposite ways making it infeasible to compute an exact solution using analytical methods. Furthermore, in the interest of robustness, error analysis is also concerned with the *propagation* of error through different stages of computation. In particular, a computed solution should not be too sensitive to input data, because if is is, any error in the input can result in a solution that is essentially useless.

Error can arise from many sources, such as

- **data error**, which accounts for error in any of the input data to a numerical method, including error in measurement or neglected components of a mathematical model,
- **discretization error**, which arises from approximating continuous functions by sets of discrete data points,
- **truncation error**, which arises from truncating a sequence of approximations that is meant to converge to the exact solution, to make computation possible, and
- **roundoff error**, which is due to the fact that computers represent real numbers approximately, in a fixed amount of storage in memory.

We will see that in some cases, these errors can be surprisingly large, so one must be careful when designing and implementing numerical methods. Section 2.1 will introduce fundamental concepts of error analysis that will be used throughout this book, and Section 2.2 will discuss computer arithmetic and roundoff error in detail.

1.1.2 Systems of Linear Equations

In Chapter 3, we will learn about how to solve a system of linear equations

$$a_{11}x_1 + a_{12}x_2 + \cdots + a_{1n}x_n = b_1$$
$$a_{21}x_1 + a_{22}x_2 + \cdots + a_{2n}x_n = b_2$$
$$\vdots$$
$$a_{n1}x_1 + a_{n2}x_2 + \cdots + a_{nn}x_n = b_n,$$

which can be more conveniently written in matrix-vector form

$$A\mathbf{x} = \mathbf{b},$$

where A is an $n \times n$ matrix, because the system has n equations (corresponding to *rows* of A) and n unknowns (corresponding to *columns*).

To solve a general system with n equations and unknowns, we can use **Gaussian elimination** to reduce the system to one with an **upper-triangular** matrix, which is easily solved. In some cases, this process requires **pivoting**, which entails interchanging of rows or columns of the matrix A. Gaussian elimination with pivoting can be used not only to solve a system of equations, but also to compute the *inverse* of a matrix, even though this is not normally practical. It can also be used to efficiently compute the determinant of a matrix.

Gaussian elimination with pivoting can be viewed as a process of factorizing the matrix A. Specifically, it achieves the decomposition

$$PA = LU,$$

where P is a **permutation matrix** that describes any row interchanges, L is a **lower-triangular** matrix, and U is an upper-triangular matrix. This decomposition, called the LU **Decomposition**, is particularly useful for solving $A\mathbf{x} = \mathbf{b}$ when the right-hand side vector $\mathbf{b}$ varies. We will see that for certain special types of matrices, variations of the general approach to solving $A\mathbf{x} = \mathbf{b}$ can lead to improved efficiency.

In Chapter 4, we will consider systems of equations for which the number of equations, m, is greater than the number of unknowns, n. This is the **least-squares** problem, which is reduced to a system with n equations and unknowns,

$$A^T A\mathbf{x} = A^T\mathbf{b},$$

called the **normal equations**. While this system can be solved directly using methods discussed above, this can be problematic due to sensitivity to roundoff error. We therefore explore other approaches based on **orthogonalization** of the columns of A.

Gaussian elimination and related methods are called **direct methods** for solving $A\mathbf{x} = \mathbf{b}$, because they compute the exact solution (up to roundoff error, which can be significant in some cases) in a fixed number of arithmetic operations that depends on n. However, such methods are often not practical, especially when A is very large, or when it is **sparse**, meaning that most of its entries are equal to zero. Therefore, in Chapter 5 we also consider **iterative methods**. Two general classes of iterative methods are:

- **stationary iterative methods**, in which the solution is characterized as a **fixed point**, or **stationary point**, of some function. These methods rely primarily on **splittings** of A to obtain a system of equations that can be solved rapidly in each iteration.
- **non-stationary methods**, which tend to rely on matrix-vector multiplication in each iteration and a judicious choice of **search direction** and **line search** to compute each iterate from the previous one.

Another fundamental problem from linear algebra, covered in Chapter 6, is the solution of the **eigenvalue problem**

$$A\mathbf{x} = \lambda\mathbf{x},$$

where the scalar λ is called an **eigenvalue** of A, and the nonzero vector $\mathbf{x}$ is called an **eigenvector**. This problem has many applications throughout applied mathematics, including the solution of differential equations and statistics. We will see that the tools developed for efficient and robust solution of least squares problems are useful for the eigenvalue problem as well.

1.1.3 Polynomial Interpolation and Approximation

Polynomials are the easiest functions to work with, because it is possible to evaluate them, as well as perform operations from calculus on them, with great efficiency. For this reason, more complicated functions, or functions that are represented only by values on a discrete set of points in their domain, are often approximated by polynomials.

Such an approximation can be computed in various ways. In Chapter 7 we consider **interpolation**, in which we construct a polynomial that agrees with the given data at selected points. While interpolation methods are efficient, they must be used carefully, because it is not necessarily true that a polynomial that agrees with a given function at certain points is a good approximation of the function elsewhere in its domain.

One remedy for this is to use **piecewise** polynomial interpolation, covered in Section 7.6. A low-degree polynomial, typically linear or cubic, is used to approximate data only on a given subdomain, and these polynomial "pieces" are "glued" together to obtain a piecewise polynomial approximation. This approach is also efficient, and tends to be more robust than standard polynomial interpolation, but there are disadvantages, such as the fact that a piecewise polynomial only has very few continuous derivatives.

An alternative to polynomial interpolation, whether piecewise or not, is **polynomial approximation**, covered in Chapter 8. The goal is to find a polynomial of a specified degree that, in some sense, best fits given data. For example, it is not possible to exactly fit a large number of points with a low-degree polynomial, but such an approximate fit can be more useful than a polynomial that can fit the given data exactly but still fail to capture the overall behavior of the data. This is illustrated in Figure 1.1. We also investigate approximation by rational functions (Section 8.3) and Fourier series (Section 8.4).

1.1.4 Numerical Differentiation and Integration

It is often necessary to approximate derivatives or integrals of functions that are represented only by values at a discrete set of points, thus making differentiation or integration rules impossible to use directly. Even when this is not the case, derivatives or integrals produced by differentiation or integration rules can often be very complicated functions, making their computation and evaluation computationally expensive.

While there are many software tools, such as Mathematica or Maple, that can compute derivatives or integrals symbolically using such rules, they are inherently unreliable because they require detection of patterns in whatever data structure is used to represent the function being differentiated or integrated, and it is not as easy to implement software that performs this kind of task effectively as it is for a person to learn how to do so through observation and intuition.

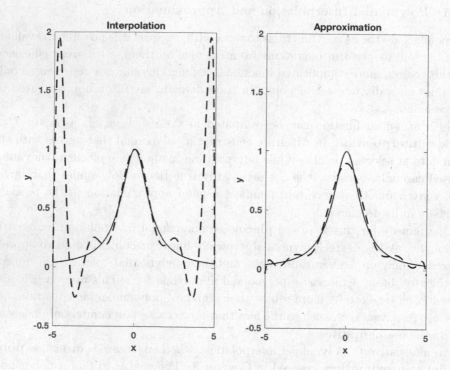

Fig. 1.1 The dashed curves demonstrate polynomial interpolation (left plot) and least-squares approximation (right plot) applied to $f(x) = 1/(1 + x^2)$ (solid curves).

Therefore, it is important to have methods for evaluating derivatives and integrals that are insensitive to the complexity of the function being acted upon. Numerical techniques for these operations, covered in Chapter 9, make use of polynomial interpolation by (implicitly) constructing a polynomial interpolant that fits the given data, and then applying differentiation or integration rules to the polynomial. We will see that by choosing the method of polynomial approximation judiciously, accurate results can be obtained with far greater efficiency than one might expect.

As an example, consider the definite integral

$$\int_0^1 \frac{1}{x^2 - 5x + 6}\, dx.$$

Evaluating this integral exactly entails factoring the denominator, which is simple in this case but not so in general, and then applying partial fraction decomposition to obtain an antiderivative, which is then evaluated at the limits. Alternatively, simply computing

$$\frac{1}{12}[f(0) + 4f(1/4) + 2f(1/2) + 4f(3/4) + f(1)],$$

where $f(x)$ is the integrand, yields an approximation with 0.01% error (that is, the error is 10^{-4}). While the former approach is less tedious to carry out by hand, at least if one has a calculator, clearly the latter approach is the far more practical use of computational resources.

1.1.5 Nonlinear Equations

The vast majority of equations, especially nonlinear equations, cannot be solved using analytical techniques such as algebraic manipulations or knowledge of trigonometric functions. For example, while the equations

$$x^2 - 5x + 6 = 0, \quad \cos x = \frac{1}{2}$$

can easily be solved to obtain exact solutions, these slightly different equations

$$x^2 - 5xe^x + 6 = 0, \quad x \cos x = \frac{1}{2}$$

cannot be solved using analytical methods.

Therefore, **iterative methods** must instead be used to obtain an approximate solution. In Chapter 10, we will study a variety of such methods, which have distinct advantages and disadvantages. For example, some methods are guaranteed to produce a solution under reasonable assumptions, but they might do so slowly. On the other hand, other methods may produce a sequence of iterates that quickly converge to the solution, but may be unreliable for some problems.

After learning how to solve nonlinear equations of the form $f(x) = 0$ using iterative methods such as **Newton's Method** (see Section 10.4), in Section 11.1 we will learn how to generalize such methods to solve systems of nonlinear equations of the form $\mathbf{F}(\mathbf{x}) = \mathbf{0}$, where $\mathbf{F} : D \subseteq \mathbb{R}^n \to \mathbb{R}^n$. In particular, for Newton's Method, computing $x^{(k+1)} = x^{(k)} - f(x^{(k)})/f'(x^{(k)})$ in the single-variable case is generalized to computing $\mathbf{x}^{(k+1)} = \mathbf{x}^{(k)} - [J_\mathbf{F}(\mathbf{x}^{(k)})]^{-1}\mathbf{F}(\mathbf{x}^{(k)})$, where $J_\mathbf{F}(\mathbf{x}^{(k)})$ is the **Jacobian matrix** of $\mathbf{F}$ evaluated at $\mathbf{x}^{(k)}$. This entails solving a system of linear equations, as in Chapter 3.

1.1.6 Optimization

We then turn our attention to the very important problem of **optimization**, in which we seek to minimize or maximize a given function, known as an **objective function**. Methods for the solution of nonlinear equations play an essential role in optimization techniques that require finding critical points of the objective function, as the derivative or gradient of the objective function is equal to zero at such points. In Section 11.2, we will examine a variety of methods based on this simple idea, as designing an efficient and robust method for optimization is far from simple. Our exploration of optimization concludes in Section 11.3 with relatively simple techniques that do not require any information about derivatives of the objective function. It will be seen that these techniques have natural analogues to certain methods for solving nonlinear equations.

1.1.7 Initial Value Problems

Next, in Chapter 12, we study various techniques for solving an **initial value problem**, which consists of an **ordinary differential equation (ODE)**

$$\frac{dy}{dt} = f(t, y), \quad t_0 < t \le T,$$

and an **initial condition**

$$y(t_0) = y_0.$$

Unlike analytical methods for solving such problems, that are used to find the exact solution in the form of a function $y(t)$, numerical methods typically compute values $y_1, y_2, y_3, \ldots$ that approximate $y(t)$ at discrete time values $t_1, t_2, t_3, \ldots$. At each time t_{n+1}, for $n \ge 0$, the value of the solution is approximated using its values at previous times.

We will learn about two general classes of methods: **one-step** methods (Section 12.2), which are derived using Taylor expansion and compute y_{n+1} only from y_n, and **multistep** methods (Section 12.3), which are usually based on polynomial interpolation and compute y_{n+1} from $y_n, y_{n-1}, \ldots, y_{n-m+1}$, where m is the number of steps in the method. Either type of method can be **explicit**, in which y_{n+1} can be described in terms of an explicit formula (the use of which is informally known as "plug-and-chug"), or **implicit**, in which y_{n+1} is described implicitly using an equation, usually nonlinear, that must be solved during each time step.

The difference between consecutive times t_n and t_{n+1}, called the **time step**, need not be uniform; in Section 12.5 we will learn about how it can be varied to achieve a desired level of accuracy as efficiently as possible. In Section 12.6 we will learn about how the methods used for the first-order initial-value problem described above can be generalized to solve higher-order equations, as well as systems of equations.

One key issue with time-stepping methods is **stability**. If the time step is not chosen to be sufficiently small, the computed solution can grow without bound, even if the exact solution is bounded. Generally, the need for stability imposes a more severe restriction on the size of the time step for explicit methods, which is why implicit methods are commonly used, even though they tend to require more computational effort per time step. Certain systems of differential equations can require an extraordinarily small time step to be solved by explicit methods; such systems are said to be **stiff** (see Section 12.4.4).

1.1.8 Boundary Value Problems

We then discuss solution methods for the **two-point boundary value problem**

$$y'' = f(x, y, y'), \quad a < x < b,$$

with **boundary conditions**

$$y(a) = \alpha, \quad y(b) = \beta.$$

One approach, the **Shooting Method** (Section 13.1), transforms this boundary-value problem into an initial-value problem so that methods for such problems from Chapter 12 can then be used. However, it is necessary to find the correct initial values so that the boundary condition at $x = b$ is satisfied. An alternative approach is to discretize y'' and y' using **finite differences** (Section 13.2), the approximation schemes covered in Chapter 9, to obtain a system of equations to solve for an approximation of $y(x)$; this system can be linear or nonlinear. Section 13.3 presents **collocation**, in which the solution is represented as a linear combination of functions, and requiring the ODE to be satisfied at selected points, called **collocation points**, yields a system of equations for the unknown coefficients. We conclude in Section 13.4 with the **Finite Element Method**, which, in a sense, treats the boundary value problem as a continuous least-squares problem as in Section 8.2.

1.1.9 Partial Differential Equations

We conclude this book with an introduction in Chapter 14 to numerical methods for the solution of **partial differential equations (PDEs)**. Techniques for two-point boundary value problems are generalized to higher spatial dimensions for the purpose of solving equations such as **Laplace's equation**. For time-dependent PDEs, such as the **heat equation** or the **wave equation**, techniques for boundary-value problems are combined with time-stepping techniques for initial value problems.

In Chapter 14, we first examine **finite-difference methods**, similar to those used for boundary-value problems, except applied to all partial derivatives with respect to both spatial and temporal variables. Then, in Section 14.5, we provide an introduction to the **Finite Element Method**, that is more conducive to solving PDEs on non-rectangular domains. We conclude in Section 14.6 with an overview of **spectral methods**, which yield extraordinary accuracy for PDEs with smooth solutions.

1.2 Getting Started with MATLAB

Throughout this book, we will make use of MATLAB to experiment with concepts and implement numerical methods. MATLAB is commercial software, originally developed by Cleve Moler in 1982 [Moler (1982)] and currently sold by The Mathworks. It can be purchased and downloaded from mathworks.com. As of this writing, the student version can be obtained for approximately $50, whereas academic and industrial licenses are much more expensive. For any license, "toolboxes" can be purchased in order to obtain additional functionality, but for the tasks performed in this book, the core product will be sufficient.

As an alternative, one can instead use Octave, a free application which uses the same programming language as MATLAB, with only minor differences. It can be obtained from gnu.org. Its user interface is not as "friendly" as that of MATLAB, but it has improved significantly in its most recent versions. In this book, examples will feature only MATLAB, but the code will also work in Octave, generally without modification. Other software alternatives are discussed in Section 1.3.

Figure 1.2 shows MATLAB when it is launched. The large window on the right is the **command window**, in which commands can be entered interactively at the >> prompt. On the left, the top window lists the files in the **current working directory**. By default, this directory is a subdirectory of your Documents directory (or My Documents, in some versions of Windows) that is named MATLAB. It is important to keep track of your current working directory, because that is the first directory in which MATLAB looks for **M-files**, or files that contain MATLAB code. These files will be discussed in detail later in this section.

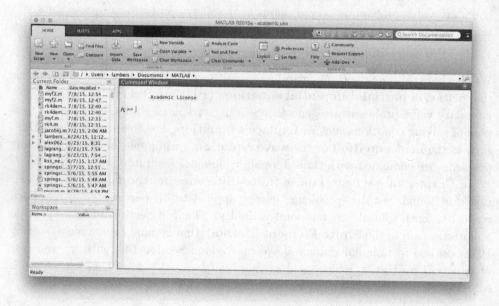

Fig. 1.2   Screen shot of MATLAB at startup in Mac OS X

We will now present basic commands, operations and programming constructs in MATLAB. Before we begin, we will use the `diary` command to save all subsequent commands, and their output, to a text file. Once a command is typed at the prompt, simply hit the Enter key to execute the command.

```
>> diary tutorial.txt
```

By default, this output is saved to a file that is named `diary` in the current working directory, but we will supply our own filename as an argument, to make the saved file easier to open later in a text editor.

1.2.1 Obtaining Help

Naturally, when learning new software, it is important to be able to obtain help. MATLAB includes the `help` command for this purpose. Try the following commands. You will observe that MATLAB offers help on individual commands, functions or operators, such as `lu`, and also help pages on various categories of same, such as `ops` for operators.

```
>> help
>> help ops
>> help lu
```

1.2.2 Basic Mathematical Operations

Next, we will perform basic mathematical operations. Try these commands, and observe the output. Note that the output has been abbreviated in this book by removing blank lines.

```
>> a=3+4
a =
     7
>> b=sqrt(a)
b =
    2.6458
>> format long
>> c=exp(a)
c =
     1.096633158428459e+003
>> d = 2*a
d =
    14
>> d=2b
 d=2b
Error: Unexpected MATLAB expression.
Did you mean:
>> d = 2*b
d =
    5.291502622129181
>> a=b^2
a =
    7.0000
```

As can be seen from these statements, arithmetic operations and standard mathematical functions can readily be performed, so MATLAB could be used as a "desk calculator" in which results of expressions can be stored in **variables**, such as a, b, c, and d in the preceding example. Also, note that once a command is executed, the output displayed is the variable name, followed by its value. This is typical behavior in MATLAB, so for the rest of this tutorial, the output will not be displayed in the text in most instances.

By default, numbers are displayed using five significant digits. The command `format long` causes numbers to be displayed using 15 significant digits instead. Formatting will be discussed in more detail later in this section. It is also very important to keep in mind that unlike in written mathematical expressions, multiplication must be indicated explicitly using a *. Note that an erroneous expression in MATLAB generates an appropriate error message, and at least in some cases, a suggestion for correcting the expression.

Exploration 1.2.1 At the prompt, type `help elfun` and familiarize yourself with the list of elementary math functions available in MATLAB. Then, enter and execute statements at the prompt that perform the following tasks:

1. Store the values 1, −5, and 6 in the variables a, b and c, respectively.

2. Use the quadratic formula, typed at the MATLAB prompt, to compute the two roots of the quadratic equation $ax^2 + bx + c = 0$. Store the roots in the variables x1 and x2. Use your knowledge of the roots of this equation, that can easily be computed by hand, to confirm that you typed in the expressions correctly. Be careful with parentheses.

3. Evaluate the Gaussian function
$$\frac{1}{\sqrt{2\pi}} e^{-\frac{1}{2}x^2}$$
at $x = $ a, b, c. To obtain an expression for π, solve the equation $\tan(\pi/4) = 1$ by hand and use an elementary math function to implement this expression in MATLAB. *Note:* the value of π is stored in the predefined variable `pi`, but the goal of this exploration is to learn to use the elementary math functions.

1.2.3 Basic Matrix Operations

In MATLAB, which is short for "matrix laboratory", it is particularly easy to work with vectors and matrices. We will now see this for ourselves by constructing and working with some simple matrices. Try the following commands.

```
>> A=[ 1 0; 0 2 ]
>> B=[ 5 7; 9 10 ]
>> A+B
>> 2*ans
```

```
>> C=A+B
>> 4*A
>> C=A-B
>> C=A*B
>> w=[ 4; 5; 6 ]
```

Note that semicolons are used to separate rows of a matrix, while entries within a row are separated by spaces. Entries within a row can also be separated by commas, which removes ambiguity.

As we can see, matrix arithmetic is easily performed. However, what happens if we attempt an operation that, mathematically, does not make sense? Consider the following example, in which an attempt is made to multiply a 2×2 matrix by a 3×1 vector.

```
>> A*w
??? Error using ==> mtimes
Inner matrix dimensions must agree.
```

Since this operation is invalid, MATLAB does not attempt to perform the operation and instead displays an error message. The function name mtimes refers to the function that implements the matrix multiplication operator, represented in the above command by *.

Exploration 1.2.2 Let

$$A = \begin{bmatrix} 2 & -4 \\ 3 & -5 \\ 7 & 0 \end{bmatrix}, \quad B = \begin{bmatrix} -5 & 1 & 4 \\ 4 & 6 & 8 \\ 10 & 5 & 0 \end{bmatrix}, \quad C = \begin{bmatrix} -8 & 0 \\ 2 & -3 \\ -1 & 4 \end{bmatrix}.$$

Carry out the computation $D = A + 3BC$ by hand, and then use MATLAB to check your work. *Note:* You may need to review Section B.6 about matrix-matrix multiplication and Section B.7.1 about matrix addition and scalar multiplication.

1.2.4 Storage of Variables

Let's examine the variables that we have created so far (exclusive of any explorations). The whos command is useful for this purpose.

```
>> whos
  Name      Size          Bytes  Class     Attributes

  A         2x2              32  double
  B         2x2              32  double
  C         2x2              32  double
  a         1x1               8  double
```

ans	2x2	32	double
b	1x1	8	double
c	1x1	8	double
w	3x1	24	double

Note that each number, such as a, or each entry of a matrix, occupies 8 bytes of storage, which is the amount of memory allocated to a double-precision floating-point number. This system of representing real numbers will be discussed further in Section 2.2. Also, note the variable ans, which was not explicitly created by any of the commands that we have entered. It is a special variable that is assigned the most recent expression that is not already assigned to a variable. In this case, the value of ans is the output of the operation 4*A, since that was not assigned to any variable.

1.2.5   Complex Numbers

MATLAB can also work with complex numbers. The following command creates a vector with one real element and one complex element.

```
>> z=[ 6; 3+4i ]
```

Now run the whos command again. Note that in the Attributes column, it is indicated that z is complex. Also, note that it states that z occupies 32 bytes, even though it has only two elements. This is because each element of z has a real part and an imaginary part, and each part occupies 8 bytes. It is important to note that if a single element of a vector or matrix is complex, then the entire vector or matrix is considered complex. This can result in wasted storage if imaginary parts are supposed to be zero, but in fact are small, nonzero numbers due to roundoff error (which will be discussed in Section 2.2).

The real and imag functions can be used to extract the real and imaginary parts, respectively, of a complex scalar, vector, or matrix. The values output by these functions are stored as real numbers.

```
>> y=real(z)
>> y=imag(z)
```

We call attention to the use of a MATLAB function in the above statements. Functions have zero or more **input arguments**, and zero or more **output arguments**. In both of the above statements, the input argument is z, and the output argument is y. Input arguments are enclosed in parentheses following the function name, with multiple arguments separated by commas. Output arguments are followed by an equals sign and then the function name. A single output argument, as in this case, does not require any delimiters, while multiple output arguments must be separated by commas and enclosed in square brackets. We will see more examples of function usage later in this tutorial.

Exploration 1.2.3 Use MATLAB to compute the (complex) roots of the quadratic equation $x^2 + 4x + 9 = 0$, as in Exploration 1.2.1, and then substitute them into the equation to verify the correctness of the roots.

Exploration 1.2.4 At the prompt, set the variable **theta** equal to a (radian) angle θ of your choosing. Then, form the complex number $z = c + is$, where $c = \cos\theta$ and $s = \sin\theta$. Compare to the result that you obtain by computing $e^{i\theta}$. *Note:* in place of i, use 1i for the imaginary unit $i = \sqrt{-1}$, as it is more efficient.

1.2.6 Creating Special Vectors and Matrices

It can be very tedious to create matrices entry-by-entry, as we have done so far. Fortunately, MATLAB has several functions that can be used to easily create certain matrices of interest. Try the following commands to learn what these functions do. In particular, note the behavior when only one input argument is given, instead of two.

```
>> E=ones(6,5)
>> E=ones(3)
>> Z=zeros(3,4)
>> Z=zeros(1,6)
>> R=rand(3,2)
```

As the names suggest, **ones** creates a matrix with all entries equal to one, **zeros** creates a matrix with all entries equal to zero, and **rand** creates a matrix with random entries. More precisely, the entries are random numbers that are uniformly distributed on $[0, 1]$.

Exploration 1.2.5 What if we want the entries of a matrix to be random numbers that are distributed within a different interval, such as $[-1, 1]$? Create such a matrix, of size 3×2, using matrix arithmetic that we have seen, and the **ones** function.

In many situations, it is helpful to have a vector of equally spaced values. For example, if we want a vector consisting of the integers from 1 to 10, inclusive, we can create it using the statement

```
>> z=[ 1 2 3 4 5 6 7 8 9 10 ]
```

However, this can be very tedious if a vector with many more entries is needed. Imagine creating a vector with all of the integers from 1 to 1000! Fortunately, this can easily be accomplished using the **colon operator**. Try the following commands to see how this operator behaves.

```
>> z=1:10
>> z=1:2:10
>> z=10:-2:1
>> z=1:-2:10
```

It should be noted that when the colon operator has three operands, the second operand, that determines spacing between entries, need not be an integer.

> **Exploration 1.2.6** Use the colon operator to create a vector of real numbers between 0 and 1, inclusive, with spacing 0.01.

> **Exploration 1.2.7** Use the `help` command to learn about the `linspace` function. How would you use this function to create the vectors `1:10`, `1:2:10` or `1:0.01:10`?

1.2.7 Suppressing Output

Suppose that we create a random 3×3 matrix at the prompt. The output will look something like this:

```
>> A=rand(3)
A =
      0.5667       0.8350       0.0031
      0.1063       0.5301       0.2453
      0.0798       0.3264       0.0194
```

We see that the matrix is printed out. Now, try the same statement, but instead create a 50×50 matrix, or an even larger one. Needless to say, such a statement produces a substantial amount of output.

This is problematic for several reasons. First, such output may not be necessary, and therefore the time spent by the operating system producing it, which can be substantial, is wasted. Second, this output is not conducive to "user-friendliness" of an application built in MATLAB. Fortunately, this output can be suppressed by adding a semicolon to the end of the statement:

```
>> A=rand(50);
>>
```

In general, any statement that evaluates an arithmetic or logical expression produces output, if it is not concluded with a semicolon. Specifically, the value of any expression that is computed in that statement is displayed, along with its variable name (or `ans`, if there is no variable associated with the expression). It is interesting to note that in MATLAB, semicolons at the end of statements are optional, unlike languages such as C, C++ or Java, in which they are required.

In most cases, suppressing output is the desired behavior, for the reasons given above. However, omitting semicolons can be useful when writing and debugging new code, because seeing intermediate results of a computation can expose coding errors, known informally as "bugs". Once the code is working, then semicolons can be added to suppress superfluous output.

1.2.8 Building Matrices

We have seen how to create matrices either entry-by-entry, as in Section 1.2.3, or with various helper functions, as in Section 1.2.6. Here, we demonstrate how to easily create larger matrices from smaller ones. First, we create the following matrices:

```
>> A=[ 2 3; 1 4 ];
>> B=[ -1 4 5; 2 -6 7 ];
>> C=[ -1 1; 2 -3; 5 -4 ];
```

Now, try these statements:

```
>> X=[ A B ];
>> Y=[ A; C ];
```

We see that the first statement concatenates matrices horizontally, while the second concatenates them vertically.

Exploration 1.2.8 With the matrices A, B, C above, try these statements:

```
>> Z=[ A; B ];
>> W=[ A C ];
```

What happens with these statements? In general, when can matrices be concatenated horizontally, or vertically?

An empty matrix can be constructed using an empty pair of square brackets:

```
>> x=[]
```

Then, entries or matrices can be concatenated onto the empty matrix. For example, here is how we can iteratively build up a row vector from "nothing":

```
>> x=[ x 1 ]
x =
    1
>> x=[ x 2 ]
x =
    1    2
>> x=[ x 3 4 5 ]
x =
    1    2    3    4    5
```

> **Exploration 1.2.9** How would the preceding statements be modified to
> build a column vector from an empty matrix, rather than a row vector?

While this is a convenient way to iteratively build a matrix (or a vector, if there
is only one row or column), it is important to note that it is also inefficient, as the
memory associated with the matrix must be deallocated and then reallocated to
make room for new entries. This adds substantial overhead that can be avoided if
the entire matrix is **preallocated**, once, using the `zeros` function.

1.2.9 Accessing and Changing Entries of Matrices

Now that we know how to create matrices, we need to be able to work with individual
entries of matrices, including accessing them and modifying them. Individual entries
are accessed using the following notation:

```
>> A=[ 2 3; 1 4 ];
>> A(1,2)
ans =
     3
```

In general, a_{ij} is accessed using the syntax `A(i,j)`. It's important to note that
MATLAB uses "1-based indexing", meaning that the indices `i` and `j` must assume
values in the ranges $1, 2, \ldots, m$ and $1, 2, \ldots, n$, respectively, where A is $m \times n$. This
contrasts with languages such as C++, which use 0-based indexing.

In many instances, it is helpful to be able to extract submatrices, such as a row,
column, or block. Try these statements and describe the portion of A that each one
extracts:

```
>> A=rand(3)
>> A(1,1:3)
>> A(1:2,1)
>> A(1:2,1:2)
>> A(:,3)
>> A(1,:)
>> A(:,end)
>> A(2:end,1)
>> A([ 1 3 ],:)
>> A(:,[ 3 2 ])
>> B=rand(5)
>> B([ 1 4 3 ],[ 5 2 ])
```

It can be seen from these statements that row vectors can be used as indices. That
is, if A is $m \times n$, and if `p` is a row vector of length k and `q` is a row vector of length

l, where $k \leq m$ and $l \leq n$, then the expression `A(p,q)` returns the $k \times l$ submatrix

$$\begin{bmatrix} a_{p_1,q_1} & a_{p_1,q_2} & \cdots & a_{p_1,q_l} \\ a_{p_2,q_1} & a_{p_2,q_2} & \cdots & a_{p_2,q_l} \\ \vdots & & & \vdots \\ a_{p_k,q_1} & a_{p_k,q_2} & \cdots & a_{p_k,q_l} \end{bmatrix}.$$

Furthermore, using a : by itself in place of a row vector is a shorthand for all possible indices, which is convenient for extracting entire rows or columns. Finally, **end**, when used as an index, refers to the largest valid index.

Exploration 1.2.10 Given a matrix A, write a MATLAB expression that represents A with its rows and columns both in reverse order, using the : operator.

Exploration 1.2.11 Given a matrix A, and two indices i and j, with $i \neq j$, write a MATLAB statement that interchanges rows i and j of A (that is, the statement overwrites these rows of A with the interchanged rows).

1.2.10 Transpose Operators

We now know how to create *row* vectors with equally spaced values, but what if we would rather have a *column* vector? This is just one of many instances in which we need to be able to compute the **transpose** of a matrix (see Section B.7.2) in MATLAB. Fortunately, this is easily accomplished, using the single quote as an operator. For example, this statement

```
>> z=(0:0.1:1)'
```

has the desired effect. However, one should not simply conclude that the single quote is the transpose operator, or they could be in for an unpleasant surprise when working with complex-valued matrices. Try these commands to see why:

```
>> z=[ 6; 3+4i ]
>> z'
>> z.'
```

We can see that the single quote is an operator that takes the **Hermitian transpose** of a matrix A, commonly denoted by A^H: it is the transpose *and* complex conjugate of A. That is, $A^H = \overline{A^T}$, where the bar denotes complex conjugation. Meanwhile, the dot followed by the single quote is the transpose operator.

Either operator can be used to take the transpose for matrices with real entries, but one must be more careful when working with complex entries. That said, why is the "default" behavior, represented by the simpler single quote operator, the Hermitian transpose rather than the transpose? This is because in general, results or techniques established for real matrices, that make use of the transpose, do not generalize to the complex case unless the Hermitian transpose is used instead.

Exploration 1.2.12 The ℓ_2-**norm**, or simply 2-norm, of a vector $\mathbf{x} \in \mathbb{C}^n$ is defined in Section B.13.1 by $\|\mathbf{x}\|_2 = \sqrt{\mathbf{x}^H \mathbf{x}}$. As $\|\mathbf{x}\|_2$ must be a real, positive number for any nonzero vector $\mathbf{x}$, justify this definition by finding a nonzero vector $\mathbf{x}$ such that $\mathbf{x}^T \mathbf{x}$, as opposed to $\mathbf{x}^H \mathbf{x}$, is a) negative, b) zero, or c) imaginary.

1.2.11 Conditional Expressions

We have already learned about arithmetic operators such as + or *. Here, we learn about **relational** and **logical** operators, which are featured in other programming languages as well. Help on these operators can be obtained by typing `help ops` at the prompt. The relational operators are:

==	equal to
~=	not equal to
<	less than
<=	less than or equal to
>	greater than
>=	greater than or equal to

Note that the "not equal" operator is different than that used in other languages such as C++ or Java. Evaluate the following relational expressions at the prompt:

```
>> 4>3
ans =
  logical
  1
>> 1==0
ans =
  logical
  0
```

Note that the value of a conditional expression is 1 if the expression is true, and 0 if it is false. Also note that these values are considered to be of type `logical`, and therefore occupies only one byte, rather than the eight bytes occupied by a real number.

The logical operators are

&&	short-circuit logical AND
||	short-circuit logical OR
&	element-wise logical AND
|	element-wise logical OR
~	logical NOT

The short-circuit operators can only be used with operands that are scalars (that is, not a vector or matrix). **Short-circuit evaluation** is an efficient approach to evaluating expressions involving several AND or OR operators. For example, x && y && z is false if x is false, in which case y and z are not evaluated at all. Similarly, x || y || z is true if x is true, and again y and z are not evaluated at all.

The element-wise operators are used to efficiently apply logical operators to corresponding elements of vectors or matrices of the same size. Consider the following example:

```
>> x=rand(3,1)
x =
    0.3705
    0.6224
    0.9976
>> y=rand(3,1)
y =
    0.5173
    0.9905
    0.2265
>> (x<1/2)|(y<1/2)
ans =
  3x1 logical array
   1
   0
   1
```

The last statement applies both the < and | operator in an element-wise manner. The result of x<1/2 is a 3×1 logical vector with elements $1, 0, 0$, as only the first element of x is less than $1/2$. Similarly, y<1/2 is a logical vector with elements $0, 0, 1$. The result of the element-wise OR operator | applied to these logical vectors is also a logical vector. Each element of the result is the logical OR of corresponding elements of these vectors.

> **Exploration 1.2.13** Let x be a column vector of 10 random numbers between -10 and 10. Construct a logical column vector consisting of 10 elements, in which each element is true if and only if the corresponding element of x is in the interval $(-1, 1)$.

1.2.12 Script M-files

We are about to learn some essential programming language constructs, that MATLAB shares with many other programming languages. First, though, we will learn how to write a **script** in MATLAB. Scripts are very useful for the following reasons:

- Some MATLAB statements, such as the programming constructs we are about to discuss, are quite complicated and span several lines. Typing them at the command window prompt can be very tedious, and if a mistake is made, then the entire construct must be retyped.

- It frequently occurs that a sequence of commands needs to be executed several times, with no or minor changes. It can be very tedious and inefficient to repeatedly type out such command sequences several times, even if MATLAB's history features (such as using the arrow keys to scroll through past commands) are used.

A script can be written in a plain text file, called an **M-file**, which is a file that has a .m extension. An M-file can be written in any text editor, or in MATLAB's own built-in editor. To create a new M-file or edit an existing M-file, one can use the `edit` command at the prompt:

```
>> edit signofx
```

If no extension is given, a .m extension is assumed. If the file does not exist in the current working directory or in MATLAB's **search path**, which is a list of directories in which MATLAB looks for M-files, MATLAB will ask if the file should be created (this behavior can be changed, if desired). The preceding command opens the M-file `signofx.m` in the default editor, which will be an empty file if it did not already exist.

1.2.13   if Statements

The first programming language construct we will learn is an `if` statement, which is used to perform a different task based on the result of a given conditional expression, that is either true or false.

In the file `signofx.m` that was just opened in the editor, type in the following code, that displays an appropriate message according to whether the (numeric) value of a variable x is positive, negative, or zero.

```
x=2*rand(1)-1;
disp(x)
if x>0
    disp('x is positive')
elseif x<0
    disp('x is negative')
else
    disp('x is zero')
end
```

Once you save, the file `signofx.m` will be written to the current working directory. To execute a script M-file, simply type the name of the file (without the .m extension) at the prompt.

```
>> signofx
   -0.8169
x is negative
```

The code in this script first assigns a value to x, and then displays this value using the disp function. Note that the output produced by disp differs from statements that produce output due to the lack of a semicolon, as only the value is displayed, not the name of the variable or an equals sign. For this reason, the use of disp is often the preferred approach to producing output, as the output can be given a more professional appearance. This will be illustrated in later examples.

Next, the conditional expression x>0 is evaluated. Then, the if statement containing this conditional expression is executed according to the result of the evaluation. If the conditional expression is true, then the statements immediately following it are executed. In this example, that means the statement disp('x is positive') would be executed.

Also note the use of the keywords else and elseif. These are used to provide alternative conditions under which different code can be executed, if the original condition in the if statement turns out to be false. If any conditions paired with the elseif keyword also turn out to be false, then the code following the else keyword, if any, is executed. Finally, note that the entire if statement is concluded with the end keyword.

For a more advanced example, type the following code into a script file named entermonthyear.m. This code computes and displays the number of days in a given month, while taking leap years into account. As this example demonstrates, if statements can be nested within one another.

```
% entermonthyear - script that asks the user to provide a month and
% year, and displays the number of days in that month

% Prompt user for month and year
month=input('Enter the month (1-12): ');
year=input('Enter the 4-digit year: ');
% For displaying the month by day, we construct an array of strings
% containing the names of all the months, in numerical order.
% This is a 12-by-9 matrix, since the longest month name (September)
% contains 9 letters.  Each row must contain the same number of
% columns, so other month names must be padded to 9 characters.
months=[ 'January  '; 'February '; 'March    '; 'April    '; ...
         'May      '; 'June     '; 'July     '; 'August   '; ...
         'September'; 'October  '; 'November '; 'December ' ];
% extract the name of the month indicated by the user
monthname=months(month,:);
% remove trailing blanks
```

```
monthname=deblank(monthname);
if month==2
    % month is February
    if rem(year,4)==0
        % leap year
        days=29;
    else
        % non-leap year
        days=28;
    end
elseif month==4 || month==6 || month==9 || month==11
    % "30 days hath April, June, September and November..."
    days=30;
else
    % "...and all the rest have 31"
    days=31;
end
% display number of days in the given month
disp([ monthname ', ' num2str(year) ' has ' num2str(days) ' days.' ])
```

Can you figure out how the nested if statements work, based on your knowledge of what the result should be? We point out the use of ... to continue a long line onto the next line. Also, we note that in this script, most of the statements are terminated with semicolons.

This M-file includes **comments**, which are preceded by a percent sign (%). Once a % is entered on a line, the rest of that line is ignored. This is very useful for documenting code so that a reader can understand what it is doing. The importance of documenting one's code cannot be overstated. In fact, it is good practice to write the documentation *before* the code, so that the process of writing code is informed with a clearer idea of the task at hand.

This script features some new functions that can be useful in many situations:

- deblank(s): returns a new string variable that is the same as s, except that any "white space" (spaces, tabs, or newlines) at the end of s is removed
- rem(a,b): returns the remainder after dividing a by b
- num2str(x): returns a string variable based on formatting the number x as text

We now execute the script in this M-file:

```
>> entermonthyear
Enter the month (1-12): 5
Enter the 4-digit year: 2001
May, 2001 has 31 days.
```

Observe that each `input` statement displays the prompt that is passed as an argument to the `input` function, and then waits for the user to type input and then hit the Enter key. Then, the number of days in the specified month is determined, and a text string is cobbled together from text and variables in the concluding `disp` statement so that the result can be reported in a visually appealing manner.

Exploration 1.2.14 Write a MATLAB script `dayinyear` that uses the code in `entermonthyear` to perform the following task: prompt the user to enter a month (by number, 1-12), a 4-digit year, and a day (1-n), where n is the number of days in the given month. The prompt for the day must include the appropriate range of days. Then, print out the number of the day (1-366) within the year; that is, the number of days since December 31 of the previous year. For example, the result for January 1 would be 1, the result for February 15 would be 46, and the result for June 10 would be 161, unless the given year is a leap year, in which case it would be 162. Use `if` statements and the number of days in each month to compute this value.

1.2.14 `for` Loops

The next programming construct is the `for` loop, which is used to carry out an iterative process that runs for a predetermined number of iterations. To illustrate `for` loops, we consider the task of computing the mean of a list of numbers, stored in a vector x.

Type this code into a script file called `computemean.m`:

```
x=rand(10,1);
disp('x is:')
disp(x)
m=0;
for i=1:10
    m=m+x(i);
end
m=m/10;
disp('The mean of x is:')
disp(m)
```

Since the vector x has 10 elements, the mean of the elements of x is given by the formula

$$m = \frac{1}{10} \sum_{i=1}^{10} x_i.$$

To "count from 1 to 10" in the process of accessing all of the elements of x for the purpose of adding them, we use a `for` statement. Note the syntax for a `for` statement: the keyword `for` is followed by an **index variable**, such as i in the preceding example, and that variable is assigned a value. Then the body of the loop is given, followed by the keyword end.

Now run this script, just like in the previous example.

```
>> computemean
x is:
    0.3148
    0.7267
    0.5158
    0.7906
    0.2045
    0.6781
    0.0525
    0.8012
    0.6786
    0.9460


The mean of x is:
    0.5709
```

The loop performed 10 iterations to compute the sum of the elements of x, which is approximately 5.7088. Then, this sum is divided by 10, the number of elements, to obtain the mean.

What does a for loop actually do? During the ith iteration, the index variable i is set equal to the ith *column* of the expression that is assigned to it by the for statement. Then, the index variable retains this value throughout the body of the loop (unless the index variable is changed within the body of the loop, which is ill-advised, and sometimes done by mistake!), until the iteration is completed. Then, the index variable is assigned the next column for the next iteration. In most cases, such as in this example, the index variable is simply used as a counter, in which case assigning to it a row vector of values, created using the colon operator, yields the desired behavior.

Exploration 1.2.15 Modify the script computemean.m to write a new script computevariance.m that computes the (population) variance of x, given by

$$v = \frac{1}{10} \sum_{i=1}^{10} (x_i - m)^2,$$

where m is the mean of x.

For our next example, we use an if statement *inside* a for statement. The following script computes the maximum of a list of numbers, again stored in a vector x. We will call this script findmax.m:

```
x=rand(10,1);
disp('x is:')
disp(x)
xmax=x(1);
for i=2:10
    if x(i)>xmax
        xmax=x(i);
    end
end
disp('The max of x is:')
disp(xmax)
```

The body of a `for` statement can contain any number of statements, and of any kind, just as each clause of an `if` statement can.

Note that in this example, the `for` statement only needs to execute nine times, because the maximum element of x is initially assumed to be its first element, and then the `for` loop is used to examine the remaining nine elements of x to check whether any of them are greater than all previously examined elements. The variable `xmax` stores the maximum value found thus far.

> **Exploration 1.2.16** Modify the script `findmax.m` to obtain a new script `findmaxpos.m` that not only finds the maximum element of the vector x, but also determines and displays the *index* `imax`, where $1 \leq \text{imax} \leq 10$, at which the maximum value is found.

For a more advanced example, we illustrate the use of nested `for` statements, as well as index variables whose upper and lower bounds are not constants, but rather variables. We examine the script file `gausselim.m`:

```
% gausselim - script that performs Gaussian elimination on a random
% 40-by-40 matrix
m=40;
n=40;
% generate random matrix
A=rand(m,n);
% display it
disp('Before elimination:')
disp(A)
for j=1:n-1
    % use elementary row operations to zero all elements in
    % column j below the diagonal
    for i=j+1:m
        mult=A(i,j)/A(j,j);
        % subtract mult * row j from row i
        for k=j:n
```

```
            A(i,k)=A(i,k)-mult*A(j,k);
        end
        % equivalent code:
        %A(i,j:n)=A(i,j:n)-mult*A(j,j:n);
    end
end
% display updated matrix
disp('After elimination:')
disp(A)
```

The script displays a randomly generated matrix A, then performs Gaussian elimination (see Section 3.1) on A to obtain an **upper triangular matrix** (see Section
B.10), and then displays the final result. Note that while the outermost loop, with
index variable j, executes n−1 times, where the matrix A is m×n, the next innermost
loop, with index variable j, executes m − j times, due to the variable starting value
of j + 1. Similarly, the innermost loop, with index variable k, executes n − j + 1
times.

Note the commented-out statement A(i,j:n)=A(i,j:n)-mult*A(j,j:n). This
statement can be used *instead* of the innermost for loop that begins with for k=j:n.
This is because each iteration of this loop does not depend on the result of previous
iterations, so the relevant portion (columns j through n) of the jth row of A, scaled
by mult, can be subtracted from the same portion of the ith row of A in a single
operation. In other words, this statement is an *implicit* for loop, because it must
still be carried out through an iteration, but that iteration is implied by the range
of indices j:n.

Exploration 1.2.17 An upper triangular matrix U has the property that
$u_{ij} = 0$ whenever $i > j$; that is, the entire "lower triangle" of U, consisting
of all entries below the main diagonal, must be zero (see Section B.10).
Examine the matrix A produced by the script **gausselim** above. Can you
explain why some subdiagonal entries are nonzero?

Exploration 1.2.18 Write a script **matadd** that creates two $m \times n$ matrices
A and B, and uses for loops to compute the sum $C = A + B$ using the
definition of matrix addition in Section B.7.1.

Exploration 1.2.19 Write a script **matmult** that creates two matrices A
and B, where A is $m \times n$ and B is $n \times p$, and uses for loops to compute
the product $C = AB$ using the definition of matrix-matrix multiplication
in Section B.6.

1.2.15 while Loops

Next, we introduce the while loop, which, like a for loop, is also used to implement an iterative process, but is controlled by a conditional expression rather than a predetermined set of values such as 1:n. A while loop executes as long as the condition in the while statement is true.

Type the following code into the script file guessnumber.m:

```
number=ceil(10*rand(1));
disp('I am thinking of a number between 1 and 10.')
guess=input('What is it? Enter your guess: ');
while number~=guess
    guess=input('Your guess is wrong! Try again: ');
end
disp('You guessed right!')
```

Then save and run the script as shown in this sample output:

```
>> guessnumber
I am thinking of a number between 1 and 10.
What is it? Enter your guess: 3
Your guess is wrong! Try again: 7
Your guess is wrong! Try again: 4
Your guess is wrong! Try again: 8
Your guess is wrong! Try again: 9
Your guess is wrong! Try again: 5
Your guess is wrong! Try again: 10
Your guess is wrong! Try again: 1
You guessed right!
```

Can you see how the while loop behaves, based on this output? The code first chooses a random integer between 1 and 10. This is accomplished by using rand to compute a random real number between 0 and 1, multiplying by 10, and then using the ceil function, short for "ceiling", to round this number *upward* to the closest integer that is greater than or equal to it. Then, the input function is used to obtain a number from the user that is a guess of the randomly chosen number.

The while statement checks whether the user's guess is equal to the randomly chosen number. If they are equal, then the loop does not execute at all, and control flows to the statement following the while loop, that displays the message 'You guessed right!' If they are not equal, then the statement in the body of the while loop, that solicits a new guess from the user, is executed. Then control flows to the top of the while loop, that checks the condition again using the new value of guess. The loop will continue in this manner until number and guess are equal, at which time control will flow to the disp statement following the while loop.

Generally, a while loop executes all of the statements in its body (that is, the statements between the keyword while followed by its conditional expression, and the matching end) as long as the conditional expression following the while keyword is true. If the condition is always true, then the loop will execute indefinitely, unless some statement in the body of the loop causes the loop to terminate, as the next example illustrates.

The following script, saved in the file newtonsqrt.m, illustrates the use of a while loop.

```
% newtonsqrt - script that uses Newton's Method to compute the
% square root of 2

% choose initial iterate
x=1;
% announce what we are doing
disp('Computing the square root of 2...')
% iterate until convergence.  we will test for convergence inside
% the loop and use the break keyword to exit, so we can use a
% loop condition that's always true
while true
    % save previous iterate for convergence test
    oldx=x;
    % compute new iterate
    x=x/2+1/x;
    % display new iterate
    disp(x)
    % if relative difference in iterates is < 10^(-15), exit the loop
    if abs(x-oldx)<10^(-15)*abs(x)
        break;
    end
end
% display result and verify that it really is the square root of 2
disp('The square root of 2 is:')
x
disp('x^2 is:')
disp(x^2)
```

Note the use of the expression true in the while statement. The value of the predefined variable true is 1, while the value of false is 0, following the convention used in many programming languages that a nonzero number is interpreted as the logical value "true", while zero is interpreted as "false". It follows that this particular while statement is an infinite loop, since the value of true will never be false. However, this loop will exit when the condition in the enclosed if statement is true, due to the break statement. A break statement causes the enclosing for or `while` loop to immediately exit.

This particular `while` loop computes an approximation of $\sqrt{2}$ such that the *relative* difference between each new iterate `x` and the previous iterate `oldx` is less than 10^{-15}. The actual process used to obtain this approximation of $\sqrt{2}$ is obtained using **Newton's Method**, which will be discussed in Section 10.4. Go ahead and run this script by typing its name, `newtonsqrt`, at the prompt. The code will display each iteration as the approximation is improved until it is sufficiently accurate. Note that convergence to $\sqrt{2}$ is quite rapid! This convergence will be explored further in Chapter 10.

> **Exploration 1.2.20** Modify the script `newtonsqrt` so that it does not use an "infinite loop" caused by the condition `while true`, or a `break` statement. Instead, the `while` statement must use the condition given in the `if` statement (or really, its logical negation).

While an "infinite loop" may be convenient in some cases, ideally loops should be written in such a way that they are guaranteed to terminate after a finite number of iterations. For example, a maximum number of iterations can be imposed.

> **Exploration 1.2.21** Modify the script from Exploration 1.2.20 so that it performs a maximum of ten iterations, enforced by checking the value of a variable `niter` that keeps track of the number of iterations. Use the `&&` operator to perform a logical AND of two conditional expressions. Type `help ops` and consult the "logical operators" section for more information.

1.2.16 Function M-files

In addition to script M-files, MATLAB also uses **function M-files**. A function M-file is also a text file with a `.m` extension, but unlike a script M-file, that simply includes a sequence of commands that could have instead been typed at the prompt, the purpose of a function M-file is to extend the capability of MATLAB by defining a new function, that can then be used by other code.

The following function M-file, called `converttemp.m`, illustrates function definition.

```
function tc=converttemp(tf)
% converts function input 'tf' of temperatures in Fahrenheit to
% function output 'tc' of temperatures in Celsius
temp=tf-32;
tc=temp*5/9;
```

Note that a function definition begins with the keyword `function`; this is how MATLAB distinguishes a script M-file from a function M-file (though in a function M-file, comments can still precede the use of `function`).

After the keyword `function`, the **output arguments** of the function are specified. In this function, there is only one output argument, `tc`, which represents the Celsius temperature. If there were more than one, then they would be enclosed in square brackets, and in a comma-separated list. After the output arguments, there is a = sign, then the function name, which should match the name of the M-file aside from the `.m` extension. Finally, if there are any **input arguments**, then they are listed after the function name, separated by commas and enclosed in parentheses.

After this first line, all subsequent code is considered the body of the function–the statements that are executed when the function is called. The only exception is that other functions can be defined within a function M-file, but they are "helper" functions, that can only be called by code within the same M-file. Helper functions must appear *after* the function for which the M-file is named.

Type in the above code for `converttemp` into a file `converttemp.m` in the current working directory. Then, it can be executed as follows:

```
>> tc=converttemp(212)
tc =
   100
```

If `tc=` had been omitted, then the output value 100 would have been assigned to the special variable `ans`, described in Section 1.2.4.

Note that the definition of `converttemp` uses a variable `temp`. Here, it should be emphasized that all variables defined within a function, including input and output arguments, are only defined within the function itself. If a variable inside a function, such as `temp`, happens to have the same name as another variable defined in the top-level workspace (the memory space used by variables defined outside of any function), or in another function, then this other variable is completely independent of the one that is internal to the function. Consider the following example:

```
>> temp=32
temp =
    32
>> tfreeze=converttemp(temp)
tfreeze =
    0
>> temp
temp =
    32
```

Inside `converttemp`, `temp` is set equal to zero by the subtraction of 32, but the `temp` in the top-level workspace retains its value of 32.

Comments included at the top of an M-file (whether script or function) are assumed by MATLAB to provide documentation of the M-file. As such, these comments are displayed by the `help` command, as applied to that function. Try the following commands:

```
>> help converttemp
>> help newtonsqrt
```

> **Exploration 1.2.22** The MATLAB functions sind and cosd are the same as sin and cos, except they assume that the input argument is given in degrees rather than radians. Write your own versions of sind and cosd (with different names) that rely on sin and cos.

We now illustrate other important aspects of functions, using the following M-file, which is called vecangle2.m:

```
% vecangle2 - function that computes the angle between two given
% vectors in both degrees and radians.  We use the formula
% x'*y = ||x||_2 ||y||_2 cos(theta), where theta is the angle
% between x and y
function [anglerad,angledeg]=vecangle2(x,y)
n=length(x);
if n~=length(y)
    error('vector lengths must agree')
end
% compute needed quantities for above formula
dotprod=x'*y;
xnorm=norm(x);
ynorm=norm(y);
% obtain cos(angle)
cosangle=dotprod/(xnorm*ynorm);
% use inverse cosine to obtain angle in radians
anglerad=acos(cosangle);
% if angle in degrees is desired (that is, two output arguments are
% specified), then convert to degrees.  Otherwise, don't bother
if nargout==2
    angledeg=anglerad*180/pi;
end
```

As described in the comments, the purpose of this function is to compute the angle between two vectors in n-dimensional space, in both radians and degrees. Note that this function accepts multiple input arguments and returns multiple output arguments. The way in which this function is called is similar to how it is defined. For example, try this command:

```
>> [arad,adeg]=vecangle2(rand(5,1),rand(5,1))
```

It is important that code include error-checking, especially if it might be used by other people. To that end, the first task performed in this function is to check whether the input arguments x and y have the same length, using the length function that returns the number of elements of a vector or matrix. If they do not have the same length, then the `error` function is used to immediately exit the function `vecangle2` and display an informative error message.

Note the use of the variable `nargout` at the end of the function definition. The function `vecangle2` is defined to have two output arguments, but `nargout` is the number of output arguments that are actually specified when the function is called. Similarly, `nargin` is the number of input arguments that are specified.

These variables allow functions to behave more flexibly and more efficiently. In this case, the angle between the vectors is only converted to degrees if the user specified both output arguments, thus making `nargout` equal to 2. Otherwise, it is assumed that the user only wanted the angle in radians, so the conversion is never performed. MATLAB typically provides several interfaces to its functions, and uses `nargin` and `nargout` to determine which interface is being used. These multiple interfaces are described in the `help` pages for such functions.

Exploration 1.2.23 Try calling `vecangle2` in various ways, with different numbers of input and output arguments, and with vectors of either the same or different lengths. Observe the behavior of MATLAB in each case.

1.2.17   Graphics

Next, we learn some basic graphics operations. We begin by plotting the graph of the function $y = x^2$ on $[-1, 1]$. Start by creating a vector `x`, of equally spaced values between -1 and 1, using the colon operator:

```
>> x=(-1:0.001:1)';
```

Then, create a vector `y` that contains the squares of the values in `x`. We can do this using a `for` loop, but there is an easier way, using the exponentiation operator `^`. However, we cannot square a vector as we would a number:

```
>> y=x^2;
Error using  ^
One argument must be a square matrix and the other must be a scalar.
Use POWER (.^) for elementwise power.
```

Instead, we must use the element-wise exponentiation operator, `.^`:

```
>> y=x.^2;
```

Multiplication and division also have element-wise versions, denoted with the preceding dot.

The `plot` function, in one of its simplest forms, takes two input arguments that are vectors, that must have the same length. The first input argument contains x-values, and the second input argument contains y-values. The command `plot(x,y)` creates a new figure window (if none already exists) and plots y versus x in a set of axes contained in the figure window. Try plotting the graph of the function $y = x^2$ on $[-1, 1]$ using this command.

Note that by default, `plot` appears to produce a solid blue curve. In reality, it is not a curve; it simply "connects the dots" using solid blue line segments, but if the segments are small enough, the resulting piecewise linear function resembles a smooth curve. But what if we want to plot curves using different colors, different line styles, or different symbols at each point?

Use the `help` command to view the help page for `plot`, which lists the specifications for different colors, line styles, and marker styles (which are used at each point that is plotted). The optional third argument to the `plot` function is used to specify these colors and styles. They can be mixed together; for example, the third argument `'r--'` plots a dashed red curve. Experiment with these different colors and styles, and with different functions.

MATLAB provides several functions that can be used to produce more sophisticated plots. It is recommended that you view the help pages for these functions, and also experiment with their usage.

- `hold` is used to specify that subsequent `plot` commands should be superimposed on the same set of axes, rather than the default behavior in which the current axes are cleared with each new `plot` command.
- `subplot` is used to divide a figure window into an $m \times n$ matrix of axes, and specify which set of axes should be used for subsequent `plot` commands.
- `xlabel` and `ylabel` are used to label the horizontal and vertical axes, respectively, with given text.
- `title` is used to place given text at the top of a set of axes.
- `legend` is used to place a legend within a set of axes, so that the curves displayed on the axes can be labeled with given text.
- `gtext` is used to place given text at an arbitrary point within a figure window, indicated by clicking the mouse at the desired point.

Exploration 1.2.24 Reproduce the plot shown in Figure 1.3 using the commands discussed in this section, with the colors of your choice.

Finally, it is essential to be able to save a figure so that it can be printed or included in a document. In the figure window, go to the File menu and choose "Save" or "Save As". You will see that the figure can be saved in a variety of standard image formats, such as JPEG or Windows bitmap (BMP).

Another format is "Matlab Figure (*.fig)". It is strongly recommended that you save your figure in this format, as well as the desired image format. Then, if you need to go back and change something about the figure after you have already closed the figure window, you can simply use the `open` function, with the `.fig` filename as its input argument, to reopen the figure window. Otherwise, you would have to recreate the entire figure from scratch.

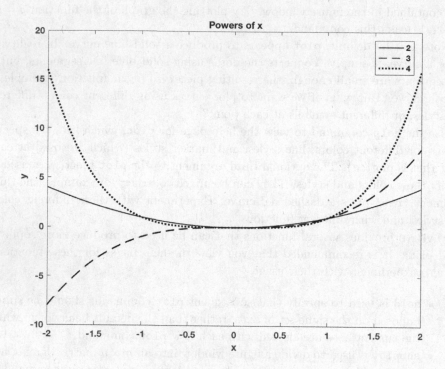

Fig. 1.3   Figure for Exploration 1.2.24

1.2.18   Polynomial Functions

MATLAB provides several functions for working with polynomials. A polynomial is represented within MATLAB using a *row* vector of coefficients, \with the highest degree coefficient listed first. For example, $f(x) = x^4 - 3x^2 + 2x + 5$ is represented by the vector [1 0 -3 2 5]. Note that the zero for the second element is necessary, or this vector would be interpreted as $x^3 - 3x^2 + 2x + 5$. The following functions work with polynomials in this format:

- r=roots(p) returns a column vector r consisting of the roots of the polynomial represented by p.
- p=poly(r) is, in a sense, an inverse of roots. This function produces a row vector p that represents the **monic** polynomial (that is, with leading coefficient 1) whose roots are the entries of the vector r.
- y=polyval(p,x) evaluates the polynomial represented by p at all of the entries of x (which can be a scalar, vector or matrix) and returns the resulting values in y.
- q=polyder(p) computes the coefficients of the polynomial q that is the derivative of the polynomial p.

- q=polyint(p) computes the coefficients of the polynomial q that is the antiderivative, or indefinite integral, of the polynomial p. A constant of integration of zero is assumed.
- r=conv(p,q) computes the coefficients of the polynomial r that is the product of the polynomials p and q.

It is recommended that you experiment with these functions in order to get used to working with them.

> **Exploration 1.2.25** Write a function graphpoly(p,a,b) that accepts as inputs a row vector p and numbers a and b, and plots the graph of the polynomial represented by p on the interval $[a, b]$. Then, on the same graph, it must plot any *real* roots of the polynomial that are in $[a, b]$ on the x-axis using red circles. Use the help page for the plot function to determine how to produce red circles. *Hint:* Use the imag function to determine whether a root is real or complex.

1.2.19 Number Formatting

By default, numbers are displayed in "short" format, which uses 5 significant digits. The format command, with one of MATLAB's predefined formats, can be used to change how numbers are displayed. For example, type format long at the prompt, and afterwards, numbers are displayed using 15 digits. Here is how 1/3 is displayed using various formats:

short	0.33333
short e	3.33333e-01
long	0.333333333333333
long e	3.33333333333333e-01
bank	0.33
hex	3fd5555555555555
rat	1/3

> **Exploration 1.2.26** The MATLAB function hilb, given an argument n, generates a $n \times n$ **Hilbert matrix**. Use this function to obtain and display a 5×5 Hilbert matrix using the various formats listed above. Use format rat to obtain a formula for each entry h_{ij} of a Hilbert matrix H.

1.2.20 Anonymous Functions

Often, it is desirable to use functions in MATLAB that compute relatively simple expressions, but it is tedious to make a single small function M-file for each such function. Instead, very simple functions can be defined as **anonymous functions**, using syntax such as the following:

```
>> f=@(x)(exp(sin(2*x)));
>> f(pi/4)
ans =
          2.7183
```

The variable f is called a **function handle**. If an anonymous function takes more than one argument, it is important to specify which argument is first, which is second, and so on. For example, to construct an anonymous function for $f(x, y) = \sqrt{x^2 + y^2}$, it is best to proceed as follows:

```
>> f=@(x,y)(sqrt(x^2+y^2));
>> f(2,1)
ans =
          2.2361
```

Anonymous functions are particularly helpful when it is necessary to pass a function f as a *parameter* to another function g, but g assumes that f has fewer input parameters than it actually accepts. An anonymous function can be used to fill in values for the extra parameters of f before it is passed to g.

This will be particularly helpful when using MATLAB's functions for solving ordinary differential equations, as they expect the time derivative f in the ODE $dy/dt = f(t, y)$ to be a function of only two arguments. If the function f that computes the time derivative has additional input arguments, one can use, for example, the anonymous function @(t,y)f(t,y,a,b,c) where a, b and c are the additional input arguments, whose values must be known.

Exploration 1.2.27 Modify your function graphpoly from Exploration 1.2.25 to create a function graphfunc(f,a,b) that accepts as input a function handle f and plots its graph on the interval [a, b]. Unlike graphpoly, it does not need to plot the roots of f.

It is worth noting that the function graphfunc from the preceding exploration duplicates one particular usage of the existing MATLAB function fplot. Consult the help page for fplot for more information about what this function does.

1.2.21 Other Helpful Functions

Here, we provide a list of MATLAB functions that tend to be very useful, as they perform fundamental tasks:

- [m,n]=size(A) returns the dimensions of the matrix A and stores them in the variables m and n. Specifically, m is the number of rows in A, and n is the number of columns.
- n=length(v) is a simplification of size for vectors. If v is a (row or column) vector, then n is the number of elements in v.

- n=numel(A) returns the total number of elements in A (that is, number of rows times the number of columns, for a matrix) and stores it in n.
- B=reshape(A,m,n) rearranges the elements of A and stores them in a m × n matrix B. It is required that the total number of elements in A is mn. This is useful, for example, to reshape a matrix into a vector, or vice versa. It is important to note that elements of A are placed into B *column-wise*. That is, if reshape is being used to reshape a matrix A into a vector B, the first column of A is stored in B first, and then the second column, and so on.
- y=sum(x) adds all of the elements of the vector x and stores the sum in y.
- y=prod(x) is similar to sum, in that the product of the elements in x is computed and stored in y.
- B=repmat(A,m,n) creates a new matrix B by creating a m × n "tiling" of A. As a result, if A is $p \times q$, then B is $mp \times nq$.

Try these functions on random vectors or matrices to check your understanding of their behavior.

Exploration 1.2.28 The behavior of the sum function was described for the case where its input argument x is a vector, but what if x is a matrix? How can you get sum to produce the "opposite" of this default behavior? Consult the help page for sum to answer this question.

Exploration 1.2.29 Write a MATLAB script dayofweek that extends your work from dayinyear from Exploration 1.2.14 to print the day of the week corresponding to a given date. Some of the functions listed above can be helpful. Use the fact that December 31, 1999 was a Friday. *Hint:* Type help elfun to learn about the various rounding and remainder functions, such as mod, round, floor, ceil, etc. Make sure that your script works even if the given date is before December 31, 1999.

1.2.22 Saving and Loading Data

It is natural to have to end a MATLAB session, and then start a new session in which the data from the previous session needs to be re-used. Also, one might wish to send their data to another user. Fortunately, MATLAB supports saving variables to a file and loading them into a workspace.

The save command saves all variables in the current workspace (whether it is the top-level workspace, or the scope of a function) into a file. By default, this file is given a .mat extension. For example, save mydata saves all variables in a file called mydata.mat in the current working directory. Similarly, the load command loads data from a given file.

1.2.23   Measuring Efficiency

With efficiency being one of the essential criteria for any numerical method, it is desirable to be able to time operations in MATLAB. For this purpose, one can use the commands `tic` and `toc`. Informally, the `tic` command "starts the clock", and the `toc` command "stops" it. Use of `toc` by itself causes the time, in seconds, since the last `tic` command to be displayed; alternatively, the time can be saved into a variable. The following commands illustrate the use of `tic` and `toc`.

```
>> A=rand(100);
>> B=rand(100);
>> tic; C=A*B; toc
Elapsed time is 0.030602 seconds.
>> tic; C=A*B; toc
Elapsed time is 0.002927 seconds.
>> tic; C=A*B; multtime=toc
multtime =
    0.0016
```

It is worth noting that the time required to multiply the same two matrices in the above example decreases. This is typical behavior, as MATLAB optimizes code on the fly, which has the effect of subsequent execution of the same code tending to be more efficient. In several exercises throughout this book, examination of some algorithm's efficiency is required, for which the `tic` and `toc` commands should be used.

1.3   Additional Resources

The definition of numerical analysis, and how it has evolved over time, is discussed at length in the appendix of [Trefethen and Bau (1997)]. While on the subject of the evolution of the field, it is important to note that although numerical analysis, as we know it, is concerned with the development of effective numerical methods to be executed by computers, the field itself predates the invention of digital computers by centuries. This is because mathematicians have always needed techniques for obtaining approximate solutions, whether on paper or by computer. The early history of numerical analysis is covered in [Goldstine (1977)], while more recent history is presented in [Nash (1990)].

The web site for MathWorks,

http://www.mathworks.com

has extensive resources to aid in learning MATLAB, including tutorials and detailed documentation. Octave can be obtained from the GNU web site at

https://www.gnu.org/software/octave/

This site also has extensive documentation, as does the Octave Wiki at

`http://wiki.octave.org/GNU_Octave_Wiki`

While this book works with the language of MATLAB, other languages that are widely used within the scientific computing community are Python, for which resources can be found at

`http://www.python.org`

and Julia, which can be obtained from

`http://julialang.org`

Chapter 2

Understanding Error

As numerical analysis is all about methods that produce approximate solutions of mathematical problems, one could almost claim that numerical analysis is, essentially, the study of errors. In fact, several books, such as [Atkinson (1989); Conte and de Boor (1972); Dahlquist and Björck (1974)], among others, have essentially stated as much in their definitions of numerical analysis. While this does not capture the entire essence of numerical analysis, there is no denying that a thorough understanding of error is an essential ingredient for all aspects–design, analysis, and implementation–of the development of numerical methods.

As such, the goal of this chapter is to provide a foundation for this understanding of error. Section 2.1 introduces concepts and terminology that will be used throughout the book for the purpose of analyzing the error inherent in numerical methods. We will learn that this analysis yields insight into not only the accuracy of these methods, but also their efficiency and robustness. It is recommended that the reader review Taylor's Theorem in Section A.6 before proceeding, as Taylor expansion will play a significant role. While this section considers various types of error, the focus of Section 2.2 will be on one type of error in particular: roundoff error due to the representation of numbers used by computers.

2.1 Error Analysis

Mathematical problems arising from scientific applications present a wide variety of difficulties that prevent us from solving them exactly. This has led to a similarly wide variety of techniques for computing approximations to quantities occurring in such problems in order to obtain approximate solutions. In this chapter, we will describe the types of approximations that can be made, and learn some basic techniques for analyzing the accuracy of these approximations.

2.1.1 Sources of Error

Suppose that we are attempting to solve a particular instance of a problem arising from a mathematical model of a scientific application. We say that such a problem is **well-posed** if it meets the following criteria, due to Hadamard:

- A solution of the problem exists.
- The solution is unique.
- A small perturbation in the problem data results in a commensurately small perturbation in the solution; that is, the solution *depends continuously on the data.*

By the first condition, the process of solving a well-posed problem can be seen to be equivalent to the evaluation of some function f at some known value x, where x represents the problem data. Since, in many cases, knowledge of the function f is limited, the task of computing $f(x)$ can be viewed, at least conceptually, as the execution of some (possibly infinite) sequence of steps that solves the underlying problem, given the data x. The goal in numerical analysis is to develop a *finite* sequence of steps, i.e., an **algorithm**, for computing an approximation of the value $f(x)$.

There are two general types of error that occur in the process of computing this approximation of $f(x)$:

1. **data error** is the error in the data x. In reality, numerical analysis involves solving a problem with *approximate* data $\hat{x}$. The exact data is often unavailable because it must be obtained by measurements or other computations that fail to be exact, due to limited precision. In addition, data may be altered in order to simplify the solution process. For example, a coefficient of a differential equation may be assumed constant, or a term of an equation may be neglected altogether.

2. **computational error** refers to the error that occurs when attempting to compute $f(\hat{x})$. Effectively, we must approximate $f(\hat{x})$ by the quantity $\hat{f}(\hat{x})$, where $\hat{f}$ is a function that approximates f. This approximation may be the result of:

   - **discretization**, which results from the replacement of continuous functions or processes by discrete analogues, such as the replacement of a derivative by a difference quotient, or an integral by a Riemann sum. This kind of error will be discussed in 2.1.5, but will be covered in greater depth in several later chapters.
   - **truncation**, which occurs when it is not possible to evaluate f exactly using a finite sequence of steps, and therefore a finite sequence that evaluates f approximately must be used instead. This particular source of computational error will be discussed in Section 2.1.5.
   - **roundoff error**, a consequence of only having limited space in which to store real numbers (that is, each number has finite **precision**). This source of error will discussed in Section 2.2.

Exploration 2.1.1 Consider the process of computing $\cos(\pi/4)$ using a calculator or computer. Indicate sources of data error and computational error, including both truncation and roundoff error. Keep in mind that a calculator or computer can only perform addition, subtraction, multiplication and division. Which types of functions can be evaluated exactly using only these operations? How can a function such as cosine be approximated using such a function?

2.1.2 Error Measurement

Now that we have been introduced to some specific types of errors that can occur during computation, we introduce useful terminology for discussing such errors. Suppose that a real number $\hat{y}$ is an approximation of some real number y. For instance, $\hat{y}$ may be the partial sum of an infinite series that converges to y, or $\hat{y}$ may be the result of a sequence of arithmetic operations performed using finite-precision arithmetic, where y is the result of the same operations performed using exact arithmetic.

Exploration 2.1.2 Consider the following approximations: a) $y = 0.5, \hat{y} = 0.75$; b) $y = 1327.5, \hat{y} = 1328$. Which approximation do you believe is the more accurate one? Explain your reasoning.

The ambiguity in Exploration 2.1.2 leads to the following definitions.

Definition 2.1.1 (Absolute Error, Relative Error) Let $\hat{y}$ be a real number that is an approximation of the real number y. The **absolute error** in $\hat{y}$ is

$$E_{\text{abs}} = \hat{y} - y.$$

The **relative error** in $\hat{y}$ is

$$E_{\text{rel}} = \frac{\hat{y} - y}{y},$$

provided that y is nonzero.

The absolute error is the most natural measure of the accuracy of an approximation, but it can be misleading. Even if the absolute error is rather large in magnitude, the approximation may still be quite accurate if the exact value y is also large in magnitude. For this reason, it is generally preferable to measure accuracy in terms of relative error, unless the exact value is quite small.

The magnitude of the relative error in $\hat{y}$ can be interpreted as a percentage of $|y|$. For example, if the relative error is greater than 1 in magnitude, then $\hat{y}$ can be considered completely erroneous, since the error is larger in magnitude than the exact value, and $\hat{y} = 0$ would actually be more accurate. Another useful interpretation of the relative error concerns **significant digits**, which are all digits excluding leading zeros. Specifically, if the relative error is at most 10^{-p}, then the decimal representation of $\hat{y}$ has at least p correct significant digits. Relative error also provides the following convenient relationship between y and $\hat{y}$.

Exploration 2.1.3 Show that if the relative error in $\hat{y}$ is r, then

$$\hat{y} = y(1 + r).$$

It should be noted that the absolute error and relative error are often defined using absolute value; that is,

$$E_{\text{abs}} = |\hat{y} - y|, \quad E_{\text{rel}} = \left| \frac{\hat{y} - y}{y} \right|.$$

This definition is preferable when one is only interested in the magnitude of the error, which is often the case. If the sign, or direction, of the error is also of interest, then Definition 2.1.1 must be used.

Example 2.1.2 If we add the numbers 4.567×10^{-1} and 8.530×10^{-3}, we obtain the exact result

$$x = 4.567 \times 10^{-1} + 0.008530 \times 10^{-1} = 4.6523 \times 10^{-1}.$$

To maintain four significant digits, this result is rounded to

$$\hat{x} = 4.652 \times 10^{-1}.$$

The absolute error in this computation is

$$E_{\text{abs}} = \hat{x} - x = 0.4652 - 0.46523 = -3 \times 10^{-5},$$

while the relative error is

$$E_{\text{rel}} = \frac{\hat{x} - x}{x} = \frac{0.4652 - 0.46523}{0.46523} \approx -6.5756 \times 10^{-5}.$$

Now, suppose that we multiply 4.567×10^3 and 8.530×10^{-3}. The exact result is

$$x = (4.567 \times 10^3) \times (8.530 \times 10^{-3}) = 3.895651 \times 10^1 = 38.95651,$$

which is rounded to

$$\hat{x} = 3.896 \times 10^1 = 38.96.$$

The absolute error in this computation is

$$E_{\text{abs}} = \hat{x} - x = 38.96 - 38.95651 = 3.49 \times 10^{-3},$$

while the relative error is

$$E_{\text{rel}} = \frac{\hat{x} - x}{x} = \frac{38.96 - 38.95651}{38.95651} \approx 8.9587 \times 10^{-5}.$$

We see that in this case, the relative error is smaller than the absolute error, because the exact result is larger than 1 in magnitude, whereas in the previous operation, the relative error was larger in magnitude, because the exact result is smaller than 1 in magnitude. □

Example 2.1.3 Suppose that the exact value of a computation is supposed to be 10^{16}, and an approximation of 1.001×10^{16} is obtained. Then the absolute error in this approximation is

$$E_{abs} = 1.001 \times 10^{16} - 10^{16} = 10^{13},$$

which suggests the computation is horribly inaccurate because this error is very large. However, the relative error is

$$E_{rel} = \frac{1.001 \times 10^{16} - 10^{16}}{10^{16}} = 10^{-3},$$

which suggests that the computation is fairly accurate, because by this measure, the error is one-tenth of one percent. This example, although an extreme case, illustrates why the absolute error can be a misleading measure of error. $\square$

Exploration 2.1.4 What are the absolute and relative errors in the approximations 3.14 and 22/7 of π? Round your answers to three significant digits.

2.1.3 Forward and Backward Error

Suppose that we compute an approximation $\hat{y} = \hat{f}(x)$ of the value $y = f(x)$ for a given function f and given problem data x. Before we can analyze the accuracy of this approximation, we must have a precisely defined notion of error in such an approximation. We now provide this precise definition.

Definition 2.1.4 (Forward Error) Let x be a real number and let $f : \mathbb{R} \to \mathbb{R}$ be a function. If $\hat{y}$ is a real number that is an approximation of $y = f(x)$, then the **forward error** in $\hat{y}$ is the difference $\Delta y = \hat{y} - y$. If $y \neq 0$, then the **relative forward error** in $\hat{y}$ is defined by

$$\frac{\Delta y}{y} = \frac{\hat{y} - y}{y}.$$

Clearly, our primary goal in error analysis is to obtain an estimate of the forward error Δy. Unfortunately, it can be difficult to obtain this estimate directly.

An alternative approach is to instead view the computed value $\hat{y}$ as the *exact* solution of a problem with modified data; i.e., $\hat{y} = f(\hat{x})$ where $\hat{x}$ is a perturbation of x.

Definition 2.1.5 (Backward Error) Let x be a real number and let $f : \mathbb{R} \to \mathbb{R}$ be a function. Suppose that the real number $\hat{y}$ is an approximation to $y = f(x)$, and that $\hat{y}$ is in the range of f; that is, $\hat{y} = f(\hat{x})$ for some real number $\hat{x}$. Then, the quantity $\Delta x = \hat{x} - x$ is the **backward error** in $\hat{y}$. If $x \neq 0$, then the **relative backward error** in $\hat{y}$ is defined by

$$\frac{\Delta x}{x} = \frac{\hat{x} - x}{x}.$$

The process of estimating Δx is known as **backward error analysis**. As we will see, this estimate of the backward error, in conjunction with knowledge of f, can be used to estimate the forward error.

Example 2.1.6 Let $f(x) = \sin x$ and $x = \pi/4$. Then, $y = \sin(\pi/4)$, to 4 digits, is $\hat{y} = 0.7071$. This approximation has a forward error of $\Delta y = 0.7071 - \sin(\pi/4) \approx -6.7812 \times 10^{-6}$. To obtain the backward error, we note that $\sin(0.78538857) = 0.7071$, which means $\hat{x} = 0.78538857$. We conclude that the backward error is $\Delta x = \hat{x} - x = 0.78538857 - \pi/4 \approx -9.59 \times 10^{-6}$. □

Example 2.1.7 Let $f(x) = e^x$ and $x = 5$. Then, $y = e^5$, to 4 decimal places, is $\hat{y} = 148.4132$. This approximation has a forward error of $\Delta y = 148.4132 - e^5 \approx 4.0897 \times 10^{-5}$. To obtain the backward error, we note that $e^{5.0000002756} = 148.4132$, which means $\hat{x} = 5.0000002756$. We conclude that the backward error is $\Delta x = \hat{x} - x = 5.0000002756 - 5 \approx 2.7556 \times 10^{-7}$. □

Exploration 2.1.5 Consider the second-degree Taylor polynomial (see Section A.6), with center $x_0 = 0$, for $f(x) = \cos x$,

$$p_2(x) = 1 - \frac{x^2}{2}.$$

Let $x = \pi/4$, and suppose we use $\hat{y} = p_2(x)$ as an approximation of $y = \cos(x)$. What are the forward and backward errors of this approximation? What are relative forward and relative backward errors? Repeat this for the 4th-degree Taylor polynomial for $f(x)$.

As will be discussed in Section 2.2, finite-precision arithmetic does not follow the laws of real arithmetic. This tends to make forward error analysis difficult. Other sources of error in the approximation $\hat{y} = \hat{f}(x)$ of $y = f(x)$ are similarly difficult to quantify. In backward error analysis, however, exact arithmetic is employed, as well as $f(x)$ rather than $\hat{f}(x)$, since it is assumed that the computed result is the exact solution of a "nearby" problem. This is one reason why backward error analysis is sometimes preferred for estimating the forward error Δy.

Exploration 2.1.6 Let $x_0 = 1$, and $f(x) = e^x$. If the magnitude of the forward error in computing $f(x_0)$, given by $|\hat{f}(x_0) - f(x_0)|$, is 0.01, then estimate the magnitude of the backward error.

Exploration 2.1.7 For a general function $f(x)$, explain when the magnitude of the forward error is greater than, or less than, that of the backward error. Assume f is differentiable near x and use calculus to explain your reasoning.

Exploration 2.1.8 Given the conditions for a problem to be well-posed, which is more desirable: for the forward error to be larger, smaller, or of a similar magnitude to the backward error? Explain your reasoning.

2.1.4 Conditioning and Stability

The goal of error analysis is to obtain an estimate of the forward relative error $(f(\hat{x}) - f(x))/f(x)$, but it is often easier to instead estimate the relative backward error $(\hat{x} - x)/x$. Therefore, it is necessary to be able to estimate the forward error in terms of the backward error. The following definition addresses this need.

Definition 2.1.8 (Condition Number) Let x be a real number and let $f : \mathbb{R} \to \mathbb{R}$ be a function. The **absolute condition number**, denoted by κ_{abs}, is the ratio of the magnitude of the forward error to the magnitude of the backward error,

$$\kappa_{\text{abs}} = \frac{|f(\hat{x}) - f(x)|}{|\hat{x} - x|} = \frac{|\Delta y|}{|\Delta x|}.$$

If $f(x) \neq 0$, then the **relative condition number** of the problem of computing $y = f(x)$, denoted by κ_{rel}, is the ratio of the magnitude of the relative forward error to the magnitude of the relative backward error,

$$\kappa_{\text{rel}} = \frac{|(f(\hat{x}) - f(x))/f(x)|}{|(\hat{x} - x)/x|} = \frac{|\Delta y/y|}{|\Delta x/x|}.$$

Either condition number is a measure of the change in the solution due to a change in the data. Since the relative condition number tends to be a more reliable measure of this change, it is sometimes referred to as simply the **condition number**.

If the condition number is large, e.g. much greater than 1, then a small change in the input data x can cause a disproportionately large change in the solution, or output, $f(x)$, and the problem is said to be **ill-conditioned**. If the condition number is small, then the problem is said to be **well-conditioned**.

Since the condition number, as defined above, depends on knowledge of the exact solution $f(x)$, it is necessary to estimate the condition number in order to estimate the relative forward error. To that end, we assume, for simplicity, that $f : \mathbb{R} \to \mathbb{R}$ is differentiable and obtain

$$\kappa_{\text{rel}} = \frac{|x\Delta y|}{|y\Delta x|} = \frac{|x(f(x + \Delta x) - f(x))|}{|f(x)\Delta x|} \approx \frac{|xf'(x)\Delta x|}{|f(x)\Delta x|} \approx \left| \frac{xf'(x)}{f(x)} \right|.$$

Therefore, if we can estimate the backward error Δx, and if we can bound f and f' near x, we can then bound the (relative) condition number and obtain an estimate of the relative forward error. Of course, the relative condition number is undefined if the exact value $f(x)$ is zero. In this case, we can instead use the absolute condition number. Using the same approach as before, the absolute condition number can be estimated using the derivative of f. Specifically, we have $\kappa_{\text{abs}} \approx |f'(x)|$.

> **Exploration 2.1.9** Let $f(x) = e^x$, $g(x) = e^{-x}$, and $x_0 = 2$. Suppose that the relative backward error in x_0 satisfies $|\Delta x_0 / x_0| = |\hat{x}_0 - x_0| / |x_0| \le 10^{-2}$. Use the condition number to estimate the relative forward error in $f(x_0)$ and $g(x_0)$. Use MATLAB or a calculator to experimentally confirm that this estimate is valid, by evaluating $f(x)$ and $g(x)$ at selected points and comparing values.

The condition number of a function f depends on, among other things, the absolute forward error $f(\hat{x}) - f(x)$. However, an algorithm for evaluating $f(x)$ actually evaluates a function $\hat{f}$ that approximates f, producing an approximation $\hat{y} = \hat{f}(x)$ to the exact solution $y = f(x)$. In our definition of backward error, we have assumed that $\hat{f}(x) = f(\hat{x})$ for some $\hat{x}$ that is close to x; i.e., our approximate solution of the original problem is the exact solution of a nearby problem. This assumption has allowed us to define the condition number of f independently of any approximation $\hat{f}$. This independence is necessary, because the sensitivity of a problem depends solely on the problem itself, and not any algorithm that may be used to approximately solve it.

> **Exploration 2.1.10** Use the relative forward and backward errors computed in Exploration 2.1.5 to estimate the relative condition number of $f(x) = \cos x$ at $x_0 = \pi/4$. Compare these estimates to the condition number of $f(x)$ at x_0 given by (2.1.4). At what value of x_0 would the condition number of $f(x)$ be particularly large?

> **Exploration 2.1.11** Let $f(x)$ be a function that is one-to-one. Then, solving the equation $f(x) = c$ for some c in the range of f is equivalent to computing $x = f^{-1}(c)$. Using (2.1.4), what is an estimate of the condition number of the problem of solving $f(x) = c$?

Is it always reasonable to assume that any approximate solution is the exact solution of a nearby problem? Unfortunately, it is not. It is possible that even if the problem is relatively insensitive to perturbations in input data (that is, the problem is well-conditioned), an algorithm used to solve it may still be unreasonably sensitive to such perturbations. This leads to the concept of a **stable algorithm**: an algorithm applied to a given problem with given data x is said to be **stable** if it computes an approximate solution that is the exact solution of the same problem with data $\hat{x}$, where $\hat{x}$ is a small perturbation of x.

It can be shown that if a problem is well-conditioned, and if we have a stable algorithm for solving it, then the computed solution can be considered accurate, in the sense that the relative error in the computed solution is small. On the other hand, a stable algorithm applied to an ill-conditioned problem cannot be expected to produce an accurate solution.

Example 2.1.9 This example will illustrate the last point made above. To solve a system of linear equations $A\mathbf{x} = \mathbf{b}$ in MATLAB, we can use the \ operator:

$$\mathbf{x} = A\backslash\mathbf{b}$$

Enter the following matrix and column vectors in MATLAB, as shown. Recall from Section 1.2.3 that a semicolon (;) separates rows.

```
>> A=[ 0.6169 0.4798; 0.4925 0.3830 ];
>> b1=[ 0.7815; 0.6239 ];
>> b2=[ 0.7753; 0.6317 ];
```

Then, solve the systems $A*x1 = b1$ and $A*x2 = b2$. Note that b1 and b2 are not very different, but what about the solutions x1 and x2? The algorithm implemented by the \ operator is stable, but what can be said about the *conditioning* of the problem of solving $A\mathbf{x} = \mathbf{b}$ for this matrix A? The conditioning of systems of linear equations will be studied in depth in Section 3.4.1. □

Example 2.1.10 Consider the matrix

$$A = \begin{bmatrix} 10^{-10} & 0.9293 & 0.2511 \\ 0.8143 & 0.3500 & 0.6160 \\ 0.2435 & 0.1966 & 0.4733 \end{bmatrix}.$$

Using the MATLAB function cond, it can be verified that the problem of solving the system of linear equations $A\mathbf{x} = \mathbf{b}$ is very well-conditioned. However, if this system is solved using Gaussian elimination without **pivoting** (see Section 3.2.5), which is not stable for general matrices, the computed solution may still be inaccurate. Gaussian elimination *with* pivoting, on the other hand, is generally a stable algorithm. □

2.1.5 Convergence

Many algorithms in numerical analysis are *iterative methods* that produce a sequence $\{\alpha_n\}$ of approximate solutions which, ideally, converges to a limit α that is the exact solution as n approaches ∞. Because we can only perform a finite number of iterations, we cannot obtain the exact solution, and we have introduced computational error.

If our iterative method is properly designed, then this computational error will approach zero as n approaches ∞. However, it is important that we obtain a sufficiently accurate approximate solution using as few computations as possible. Therefore, it is not practical to simply perform enough iterations so that the computational error is determined to be sufficiently small, because it is possible that another method may yield comparable accuracy with less computational effort.

The total computational effort of an iterative method depends on both the effort per iteration and the number of iterations performed. Therefore, in order to determine the amount of computation that is needed to attain a given accuracy, we must be able to measure the error in α_n as a function of n. The more rapidly this function approaches zero as n approaches ∞, the more rapidly the sequence of approximations $\{\alpha_n\}$ converges to the exact solution α, and as a result, fewer iterations are needed to achieve a desired accuracy. We now introduce some terminology that will aid in the discussion of the convergence behavior of iterative methods.

Definition 2.1.11 (Big-O Notation, $n \to \infty$) Let f and g be two functions defined on a domain $D \subseteq \mathbb{R}$ that is not bounded above. We write that $f(n) = O(g(n))$ if there exists a positive constant c such that

$$|f(n)| \le c|g(n)|, \quad n \ge n_0,$$

for some $n_0 \in D$.

As sequences are functions defined on $\mathbb{N}$, the domain of the natural numbers, we can apply big-O notation to sequences. Therefore, this notation is useful to describe the rate at which a sequence of computations converges to a limit.

Definition 2.1.12 (Rate of Convergence, $n \to \infty$) Let $\{\alpha_n\}_{n=1}^{\infty}$ and $\{\beta_n\}_{n=1}^{\infty}$ be sequences that satisfy

$$\lim_{n \to \infty} \alpha_n = \alpha, \quad \lim_{n \to \infty} \beta_n = 0,$$

where α is a real number. We say that $\{\alpha_n\}$ converges to α with **rate of convergence** $O(\beta_n)$ if $\alpha_n - \alpha = O(\beta_n)$.

We say that an iterative method converges rapidly, in some sense, if it produces a sequence of approximate solutions $\{\alpha_n\}$ whose rate of convergence is $O(\beta_n)$, where the terms of the sequence $\{\beta_n\}$ approach zero rapidly as n approaches ∞. Intuitively, if two iterative methods for solving the same problem perform a comparable amount of computation during each iteration, but one method exhibits a faster rate of convergence, then that method should be used because it will require less overall computational effort to obtain an approximate solution that is sufficiently accurate, due to fewer iterations being performed.

Example 2.1.13 Consider the sequence $\{\alpha_n\}_{n=1}^{\infty}$ defined by

$$\alpha_n = \frac{n+1}{n+2}, \quad n = 1, 2, \ldots$$

Then, we have

$$\lim_{n \to \infty} \alpha_n = \lim_{n \to \infty} \frac{n+1}{n+2} \frac{1/n}{1/n} = \lim_{n \to \infty} \frac{1+1/n}{1+2/n} = \frac{1 + \lim_{n \to \infty} 1/n}{1 + \lim_{n \to \infty} 2/n} = 1.$$

That is, the sequence $\{\alpha_n\}$ converges to $\alpha = 1$. To determine the rate of convergence, we note that

$$\alpha_n - \alpha = \frac{n+1}{n+2} - 1 = \frac{n+1}{n+2} - \frac{n+2}{n+2} = \frac{-1}{n+2},$$

and since

$$\left| \frac{-1}{n+2} \right| \le \left| \frac{1}{n} \right|.$$

for any positive integer n, it follows that

$$\alpha_n = \alpha + O\left(\frac{1}{n}\right).$$

On the other hand, consider the sequence $\{\alpha_n\}_{n=1}^{\infty}$ defined by

$$\alpha_n = \frac{2n^2 + 4n}{n^2 + 2n + 1}, \quad n = 1, 2, \ldots$$

Then, we have

$$\lim_{n \to \infty} \alpha_n = \lim_{n \to \infty} \frac{2n^2 + 4n}{n^2 + 2n + 1} \frac{1/n^2}{1/n^2} = \frac{2 + \lim_{n \to \infty} 4/n}{1 + \lim_{n \to \infty}(2/n + 1/n^2)} = 2.$$

That is, the sequence $\{\alpha_n\}$ converges to $\alpha = 2$. To determine the rate of convergence, we note that

$$\alpha_n - \alpha = \frac{2n^2 + 4n}{n^2 + 2n + 1} - 2 = \frac{2n^2 + 4n}{n^2 + 2n + 1} - \frac{2n^2 + 4n + 2}{n^2 + 2n + 1} = \frac{-2}{n^2 + 2n + 1},$$

and since

$$\left|\frac{-2}{n^2 + 2n + 1}\right| = \left|\frac{2}{(n+1)^2}\right| \leq \left|\frac{2}{n^2}\right|$$

for any positive integer n, it follows that

$$\alpha_n = \alpha + O\left(\frac{1}{n^2}\right).$$

□

It is worth noting that Big-O notation is also useful for describing the **runtime complexity** of an algorithm in terms of the size of its input. In this context, we are not examining a convergent sequence, but rather describing the efficiency of an algorithm in a concise way.

Example 2.1.14 The number of arithmetic operations required to perform multiplication of two $n \times n$ real matrices, in the most straightforward manner presented in Section B.6, is $f(n) = 2n^3 - n^2$. For $n \geq 1$, we have

$$|2n^3 - n^2| \leq |2n^3| + |n^2| \leq |2n^3| + |n^3| \leq 3n^3.$$

We conclude that $f(n)$ is $O(n^3)$. □

We can also use big-O notation to describe the rate of convergence of a function as its argument approaches zero, rather than infinity.

Definition 2.1.15 (Big-O Notation, $h \to 0$) Let f and g be two functions defined on a domain $D \subseteq \mathbb{R}$ that contains the origin. We write that $f(h) = O(g(h))$ if there exists a positive constant c such that

$$|f(h)| \leq c|g(h)|, \quad h \leq h_0,$$

for some $h_0 \in D$.

Definition 2.1.16 (Rate of Convergence, $h \to 0$) Let $f(h)$ and $g(h)$ be functions that satisfy

$$\lim_{h \to 0} f(h) = f_0, \quad \lim_{h \to 0} g(h) = 0,$$

where f_0 is a real number. We say that $f(h)$ converges to f_0 with **rate of convergence** $O(g(h))$ if $f(h) - f_0 = O(g(h))$.

Example 2.1.17 Consider the function $f(h) = 1 + 2h$. Since this function is continuous for all h, we have

$$\lim_{h \to 0} f(h) = f(0) = f_0 = 1.$$

It follows that

$$f(h) - f_0 = (1 + 2h) - 1 = 2h = O(h),$$

so we can conclude that as $h \to 0$, $1 + 2h$ converges to 1 of order $O(h)$. $\square$

Example 2.1.18 Consider the function $f(h) = 1 + 4h + 2h^2$. Since this function is continuous for all h, we have

$$\lim_{h \to 0} f(h) = f(0) = f_0 = 1.$$

It follows that

$$f(h) - f_0 = (1 + 4h + 2h^2) - 1 = 4h + 2h^2.$$

To determine the rate of convergence as $h \to 0$, we consider h in the interval $[-1, 1]$. In this interval, $|h^2| \le |h|$. It follows that

$$|4h + 2h^2| \le |4h| + |2h^2|$$
$$\le |4h| + |2h|$$
$$\le 6|h|.$$

Since there exists a constant C (namely, 6) such that $|4h + 2h^2| \le C|h|$ for h satisfying $|h| \le h_0$ for some h_0 (namely, 1), we can conclude that as $h \to 0$, $1 + 4h + 2h^2$ converges to 1 of order $O(h)$. $\square$

In general, when $f(h)$ denotes an approximation that depends on h, and

$$f_0 = \lim_{h \to 0} f(h)$$

denotes the exact value, $f(h) - f_0$ represents the absolute error in the approximation $f(h)$. When this error is a polynomial in h, as in this example and the previous example, the rate of convergence is $O(h^k)$ where k is the smallest exponent of h in the error. This is because as $h \to 0$, the smallest power of h approaches zero more slowly than higher powers, thereby making the dominant contribution to the error.

By contrast, when determining the rate of convergence of a sequence $\{\alpha_n\}$ as $n \to \infty$, the *highest* power of n determines the rate of convergence. As powers of n are negative if convergence occurs at all as $n \to \infty$, and powers of h are positive if convergence occurs at all as $h \to 0$, it can be said that for either type of convergence, it is the exponent that is closest to zero that determines the rate of convergence.

Example 2.1.19 Consider the function $f(h) = \cos h$. Since this function is continuous for all h, we have

$$\lim_{h \to 0} f(h) = f(0) = 1.$$

Using **Taylor's Theorem** (Theorem A.6.1), with center $h_0 = 0$, we obtain

$$f(h) = f(0) + f'(0)h + \frac{f''(\xi(h))}{2}h^2,$$

where $\xi(h)$ is between 0 and h. Substituting $f(h) = \cos h$ into the above, we obtain

$$\cos h = 1 - \frac{\cos \xi(h)}{2}h^2.$$

Because $|\cos x| \leq 1$ for all x, we have

$$|\cos h - 1| = \left| -\frac{\cos \xi(h)}{2}h^2 \right| \leq \frac{1}{2}h^2,$$

so we can conclude that as $h \to 0$, $\cos h$ converges to 1 with order $O(h^2)$. □

Exploration 2.1.12 Determine the rate of convergence of

$$\lim_{h \to 0} e^h - h - \frac{1}{2}h^2 = 1.$$

Example 2.1.20 We will study the convergence of the **centered difference** approximation (see Section 9.1) of the second derivative,

$$f''(x_0) \approx \frac{f(x_0 + h) - 2f(x_0) + f(x_0 - h)}{h^2}.$$

We assume that $f \in C^4[x_0 - h, x_0 + h]$. That is, f belongs to the space of functions that have at least four continuous derivatives on the interval $[x_0 - h, x_0 + h]$ (see Section B.15 for an overview of function spaces).

Using Taylor's Theorem with $n = 3$ and center x_0, we obtain the expansions

$$f(x_0 + h) = f(x_0) + f'(x_0)h + \frac{1}{2}f''(x_0)h^2 + \frac{1}{6}f'''(x_0)h^3 + \frac{1}{24}f^{(4)}(\xi_1)h^4$$

$$f(x_0 - h) = f(x_0) - f'(x_0)h + \frac{1}{2}f''(x_0)h^2 - \frac{1}{6}f'''(x_0)h^3 + \frac{1}{24}f^{(4)}(\xi_2)h^4$$

that each consist of a third-degree Taylor polynomial in h, and a remainder term that is $O(h^4)$. The remainders include the fourth derivative of $f(x)$ evaluated at unknown points $\xi_1 \in (x_0, x_0 + h)$ and $\xi_2 \in (x_0 - h, x_0)$.

Adding these expansions and rearranging yields

$$f''(x_0) = \frac{f(x_0 + h) - 2f(x_0) + f(x_0 - h)}{h^2} - \frac{1}{12}h^2\frac{1}{2}[f^{(4)}(\xi_1) + f^{(4)}(\xi_2)].$$

Because $f^{(4)}$ is continuous on $[x_0 - h, x_0 + h]$, we can apply the **Intermediate Value Theorem** (Theorem A.1.8) to conclude that in the interval (ξ_1, ξ_2), there exists a point ξ such that $f^{(4)}(\xi)$ is equal to the average of $f^{(4)}(\xi_1)$ and $f^{(4)}(\xi_2)$, which must lie between these two values.

We therefore have

$$f''(x_0) = \frac{f(x_0 + h) - 2f(x_0) + f(x_0 - h)}{h^2} - \frac{1}{12}h^2 f^{(4)}(\xi).$$

We conclude that the error is $O(h^2)$ because

$$\left| -\frac{1}{12}h^2 f^{(4)}(\xi) \right| \le \frac{1}{12} M h^2$$

where $|f^{(4)}(x)| \le M$ on $[x_0 - h, x_0 + h]$. $\square$

Try using the approach from the preceding example with other formulas for approximating derivatives.

Exploration 2.1.13 Use Taylor's Theorem to derive the error terms in the formulas

$$f'(x_0) \approx \frac{f(x_0 + h) - f(x_0)}{h}, \quad \text{error} = -\frac{h}{2}f''(\xi)$$

and

$$f'(x_0) \approx \frac{f(x_0 + h) - f(x_0 - h)}{2h}, \quad \text{error} = -\frac{h^2}{6}f'''(\xi).$$

Exploration 2.1.14 Use both of the formulas from Exploration 2.1.13 to compute approximations of $f'(x_0)$, with $f(x) = \sin x$, $x_0 = 1$, and $h = 10^{-1}, 10^{-2}, 10^{-3}$, and then $h = 10^{-14}$. What should the result be? What happens, and can you explain *why*?

If you can't explain what happens for the smallest value of h, fortunately this will be addressed in Section 2.2.

2.1.6 Concept Check

1. For each of the following questions, answer true or false, with reasons:

    (a) A more accurate algorithm for solving an ill-conditioned problem will make the problem better conditioned.

    (b) The choice of algorithm for solving a given problem does not affect the sensitivity of error.

    (c) A stable algorithm applied to a well-conditioned problem will produce an accurate solution.

2. What does it mean for a problem to be well-posed?
3. What are the two broad categories of error in numerical computation?
4. Explain the difference between truncation error and roundoff error.
5. When should error be measured using absolute error or relative error?
6. What does it mean for a problem to be well-conditioned?

7. Explain the distinction between a well-conditioned problem and a stable algorithm.
8. Explain the distinction between forward error and backward error. How is backward error useful?
9. What does it mean to say that a sequence $\{a_n\}$ is $O(b_n)$?
10. What is the significance of the rate of convergence?
11. Explain the difference between how the rate of convergence is determined for a sequence $\{a_n\}$ as $n \to \infty$ and a function $f(h)$ as $h \to 0$.
12. What is the role that Taylor expansion plays in determining the rate of convergence?

2.2 Computer Arithmetic

In this section, we will focus on one type of error that occurs in all computation, whether performed by hand or on a computer: **roundoff error**. This error is due to the fact that in computation, real numbers can only be represented using a finite number of digits. In general, it is not possible to represent real numbers exactly with this limitation, and therefore they must be approximated by real numbers that *can* be represented using a fixed number of digits, which is called the **precision**.

> **Exploration 2.2.1** Use scientific notation to write down approximations of 100π and $10e$ consisting of five significant digits. Now multiply these approximations, retaining only five significant digits, and express the result in the same scientific notation. How accurate is this approximation of $1000e\pi$? What is the percentage error?

As Exploration 2.2.1 illustrates, arithmetic operations applied to numbers that can be represented exactly using a given precision do not necessarily produce a result that can be represented using the same precision. It follows that if a fixed precision is used, then *every* arithmetic operation introduces error into a computation.

Given that scientific computations can have several sources of error, one would think that it would be foolish to compound the problem by performing arithmetic using fixed precision. However, using a fixed precision is actually far more efficient than other options, such as arbitrary-precision arithmetic or symbolic computation. Furthermore, as long as computations are performed carefully, sufficient accuracy can still be achieved.

2.2.1 Floating-Point Representation

We now describe a general system for representing real numbers on a computer.

> **Definition 2.2.1 (Floating-point Number System)** Given integers $\beta > 1$, $p \geq 1$, L, and $U \geq L$, a **floating-point number system** $\mathbb{F}$ is defined to be the set of all real numbers of the form
>
> $$x = \pm m\beta^E.$$
>
> The number m is the **mantissa** of x, and has the form
>
> $$m = \left(\sum_{j=0}^{p-1} d_j \beta^{-j} \right), \tag{2.1}$$
>
> where each digit d_j, $j = 0, \ldots, p-1$ is an integer satisfying $0 \leq d_i \leq \beta - 1$. The number E is called the **exponent** of x, and it is an integer satisfying $L \leq E \leq U$. The integer p is called the **precision** of $\mathbb{F}$, and β is called the **base** of $\mathbb{F}$.

The term "floating-point" comes from the fact that as a number $x \in \mathbb{F}$ is multiplied by or divided by a power of β, the mantissa does not change, only the exponent. As a result, the decimal point (or whatever it should be called, if $\beta \neq 10$) shifts, or "floats," to account for the changing exponent. Nearly all computers use a **binary** floating-point system, in which $\beta = 2$.

Example 2.2.2 Let $x = -117$. Then, in a floating-point number system with base $\beta = 10$, x is represented as

$$x = -(1.17)10^2,$$

where 1.17 is the mantissa and 2 is the exponent. If the base $\beta = 2$, then we have

$$x = -(1.110101)2^6,$$

where 1.110101 is the mantissa and 6 is the exponent. The mantissa should be interpreted as a string of binary digits, rather than decimal digits; that is,

$$
\begin{aligned}
1.110101 &= 1 \cdot 2^0 + 1 \cdot 2^{-1} + 1 \cdot 2^{-2} + 0 \cdot 2^{-3} + 1 \cdot 2^{-4} + 0 \cdot 2^{-5} + 1 \cdot 2^{-6} \\
&= 1 + \frac{1}{2} + \frac{1}{4} + \frac{1}{16} + \frac{1}{64} \\
&= \frac{117}{64} \\
&= \frac{117}{2^6}.
\end{aligned}
$$

□

We now examine how to represent real numbers in a floating-point number system. For simplicity, we first consider the case of a decimal system, in which $\beta = 10$. Then, the floating-point representation of a number is essentially *scientific notation*, in which the significant digits of a number are written separately from its magnitude.

Example 2.2.3 We write $x = -1234.56789$ as $x = -1.23456789 \times 10^3$, so the exponent is $E = 3$, and the mantissa is $m = 1.23456789$. To represent this number exactly in a floating-point number system with $\beta = 10$, we need at least nine digits in the mantissa; that is, we must have $p \geq 9$. $\square$

For now, we will assume that the mantissa m, as defined in (2.1), has the property that the leading digit, d_0, must be nonzero. This ensures that each number represented by a floating-point number system has a unique representation. This choice will be discussed in greater detail later in this section, when we introduce **normalized** and **denormalized** floating-point numbers.

Exploration 2.2.2 What is the largest possible mantissa in a system with $\beta = 10$ and $p = 5$? What if $\beta = 2$ and $p = 10$? Express this value as both a sequence of binary digits, and as a decimal.

Exploration 2.2.3 Show that if $d_0 \neq 0$, the mantissa m of a floating-point number must satisfy $1 \leq m < \beta$, where β is the base of the floating-point number system in question. Based on Exploration 2.2.2, can you provide a formula for the largest mantissa?

Exploration 2.2.4 Assume that $d_0 \neq 0$. What is the exponent for the following numbers in the given floating-point number systems? *Hint:* Use the result of Exploration 2.2.3.

1. $x = 234597.34$, $\beta = 10$
2. $x = 0.00000178$, $\beta = 10$
3. $x = 998$, $\beta = 2$
4. $x = 0.01$, $\beta = 2$

Exploration 2.2.5 Show that if $d_0 \neq 0$, then the exponent E of a floating-point number x is given by $E = \lfloor \log_\beta |x| \rfloor$, where β is the base of the floating-point number system, and $\lfloor x \rfloor$ is the greatest integer that is less than or equal to x ($\lfloor x \rfloor$ is known as the **greatest integer function** or the **floor function**; it is implemented by the floor function in MATLAB). *Hint:* Work with the representation $x = \pm m\beta^E$.

Example 2.2.4 Consider the base of the natural logarithm, e, rounded to five significant digits: $e \approx 2.7183$. To represent this number in a binary floating-point number system (that is, $\beta = 2$), we first obtain the exponent. Taking $\log_2 2.7183$ and rounding down, we obtain $E = \lfloor 1.4427 \rfloor = 1$. That is, $2.7183 = m \times 2^1$, where $m = 2.7183/2 = 1.35915$ is the mantissa.

To obtain a binary representation of the mantissa, we first take $d_0 = 1$; by requiring $d_0 \neq 0$, in a binary system we must have $d_0 = 1$ because that is the only remaining option. Subtracting off 1, we focus on representing 0.35915 using binary digits. We repeat the following process: multiply by 2, take the digit to the left of the decimal point, and subtract. The result of this process is:

1. $2 \times 0.35915 = 0.7183 \Longrightarrow d_1 = 0$
2. $2 \times 0.7183 = 1.4366 \Longrightarrow d_2 = 1$, then subtract 1
3. $2 \times 0.4366 = 0.8732 \Longrightarrow d_3 = 0$
4. $2 \times 0.8732 = 1.7464 \Longrightarrow d_4 = 1$, then subtract 1
5. $2 \times 0.7464 = 1.4928 \Longrightarrow d_5 = 1$, then subtract 1
6. $2 \times 0.4928 = 0.9856 \Longrightarrow d_6 = 0$
7. $2 \times 0.9856 = 1.9712 \Longrightarrow d_7 = 1$, then subtract 1
8. $2 \times 0.9712 = 1.9424 \Longrightarrow d_8 = 1$, then subtract 1
9. $2 \times 0.9424 = 1.8848 \Longrightarrow d_9 = 1$, then subtract 1
10. $2 \times 0.8848 = 1.7696 \Longrightarrow d_{10} = 1$, then subtract 1
11. $2 \times 0.7696 = 1.5392 \Longrightarrow d_{11} = 1$, then subtract 1
12. $2 \times 0.5392 = 1.0784 \Longrightarrow d_{12} = 1$, then subtract 1
13. $2 \times 0.0784 = 0.1568 \Longrightarrow d_{13} = 0$
14. $2 \times 0.1568 = 0.3136 \Longrightarrow d_{14} = 0$
15. $2 \times 0.3136 = 0.6272 \Longrightarrow d_{15} = 0$
16. $2 \times 0.6272 = 1.2544 \Longrightarrow d_{16} = 1$, then subtract 1
17. $2 \times 0.2544 = 0.5088 \Longrightarrow d_{17} = 0$
18. $2 \times 0.5088 = 1.0176 \Longrightarrow d_{18} = 1$, then subtract 1
19. $2 \times 0.0176 = 0.0352 \Longrightarrow d_{19} = 0$

Limiting our floating-point number system to a precision of $p = 20$ digits, our mantissa is, in binary, 1.01011011111110001010, which, in decimal, is equal to

$$m = 1 + \frac{1}{2^2} + \frac{1}{2^4} + \frac{1}{2^5} + \frac{1}{2^7} + \frac{1}{2^8} + \frac{1}{2^9} + \frac{1}{2^{10}} + \frac{1}{2^{11}} + \frac{1}{2^{12}} + \frac{1}{2^{16}} + \frac{1}{2^{18}}$$
$$= 1.359149932861328.$$

It follows that our floating-point representation of 2.7183 is

$$2 \times 1.359149932861328 = 2.718299865722656.$$

We see that even with 20 binary digits of precision, we cannot represent 2.7183 exactly. Our floating-point number has a relative error of approximately -4.9×10^{-8}, which has a magnitude comparable to that of 2^{-24}. $\square$

Exploration 2.2.6 Using the approach from Example 2.2.4, express $x = \pi$ as accurately as possible in a floating-point number system with base $\beta = 2$ and precision $p = 12$.

Exploration 2.2.7 Let $\beta = 10$. What are the smallest values of p and U, and the largest value of L, such that the numbers 239487.234 and 0.0000000034394 can be represented exactly in a floating-point number system of base β with precision p and exponent range $[L, U]$?

2.2.1.1 *Overflow and Underflow*

A floating-point system $\mathbb{F}$ can only represent a finite subset of the real numbers. As such, it is important to know how large in magnitude a number can be and still be represented, at least approximately, by a number in $\mathbb{F}$. Similarly, it is important to know how small in magnitude a number can be and still be represented by a *nonzero* number in $\mathbb{F}$; if its magnitude is too small, then it is most accurately represented by zero.

Exploration 2.2.8 Consider a floating-point number system with $\beta = 10$, $p = 5$, $L = -20$ and $U = 20$. What are the largest and smallest positive numbers that can be represented in this system? Do *not* assume $d_0 \neq 0$. How would you describe these values for a general floating-point number system with $\beta = 10$?

The answer to Exploration 2.2.8 leads to the following formal definition, that applies to any valid base.

Definition 2.2.5 (Underflow, Overflow) Let $\mathbb{F}$ be a floating-point number system. The smallest positive number in $\mathbb{F}$ is called the **underflow level**, and it has the value

$$\text{UFL} = m_{\min}\beta^L,$$

where L is the smallest valid exponent and $m_{\min}$ is the smallest mantissa. The largest positive number in $\mathbb{F}$ is called the **overflow level**, and it has the value

$$\text{OFL} = \beta^{U+1}(1 - \beta^{-p}).$$

The value of $m_{\min}$ depends on whether floating-point numbers are **normalized** in $\mathbb{F}$; this point will be discussed shortly. The overflow level is the value obtained by setting each digit in the mantissa to $\beta - 1$ and using the largest possible value, U, for the exponent.

Exploration 2.2.9 Determine the value of OFL for a floating-point system with base $\beta = 2$, precision $p = 53$, and largest exponent $U = 1023$.

It is important to note that the real numbers that can be represented in $\mathbb{F}$ are not equally spaced along the real number line. Numbers having the same exponent are equally spaced, and the spacing between numbers in $\mathbb{F}$ decreases as their magnitude decreases.

2.2.1.2 *Normalization*

It is common to **normalize** floating-point numbers by specifying that the leading digit d_0 of the mantissa (2.1) be nonzero. In a binary system, with $\beta = 2$, this implies that the leading digit is equal to 1, and therefore need not be stored. In addition to the benefit of gaining one additional bit of precision, normalization also ensures that each floating-point number has a unique representation.

Example 2.2.6 Without normalization, the number 3.14 could also be represented as 0.314×10^1 or 0.00314×10^3, but with normalization, these two alternative representations would not be allowed, because in both cases the leading digit of the mantissa is zero. □

Exploration 2.2.10 Consider a floating-point number system with $\beta = 10$, $p = 10$, $L = -20$ and $U = 20$. Write down three representations of the closest representable number to 100π.

Exploration 2.2.11 How does the answer to Exploration 2.2.8 change if numbers are normalized?

One drawback of normalization is that fewer numbers near zero can be represented exactly than if normalization is not used. One workaround is a practice called **gradual underflow**, in which the leading digit of the mantissa is allowed to be zero when the exponent is equal to L, thus allowing smaller values of the mantissa. In such a system, the number UFL is equal to β^{L-p+1}, whereas in a normalized system, UFL $= \beta^L$.

Exploration 2.2.12 Determine the value of UFL for a floating-point system with base $\beta = 2$, precision $p = 53$, and smallest exponent $L = -1022$, both with and without normalization.

2.2.1.3 *Rounding*

A number that can be represented exactly in a floating-point system is called a **machine number**. Since only finitely many real numbers are machine numbers, it is necessary to determine how non-machine numbers are to be approximated by machine numbers. The process of choosing a machine number to approximate a non-machine number is called **rounding**, and the error introduced by such an approximation is called **roundoff error**. Given a real number x, the machine number obtained by rounding x is denoted by fl(x).

In most floating-point systems, rounding is achieved by one of two strategies:

- **chopping**, or **rounding to zero**, is the simplest strategy, in which the base-β expansion of a number is truncated after the first p digits. As a result, fl(x) is the unique machine number between 0 and x that is nearest to x.
- **rounding to nearest** sets fl(x) to be the machine number that is closest to x in absolute value; if two numbers satisfy this property, then an appropriate

tie-breaking rule must be used, such as setting fl(x) equal to the choice whose last digit is even.

Example 2.2.7 Suppose we are using a floating-point system with $\beta = 10$ (decimal), with $p = 4$ significant digits. Then, if we use chopping, or rounding to zero, we have fl($2/3$) = 0.6666, whereas if we use rounding to nearest, then we have fl($2/3$) = 0.6667. □

Example 2.2.8 When **rounding to even** in decimal, 88.5 is rounded to 88, not 89, so that the last digit is even, while 89.5 is rounded to 90, again to make the last digit even. □

Rounding to even, as described in the preceding example, is preferable to consistently breaking a tie by rounding up, because it is statistically unbiased.

2.2.1.4 *Machine Precision*

In error analysis, it is necessary to estimate error incurred in each step of a computation. As such, it is desirable to know an upper bound for the relative error introduced by rounding. This leads to the following definition.

Definition 2.2.9 (Machine Precision) Let $\mathbb{F}$ be a floating-point number system. The **unit roundoff** or **machine precision**, denoted by $\mathbf{u}$, is the real number that satisfies

$$\left| \frac{fl(x) - x}{x} \right| \leq \mathbf{u}$$

for any real number x such that UFL $< x <$ OFL.

The value of $\mathbf{u}$ depends on the rounding strategy that is used. If rounding toward zero is used, then $\mathbf{u} = \beta^{1-p}$, whereas if rounding to nearest is used, $\mathbf{u} = \frac{1}{2}\beta^{1-p}$.

An alternative, but not equivalent, definition of $\mathbf{u}$ is that it is the smallest positive number such that

$$fl\,(1 + \mathbf{u}) > 1.$$

That is, $\mathbf{u}$ is the distance from 1 to the next larger representable number. The value of $\mathbf{u}$ using this definition is β^{1-p}. Regardless of which definition is used, it is important to avoid confusing $\mathbf{u}$ with the underflow level UFL. The unit roundoff is determined by the number of digits in the mantissa, whereas the underflow level is determined by the range of allowed exponents.

Exploration 2.2.13 Based on Definition 2.2.9, the value of $\mathbf{u}$ in IEEE double-precision floating-point arithmetic is $\mathbf{u} = 2^{-53}$. Explain why fl($1 + 2^{-53}$) = 1, rather than a number greater than one.

> **Exploration 2.2.14** The MATLAB predefined variable eps is equal to 2^{-52}, the distance from 1 to the next larger representable number (that is, the second definition of **u** given above). However, it is actually possible to find a number x smaller than eps such that $\mathrm{fl}(1 + x) > 1$. Find such a number by experimenting with powers of 2 in MATLAB.

In analysis of roundoff error, it is assumed that $\mathrm{fl}(x \text{ op } y) = (x \text{ op } y)(1 + \delta)$, where op is an arithmetic operation and δ is an unknown constant satisfying $|\delta| \leq \mathbf{u}$. From this assumption, it can be seen that the relative error in $\mathrm{fl}(x \text{ op } y)$ is δ (recall Exploration 2.1.3). In the case of addition, the relative backward error in each operand is also δ.

2.2.1.5 *The IEEE Floating-Point Standard*

Generally, computers conform to the **IEEE standard** for floating-point arithmetic. The standard specifies, among other things, how floating-point numbers are to be represented in memory. Two representations prescribed in the standard are **single-precision** and **double-precision**.

Under the standard, single-precision floating-point numbers occupy 4 bytes in memory, with 23 bits used for the mantissa, 8 for the exponent, and one for the sign. IEEE double-precision floating-point numbers occupy eight bytes in memory, with 52 bits used for the mantissa, 11 for the exponent, and one for the sign. Due to normalization, it follows that in the IEEE floating-point standard, $p = 24$ for single precision, and $p = 53$ for double precision, even though only 23 and 52 bits, respectively, are used to store mantissas.

Example 2.2.10 The following table summarizes the main aspects of a general floating-point system and a double-precision floating-point system that uses a 52-bit mantissa and 11-bit exponent. For both systems, we assume that rounding to nearest is used, and that normalization is used. $\square$

	General	Double Precision
Form of machine number	$\pm m \beta^E$	$\pm 1.d_1 d_2 \cdots d_{52} 2^E$
Precision	p	53
Exponent range	$L \leq E \leq U$	$-1023 \leq E \leq 1024$
UFL (Underflow Level)	β^L	2^{-1023}
OFL (Overflow Level)	$\beta^{U+1}(1 - \beta^{-p})$	$2^{1025}(1 - 2^{-53})$
u	$\frac{1}{2}\beta^{1-p}$	2^{-53}

> **Exploration 2.2.15** Are the values for UFL and OFL given in the table above the actual values used in the IEEE double-precision floating point system? Experiment with powers of 2 in MATLAB to find out. What are the largest and smallest positive numbers you can represent? Can you explain any discrepancies between these values and the ones in the table?

> **Exploration 2.2.16** Once you have determined the value of U in the IEEE double-precision floating-point system from Exploration 2.2.15, explain why the MATLAB expression for $2^{U+1}(1 - 2^{-p})$, if typed at the prompt without algebraic modification, is represented as `Inf`. How can the expression be modified so that it is a finite representable number? Verify this modification in MATLAB.

One can gain insight into the representation of IEEE double-precision floating-point numbers using `format hex` in MATLAB (see Section 1.2.19). Consider the following output:

```
>> format hex
>> 4075/64
ans =
   404fd60000000000
>> format long
>> ans
ans =
  63.671875000000000
```

The hexadecimal, or base 16, representation of the number $4075/64 = 63.671875$ can readily be converted to binary, because $16 = 2^4$ and therefore each hexadecimal digit corresponds to four binary digits.

For convenience, we use the following conversion table,

Decimal	Hexadecimal	Binary	Decimal	Hexadecimal	Binary
0	0	0000	8	8	1000
1	1	0001	9	9	1001
2	2	0010	10	a	1010
3	3	0011	11	b	1011
4	4	0100	12	c	1100
5	5	0101	13	d	1101
6	6	0110	14	e	1110
7	7	0111	15	f	1111

where each binary number should be interpreted in the same way as a number in decimal is naturally interpreted, only with base 2 instead of base 10. That is, the rightmost digit is multiplied by $2^0 = 1$, and then added to the next digit to the left times $2^1 = 2$, plus the next digit to the left times $2^2 = 4$, and so on. For example,

$$13 = 1101 = 1 \times 2^3 + 1 \times 2^2 + 0 \times 2^1 + 1 \times 2^0 = 8 + 4 + 0 + 1.$$

Using this table, the binary representation of `404fd60000000000` is

0100 0000 0100 1111 1101 0110 0000 0000 0000 0000 0000 0000 0000

0000 0000 0000

where spaces are introduced to separate sequences of binary digits corresponding to the original hexadecimal digits.

Now, we interpret this sequence of 64 binary digits, or bits, in terms of the IEEE double-precision floating-point standard. Reading from left to right, the first bit, 0, is the sign bit, which indicates that the number is positive; a sign bit of 1 is used for a negative number. The next 11 bits, 10000000100, represents the unsigned integer $2^{10} + 2^2 = 1028$. We subtract 1023 from this value to obtain the exponent $E = 5$.

Finally, the twelve leading binary digits of the mantissa are 111111010110, followed by forty digits that are zero. Because the number is normalized, there is an implied leading digit $d_0 = 1$. The bits of the mantissa that are explicitly stored correspond to the digits $d_1, d_2, \ldots, d_{p-1}$ in Definition 2.2.1, with $p = 53$. It follows that the value represented by the mantissa is

$$\sum_{j=0}^{52} d_j 2^{-j} = 1 + \frac{1}{2} + \frac{1}{4} + \frac{1}{8} + \frac{1}{16} + \frac{1}{32} + \frac{1}{64} + \frac{1}{256} + \frac{1}{1024} + \frac{1}{2048} = \frac{4075}{2048}.$$

Therefore, the number represented is

$$(-1)^0 \times 2^5 \times \frac{4075}{2048} = \frac{4075}{64},$$

where the three numbers multiplied on the left side account for the sign bit 0, the exponent 5, and mantissa $4075/2048$, respectively.

2.2.2   Issues with Floating-Point Arithmetic

We now discuss the various issues that arise when performing **floating-point arithmetic**, or **finite-precision arithmetic**, which approximates arithmetic operations on real numbers.

2.2.2.1   *Loss of Precision*

When adding or subtracting floating-point numbers, it is necessary to shift one of the operands so that both operands have the same exponent, before adding or subtracting the mantissas. As a result, digits of precision are lost in the operand that is smaller in magnitude, and the result of the operation cannot be represented using a machine number. In fact, if x is the smaller operand and y is the larger operand, and $|x| < |y|\mathbf{u}$, then the result of the operation will simply be y (or $-y$, if y is to be subtracted from x), since the entire value of x is lost in rounding the result.

Example 2.2.11 Consider a floating-point system with $\beta = 10$, $p = 10$, $L = -20$ and $U = 20$. If $x = 2 \times 10^4$ and $y = 3 \times 10^{-10}$, then the exact value of $x + y$ is 20000.0000000003, but that cannot be represented in this floating-point system because 15 digits are needed to represent this number but only $p = 10$ digits are available. The floating-point representation of the sum would simply be x, because $|y| < |x|\mathbf{u}$, where $\mathbf{u} = \frac{1}{2}10^{-10}$ is the machine precision. $\square$

Exploration 2.2.17 Consider the evaluation of the summation

$$\sum_{i=1}^{n} x_i,$$

where each term x_i is positive. Will the sum be computed more accurately in floating-point arithmetic if the numbers are added in order from smallest to largest, or largest to smallest? Justify your answer.

In multiplication or division, the operands need not be shifted, but the mantissas, when multiplied or divided, cannot necessarily be represented using only p digits of precision. The product of two mantissas requires $2p$ digits to be represented exactly, while the quotient of two mantissas could conceivably require infinitely many digits.

2.2.2.2 *Violation of Arithmetic Rules*

Because floating-point arithmetic operations are not exact, they do not follow all of the laws of real arithmetic. In particular, while floating-point arithmetic is still commutative, it is not associative; i.e., $x + (y + z) \neq (x + y) + z$ in floating-point arithmetic.

Exploration 2.2.18 In MATLAB, generate three random numbers x, y and z, and compute $x + (y + z)$ and $(x + y) + z$. Do they agree? Try this a few times with different random numbers, of different magnitudes, and observe what happens. What would have to be true of x, y and z to cause a significant deviation?

2.2.2.3 *Overflow and Underflow*

When multiplying or dividing, overflow or underflow may occur depending on the exponents of the operands, since their sum or difference may lie outside of the interval $[L, U]$.

Example 2.2.12 Consider a floating-point system with $\beta = 10$, $L = -20$ and $U = 20$. Then the numbers $x = 4 \times 10^{18}$ and $y = -2 \times 10^{-17}$ can be represented exactly in this system, but computing $x/y = -2 \times 10^{35}$ would cause overflow because the exponent of 35 is greater than U. Similarly, computing $y^2 = 4 \times 10^{-34}$ would cause underflow, because the exponent of -34 is less than L. This value would therefore be rounded to zero. □

Exploration 2.2.19 Consider the formula $z = \sqrt{x^2 + y^2}$. Explain how overflow can occur in computing z, even if x, y and z all have magnitudes that can be represented. How can this formula be rewritten so that overflow does not occur?

Exploration 2.2.20 Let $\mathbf{x} \in \mathbb{R}^n$. Use the result of Exploration 2.2.19 to describe an algorithm for computing

$$\|\mathbf{x}\|_2 = \left(\sum_{i=1}^{n} x_i^2 \right)^{1/2}$$

in such a way as to prevent overflow.

2.2.2.4 *Cancellation*

Subtraction of floating-point numbers presents a unique difficulty, in addition to the rounding error previously discussed.

Exploration 2.2.21 When performing the subtraction

$$\begin{array}{r} 0.345769258233 \\ -0.345769258174 \\ \hline \end{array}$$

how many significant digits are included in the result? In general, if the two numbers being subtracted have p significant digits, and the first (most significant, or left to right) m digits agree, how many significant digits, at most, can the result have?

In an extreme case, if the two operands differ by less than $\mathbf{u}$, then the result contains no correct digits; it consists entirely of roundoff error from previous computations. This phenomenon of losing precision due to subtraction of nearly equal numbers is known as **catastrophic cancellation**, or **cancellation error**. Unfortunately, the term "catastrophic" is not an exaggeration, as failure to take roundoff error into account led to the deaths of 28 American soldiers during the 1991 Gulf War. Details of this tragedy, including the steps taken to prevent a reoccurrence, can be found at http://www-users.math.umn.edu/~arnold/disasters/patriot.html.

Because of the highly detrimental effect of cancellation error, it is important to ensure that no steps in a computation compute small values by subtracting relatively large operands. Often, computations can be rearranged to avoid this risky practice.

Example 2.2.13 Consider the quadratic equation

$$ax^2 + bx + c = 0,$$

which has the solutions

$$x_1 = \frac{-b + \sqrt{b^2 - 4ac}}{2a}, \quad x_2 = \frac{-b - \sqrt{b^2 - 4ac}}{2a}.$$

Suppose that $b > 0$. Then, in computing x_1, we encounter catastrophic cancellation if b is much larger than a and c, because this implies that $\sqrt{b^2 - 4ac} \approx b$ and as a result we are subtracting two numbers that are nearly equal in computing the numerator. On the other hand, if $b < 0$, we encounter this same difficulty in computing x_2.

Suppose that we use 4-digit rounding arithmetic to compute the roots of the equation

$$x^2 + 10,000x + 1 = 0.$$

Then, we obtain $x_1 = 0$ and $x_2 = -10,000$. Clearly, x_1 is incorrect because if we substitute $x = 0$ into the equation then we obtain the contradiction $1 = 0$. In fact, if we use 7-digit rounding arithmetic then we obtain the same result. Only if we use at least 8 digits of precision do we obtain roots that are reasonably correct,

$$x_1 \approx -1 \times 10^{-4}, \quad x_2 \approx -9.9999999 \times 10^3.$$

☐

Exploration 2.2.22 Use an algebraic manipulation commonly used on expressions involving square roots to obtain a new quadratic formula that circumvents the cancellation error exhibited in Example 2.2.13. Try this formula on the quadratic equation featured in this example, with 4-digit rounding arithmetic.

As the following explorations illustrate, cancellation error can occur in a variety of situations, so it is a phenomenon that one should always be aware of when designing numerical algorithms.

Exploration 2.2.23 Use the MATLAB function **randn** to generate 1000 normally distributed random numbers with mean 1000 and standard deviation 0.1. Then, use these formulas to compute the variance:

1. $\frac{1}{n}\sum_{i=1}^{n}(x_i - \bar{x})^2$
2. $\left(\frac{1}{n}\sum_{i=1}^{n}x_i^2\right) - \bar{x}^2$

where $\bar{x}$ is the mean. How do the results differ? Which formula is more susceptible to issues with floating-point arithmetic, and why?

Exploration 2.2.24 Recall Exploration 2.1.14, in which two approximations of the derivative were tested using various values of the spacing h. In light of the discussion in this section, explain the behavior for the case of $h = 10^{-14}$.

Exploration 2.2.25 Consider the expression

$$\frac{1}{1-x} - \frac{1}{1+x}.$$

For which values of x can the evaluation of this expression, as written, exhibit catastrophic cancellation? Rewrite the expression to address the issue.

Exploration 2.2.26 Assume that $x \approx y$, and that $|x|, |y| \gg 1$. Consider the equation

$$x^2 - y^2 = (x-y)(x+y).$$

Which side of this expression is evaluated more accurately with floating-point arithmetic? Justify your answer.

2.2.3 Concept Check

1. For each of the following questions, answer true or false, with reasons:

   (a) Floating-point numbers are uniformly distributed on the real number line.

   (b) If two real numbers can be represented exactly in a floating-point number system, then the result of an arithmetic operation on these numbers can also be represented exactly.

   (c) Floating-point addition is commutative.

   (d) Floating-point addition is associative.

   (e) IEEE double-precision floating-point numbers have a mantissa that is twice the size of IEEE single-precision floating-point numbers.

2. Describe the four numbers (β, p, L, U) that define a floating-point number system.

3. Define underflow and overflow. Which one is more problematic, and why?

4. Of the four basic arithmetic operations (addition, subtraction, multiplication and division), which ones can *not* produce overflow when applied to two positive numbers?

5. What is the benefit of normalizing floating-point numbers? What is a good reason not to normalize them? Explain how the use of gradual underflow yields a "best-of-both-worlds" situation.

6. Explain the difference between chopping and rounding. What is the benefit of rounding to even?

7. What is cancellation error, and under what circumstances does it arise?

8. Give an example of a number that can be represented exactly in decimal, but not in binary.

9. Explain the distinction between the unit roundoff **u** and the underflow level UFL.

2.3 Additional Resources

The concept of backward error analysis was introduced by James Wilkinson in the context of numerical linear algebra; see for example [Wilkinson (1960, 1963)]. An early treatment of the notion of conditioning can be found in [Rice (1966)]. "Big-O" notation, and variations thereof, are covered extensively in [Graham et al. (1994)]. In this chapter, this notation was used primarily to discuss convergence, but in later chapters we will also use it to discuss the *complexity*, or, informally, "running time" of algorithms, as in Example 2.1.14. Complexity theory, as applied to algorithms in general, is presented in [Blum et al. (2001)].

The IEEE floating-point standard is published in [IEEE (1987)]; a tutorial is also provided in [Goldberg (1991)]. The web page of William Kahan is an excellent resource for information about floating-point arithmetic:

http://people.eecs.berkeley.edu/~wkahan/

An accessible treatment of floating-point architecture, programming language support, and the concepts of conditioning, stability and cancellation is provided by [Overton (2001)]. A thorough error analysis of roundoff error and stability of numerical algorithms, along with historical perspective, is given in [Higham (2003)].

Books that focus on the hardware aspects of computer arithmetic are [Koren (1993); Omondi (1994); Parhami (1999)]. Beyond the scope of this book is **interval arithmetic** [Alefeld and Herzberger (1983); Jaulin et al. (2001); Moore et al. (2009)], in which arithmetic operations are performed not on numbers, but intervals that account for measurement and rounding errors.

2.4 Exercises

1. Let $f(x) = x - y$, where y is a fixed constant. Use the relative condition number of $f(x)$ to explain why subtraction suffers from cancellation error.

2. Let $x, y > 0$ be such that $x \approx y$. It follows that $\ln(x) - \ln(y)$ is susceptible to cancellation error. Use the relative condition number to explain why the equivalent expression $\ln(x/y)$ is also problematic, even though it does not involve subtraction. *Hint:* Use the same approach as in Exercise 1.

3. Let $n = 100$. In MATLAB, form a vector of $n + 1$ equally spaced points $x_0, x_1, x_2, \ldots, x_n$ on $[0, 1]$, with spacing $h = 1/n$, in two different ways:

   (a) $x_0 = 0$, $x_i = x_{i-1} + h$, $i = 1, 2, \ldots, n$
   (b) $x_i = ih$, $i = 0, 1, 2, \ldots, n$

   Which approach is more accurate, and why?

4. Find the IEEE double-precision floating-point representation of 0.1. How would the representation change if chopping was used instead of rounding? *Hint:* Use **format hex** in MATLAB to display this value in hexadecimal.

5. Write a MATLAB function [x1,x2]=quadroots(a,b,c) based on Example

2.2.13 to compute the roots x1 and x2 of the quadratic equation $ax^2 + bx + c = 0$ in such a way as to avoid cancellation error.

6. Graph the function $f(x) = (x - 1)^6$ on the interval $[0.995, 1.005]$, using at least 100 equally spaced points, in two different ways: a) by evaluating $(x-1)^6$, exactly as written, for each x-value, and b) by expanding $(x-1)^6$. Use one of the functions in Section 1.2.18 to help with this. Explain the discrepancy in the graphs.

7. Consider a floating-point number system with $\beta = 2$ in which all numbers are normalized. Describe the circumstances in which $\mathbf{u} <$ UFL. Your answer should be a relationship between p and L. How does your answer change if denormalized numbers are used? In both cases, assume rounding is used rather than chopping.

8. Write a MATLAB function [s,m,e]=dissectfp(x) that accepts as input a double-precision floating-point number x and returns as output the components of the floating-point representation of x: the sign s, the mantissa m and the exponent e. s is 1 if x is positive; otherwise it is -1. The mantissa x consists of the binary digits $d_0, d_1, \ldots, d_{p-1}$, from Definition 2.2.1, where $p = 53$, that form a double-precision floating-point number x_0 where $0 \leq x_0 < 2$. The exponent e must be an integer in the interval $[-1022, 1023]$. Make sure to account for denormalized numbers.

9. The MATLAB function round, when used with a second input argument n, rounds a number x (specified as the first input) to n digits, whether they are the n most significant digits, or the first n digits to the right of the decimal point. Write functions roundsig and rounddec that perform rounding in these manners. Compare your results to those returned by round with the appropriate usage. Consult the help page for round as needed.

10. As a follow-up to Exploration 2.2.17, consider the following algorithm, due to W. Kahan [Kahan (1965)] and called **compensated summation**, for computing the sum of n numbers $x_1, x_2, \ldots, x_n$:

$s = x_1$
$c = 0$
for $i = 2, \ldots, n$ **do**
     $y = x_i - c$
     $t = s + y$
     $c = (t - s) - y$
     $s = t$
end for

Explain why this is a more efficient approach to computing the sum than the most accurate approach found in Exploration 2.2.17. Assume that the elements of x are not arranged in any particular order.

PART II
Numerical Linear Algebra

Chapter 3

Direct Methods for Linear Systems

Let A be an $n \times n$ matrix; alternatively, we say $A \in \mathbb{R}^{n \times n}$ if the entries of A are assumed to be real numbers. In this chapter, we discuss methods for solving the *system of linear equations* $A\mathbf{x} = \mathbf{b}$, where A is a nonsingular (that is, invertible) matrix, $\mathbf{x}$ is an unknown n-vector, and $\mathbf{b}$ is an n-vector. The need to solve such a system of linear equations, or *linear system*, arises in a number of applications, such as data fitting and the solution of linear differential equations.

Section 3.1 develops an algorithm for **Gaussian Elimination**, to reduce the system $A\mathbf{x} = \mathbf{b}$ to an equivalent system that can easily be solved. Section 3.2 introduces the **LU Decomposition**, which makes Gaussian Elimination more practical for coding purposes. Section 3.3 examines special cases of $A\mathbf{x} = \mathbf{b}$, and how Gaussian Elimination can be made more efficient in such cases. Finally, Section 3.4 presents error analysis and approaches to improving accuracy and problem *conditioning*, which was discussed in Section 2.1.4. It is recommended that the reader review Appendix B to ensure having a sufficient linear algebra background.

3.1 Gaussian Elimination

To develop an algorithm for solving a very general problem such as $A\mathbf{x} = \mathbf{b}$, it is recommended to proceed by posing the following questions:

- Are there cases in which this problem is easy to solve?
- Is it possible to extend our method of solving the problem in these "easy" cases to more general cases?

This approach will be a recurring theme in this book. We begin with an examination of some easy cases.

3.1.1 Triangular Systems

The basic idea behind methods for solving a system of linear equations is to reduce them to equations involving a single unknown, because such equations are trivial to solve. There are three types of *triangular* systems that feature such trivial equations:

- **diagonal**, where $a_{ij} = 0$ for $i \neq j$,
- **upper triangular**, where $a_{ij} = 0$ for $i > j$, and
- **lower triangular**, where $a_{ij} = 0$ for $i < j$.

In this section we will look at how each type of system can be solved efficiently, and analyze the computational cost in each case.

3.1.1.1 *Diagonal Systems*

We begin with the type of system that is simplest to solve, that has the form

$$Ax = \begin{bmatrix} a_{11} & 0 & 0 & \cdots & 0 \\ 0 & a_{22} & 0 & \cdots & 0 \\ 0 & 0 & a_{33} & \cdots & 0 \\ \vdots & \vdots & \vdots & \ddots & \vdots \\ 0 & 0 & 0 & \cdots & a_{nn} \end{bmatrix} \begin{bmatrix} x_1 \\ x_2 \\ x_3 \\ \vdots \\ x_n \end{bmatrix} = \begin{bmatrix} b_1 \\ b_2 \\ b_3 \\ \vdots \\ b_n \end{bmatrix} = b. \tag{3.1}$$

The matrix A is a diagonal matrix, so named because the only nonzero entries in A lie on the **main diagonal** consisting of the entries $a_{11}, a_{22}, \ldots, a_{nn}$. It follows that in this system, each equation only has one unknown, so we can solve each equation independently of the other equations. The equations we need to solve are

$$a_{11}x_1 = b_1$$
$$a_{22}x_2 = b_2$$
$$\vdots$$
$$a_{nn}x_n = b_n.$$

Exploration 3.1.1 Write a MATLAB script to solve the following diagonal system of equations:

$$\begin{aligned} 3x & = 4 \\ 2y & = 8 \\ 7z & = 21 \end{aligned}$$

How many floating-point arithmetic operations does this script perform?

We can see that each component of the solution is found by $x_i = b_i / a_{ii}$, provided that $a_{ii} \neq 0$, for $i = 1, 2, \ldots, n$. Thus we have the following algorithm:

Algorithm 3.1.1 (Solution of Diagonal System) Given an $n \times n$ diagonal, nonsingular matrix A and n-vector b, the following algorithm computes the solution x of the system $Ax = b$.

for $i = 1, 2, \ldots, n$ **do**
    $x_i = b_i / a_{ii}$
end for

> **Exploration 3.1.2** Write a MATLAB function
>
> $$x\texttt{=solvediag(A,b)}$$
>
> to solve the diagonal system of equations (3.1). How many floating-point operations does this function perform, in terms of n?

It does not matter which equation we start with in solving this system, since they can all be solved independently of each other, or in parallel. The ordering of equations will become relevant as we consider more general systems of equations.

3.1.1.2 *Upper Triangular Systems*

Next, we will study an upper triangular system, which has a matrix of the form

$$A = \begin{bmatrix} a_{11} & a_{12} & a_{13} & \cdots & a_{1n} \\ & a_{22} & a_{23} & \cdots & a_{2n} \\ & & \ddots & \ddots & \vdots \\ & & & \ddots & \vdots \\ & & & & a_{nn} \end{bmatrix}.$$

This matrix is upper triangular, so named because its nonzero entries are confined to the "upper triangle" that includes the main diagonal and all entries "above" it. To understand the solution process, let us take a look at a system of n equations in upper triangular form, where $n = 3$.

Example 3.1.2 We will solve the system

$$x_1 + 2x_2 + x_3 = 5,$$
$$-4x_2 + x_3 = 2,$$
$$-2x_3 = 4$$

for the unknowns x_1, x_2 and x_3. Because the third equation is a linear equation in only x_3, it can easily be solved by dividing by the coefficient of x_3.

$$\begin{aligned} x_1 + 2x_2 + x_3 &= 5, \\ -4x_2 + x_3 &= 2, \\ -2x_3 &= 4. \end{aligned} \qquad \rightarrow \qquad \begin{aligned} x_1 + 2x_2 + x_3 &= 5, \\ -4x_2 + x_3 &= 2, \\ x_3 &= -2. \end{aligned}$$

Then, we can substitute this value into the second equation. After moving the term involving x_3 to the right side, can solve for x_2, again by dividing by its coefficient.

$$\begin{aligned} x_1 + 2x_2 + x_3 &= 5, \\ -4x_2 + -2 &= 2, \\ x_3 &= -2. \end{aligned} \qquad \rightarrow \qquad \begin{aligned} x_1 + 2x_2 + x_3 &= 5, \\ x_2 &= -1, \\ x_3 &= -2. \end{aligned}$$

Finally, we substitute the values of x_2 and x_3 into the first equation, and move those terms to the right side to obtain x_1.

$$x_1 + 2(-1) + -2 = 5,$$
$$x_2 = -1,$$
$$x_3 = -2.$$

$\rightarrow$

$$x_1 = 9,$$
$$x_2 = -1,$$
$$x_3 = -2.$$

□

Exploration 3.1.3 Write a MATLAB script to solve the following triangular system:

$$2x + y - 3z = -10$$
$$-2y + z = -2$$
$$3z = 6$$

How many floating-point operations does this script perform?

Based on the preceding example, we now examine the process of solving a general 3×3 upper triangular system of the form

$$a_{11}x_1 + a_{12}x_2 + a_{13}x_3 = b_1$$
$$a_{22}x_2 + a_{23}x_3 = b_2$$
$$a_{33}x_3 = b_3.$$

We can now simply solve for x_3 as follows:

$$a_{33}x_3 = b_3$$
$$\implies \quad x_3 = b_3/a_{33}$$

Now that we have found x_3, we substitute it into the previous equation to solve for the unknown in that equation, which yields

$$a_{22}x_2 + a_{23}x_3 = b_2$$
$$\implies \quad x_2 = (b_2 - a_{23}x_3)/a_{22}.$$

Similarly, we substitute x_2 and x_3 into the first equation, and again we have a linear equation with only one unknown.

$$a_{11}x_1 + a_{12}x_2 + a_{13}x_3 = b_1$$
$$\implies \quad x_1 = (b_1 - a_{12}x_2 - a_{13}x_3)/a_{11}$$

Thus the solution process is complete.

Exploration 3.1.4 Modify your script from Exploration 3.1.3 to solve a general upper triangular system of three equations in three unknowns x, y and z. Assume that the coefficients of the system are stored in a 3×3 matrix U, and that the right-hand sides of the equations are stored in a 3-vector **b**. Test your script on a randomly generated U and **b**, using MATLAB's backslash operator to check your solution.

The process we have just described is known as **back substitution**, since it proceeds backwards from the last equation. Just by looking at the above 3×3 case we can see a pattern emerging, and this pattern can be described in terms of an algorithm that can be used to solve this type of triangular system.

Algorithm 3.1.3 (Back Substitution) Given an $n \times n$ nonsingular upper triangular matrix U and n-vector $\mathbf{y}$, the following algorithm computes the solution $\mathbf{x}$ of the system of linear equations $U\mathbf{x} = \mathbf{y}$.

for $i = n, n - 1, \ldots, 1$ do
$\quad x_i = y_i$
$\quad$ for $j = i + 1, i + 2, \ldots, n$ do
$\quad\quad x_i = x_i - u_{ij} x_j$
$\quad$ end for
$\quad x_i = x_i / u_{ii}$
end for

If we examine the nested loops in the above algorithm, we can see that there are two levels of nesting involved, and each loop can perform up to $O(n)$ iterations. Therefore, it takes $O(n^2)$ floating-point operations to solve this kind of system. In fact, it can be shown that the number of operations required is *exactly* n^2.

Exploration 3.1.5 Generalize your script from Exploration 3.1.4 to an implementation of Algorithm 3.1.3 for solving an upper triangular system of n equations in n unknowns $x_1, x_2, \ldots, x_n$. Convert your script to a function x=backsub(U,y) that accepts as input arguments an $n \times n$ matrix U and n-vector y, and returns as output a column vector x consisting of the values of the unknowns $x_1, x_2, \ldots, x_n$.

3.1.1.3 *Lower Triangular Systems*

The last type of system we will look at is a lower triangular system, which has a matrix of the form

$$
A = \begin{bmatrix}
a_{11} & & & & \\
a_{21} & a_{22} & & & \\
a_{31} & a_{32} & \ddots & & \\
\vdots & \vdots & \ddots & \ddots & \\
a_{n1} & a_{n2} & \cdots & a_{n,n-1} & a_{nn}
\end{bmatrix}.
$$

This type of system can be solved using a similar approach as with upper triangular systems, using a method called **forward substitution**. This algorithm will be discussed later in this chapter. Given the structure of A, we can see that forward substitution proceeds like back substitution, except that it would iterate over the equations in the opposite order. Like back substitution, forward substitution requires $O(n^2)$ floating-point operations.

3.1.2   Row Operations

Now that we have looked at the simplest cases, we want to switch our focus to solving a general square nonsingular system $A\mathbf{x} = \mathbf{b}$. The approach that we use is to first reduce our original system to one of the types of systems discussed in Section 3.1.1, since we now know an efficient way to solve these systems.

Such a reduction is achieved by manipulating the equations in the system in such a way that the solution does not change, but unknowns are eliminated from selected equations until, finally, we obtain an equation involving only a single unknown. These manipulations are called **elementary row operations**, and they are defined as follows. We denote by R_i the ith equation, or row, of the system, and s is a scalar.

- Multiplying an equation by a scalar $(R_i \leftarrow sR_i)$
- Reordering the equations by interchanging the ith and jth equations in the system $(R_i \leftrightarrow R_j)$
- Replacing equation i by the difference of itself and a multiple of equation j $(R_i \leftarrow R_i - sR_j)$

Exploration 3.1.6 Prove that the following row operations do not change the solution set.

- Multiplying an equation by a scalar.
- Replacing equation i by the difference of itself and a multiple of equation j.

The third operation is by far the most useful. We will now demonstrate how it can be used to reduce a system of equations to a form in which it can easily be solved.

Example 3.1.4 Consider the system of linear equations

$$x_1 + 2x_2 + x_3 = 5,$$
$$3x_1 + 2x_2 + 4x_3 = 17,$$
$$4x_1 + 4x_2 + 3x_3 = 26.$$

First, we eliminate x_1 from the second equation by subtracting 3 times the first equation from the second. This yields the equivalent system

$$x_1 + 2x_2 + x_3 = 5,$$
$$-4x_2 + x_3 = 2,$$
$$4x_1 + 4x_2 + 3x_3 = 26.$$

Next, we subtract 4 times the first equation from the third, to eliminate x_1 from the third equation as well:

$$x_2 + 2x_2 + x_3 = 5,$$
$$-4x_2 + x_3 = 2,$$
$$-4x_2 - x_3 = 6.$$

Then, we eliminate x_2 from the third equation by subtracting the second equation from it, which yields the system

$$x_1 + 2x_2 + x_3 = 5,$$
$$-4x_2 + x_3 = 2,$$
$$-2x_3 = 4.$$

This system is in upper triangular form, because the third equation depends only on x_3, and the second equation depends on only x_2 and x_3. It was solved using back substitution in Example 3.1.2. □

Performing row operations on the system $Ax = b$ can be accomplished by performing them on the **augmented matrix**

$$\begin{bmatrix} A\,\mathbf{b} \end{bmatrix} = \begin{bmatrix} a_{11} & a_{12} & \cdots & a_{1n} & b_1 \\ a_{21} & a_{22} & \cdots & a_{2n} & b_2 \\ \vdots & & & \vdots & \vdots \\ a_{n1} & a_{n2} & \cdots & a_{nn} & b_n \end{bmatrix}.$$

By working with the augmented matrix instead of the original system, there is no need to continually rewrite the unknowns or arithmetic operators. Once the augmented matrix is reduced to upper triangular form, also known as **row-echelon form**, the corresponding system of linear equations can be solved by back substitution, as before.

The process of eliminating variables from the equations, or, equivalently, zeroing entries of the corresponding matrix, in order to reduce the system to upper triangular form is called **Gaussian Elimination**. We will now step through an example as we discuss the steps of the Gaussian Elimination algorithm.

Example 3.1.5 Consider the system of linear equations

$$2x_1 + 5x_2 + 5x_3 + 3x_4 = 4$$
$$3x_1 + 3x_2 + x_3 + 3x_4 = 3$$
$$x_1 + 4x_2 + x_3 + 2x_4 = 3$$
$$5x_1 + 5x_2 + x_3 + 2x_4 = 4.$$

This system can be represented by the coefficient matrix A and right-hand side vector $\mathbf{b}$, as follows:

$$A = \begin{bmatrix} 2 & 5 & 5 & 3 \\ 3 & 3 & 1 & 3 \\ 1 & 4 & 1 & 2 \\ 5 & 5 & 1 & 2 \end{bmatrix}, \quad \mathbf{b} = \begin{bmatrix} 4 \\ 3 \\ 3 \\ 4 \end{bmatrix}.$$

To perform row operations to reduce this system to upper triangular form, we define the augmented matrix

$$\tilde{A} = [\, A \; \mathbf{b} \,] = \begin{bmatrix} 2 & 5 & 5 & 3 & 4 \\ 3 & 3 & 1 & 3 & 3 \\ 1 & 4 & 1 & 2 & 3 \\ 5 & 5 & 1 & 2 & 4 \end{bmatrix}.$$

The first elementary row operation entails subtracting 3/2 times the first row from the second in order to eliminate x_1 from the second equation. For convenience, we overwrite $\tilde{A}$, and therefore A and $\mathbf{b}$, with the result of this row operation:

$$\tilde{A} = \begin{bmatrix} 2 & 5 & 5 & 3 & 4 \\ 0 & -9/2 & -13/2 & -3/2 & -3 \\ 1 & 4 & 1 & 2 & 3 \\ 5 & 5 & 1 & 2 & 4 \end{bmatrix}.$$

Next, we eliminate x_1 from the third equation by subtracting 1/2 times the first row from the third:

$$\tilde{A} = \begin{bmatrix} 2 & 5 & 5 & 3 & 4 \\ 0 & -9/2 & -13/2 & -3/2 & -3 \\ 0 & 3/2 & -3/2 & 1/2 & 1 \\ 5 & 5 & 1 & 2 & 4 \end{bmatrix}.$$

Then, we complete the elimination of x_1 by subtracting 5/2 times the first row from the fourth:

$$\tilde{A} = \begin{bmatrix} 2 & 5 & 5 & 3 & 4 \\ 0 & -9/2 & -13/2 & -3/2 & -3 \\ 0 & 3/2 & -3/2 & 1/2 & 1 \\ 0 & -15/2 & -23/2 & -11/2 & -6 \end{bmatrix}.$$

We now need to eliminate x_2 from the third and fourth equations. This is accomplished by subtracting $(3/2)/(-9/2) = -1/3$ times the second row from the third, which yields

$$\tilde{A} = \begin{bmatrix} 2 & 5 & 5 & 3 & 4 \\ 0 & -9/2 & -13/2 & -3/2 & -3 \\ 0 & 0 & -11/3 & 0 & 0 \\ 0 & -15/2 & -23/2 & -11/2 & -6 \end{bmatrix},$$

and $15/9 = 5/3$ times the second row from the fourth, which yields

$$\tilde{A} = \begin{bmatrix} 2 & 5 & 5 & 3 & 4 \\ 0 & -9/2 & -13/2 & -3/2 & -3 \\ 0 & 0 & -11/3 & 0 & 0 \\ 0 & 0 & -2/3 & -3 & -1 \end{bmatrix}.$$

Finally, we subtract $(-2/3)/(-11/3) = 2/11$ times the third row from the fourth to obtain the augmented matrix of an upper triangular system,

$$\tilde{A} = \begin{bmatrix} 2 & 5 & 5 & 3 & 4 \\ 0 & -9/2 & -13/2 & -3/2 & -3 \\ 0 & 0 & -11/3 & 0 & 0 \\ 0 & 0 & 0 & -3 & -1 \end{bmatrix}.$$

Now, we can perform back substitution on the corresponding system,

$$2x_1 + 5x_2 + 5x_3 + 3x_4 = 4,$$
$$-\frac{9}{2}x_2 - \frac{13}{2}x_3 - \frac{3}{2}x_4 = -3,$$
$$-\frac{11}{3}x_3 + 0x_4 = 0,$$
$$-3x_4 = -1,$$

to obtain the solution, which yields $x_4 = 1/3$, $x_3 = 0$, $x_2 = 5/9$, and $x_1 = 1/9$. $\square$

Let us review the steps in the preceding example. Note that we started by eliminating the first entry of the second row, $a_{21} = 3$. Why did it work to multiply the first row by $3/2$, and subtract it from the second? The first entry of the first row is $a_{11} = 2$. By dividing by 2, we normalize the first row so that its first entry is one. Then, by multiplying by the entry we wish to eliminate, $a_{21} = 3$, the first row is scaled so that its first entry has this value as well, so that subtracting the scaled first row from the second row eliminates a_{21}. Next, we eliminate everything else below the main diagonal in the first column. In other words, while working in this column, we eliminate $a_{21}, a_{31}, \ldots, a_{n1}$, where $2, 3, \ldots n$ refer to the row numbers. In each case, the multiple of the first row that is subtracted is the ratio of the entry we wish to eliminate to the first entry of the row we use to perform the elimination. That is, the multiples are a_{21}/a_{11}, a_{31}/a_{11}, and so on.

Now, hold that thought while we discuss what happens next. After that is completed, we move on to column number two, and eliminate everything below the main diagonal in that column. While working in column two, we eliminate $a_{32}, a_{42}, \ldots, a_{n2}$, where $3, 4, \ldots n$ refer to the row numbers. To eliminate $a_{32} = 3/2$, because we are subtracting a multiple of the *second* row from the third row (using the first row would undo what we've accomplished in the first column!), the multiple we need is obtained by dividing the second row by its first *nonzero* entry, a_{22}, to make the first entry equal to one, and then multiplying by the entry we wish to eliminate, a_{32}, so that subtracting the scaled second row and third row will leave $a_{32} = 0$. That is, the multiple is a_{32}/a_{22}. What multiple of the second row should be subtracted from the fourth row to eliminate a_{42}? What multiples should we use when we move to the third column to eliminate entries below the diagonal there?

A pattern can be seen from what we have already discussed, but now let us think about what happens when we get to the last column. The last column would be column n, and we would eliminate, as before, everything below the main diagonal.

But, there are no entries below the main diagonal in the last column! From this, we see that we need to examine all columns except the last. That is, we examine the jth column, for $j = 1, \ldots, n-1$. Within each column, we need to eliminate entries a_{ij}, where $i = j+1, j+2, \ldots, n$ (that is, entries below the diagonal entry). Finally, eliminating a_{ij} entails performing a row operation in which a multiple, which we denote by m_{ij}, of row j is subtracted from row i, thus affecting entries a_{ik} in row i, where $k = j+1, \ldots, n$. From these indexing patterns, and a pattern in the multiple used for each row operation, we can finally formulate a general algorithm.

Algorithm 3.1.6 (**Gaussian Elimination**) Given an $n \times n$ matrix A and n-vector $\mathbf{b}$, the following algorithm reduces the system of linear equations $A\mathbf{x} = \mathbf{b}$ to an equivalent upper triangular system. A and $\mathbf{b}$ are overwritten with the coefficients and right-hand side, respectively, of the reduced system.

for $j = 1, 2, \ldots, n-1$ **do**
    for $i = j+1, j+2, \ldots, n$ **do**
        $m_{ij} = a_{ij}/a_{jj}$
        for $k = j+1, j+2, \ldots, n$ **do**
            $a_{ik} = a_{ik} - m_{ij}a_{jk}$
        end for
        $b_i = b_i - m_{ij}b_j$
    end for
end for

Exploration 3.1.7 Write a MATLAB function

$$[\text{A},\text{b}]=\text{gausselim(A,b)}$$

that implements Algorithm 3.1.6.

Exploration 3.1.8 Based on the reasoning used to figure out the algorithmic complexity of back substitution (that is, to determine that it performs $O(n^2)$ floating-point operations), what is the complexity of Gaussian Elimination?

Exploration 3.1.9 Modify your function `gausselim` from Exploration 3.1.7 to return a third output argument, `opcount`, that is equal to the number of floating-point operations performed by Gaussian elimination. Obtain operation counts for matrices of various sizes and plot them against the matrix size. If the operation count is approximately Cn^p, where C and p are constants and the matrix is $n \times n$, can you figure out the values of C and p? Does the value of p match what you obtained in Exploration 3.1.8?

Based on Exploration 3.1.9, this algorithm can be quite expensive if n is large. Later, we will discuss alternative approaches that are more efficient for certain kinds of systems, but Gaussian Elimination remains the most generally applicable method of solving systems of linear equations.

In the above algorithm, the number m_{ij} is called a **multiplier**. It is the number by which row j is multiplied before subtracting it from row i, in order to eliminate the unknown x_j from the ith equation. Note that this algorithm is applied to the augmented matrix, as the elements of the vector **b** are updated by the row operations as well.

It should be noted that in Gaussian Elimination as described in Algorithm 3.1.6, each entry below the main diagonal is never explicitly zeroed, because that computation is unnecessary. It is only necessary to update entries of the matrix that are involved in subsequent row operations or the solution of the resulting upper triangular system.

3.1.3 Concept Check

1. What are the different types of triangular systems of linear equations? How are they defined in terms of the entries of the coefficient matrix?
2. What is the purpose for studying the solution of triangular systems of equations?
3. Describe the process of back substitution. For what kind of systems of linear equations is it useful?
4. For which systems is forward substitution useful? Based on your understanding of back substitution, how would you describe forward substitution?
5. What is the algorithmic complexity of the solution of upper triangular, lower triangular, and diagonal systems of linear equations? For each type of system, your answer should have the form $O(n^p)$ where the matrix is $n \times n$.
6. What are the three types of elementary row operations?
7. What is Gaussian Elimination, and what role does it play in the solution of the linear system $A\mathbf{x} = \mathbf{b}$?
8. What is a multiplier?

3.2 The LU Decomposition

We have learned how to solve a system of linear equations $A\mathbf{x} = \mathbf{b}$ by applying Gaussian Elimination to the augmented matrix $\tilde{A} = \begin{bmatrix} A\,\mathbf{b} \end{bmatrix}$, and then performing back substitution on the resulting upper triangular matrix. However, this approach is not practical if the right-hand side **b** of the system is changed, while A is not. This is due to the fact that the choice of **b** has no effect on the row operations needed to reduce A to upper triangular form. Therefore, it is desirable to instead apply these row operations to A only once, and then "store" them in some way in order to apply them to any number of right-hand sides. This will be useful in Chapter 12 when using **implicit time-stepping methods** for solving systems of ordinary differential equations.

3.2.1 Derivation of the LU Decomposition

To figure out how to perform Gaussian Elimination in such a way as to conveniently store row operations, we examine the process from a different perspective–matrix multiplication, rather than row operations.

3.2.1.1 *Elementary row Matrices*

To accomplish this, we first note that subtracting m_{ij} times row j from row i to eliminate a_{ij} is equivalent to multiplying A on the left by the matrix

$$M_{ij} = \begin{bmatrix} 1 & 0 & \cdots & \cdots & \cdots & \cdots & \cdots & 0 \\ 0 & 1 & 0 & & & & & 0 \\ \vdots & \ddots & & \ddots & \ddots & & & \vdots \\ \vdots & & & & \ddots & \ddots & & \vdots \\ \vdots & & -m_{ij} & & \ddots & \ddots & \ddots & \vdots \\ \vdots & & & & & \ddots & \ddots & \vdots \\ \vdots & & & & & 0 & 1 & 0 \\ 0 & \cdots & \cdots & \cdots & \cdots & \cdots & 0 & 1 \end{bmatrix},$$

where the entry $-m_{ij}$ is in row i, column j. Each such matrix M_{ij} is an example of an **elementary row matrix**, which is a matrix that results from applying a elementary row operation to the identity matrix I (see Section B.8).

Example 3.2.1 The matrix

$$A = \begin{bmatrix} 1 & 2 & 1 & -1 \\ 3 & 2 & 4 & 4 \\ 4 & 4 & 3 & 4 \\ 2 & 0 & 1 & 5 \end{bmatrix}$$

can be reduced to the upper triangular matrix

$$U = \begin{bmatrix} 1 & 2 & 1 & -1 \\ 0 & -4 & 1 & 7 \\ 0 & 0 & -2 & 1 \\ 0 & 0 & 0 & -1 \end{bmatrix}$$

by performing the following row operations, or, equivalently, multiplying by the following elementary row matrices.

• M_{21}: Subtracting three times the first row from the second

- M_{31}: Subtracting four times the first row from the third
- M_{41}: Subtracting two times the first row from the fourth
- M_{32}: Subtracting the second row from the third
- M_{42}: Subtracting the second row from the fourth
- M_{43}: Subtracting the third row from the fourth

The elementary row matrices themselves are as follows.

$$M_{21} = \begin{bmatrix} 1 & 0 & 0 & 0 \\ -3 & 1 & 0 & 0 \\ 0 & 0 & 1 & 0 \\ 0 & 0 & 0 & 1 \end{bmatrix}, \quad M_{31} = \begin{bmatrix} 1 & 0 & 0 & 0 \\ 0 & 1 & 0 & 0 \\ -4 & 0 & 1 & 0 \\ 0 & 0 & 0 & 1 \end{bmatrix},$$

$$M_{41} = \begin{bmatrix} 1 & 0 & 0 & 0 \\ 0 & 1 & 0 & 0 \\ 0 & 0 & 1 & 0 \\ -2 & 0 & 0 & 1 \end{bmatrix}, \quad M_{32} = \begin{bmatrix} 1 & 0 & 0 & 0 \\ 0 & 1 & 0 & 0 \\ 0 & -1 & 1 & 0 \\ 0 & 0 & 0 & 1 \end{bmatrix},$$

$$M_{42} = \begin{bmatrix} 1 & 0 & 0 & 0 \\ 0 & 1 & 0 & 0 \\ 0 & 0 & 1 & 0 \\ 0 & -1 & 0 & 1 \end{bmatrix}, \quad M_{43} = \begin{bmatrix} 1 & 0 & 0 & 0 \\ 0 & 1 & 0 & 0 \\ 0 & 0 & 1 & 0 \\ 0 & 0 & -1 & 1 \end{bmatrix}.$$

Exploration 3.2.1 (a) Compute the matrix products $M^{(1)} = M_{41}M_{31}M_{21}$ and $M^{(2)} = M_{42}M_{32}$. What patterns do you observe in these products?

(b) Compute the inverse of each elementary row matrix in this example, by solving the equation $M_{ij}X = I$ directly for X. Use this result and the result of part (a), along with properties of the inverse of a matrix listed in Section B.9, to immediately write down $[M^{(j)}]^{-1}$ for $j = 1, 2, 3$.

(c) Use the order in which the row operations are applied to write down the relationship between A and U in terms of $M^{(1)}$, $M^{(2)}$ and $M^{(3)} = M_{43}$. Make sure all matrices are multiplied in the correct order!

(d) Use the preceding results to obtain a formula for, and then the entries of, the matrix L in the matrix equation $A = LU$. What structure (that is, pattern of zero and nonzero entries) does the matrix L have? How do the entries of L relate to the row operations used to reduce A to upper triangular form?

Let $\mathbf{b} = \begin{bmatrix} 5 & 1 & 22 & 15 \end{bmatrix}^T$. Applying the same row operations to $\mathbf{b}$ yields the modified right-hand side

$$\mathbf{y} = \begin{bmatrix} 5 \\ 1 \\ 1 \\ 3 \end{bmatrix}.$$

Exploration 3.2.2 How are the vectors $\mathbf{b}$ and $\mathbf{y}$ related, in terms of the matrix L from Exploration 3.2.1(d)?

We then use back substitution (Algorithm 3.1.3) to solve the system $U\mathbf{x} = \mathbf{y}$:

$$x_4 = y_4/u_{44} = 3/(-1) = -3,$$
$$x_3 = (y_3 - u_{34}x_4)/u_{33} = (1 - 1(-3))/(-2) = -2,$$
$$x_2 = (y_2 - u_{23}x_3 - u_{24}x_4)/u_{22} = (1 - 1(-2) - 7(-3))/(-4) = -6,$$
$$x_1 = (y_1 - u_{12}x_2 - u_{13}x_3 - u_{14}x_4)/u_{11}$$
$$= (5 - 2(-6) - 1(-2) + 1(-3))/1 = 16.$$

□

In general, if we let $A^{(1)} = A$ and let $A^{(k+1)}$ be the matrix obtained by eliminating entries of column k in $A^{(k)}$, then we have, for $k = 1, 2, \ldots, n-1$,

$$A^{(k+1)} = M^{(k)} A^{(k)}$$

where

$$M^{(k)} = \begin{bmatrix} 1 & 0 & \cdots & & & \cdots & \cdots & \cdots & \cdots & 0 \\ 0 & 1 & 0 & & & & & & & 0 \\ \vdots & \ddots & \ddots & \ddots & & & & & & \vdots \\ \vdots & & 0 & \ddots & \ddots & & & & & \vdots \\ \vdots & & \vdots & -m_{k+1,k} & \ddots & \ddots & & & & \vdots \\ \vdots & & \vdots & 0 & \ddots & \ddots & & & & \vdots \\ \vdots & & \vdots & \vdots & \vdots & \ddots & \ddots & \ddots & & \vdots \\ \vdots & & \vdots & \vdots & \vdots & & \ddots & 1 & 0 \\ 0 & \cdots & 0 & -m_{nk} & 0 & \cdots & \cdots & \cdots & 0 & 1 \end{bmatrix}, \tag{3.2}$$

with the entries $-m_{k+1,k}, \ldots, -m_{nk}$ occupying column k. It follows that the matrix

$$U = A^{(n)} = M^{(n-1)} A^{(n-1)} = M^{(n-1)} M^{(n-2)} \cdots M^{(1)} A \tag{3.3}$$

is upper triangular, and the vector

$$\mathbf{y} = M^{(n-1)} M^{(n-2)} \cdots M^{(1)} \mathbf{b},$$

being the result of applying the same row operations to $\mathbf{b}$, is the right-hand side for the upper triangular system that is to be solved by back substitution.

3.2.1.2 *Unit Lower Triangular Matrices*

We have previously learned about upper triangular matrices that result from Gaussian Elimination. Recall that an $m \times n$ matrix A is upper triangular if $a_{ij} = 0$ whenever $i > j$. This means that all entries *below* the main diagonal are equal to zero. A system of linear equations of the form $U\mathbf{x} = \mathbf{y}$, where U is an $n \times n$ nonsingular upper triangular matrix, can be solved by back substitution.

Exploration 3.2.3 Prove that an upper triangular matrix is nonsingular if and only if all of its diagonal entries are nonzero.

Similarly, a matrix L is lower triangular if all of its entries *above* the main diagonal, that is, entries ℓ_{ij} for which $i < j$, are equal to zero. We will see that a system of equations of the form $L\mathbf{y} = \mathbf{b}$, where L is an $n \times n$ nonsingular lower triangular matrix, can be solved using a process similar to back substitution, called **forward substitution**. As with upper triangular matrices, a lower triangular matrix is nonsingular if and only if all of its diagonal entries are nonzero.

Exploration 3.2.4 Prove the following useful properties for triangular matrices:

(a) The product of two upper/lower triangular matrices is upper/lower triangular.

(b) The inverse of a nonsingular upper/lower triangular matrix is upper/lower triangular. *Hint:* What system of equations does a column of the inverse solve, and how can that system be solved in the case of a triangular matrix?

Now that we have proven these properties, we can say that matrix multiplication and inversion preserve triangularity.

We note that each matrix $M^{(k)}$ from (3.2), for $k = 1, 2, \ldots, n-1$, is not only a lower triangular matrix, but a **unit lower triangular** matrix, because all of its diagonal entries are equal to 1. Next, we note two important properties of unit lower/upper triangular matrices:

- The product of two unit lower/upper triangular matrices is unit lower/upper triangular.
- A unit lower/upper triangular matrix is nonsingular, and its inverse is unit lower/upper triangular.

Exploration 3.2.5 A lower triangular matrix is not necessarily nonsingular. How do we know that a *unit* lower triangular matrix must be nonsingular?

In fact, the inverse of each $M^{(k)}$ is easily computed. We have

$$
L^{(k)} = [M^{(k)}]^{-1} =
\begin{bmatrix}
1 & 0 & \cdots & & \cdots & & \cdots\cdots\cdots\cdots\cdots\cdots & 0 \\
0 & 1 & 0 & & & & & 0 \\
\vdots & \ddots & \ddots & & \ddots & & & \vdots \\
\vdots & & 0 & & \ddots & & \ddots & \vdots \\
\vdots & & \vdots & m_{k+1,k} & & \ddots & \ddots & \vdots \\
\vdots & & \vdots & & 0 & & \ddots & \ddots & \vdots \\
\vdots & & \vdots & & \vdots & & \ddots & \ddots & \ddots & \vdots \\
\vdots & & \vdots & & \vdots & & \vdots & & \ddots & 1 & 0 \\
0 & \cdots & 0 & & m_{nk} & & 0 & \cdots & \cdots & 0 & 1
\end{bmatrix}.
\tag{3.4}
$$

Exploration 3.2.6 Prove (3.4).

It follows that if we define $M = M^{(n-1)} \cdots M^{(1)}$, then M is unit lower triangular, and $MA = U$, where U is upper triangular. We then have $A = M^{-1}U = LU$, where

$$
L = L^{(1)} \cdots L^{(n-1)} = [M^{(1)}]^{-1} \cdots [M^{(n-1)}]^{-1}
$$

is also unit lower triangular. Furthermore, from the structure of each matrix $L^{(k)}$, it can readily be determined that

$$
L =
\begin{bmatrix}
1 & 0 & \cdots & & \cdots & 0 \\
m_{21} & 1 & 0 & & & \vdots \\
\vdots & m_{32} & \ddots & & \ddots & \vdots \\
\vdots & & \ddots & & 1 & 0 \\
m_{n1} & m_{n2} & \cdots & & m_{n,n-1} & 1
\end{bmatrix}.
\tag{3.5}
$$

That is, L stores all of the multipliers used during Gaussian Elimination. The factorization of A that we have obtained,

$$
A = LU,
$$

is called the **LU Decomposition**, or **LU Factorization**, of A.

Exploration 3.2.7 Express each matrix $L^{(k)}$ from (3.4) as an **outer product update** (see Section B.7.3) of the form $L^{(k)} = I + \mathbf{u}\mathbf{v}^T$. What are the vectors $\mathbf{u}$ and $\mathbf{v}$?

Exploration 3.2.8 Use the result of Exploration 3.2.7 to prove (3.5).

3.2.2 Solution of $Ax = b$

Once the LU Decomposition $A = LU$ has been computed, we can solve the system $Ax = b$ by first noting that if x is the solution, then

$$Ax = LUx = b.$$

Therefore, we can obtain x by first solving the system

$$Ly = b,$$

and then solving

$$Ux = y.$$

Then, if b should change, only these two triangular systems need to be solved in order to obtain the new solution; the LU Decomposition does not need to be recomputed.

The system $Ux = y$ can be solved by back substitution, since U is upper triangular. To solve $Ly = b$, we can use **forward substitution**, since L is unit lower triangular.

Algorithm 3.2.2 (Forward Substitution) Given an $n \times n$ unit lower triangular matrix L and an n-vector b, the following algorithm computes an n-vector y such that $Ly = b$.

for $i = 1, 2, \ldots, n$ **do**
  $y_i = b_i$
  for $j = 1, 2, \ldots, i - 1$ **do**
    $y_i = y_i - \ell_{ij} y_j$
  end for
end for

Exploration 3.2.9 Write a MATLAB function

$$y=\texttt{forwsub(L,b)}$$

that implements Algorithm 3.2.2 for forward substitution. Try your function on the following unit lower triangular matrix.

$$L = \begin{bmatrix} 1 & 0 & 0 & 0 \\ 2 & 1 & 0 & 0 \\ 3 & 3 & 1 & 0 \\ 4 & 6 & 4 & 1 \end{bmatrix}$$

Like back substitution, this algorithm requires $O(n^2)$ floating-point operations. Unlike back substitution, there is no division of the ith component of the solution by a diagonal entry of the matrix, but this is only because in this context, L is *unit* lower triangular, so $\ell_{ii} = 1$. When applying forward substitution to a general lower triangular matrix, such a division is required.

> **Exploration 3.2.10** Carry out, by hand, the LU Decomposition of the matrix
> $$A = \begin{bmatrix} 5 & 0 & 2 & -5 \\ -1 & -5 & 3 & 2 \\ 2 & -2 & 1 & -3 \\ 3 & 3 & -1 & 0 \end{bmatrix}.$$

Because both forward and back substitution require only $O(n^2)$ operations, whereas Gaussian Elimination requires $O(n^3)$ operations (see Exercise 6), changes in the right-hand side **b** are best handled by computing the factors L and U once, and storing them, rather than performing Gaussian Elimination on the entire system $A\mathbf{x} = \mathbf{b}$ for each choice of **b**. This storage of L and U can be accomplished quite efficiently, because L is a unit lower triangular matrix. It follows from this that L and U can be stored in a single $n \times n$ matrix by storing U in the upper triangular part, and the multipliers m_{ij} in the strictly lower triangular part (that is, the part below the main diagonal). Once the LU Decomposition has been computed, the appropriate entries of this matrix, that contains the "interesting" parts of both L and U, can be accessed by the algorithms for forward and back substitution when solving $Ly = \mathbf{b}$ and $U\mathbf{x} = \mathbf{y}$, respectively.

> **Exploration 3.2.11** Modify your function `gausselim` from Exploration 3.1.7 so that it is called as follows: `LU=gausselim(A)`. That is, it no longer takes a vector **b** as an input argument, nor does it return it as an output argument. The entries of the output argument `LU` are set as follows: for $i > j$, the (i,j) entry is equal to the multiplier m_{ij} from Algorithm 3.1.6, and for $i \leq j$, the (i,j) entry is equal to u_{ij}, where U is the upper triangular matrix that is the final result of Gaussian Elimination.

3.2.3 Existence and Uniqueness

Not every nonsingular $n \times n$ matrix A has an LU Decomposition. For example, if $a_{11} = 0$, then the multipliers $m_{i1} = a_{i1}/a_{11}$, for $i = 2, 3, \ldots, n$, are not defined, so no multiple of the first row can be subtracted from the other rows to eliminate subdiagonal entries in the first column. That is, Gaussian Elimination can break down. Even if $a_{11} \neq 0$, it can happen that the (j, j) entry of $A^{(j)}$ is zero, in which case a similar breakdown occurs. When this is the case, the LU Decomposition of A does not exist. This will be addressed by **pivoting**, resulting in a modification of the LU Decomposition.

It can be shown that the LU Decomposition of an $n \times n$ matrix A *does* exist if and only if the **leading principal submatrices** of A, defined by

$$[A]_{1:k,1:k} = \begin{bmatrix} a_{11} & \cdots & a_{1k} \\ \vdots & \ddots & \vdots \\ a_{k1} & \cdots & a_{kk} \end{bmatrix}, \quad k = 1, 2, \ldots, n,$$

are all nonsingular. Furthermore, when the LU Decomposition exists, it is unique.

Exploration 3.2.12 Prove that the matrix

$$A = \begin{bmatrix} -1 & 2 & 5 \\ 2 & -4 & 5 \\ -1 & 0 & 5 \end{bmatrix}$$

does not have an LU Decomposition, without actually trying to compute it. If Gaussian elimination was performed on this matrix, at what point would it fail?

Exploration 3.2.13 Prove, by contradiction, that if the LU Decomposition exists, it is unique.

3.2.4 Practical Computation of Determinants

Computing the determinant of an $n \times n$ matrix A using its definition, as given in Section B.11, requires a number of floating-point operations that is $O(n!)$. A more practical method for computing the determinant can be obtained using its properties, some of which we recall from Section B.11:

- If $\tilde{A}$ is obtained from A by adding or subtracting a multiple of a row of A to another row, then $\det(\tilde{A}) = \det(A)$.
- If B is an $n \times n$ matrix, then $\det(AB) = \det(A)\det(B)$.
- If A is a triangular matrix (either upper or lower), then $\det(A) = \prod_{i=1}^{n} a_{ii}$.

It follows from these properties that if Gaussian Elimination is used to reduce A to an upper triangular matrix U, then $\det(A) = \det(U)$, where U is the resulting upper triangular matrix, because the elementary row operations needed to reduce A to U do not change the determinant. Because U is upper triangular, $\det(U)$, being the product of its diagonal entries, can be computed in $n - 1$ multiplications. It follows that the determinant of any matrix can be computed in $O(n^3)$ operations, since the LU Decomposition must be computed first. Still, this is a great improvement over the $O(n!)$ operations required when using the definition of the determinant.

It can also be seen that $\det(A) = \det(U)$ by noting that if $A = LU$, then $\det(A) = \det(L)\det(U)$, by one of the abovementioned properties, but $\det(L) = 1$, because L is a unit lower triangular matrix. It follows from the fact that L is lower triangular that $\det(L)$ is the product of its diagonal entries, and it follows from the fact that L is *unit* lower triangular that all of its diagonal entries are equal to 1.

3.2.5 Pivoting

During Gaussian Elimination, it is necessary to interchange rows of the augmented matrix whenever the diagonal entry of the column currently being processed, known as the **pivot element**, is equal to zero. However, if we examine the main step in Gaussian Elimination,

$$a_{ik}^{(j+1)} = a_{ik}^{(j)} - m_{ij} a_{jk}^{(j)}, \qquad (3.6)$$

we can see that any roundoff error in the computation of $a_{jk}^{(j)}$ is amplified by m_{ij}. Because the multipliers can be arbitrarily large, it follows from the previous analysis that the error in the computed solution can be arbitrarily large, meaning that Gaussian Elimination is an **unstable algorithm** (see the discussion of stable algorithms in Section 2.1.4). Therefore, it is helpful if it can be ensured that the multipliers are small. This can be accomplished by performing row interchanges, or **pivoting**, even when it is not absolutely necessary to do so for elimination to proceed.

3.2.5.1 _Partial Pivoting_

One approach to pivoting is called **partial pivoting**. When eliminating entries in column j, we seek the largest entry in column j, on or below the main diagonal, and then interchange that entry's row with row j. That is, we find an integer p, $j \le p \le n$, such that

$$|a_{pj}^{(j)}| = \max_{j \le i \le n} |a_{ij}^{(j)}|.$$

Then, we interchange rows p and j.

In view of the definition of the multiplier, $m_{ij} = a_{ij}^{(j)}/a_{jj}^{(j)}$, it follows that $|m_{ij}| \le 1$ for $j = 1, \ldots, n-1$ and $i = j+1, \ldots, n$. Furthermore, while pivoting in this manner requires $O(n^2)$ comparisons to determine the appropriate row interchanges, that extra expense is negligible compared to the overall cost of Gaussian Elimination, which is $O(n^3)$, and therefore is outweighed by the potential reduction in roundoff error.

We note that when partial pivoting is used, even though the entries of L satisfy $|\ell_{ij}| \le 1$, the entries of U can still grow substantially. We now illustrate this with a worst-case scenario.

Exploration 3.2.14 Carry out Gaussian Elimination with partial pivoting on the matrix

$$A = \begin{bmatrix} 1 & 0 & 0 & 0 & 1 \\ -1 & 1 & 0 & 0 & 1 \\ -1 & -1 & 1 & 0 & 1 \\ -1 & -1 & -1 & 1 & 1 \\ -1 & -1 & -1 & -1 & 1 \end{bmatrix}.$$

How do the entries of U grow?

3.2.5.2 *Complete Pivoting*

While partial pivoting helps to control the propagation of roundoff error, loss of significant digits can still result if, in the main step of Gaussian Elimination (3.6), $m_{ij}a_{jk}^{(j)}$ is much larger in magnitude than $a_{ik}^{(j)}$. Even though m_{ij} is not large, this can still occur if $a_{jk}^{(j)}$ is particularly large.

Complete pivoting entails finding integers p and q such that

$$|a_{pq}| = \max_{j \leq i \leq n, j \leq q \leq n} |a_{ij}|,$$

and then using both row *and column* interchanges to move a_{pq} into the pivot position in row j and column j. It has been proven [Wilkinson (1961)] that this is an effective strategy for ensuring that Gaussian Elimination is a **stable algorithm**, meaning that the computed solution is the exact solution of a nearby system, as discussed in Section 2.1.4. Unfortunately, complete pivoting requires $O(n^3)$ comparisons, which is why partial pivoting is generally still preferred.

> **Exploration 3.2.15** Perform Gaussian elimination with complete pivoting on the matrix A from Exploration 3.2.14. Does the same growth in the entries of U occur? Explain any difference in behavior.

3.2.5.3 *The LU Decomposition with Partial Pivoting*

Suppose that partial pivoting is performed during Gaussian Elimination. Then, if row j is interchanged with row p, for $p > j$, before entries in column j are eliminated, the matrix $A^{(j)}$ is effectively multiplied on the left by a **permutation matrix** $P^{(j)}$. A permutation matrix is a matrix obtained by permuting the rows (or columns) of the identity matrix I. In $P^{(j)}$, rows j and p of I are interchanged, so that multiplying $A^{(j)}$ on the left by $P^{(j)}$ interchanges these rows of $A^{(j)}$. It follows that the process of Gaussian Elimination with partial pivoting can be described in terms of matrix multiplications, just as it can when no pivoting is performed.

Before proceeding, we discuss an important property of permutation matrices. Let $i_1, i_2, \ldots, i_n$ be a permutation of the indices $1, 2, \ldots, n$. Then, let the permutation matrix P be defined by $p_{i_k,k} = 1$ for $k = 1, 2, \ldots, n$, with all other entries equal to zero.

> **Exploration 3.2.16** Show that for $k = 1, 2, \ldots, n$, row i_k of PA is equal to row k of A.

> **Exploration 3.2.17** Show that $P^{-1} = P^T$.

We then have $P^T P = I$. That is, P is an **orthogonal matrix**. In general, any matrix $Q \in \mathbb{R}^{n \times n}$ is said to be orthogonal if $Q^T Q = I$. The term comes from the fact that if $Q = [\, \mathbf{q}_1 \cdots \mathbf{q}_n \,]$, then $\mathbf{q}_i^T \mathbf{q}_j = 0$ for $i \neq j$. That is, the columns of Q are orthogonal to one another. Furthermore, from $\mathbf{q}_i^T \mathbf{q}_i = 1$, we see that the columns actually form an **orthonormal** set.

Exploration 3.2.18 Show that the *rows* of an orthogonal matrix Q are also orthonormal.

Orthogonal matrices will play an essential role in later chapters.

Example 3.2.3 Let

$$A = \begin{bmatrix} 1 & 4 & 7 \\ 2 & 8 & 5 \\ 3 & 6 & 9 \end{bmatrix}.$$

Applying Gaussian Elimination to A, we subtract twice the first row from the second, and three times the first row from the third, to obtain

$$A^{(2)} = \begin{bmatrix} 1 & 4 & 7 \\ 0 & 0 & -9 \\ 0 & -6 & -12 \end{bmatrix}.$$

At this point, Gaussian Elimination breaks down, because the multiplier $m_{32} = a_{32}/a_{22} = -6/0$ is undefined.

Therefore, we must interchange the second and third rows, which yields the upper triangular matrix

$$U = A^{(3)} = P^{(2)}A^{(2)} = \begin{bmatrix} 1 & 4 & 7 \\ 0 & -6 & -12 \\ 0 & 0 & -9 \end{bmatrix},$$

where $P^{(2)}$ is the *permutation matrix*

$$P^{(2)} = \begin{bmatrix} 1 & 0 & 0 \\ 0 & 0 & 1 \\ 0 & 1 & 0 \end{bmatrix}$$

obtained by interchanging the second and third rows of the identity matrix.

It follows that we have computed the factorization

$$PA = LU,$$

or

$$\begin{bmatrix} 1 & 0 & 0 \\ 0 & 0 & 1 \\ 0 & 1 & 0 \end{bmatrix} \begin{bmatrix} 1 & 4 & 7 \\ 2 & 8 & 5 \\ 3 & 6 & 9 \end{bmatrix} = \begin{bmatrix} 1 & 0 & 0 \\ 3 & 1 & 0 \\ 2 & 0 & 1 \end{bmatrix} \begin{bmatrix} 1 & 4 & 7 \\ 0 & -6 & -12 \\ 0 & 0 & -9 \end{bmatrix}.$$

It can be seen in advance that A does not have an LU Decomposition because the second leading principal submatrix of A, expressed in MATLAB-like syntax as $A(1 : 2, 1 : 2)$, is a singular matrix. $\square$

In the preceding example, Gaussian Elimination produces

$$P^{(2)}M^{(1)}A = U,$$

where

$$M^{(1)} = \begin{bmatrix} 1 & 0 & 0 \\ -2 & 1 & 0 \\ -3 & 0 & 1 \end{bmatrix}.$$

This equation does not lead directly to a factorization of the form $PA = LU$, because the matrices involved are not in an order that is conducive to such a factorization. Ideally, we would like to apply all of the row interchanges to A "up front", so that the resulting matrix PA has an LU Decomposition.

To achieve such a factorization, we use the fact that a permutation matrix is orthogonal to obtain

$$P^{(2)} M^{(1)} [P^{(2)}]^T P^{(2)} A = U,$$

which can be rewritten as

$$\tilde{M}^{(1)} P^{(2)} A = U, \quad \tilde{M}^{(1)} = P^{(2)} M^{(1)} [P^{(2)}]^T = \begin{bmatrix} 1 & 0 & 0 \\ -3 & 1 & 0 \\ -2 & 0 & 1 \end{bmatrix}.$$

We then have $PA = LU$, where $P = P^{(2)}$ and $L = [\tilde{M}^{(1)}]^{-1}$.

Exploration 3.2.19   (a) In the general case of performing Gaussian Elimination with pivoting on an $n \times n$ matrix, find the order in which the permutation matrices $P^{(j)}$ and the multiplier matrices $M^{(j)}$ should be multiplied with A to produce U.

(b) Express the result of part (a) in the form

$$[\tilde{M}^{(n-1)} \tilde{M}^{(n-2)} \cdots \tilde{M}^{(1)}][P^{(n-1)} P^{(n-2)} \cdots P^{(1)}] A = U. \qquad (3.7)$$

What is the formula for $\tilde{M}^{(j)}$, for $j = 1, 2, \ldots, n-1$?

From (3.7), we obtain the decomposition $PA = LU$ where

$$P = P^{(n-1)} P^{(n-2)} \cdots P^{(1)}, \quad L = [\tilde{M}^{(1)}]^{-1} [\tilde{M}^{(2)}]^{-1} \cdots [\tilde{M}^{(n-1)}]^{-1}.$$

Fortunately, it is easier to obtain L than Exploration 3.2.19 suggests. As Gaussian Elimination stores the multipliers m_{ij} in the lower triangle of A while reducing the rest of the matrix to U, it is sufficient to apply any row interchange to the *entirety* of the rows–including the portion containing multipliers. This is because $\tilde{M}^{(j)}$ is obtained by applying permutations $P^{(i)}$ to $M^{(j)}$, for $i > j$–row interchanges that occur *after* the multipliers from column j are stored in $M^{(j)}$.

Once the LU Decomposition $PA = LU$ has been computed, we can solve the system $Ax = b$ by first noting that if $\mathbf{x}$ is the solution, then

$$PA\mathbf{x} = LU\mathbf{x} = P\mathbf{b}.$$

Therefore, we can obtain $\mathbf{x}$ by first solving the system $L\mathbf{y} = P\mathbf{b}$, and then solving $U\mathbf{x} = \mathbf{y}$. Then, if $\mathbf{b}$ should change, only these two triangular systems need to be solved in order to obtain the solution. As in the case of Gaussian Elimination without pivoting, the LU Decomposition does not need to be recomputed.

Exploration 3.2.20 Write a MATLAB function

$$x=solveAxb(A,b)$$

that accepts as input arguments an $n \times n$ matrix A and column n-vector b, and returns as output a column n-vector x such that $A * x = b$. The solution must be computed by performing Gaussian Elimination with partial pivoting to obtain the LU Decomposition of PA, where P is a permutation matrix, followed by forward and back substitution. Use your implementations of algorithms from previous explorations to simplify your task. If the matrix A is found to be singular, or if the input arguments do not have compatible dimensions, then display an appropriate error message and abort. Consult the help page for the MATLAB function **error** for this purpose.

3.2.5.4 *Practical Computation of Determinants, Revisited*

When Gaussian Elimination is used without pivoting to obtain the factorization $A = LU$, we have $\det(A) = \det(U)$, because $\det(L) = 1$ due to L being unit lower triangular. When pivoting is used, we have the factorization $PA = LU$, where P is a permutation matrix. Because a permutation matrix is **orthogonal**; that is, $P^T P = I$, and $\det(A) = \det(A^T)$ for any square matrix, it follows that $\det(P)^2 = 1$, or $\det(P) = \pm 1$. Therefore, $\det(A) = \pm \det(U)$, where the sign depends on the sign of $\det(P)$.

To determine this sign, we note that when two rows (or columns) of A are interchanged, the sign of the determinant changes (see Section B.11). Therefore, $\det(P) = (-1)^p$, where p is the number of row interchanges that are performed during Gaussian Elimination. The number p is known as the *sign* of the permutation represented by P that determines the final ordering of the rows. We conclude that $\det(A) = (-1)^p \det(U)$.

Exploration 3.2.21 Write a MATLAB function

$$d=mydet(A)$$

that computes the determinant of an $n \times n$ matrix A that is passed as an input argument, using the LU Decomposition and partial pivoting.

3.2.6 Concept Check

1. What is the LU Decomposition of a matrix A? How are the matrices L and U related to Gaussian Elimination?
2. What is a unit lower triangular matrix, and what properties does it have?
3. Does the LU Decomposition always exist? If it does exist, is it unique?

4. Why is pivoting necessary in the numerical solution of linear systems, besides avoiding division by zero?
5. Explain the main ideas behind partial pivoting and complete pivoting.
6. What are permutation matrices, and how are they used in the solution of $A\mathbf{x} = \mathbf{b}$?
7. How is the determinant of A computed after its LU Decomposition is obtained? How is this affected by pivoting?

3.3 Special Matrices

For a general system of linear equations $A\mathbf{x} = \mathbf{b}$, in which A is $n \times n$ and nonsingular, Gaussian Elimination with partial pivoting is most often the method of choice. However, for certain categories of matrices, more efficient methods are available. In this section, we consider a few such categories.

3.3.1 Banded Matrices

The first type of special matrix we will discuss is called a **banded matrix**. An $n \times n$ matrix A is said to have **upper bandwidth** p if $a_{ij} = 0$ whenever $j - i > p$. We consider an example so that we can visualize this scenario.

$$B = \begin{bmatrix} -2 & 1 & 5 & 0 & 0 \\ 1 & 1 & 8 & 4 & 0 \\ 0 & 4 & 3 & -2 & 7 \\ 0 & 0 & 11 & 1 & 1 \\ 0 & 0 & 0 & 3 & -2 \end{bmatrix},$$

The above matrix has an upper bandwidth of $p = 2$ since $a_{ij} = 0$ whenever $j - i > 2$; see, for example, that $a_{14} = 0$. Similarly, A has **lower bandwidth** q if $a_{ij} = 0$ whenever $i - j > q$. The above matrix has a lower bandwidth of $q = 1$ since $a_{ij} = 0$ whenever $i - j > 1$; see, for example, that $a_{31} = 0$.

A matrix that has upper bandwidth p and lower bandwidth q is said to have **bandwidth** $w = p + q + 1$. Any $n \times n$ matrix A has a bandwidth $w \le 2n - 1$. If $w < 2n - 1$, then A is said to be **banded**. However, cases in which the bandwidth is $O(1)$, such as when A is a **tridiagonal** matrix for which $p = q = 1$, are of particular interest because for such matrices, Gaussian Elimination, forward substitution and back substitution are much more efficient than in the general case.

Example 3.3.1 The matrix

$$A = \begin{bmatrix} -2 & 1 & 0 & 0 & 0 \\ 1 & -2 & 1 & 0 & 0 \\ 0 & 1 & -2 & 1 & 0 \\ 0 & 0 & 1 & -2 & 1 \\ 0 & 0 & 0 & 1 & -2 \end{bmatrix},$$

which arises from the discretization of the second derivative operator (see Section 14.2.1), is banded with lower bandwidth and upper bandwidth 1, and total bandwidth 3. That is, A is tridiagonal. Its LU Decomposition is

$$
\begin{bmatrix}
-2 & 1 & 0 & 0 & 0 \\
1 & -2 & 1 & 0 & 0 \\
0 & 1 & -2 & 1 & 0 \\
0 & 0 & 1 & -2 & 1 \\
0 & 0 & 0 & 1 & -2
\end{bmatrix} =
$$

$$
\begin{bmatrix}
1 & 0 & 0 & 0 & 0 \\
-\frac{1}{2} & 1 & 0 & 0 & 0 \\
0 & -\frac{2}{3} & 1 & 0 & 0 \\
0 & 0 & -\frac{3}{4} & 1 & 0 \\
0 & 0 & 0 & -\frac{4}{5} & 1
\end{bmatrix}
\begin{bmatrix}
-2 & 1 & 0 & 0 & 0 \\
0 & -\frac{3}{2} & 1 & 0 & 0 \\
0 & 0 & -\frac{4}{3} & 1 & 0 \\
0 & 0 & 0 & -\frac{5}{4} & 1 \\
0 & 0 & 0 & 0 & -\frac{6}{5}
\end{bmatrix}.
$$

We see that L has lower bandwidth 1, and U has upper bandwidth 1. $\square$

Exploration 3.3.1   (a) If A has lower bandwidth q, and $A = LU$ is the LU Decomposition of A (without pivoting), then what is the lower bandwidth of the lower triangular matrix L?

(b) How many entries, at most, need to be eliminated per column?

(c) If A has upper bandwidth p, and $A = LU$ is the LU Decomposition of A (without pivoting), then what is the upper bandwidth of the upper triangular matrix U?

(d) How many entries, at most, need to be updated per row by each row operation?

Exploration 3.3.2 If A has $O(1)$ bandwidth, then approximately how many floating-point operations do Gaussian Elimination, forward substitution and back substitution require? Use big-O notation to express your answers.

Exploration 3.3.3   (a) Write a MATLAB function

        [L,U]=tridiaglu(A)

to find the LU Decomposition of a tridiagonal matrix.

(b) Now, write a MATLAB function [L,U]=bandedlu(A,p,q) that does the same for any banded matrix with upper bandwidth p and lower bandwidth q.

(c) Approximately how many floating-point operations does each function require? Your answer for part (b) should be in terms of p and q. You may use big-O notation.

When a matrix A is banded with bandwidth w, it is wasteful to store it in the traditional 2-dimensional array. Instead, it is much more efficient to store the entries of A in w vectors of length at most n. Then, the algorithms for Gaussian Elimination, forward substitution and back substitution can be modified appropriately to work with these vectors. For example, to perform Gaussian Elimination on a tridiagonal matrix, we can proceed as in the following algorithm. We assume that the main diagonal of A is stored in the vector $\mathbf{a}$, the **subdiagonal** (entries $a_{j+1,j}$) is stored in the vector $\mathbf{l}$, and the **superdiagonal** (entries $a_{j,j+1}$) is stored in the vector $\mathbf{u}$.

for $j = 1, 2, \ldots, n-1$ **do**
 $l_j = l_j / a_j$
 $a_{j+1} = a_{j+1} - l_j u_j$
end for

After Gaussian Elimination, the components of the vector $\mathbf{l}$ are the subdiagonal entries of L in the LU Decomposition of A (that is, the multipliers), and the components of the vector $\mathbf{u}$ are the superdiagonal entries of U.

Notice that this algorithm is much simpler than Gaussian Elimination (Algorithm 3.1.6) for a general matrix. This is because the number of operations for solving a tridiagonal system is significantly reduced. Once it has completed, we can use these updated vectors to solve the system $A\mathbf{x} = \mathbf{b}$ using forward and back substitution as follows:

$y_1 = b_1$
for $i = 2, 3, \ldots, n$ **do**
 $y_i = b_i - l_{i-1} y_{i-1}$
end for
$x_n = y_n / a_n$
for $i = n-1, n-2, \ldots, 1$ **do**
 $x_i = (y_i - u_i x_{i+1}) / a_i$
end for

Exploration 3.3.4 Write a MATLAB function

$$\texttt{x=tridiagsolve(l,a,u,b)}$$

that accepts as input three vectors $\mathbf{l}$, $\mathbf{a}$, and $\mathbf{u}$ that represent the subdiagonal, diagonal, and superdiagonal, respectively, of an $n \times n$ tridiagonal matrix A, and a column n-vector $\mathbf{b}$. The function returns a column n-vector $\mathbf{x}$ that is the solution of $A\mathbf{x} = \mathbf{b}$.

Pivoting can cause difficulties for banded systems because it can cause **fill-in**: the introduction of nonzero entries outside of the band. For this reason, when pivoting is necessary, pivoting schemes that offer more flexibility than partial pivoting are typically used. The resulting trade-off is that the entries of L are permitted to be somewhat larger, but the **sparsity** (that is, the occurrence of zero entries) of A is preserved to a greater extent.

3.3.2   Symmetric Matrices

As defined in Section B.7.2, an $n \times n$ matrix A is **symmetric** if $A = A^T$. We will see that symmetry allows more efficient solution of $A\mathbf{x} = \mathbf{b}$.

3.3.2.1   *The LDL^T Factorization*

Suppose that A is a nonsingular $n \times n$ matrix that has an LU Decomposition $A = LU$. If we define the diagonal matrix D by

$$D = \begin{bmatrix} u_{11} & 0 & \cdots & 0 \\ 0 & u_{22} & \ddots & 0 \\ \vdots & \ddots & \ddots & \vdots \\ 0 & \cdots & 0 & u_{nn} \end{bmatrix} = \mathrm{diag}(u_{11}, u_{22}, \ldots, u_{nn}),$$

then D is also nonsingular, and then the matrix $D^{-1}U$ has entries

$$[D^{-1}U]_{ij} = \frac{u_{ij}}{u_{ii}}, \quad i, j = 1, 2, \ldots, n.$$

The diagonal entries of this matrix are equal to one, and therefore $D^{-1}U$ is **unit upper triangular**.

Therefore, if we define the matrix M by $M^T = D^{-1}U$, then we have the factorization

$$A = LU = LDD^{-1}U = LDM^T,$$

where *both* L and M are unit lower triangular, and D is diagonal. This is called the LDM^T Factorization of A. Because of the close connection between the LDM^T factorization and the LU Decomposition, the LDM^T Factorization is not normally used in practice for solving the system $A\mathbf{x} = \mathbf{b}$ for a general nonsingular matrix A. However, this factorization becomes much more interesting when A is symmetric.

If $A = A^T$, then $LDM^T = (LDM^T)^T = MD^TL^T = MDL^T$, because D, being a diagonal matrix, is also symmetric. Because L and M, being unit lower triangular, are nonsingular, it follows that

$$M^{-1}LD = DL^TM^{-T} = D(M^{-1}L)^T.$$

The matrix $M^{-1}L$ is unit lower triangular. Therefore, the above equation states that a lower triangular matrix is equal to an upper triangular matrix, which implies that both matrices must be diagonal. It follows that $M^{-1}L = I$, because its diagonal entries are already known to be equal to one. We conclude that $L = M$, and thus we have the **LDL^T Factorization**

$$A = LDL^T. \tag{3.8}$$

We now examine this factorization of a symmetric matrix

Example 3.3.2 Consider Gaussian Elimination applied to the symmetric matrix

$$S = \begin{bmatrix} 4 & 2 \\ 2 & 3 \end{bmatrix}.$$

The only row operation required would be to subtract half of row 1 from row 2

$$S = LU = \begin{bmatrix} 1 & 0 \\ \frac{1}{2} & 1 \end{bmatrix} \cdot \begin{bmatrix} 4 & 2 \\ 0 & 2 \end{bmatrix}$$

There is no symmetry in this factorization, but this can be fixed by applying the same row operation as a *column* operation. This yields

$$S = LDL^T = \begin{bmatrix} 1 & 0 \\ \frac{1}{2} & 1 \end{bmatrix} \cdot \begin{bmatrix} 4 & 0 \\ 0 & 2 \end{bmatrix} \begin{bmatrix} 1 & \frac{1}{2} \\ 0 & 1 \end{bmatrix}$$

$\square$

From the above example we see that whenever we have a symmetric matrix A we have the factorization $A = LDU$ where L is lower unit triangular, D is diagonal, and U is upper unit triangular. By symmetry, we have $U = L^T$, and therefore we have the factorization $A = LDL^T$. This factorization is quite economical, compared to the LU and LDM^T Factorizations, because only $n(n+1)/2$ entries are needed to represent L and D. Once these factors are obtained, we can solve $Ax = b$ by solving the simple systems

$$Ly = b, \quad Dz = y, \quad L^T x = z,$$

using forward substitution, simple divisions, and back substitution, respectively.

The LDL^T Factorization can be obtained by performing Gaussian Elimination, but this is not efficient, because Gaussian Elimination requires performing operations on entire rows of A, which does not exploit symmetry. This can be addressed by omitting updates of the upper triangular portion of A, as they do not influence the computation of L and D. An alternative approach, that is equally efficient in terms of the number of floating-point operations, but more desirable overall due to its use of vector operations, involves computing L column-by-column. If we multiply both sides of the matrix equation $A = LDL^T$ by the standard basis vector e_j (see Section B.4) to extract the jth column of each side, we obtain

$$a_j = \sum_{k=1}^{j} \ell_k v_{kj},$$

where

$$A = \begin{bmatrix} a_1 & \cdots & a_n \end{bmatrix}, \quad L = \begin{bmatrix} \ell_1 & \cdots & \ell_n \end{bmatrix}$$

are column partitions of A and L, respectively, and $v_j = DL^T e_j$.

Suppose that columns $1, 2, \ldots, j-1$ of L, as well as $d_{11}, d_{22}, \ldots, d_{j-1,j-1}$, the first $j-1$ diagonal entries of D, have already been computed. Then, we can compute

$v_{kj} = d_{kk}\ell_{jk}$ for $k = 1, 2, \ldots, j-1$, because these quantities depend on entries of L and D that are available. It follows that

$$\mathbf{a}_j - \sum_{k=1}^{j-1} \ell_k v_{kj} = \ell_j v_{jj} = \ell_j d_{jj} \ell_{jj}.$$

However, $\ell_{jj} = 1$, which means that we can obtain d_{jj} from the jth component of the vector

$$\mathbf{u}_j = \mathbf{a}_j - \sum_{k=1}^{j-1} \ell_k v_{kj},$$

and then obtain the "interesting" portion of the new column ℓ_j, that is, entries $j+1, \ldots, n$, by computing $\ell_j = \mathbf{u}_j / d_{jj}$. The remainder of this column is zero, because L is lower triangular. The entire algorithm proceeds as follows:

Algorithm 3.3.3 (LDL^T Factorization) Given a symmetric matrix A, the following algorithm computes the factorization $A = LDL^T$, where L is a unit lower triangular matrix and D is a diagonal matrix.

$L = 0$
$D = 0$
for $j = 1, 2, \ldots, n$ **do**
    for $k = 1, 2, \ldots, j-1$ **do**
        $v_{kj} = d_{kk}\ell_{jk}$
    end for
    $\mathbf{u}_j = A(j:n, j)$
    for $k = 1, 2, \ldots, j-1$ **do**
        $\mathbf{u}_j = \mathbf{u}_j - L(j:n, k)v_{kj}$
    end for
    $d_{jj} = u_{1j}$
    $L(j:n, j) = \mathbf{u}_j / d_{jj}$
end for

Exploration 3.3.5 Write a MATLAB function

```
[L,D]=ldlt(A)
```

that implements Algorithm 3.3.3. This algorithm should take the random symmetric matrix A as input, and return L and D as output. Check to see if A = L * D * L', aside from roundoff error. How can the output be stored more economically?

This algorithm requires approximately $\frac{1}{3}n^3$ floating-point operations, which is approximately half as many as Gaussian Elimination. If pivoting is required, then it is necessary to interchange both rows *and* columns to preserve symmetry. However, we will soon see that for an important class of symmetric matrices, pivoting is unnecessary.

> **Exploration 3.3.6** Using the same approach as in Exploration 3.1.9, verify that Algorithm 3.3.3 requires $\frac{1}{3}n^3 + O(n^2)$ operations.

3.3.3 Symmetric Positive Definite Matrices

Now, we consider the special case of a symmetric matrix A in which all the entries of D, where $A = LDL^T$, are positive.

3.3.3.1 *Properties*

A real, $n \times n$ symmetric matrix A is **symmetric positive definite**, or simply **positive definite**, if for any nonzero vector $\mathbf{x}$,

$$\mathbf{x}^T A \mathbf{x} > 0.$$

A symmetric positive definite matrix is the generalization to $n \times n$ matrices of a positive number. It should be noted that only a symmetric matrix can be called positive definite. A nonsymmetric matrix A can still satisfy $\mathbf{x}^T A \mathbf{x} > 0$ for any nonzero vector $\mathbf{x}$, but it is still not considered positive definite.

If A is symmetric positive definite, then it has the following properties:

- A is nonsingular; in fact, $\det(A) > 0$.
- All of the diagonal entries of A are positive.
- The largest entry of the matrix lies on the diagonal.
- All of the eigenvalues of A are positive.

> **Exploration 3.3.7** Show that if matrices A and B are positive definite, then $A + B$ is positive definite.

> **Exploration 3.3.8** Use the definition of a symmetric positive definite matrix to prove that if A is symmetric positive definite, which implies A is nonsingular, then A^{-1} is also symmetric positive definite.

While the above properties can be used to quickly conclude that a matrix is *not* symmetric positive definite, in general it is not easy to determine whether a given $n \times n$ symmetric matrix A *is* positive definite. One approach is to check the matrices

$$A_k = \begin{bmatrix} a_{11} & a_{12} & \cdots & a_{1k} \\ a_{21} & a_{22} & \cdots & a_{2k} \\ \vdots & \vdots & & \vdots \\ a_{k1} & a_{k2} & \cdots & a_{kk} \end{bmatrix}, \quad k = 1, 2, \ldots, n, \tag{3.9}$$

which are the **leading principal submatrices** of A. Recall these submatrices were used in Section 3.2.3 to discuss the existence of the LU Decomposition.

Exploration 3.3.9 Let A be a symmetric matrix such that the factorization $A = LDL^T$ exists, where L is unit lower triangular and D is a diagonal matrix with nonzero diagonal entries. Use induction to show that

$$d_{11} = a_{11}, \quad d_{jj} = \frac{\det(A_j)}{\det(A_{j-1})}, \quad j = 2, 3, \ldots, n$$

where A_j is a leading principal submatrix of A, as defined in (3.9).

Exploration 3.3.10 Use the result of Exploration 3.3.9 to prove that A is symmetric positive definite if and only if all of its leading principal submatrices $A_1, A_2, \ldots, A_n$ satisfy $\det(A_k) > 0$ for $k = 1, 2, \ldots, n$. *Hint:* Various properties of determinants from Section B.11 will be helpful.

There are other classes of matrices with similar properties. They are defined as follows: **negative definite**, where $\mathbf{x}^T A \mathbf{x} < 0$; **positive semi-definite**, where $\mathbf{x}^T A \mathbf{x} \geq 0$; and **negative semi-definite**, where $\mathbf{x}^T A \mathbf{x} \leq 0$.

Exploration 3.3.11 Find the values of c for which the following matrix is

    (a) positive definite
    (b) positive semi-definite
    (c) negative definite
    (d) negative semi-definite

$$\begin{bmatrix} 3 & -1 & c \\ -1 & 3 & -1 \\ c & -1 & 3 \end{bmatrix}$$

One desirable property of symmetric positive definite matrices is that Gaussian Elimination can be performed on them without pivoting, and all pivot elements are positive. Furthermore, Gaussian Elimination applied to such matrices is robust with respect to the accumulation of roundoff error. However, Gaussian Elimination is not the most practical approach to solving systems of linear equations involving symmetric positive definite matrices, in terms of the number of floating-point operations that are required.

3.3.3.2 *The Cholesky Factorization*

Instead, it is preferable to compute the **Cholesky Factorization** of A,

$$A = GG^T,$$

where G is a lower triangular matrix with positive diagonal entries. Because A is factored into two matrices that are the transpose of one another, the process of computing the Cholesky Factorization requires about half as many floating-point operations as that of the LU Decomposition.

The algorithm for computing the Cholesky Factorization can be derived by matching entries of GG^T with those of A. This yields the following relation between the entries of G and A,

$$a_{ik} = \sum_{j=1}^{k} g_{ij} g_{kj}, \quad i, k = 1, 2, \ldots, n, \quad i \geq k. \tag{3.10}$$

This relation can be used to obtain formulas for the entries of G, when entries of A are examined in the right order.

Example 3.3.4 Let

$$A = \begin{bmatrix} 9 & -3 & 3 & 9 \\ -3 & 17 & -1 & -7 \\ 3 & -1 & 17 & 15 \\ 9 & -7 & 15 & 44 \end{bmatrix}.$$

A is a symmetric positive definite matrix. To compute its Cholesky Factorization $A = GG^T$, we equate entries of A to those of GG^T, which yields the matrix equation

$$\begin{bmatrix} a_{11} & a_{12} & a_{13} & a_{14} \\ a_{21} & a_{22} & a_{23} & a_{24} \\ a_{31} & a_{32} & a_{33} & a_{34} \\ a_{41} & a_{42} & a_{43} & a_{44} \end{bmatrix} = \begin{bmatrix} g_{11} & 0 & 0 & 0 \\ g_{21} & g_{22} & 0 & 0 \\ g_{31} & g_{32} & g_{33} & 0 \\ g_{41} & g_{42} & g_{43} & g_{44} \end{bmatrix} \begin{bmatrix} g_{11} & g_{21} & g_{31} & g_{41} \\ 0 & g_{22} & g_{32} & g_{42} \\ 0 & 0 & g_{33} & g_{43} \\ 0 & 0 & 0 & g_{44} \end{bmatrix},$$

and the equivalent scalar equations

$$a_{11} = g_{11}^2,$$
$$a_{21} = g_{21} g_{11},$$
$$a_{31} = g_{31} g_{11},$$
$$a_{41} = g_{41} g_{11},$$
$$a_{22} = g_{21}^2 + g_{22}^2,$$
$$a_{32} = g_{31} g_{21} + g_{32} g_{22},$$
$$a_{42} = g_{41} g_{21} + g_{42} g_{22},$$
$$a_{33} = g_{31}^2 + g_{32}^2 + g_{33}^2,$$
$$a_{43} = g_{41} g_{31} + g_{42} g_{32} + g_{43} g_{33},$$
$$a_{44} = g_{41}^2 + g_{42}^2 + g_{43}^2 + g_{44}^2.$$

We compute the nonzero entries of G one column at a time. For the first column, we have

$$g_{11} = \sqrt{a_{11}} = \sqrt{9} = 3,$$
$$g_{21} = a_{21}/g_{11} = -3/3 = -1,$$
$$g_{31} = a_{31}/g_{11} = 3/3 = 1,$$
$$g_{41} = a_{41}/g_{11} = 9/3 = 3.$$

Before proceeding to the next column, we first subtract all contributions to the remaining entries of A from the entries of the first column of G. That is, we update A as follows:

$$a_{22} = a_{22} - g_{21}^2 = 17 - (-1)^2 = 16,$$
$$a_{32} = a_{32} - g_{31}g_{21} = -1 - (1)(-1) = 0,$$
$$a_{42} = a_{42} - g_{41}g_{21} = -7 - (3)(-1) = -4,$$
$$a_{33} = a_{33} - g_{31}^2 = 17 - 1^2 = 16,$$
$$a_{43} = a_{43} - g_{41}g_{31} = 15 - (3)(1) = 12,$$
$$a_{44} = a_{44} - g_{41}^2 = 44 - 3^2 = 35.$$

Now, we can compute the nonzero entries of the second column of G just as for the first column:

$$g_{22} = \sqrt{a_{22}} = \sqrt{16} = 4,$$
$$g_{32} = a_{32}/g_{22} = 0/4 = 0,$$
$$g_{42} = a_{42}/g_{22} = -4/4 = -1.$$

We then remove the contributions from G's second column to the remaining entries of A:

$$a_{33} = a_{33} - g_{32}^2 = 16 - 0^2 = 16,$$
$$a_{43} = a_{43} - g_{42}g_{32} = 12 - (-1)(0) = 12,$$
$$a_{44} = a_{44} - g_{42}^2 = 35 - (-1)^2 = 34.$$

The nonzero portion of the third column of G is then computed as follows:

$$g_{33} = \sqrt{a_{33}} = \sqrt{16} = 4,$$
$$g_{43} = a_{43}/g_{43} = 12/4 = 3.$$

Finally, we compute g_{44}:

$$a_{44} = a_{44} - g_{43}^2 = 34 - 3^2 = 25, \quad g_{44} = \sqrt{a_{44}} = \sqrt{25} = 5.$$

Thus the complete Cholesky Factorization of A is

$$\begin{bmatrix} 9 & -3 & 3 & 9 \\ -3 & 17 & -1 & -7 \\ 3 & -1 & 17 & 15 \\ 9 & -7 & 15 & 44 \end{bmatrix} = \begin{bmatrix} 3 & 0 & 0 & 0 \\ -1 & 4 & 0 & 0 \\ 1 & 0 & 4 & 0 \\ 3 & -1 & 3 & 5 \end{bmatrix} \begin{bmatrix} 3 & -1 & 1 & 3 \\ 0 & 4 & 0 & -1 \\ 0 & 0 & 4 & 3 \\ 0 & 0 & 0 & 5 \end{bmatrix}.$$

□

This approach to the Cholesky Factorization arises from the **outer product** perspective of matrix multiplication, discussed in Section B.7.6. From this perspective, we see that $A = GG^T$ is equivalent to

$$A = \sum_{j=1}^{n} \mathbf{g}_j \mathbf{g}_j^T, \tag{3.11}$$

where $G = \begin{bmatrix} \mathbf{g}_1 \cdots \mathbf{g}_n \end{bmatrix}$ is a column partition of G. Equation (3.11) suggests that we can first compute $\mathbf{g}_1$, as in Example 3.3.4, and then subtract $\mathbf{g}_1 \mathbf{g}_1^T$ from A. Because G is lower triangular, only $\mathbf{g}_1$ can make any contribution to the first row or column of A. Therefore, once $\mathbf{g}_1$ has been computed and $\mathbf{g}_1 \mathbf{g}_1^T$ subtracted off from A, we can focus on the lower right $(n-1) \times (n-1)$ block of A, and use its first column to obtain $\mathbf{g}_2$, and so on. This leads to the following algorithm.

Algorithm 3.3.5 (Cholesky Factorization) Given a $A \in \mathbb{R}^{n \times n}$ that is symmetric positive definite matrix, the following algorithm computes the Cholesky Factorization $A = GG^T$, where G is lower triangular with positive diagonal entries.

for $j = 1, 2, \ldots, n$ do
  $g_{jj} = \sqrt{a_{jj}}$
  for $i = j + 1, j + 2, \ldots, n$ do
    $g_{ij} = a_{ij}/g_{jj}$
    for $k = j + 1, \ldots, i$ do
      $a_{ik} = a_{ik} - g_{ij}g_{kj}$
    end for
  end for
end for

For each j, the innermost loop subtracts the matrix $\mathbf{g}_j \mathbf{g}_j^T$ from A. Equivalently, it subtracts off all terms but the last (corresponding to $j = k$) in (3.10). Therefore, in the outermost loop of Algorithm 3.3.5, for each j, the contributions of all columns $\mathbf{g}_\ell$ of G, where $\ell < j$, have already been subtracted from A. This allows column j of G to easily be computed by the steps in the outer loops, which account for the last term in the summation for a_{ik} in (3.10), in which $j = k$.

If A is *not* symmetric positive definite, then the algorithm will break down, because it will attempt to compute g_{jj}, for some j, by taking the square root of a negative number, or divide by a zero g_{jj}.

Example 3.3.6 The matrix

$$A = \begin{bmatrix} 4 & 3 \\ 3 & 2 \end{bmatrix}$$

is symmetric but not positive definite, because $\det(A) = 4(2) - 3(3) = -1 < 0$. If we attempt to compute the Cholesky Factorization $A = GG^T$, we have

$$g_{11} = \sqrt{a_{11}} = \sqrt{4} = 2,$$
$$g_{21} = a_{21}/g_{11} = 3/2,$$
$$a_{22} = a_{22} - g_{21}^2 = 2 - 9/4 = -1/4,$$
$$g_{22} = \sqrt{a_{22}} = \sqrt{-1/4},$$

and the algorithm breaks down. $\square$

In fact, due to the expense involved in computing determinants, the Cholesky Factorization is also an efficient method for checking whether a symmetric matrix is also positive definite.

Exploration 3.3.12 Write a MATLAB function

$$[G,isposdef]=cholesky(A)$$

that implements Algorithm 3.3.5. The second output argument `isposdef` is a logical variable that has the value `true` if A is in fact symmetric positive definite. Your function should ensure that this variable is set to `false` if the Cholesky Factorization should break down for any reason.

Exploration 3.3.13 Use the approach from Exploration 3.1.9 to estimate the number of floating-point operations required by your function `cholesky` from Exploration 3.3.12. Also, use the MATLAB commands `tic` and `toc` to determine the execution time for matrices of different sizes. As n increases, where A is $n \times n$, is the growth in execution time consistent with the increase in the number of floating-point operations?

Once the Cholesky factor G of A is computed, a system $A\mathbf{x} = \mathbf{b}$ can be solved by first solving $G\mathbf{y} = \mathbf{b}$ by forward substitution, and then solving $G^T\mathbf{x} = \mathbf{y}$ by back substitution. This is similar to the process of solving $A\mathbf{x} = \mathbf{b}$ using the LDL^T factorization, except that there is no diagonal system to solve. In fact, the LDL^T factorization is also known as the "square-root-free Cholesky Factorization", since it computes factors that are similar in structure to the Cholesky factors, but without computing any square roots. Specifically, if $A = GG^T$ is the Cholesky Factorization of A, then $G = LD^{1/2}$. As with the LU Decomposition, the Cholesky Factorization is unique, because the diagonal is required to be positive.

The MATLAB function `chol` computes the Cholesky Factorization of a given matrix A. It produces an error message if it is determined that A is not symmetric positive definite. One key difference between `chol` and the `cholesky` function from Exploration 3.3.12 is that `chol` returns an *upper* triangular matrix R such that $A = R^T R$. That is, $R = G^T$, where G is the Cholesky factor as described in this section.

Exploration 3.3.14 Prove that a square matrix A is symmetric positive definite if and only if it has a Cholesky Factorization $A = GG^T$, where G is lower triangular and has positive diagonal entries. *Hint:* Use the LU Decomposition.

3.3.4   Concept Check

1. Define the terms upper bandwidth and lower bandwidth. What does it mean for a matrix to be banded?

2. What is the computational expense of Gaussian Elimination when applied to a banded matrix?
3. What is the structure of the matrices L and U in the LU Decomposition of a banded matrix?
4. How should a banded matrix be stored in memory?
5. What is fill-in, and why is it problematic?
6. What is the LDL^T factorization, and when is it useful? What is its advantage over other factorizations?
7. What is a symmetric positive definite matrix, and what are its properties?
8. What is the Cholesky Factorization, and when does it exist? How is it beneficial for solving linear systems?

3.4   Estimating and Improving Accuracy

In this section, we investigate the sensitivity of the problem of solving $Ax = b$, and discover that it can be **ill-conditioned**, in the sense of Definition 2.1.8. We will then explore various approaches to dealing with this issue.

3.4.1   The Condition Number

To measure the sensitivity of the system $Ax = b$, where $A \in \mathbb{R}^{n \times n}$, we follow the approach from [Golub and van Loan (2012)] and consider the parameterized system

$$(A + \epsilon E)x(\epsilon) = b + \epsilon e, \tag{3.12}$$

where $E \in \mathbb{R}^{n \times n}$ and $e \in \mathbb{R}^n$. Thus $x(0)$ is the exact solution x of $Ax = b$.

> **Exploration 3.4.1** Show that if $\|\epsilon A^{-1} E\| < 1$ for some natural matrix norm, then $A + \epsilon E$ is nonsingular. *Hint:* Use the properties of singular and nonsingular matrices from Section B.9.

Using **backward error analysis**, as described in Section 2.1.3, we can think of $x(\epsilon)$ as the computed solution of $Ax = b$. As such, the perturbations ϵE and ϵe of A and b, respectively, together represent the **backward error** in the computed solution (see Definition 2.1.5).

> **Exploration 3.4.2** Show that the forward absolute error $x(\epsilon) - x$ is given by
> $$x(\epsilon) - x = \epsilon A^{-1}(e - Ex(\epsilon)).$$

For convenience, we let $r = \|\epsilon A^{-1} E\|$ for some natural matrix norm (see Section B.13), and assume ϵ is sufficiently small so that $r < 1$. Using norms to measure the relative forward error in $\mathbf{x}$, we obtain

$$\frac{\|\mathbf{x}(\epsilon) - \mathbf{x}\|}{\|\mathbf{x}\|} = |\epsilon| \frac{\|A^{-1}(\mathbf{e} - E\mathbf{x}(\epsilon))\|}{\|\mathbf{x}\|}$$

$$= |\epsilon| \frac{\|A^{-1}(\mathbf{e} - E\mathbf{x}) - A^{-1}(\epsilon)E(\mathbf{x}(\epsilon) - \mathbf{x})\|}{\|\mathbf{x}\|}$$

$$\leq |\epsilon| \frac{\|A^{-1}(\mathbf{e} - E\mathbf{x})\|}{\|\mathbf{x}\|} + r \frac{\|\mathbf{x}(\epsilon) - \mathbf{x}\|}{\|\mathbf{x}\|}$$

$$\leq \frac{1}{1-r} |\epsilon| \|A^{-1}\| \left(\frac{\|\mathbf{e}\|}{\|\mathbf{x}\|} + \|E\| \right).$$

Multiplying and dividing by $\|A\|$, and using $A\mathbf{x} = \mathbf{b}$ to obtain $\|\mathbf{b}\| \leq \|A\|\|\mathbf{x}\|$, yields

$$\frac{\|\mathbf{x}(\epsilon) - \mathbf{x}\|}{\|\mathbf{x}\|} \leq \frac{1}{1-r} \kappa(A) |\epsilon| \left(\frac{\|\mathbf{e}\|}{\|\mathbf{b}\|} + \frac{\|E\|}{\|A\|} \right),$$

where

$$\kappa(A) = \|A\|\|A^{-1}\|$$

is called the **condition number** of A. We note that this definition depends on the matrix norm used, so for this reason we indicate the norm with a subscript. For example, we define $\kappa_2(A) = \|A\|_2\|A^{-1}\|_2$.

We can see that the relative errors in A and $\mathbf{b}$ can be amplified by a factor of $\kappa(A)$ in the error in the solution. Therefore, if $\kappa(A)$ is large, the problem $A\mathbf{x} = \mathbf{b}$ can be quite sensitive to perturbations in A and $\mathbf{b}$. In this case, we say that A is **ill-conditioned**; otherwise, we say that A is **well-conditioned**. The MATLAB function cond can be used to compute the condition number of a given matrix. By default, the ℓ_2-norm is used, but a different norm can be specified.

Exploration 3.4.3 As will be seen in Part V, the $n \times n$ matrix

$$A = \begin{bmatrix} -2 & 1 & & & \\ 1 & -2 & 1 & & \\ & \ddots & \ddots & \ddots & \\ & & 1 & -2 & 1 \\ & & & 1 & -2 \end{bmatrix}$$

is particularly useful when solving differential equations. Form this matrix for values of n ranging over several orders of magnitude (for example, powers of 2 or 10) and compute $\kappa_2(A)$ for each. Plot $\kappa_2(A)$ versus n, using logarithmic scales on both axes.

Exploration 3.4.4 Explain how the definition of the condition number of A is consistent with the definition of the condition number κ_{rel} given in Definition 2.1.8, applied to the problem of solving $A\mathbf{x} = \mathbf{b}$.

Exploration 3.4.5 Let $A \in \mathbb{R}^{n \times n}$ be nonsingular. Use the fact that for any nonzero $\mathbf{y} \in \mathbb{R}^n$, there exists $\mathbf{x} \in \mathbb{R}^n$ such that $A\mathbf{x} = \mathbf{y}$ to show that

$$\|A^{-1}\|_2 = \frac{1}{\min_{\|\mathbf{x}\|_2 = 1} \|A\mathbf{x}\|_2}.$$

What formula does this result yield for $\kappa_2(A)$?

It follows from the preceding exploration that $\kappa_2(A)$ is a measure of the elongation of the hyperellipsoid $\{A\mathbf{x} \mid \|\mathbf{x}\|_2 = 1\}$.

Example 3.4.1 The matrices

$$A_1 = \begin{bmatrix} 0.45368292 & 0.19382865 \\ 0.70364726 & 0.52104011 \end{bmatrix}, \quad A_2 = \begin{bmatrix} 0.40563526 & 0.26686200 \\ 0.73033346 & 0.48047658 \end{bmatrix}$$

do not appear to be very different from one another, but $\kappa_2(A_1) \approx 10$ while $\kappa_2(A_2) \approx 1.869 \times 10^8$. That is, A_1 is well-conditioned while A_2 is ill-conditioned.

To illustrate the ill-conditioned nature of A_2, we solve the two systems of equations $A_2\mathbf{x}_1 = \mathbf{b}_1$ and $A_2\mathbf{x}_2 = \mathbf{b}_2$ for the unknown vectors $\mathbf{x}_1$ and $\mathbf{x}_2$, where

$$\mathbf{b}_1 = \begin{bmatrix} 0.48554638 \\ 0.87421091 \end{bmatrix}, \quad \mathbf{b}_2 = \begin{bmatrix} 0.39812529 \\ 0.92276554 \end{bmatrix}.$$

These vectors differ from one another by roughly 10%, but the solutions

$$\mathbf{x}_1 = \begin{bmatrix} 1.45237206 \\ -0.38816669 \end{bmatrix}, \quad \mathbf{x}_2 = \begin{bmatrix} 1.0272524 \times 10^7 \\ -1.5614428 \times 10^7 \end{bmatrix}$$

differ by several orders of magnitude, because of the sensitivity of A_2 to perturbations. $\square$

From [Kahan (1966)] we have, for any ℓ_p-norm,

$$\frac{1}{\kappa_p(A)} = \min_{A + \Delta A \text{ singular}} \frac{\|\Delta A\|_p}{\|A\|_p}.$$

That is, in any ℓ_p-norm, $\kappa_p(A)$ measures the relative distance in that norm from A to the set of singular matrices.

Because $\det(A) = 0$ if and only if A is singular, it would appear that the determinant could be used to measure the distance from A to the nearest singular matrix. However, this is generally not the case. It is possible for a matrix to have a relatively large determinant, but be very close to a singular matrix, or for a matrix to have a relatively small determinant, but not be nearly singular. In other words, there is not necessarily correlation between $\det(A)$ and the condition number of A.

Example 3.4.2 Let

$$A = \begin{bmatrix} 1 & -1 & -1 & -1 & -1 & -1 & -1 & -1 & -1 & -1 \\ 0 & 1 & -1 & -1 & -1 & -1 & -1 & -1 & -1 & -1 \\ 0 & 0 & 1 & -1 & -1 & -1 & -1 & -1 & -1 & -1 \\ 0 & 0 & 0 & 1 & -1 & -1 & -1 & -1 & -1 & -1 \\ 0 & 0 & 0 & 0 & 1 & -1 & -1 & -1 & -1 & -1 \\ 0 & 0 & 0 & 0 & 0 & 1 & -1 & -1 & -1 & -1 \\ 0 & 0 & 0 & 0 & 0 & 0 & 1 & -1 & -1 & -1 \\ 0 & 0 & 0 & 0 & 0 & 0 & 0 & 1 & -1 & -1 \\ 0 & 0 & 0 & 0 & 0 & 0 & 0 & 0 & 1 & -1 \\ 0 & 0 & 0 & 0 & 0 & 0 & 0 & 0 & 0 & 1 \end{bmatrix}.$$

Then $\det(A) = 1$, but $\kappa_2(A) \approx 1,918$. That is, A is quite close to a singular matrix, even though $\det(A)$ is not near zero. In fact, there exists a singular matrix $\tilde{A}$ such that $\|A - \tilde{A}\|_2 \approx 0.0029$. That is, a matrix whose entries are equal to those of A to within two decimal places is singular. $\square$

Exploration 3.4.6 Find a matrix A such that $\det(A)$ is very small, but A is very well-conditioned. *Hint:* How are $\det(A)$ and $\kappa(A)$ affected by scaling A?

Exploration 3.4.7 Let $A \in \mathbb{R}^{2 \times 2}$. Show that there actually is a strong correlation between $\det(A)$ and $\kappa_F(A) = \|A\|_F \|A^{-1}\|_F$, where $\|A\|_F$ is the **Frobenius norm** of A, defined in (B.6).

3.4.2 Iterative Refinement

Although we have learned about solving a system of linear equations $A\mathbf{x} = \mathbf{b}$, we have yet to discuss methods of estimating the error in a computed solution $\tilde{\mathbf{x}}$. A simple approach to judging the accuracy of $\tilde{\mathbf{x}}$ is to compute the **residual** vector $\mathbf{r} = \mathbf{b} - A\tilde{\mathbf{x}}$, and then compute the magnitude of $\mathbf{r}$ using any vector norm. However, this approach can be misleading, as a small residual does not necessarily imply that the *error* in the solution, which is $\mathbf{e} = \mathbf{x} - \tilde{\mathbf{x}}$, is small.

Exploration 3.4.8 Prove the following:

(a) $A\mathbf{e} = \mathbf{r}$

(b) $\dfrac{\|\mathbf{e}\|}{\|\mathbf{x}\|} \leq \kappa(A) \dfrac{\|\mathbf{r}\|}{\|\mathbf{b}\|}$

Therefore, if A is ill-conditioned, it is possible for the residual to be small, and the error to still be large.

We can exploit the relationship between the error $\mathbf{e}$ and the residual $\mathbf{r}$, $A\mathbf{e} = \mathbf{r}$, to obtain an estimate of the error, $\tilde{\mathbf{e}}$, by solving the system $A\mathbf{e} = \mathbf{r}$ in the same manner in which we obtained $\tilde{\mathbf{x}}$ by attempting to solve $A\mathbf{x} = \mathbf{b}$. Since $\tilde{\mathbf{e}}$ is an estimate of the error $\mathbf{e} = \mathbf{x} - \tilde{\mathbf{x}}$ in $\tilde{\mathbf{x}}$, it follows that $\tilde{\mathbf{x}} + \tilde{\mathbf{e}}$ is a more accurate approximation of $\mathbf{x}$ than $\tilde{\mathbf{x}}$ is. This is the basic idea behind **iterative refinement**, also known as **iterative improvement** or **residual correction**. The algorithm is as follows:

Algorithm 3.4.3 (Iterative refinement) Given a nonsingular matrix $A \in \mathbb{R}^{n \times n}$ and $\mathbf{b} \in \mathbb{R}^n$, the following algorithm computes the solution of $A\mathbf{x} = \mathbf{b}$.

$\tilde{\mathbf{x}}^{(0)} = \mathbf{0}$
$\mathbf{r}^{(0)} = \mathbf{b}$
for $k = 0, 1, 2, \ldots$ **do** until convergence
    Solve $A\tilde{\mathbf{e}}^{(k)} = \mathbf{r}^{(k)}$
    $\tilde{\mathbf{x}}^{(k+1)} = \tilde{\mathbf{x}}^{(k)} + \tilde{\mathbf{e}}^{(k)}$
    $\mathbf{r}^{(k+1)} = \mathbf{r}^{(k)} - A\tilde{\mathbf{e}}^{(k)}$
end for

A test convergence could be, for example, checking whether the norm of the error estimate $\mathbf{e}^{(k)}$ is less than some tolerance.

The algorithm repeatedly applies the relationship $A\mathbf{e} = \mathbf{r}$, where $\mathbf{e}$ is the error and $\mathbf{r}$ is the residual, to update the computed solution with an estimate of its error. For this algorithm to be effective, it is important that the residual $\tilde{\mathbf{r}}^{(k)}$ be computed as accurately as possible, for example using higher-precision arithmetic than for the rest of the computation. It is shown in [Moler (1967)] that if the vector $\mathbf{r}^{(k)}$ is computed using extended precision, then $\mathbf{x}^{(k)}$ converges to a solution where almost all digits are correct in the original precision when A is well-conditioned relative to machine precision.

Exploration 3.4.9 Write a MATLAB function

$$x = \texttt{iterrefine(A,b,TOL)}$$

that implements Algorithm 3.4.3.

Exploration 3.4.10 Modify your function `iterrefine` from Exploration 3.4.9 to use single-precision floating-point numbers, except in the computation of the residual. Use the MATLAB functions `single` and `double` to convert numbers to single- and double-precision, respectively. Test your function with an error tolerance of 2^{-24}, which is the value of the unit roundoff $\mathbf{u}$ for the IEEE single precision floating-point number system. Is convergence achieved?

> **Exploration 3.4.11** Note that in Algorithm 3.4.3, the new residual $\mathbf{r}^{(k+1)}$ is computed using the formula $\mathbf{r}^{(k+1)} = \mathbf{r}^{(k)} - A\tilde{\mathbf{e}}^{(k)}$, rather than the definition $\mathbf{r}^{(k+1)} = \mathbf{b} - A\mathbf{x}^{(k+1)}$. Show that these formulas are equivalent. Why is the first formula more desirable, as far as floating-point arithmetic is concerned?

3.4.3   Scaling and Equilibration*

As we have seen, the bounds for the error depend on $\kappa(A) = \|A\|\|A^{-1}\|$. Fortunately, in at least some cases we can re-scale the equations so that the condition number is reduced. We replace the system

$$Ax = b$$

by the equivalent system

$$DAx = Db$$

or possibly

$$DAEy = Db$$

where D and E are diagonal matrices and $\mathbf{y} = E^{-1}\mathbf{x}$.

Suppose A is **symmetric positive definite**, as discussed in Section 3.3.3. We want to replace A by DAD, that is, replace a_{ij} by $d_i d_j a_{ij}$, so that $\kappa(DAD)$ is minimized. It turns out that for a class of matrices, such minimization is possible. A matrix A is said to have **Property A** if there exists a permutation matrix Π such that

$$\Pi A \Pi^T = \begin{bmatrix} D_1 & F \\ G & D_2 \end{bmatrix},$$

where D_1 and D_2 are diagonal matrices. For example, all **tridiagonal** matrices (see Section 3.3.1) have Property A. We will encounter such matrices in Part V when solving differential equations.

Example 3.4.4 Suppose

$$A = \begin{bmatrix} 50 & 7 \\ 7 & 1 \end{bmatrix}.$$

Then $\kappa(A) \approx 2599$. However,

$$DAD = \begin{bmatrix} \frac{1}{\sqrt{50}} & 0 \\ 0 & 1 \end{bmatrix} \begin{bmatrix} 50 & 7 \\ 7 & 1 \end{bmatrix} \begin{bmatrix} \frac{1}{\sqrt{50}} & 0 \\ 0 & 1 \end{bmatrix} = \begin{bmatrix} 1 & \frac{7}{\sqrt{50}} \\ \frac{7}{\sqrt{50}} & 1 \end{bmatrix}$$

which yields $\kappa(DAD) \approx 198$.

In general, the optimal scaling of symmetric positive definite matrices with Property A is that which makes the diagonal blocks D_1 and D_2 equal to the identity matrix [Golub and Varah (1974)].

For more general matrices, one scaling strategy is called **equilibration**. The idea is to set $A^{(0)} = A$ and compute $A^{(1/2)} = D^{(1)} A^{(0)} = (d_i^{(1)} a_{ij})$, choosing the diagonal matrix $D^{(1)}$ so that $d_i^{(1)} \sum_{j=1}^{n} |a_{ij}^{(0)}| = 1$. That is, all *row sums* of $|D^{(1)} A^{(0)}|$ are equal to one. Then, we compute $A^{(1)} = A^{(1/2)} E^{(1)} = (a_{ij}^{(1/2)} e_j^{(1)})$, choosing each diagonal entry of the diagonal matrix $E^{(1)}$ so that $e_j^{(1)} \sum_{i=1}^{n} |a_{ij}^{(1/2)}| = 1$. That is, all *column sums* of $|A^{(1/2)} E^{(1)}|$ are equal to one. We then repeat this process, which yields

$$A^{(k+1/2)} = D^{(k+1)} A^{(k)},$$
$$A^{(k+1)} = A^{(k+1/2)} E^{(k+1)}.$$

Under very general conditions, the $A^{(k)}$ converge to a matrix DAE whose row and column sums are all equal. Optimal scaling is discussed in [Bauer (1963)].

3.4.4   Concept Check

1. What is the condition number of a square nonsingular matrix A, and what is its significance for solving $A\mathbf{x} = \mathbf{b}$?
2. What is a geometric interpretation of the condition number?
3. Which measure is more appropriate for determining whether a matrix is nearly singular: the determinant, or the condition number?
4. Briefly describe, in words, the process of iterative refinement. When is it most effective?
5. What is the benefit of scaling the equations in the system $A\mathbf{x} = \mathbf{b}$, even though it does not change the solution?

3.5   Additional Resources

The reader seeking a comprehensive background in linear algebra is referred to the text [Strang (2006)]. Advanced matrix theory is presented in [Horn and Johnson (1985)]. An encyclopedic reference on matrix computations is provided by [Golub and van Loan (2012)]; other excellent textbooks in this area are, among others, [Demmel (1997); Trefethen and Bau (1997)].

In the early days of electronic computers, there was great concern that Gaussian Elimination would not be practical for the solution of large-scale systems of equations due to the limitations of computer arithmetic. In particular, [Hotelling (1943)] warned of exponential accumulation of error. The papers [Turing (1948); von Neumann and Goldstine (1947)] included rigorous analysis of roundoff error and concluded that such exponential growth would only occur in exceptional cases and that Gaussian Elimination was indeed practical. This was confirmed through an analysis in [Wilkinson (1961)]. More accessible presentations of this analysis can be found in [Demmel (1997); Golub and van Loan (2012); Trefethen and Bau (1997)]. Related work covers stability [Higham (2003)] and condition number estimation [Higham (1987)].

Gaussian Elimination, including pivoting, may appear to be a straightforward process. However, even after its acceptance as a general-purpose algorithm, it has still been investigated extensively due to its unusual behavior in certain cases. More recent papers discuss its occasional instability [Foster (1994); Wright (1993)], its general stability [Trefethen (1985); Trefethen and Schreiber (1990)], its sub-optimality [Strassen (1969)] and lack of parallelism [Vavasis (1989)].

Beyond the scope of this book are implementation details related to vectorization, parallelization, blocking algorithms, and other aspects related to hardware, which are discussed in [Demmel (1997); Demmel, et al. (1993); Dongarra, et al. (1998, 1984); Ortega (1988)]. Also not covered are direct methods for sparse matrices [Davis (2006)].

LINPACK is a software package for solving linear systems that is available from Netlib; its documentation is in [Dongarra, et al. (1979)]. It has essentially been superseded by LAPACK, the user guide for which is [Anderson, et al. (1999)]. Both build on BLAS (Basic Linear Algebra Subprograms) which provides architecture-specific implementations of fundamental matrix computations. Helpful references for BLAS are [Dongarra, et al. (1990, 1988); Lawson, et al. (1979)].

3.6   Exercises

1. Consider the matrix

$$A = \begin{bmatrix} 1 & 1+\epsilon \\ 1-\epsilon & 1 \end{bmatrix},$$

where $0 < \epsilon \ll 1$.

   (a) How small can ϵ be for $\mathrm{fl}(\det(A))$ to be exactly zero in double-precision floating point arithmetic?

   (b) Compute the LU Decomposition of A. How small can ϵ be for the double-precision floating-point representation of U to be singular?

   Verify your answers using MATLAB.

2. Let

$$A = \begin{bmatrix} 1/3 & 3/4 \\ 5/3 & 15/4 \end{bmatrix}, \quad \mathbf{b} = \begin{bmatrix} 5/12 \\ 25/12 \end{bmatrix}.$$

   (a) Show that A is singular in exact arithmetic.

   (b) Show that in exact arithmetic, $A\mathbf{x} = \mathbf{b}$ has infinitely many solutions. Describe all such solutions.

   (c) If Gaussian Elimination with partial pivoting is used to compute the factorization $PA = LU$, then at what point would the algorithm fail?

(d) What happens if you use the backslash operator to solve $A\mathbf{x} = \mathbf{b}$, or $U\mathbf{x} = \mathbf{y}$ where $PA = LU$ is the factorization obtained using the MATLAB function lu, and $L\mathbf{y} = P\mathbf{b}$? Are these solutions consistent with the solution set obtained in part 2b?

(e) What is the condition number of A as reported by cond(A)? Based on this result, how many digits of accuracy would you expect in a computed solution?

3. Consider a system of equations of the form

$$\begin{bmatrix} A_1 & 0 \\ B_2 & A_2 \end{bmatrix} \begin{bmatrix} \mathbf{x} \\ \mathbf{y} \end{bmatrix} = \begin{bmatrix} \mathbf{b} \\ \mathbf{c} \end{bmatrix},$$

where A_1 is $m \times m$, A_2 is $n \times n$, $\mathbf{x}$ and $\mathbf{b}$ are m-vectors, and $\mathbf{y}$ and $\mathbf{c}$ are n-vectors. We say that the coefficient matrix in this system is **block lower triangular**. We will assume that A_1 and A_2 are nonsingular.

(a) Describe an algorithm for solving this system by solving two systems with coefficient matrices A_1 and A_2. How many floating-point operations would be required?

(b) How many floating-point operations would be needed to solve this system by simply applying Gaussian Elimination to the entire matrix? For simplicity, we assume that no pivoting is necessary. How does this approach compare to the approach in part 3a?

4. A $n \times n$ matrix A is said to be **strictly column diagonally dominant** if it satisfies the conditions

$$|a_{jj}| > \sum_{i=1, i \neq j}^{n} |a_{ij}|, \quad j = 1, 2, \dots, n.$$

Suppose that Gaussian Elimination with partial pivoting is applied to A. Prove that no pivoting is actually performed. *Hint:* What can be said about the multipliers for such a matrix?

5. Let A be an $n \times n$ nonsingular matrix, and let $\mathbf{u}$ and $\mathbf{v}$ be n-vectors. Explain how each of the following expressions can be computed *without* explicitly computing A^{-1}. Approximately how many floating-point operations are required in each case?

(a) $c = \mathbf{u}^T A^{-1} \mathbf{v}$

(b) $\mathbf{y}^T = \mathbf{v}^T A^{-1}$

6. Use the summation formula

$$\sum_{i=1}^{n} i^2 = \frac{n(n+1)(2n+1)}{6}$$

to prove that the number of floating-point operations required by Gaussian Elimination is $\frac{2}{3}n^3 + O(n^2)$. Use this result, and the number of floating-point operations required by forward and back substitution, to explain why it is far more efficient to solve $A\mathbf{x} = \mathbf{b}$ using Gaussian Elimination and forward and back substitution than to compute A^{-1} explicitly and then obtain $\mathbf{x} = A^{-1}\mathbf{b}$ through matrix-vector multiplication.

7. Let A be an $n \times n$ symmetric positive definite matrix, and assume that its Cholesky Factorization $A = GG^T$ is known. Now, suppose that a row and column are added to A, creating a new matrix

$$B = \begin{bmatrix} A & \mathbf{v} \\ \mathbf{v}^T & \beta \end{bmatrix}.$$

(a) Derive a sufficient condition for B being positive definite, in terms of β and $\|\mathbf{v}\|_2$. *Hint:* The discussion of the **Cauchy-Schwarz inequality** in Section B.13.1 will be helpful.

(b) Assuming B is positive definite, show how the Cholesky Factorization of B can be obtained efficiently from that of A. Approximately how many floating-point operations are required?

8. Suppose an $n \times n$ banded matrix A has upper and lower bandwidth both equal to p. If Gaussian Elimination with partial pivoting is applied to A, what is the maximum possible bandwidth of L and U? Justify your answer.

9. The MATLAB statement H=hilb(n) computes and stores in H the $n \times n$ **Hilbert matrix**, which has entries

$$h_{ij} = \frac{1}{i + j - 1}, \quad i, j = 1, 2, \ldots, n.$$

This matrix is known to be highly ill-conditioned, even for small values of n. Let x=ones(n,1) and let b=H*x. How large does n have to be for the computed solution x=A\b to have relative error greater than 100 percent?

10. Modify Algorithm 3.1.3 (back substitution) so that the outer loop uses the column index j and the inner loop uses the row index i. Test your algorithm in MATLAB to ensure that it is correct. Which version of back substitution is faster? Use a large matrix U to check.

11. A complex $n \times n$ matrix A is **Hermitian** if $A^H = A$, where $A^H = \overline{A^T}$ is the **Hermitian transpose** of A. That is, A^H is the *complex conjugate* of the transpose of A. We say that a Hermitian matrix A is **positive definite** if

$$\mathbf{x}^H A \mathbf{x} > 0$$

for any nonzero complex vector $\mathbf{x}$. Modify Algorithm 3.3.5 for the Cholesky factorization of a real symmetric positive definite matrix so that it computes the factorization $A = GG^H$ of a complex Hermitian matrix A.

12. A matrix H is said to be **upper Hessenberg** if $h_{ij} = 0$ for $i > j + 1$. That is, all entries of H below the subdiagonal are zero. How many floating-point operations are required to compute the LU Decomposition of H? What is the bandwidth of L? Consider the cases of no pivoting and partial pivoting.

Chapter 4

Least Squares Problems

Suppose we have m data points $(x_1, y_1), (x_2, y_2), \ldots, (x_m, y_m) \in \mathbb{R}^2$ that we wish to describe using a set of real-valued functions $\varphi_1(x)$, $\varphi_2(x)$, $\ldots$, $\varphi_n(x)$, where $n \leq m$. Taking a linear combination of these functions leads to the system of linear equations

$$\sum_{j=1}^{n} c_i \varphi_j(x_i) = y_i, \quad i = 1, 2, \ldots, m,$$

which can be written in matrix-vector form $\Phi c = y$, where $\Phi \in \mathbb{R}^{m \times n}$, $y \in \mathbb{R}^m$ and $c \in \mathbb{R}^n$. If $m = n$ and Φ is invertible, then the system can be solved using the techniques of Chapter 3. In this chapter, we consider the case $m \geq n$, in which this system may not have an exact solution, and therefore we seek a solution that, in some sense, comes "as close as possible" to solving the system.

In Section 4.1 we consider systems of the form $Ax = b$, where $A \in \mathbb{R}^{m \times n}$, $m \geq n$, and $\text{rank}(A) = n$. Such systems lead to the **full-rank least squares problem**, which has a unique solution. To obtain a numerically stable method for solving this problem, we introduce the **QR Factorization** in Section 4.2. In Section 4.3, we consider the case in which $\text{rank}(A) < n$, which leads to infinitely many solutions. Section 4.4 introduces the **Singular Value Decomposition (SVD)**, which is extremely useful for solving and analyzing least squares problems, and has several other applications of interest. Finally, in Section 4.5, we examine variations on these least squares problems, including the imposition of constraints.

4.1 The Full Rank Least Squares Problem

Given $A \in \mathbb{R}^{m \times n}$, with $m \geq n$, and $b \in \mathbb{R}^m$, we consider the system of equations $Ax = b$, in the case where A has **full column rank** (that is, $\text{rank}(A) = n$, the largest value possible for an $m \times n$ matrix if $m \geq n$). There exists a unique solution to this system if and only if b is in the **range** of A (see Section B.5.3). For example, there exists a unique solution in the case of

$$A = \begin{bmatrix} 0 & 1 \\ 1 & 0 \\ 0 & 0 \end{bmatrix}, \quad b = \begin{bmatrix} 1 \\ 1 \\ 0 \end{bmatrix},$$

but not if $\mathbf{b} = \begin{bmatrix} 1 & 1 & 1 \end{bmatrix}^T$. In such cases, when $\mathbf{b}$ is not in the **range** of A, we instead seek to minimize $\|\mathbf{b} - A\mathbf{x}\|_p$ for some p, where $\|\cdot\|_p$ is the ℓ_p-**norm** defined in Section B.13.1. Recall from Section 3.4.2 that the vector $\mathbf{r} = \mathbf{b} - A\mathbf{x}$ is known as the **residual** vector.

Different norms give different solutions. If $p = 1$ or $p = \infty$, then the function we seek to minimize, $f(x) = \|\mathbf{b} - A\mathbf{x}\|_p$ is not differentiable, so we cannot use standard minimization techniques from calculus. However, if $p = 2$, $f(x)$ is differentiable, and thus the problem is more tractable. Because the ℓ_2-norm of a vector includes the sum of squares of its components, the problem of minimizing the ℓ_2-norm of the residual is called a **least squares problem**. We now consider two methods for solving this problem in the case where A has full column rank.

4.1.1   Derivation of the Normal Equations

The first method is to define the function $\phi : \mathbb{R}^n \to \mathbb{R}$ by $\phi(\mathbf{x}) = \frac{1}{2}\|\mathbf{b} - A\mathbf{x}\|_2^2$, which is a differentiable function of $\mathbf{x}$, and then use our knowledge of multivariable calculus to find $\mathbf{x} \in \mathbb{R}^n$ that minimizes ϕ. To help us to characterize the minimum of this function, we first compute the gradient of simpler functions.

Exploration 4.1.1 Let $\mathbf{c} \in \mathbb{R}^n$ and $B \in \mathbb{R}^{n \times n}$. We define $\psi : \mathbb{R}^n \to \mathbb{R}$ and $\varphi : \mathbb{R}^n \to \mathbb{R}$ by

$$\psi(\mathbf{x}) = \mathbf{c}^T \mathbf{x}, \quad \varphi(\mathbf{x}) = \mathbf{x}^T B \mathbf{x}.$$

Use the formula for matrix-vector multiplication to show that

$$\nabla \psi(\mathbf{x}) = \mathbf{c}, \quad \nabla \varphi(\mathbf{x}) = (B + B^T)\mathbf{x}.$$

Exploration 4.1.2 Use the result of Exploration 4.1.1, and the properties of inner products, to show that

$$\nabla \left(\frac{1}{2} \|\mathbf{b} - A\mathbf{x}\|_2^2 \right) = A^T A \mathbf{x} - A^T \mathbf{b}.$$

Hint: First, show that $A^T A$ is symmetric.

The **Hessian** of a twice differentiable function $\varphi : \mathbb{R}^n \to \mathbb{R}$, denoted by $H_\varphi(\mathbf{x})$, is the matrix with entries

$$h_{ij} = \frac{\partial^2 \varphi}{\partial x_i \partial x_j}.$$

Because mixed second partial derivatives satisfy

$$\frac{\partial^2 \varphi}{\partial x_i \partial x_j} = \frac{\partial^2 \varphi}{\partial x_j \partial x_i}$$

as long as they are continuous, the Hessian is symmetric under these assumptions. In the case of $\varphi(\mathbf{x}) = \mathbf{x}^T B \mathbf{x}$, the gradient of which is $\nabla \varphi(\mathbf{x}) = (B + B^T)\mathbf{x}$, the

Hessian is $H_\varphi(\mathbf{x}) = B + B^T$. It follows from the result of Exploration 4.1.2 that the Hessian of $\phi(\mathbf{x}) = \frac{1}{2}\|\mathbf{b} - A\mathbf{x}\|_2^2$ is $A^T A$.

Recall that A is $m \times n$, with $m \geq n$ and $\operatorname{rank}(A) = n$. Then, if $\mathbf{x} \neq \mathbf{0}$, it follows from the linear independence of A's columns that $A\mathbf{x} \neq \mathbf{0}$. We then have

$$\mathbf{x}^T A^T A\mathbf{x} = (A\mathbf{x})^T A\mathbf{x} = \|A\mathbf{x}\|_2^2 > 0,$$

since the norm of a nonzero vector must be positive. It follows that $A^T A$ is not only symmetric, but **positive definite** as well (see Section 3.3.3). Therefore, from multivariable calculus, the Hessian of $\phi(\mathbf{x})$ is positive definite, which means that a critical point $\mathbf{x}$, which is a solution to the equations $A^T A\mathbf{x} - A^T \mathbf{b} = \mathbf{0}$, is a local *minimum*.

In summary, we can find a local minimum of $\phi(\mathbf{x})$ by noting that $\nabla\phi(\mathbf{x}) = A^T A\mathbf{x} - A^T \mathbf{b}$, which means that $\nabla\phi(\mathbf{x}) = \mathbf{0}$ if and only if $\mathbf{x}$ is a solution of

$$A^T A\mathbf{x} = A^T \mathbf{b}. \tag{4.1}$$

This system of linear equations is called the **normal equations**, and were used by Gauss to solve the least squares problem [Stigler (1981)]. We note that $A^T A$ is $n \times n$, so if $m \gg n$, the system of normal equations is a much smaller system to solve than $A\mathbf{x} = \mathbf{b}$. Furthermore, we can use the Cholesky Factorization (see Section 3.3.3.2) to solve for $\mathbf{x}$, as $A^T A$ is symmetric positive definite, and therefore nonsingular. It follows that the normal equations have a unique solution $\mathbf{x}^*$. A multivariable Taylor expansion

$$\phi(\mathbf{x}) = \phi(\mathbf{x}^*) + \nabla\phi(\mathbf{x}^*)^T(\mathbf{x} - \mathbf{x}^*) + \frac{1}{2}(\mathbf{x} - \mathbf{x}^*)^T H_\phi(\mathbf{x}^*)(\mathbf{x} - \mathbf{x}^*)$$

$$= \phi(\mathbf{x}^*) + \frac{1}{2}(\mathbf{x} - \mathbf{x}^*)^T A^T A(\mathbf{x} - \mathbf{x}^*)$$

shows that $\mathbf{x}^*$, which is the only critical point of ϕ, is not only a local minimizer but also a *global* minimizer. We conclude that $\mathbf{x}^*$ is the unique solution of the full rank least squares problem.

Exploration 4.1.3 Use the normal equations to solve the full-rank least squares problem of minimizing $\|A\mathbf{c} - \mathbf{y}\|_2$, where

$$A = \begin{bmatrix} 1 & x_1 \\ 1 & x_2 \\ \vdots & \vdots \\ 1 & x_m \end{bmatrix}, \quad \mathbf{y} = \begin{bmatrix} y_1 \\ y_2 \\ \vdots \\ y_m \end{bmatrix}.$$

The solution $\mathbf{c} = \begin{bmatrix} b & m \end{bmatrix}^T$ contains the coefficients of the linear function $y = mx + b$ that best fits the data, in the least squares sense. This kind of data fitting problem will be revisited in Section 8.1.

4.1.2   The Condition Number of $A^T A$

Because the coefficient matrix of the normal equations is $A^T A$, it is important to understand its condition number. When A is $n \times n$ and invertible,

$$\kappa_2(A) = \|A\|_2 \|A^{-1}\|_2.$$

We now develop a formula for the condition number of $A^T A$.

Exploration 4.1.4 Show that $A^T A$ and AA^T have the same eigenvalues. *Hint:* If $\mathbf{x}$ is an eigenvector of $A^T A$ with eigenvalue λ, what is an eigenvector of AA^T corresponding to λ?

Exploration 4.1.5 Let $A \in \mathbb{R}^{n \times n}$. Show that if λ is an eigenvalue of A, then λ^2 is an eigenvalue of A^2, and if A is also invertible, that $\lambda \neq 0$ and λ^{-1} is an eigenvalue of A^{-1}.

Exploration 4.1.6 Let $A \in \mathbb{R}^{n \times n}$ be invertible. Use the results of Explorations 4.1.4 and 4.1.5 to show that

$$\kappa_2(A) = \sqrt{\frac{\lambda_{\max}(A^T A)}{\lambda_{\min}(A^T A)}}.$$

If A is $m \times n$ with $m > n$ and $\operatorname{rank}(A) = n$, A^{-1} does not exist, but the quantity $\sqrt{\frac{\lambda_{\max}(A^T A)}{\lambda_{\min}(A^T A)}}$ from Exploration 4.1.6 is still defined and an appropriate measure of the sensitivity of the least squares problem to perturbations in the data, so we denote this ratio by $\kappa_2(A)$ in this case as well.

Exploration 4.1.7 Use the result of Exploration 4.1.5 to show that

$$\kappa_2(A^T A) = \kappa_2(A)^2.$$

We see that the normal equations can be very badly conditioned, especially if the columns of A are nearly linearly dependent. Therefore, it is worthwhile to consider an alternative approach to the full-rank least squares problem.

4.1.3   The QR Factorization

The second approach to the full-rank least squares problem is to take advantage of the fact that the vector ℓ_2-norm is invariant under orthogonal transformations. Recall from Section 3.2.5.4 that a matrix $Q \in \mathbb{R}^{n \times n}$ is **orthogonal** if $Q^T Q = I$.

Exploration 4.1.8 Let $\mathbf{x} \in \mathbb{R}^n$, and let $Q \in \mathbb{R}^{n \times n}$ be an orthogonal matrix. Prove that $\|\mathbf{x}\|_2 = \|Q\mathbf{x}\|_2$.

We therefore seek an orthogonal matrix $Q \in \mathbb{R}^{m \times m}$ such that the transformed problem

$$\min \|\mathbf{b} - A\mathbf{x}\|_2 = \min \|Q^T(\mathbf{b} - A\mathbf{x})\|_2$$

is "easy" to solve. Let

$$A = QR = \begin{bmatrix} Q_1 & Q_2 \end{bmatrix} \begin{bmatrix} R_1 \\ 0 \end{bmatrix} = Q_1 R_1, \tag{4.2}$$

where Q_1 is $m \times n$ and R_1 is $n \times n$ and nonsingular, since $\text{rank}(A) = n$. Then, because Q is orthogonal, $Q^T A = R$ and

$$\min_{\mathbf{x} \in \mathbb{R}^n} \|\mathbf{b} - A\mathbf{x}\|_2 = \min_{\mathbf{x} \in \mathbb{R}^n} \|Q^T(\mathbf{b} - A\mathbf{x})\|_2$$

$$= \min_{\mathbf{x} \in \mathbb{R}^n} \|Q^T \mathbf{b} - (Q^T A)\mathbf{x}\|_2$$

$$= \min_{\mathbf{x} \in \mathbb{R}^n} \left\| Q^T \mathbf{b} - \begin{bmatrix} R_1 \\ 0 \end{bmatrix} \mathbf{x} \right\|_2.$$

If we partition

$$Q^T \mathbf{b} = \begin{bmatrix} \mathbf{c} \\ \mathbf{d} \end{bmatrix},$$

where $\mathbf{c}$ is an n-vector, then

$$\min_{\mathbf{x} \in \mathbb{R}^n} \|\mathbf{b} - A\mathbf{x}\|_2^2 = \min_{\mathbf{x} \in \mathbb{R}^n} \left\| \begin{bmatrix} \mathbf{c} \\ \mathbf{d} \end{bmatrix} - \begin{bmatrix} R_1 \\ 0 \end{bmatrix} \mathbf{x} \right\|_2^2 = \min_{\mathbf{x} \in \mathbb{R}^n} \|\mathbf{c} - R_1 \mathbf{x}\|_2^2 + \|\mathbf{d}\|_2^2.$$

Therefore, the minimum is achieved by the vector $\mathbf{x}$ such that $R_1 \mathbf{x} = \mathbf{c}$ and therefore

$$\min_{\mathbf{x} \in \mathbb{R}^n} \|\mathbf{b} - A\mathbf{x}\|_2 = \|\mathbf{d}\|_2.$$

Thus we seek a factorization of the form $A = QR$ where Q is orthogonal, and R is upper triangular, so that $R_1 \mathbf{x} = \mathbf{c}$ is easily solved via back substitution. This factorization is called the **QR Factorization** of A.

> **Exploration 4.1.9** Why is it not feasible to reduce A to upper triangular form using elementary row operations, as in Gaussian Elimination?

> **Exploration 4.1.10** Show that R_1^T is the Cholesky factor of $A^T A$, provided that the diagonal entries of R_1 are positive. If they are not, how could it be "fixed" so that they are?

4.1.4 Perturbation Theory*

We now investigate the sensitivity of the full-rank least squares problem, using **backward error analysis**, introduced in Section 2.1. That is, we assume that our computed solution $\hat{\mathbf{x}} = \mathbf{x} + \Delta \mathbf{x}$ is the exact solution of a "nearby" system of normal equations. Specifically,

$$(A + \Delta A)^T (A + \Delta A)(\hat{\mathbf{x}} + \Delta \mathbf{x}) = (A + \Delta A)^T (\mathbf{b} + \Delta \mathbf{b}) \tag{4.3}$$

where ΔA and $\Delta \mathbf{b}$ are perturbations of A and $\mathbf{b}$, respectively. We define

$$\epsilon = \max\{\|\Delta A\|_2/\|A\|_2, \|\Delta \mathbf{b}\|_2/\|\mathbf{b}\|_2\}$$

to quantify the size of these perturbations. We also assume $\epsilon\kappa_2(A) < 1$ to ensure that $A + \Delta A$ has full column rank. Expanding, we find that the error $\Delta\mathbf{x}$ in $\mathbf{x}$ is given by

$$\Delta\mathbf{x} = (A^T A)^{-1}\left[A^T \Delta\mathbf{b} + \Delta A^T \mathbf{r} - A^T \Delta A\mathbf{x}\right] + O(\epsilon^2) \qquad (4.4)$$

where, as before, $\mathbf{r} = \mathbf{b} - A\mathbf{x}$.

Taking 2-norms of both sides and dividing by $\|\mathbf{x}\|_2$ yields

$$\frac{\|\Delta\mathbf{x}\|_2}{\|\mathbf{x}\|_2} \le \|(A^T A)^{-1} A^T\|_2\left[\frac{\|\Delta\mathbf{b}\|_2}{\|\mathbf{x}\|_2} + \|\Delta A\|_2\right] + \|(A^T A)^{-1}\|_2\|\Delta A\|_2\frac{\|\mathbf{r}\|_2}{\|\mathbf{x}\|_2} + O(\epsilon^2).$$
$$(4.5)$$

We need to compute $\|(A^T A)^{-1} A^T\|_2$. From (B.5) in Section B.13.2, we know that

$$\|(A^T A)^{-1} A^T\|_2 = \sqrt{\lambda_{\max}(A(A^T A)^{-2} A^T)}.$$

To obtain this largest eigenvalue, we let $\mathbf{v}$ be an eigenvector of $A^T A$; that is, $A^T A\mathbf{v} = \lambda\mathbf{v}$ for some scalar λ, which we know is real and positive since $A^T A$ is symmetric positive definite. It follows that $A\mathbf{v}$ is an eigenvector of $A(A^T A)^{-2} A^T$ with eigenvalue λ^{-1}, and therefore

$$\|(A^T A)^{-1} A^T\|_2 = \frac{1}{\lambda_{\min}(A^T A)}.$$

For convenience, and consistency with notation to be used later in this chapter, we define

$$\sigma_1 = \sqrt{\lambda_{\max}(A^T A)}, \quad \sigma_n = \sqrt{\lambda_{\min}(A^T A)}.$$

Then we have

$$\|A\|_2 = \sigma_1, \quad \|(A^T A)^{-1} A^T\|_2 = \frac{1}{\sigma_n}, \quad \kappa_2(A) = \frac{\sigma_1}{\sigma_n}, \quad \|(A^T A)^{-1}\|_2 = \frac{1}{\sigma_n^2}.$$

Returning to (4.5), we now have

$$\frac{\|\Delta\mathbf{x}\|_2}{\|\mathbf{x}\|_2} \le \frac{1}{\sigma_n}\left[\frac{\|\Delta\mathbf{b}\|_2}{\|\mathbf{x}\|_2} + \|\Delta A\|_2\right] + \frac{1}{\sigma_n^2}\|\Delta A\|_2\frac{\|\mathbf{r}\|_2}{\|\mathbf{x}\|_2} + O(\epsilon^2).$$

Using the fact that $\|A\mathbf{x}\|_2 \le \|A\|_2\|\mathbf{x}\|_2$, we obtain

$$\frac{\|\Delta\mathbf{x}\|_2}{\|\mathbf{x}\|_2} \le \frac{1}{\sigma_n}\left[\sigma_1\frac{\|\Delta\mathbf{b}\|_2}{\|A\mathbf{x}\|_2} + \|\Delta A\|_2\right] + \frac{\sigma_1}{\sigma_n^2}\|\Delta A\|_2\frac{\|\mathbf{r}\|_2}{\|A\mathbf{x}\|_2} + O(\epsilon^2).$$

Because $A\mathbf{x}$ and $\mathbf{r}$ are orthogonal and sum to $\mathbf{b}$, we have, for some θ,

$$\sin\theta = \frac{\|\mathbf{r}\|_2}{\|\mathbf{b}\|_2}, \quad \cos\theta = \frac{\|A\mathbf{x}\|_2}{\|\mathbf{b}\|_2}.$$

We then have

$$\frac{\|\Delta\mathbf{x}\|_2}{\|\mathbf{x}\|_2} \le \frac{1}{\sigma_n}\left[\sigma_1\frac{\|\Delta\mathbf{b}\|_2}{\|\mathbf{b}\|_2}\frac{\|\mathbf{b}\|_2}{\|A\mathbf{x}\|_2} + \|A\|_2\frac{\|\Delta A\|_2}{\|A\|_2}\right] + \frac{\sigma_1}{\sigma_n^2}\|A\|_2\frac{\|\Delta A\|_2}{\|A\|_2}\frac{\|\mathbf{r}\|_2}{\|A\mathbf{x}\|_2} + O(\epsilon^2)$$

$$\le \frac{\sigma_1}{\sigma_n}\left[\epsilon\frac{\|\mathbf{b}\|_2}{\|A\mathbf{x}\|_2} + \epsilon\right] + \frac{\sigma_1^2}{\sigma_n^2}\epsilon\frac{\|\mathbf{r}\|_2}{\|A\mathbf{x}\|_2} + O(\epsilon^2)$$

$$\le \epsilon\kappa_2(A)\sec\theta + \epsilon\kappa_2(A) + \epsilon\kappa_2(A)^2\tan\theta + O(\epsilon^2). \qquad (4.6)$$

These next explorations lead to a similar result pertaining to the sensitivity of the residual.

Exploration 4.1.11 A $m \times m$ matrix P is a **projection** if $P = P^T$ and $P^2 = P$.

    (a) Prove that if P is a projection, then $\|P\|_2 = 1$.
    (b) Prove that if P is a projection, then $I - P$ is also a projection.

Hint: What are the eigenvalues of P?

Exploration 4.1.12 Let $A \in \mathbb{R}^{m \times n}$ with $m \geq n$ and rank$(A) = n$. Prove that $P = A(A^T A)^{-1} A^T$ is a projection.

Exploration 4.1.13 Let $\mathbf{r} = \mathbf{b} - A\mathbf{x}$ and let $\mathbf{r} + \Delta\mathbf{r}$ be the residual of the perturbed full-rank least squares problem (4.3). Use the results of Explorations 4.1.11 and 4.1.12 to prove

$$\frac{\|\Delta\mathbf{r}\|_2}{\|\mathbf{b}\|_2} \leq \epsilon[1 + 2\kappa_2(A)] + O(\epsilon^2). \qquad (4.7)$$

Hint: Use a similar approach as the one used to establish the bound on $\|\Delta\mathbf{x}\|_2/\|\mathbf{x}\|_2$. Also, (4.4) will be helpful.

It follows from (4.6) and (4.7) that even if the perturbation in the residual is small, the perturbation in the solution can still be large.

4.1.5 Concept Check

    1. What is the residual vector of a system of linear equations $A\mathbf{x} = \mathbf{b}$?
    2. Why is the ℓ_2-norm of the residual minimized, instead of some other norm?
    3. What is the least squares problem?
    4. What are the normal equations? What difficulty arises when solving them?
    5. What is the QR Factorization and how is it helpful with solving least squares problems?
    6. How sensitive are the solution and residual of the full-rank least squares problem to perturbations?

4.2 Computing the QR Factorization

Let $A \in \mathbb{R}^{m \times n}$ matrix with full column rank. The **QR Factorization** of A is a decomposition $A = QR$, where Q is an $m \times m$ orthogonal matrix and R is an $m \times n$ upper triangular matrix. There are three ways to compute this decomposition:

    1. Using **Givens rotations**, also known as **Jacobi rotations**, used by Givens and originally invented by Jacobi for use with in solving the symmetric eigenvalue problem in 1846 [Jacobi (1846)].

2. Using **Householder reflections**, also known as **Householder transformations**, developed by Householder.

3. A third, less frequently used approach known as **Gram-Schmidt Orthogonalization**.

4.2.1 Givens Rotations

We have seen how elementary row operations can be used to reduce a matrix to upper triangular form, resulting in the LU Decomposition $PA = LU$. To compute the factorization $A = QR$, we can use a similar approach, in which (non-elementary) row operations are applied to A to reduce A to upper triangular form. If each such row operation, designed to zero a_{ij} where $i > j$, can be implemented through pre-multiplication by an orthogonal matrix, then the accumulation of these row operations is implemented through pre-multiplication by the product of these orthogonal matrices, which is itself the orthogonal matrix Q^T.

We illustrate the process in the case where A is a 2×2 matrix, for which we need only zero a_{21}. The QR Factorization computes $Q^T A = R$, or

$$\begin{bmatrix} c & -s \\ s & c \end{bmatrix}^T \begin{bmatrix} a_{11} & a_{12} \\ a_{21} & a_{22} \end{bmatrix} = \begin{bmatrix} r_{11} & r_{12} \\ 0 & r_{22} \end{bmatrix},$$

where $c^2 + s^2 = 1$ to ensure Q is orthogonal. From the relationship $-sa_{11} + ca_{21} = 0$ we obtain

$$c^2 a_{21}^2 = s^2 a_{11}^2 = (1 - c^2) a_{11}^2$$

which yields

$$c = \pm \frac{a_{11}}{\sqrt{a_{21}^2 + a_{11}^2}}.$$

It is conventional to choose the $+$ sign. Then, we obtain

$$s^2 = 1 - c^2 = 1 - \frac{a_{11}^2}{a_{21}^2 + a_{11}^2} = \frac{a_{21}^2}{a_{21}^2 + a_{11}^2},$$

or

$$s = \pm \frac{a_{21}}{\sqrt{a_{21}^2 + a_{11}^2}}.$$

Again, we choose the $+$ sign. As a result, we have

$$r_{11} = a_{11} \frac{a_{11}}{\sqrt{a_{21}^2 + a_{11}^2}} + a_{21} \frac{a_{21}}{\sqrt{a_{21}^2 + a_{11}^2}} = \sqrt{a_{21}^2 + a_{11}^2}.$$

The matrix

$$Q = \begin{bmatrix} c & -s \\ s & c \end{bmatrix}$$

is called a **Givens rotation**. It is called a rotation because it is orthogonal, and therefore length-preserving, and also because there is an angle θ such that $\sin \theta = s$

and $\cos\theta = c$. The effect of pre-multiplying a vector by Q^T is to rotate the vector *clockwise* through the angle θ. In particular, if $a = r\cos\theta$ and $b = r\sin\theta$ for some angle θ, then

$$\begin{bmatrix} c & -s \\ s & c \end{bmatrix}^T \begin{bmatrix} a \\ b \end{bmatrix} = \begin{bmatrix} r \\ 0 \end{bmatrix}, \quad r = \sqrt{a^2 + b^2}.$$

That is, the point (a, b) is rotated to the positive x-axis, effectively "undoing" the counterclockwise rotation by θ inherent in polar coordinates.

Exploration 4.2.1 Verify that the product of two 2×2 Givens rotations, with angles θ_1 and θ_2, is also a Givens rotation with angle $\theta_1 + \theta_2$.

Now, to see how Givens rotations can be used to zero entries of an $m \times n$ matrix A, suppose that we have the vector

$$\begin{bmatrix} \times \\ \vdots \\ \times \\ a \\ \times \\ \vdots \\ \times \\ b \\ \times \\ \vdots \\ \times \end{bmatrix}$$

that is a column of A. Then

$$\begin{bmatrix} 1 \\ & \ddots \\ & & 1 \\ & & & c & & & s \\ & & & & 1 \\ & & & & & \ddots \\ & & & -s & & & c \\ & & & & & & & 1 \\ & & & & & & & & \ddots \\ & & & & & & & & & 1 \end{bmatrix} \begin{bmatrix} \times \\ \vdots \\ \times \\ a \\ \times \\ \vdots \\ \times \\ b \\ \times \\ \vdots \\ \times \end{bmatrix} = \begin{bmatrix} \times \\ \vdots \\ \times \\ r \\ \times \\ \vdots \\ \times \\ 0 \\ \times \\ \vdots \\ \times \end{bmatrix}.$$

So, to transform A into an upper triangular matrix R, we can find a product of rotations Q such that $Q^T A = R$. It is easy to see that $O(mn)$ rotations are required. Each rotation takes $O(n)$ floating-point operations, so the entire process of computing the QR Factorization requires $O(mn^2)$ operations.

It is important to note that the straightforward approach to computing the entries c and s of the Givens rotation,

$$c = \frac{a}{\sqrt{a^2 + b^2}}, \quad s = \frac{b}{\sqrt{a^2 + b^2}},$$

is not always advisable, because in floating-point arithmetic, the computation of $\sqrt{a^2 + b^2}$ could overflow (see Exploration 2.2.19). To get around this problem, suppose that $|b| \geq |a|$. Then, we can instead compute

$$t = \frac{a}{b}, \quad s = \frac{1}{\sqrt{1 + t^2}}, \quad c = st, \tag{4.8}$$

which is guaranteed not to overflow since the only number that is squared is at most one in magnitude. Similarly, if $|a| \geq |b|$, then we compute

$$t = \frac{b}{a}, \quad c = \frac{1}{\sqrt{1 + t^2}}, \quad s = ct. \tag{4.9}$$

Exploration 4.2.2 Write a MATLAB function

$$[\texttt{c,s}]=\texttt{givens(a,b)}$$

that uses (4.8) and (4.9) to compute c and s such that

$$\begin{bmatrix} c & -s \\ s & c \end{bmatrix}^T \begin{bmatrix} a \\ b \end{bmatrix} = \begin{bmatrix} r \\ 0 \end{bmatrix}, \quad r = \sqrt{a^2 + b^2}.$$

Example 4.2.1 We illustrate how Givens rotations can be used to compute the QR Factorization of

$$A = \begin{bmatrix} 0.8147 & 0.0975 & 0.1576 \\ 0.9058 & 0.2785 & 0.9706 \\ 0.1270 & 0.5469 & 0.9572 \\ 0.9134 & 0.9575 & 0.4854 \\ 0.6324 & 0.9649 & 0.8003 \end{bmatrix}.$$

First, we compute a Givens rotation that, when applied to a_{41} and a_{51}, zeros a_{51}:

$$\begin{bmatrix} 0.8222 & -0.5692 \\ 0.5692 & 0.8222 \end{bmatrix}^T \begin{bmatrix} 0.9134 \\ 0.6324 \end{bmatrix} = \begin{bmatrix} 1.1109 \\ 0 \end{bmatrix}.$$

Applying this rotation to rows 4 and 5 yields

$$\begin{bmatrix} 1 & 0 & 0 & 0 & 0 \\ 0 & 1 & 0 & 0 & 0 \\ 0 & 0 & 1 & 0 & 0 \\ 0 & 0 & 0 & 0.8222 & -0.5692 \\ 0 & 0 & 0 & 0.5692 & 0.8222 \end{bmatrix}^T \begin{bmatrix} 0.8147 & 0.0975 & 0.1576 \\ 0.9058 & 0.2785 & 0.9706 \\ 0.1270 & 0.5469 & 0.9572 \\ 0.9134 & 0.9575 & 0.4854 \\ 0.6324 & 0.9649 & 0.8003 \end{bmatrix} =$$

$$\begin{bmatrix} 0.8147 & 0.0975 & 0.1576 \\ 0.9058 & 0.2785 & 0.9706 \\ 0.1270 & 0.5469 & 0.9572 \\ 1.1109 & 1.3365 & 0.8546 \\ 0 & 0.2483 & 0.3817 \end{bmatrix}.$$

Next, we compute a Givens rotation that, when applied to a_{31} and a_{41}, zeros a_{41}:

$$\begin{bmatrix} 0.1136 & -0.9935 \\ 0.9935 & 0.1136 \end{bmatrix}^T \begin{bmatrix} 0.1270 \\ 1.1109 \end{bmatrix} = \begin{bmatrix} 1.1181 \\ 0 \end{bmatrix}.$$

Applying this rotation to rows 3 and 4 yields

$$\begin{bmatrix} 1 & 0 & 0 & 0 & 0 \\ 0 & 1 & 0 & 0 & 0 \\ 0 & 0 & 0.1136 & -0.9935 & 0 \\ 0 & 0 & 0.9935 & 0.1136 & 0 \\ 0 & 0 & 0 & 0 & 1 \end{bmatrix}^T \begin{bmatrix} 0.8147 & 0.0975 & 0.1576 \\ 0.9058 & 0.2785 & 0.9706 \\ 0.1270 & 0.5469 & 0.9572 \\ 1.1109 & 1.3365 & 0.8546 \\ 0 & 0.2483 & 0.3817 \end{bmatrix} =$$

$$\begin{bmatrix} 0.8147 & 0.0975 & 0.1576 \\ 0.9058 & 0.2785 & 0.9706 \\ 1.1181 & 1.3899 & 0.9578 \\ 0 & -0.3916 & -0.8539 \\ 0 & 0.2483 & 0.3817 \end{bmatrix}.$$

Next, we compute a Givens rotation that, when applied to a_{21} and a_{31}, zeros a_{31}:

$$\begin{bmatrix} 0.6295 & -0.7770 \\ 0.7770 & 0.6295 \end{bmatrix}^T \begin{bmatrix} 0.9058 \\ 1.1181 \end{bmatrix} = \begin{bmatrix} 1.4390 \\ 0 \end{bmatrix}.$$

Applying this rotation to rows 2 and 3 yields

$$\begin{bmatrix} 1 & 0 & 0 & 0 & 0 \\ 0 & 0.6295 & -0.7770 & 0 & 0 \\ 0 & 0.7770 & 0.6295 & 0 & 0 \\ 0 & 0 & 0 & 1 & 0 \\ 0 & 0 & 0 & 0 & 1 \end{bmatrix}^T \begin{bmatrix} 0.8147 & 0.0975 & 0.1576 \\ 0.9058 & 0.2785 & 0.9706 \\ 1.1181 & 1.3899 & 0.9578 \\ 0 & -0.3916 & -0.8539 \\ 0 & 0.2483 & 0.3817 \end{bmatrix} =$$

$$\begin{bmatrix} 0.8147 & 0.0975 & 0.1576 \\ 1.4390 & 1.2553 & 1.3552 \\ 0 & 0.6585 & -0.1513 \\ 0 & -0.3916 & -0.8539 \\ 0 & 0.2483 & 0.3817 \end{bmatrix}.$$

To complete the first column, we compute a Givens rotation that, when applied to a_{11} and a_{21}, zeros a_{21}:

$$\begin{bmatrix} 0.4927 & -0.8702 \\ 0.8702 & 0.4927 \end{bmatrix}^T \begin{bmatrix} 0.8147 \\ 1.4390 \end{bmatrix} = \begin{bmatrix} 1.6536 \\ 0 \end{bmatrix}.$$

Applying this rotation to rows 1 and 2 yields

$$
\begin{bmatrix}
0.4927 & -0.8702 & 0 & 0 & 0 \\
0.8702 & 0.4927 & 0 & 0 & 0 \\
0 & & 0 & 1 & 0 & 0 \\
0 & & 0 & 0 & 1 & 0 \\
0 & & 0 & 0 & 0 & 1
\end{bmatrix}^T
\begin{bmatrix}
0.8147 & 0.0975 & 0.1576 \\
1.4390 & 1.2553 & 1.3552 \\
0 & 0.6585 & -0.1513 \\
0 & -0.3916 & -0.8539 \\
0 & 0.2483 & 0.3817
\end{bmatrix}
=
$$

$$
\begin{bmatrix}
1.6536 & 1.1405 & 1.2569 \\
0 & 0.5336 & 0.5305 \\
0 & 0.6585 & -0.1513 \\
0 & -0.3916 & -0.8539 \\
0 & 0.2483 & 0.3817
\end{bmatrix}.
$$

Moving to the second column, we compute a Givens rotation that, when applied to a_{42} and a_{52}, zeros a_{52}:

$$
\begin{bmatrix}
0.8445 & 0.5355 \\
-0.5355 & 0.8445
\end{bmatrix}^T
\begin{bmatrix}
-0.3916 \\
0.2483
\end{bmatrix}
=
\begin{bmatrix}
0.4636 \\
0
\end{bmatrix}.
$$

Applying this rotation to rows 4 and 5 yields

$$
\begin{bmatrix}
1 & 0 & 0 & 0 & 0 \\
0 & 1 & 0 & 0 & 0 \\
0 & 0 & 1 & 0 & 0 \\
0 & 0 & 0 & 0.8445 & 0.5355 \\
0 & 0 & 0 & -0.5355 & 0.8445
\end{bmatrix}^T
\begin{bmatrix}
1.6536 & 1.1405 & 1.2569 \\
0 & 0.5336 & 0.5305 \\
0 & 0.6585 & -0.1513 \\
0 & -0.3916 & -0.8539 \\
0 & 0.2483 & 0.3817
\end{bmatrix}
=
$$

$$
\begin{bmatrix}
1.6536 & 1.1405 & 1.2569 \\
0 & 0.5336 & 0.5305 \\
0 & 0.6585 & -0.1513 \\
0 & -0.4636 & -0.9256 \\
0 & 0 & -0.1349
\end{bmatrix}.
$$

This rotation does not change the first column, because both of the entries of the first column that would be affected are already equal to zero. Next, we compute a Givens rotation that, when applied to a_{32} and a_{42}, zeros a_{42}:

$$
\begin{bmatrix}
0.8177 & 0.5757 \\
-0.5757 & 0.8177
\end{bmatrix}^T
\begin{bmatrix}
0.6585 \\
-0.4636
\end{bmatrix}
=
\begin{bmatrix}
0.8054 \\
0
\end{bmatrix}.
$$

Applying this rotation to rows 3 and 4 yields

$$
\begin{bmatrix}
1 & 0 & 0 & 0 & 0 \\
0 & 1 & 0 & 0 & 0 \\
0 & 0 & 0.8177 & 0.5757 & 0 \\
0 & 0 & -0.5757 & 0.8177 & 0 \\
0 & 0 & 0 & 0 & 1
\end{bmatrix}^{T}
\begin{bmatrix}
1.6536 & 1.1405 & 1.2569 \\
0 & 0.5336 & 0.5305 \\
0 & 0.6585 & -0.1513 \\
0 & -0.4636 & -0.9256 \\
0 & 0 & -0.1349
\end{bmatrix} =
$$

$$
\begin{bmatrix}
1.6536 & 1.1405 & 1.2569 \\
0 & 0.5336 & 0.5305 \\
0 & 0.8054 & 0.4091 \\
0 & 0 & -0.8439 \\
0 & 0 & -0.1349
\end{bmatrix}.
$$

Next, we compute a Givens rotation that, when applied to a_{22} and a_{32}, zeros a_{32}:

$$
\begin{bmatrix}
0.5523 & -0.8336 \\
0.8336 & 0.5523
\end{bmatrix}^{T}
\begin{bmatrix}
0.5336 \\
0.8054
\end{bmatrix} =
\begin{bmatrix}
0.9661 \\
0
\end{bmatrix}.
$$

Applying this rotation to rows 3 and 4 yields

$$
\begin{bmatrix}
1 & 0 & 0 & 0 & 0 \\
0 & 0.5523 & -0.8336 & 0 & 0 \\
0 & 0.8336 & 0.5523 & 0 & 0 \\
0 & 0 & 0 & 1 & 0 \\
0 & 0 & 0 & 0 & 1
\end{bmatrix}^{T}
\begin{bmatrix}
1.6536 & 1.1405 & 1.2569 \\
0 & 0.5336 & 0.5305 \\
0 & 0.8054 & 0.4091 \\
0 & 0 & -0.8439 \\
0 & 0 & -0.1349
\end{bmatrix} =
$$

$$
\begin{bmatrix}
1.6536 & 1.1405 & 1.2569 \\
0 & 0.9661 & 0.6341 \\
0 & 0 & -0.2163 \\
0 & 0 & -0.8439 \\
0 & 0 & -0.1349
\end{bmatrix}.
$$

Moving to the third column, we compute a Givens rotation that, when applied to a_{43} and a_{53}, zeros a_{53}:

$$
\begin{bmatrix}
0.9875 & -0.1579 \\
0.1579 & 0.9875
\end{bmatrix}^{T}
\begin{bmatrix}
-0.8439 \\
-0.1349
\end{bmatrix} =
\begin{bmatrix}
0.8546 \\
0
\end{bmatrix}.
$$

Applying this rotation to rows 4 and 5 yields

$$
\begin{bmatrix}
1 & 0 & 0 & 0 & 0 \\
0 & 1 & 0 & 0 & 0 \\
0 & 0 & 1 & 0 & 0 \\
0 & 0 & 0 & 0.9875 & -0.1579 \\
0 & 0 & 0 & 0.1579 & 0.9875
\end{bmatrix}^{T}
\begin{bmatrix}
1.6536 & 1.1405 & 1.2569 \\
0 & 0.9661 & 0.6341 \\
0 & 0 & -0.2163 \\
0 & 0 & -0.8439 \\
0 & 0 & -0.1349
\end{bmatrix} =
$$

$$
\begin{bmatrix}
1.6536 & 1.1405 & 1.2569 \\
0 & 0.9661 & 0.6341 \\
0 & 0 & -0.2163 \\
0 & 0 & -0.8546 \\
0 & 0 & 0
\end{bmatrix}.
$$

Finally, we compute a Givens rotation that, when applied to a_{33} and a_{43}, zeros a_{43}:

$$\begin{bmatrix} 0.2453 & -0.9694 \\ 0.9694 & 0.2453 \end{bmatrix}^T \begin{bmatrix} -0.2163 \\ -0.8546 \end{bmatrix} = \begin{bmatrix} 0.8816 \\ 0 \end{bmatrix}.$$

Applying this rotation to rows 3 and 4 yields

$$\begin{bmatrix} 1 & 0 & 0 & 0 & 0 \\ 0 & 1 & 0 & 0 & 0 \\ 0 & 0 & 0.2453 & -0.9694 & 0 \\ 0 & 0 & 0.9694 & 0.2453 & 0 \\ 0 & 0 & 0 & 0 & 1 \end{bmatrix}^T \begin{bmatrix} 1.6536 & 1.1405 & 1.2569 \\ 0 & 0.9661 & 0.6341 \\ 0 & 0 & -0.2163 \\ 0 & 0 & -0.8546 \\ 0 & 0 & 0 \end{bmatrix} =$$

$$\begin{bmatrix} 1.6536 & 1.1405 & 1.2569 \\ 0 & 0.9661 & 0.6341 \\ 0 & 0 & -0.8816 \\ 0 & 0 & 0 \\ 0 & 0 & 0 \end{bmatrix} = R.$$

Applying these Givens rotations, in the same order, to the *columns* of the identity matrix yields the orthogonal matrix $Q = H_1 H_2 H_3$ such that $Q^T A = R$ is upper triangular. $\square$

Now, we can describe the entire algorithm for computing the QR Factorization of $A \in \mathbb{R}^{m \times n}$ using Givens rotations. Let $\mathbf{v} \in \mathbb{R}^n$ and let $v_i = a$ and $v_j = b$, with $j > i$. We compute $[c, s] = \texttt{givens}(a, b)$, where $\texttt{givens}$ is the function implemented in Exploration 4.2.2. We denote by $G(i, j, c, s)$ be the $m \times m$ Givens rotation matrix that rotates the ith and jth elements of the vector $\mathbf{v}$ clockwise by the angle θ such that $\cos \theta = c$ and $\sin \theta = s$. Then, in the updated vector $\mathbf{u} = G(i, j, c, s)^T \mathbf{v}$, $u_i = r = \sqrt{a^2 + b^2}$ and $u_j = 0$.

Based on Example 4.2.1, the QR Factorization of an $m \times n$ matrix A is then computed as follows, using such Givens rotations.

Algorithm 4.2.2 (QR Factorization via Givens rotations) Let $m \geq n$ and let $A \in \mathbb{R}^{m \times n}$ have full column rank. The following algorithm uses Givens rotations to compute the QR Factorization $A = QR$, where $Q \in \mathbb{R}^{m \times m}$ is orthogonal and $R \in \mathbb{R}^{m \times n}$ is upper triangular.

$Q = I$
$R = A$
for $j = 1, 2, \ldots, n$ **do**
    for $i = m, m - 1, \ldots, j + 1$ **do**
        $[c, s] = \texttt{givens}(r_{i-1,j}, r_{ij})$
        $R = G(i - 1, i, c, s)^T R$
        $Q = QG(i - 1, i, c, s)$
    end for
end for

Note that the matrix Q is accumulated by *column* rotations of the identity matrix, because the matrix by which R is multiplied to reduce R to upper triangular form, a product of *row* rotations, is Q^T. We also note that in a practical implementation, the matrix $G(i, j, c, s)$ is not formed explicitly; rather, rows i and j of R are modified to compute $G(i, j, c, s)^T R$, or columns i and j of Q to compute $QG(i, j, c, s)$.

Exploration 4.2.3 Write a MATLAB function `[Q,R]=givensqr(A)` that implements Algorithm 4.2.2.

We showed how to construct Givens rotations in order to rotate two elements of a column vector so that one element would be zero, and that approximately $mn - n^2/2$ such rotations could be used to transform A into an upper triangular matrix R. Because each rotation only modifies two rows of A, it is possible to interchange the order of rotations that affect different rows, and thus apply sets of rotations in parallel. This is the main reason why Givens rotations can be preferable to other approaches. Other reasons are that they are easy to use when the QR Factorization needs to be updated as a result of adding a row to A or deleting a column of A. They are also more efficient when A is sparse.

Exploration 4.2.4 Write MATLAB functions

$$[Q1,R1]=\texttt{qrinscolumn(Q,R,j,x)}$$
$$[Q1,R1]=\texttt{qrdelcolumn(Q,R,j)}$$

that update the QR Factorization when a column is inserted into or deleted from the matrix A such that $A = \texttt{QR}$. For `qrinscolumn`, the input argument `j` is the index of the column *after* which the new column `x` is to be inserted, and for `qrdelcolumn`, `j` is the index of the column to be deleted. *Hint:* Determine the product of $\texttt{Q}^T$ and the updated matrix, and use as few Givens rotations as necessary to "fix" the result so that it is upper triangular.

4.2.2 Householder Reflections

Givens rotations are an ideal approach for computing the QR Factorization of a matrix A when it is necessary to be selective regarding which entries of A need to be zeroed. However, for the general case, it is not the most efficient. To achieve greater efficiency, we use orthogonal transformations that introduce more zeros, which are **Householder reflections**. Consider a matrix of the form $P = I - \tau u u^T$, where $u \neq 0$ and τ is a nonzero constant. It is clear that P is a symmetric **rank-one update** of I (see Section B.7.3). Can we choose τ so that P is also orthogonal?

From the desired relation $P^T P = I$ we obtain

$$P^T P = (I - \tau \mathbf{u}\mathbf{u}^T)^T (I - \tau \mathbf{u}\mathbf{u}^T)$$
$$= I - 2\tau \mathbf{u}\mathbf{u}^T + \tau^2 \mathbf{u}\mathbf{u}^T \mathbf{u}\mathbf{u}^T$$
$$= I - 2\tau \mathbf{u}\mathbf{u}^T + \tau^2 (\mathbf{u}^T \mathbf{u})\mathbf{u}\mathbf{u}^T$$
$$= I + \tau(\tau \mathbf{u}^T \mathbf{u} - 2)\mathbf{u}\mathbf{u}^T.$$

It follows that if $\tau = 2/\mathbf{u}^T \mathbf{u}$, then $P^T P = I$ for any nonzero $\mathbf{u}$. Without loss of generality, we can stipulate that $\mathbf{u}^T \mathbf{u} = 1$, and therefore P takes the form $P = I - 2\mathbf{v}\mathbf{v}^T$, where $\mathbf{v}^T \mathbf{v} = 1$.

Why is the matrix P called a reflection? This is because for any nonzero vector $\mathbf{x}$, $P\mathbf{x}$ is the reflection of $\mathbf{x}$ across the hyperplane that is normal to $\mathbf{v}$. To see this, we consider the 2×2 case and set $\mathbf{v} = \begin{bmatrix} 1 & 0 \end{bmatrix}^T$ and $\mathbf{x} = \begin{bmatrix} 1 & 2 \end{bmatrix}^T$. Then

$$P = I - 2\mathbf{v}\mathbf{v}^T$$
$$= I - 2 \begin{bmatrix} 1 \\ 0 \end{bmatrix} \begin{bmatrix} 1 & 0 \end{bmatrix}$$
$$= \begin{bmatrix} 1 & 0 \\ 0 & 1 \end{bmatrix} - 2 \begin{bmatrix} 1 & 0 \\ 0 & 0 \end{bmatrix}$$
$$= \begin{bmatrix} -1 & 0 \\ 0 & 1 \end{bmatrix}$$

Therefore

$$P\mathbf{x} = \begin{bmatrix} -1 & 0 \\ 0 & 1 \end{bmatrix} \begin{bmatrix} 1 \\ 2 \end{bmatrix} = \begin{bmatrix} -1 \\ 2 \end{bmatrix}.$$

On the other hand, if $\mathbf{y} = \begin{bmatrix} 2 & -1 \end{bmatrix}$, then $\mathbf{v}^T \mathbf{y} = 0$ and we obtain $P\mathbf{y} = \mathbf{y}$. That is, $P\mathbf{x}$, for an arbitrary vector $\mathbf{x}$, preserves all components that are orthogonal to $\mathbf{v}$, and negates the component in the direction of $\mathbf{v}$.

Now, let $\mathbf{x} \in \mathbb{R}^n$. We wish to construct $P \in \mathbb{R}^{n \times n}$, of the form $P = I - 2\mathbf{v}\mathbf{v}^T$ where $\mathbf{v}^T \mathbf{v} = 1$, so that $P\mathbf{x} = \alpha \mathbf{e}_1$ for some α, where $\mathbf{e}_1 = \begin{bmatrix} 1 & 0 & \cdots & 0 \end{bmatrix}^T$. From the relations

$$\|P\mathbf{x}\|_2 = \|\mathbf{x}\|_2, \quad \|\alpha \mathbf{e}_1\|_2 = |\alpha| \|\mathbf{e}_1\|_2 = |\alpha|,$$

we obtain $\alpha = \pm \|\mathbf{x}\|_2$. To determine P, we begin with the equation

$$P\mathbf{x} = (I - 2\mathbf{v}\mathbf{v}^T)\mathbf{x} = \mathbf{x} - 2\mathbf{v}\mathbf{v}^T \mathbf{x} = \alpha \mathbf{e}_1.$$

Rearranging, we obtain

$$\frac{1}{2}(\mathbf{x} - \alpha \mathbf{e}_1) = (\mathbf{v}^T \mathbf{x})\mathbf{v}.$$

It follows that the vector $\mathbf{v}$, which is a unit vector, must be a scalar multiple of $\mathbf{x} - \alpha \mathbf{e}_1$. Therefore, $\mathbf{v}$ is defined by the equations

$$
\begin{aligned}
v_1 &= \frac{x_1 - \alpha}{\|\mathbf{x} - \alpha \mathbf{e}_1\|_2} \\
&= \frac{x_1 - \alpha}{\sqrt{\|\mathbf{x}\|_2^2 - 2\alpha x_1 + \alpha^2}} \\
&= -\frac{\alpha - x_1}{\sqrt{2\alpha(\alpha - x_1)}} \\
&= -\operatorname{sgn}(\alpha)\sqrt{\frac{\alpha - x_1}{2\alpha}}, \tag{4.10}
\end{aligned}
$$

$$
\begin{aligned}
v_2 &= \frac{x_2}{\sqrt{2\alpha(\alpha - x_1)}} \\
&= -\frac{x_2}{2\alpha v_1}, \tag{4.11}
\end{aligned}
$$

$$
\vdots
$$

$$
v_n = -\frac{x_n}{2\alpha v_1}. \tag{4.12}
$$

To avoid catastrophic cancellation (see Section 2.2.2.4), it is best to choose the sign of α so that it has the opposite sign of x_1. It can be seen that the computation of $\mathbf{v}$ requires about n floating-point operations.

Note that the matrix P is *never* formed explicitly. For any vector $\mathbf{b}$, the product $P\mathbf{b}$ can be computed as follows:

$$
P\mathbf{b} = (I - 2\mathbf{v}\mathbf{v}^T)\mathbf{b} = \mathbf{b} - 2(\mathbf{v}^T\mathbf{b})\mathbf{v}.
$$

This process requires only $4n$ operations. It is easy to see that we can represent P simply by storing only $\mathbf{v}$.

> **Exploration 4.2.5** Let $\mathbf{a}, \mathbf{b} \in \mathbb{R}^m$, with $\|\mathbf{a}\|_2 = \|\mathbf{b}\|_2$. Find a Householder reflection P such that $P\mathbf{a} = \mathbf{b}$.

Now, suppose that that $\mathbf{x} = \mathbf{a}_1$ is the first column of a matrix A. Then we construct a Householder reflection $H_1 = I - 2\mathbf{v}_1\mathbf{v}_1^T$ such that $H\mathbf{x} = \alpha \mathbf{e}_1$, and we have

$$
A^{(2)} = H_1 A = \begin{bmatrix} r_{11} & r_{12} & \cdots & r_{1n} \\ 0 & & & \\ \vdots & \mathbf{a}_{2:m,2}^{(2)} & \cdots & \mathbf{a}_{2:m,n}^{(2)} \\ 0 & & & \end{bmatrix}.
$$

where we denote the constant α by r_{11}, as it is the $(1,1)$ entry of the updated matrix $A^{(2)}$. Now, we can construct $\tilde{H}_2$ such that

$$
\tilde{H}_2 \mathbf{a}_{2:m,2}^{(2)} = \begin{bmatrix} r_{22} \\ 0 \\ \vdots \\ 0 \end{bmatrix},
$$

which yields the updated matrix

$$A^{(3)} = \begin{bmatrix} 1 & 0 \\ 0 & \tilde{H}_2 \end{bmatrix} A^{(2)} = \begin{bmatrix} r_{11} & r_{12} & r_{13} & \cdots & r_{1n} \\ 0 & r_{22} & r_{23} & \cdots & r_{2n} \\ 0 & 0 & & & \\ \vdots & \vdots & \mathbf{a}_{3:m,3}^{(3)} & \cdots & \mathbf{a}_{3:m,n}^{(3)} \\ 0 & 0 & & & \end{bmatrix}.$$

Note that the first column of $A^{(2)}$ is unchanged by $\tilde{H}_2$, because $\tilde{H}_2$ only operates on rows 2 through m, which, in the first column, have zero entries. Continuing this process, we obtain

$$H_n \cdots H_1 A = A^{(n+1)} = R,$$

where, for $j = 1, 2, \ldots, n$,

$$H_j = \begin{bmatrix} I_{j-1} & 0 \\ 0 & \tilde{H}_j \end{bmatrix} \tag{4.13}$$

and R is an upper triangular matrix. We have thus factored $A = QR$, where $Q = H_1 H_2 \cdots H_n$ is an orthogonal matrix.

Example 4.2.3 We apply Householder reflections to compute the QR Factorization of the matrix from Example 4.2.1,

$$A^{(1)} = A = \begin{bmatrix} 0.8147 & 0.0975 & 0.1576 \\ 0.9058 & 0.2785 & 0.9706 \\ 0.1270 & 0.5469 & 0.9572 \\ 0.9134 & 0.9575 & 0.4854 \\ 0.6324 & 0.9649 & 0.8003 \end{bmatrix}.$$

First, we work with the first column of A,

$$\mathbf{x}_1 = \mathbf{a}_{1:5,1}^{(1)} = \begin{bmatrix} 0.8147 \\ 0.9058 \\ 0.1270 \\ 0.9134 \\ 0.6324 \end{bmatrix}, \quad \|\mathbf{x}_1\|_2 = 1.6536.$$

The corresponding Householder vector is

$$\tilde{\mathbf{v}}_1 = \mathbf{x}_1 + \|\mathbf{x}_1\|_2 \mathbf{e}_1 = \begin{bmatrix} 0.8147 \\ 0.9058 \\ 0.1270 \\ 0.9134 \\ 0.6324 \end{bmatrix} + 1.6536 \begin{bmatrix} 1.0000 \\ 0 \\ 0 \\ 0 \\ 0 \end{bmatrix} = \begin{bmatrix} 2.4684 \\ 0.9058 \\ 0.1270 \\ 0.9134 \\ 0.6324 \end{bmatrix}.$$

From this vector, we build the Householder reflection

$$c = \frac{2}{\tilde{\mathbf{v}}_1^T \tilde{\mathbf{v}}_1} = 0.2450, \quad \tilde{H}_1 = I - c\tilde{\mathbf{v}}_1 \tilde{\mathbf{v}}_1^T.$$

Applying this reflection to $A^{(1)}$ yields

$$\tilde{H}_1 A^{(1)}_{1:5,1:3} = \begin{bmatrix} -1.6536 & -1.1405 & -1.2569 \\ 0 & -0.1758 & 0.4515 \\ 0 & 0.4832 & 0.8844 \\ 0 & 0.4994 & -0.0381 \\ 0 & 0.6477 & 0.4379 \end{bmatrix},$$

$$A^{(2)} = \begin{bmatrix} -1.6536 & -1.1405 & -1.2569 \\ 0 & -0.1758 & 0.4515 \\ 0 & 0.4832 & 0.8844 \\ 0 & 0.4994 & -0.0381 \\ 0 & 0.6477 & 0.4379 \end{bmatrix}.$$

Next, we take the "interesting" portion of the second column of the updated matrix $A^{(2)}$, from rows 2 to 5:

$$\mathbf{x}_2 = \mathbf{a}^{(2)}_{2:5,2} = \begin{bmatrix} -0.1758 \\ 0.4832 \\ 0.4994 \\ 0.6477 \end{bmatrix}, \quad \|\mathbf{x}_2\|_2 = 0.9661.$$

The corresponding Householder vector is

$$\tilde{\mathbf{v}}_2 = \mathbf{x}_2 - \|\mathbf{x}_2\|_2 \mathbf{e}_1 = \begin{bmatrix} -0.1758 \\ 0.4832 \\ 0.4994 \\ 0.6477 \end{bmatrix} - 0.9661 \begin{bmatrix} 1.0000 \\ 0 \\ 0 \\ 0 \end{bmatrix} = \begin{bmatrix} -1.1419 \\ 0.4832 \\ 0.4994 \\ 0.6477 \end{bmatrix}.$$

From this vector, we build the Householder reflection

$$c = \frac{2}{\tilde{\mathbf{v}}_2^T \tilde{\mathbf{v}}_2} = 0.9065, \quad \tilde{H}_2 = I - c\tilde{\mathbf{v}}_2 \tilde{\mathbf{v}}_2^T.$$

Applying this reflection to $A^{(2)}$ yields

$$\tilde{H}_2 A^{(2)}_{2:5,2:3} = \begin{bmatrix} 0.9661 & 0.6341 \\ 0 & 0.8071 \\ 0 & -0.1179 \\ 0 & 0.3343 \end{bmatrix},$$

$$A^{(3)} = \begin{bmatrix} -1.6536 & -1.1405 & -1.2569 \\ 0 & 0.9661 & 0.6341 \\ 0 & 0 & 0.8071 \\ 0 & 0 & -0.1179 \\ 0 & 0 & 0.3343 \end{bmatrix}.$$

Finally, we take the interesting portion of the third column of $A^{(3)}$, from rows 3 to 5:

$$\mathbf{x}_3 = \mathbf{a}^{(3)}_{3:5,3} = \begin{bmatrix} 0.8071 \\ -0.1179 \\ 0.3343 \end{bmatrix}, \quad \|\mathbf{x}_3\|_2 = 0.8816.$$

The corresponding Householder vector is

$$\tilde{\mathbf{v}}_3 = \mathbf{x}_3 + \|\mathbf{x}_3\|_2 \mathbf{e}_1 = \begin{bmatrix} 0.8071 \\ -0.1179 \\ 0.3343 \end{bmatrix} + 0.8816 \begin{bmatrix} 1.0000 \\ 0 \\ 0 \end{bmatrix} = \begin{bmatrix} 1.6887 \\ -0.1179 \\ 0.3343 \end{bmatrix}.$$

From this vector, we build the Householder reflection

$$c = \frac{2}{\tilde{\mathbf{v}}_3^T \tilde{\mathbf{v}}_3} = 0.6717, \quad \tilde{H}_3 = I - c\tilde{\mathbf{v}}_3 \tilde{\mathbf{v}}_3^T.$$

Applying this reflection to $A^{(3)}$ yields

$$\tilde{H}_3 A_{3:5,3:3}^{(3)} = \begin{bmatrix} -0.8816 \\ 0 \\ 0 \end{bmatrix}, \quad A^{(4)} = \begin{bmatrix} -1.6536 & -1.1405 & -1.2569 \\ 0 & 0.9661 & 0.6341 \\ 0 & 0 & -0.8816 \\ 0 & 0 & 0 \\ 0 & 0 & 0 \end{bmatrix}.$$

Applying these same Householder reflections, in order, on the *right* of the identity matrix, yields the orthogonal matrix

$$Q = H_1 H_2 H_3 = \begin{bmatrix} -0.4927 & -0.4806 & 0.1780 & -0.6015 & -0.3644 \\ -0.5478 & -0.3583 & -0.5777 & 0.3760 & 0.3104 \\ -0.0768 & 0.4754 & -0.6343 & -0.1497 & -0.5859 \\ -0.5523 & 0.3391 & 0.4808 & 0.5071 & -0.3026 \\ -0.3824 & 0.5473 & 0.0311 & -0.4661 & 0.5796 \end{bmatrix}$$

such that

$$A^{(4)} = R = Q^T A = H_3 H_2 H_1 A = \begin{bmatrix} -1.6536 & -1.1405 & -1.2569 \\ 0 & 0.9661 & 0.6341 \\ 0 & 0 & -0.8816 \\ 0 & 0 & 0 \\ 0 & 0 & 0 \end{bmatrix}$$

is upper triangular, where

$$H_1 = \tilde{H}_1, \quad H_2 = \begin{bmatrix} 1 & \mathbf{0} \\ \mathbf{0} & \tilde{H}_2 \end{bmatrix}, \quad H_3 = \begin{bmatrix} 1 & 0 & 0 \\ 0 & 1 & 0 \\ 0 & 0 & \tilde{H}_3 \end{bmatrix},$$

are the Householder reflections from (4.13), defined in such a way that they can be applied to the *entire* matrix A. Note that for $j = 1, 2, 3$,

$$H_j = I - 2\mathbf{v}_j \mathbf{v}_j^T, \quad \mathbf{v}_j = \begin{bmatrix} \mathbf{0} \\ \tilde{\mathbf{v}}_j \end{bmatrix}, \quad \|\mathbf{v}_j\|_2 = \|\tilde{\mathbf{v}}_j\|_2 = 1,$$

where the first $j - 1$ components of $\mathbf{v}_j$ are equal to zero.

Also, note that the first $n = 3$ columns of Q are the same as those of the matrix Q that was computed in the Example 4.2.1, except for possible negation. The fourth and fifth columns are not the same, but they do span the same subspace, which is the subspace consisting of all vectors that are orthogonal to the range of A. This leads to the following definition.

Definition 4.2.4 The **orthogonal complement** of a subspace S of a vector space V is denoted by $S^\perp$ and is defined by

$$S^\perp = \{\mathbf{x} \in V \mid \mathbf{x}^T \mathbf{y} = 0 \,\forall \mathbf{y} \in S\}.$$

That is, $S^\perp$ is the subspace of V consisting of all vectors that are orthogonal to every vector in S.

Exploration 4.2.6 Let $A \in \mathbb{R}^{m \times n}$. Show that $(\text{range}(A))^\perp = \text{null}(A^T)$. That is, the orthogonal complement of the range of A is the null space of A^T. See Section B.5.3 for the definitions of the range and null space of a matrix.

Because A is a 5×3 matrix with full column rank, it follows from the form of the QR Factorization given in (4.2) that the first three columns of Q, which correspond to Q_1 in (4.2), span range(A), and therefore the fourth and fifth columns, which make up Q_2 in (4.2), span $(\text{range}(A))^\perp$, the orthogonal complement of range(A). $\square$

Note that for each $j = 1, 2, \ldots, n$, H_j is a Householder reflection, based on a vector whose first $j - 1$ components are equal to zero. Therefore, application of H_j to a matrix does not affect the first $j - 1$ rows or columns. We now present the entire algorithm. We assume the availability of a function $\mathbf{v} = \text{house}(\mathbf{x})$ that computes the n-vector $\mathbf{v}$ from the n-vector $\mathbf{x}$ using the formulas (4.10)-(4.12).

Algorithm 4.2.5 (QR Factorization via Householder reflections) Let $m \geq n$ and let $A \in \mathbb{R}^{m \times n}$ have full column rank. The following algorithm uses Householder reflections to compute the QR Factorization $A = QR$, where $Q \in \mathbb{R}^{m \times m}$ is orthogonal and $R \in \mathbb{R}^{m \times n}$ is upper triangular.

$Q = I$
$R = A$
for $j = 1, 2, \ldots, n$ **do**
    $\mathbf{v} = \text{house}(R(j:m, j))$
    $c = 2/\mathbf{v}^T \mathbf{v}$
    $R(j:m, j:n) = R(j:m, j:n) - c\mathbf{v}(\mathbf{v}^T R(j:m, j:n))$
    $Q(1:m, j:m) = Q(j:m, j:m) - cQ(j:m, j:m)\mathbf{v}\mathbf{v}^T$
end for

It is worth noting that it is not necessary to store Q as an $m \times m$ matrix, as the n vectors produced by house are sufficient to reconstruct Q if it should be needed. This saves a significant amount of storage if $m \gg n$.

Exploration 4.2.7 Write a MATLAB function

$$[Q,R]=\texttt{houseqr(A)}$$

that implements Algorithm 4.2.5. How does this function compare to your function `givensqr` from Exploration 4.2.3 in terms of efficiency? Does it produce the same matrices Q and R as `givensqr`, when applied to a square matrix A? If not, can you explain why they differ?

Exploration 4.2.8 Show that Householder QR requires about 2/3 as many floating-point operations as Givens QR. In both cases, do not include the accumulation of transformations needed to obtain Q; only consider the reduction of A to upper triangular form.

4.2.3 Gram-Schmidt Orthogonalization

Givens rotations or Householder reflections can be used to compute the "full" QR Factorization

$$A = QR = \begin{bmatrix} Q_1 & Q_2 \end{bmatrix} \begin{bmatrix} R_1 \\ 0 \end{bmatrix}$$

where Q is an $m \times m$ orthogonal matrix, and R_1 is an $n \times n$ upper triangular matrix that is nonsingular if and only if A is of full column rank (that is, $\operatorname{rank}(A) = n$).

It can be seen from the above partitions of Q and R that $A = Q_1 R_1$. Furthermore, it can be shown that if $\operatorname{rank}(A) = n$, then $\operatorname{range}(A) = \operatorname{range}(Q_1)$. In fact, for $k = 1, \ldots, n$, we have

$$\operatorname{span}\{\mathbf{a}_1, \ldots, \mathbf{a}_k\} = \operatorname{span}\{\mathbf{q}_1, \ldots, \mathbf{q}_k\}, \tag{4.14}$$

where

$$A = \begin{bmatrix} \mathbf{a}_1 & \cdots & \mathbf{a}_n \end{bmatrix}, \quad Q = \begin{bmatrix} \mathbf{q}_1 & \cdots & \mathbf{q}_m \end{bmatrix}$$

are column partitions of A and Q, respectively.

Exploration 4.2.9 Prove (4.14).

We now examine two methods for computing the "thin" or "economy-size" QR Factorization $A = Q_1 R_1$. This is sufficient for solving full-rank least-squares problems, as the least-squares solution $\mathbf{x}$ that minimizes $\|\mathbf{b} - A\mathbf{x}\|_2$ can be obtained by solving the upper triangular system $R_1 \mathbf{x} = Q_1^T \mathbf{b}$ by back substitution.

4.2.3.1 *Classical Gram-Schmidt*

Consider the "thin" QR Factorization

$$A = \begin{bmatrix} \mathbf{a}_1 & \cdots & \mathbf{a}_n \end{bmatrix} = \begin{bmatrix} \mathbf{q}_1 & \cdots & \mathbf{q}_n \end{bmatrix} \begin{bmatrix} r_{11} & \cdots & r_{1n} \\ & \ddots & \vdots \\ & & r_{nn} \end{bmatrix} = Q_1 R_1.$$

From the above matrix product we can see that $\mathbf{a}_1 = r_{11}\mathbf{q}_1$, from which it follows that

$$r_{11} = \pm\|\mathbf{a}_1\|_2, \quad \mathbf{q}_1 = \frac{1}{\|\mathbf{a}_1\|_2}\mathbf{a}_1.$$

For convenience, we choose the $+$ sign for r_{11}.

Next, by taking the inner product (see Section B.7.3) of both sides of the equation $\mathbf{a}_2 = r_{12}\mathbf{q}_1 + r_{22}\mathbf{q}_2$ with $\mathbf{q}_1$, and imposing the requirement that the columns of Q_1 form an orthonormal set, we obtain

$$r_{12} = \mathbf{q}_1^T\mathbf{a}_2, \quad r_{22} = \|\mathbf{a}_2 - r_{12}\mathbf{q}_1\|_2, \quad \mathbf{q}_2 = \frac{1}{r_{22}}(\mathbf{a}_2 - r_{12}\mathbf{q}_1).$$

In general, we use the relation

$$\mathbf{a}_k = \sum_{j=1}^{k} r_{jk}\mathbf{q}_j$$

to obtain

$$\mathbf{q}_k = \frac{1}{r_{kk}}\left(\mathbf{a}_k - \sum_{j=1}^{k-1} r_{jk}\mathbf{q}_j\right),$$

where

$$r_{kk} = \left\|\mathbf{a}_k - \sum_{j=1}^{k-1} r_{jk}\mathbf{q}_j\right\|_2, \quad r_{jk} = \mathbf{q}_j^T\mathbf{a}_k, \quad j = 1, 2, \ldots, k-1.$$

The algorithm for computing the "thin" QR Factorization via Gram-Schmidt orthogonalization is as follows.

Algorithm 4.2.6 (Classical Gram-Schmidt Orthogonalization) Let $m \geq n$ and let $A \in \mathbb{R}^{m \times n}$ have full column rank. The following algorithm uses classical Gram-Schmidt orthogonalization to compute the QR Factorization $A = Q_1R_1$, where $Q \in \mathbb{R}^{m \times n}$ has orthonormal columns and $R \in \mathbb{R}^{n \times n}$ is upper triangular and nonsingular. We assume A has the column partition $A = [\mathbf{a}_1 \cdots \mathbf{a}_n]$.

for $j = 1, 2, \ldots, n$ **do**
    $\mathbf{v} = \mathbf{a}_j$
    for $i = 1, 2, \ldots, j-1$ **do**
        $r_{ij} = \mathbf{q}_i^T\mathbf{a}_j$
        $\mathbf{v} = \mathbf{v} - r_{ij}\mathbf{q}_i$
    end for
    $r_{jj} = \|\mathbf{v}\|_2$
    $\mathbf{q}_j = \mathbf{v}/r_{jj}$
end for

Exploration 4.2.10 Write a MATLAB function

$$[Q,R]=classgs(A)$$

that implements Algorithm 4.2.6.

Unfortunately, Gram-Schmidt orthogonalization, as described, is numerically unstable. For example, if $\mathbf{a}_1$ and $\mathbf{a}_2$ are almost parallel, then $\mathbf{a}_2 - r_{12}\mathbf{q}_1$ is almost zero, and roundoff error becomes significant due to catastrophic cancellation.

4.2.3.2 *Modified Gram-Schmidt*

Modified Gram-Schmidt orthogonalization alleviates the numerical instability of Classical Gram-Schmidt. To derive this modification, we use the **outer product** perspective of matrix multiplication, as highlighted in Section B.7.6. Let $A = Q_1 R_1$ be the "thin" QR Factorization of $A \in \mathbb{R}^{m \times n}$, and let A, Q_1 and R_1 be partitioned as follows:

$$A = \begin{bmatrix} \mathbf{a}_1 \, \mathbf{a}_2 \cdots \mathbf{a}_n \end{bmatrix}, \quad Q_1 = \begin{bmatrix} \mathbf{q}_1 \, \mathbf{q}_2 \cdots \mathbf{q}_n \end{bmatrix}, \quad R_1 = \begin{bmatrix} \mathbf{r}_1^T \\ \mathbf{r}_2^T \\ \vdots \\ \mathbf{r}_n^T \end{bmatrix}.$$

Then we have

$$A = Q_1 R_1 = \sum_{j=1}^{n} \mathbf{q}_j \mathbf{r}_j^T. \tag{4.15}$$

Taking the first column of both sides of this matrix equation, we have $\mathbf{a}_1 = \mathbf{q}_1 r_{11}$, because R_1 is upper triangular. As in Classical Gram-Schmidt, we then set

$$r_{11} = \|\mathbf{a}_1\|_2, \quad \mathbf{q}_1 = \frac{\mathbf{a}_1}{r_{11}}.$$

Next, we take the inner product of both sides of (4.15) with $\mathbf{q}_1$ to obtain

$$\mathbf{q}_1^T \mathbf{a}_j = r_{1j}, \quad j = 2, 3, \ldots, n.$$

From (4.15), we then subtract off from A the contribution of $\mathbf{q}_1$, which yields

$$A^{(2)} = A - \mathbf{q}_1 \mathbf{r}_1^T = \sum_{j=2}^{n} \mathbf{q}_j \mathbf{r}_j^T. \tag{4.16}$$

Taking the *second* column of both sides of this matrix equation, and again using the upper triangularity of R_1, we obtain $\mathbf{a}_2^{(2)} = \mathbf{q}_2 r_{22}$, which yields

$$r_{22} = \|\mathbf{a}_2^{(2)}\|_2, \quad \mathbf{q}_2 = \frac{\mathbf{a}_2^{(2)}}{r_{22}}.$$

Next, we take the inner product of both sides of (4.16) with $\mathbf{q}_2$ to obtain

$$\mathbf{q}_2^T \mathbf{a}_j^{(2)} = r_{2j}, \quad j = 3, 4, \ldots, n.$$

Continuing this process, we obtain all of the columns of Q_1 and all of the entries of the upper triangle of R_1.

Note that Modified Gram-Schmidt computes the entries of R_1 row-by-row, rather than column-by-column, as Classical Gram-Schmidt does. This rearrangement of the order of operations, while mathematically equivalent to Classical Gram-Schmidt, is much more stable, numerically, because each entry of R_1 is obtained by computing an inner product of a column of Q_1 with a *modified* column of A, from which the contributions of all previous columns of Q_1 have been removed. To see why this is significant, consider the inner products

$$\mathbf{u}^T\mathbf{v}, \quad \mathbf{u}^T(\mathbf{v}+\mathbf{w}),$$

where $\mathbf{u}^T\mathbf{w} = 0$. The above inner products are equal, but suppose that $|\mathbf{u}^T\mathbf{v}| \ll \|\mathbf{w}\|$. Then $\mathbf{u}^T\mathbf{v}$ is a small number that is being computed by subtraction of potentially large numbers, which is susceptible to catastrophic cancellation.

It is shown in [Björck (1967)] that Modified Gram-Schmidt produces a matrix $\hat{Q}_1$ such that

$$\|\hat{Q}_1^T\hat{Q}_1 - I\|_2 \approx \mathbf{u}\kappa_2(A),$$

and $\hat{Q}_1$ can be computed in approximately $2mn^2$ floating-point operations, whereas with Householder QR,

$$\|\hat{Q}_1^T\hat{Q}_1 - I\|_2 \approx \mathbf{u},$$

with $\hat{Q}_1$ being computed in approximately $2mn^2 - 2n^2/3$ operations to compute R and an additional $2mn^2 - 2n^2/3$ operations to obtain Q_1, the first n columns of Q. That is, Householder QR is much less sensitive to roundoff error than Gram-Schmidt, even with modification, although Gram-Schmidt is more efficient if an explicit representation of Q_1 desired.

Exploration 4.2.11 Modify your function `classgs` from Exploration 4.2.10 so that it implements Modified Gram-Schmidt orthogonalization, resulting in a new function `[Q,R]=modgs(A)`. It turns out that only one line needs to be changed–can you figure out what that change should be?

Exploration 4.2.12 Consider the matrix

$$A = \begin{bmatrix} 0.75126707 & 0.75126707 & 0.75126706 & 0.75126706 \\ 0.25509512 & 0.25509512 & 0.25509512 & 0.25509512 \\ 0.50595705 & 0.50595706 & 0.50595705 & 0.50595706 \\ 0.69907672 & 0.69907673 & 0.69907673 & 0.69907673 \\ 0.89090326 & 0.89090326 & 0.89090326 & 0.89090326 \end{bmatrix}.$$

Use both of your functions from Explorations 4.2.10 and 4.2.11 to compute the QR Factorization of A. Can you explain any discrepancies in the results?

4.2.4   Concept Check

1. What is a Givens rotation? What geometric effect does it have on a vector?
2. What are three advantages of using Givens rotations to compute a QR Factorization, compared to other approaches?
3. What is a Householder reflection? Describe geometrically how it acts on a vector.
4. What is the difference between the "full" and "thin" QR Factorizations?
5. What is wrong with classical Gram-Schmidt, and how does modified Gram-Schmidt address the problem?

4.3   Rank-Deficient Least Squares

So far, we have assumed that the matrix A in our least-squares problem has full column rank. In this section, we drop this assumption. It follows that the normal equations no longer have a unique solution, so we must develop criteria for choosing a solution from among the infinitely many that exist. We must also modify our solution methods, which would break down in the rank-deficient case.

4.3.1   QR with Column Pivoting

When A does not have full column rank, the property

$$\text{span}\{\mathbf{a}_1, \ldots, \mathbf{a}_k\} = \text{span}\{\mathbf{q}_1, \ldots, \mathbf{q}_k\}$$

can not be expected to hold, because the first k columns of A could be linearly dependent, while the first k columns of Q, being orthonormal, must be linearly independent.

Example 4.3.1 The matrix

$$A = \begin{bmatrix} 1 & -2 & 1 \\ 2 & -4 & 0 \\ 1 & -2 & 3 \end{bmatrix}$$

has rank 2, because the first two columns are parallel, and therefore are linearly dependent, while the third column is not parallel to either of the first two. Columns 1 and 3, or columns 2 and 3, form linearly independent sets. $\square$

Therefore, in the case where $\text{rank}(A) = r < n$, we seek a decomposition of the form $A\Pi = QR$, where Π is a **permutation matrix** (see Section 3.2.5.3) chosen so that the diagonal entries of R are maximized at each stage. Specifically, suppose H_1 is a Householder reflection chosen so that

$$H_1\mathbf{a}_1 = \begin{bmatrix} r_{11} \\ 0 \\ \vdots \\ 0 \end{bmatrix}, \quad r_{11} = \|\mathbf{a}_1\|_2.$$

To maximize r_{11}, we choose Π_1 so that in the column-permuted matrix $A_1 = A\Pi_1$, we have $\|\mathbf{a}_1\|_2 \geq \|\mathbf{a}_j\|_2$ for $j \geq 2$. For Π_2, we examine the lengths of the columns of the submatrix of A obtained by removing the first row and column. It is not necessary to recompute the lengths of the columns, because we can update them by subtracting the square of the first component from the square of the total length.

This process is called **QR with column pivoting**. It yields the decomposition

$$Q \begin{bmatrix} R & S \\ 0 & 0 \end{bmatrix} = A\Pi$$

where $Q = H_1 \cdots H_r$, $\Pi = \Pi_1 \cdots \Pi_r$, and R is an upper triangular, $r \times r$ matrix. The last $m - r$ rows are necessarily zero, because every column of A is a linear combination of the first r columns of Q.

Example 4.3.2 We perform QR with column pivoting on the matrix

$$A = \begin{bmatrix} 1 & 3 & 5 & 1 \\ 2 & -1 & 2 & 1 \\ 1 & 4 & 6 & 1 \\ 4 & 5 & 10 & 1 \end{bmatrix}.$$

Computing the 2-norms of the columns yields

$$\|\mathbf{a}_1\|_2^2 = 22, \quad \|\mathbf{a}_2\|_2^2 = 51, \quad \|\mathbf{a}_3\|_2^2 = 165, \quad \|\mathbf{a}_4\|_2^2 = 4.$$

We see that the third column has the largest 2-norm. We therefore interchange the first and third columns to obtain

$$A^{(1)} = A\Pi_1 = A \begin{bmatrix} 0 & 0 & 1 & 0 \\ 0 & 1 & 0 & 0 \\ 1 & 0 & 0 & 0 \\ 0 & 0 & 0 & 1 \end{bmatrix} = \begin{bmatrix} 5 & 3 & 1 & 1 \\ 2 & -1 & 2 & 1 \\ 6 & 4 & 1 & 1 \\ 10 & 5 & 4 & 1 \end{bmatrix}.$$

We then apply a Householder reflection H_1 to $A^{(1)}$ to make the first column a multiple of $\mathbf{e}_1$, which yields

$$H_1 A^{(1)} = \begin{bmatrix} -12.8452 & -6.7729 & -4.2817 & -1.7905 \\ 0 & -2.0953 & 1.4080 & 0.6873 \\ 0 & 0.7141 & -0.7759 & 0.0618 \\ 0 & -0.4765 & 1.0402 & -0.5637 \end{bmatrix}.$$

Next, we consider the submatrix obtained by removing the first row and column of $H_1 A^{(1)}$:

$$\tilde{A}^{(2)} = \begin{bmatrix} -2.0953 & 1.4080 & 0.6873 \\ 0.7141 & -0.7759 & 0.0618 \\ -0.4765 & 1.0402 & -0.5637 \end{bmatrix}.$$

We compute the lengths of the columns, as before, except that this time, we update the lengths of the columns of A, rather than recomputing them. This yields

$$\|\tilde{\mathbf{a}}_1^{(2)}\|_2^2 = \|\mathbf{a}_2^{(1)}\|^2 - [a_{12}^{(1)}]^2 = 51 - (-6.7729)^2 = 5.1273,$$

$$\|\tilde{\mathbf{a}}_2^{(2)}\|_2^2 = \|\mathbf{a}_3^{(1)}\|^2 - [a_{13}^{(1)}]^2 = 22 - (-4.2817)^2 = 3.6667,$$

$$\|\tilde{\mathbf{a}}_3^{(2)}\|_2^2 = \|\mathbf{a}_4^{(1)}\|^2 - [a_{14}^{(1)}]^2 = 4 - (-1.7905)^2 = 0.7939.$$

The second column is the largest, so there is no need for a column interchange this time. We apply a Householder transformation $\tilde{H}_2$ to the first column of $\tilde{A}^{(2)}$ so that the updated column is a multiple of $\mathbf{e}_1$, which is equivalent to applying a 4×4 Householder reflection $H_2 = I - 2\mathbf{v}_2\mathbf{v}_2^T$, where the first component of $\mathbf{v}_2$ is zero, to the *second* column of $A^{(2)}$ so that the updated column is a linear combination of $\mathbf{e}_1$ and $\mathbf{e}_2$. This yields

$$\tilde{H}_2\tilde{A}^{(2)} = \begin{bmatrix} 2.2643 & -1.7665 & -0.4978 \\ 0 & -0.2559 & 0.2559 \\ 0 & 0.6933 & -0.6933 \end{bmatrix}.$$

Then, we consider the submatrix obtained by removing the first row and column of $H_2\tilde{A}^{(2)}$:

$$\tilde{A}^{(3)} = \begin{bmatrix} -0.2559 & 0.2559 \\ 0.6933 & -0.6933 \end{bmatrix}.$$

Both columns have the same lengths, so no column interchange is required. Applying a Householder reflection $\tilde{H}_3$ to the first column to make it a multiple of $\mathbf{e}_1$ will have the same effect on the second column, because they are parallel. We have

$$\tilde{H}_3\tilde{A}^{(3)} = \begin{bmatrix} 0.7390 & -0.7390 \\ 0 & 0 \end{bmatrix}.$$

It follows that the matrix $\tilde{A}^{(4)}$ obtained by removing the first row and column of $H_3\tilde{A}^{(3)}$ will be the zero matrix. We conclude that $\text{rank}(A) = 3$, and that A has the factorization

$$A\Pi = Q\begin{bmatrix} R & S \\ 0 & 0 \end{bmatrix},$$

where

$$\Pi = \begin{bmatrix} 0 & 0 & 1 & 0 \\ 0 & 1 & 0 & 0 \\ 1 & 0 & 0 & 0 \\ 0 & 0 & 0 & 1 \end{bmatrix},$$

$$R = \begin{bmatrix} -12.8452 & -6.7729 & -4.2817 \\ 0 & 2.2643 & -1.7665 \\ 0 & 0 & 0.7390 \end{bmatrix}, \quad S = \begin{bmatrix} -1.7905 \\ -0.4978 \\ -0.7390 \end{bmatrix},$$

and $Q = H_1H_2H_3$, where H_2 and H_3 are defined in terms of $\tilde{H}_2$ and $\tilde{H}_3$ as in (4.13), is the product of the Householder reflections used to reduce $A\Pi$ to upper triangular form. $\square$

Exploration 4.3.1 Modify your function `houseqr` from Exploration 4.2.7 so that it performs QR with column pivoting, and therefore returns a third output argument P that is a permutation matrix, corresponding to the matrix Π in the preceding discussion.

Using this decomposition, we can solve the linear least squares problem $A\mathbf{x} = \mathbf{b}$ by observing that

$$\|\mathbf{b} - A\mathbf{x}\|_2^2 = \left\| \mathbf{b} - Q \begin{bmatrix} R & S \\ 0 & 0 \end{bmatrix} \Pi^T \mathbf{x} \right\|_2^2$$

$$= \left\| Q^T \mathbf{b} - \begin{bmatrix} R & S \\ 0 & 0 \end{bmatrix} \begin{bmatrix} \mathbf{u} \\ \mathbf{v} \end{bmatrix} \right\|_2^2$$

$$= \left\| \begin{bmatrix} \mathbf{c} \\ \mathbf{d} \end{bmatrix} - \begin{bmatrix} R\mathbf{u} + S\mathbf{v} \\ 0 \end{bmatrix} \right\|_2^2$$

$$= \|\mathbf{c} - R\mathbf{u} - S\mathbf{v}\|_2^2 + \|\mathbf{d}\|_2^2,$$

where

$$Q^T \mathbf{b} = \begin{bmatrix} \mathbf{c} \\ \mathbf{d} \end{bmatrix}, \quad \Pi^T \mathbf{x} = \begin{bmatrix} \mathbf{u} \\ \mathbf{v} \end{bmatrix},$$

with $\mathbf{c}$ and $\mathbf{u}$ being r-vectors. Thus $\min \|\mathbf{b} - A\mathbf{x}\|_2^2 = \|\mathbf{d}\|_2^2$, provided that $R\mathbf{u} + S\mathbf{v} = \mathbf{c}$. A *basic solution* is obtained by choosing $\mathbf{v} = \mathbf{0}$. A second solution is to choose $\mathbf{u}$ and $\mathbf{v}$ so that $\|\mathbf{u}\|_2^2 + \|\mathbf{v}\|_2^2$ is minimized, but that solution is not easily found using this factorization.

Exploration 4.3.2 Write a MATLAB function

$$\texttt{x=rankdeflsq(A,b)}$$

that solves the rank-deficient least-squares problem of minimizing $\|\mathbf{b} - A\mathbf{x}\|_2$. Since the solution to this problem is not unique, have your function return the *basic solution* described above. Use your function from Exploration 4.3.1 to obtain the QR Factorization with column pivoting.

4.3.2 Complete Orthogonal Decomposition*

To obtain the minimum-norm least squares solution, we seek a factorization of A with a simpler form, meaning that more of the "middle" matrix is zeroed. After performing the QR Factorization with column pivoting, we have

$$A = Q \begin{bmatrix} R & S \\ 0 & 0 \end{bmatrix} \Pi^T$$

where R is upper triangular. Then

$$A^T = \Pi \begin{bmatrix} R^T & 0 \\ S^T & 0 \end{bmatrix} Q^T$$

where R^T is lower triangular. We apply Householder reflections so that

$$P_r \cdots P_2 P_1 \begin{bmatrix} R^T & 0 \\ S^T & 0 \end{bmatrix} = \begin{bmatrix} U & 0 \\ 0 & 0 \end{bmatrix},$$

where U is upper triangular. Then

$$A^T = Z \begin{bmatrix} U & 0 \\ 0 & 0 \end{bmatrix} Q^T$$

where $Z = \Pi P_1 \cdots P_r$. In other words,

$$A = Q \begin{bmatrix} L & 0 \\ 0 & 0 \end{bmatrix} Z^T \tag{4.17}$$

where $L = U^T$ is a nonsingular lower triangular matrix of size $r \times r$, and $r = \text{rank}(A)$. This is the **complete orthogonal decomposition** of A.

Exploration 4.3.3 Modify your function houseqr from Exploration 4.3.1 to create a new function [Q,L,Z]=comporth(A) that computes the complete orthogonal decomposition (4.17) of the input argument A.

This decomposition leads to the following definition.

Definition 4.3.3 *The $m \times n$ matrix X is the* **pseudo-inverse** *of an $m \times n$ matrix A if*

1. $AXA = A$
2. $XAX = X$
3. $(XA)^T = XA$
4. $(AX)^T = AX$

Given the above complete orthogonal decomposition of A, the pseudo-inverse of A, denoted A^+, is given by

$$A^+ = Z \begin{bmatrix} L^{-1} & 0 \\ 0 & 0 \end{bmatrix} Q^T. \tag{4.18}$$

Exploration 4.3.4 Verify that matrix A^+ from (4.18) satisfies all of the properties of the pseudo-inverse listed in Definition 4.3.3.

Exploration 4.3.5 Modify your function comporth from Exploration 4.3.3 to create a new function B=pseudoinv(A) that computes the pseudo-inverse of A using (4.18) and returns it in the output argument B.

The MATLAB function pinv computes the pseudo-inverse of a given matrix A.

Let $\mathcal{X} = \{\mathbf{x} \mid \|\mathbf{b} - A\mathbf{x}\|_2 = \text{minimum}\}$. If $\mathbf{x} \in \mathcal{X}$ and we desire $\|\mathbf{x}\|_2 = \text{minimum}$, then $\mathbf{x} = A^+\mathbf{b}$. To see this, note that if we write $Q = \begin{bmatrix} Q_1 & Q_2 \end{bmatrix}$ and $Z = \begin{bmatrix} Z_1 & Z_2 \end{bmatrix}$, where Q_1 and Z_1 both have r columns, then for any $\mathbf{x} \in \mathcal{X}$,

$$Z_1^T \mathbf{x} = L^{-1} Q_1^T \mathbf{b}. \tag{4.19}$$

Exploration 4.3.6 Use the normal equations and (4.17) to verify (4.19).

Furthermore, from (4.17) and (4.18), $A = Q_1 L Z_1^T$ and $A^+ = Z_1 L^{-1} Q_1^T$. Applying (4.19), we then have

$$\mathbf{x} = Z Z^T \mathbf{x} = Z_1 Z_1^T \mathbf{x} + Z_2 Z_2^T \mathbf{x} = A^+ \mathbf{b} + Z_2 Z_2^T \mathbf{x},$$

which, due to $Z_1^T Z_2 = 0$, yields

$$\|\mathbf{x}\|_2^2 = \|A^+ \mathbf{b}\|_2^2 + \|Z_2 Z_2^T \mathbf{x}\|_2^2.$$

It follows from the nonnegativity of norms that $\|A^+ \mathbf{b}\|_2 \leq \|\mathbf{x}\|_2$ for any $\mathbf{x} \in \mathcal{X}$. Furthermore, because $A^+ \mathbf{b}$ is the unique solution of

$$Z^T \mathbf{x} = \begin{bmatrix} Z_1^T \mathbf{x} \\ Z_2^T \mathbf{x} \end{bmatrix} = \begin{bmatrix} L^{-1} Q_1^T \mathbf{b} \\ 0 \end{bmatrix},$$

this minimum-norm solution is unique.

It is worth noting that

$$P = AA^+$$
$$= Q \begin{bmatrix} L & 0 \\ 0 & 0 \end{bmatrix} Z^T Z \begin{bmatrix} L^{-1} & 0 \\ 0 & 0 \end{bmatrix} Q^T$$
$$= Q \begin{bmatrix} I & 0 \\ 0 & 0 \end{bmatrix} Q^T.$$

It can then be verified directly that $P = P^T$ and $P^2 = P$. In other words, the matrix P is a **projection** (see Exploration 4.1.11). In particular, it is a projection onto the space spanned by the columns of A, i.e., the range of A. That is, $P = Q_r Q_r^T$, where Q_r is the matrix consisting of the first r columns of Q.

We can use the complete orthogonal decomposition to show that the minimum-norm solution $\mathbf{x} = A^+ \mathbf{b}$, while generally not an exact solution of $A\mathbf{x} = \mathbf{b}$, is the exact solution to a related system of equations. We write $\mathbf{b} = \mathbf{b}_1 + \mathbf{b}_2$, where

$$\mathbf{b}_1 = P\mathbf{b}, \quad \mathbf{b}_2 = (I - P)\mathbf{b}.$$

It follows that $\mathbf{b}_1$ is a linear combination of $\mathbf{q}_1, \ldots, \mathbf{q}_r$, the columns of Q that form an orthonormal basis for the range of A. From $\mathbf{x} = A^+ \mathbf{b}$ we obtain

$$A\mathbf{x} = AA^+ \mathbf{b} = P\mathbf{b} = \mathbf{b}_1.$$

Therefore, the solution to the least squares problem is also the exact solution to the system $A\mathbf{x} = P\mathbf{b}$. The **residual vector** $\mathbf{r} = \mathbf{b} - A\mathbf{x}$ can also be expressed conveniently using the projection P. We have

$$\mathbf{r} = \mathbf{b} - A\mathbf{x} = \mathbf{b} - AA^+ \mathbf{b} = \mathbf{b} - P\mathbf{b} = (I - P)\mathbf{b} = P^{\perp} \mathbf{b}.$$

That is, the residual is the projection of $\mathbf{b}$ onto the **orthogonal complement** of the range of A, which is the null space of A^T (see Exploration 4.2.6).

Exploration 4.3.7 Write a MATLAB function

```
[x,r]=minnormlsq(A,b)
```

that computes the minimum-norm least squares solution of $A * \mathbf{x} = \mathbf{b}$, using your function comporth from Exploration 4.3.3. The output arguments are the minimum-norm solution $\mathbf{x} = A^+ \mathbf{b}$ and the residual $\mathbf{r} = \mathbf{b} - AA^+ \mathbf{b}$.

4.3.3 Concept Check

1. Why are the approaches used to compute the QR Factorization described in Section 4.2 not viable in the case where A does not have full column rank?

2. Describe QR with column pivoting. What purpose does it serve, and how does it work?

3. How is the complete orthogonal decomposition computed?

4. What is the pseudo-inverse of a matrix, and how does it relate to the inverse of a square nonsingular matrix?

5. What is a projection matrix? Use a projection to describe the system of linear equations that the least squares solution satisfies exactly.

4.4 The Singular Value Decomposition

The Singular Value Decomposition (SVD) is one of the most important matrix factorizations in numerical linear algebra, due to its many applications. In this section, we establish its existence and provide an overview of its properties and these applications.

4.4.1 Existence

The matrix ℓ_2-norm can be used to obtain a highly useful decomposition of any $A \in \mathbb{R}^{m \times n}$. Let $\mathbf{x} \in \mathbb{R}^n$ be a unit ℓ_2-norm vector (that is, $\|\mathbf{x}\|_2 = 1$) such that $\|A\mathbf{x}\|_2 = \|A\|_2$. Such a vector exists, due to the definition of an induced matrix norm from Section B.13.2. If $\mathbf{z} = A\mathbf{x}$, and we define $\mathbf{y} = \mathbf{z}/\|\mathbf{z}\|_2$, then we have $A\mathbf{x} = \sigma \mathbf{y}$, with $\sigma = \|A\|_2$, and both $\mathbf{x}$ and $\mathbf{y}$ are unit vectors.

Let $V_1 \in \mathbb{R}^{n \times n}$ be a Householder reflection such that $V_1 \mathbf{x} = \mathbf{e}_1$, and $U_1 \in \mathbb{R}^{m \times m}$ be a Householder reflection such that $U_1 \mathbf{y} = \mathbf{e}_1$. These can be constructed using the approach from Exploration 4.2.5. Then we have

$$U_1^T A V_1 \mathbf{e}_1 = U_1^T A \mathbf{x} = \sigma U_1^T \mathbf{y} = \sigma \mathbf{e}_1,$$

and

$$\mathbf{e}_1^T U_1^T A V_1 = \mathbf{y}^T A V_1.$$

Let $\mathbf{z} = A^T \mathbf{y}$. Then $\mathbf{z} = c\mathbf{x} + d\mathbf{w}$, where c and d are scalars, $\|\mathbf{w}\|_2 = 1$, and $\mathbf{w}^T \mathbf{x} = 0$. It follows that

$$c = \mathbf{x}^T \mathbf{z} = (A\mathbf{x})^T \mathbf{y} = \sigma,$$

and thus $\|A^T \mathbf{y}\|_2^2 = \sigma^2 + d^2$.

Exploration 4.4.1 Let $A \in \mathbb{R}^{m \times n}$. Use (B.5) and the result of Exploration 4.1.4 to prove that $\|A\|_2 = \|A^T\|_2$.

It follows that $d = 0$, which means $A^T \mathbf{y} = \sigma \mathbf{x}$, and therefore

$$\mathbf{e}_1^T U_1^T A V_1 = \sigma \mathbf{x}^T V_1 = \sigma \mathbf{e}_1^T.$$

We then have

$$U_1^T A V_1 = \begin{bmatrix} \sigma & \mathbf{0}^T \\ \mathbf{0} & A_1 \end{bmatrix}.$$

Continuing this process on A_1, and keeping in mind that $\|A_1\|_2 \le \|A\|_2$, we obtain the decomposition

$$U^T A V = \Sigma$$

where

$$U = \begin{bmatrix} \mathbf{u}_1 & \cdots & \mathbf{u}_m \end{bmatrix} \in \mathbb{R}^{m \times m}, \quad V = \begin{bmatrix} \mathbf{v}_1 & \cdots & \mathbf{v}_n \end{bmatrix} \in \mathbb{R}^{n \times n}$$

are both orthogonal matrices, and

$$\Sigma = \mathrm{diag}(\sigma_1, \ldots, \sigma_p) \in \mathbb{R}^{m \times n}, \quad p = \min\{m, n\}$$

is a *diagonal* matrix, in which $\Sigma_{ii} = \sigma_i$ for $i = 1, 2, \ldots, p$, and $\Sigma_{ij} = 0$ for $i \ne j$. Furthermore, we have

$$\|A\|_2 = \sigma_1 \ge \sigma_2 \ge \cdots \ge \sigma_p \ge 0.$$

This decomposition of A is called the **Singular Value Decomposition (SVD)**, or **SVD**. It is more commonly written as a factorization of A,

$$A = U \Sigma V^T.$$

The MATLAB function svd computes the SVD of a given matrix.

4.4.2 Properties

The diagonal entries of Σ are the **singular values** of A. The columns of U are the **left singular vectors**, and the columns of V are the **right singular vectors**. It follows from the SVD itself that the singular values and vectors satisfy the relations

$$A\mathbf{v}_i = \sigma_i \mathbf{u}_i, \quad A^T \mathbf{u}_i = \sigma_i \mathbf{v}_i, \quad i = 1, 2, \ldots, \min\{m, n\}.$$

For convenience, we denote the ith largest singular value of A by $\sigma_i(A)$, or simply σ_i when the matrix is clear from context. The largest and smallest singular values are commonly denoted by $\sigma_{\max}(A)$ and $\sigma_{\min}(A)$, respectively.

The SVD conveys much useful information about the structure of a matrix, particularly with regard to systems of linear equations involving the matrix. Let r be the number of nonzero singular values. Then r is the rank of A, and

$$\mathrm{range}(A) = \mathrm{span}\{\mathbf{u}_1, \ldots, \mathbf{u}_r\}, \quad \mathrm{null}(A) = \mathrm{span}\{\mathbf{v}_{r+1}, \ldots, \mathbf{v}_n\}.$$

That is, the SVD yields orthonormal bases of the range and null space of A.

It follows that we can write

$$A = \sum_{i=1}^{r} \sigma_i \mathbf{u}_i \mathbf{v}_i^T.$$

This is called the **SVD expansion** of A. If $m \geq n$, then this expansion yields the "economy-size" SVD

$$A = U_1 \Sigma_1 V^T,$$

where

$$U_1 = \begin{bmatrix} \mathbf{u}_1 & \cdots & \mathbf{u}_n \end{bmatrix} \in \mathbb{R}^{m \times n}, \quad \Sigma_1 = \text{diag}(\sigma_1, \ldots, \sigma_n) \in \mathbb{R}^{n \times n}.$$

Example 4.4.1 The matrix

$$A = \begin{bmatrix} 11 & 19 & 11 \\ 9 & 21 & 9 \\ 10 & 20 & 10 \end{bmatrix}$$

has the SVD $A = U \Sigma V^T$ where

$$U = \begin{bmatrix} 1/\sqrt{3} & -1/\sqrt{2} & 1/\sqrt{6} \\ 1/\sqrt{3} & 1/\sqrt{2} & 1/\sqrt{6} \\ 1/\sqrt{3} & 0 & -\sqrt{2/3} \end{bmatrix}, \quad V = \begin{bmatrix} 1/\sqrt{6} & -1/\sqrt{3} & 1/\sqrt{2} \\ \sqrt{2/3} & 1/\sqrt{3} & 0 \\ 1/\sqrt{6} & -1/\sqrt{3} & -1/\sqrt{2} \end{bmatrix},$$

and

$$S = \begin{bmatrix} 42.4264 & 0 & 0 \\ 0 & 2.4495 & 0 \\ 0 & 0 & 0 \end{bmatrix}.$$

Let $U = \begin{bmatrix} \mathbf{u}_1 & \mathbf{u}_2 & \mathbf{u}_3 \end{bmatrix}$ and $V = \begin{bmatrix} \mathbf{v}_1 & \mathbf{v}_2 & \mathbf{v}_3 \end{bmatrix}$ be column partitions of U and V, respectively. Because there are only two nonzero singular values, we have $\text{rank}(A) = 2$. Furthermore, $\text{range}(A) = \text{span}\{\mathbf{u}_1, \mathbf{u}_2\}$, and $\text{null}(A) = \text{span}\{\mathbf{v}_3\}$. We also have

$$A = 42.4264 \mathbf{u}_1 \mathbf{v}_1^T + 2.4495 \mathbf{u}_2 \mathbf{v}_2^T.$$

□

Exploration 4.4.2 Use the MATLAB function svd to compute the SVD of the matrix

$$A = \begin{bmatrix} 10 & -2 & -2 \\ 10 & -2 & -2 \\ -14 & 10 & 10 \end{bmatrix}.$$

Use the output of this function to determine $\text{rank}(A)$ and write down orthonormal bases for the range and null space of A.

The SVD is also closely related to the ℓ_2-norm and Frobenius norm. We have

$$\|A\|_2 = \sigma_1, \quad \|A\|_F^2 = \sigma_1^2 + \sigma_2^2 + \cdots + \sigma_r^2, \tag{4.20}$$

and

$$\inf_{\mathbf{x}\neq 0} \frac{\|A\mathbf{x}\|_2}{\|\mathbf{x}\|_2} = \min_{\|\mathbf{x}\|_2=1} \|A\mathbf{x}\|_2 = \sigma_p, \quad p = \min\{m,n\}. \tag{4.21}$$

These relationships follow directly from the invariance of these norms under orthogonal transformations, a proof of which is guided by the following explorations.

Exploration 4.4.3 Let $A \in \mathbb{R}^{m\times n}$, and let $Q \in \mathbb{R}^{m\times m}$ and $Z \in \mathbb{R}^{n\times n}$ be orthogonal matrices. Use the result of Exploration 4.4.1 to prove that $\|QA\|_2 = \|AZ\|_2 = \|A\|_2$. That is, the matrix ℓ_2-norm is invariant under orthogonal transformations.

Exploration 4.4.4 The **trace** of a matrix $A \in \mathbb{R}^{m\times n}$, denoted by $\mathrm{tr}(A)$, is the sum of its diagonal entries:

$$\mathrm{tr}(A) = \sum_{i=1}^{\min\{m,n\}} a_{ii}. \tag{4.22}$$

Use (B.6) and the formula (B.2) for matrix multiplication to prove

$$\|A\|_F = \sqrt{\mathrm{tr}(A^T A)}.$$

Exploration 4.4.5 Prove that $\|A\|_F = \|A^T\|_F$.

Exploration 4.4.6 Use the result of Explorations 4.4.4 and 4.4.5 to prove that the Frobenius norm is invariant under orthogonal transformations, just as the ℓ_2-norm is. That is, prove that $\|QA\|_F = \|AZ\|_F = \|A\|_F$, where $A \in \mathbb{R}^{m\times n}$, and $Q \in \mathbb{R}^{m\times m}$ and $Z \in \mathbb{R}^{n\times n}$ are orthogonal matrices.

Exploration 4.4.7 Use the results of Explorations 4.4.3 and 4.4.6 to prove (4.20).

Exploration 4.4.8 Use the SVD to prove (4.21).

The SVD also provides a simple formula for the condition number of A.

Exploration 4.4.9 Show that

$$\kappa_2(A) = \|A\|_2\|A^{-1}\|_2 = \frac{\sigma_{\max}(A)}{\sigma_{\min}(A)}.$$

4.4.2.1 *Minimum-norm least squares solution*

The Singular Value Decomposition is very useful in studying the linear least squares problem. Suppose that we are given $\mathbf{b} \in \mathbb{R}^m$ and $A \in \mathbb{R}^{m\times n}$, with $\mathrm{rank}(A) = r$, and we wish to find $\mathbf{x}$ such that

$$\|\mathbf{b} - A\mathbf{x}\|_2 = \text{minimum}.$$

From the SVD of A, we can simplify this minimization problem as follows:

$$\|\mathbf{b} - A\mathbf{x}\|_2^2 = \|\mathbf{b} - U\Sigma V^T\mathbf{x}\|_2^2$$
$$= \|U^T\mathbf{b} - \Sigma V^T\mathbf{x}\|_2^2$$
$$= \|\mathbf{c} - \Sigma\mathbf{y}\|_2^2$$
$$= (c_1 - \sigma_1 y_1)^2 + \cdots + (c_r - \sigma_r y_r)^2 +$$
$$c_{r+1}^2 + \cdots + c_m^2$$

where $\mathbf{c} = U^T\mathbf{b}$ and $\mathbf{y} = V^T\mathbf{x}$. We see that in order to minimize $\|\mathbf{b} - A\mathbf{x}\|_2$, we must set $y_i = c_i/\sigma_i$ for $i = 1, \ldots, r$, but the unknowns y_i, for $i = r + 1, \ldots, n$, can have any value, since they do not influence $\|\mathbf{c} - \Sigma\mathbf{y}\|_2$. Therefore, if A does not have full rank, there are infinitely many solutions to the least squares problem. However, we can easily obtain the unique solution of minimum 2-norm by setting $y_{r+1} = \cdots = y_n = 0$.

In summary, the solution of minimum length to the linear least squares problem is

$$\mathbf{x} = V\mathbf{y}$$
$$= V\Sigma^+\mathbf{c}$$
$$= V\Sigma^+U^T\mathbf{b}$$
$$= A^+\mathbf{b}$$

where Σ^+ is a diagonal matrix with entries

$$\Sigma^+ = \begin{bmatrix} \sigma_1^{-1} & & & & & \\ & \ddots & & & & \\ & & \sigma_r^{-1} & & & \\ & & & 0 & & \\ & & & & \ddots & \\ & & & & & 0 \end{bmatrix}$$

and $A^+ = V\Sigma^+U^T$. The matrix A^+ is actually the **pseudo-inverse** of A, previously introduced in Section 4.3.2.

> **Exploration 4.4.10** Repeat Exploration 4.3.4 for the pseudo-inverse A^+ described in terms of the SVD.

The MATLAB function `pinv` can be used to compute the pseudo-inverse of a given matrix A. However, just as it is generally impractical to explicitly compute the inverse of a square invertible matrix, it is not advisable to explicitly compute the pseudo-inverse either.

> **Exploration 4.4.11** Write a new version of your function `minnormlsq` from Exploration 4.3.7 that uses svd instead of `comporth` to compute $\mathbf{x} = A^+*\mathbf{b}$, without explicitly forming A^+.

It is actually very easy to obtain the minimum-norm least squares solution in MAT-LAB, using the \ operator, which automatically solves this least squares problem in the case where A is rank-deficient.

4.4.3 Low-rank Approximations

The SVD has many useful applications, but one of particular interest is that the truncated SVD expansion

$$A_k = \sum_{i=1}^{k} \sigma_i \mathbf{u}_i \mathbf{v}_i^T, \qquad (4.23)$$

where $k < r = \text{rank}(A)$, is the best approximation of A by a rank-k matrix. More precisely, the distance, in the ℓ_2-norm, between A and the set of matrices of rank k is

$$\min_{\text{rank}(B)=k} \|A - B\|_2 = \|A - A_k\|_2 = \left\| \sum_{i=k+1}^{r} \sigma_i \mathbf{u}_i \mathbf{v}_i^T \right\| = \sigma_{k+1},$$

because the ℓ_2-norm of a matrix is its largest singular value, and σ_{k+1} is the largest singular value of the matrix obtained by removing the first k terms of the SVD expansion. Therefore, σ_p, where $p = \min\{m, n\}$, is the distance between A and the set of all rank-deficient matrices (which is zero when A is already rank-deficient). Because a matrix of full rank can have arbitrarily small, but still positive, singular values, it follows that the set of all full-rank matrices in $\mathbb{R}^{m \times n}$ is both open and dense.

Example 4.4.2 The best approximation of the matrix A from Example 4.4.1, which has rank two, by a matrix of rank one is obtained by taking only the first term of its SVD expansion,

$$A_1 = 42.4264 \mathbf{u}_1 \mathbf{v}_1^T = \begin{bmatrix} 10 & 20 & 10 \\ 10 & 20 & 10 \\ 10 & 20 & 10 \end{bmatrix}.$$

The absolute error in this approximation is

$$\|A - A_1\|_2 = \sigma_2 \|\mathbf{u}_2 \mathbf{v}_2^T\|_2 = \sigma_2 = 2.4495,$$

while the relative error is

$$\frac{\|A - A_1\|_2}{\|A\|_2} = \frac{2.4495}{42.4264} = 0.0577.$$

That is, the rank-one approximation of A is off by a little less than six percent. $\square$

Exploration 4.4.12 Prove that the best rank-k approximation of A defined by (4.23) is also the best rank-k approximation in the Frobenius norm.

Suppose that the entries of a matrix A have been determined experimentally, for example by measurements, and the error in the entries is determined to be bounded by some value ϵ. Then, if $\sigma_k > \epsilon \geq \sigma_{k+1}$ for some $k < \min\{m, n\}$, then ϵ is at least as large as the distance between A and the set of all matrices of rank k. Therefore, due to the uncertainty of the measurement, A could be a matrix of rank k, but not of lower rank, because $\sigma_k > \epsilon$. We say that the ϵ-**rank** of A, defined by

$$\text{rank}(A, \epsilon) = \min_{\|A-B\|_2 \leq \epsilon} \text{rank}(B),$$

is equal to k. If $k < \min\{m, n\}$, then we say that A is **numerically rank-deficient**.

4.4.4   Concept Check

1. What is the Singular Value Decomposition? Identify the singular values and left and right singular vectors.
2. How can the SVD be used to easily describe the range and null space of a matrix?
3. How do the singular values relate to the ℓ_2-norm and Frobenius norm of a matrix?
4. How is the condition number in the ℓ_2-norm expressed in terms of the singular values of A?
5. How can the SVD be used to obtain a low-rank approximation of a matrix?

4.5   Other Least Squares Problems*

So far in this chapter, we have only considered the problem of minimizing $\|\mathbf{b} - A\mathbf{x}\|_2$, with the additional criterion of minimizing $\|\mathbf{x}\|_2$ in the case where A is rank-deficient. In this section, we examine some variations on this problem.

4.5.1   Linear Constraints

Suppose that we wish to fit data as in the least squares problem, except that we are using different functions to fit the data on different subintervals. A common example is the process of fitting data using cubic splines, with a different cubic polynomial approximating data on each subinterval.

Typically, it is desired that the functions assigned to each piece form a function that is continuous on the entire interval within which the data lies. This requires that *constraints* be imposed on the functions themselves. It is also not uncommon to require that the function assembled from these pieces also has a continuous first or even second derivative, resulting in additional constraints. The result is a **least squares problem with linear constraints**, as the constraints are applied to coefficients of predetermined functions chosen as a basis for some function space, such as the space of polynomials of a given degree.

The general form of a least squares problem with linear constraints is as follows: Given $\mathbf{b} \in \mathbb{R}^n$ and $A \in \mathbb{R}^{m \times n}$, we wish to find $\mathbf{x} \in \mathbb{R}^n$ that minimizes $\|\mathbf{b} - A\mathbf{x}\|_2$, subject to the constraint $C^T\mathbf{x} = \mathbf{d}$, where $C \in \mathbb{R}^{n \times p}$ and $\mathbf{d} \in \mathbb{R}^p$ are known. This problem is usually solved using **Lagrange multipliers**. We define

$$f(x; \lambda) = \|\mathbf{b} - A\mathbf{x}\|_2^2 + 2\lambda^T (C^T\mathbf{x} - \mathbf{d}).$$

Then

$$\nabla f = 2(A^T A\mathbf{x} - A^T\mathbf{b} + C\lambda).$$

To minimize f, we can solve the system

$$\begin{bmatrix} A^T A \; C \\ C^T \;\; 0 \end{bmatrix} \begin{bmatrix} \mathbf{x} \\ \lambda \end{bmatrix} = \begin{bmatrix} A^T\mathbf{b} \\ \mathbf{d} \end{bmatrix}.$$

From $A^T A\mathbf{x} = A^T\mathbf{b} - C\lambda$, we see that $\mathbf{x}$ has the form $\mathbf{x} = \hat{\mathbf{x}} - (A^T A)^{-1} C\lambda$, where $\hat{\mathbf{x}}$ is the solution to the *unconstrained* least squares problem of minimizing $\|\mathbf{b} - A\mathbf{x}\|_2$. Then, from the constraint $C^T\mathbf{x} = \mathbf{d}$ we obtain the equation

$$C^T (A^T A)^{-1} C\lambda = C^T\hat{\mathbf{x}} - \mathbf{d}, \tag{4.24}$$

which we can then solve for λ to finally obtain $\mathbf{x}$.

The algorithm proceeds as follows:

Algorithm 4.5.1 (Least Squares with Linear Constraints) Given $A \in \mathbb{R}^{m \times n}$, $\mathbf{b} \in \mathbb{R}^m$, $C \in \mathbb{R}^{n \times p}$, and $\mathbf{d} \in \mathbb{R}^p$, the following algorithm computes $\mathbf{x} \in \mathbb{R}^n$ that minimizes $\|\mathbf{b} - A\mathbf{x}\|_2$, subject to the constraints $C^T\mathbf{x} = \mathbf{d}$.

$A = QR$ ("thin" QR Factorization)
Solve $R\hat{\mathbf{x}} = Q^T\mathbf{b}$ for $\hat{\mathbf{x}}$
Solve $R^T W = C$ for W
$W = PU$ ("thin" QR Factorization)
Solve $U^T\mathbf{y} = C^T\hat{\mathbf{x}} - \mathbf{d}$ for $\mathbf{y}$
Solve $U\lambda = \mathbf{y}$ for λ
Solve $R\mathbf{z} = W\lambda$ for $\mathbf{z}$
$\mathbf{x} = \hat{\mathbf{x}} - \mathbf{z}$

Exploration 4.5.1 Write a MATLAB function

```
x=lsqlinlagrange(A,b,C,d)
```

that implements Algorithm 4.5.1.

Exploration 4.5.2 Verify that $\mathbf{z}$ from Algorithm 4.5.1 satisfies $\mathbf{z} = (A^T A)^{-1} C\lambda$, and that λ as computed by this algorithm is the solution of (4.24).

This method is not the most practical since it has more unknowns than the unconstrained least squares problem, which is odd because the constraints should have the effect of *eliminating* unknowns, not adding them.

We now describe an alternative approach. Suppose that we compute the QR Factorization of C to obtain

$$C = QR = \begin{bmatrix} Q_1 & Q_2 \end{bmatrix} \begin{bmatrix} R_1 \\ 0 \end{bmatrix}$$

where Q_1 is $n \times p$ and R_1 is a $p \times p$ upper triangular matrix. Then the constraint $C^T \mathbf{x} = \mathbf{d}$ takes the form

$$R_1^T \mathbf{u} = \mathbf{d}, \quad \mathbf{u} = Q_1^T \mathbf{x}.$$

Then, if we let $\mathbf{v} = Q_2^T \mathbf{x}$, we have

$$\begin{aligned}
\|\mathbf{b} - A\mathbf{x}\|_2 &= \|\mathbf{b} - AQQ^T \mathbf{x}\| \\
&= \left\| \mathbf{b} - A \begin{bmatrix} Q_1 & Q_2 \end{bmatrix} \begin{bmatrix} Q_1^T \\ Q_2^T \end{bmatrix} \mathbf{x} \right\| \\
&= \|\mathbf{b} - AQ_1\mathbf{u} - A_2\mathbf{v}\|_2
\end{aligned}$$

where $A_2 = AQ_2$. Thus we can obtain $\mathbf{x}$ by the following procedure.

Algorithm 4.5.2 (Least Squares with Linear Constraints) Given $A \in \mathbb{R}^{m \times n}$, $\mathbf{b} \in \mathbb{R}^m$, $C \in \mathbb{R}^{n \times p}$, and $\mathbf{d} \in \mathbb{R}^p$, the following algorithm computes $\mathbf{x} \in \mathbb{R}^n$ that minimizes $\|\mathbf{b} - A\mathbf{x}\|_2$, subject to the constraints $C^T \mathbf{x} = \mathbf{d}$.

$C = QR$ (QR Factorization)
Solve $R_1^T \mathbf{u} = \mathbf{d}$
$\mathbf{z} = Q_1 \mathbf{u}$
$A_2 = AQ_2$
Solve $\|(\mathbf{b} - A\mathbf{z}) - A_2\mathbf{v}\|_2 = \text{minimum}$
$\mathbf{x} = \mathbf{z} + Q_2\mathbf{v}$

This approach has the advantage that there are fewer unknowns in each system that needs to be solved, and also that $\kappa(A_2) \leq \kappa(A)$. The drawback is that sparsity can be destroyed.

Exploration 4.5.3 Write a MATLAB function

$$\texttt{x=lsqlinqr(A,b,C,d)}$$

that implements Algorithm 4.5.2. Compare the performance with that of your function `lsqlinlagrange` from Exploration 4.5.1.

The MATLAB function `lsqlin` can be used to solve least squares problems with linear constraints. In addition to the equality constraints discussed in this section, one can also impose inequality constraints of the form $C\mathbf{x} \leq \mathbf{d}$, as well as upper and lower bounds on the elements of $\mathbf{x}$.

4.5.2 Quadratic Constraints

We wish to solve the problem
$$\|\mathbf{b} - A\mathbf{x}\|_2 = \text{minimum}, \quad \|\mathbf{x}\|_2 = \alpha, \quad \alpha \leq \|A^+\mathbf{b}\|_2.$$
This problem is known as **least squares with quadratic constraints**. Solving the least squares problem with quadratic constraints arises in several applications, including solving ill-conditioned linear systems by **Tikhonov regularization** [Tikhonov and Arsenin (1977)] or smoothing data [Reinsch (1967)].

To solve this problem, we again use a Lagrange multiplier. We define
$$\varphi(\mathbf{x}; \mu) = \|\mathbf{b} - A\mathbf{x}\|_2^2 + \mu(\|\mathbf{x}\|^2 - \alpha^2),$$
where $\mu \geq 0$, and seek to minimize φ. From
$$\nabla \varphi = -2A^T\mathbf{b} + 2A^T A\mathbf{x} + 2\mu\mathbf{x}$$
we obtain the system
$$(A^T A + \mu I)\mathbf{x} = A^T\mathbf{b}. \tag{4.25}$$
It can be shown that this system is at least as well-conditioned than the normal equations, if not better.

Exploration 4.5.4 Show that for $\mu \geq 0$, $\kappa_2(A^T A + \mu I) \leq \kappa_2(A^T A)$.

To solve this problem, we see that we need to compute
$$\mathbf{x} = (A^T A + \mu I)^{-1} A^T\mathbf{b}$$
where
$$\mathbf{x}^T\mathbf{x} = \mathbf{b}^T A (A^T A + \mu I)^{-2} A^T\mathbf{b} = \alpha^2.$$
If $A = U\Sigma V^T$ is the SVD of A, then we have
$$\alpha^2 = \mathbf{b}^T U\Sigma V^T (V\Sigma^T\Sigma V^T + \mu I)^{-2} V\Sigma^T U^T\mathbf{b}$$
$$= \mathbf{c}^T \Sigma (\Sigma^T\Sigma + \mu I)^{-2} \Sigma^T\mathbf{c}, \quad U^T\mathbf{b} = \mathbf{c}$$
$$= \sum_{i=1}^{r} \frac{c_i^2 \sigma_i^2}{(\sigma_i^2 + \mu)^2}$$
$$= \chi(\mu).$$
The rational function $\chi(\mu)$ has *poles* at $-\sigma_i^2$ for $i = 1, \dots, r$, so it is well-defined for $\mu \geq 0$. Furthermore, $\chi(\mu) \to 0$ as $\mu \to \infty$.

We now have the following procedure for solving this problem.

Algorithm 4.5.3 Let $A \in \mathbb{R}^{m \times n}$ with $m \geq n$, $\mathbf{b} \in \mathbb{R}^m$, and $\alpha \in \mathbb{R}$ with $\alpha > 0$. The following algorithm computes $\mathbf{x} \in \mathbb{R}^n$ that minimizes $\|\mathbf{b} - A\mathbf{x}\|_2$ subject to the constraint $\|\mathbf{x}\|_2 = \alpha$. It is assumed that $\alpha \leq \|A^+\mathbf{b}\|_2$.

$A = U\Sigma V^T$ (SVD)
$\mathbf{c} = U^T\mathbf{b}$
Solve $1/\chi(\mu) = 1/\alpha^2$ on the domain $\mu \geq 0$, using Newton's Method
    (See Algorithm 10.4.1)
$\mathbf{x} = V(\Sigma^T\Sigma + \mu I)^{-1}\Sigma^T\mathbf{c}$

The reason for solving the nonlinear equation $1/\chi(\mu) = 1/\alpha^2$, rather than $\chi(\mu) = \alpha^2$, is that $\chi(\mu)$ decays to zero as μ increases, and in such a way that $\chi'(\mu)$ also approaches zero. As will be discussed in Section 10.4, root-finding methods such as Newton's Method have difficulty with functions whose derivatives are small near a root.

Exploration 4.5.5 Verify that the final step in Algorithm 4.5.3 computes the solution $\mathbf{x}$ of the system 4.25.

Exploration 4.5.6 Throughout our discussion of the least squares problem with quadratic constraints, we have assumed $\alpha \leq \|A^{+}\mathbf{b}\|_2$. How would the problem be solved in the case $\alpha > \|A^{+}\mathbf{b}\|_2$?

4.5.3 Total Least Squares

In the ordinary least squares problem, we are solving

$$A\mathbf{x} = \mathbf{b} + \mathbf{r}, \quad \|\mathbf{r}\|_2 = \text{minimum}.$$

In the **total least squares** problem, we wish to solve

$$(A + E)\mathbf{x} = \mathbf{b} + \mathbf{r}, \quad \|E\|_F^2 + \lambda^2\|\mathbf{r}\|_2^2 = \text{minimum}.$$

From $A\mathbf{x} - \mathbf{b} + E\mathbf{x} - \mathbf{r} = \mathbf{0}$, we obtain the system

$$\begin{bmatrix} A & \lambda\mathbf{b} \end{bmatrix}\begin{bmatrix} \mathbf{x} \\ -\lambda^{-1} \end{bmatrix} + \begin{bmatrix} E & \lambda\mathbf{r} \end{bmatrix}\begin{bmatrix} \mathbf{x} \\ -\lambda^{-1} \end{bmatrix} = \mathbf{0},$$

or

$$(C + F)\mathbf{z} = \mathbf{0}, \quad C = \begin{bmatrix} A & \lambda\mathbf{b} \end{bmatrix}, \quad F = \begin{bmatrix} E & \lambda\mathbf{r} \end{bmatrix}.$$

We need the matrix $C + F$ to be rank-deficient, and we want to minimize $\|F\|_F$, as $\|F\|_F^2 = \|E\|_F^2 + \lambda^2\|\mathbf{r}\|_2^2$.

To solve this problem, we compute the SVD of $C = U\Sigma V^T$. Let C_n be the best rank-n approximation of C as defined in (4.23), and let $F = -\sigma_{n+1}\mathbf{u}_{n+1}\mathbf{v}_{n+1}^T$. Then $C + F = C_n$, and certainly rank$(C_n) < n + 1$. Before proceeding, we assume $\sigma_n(A) > \sigma_{n+1}(C)$. The following theorem, proved in [Wilkinson (1965)], is helpful.

Theorem 4.5.4 (Interlacing Property of Singular Values) Let $A \in \mathbb{R}^{m \times n}$, with $m \geq n$, have the column partition $A = \begin{bmatrix} \mathbf{a}_1 & \cdots & \mathbf{a}_n \end{bmatrix}$ and let $A_r = \begin{bmatrix} \mathbf{a}_1 & \cdots & \mathbf{a}_r \end{bmatrix}$. Then, for $r < n$,

$$\sigma_1(A_{r+1}) \geq \sigma_1(A_r) \geq \sigma_2(A_{r+1}) \geq \cdots \geq \sigma_r(A_{r+1}) \geq \sigma_r(A_r) \geq \sigma_{r+1}(A_{r+1}).$$

This ensures that $\sigma_n(C) \geq \sigma_n(A)$, and therefore $\sigma_n(C) > \sigma_{n+1}(C)$. It follows that C_n is the *unique* best rank-n approximation of C, and therefore F is the smallest matrix, in the Frobenius norm, for which rank$(C + F) < n + 1$. We also have

$$C_n\mathbf{v}_{n+1} = C\mathbf{v}_{n+1} + F\mathbf{v}_{n+1} = \sigma_{n+1}\mathbf{u}_{n+1} - \sigma_{n+1}\mathbf{u}_{n+1}\mathbf{v}_{n+1}^T\mathbf{v}_{n+1} = \mathbf{0}.$$

Exploration 4.5.7 Suppose that for a given total least squares problem, the assumption $\sigma_n(A) > \sigma_{n+1}(C)$ does not hold. How can the problem be modified so that it does hold?

For convenience, let $\mathbf{v}_{n+1} = \begin{bmatrix} \tilde{\mathbf{v}}_{n+1}^T & v_{n+1,n+1} \end{bmatrix}^T$. Then,

$$C\mathbf{v}_{n+1} = A\tilde{\mathbf{v}}_{n+1} + \lambda v_{n+1,n+1}\mathbf{b} = \sigma_{n+1}\mathbf{u}_{n+1}.$$

It follows that $v_{n+1,n+1} \neq 0$, for otherwise, $\|\tilde{\mathbf{v}}_{n+1}\|_2 = \|\mathbf{v}_{n+1}\|_2 = 1$ and

$$\sigma_{n+1}(C) = \|A\tilde{\mathbf{v}}_{n+1}\|_2 \geq \sigma_n(A),$$

which is a contradiction. We therefore have

$$\mathbf{0} = C_n\mathbf{v}_{n+1} = (C + F)\mathbf{v}_{n+1} = (A + E)\tilde{\mathbf{v}}_{n+1} + \lambda(\mathbf{b} + \mathbf{r})v_{n+1,n+1}$$

which yields the solution

$$\mathbf{x} = -\frac{1}{\lambda v_{n+1,n+1}}\tilde{\mathbf{v}}_{n+1}.$$

We now summarize the procedure for obtaining $\mathbf{x}$, as well as the perturbations E and $\mathbf{r}$.

Algorithm 4.5.5 (Total Least Squares) Let $A \in \mathbb{R}^{m \times n}$ with $m \geq n$ and $\text{rank}(A) = n$, $\mathbf{b} \in \mathbb{R}^n$ and $\lambda \in \mathbb{R}$ with $\lambda > 0$. The following algorithm computes $\mathbf{x} \in \mathbb{R}^n$ and $E \in \mathbb{R}^{m \times n}$ such that $(A + E)\mathbf{x} = \mathbf{b} + \mathbf{r}$ where $\|E\|_2^2 + \lambda^2\|\mathbf{r}\|_2^2$ is minimized, where $\mathbf{r} = \mathbf{b} - A\mathbf{x}$.

$C = \begin{bmatrix} A & \lambda\mathbf{b} \end{bmatrix}$
$C = U\Sigma V^T$ (SVD)
$\mathbf{x} = -V(1:n, n+1)/(\lambda V(n+1, n+1))$
$E = -\sigma_{n+1}U(:, n+1)V(1:n, n+1)^T$
$\mathbf{r} = -\sigma_{n+1}U(:, n+1)V(n+1, n+1)/\lambda$

We compute the residual $\mathbf{r}$ by scaling the $(n+1)$st column of U rather than by using its definition $\mathbf{r} = \mathbf{b} - A\mathbf{x}$, as the latter is susceptible to cancellation error if the residual is particularly small.

Exploration 4.5.8 Implement Algorithm 4.5.5 and test it on a random $A \in \mathbb{R}^{m \times n}$ with $m > n$, a random $\mathbf{b} \in \mathbb{R}^m$, and a chosen value of λ. How does the solution $\mathbf{x}$ obtained via total least squares differ from that computed by solving the full rank least squares problem?

4.5.4 Concept Check

1. What kind of data fitting problem leads to the least squares problem with linear constraints?
2. How do Lagrange multipliers simplify the least squares problem with linear constraints? What is a drawback to using them?

3. Briefly describe an alternate approach to least squares with linear constraints. What are the advantages and disadvantages of this approach?
4. What are some applications of the least squares problem with quadratic constraints?
5. What system of equations is solved to solve the least squares problem with quadratic constraints?
6. What is the distinction between the total least squares problem and the "ordinary" least squares problem?

4.6 Additional Resources

A more detailed presentation of techniques for solving least squares problems can be found in various textbooks devoted to numerical linear algebra, such as the "usual suspects" [Demmel (1997); Golub and van Loan (2012); Watkins (2002)]. Books devoted exclusively to least squares problems are [Björck (1996); Farebrother (1988); Lawson and Hanson (1995)]. An in-depth treatment of rank-deficient problems can be found in [Hansen (1998)]. The reader interested in learning all about total least squares is referred to [Van Huffel and Vandewalle (1991)]. References on statistical computing, such as [Kennedy and Gentle (1980); Thisted (1988)] offer a presentation of least squares problems from a statistical perspective and illustrate the importance of numerical methods for data analysis.

In this chapter, we focused exclusively on minimization of the ℓ_2-norm of residuals, but other norms are used as well. In particular, the ℓ_1-norm and ℓ_∞-norm are minimized in *linear programming*, which is presented in texts such as [Fletcher (1987); Nocedal and Wright (2006)]. In Section 8.1.3 are simple examples of *nonlinear least squares*; a comprehensive treatment can be found in [Kelley (1999)].

4.7 Exercises

1. Consider the full-rank least squares problem of minimizing $\|\mathbf{b} - A\mathbf{x}\|_2$, where

$$A = \begin{bmatrix} 1 & 1 \\ 1 & 2 \\ 1 & 3 \\ 1 & 4 \end{bmatrix}, \quad \mathbf{b} = \begin{bmatrix} 2 \\ 4 \\ 4 \\ 2 \end{bmatrix}.$$

(a) Set up the normal equations for this problem.
(b) Compute the Cholesky Factorization $A = R^T R$.
(c) Solve the problem using the Cholesky factor R. After obtaining the solution $\mathbf{x}$, compute the residual $\mathbf{r} = \mathbf{b} - A\mathbf{x}$.
(d) Repeat this process with $\mathbf{b} = \begin{bmatrix} -1 & 1 & 1 & -1 \end{bmatrix}$. Can you explain the result?

2. In this problem, we will obtain a uniqueness result for the QR Factorization.

(a) Prove that a matrix T that is upper triangular and orthogonal must also be diagonal.

(b) What are the diagonal entries of T?

(c) Use the results of parts 2a and 2b to prove that if $A \in \mathbb{R}^{m \times n}$, $m \geq n$, has full column rank, then the QR Factorization $A = QR$ is unique in the sense that there is only one such factorization in which R has positive diagonal entries.

3. In calculus one learns about the **scalar projection** of a vector $\mathbf{u}$ onto a vector $\mathbf{v}$,

$$\operatorname{comp}_{\mathbf{v}} \mathbf{u} = \frac{\mathbf{u} \cdot \mathbf{v}}{\|\mathbf{v}\|_2},$$

and the **vector projection** of $\mathbf{u}$ onto $\mathbf{v}$,

$$\operatorname{proj}_{\mathbf{v}} \mathbf{u} = \frac{\mathbf{u} \cdot \mathbf{v}}{\|\mathbf{v}\|_2^2} \mathbf{v}.$$

(a) Show that $\operatorname{proj}_{\mathbf{v}} \mathbf{u} = c\mathbf{v}$ where c is the solution of the least squares problem of minimizing $\|\mathbf{u} - c\mathbf{v}\|_2$.

(b) Show that $c = \operatorname{comp}_{\mathbf{v}} \mathbf{u}$ if and only if $\mathbf{v}$ is a unit vector.

4. Let $A \in \mathbb{R}^{m \times n}$ with $m \geq n$ and $\operatorname{rank}(A) = n$. Let $P = A(A^T A)^{-1} A^T$.

(a) Show that $P = UU^T$, where $A = U\Sigma V^T$ is the "economy-size" SVD of A.

(b) Show that $A^+ = (A^T A)^{-1} A^T$.

(c) Show that for the general case of $\operatorname{rank}(A) = r$, $A^+ = (A_r^T A_r)^{-1} A^T$, where A_r is defined as in (4.23).

5. Let S be a subspace of $\mathbb{R}^m$, with $\dim(S) = k$. Let $Q \in \mathbb{R}^{m \times k}$ have columns that form an orthonormal basis for S.

(a) Prove that $P = QQ^T$ is a projection onto S. That is, prove that P is a projection, and that $\operatorname{range}(P) = S$.

(b) Let $\mathbf{b} \in \mathbb{R}^m$. Prove that $P\mathbf{b}$ is the "best approximation" of $\mathbf{b}$ by a vector in S; that is, $\|P\mathbf{b} - \mathbf{b}\|_2 \leq \|\mathbf{v} - \mathbf{b}\|_2$ for all $\mathbf{v} \in S$.

6. Let $A \in \mathbb{R}^{m \times n}$ have the form $A = R + \mathbf{u}\mathbf{v}^T$, where R is upper triangular. Develop an algorithm for computing the QR Factorization of A that requires only $O(mn + n^2)$ floating-point operations.

7. Modify your function `givensqr` from Exploration 4.2.3 so that it returns a single output R that contains R from the QR Factorization of the input matrix A in its upper triangle, and in its lower triangle, the angles of the Givens rotations used to zero the subdiagonal entries of A. That is, when $i > j$, the (i, j) entry of R contains the angle θ such that the Givens rotation $G(i - 1, i, c, s)^T$ zeros a_{ij}, where $c = \cos\theta$ and $s = \sin\theta$. Then write a function Q=makeQ(R) that uses these subdiagonal entries of R to construct Q.

8. Let

$$A = \begin{bmatrix} 1 & 1 \\ \epsilon & 0 \\ 0 & \epsilon \end{bmatrix}$$

where $\epsilon > 0$.

(a) Show that if ϵ is sufficiently small, $B = A^T A$ is numerically singular, even though A still has full rank. What is the smallest ϵ so that B is numerically nonsingular in the IEEE double-precision floating-point system?

(b) Choose a random vector $\mathbf{b} \in \mathbb{R}^3$ and ϵ to be the smallest value such that B is still numerically nonsingular. Solve the full-rank least square problem of minimizing $\|\mathbf{b} - A\mathbf{x}\|_2$ using both the normal equations and QR Factorization. What are your observations?

9. Use the summation formulas

$$\sum_{i=1}^{n} i = \frac{n(n+1)}{2}, \quad \sum_{i=1}^{n} i^2 = \frac{n(n+1)(2n+1)}{6}$$

to show that the number of floating-point operations required to reduce $A \in \mathbb{R}^{m \times n}$, $m \geq n$, to upper triangular form using Householder reflections is $2mn^2 - 2n^3/3$ plus lower-degree terms.

10. Let

$$A = \begin{bmatrix} 36/65 & 3/13 \\ 48/65 & 4/13 \end{bmatrix}.$$

(a) Show that $A^+ = A^T$.

(b) Round each entry of A to 4 decimal places and compute A^+ using MATLAB. Explain your observations using the SVDs of both matrices.

Chapter 5

Iterative Methods for Linear Systems

Given a system of n linear equations in n unknowns, described by the matrix-vector equation

$$A\mathbf{x} = \mathbf{b},$$

where A is an invertible $n \times n$ matrix, we learned in Chapter 3 that we can obtain the solution using a **direct method**, such as Gaussian Elimination in conjunction with forward and back substitution. However, there are several drawbacks to this approach:

- If we have an approximation to the solution $\mathbf{x}$, a direct method does not provide any means of taking advantage of this information to reduce the amount of computation required.
- If we only require an approximate solution, rather than the exact solution except for roundoff error, it is not possible to terminate the algorithm for a direct method early in order to obtain such an approximation.
- If the matrix A is **sparse**, meaning that most of the entries are zero, Gaussian Elimination or similar methods can cause **fill-in**, which is the introduction of new nonzero entries in the matrix, thus reducing efficiency.
- In some cases, A may not be represented as a two-dimensional array or a set of vectors; instead, it might be represented implicitly by a function that returns as output the matrix-vector product $A\mathbf{x}$, given the vector $\mathbf{x}$ as input. A direct method is not practical for such a representation, because the individual entries of A are not readily available.

For this reason, it is worthwhile to consider alternative approaches to solving $A\mathbf{x} = \mathbf{b}$, such as **iterative methods**. In particular, an iterative method based on matrix-vector multiplication would address all of these drawbacks of direct methods.

There are two general classes of iterative methods: **stationary iterative methods**, covered in Section 5.1, and **non-stationary iterative methods**, to which the remainder of this chapter is devoted. Either type of method accepts as input an initial guess $\mathbf{x}^{(0)}$ (usually chosen to be the zero vector, in the absence of additional information) and computes a sequence of iterates $\mathbf{x}^{(1)}$, $\mathbf{x}^{(2)}$, $\mathbf{x}^{(3)}$, ..., that, hopefully, converges to the solution $\mathbf{x}$. A stationary iterative method has the form

$$\mathbf{x}^{(k+1)} = \mathbf{g}(\mathbf{x}^{(k)}),$$

for some function $\mathbf{g} : \mathbb{R}^n \to \mathbb{R}^n$. The solution $\mathbf{x}$ is a **fixed point**, or **stationary point**, of $\mathbf{g}$. In other words, a stationary iterative method is one in which **Fixed Point Iteration**, which is covered in Section 11.1.1 in the context of solving systems of nonlinear equations, is used to obtain the solution.

Prior to reading this chapter, it is recommended that the reader review sections of Appendix B that cover the basics of eigenvalues (Section B.12), norms (Section B.13), and convergence of sequences of vectors (Section B.14).

5.1 Stationary Iterative Methods

To construct a suitable function $\mathbf{g}$, we compute a **splitting** of the matrix $A = M - N$, where M is nonsingular. Then, the solution $\mathbf{x}$ satisfies

$$M\mathbf{x} = N\mathbf{x} + \mathbf{b},$$

or

$$\mathbf{x} = M^{-1}(N\mathbf{x} + \mathbf{b}).$$

We therefore define

$$\mathbf{g}(\mathbf{x}) = M^{-1}(N\mathbf{x} + \mathbf{b}),$$

so that the iteration takes the form

$$M\mathbf{x}^{(k+1)} = N\mathbf{x}^{(k)} + \mathbf{b}. \tag{5.1}$$

It follows that for the sake of efficiency, the splitting $A = M - N$ should be chosen so that the system $M\mathbf{y} = \mathbf{c}$ is easily solved.

5.1.1 Convergence Analysis

Before we describe specific splittings, we examine the convergence of this type of iteration. Using the fact that $\mathbf{x}$ is the exact solution of $A\mathbf{x} = \mathbf{b}$, we obtain

$$M(\mathbf{x}^{(k+1)} - \mathbf{x}) = N(\mathbf{x}^{(k)} - \mathbf{x}) + \mathbf{b} - \mathbf{b},$$

which yields

$$\mathbf{x}^{(k+1)} - \mathbf{x} = M^{-1}N(\mathbf{x}^{(k)} - \mathbf{x})$$

and

$$\mathbf{x}^{(k)} - \mathbf{x} = (M^{-1}N)^k(\mathbf{x}^{(0)} - \mathbf{x}).$$

We see that the error is multiplied by $T = M^{-1}N$ from one iteration to the next. The matrix T is called the **iteration matrix** for the iterative method (5.1).

Taking norms of both sides, and using the properties of natural matrix norms, yields

$$\|\mathbf{x}^{(k)} - \mathbf{x}\| \le \|M^{-1}N\|^k \|(\mathbf{x}^{(0)} - \mathbf{x})\|.$$

It follows that if $\|M^{-1}N\| < 1$ for any natural norm, the iteration converges. However, this condition is *sufficient*, not *necessary*.

Exploration 5.1.1 Consider the matrix

$$A = \begin{bmatrix} 9/10 & 1 \\ 0 & 9/10 \end{bmatrix}.$$

Let $T = M^{-1}N$ where $A = M - N$ and

$$M = \begin{bmatrix} 9/10 & 0 \\ 0 & 9/10 \end{bmatrix}.$$

Compute $\|T\|_\infty$. What do you expect to happen if you compute $T^k\mathbf{x}$, for any nonzero vector $\mathbf{x}$, and increasing values of k? What actually happens? Can you explain the behavior?

The error after each iteration is obtained by multiplying the error from the previous iteration by $T = M^{-1}N$. Therefore, as discussed in Section B.14, in order for the error to converge to the zero vector, for any choice of the initial guess $\mathbf{x}^{(0)}$, we must have $\rho(T) < 1$, where $\rho(T)$ is the **spectral radius** of T, which is the absolute value of the largest eigenvalue of T.

5.1.2 The Jacobi Method

We now discuss some basic stationary iterative methods. For convenience, we write

$$A = D + L + U,$$

where D is a diagonal matrix whose diagonal entries are the diagonal entries of A, L is a strictly lower triangular matrix defined by

$$\ell_{ij} = \begin{cases} a_{ij} & i > j \\ 0 & i \le j \end{cases},$$

and U is a strictly upper triangular matrix that is similarly defined: $u_{ij} = a_{ij}$ if $i < j$, and 0 otherwise.

The **Jacobi Method** is defined by the splitting

$$A = M - N, \quad M = D, \quad N = -(L + U).$$

That is,

$$\mathbf{x}^{(k+1)} = D^{-1}[-(L+U)\mathbf{x}^{(k)} + \mathbf{b}].$$

If we write each row of this vector equation individually, we obtain

$$x_i^{(k+1)} = \frac{1}{a_{ii}} \left(b_i - \sum_{j \ne i} a_{ij} x_j^{(k)} \right). \tag{5.2}$$

This description of the Jacobi method is helpful for its practical implementation, but it also reveals how the method can be improved. If the components of $\mathbf{x}^{(k+1)}$ are computed in order, then the computation of $x_i^{(k+1)}$ uses components $1, 2, \ldots, i-1$ of $\mathbf{x}^{(k)}$, even though these components of $\mathbf{x}^{(k+1)}$ have already been computed.

Exploration 5.1.2 Write a MATLAB function

$$[\text{x,niter}]=\text{jacobi(A,b,tol)}$$

that implements the Jacobi Method (5.2). Have the iteration terminate when $\|\mathbf{x}^{(k+1)} - \mathbf{x}^{(k)}\|_\infty < \text{tol}\|\mathbf{x}^{(k)}\|_\infty$. The second output argument `niter` is the number of iterations.

Exploration 5.1.3 A matrix $A \in \mathbb{R}^{n \times n}$ is **strictly row diagonally dominant** if it satisfies the conditions

$$|a_{ii}| > \sum_{j=1, j \neq i}^{n} |a_{ij}|.$$

Prove that if A is strictly row diagonally dominant, then the Jacobi Method is guaranteed to converge to the solution of $Ax = \mathbf{b}$.

Exploration 5.1.4 Solve the linear system $Ax = \mathbf{b}$ by using the Jacobi method, where

$$A = \begin{bmatrix} 4 & 1 & -1 \\ 2 & 7 & 1 \\ 1 & -3 & 12 \end{bmatrix}, \quad \mathbf{b} = \begin{bmatrix} 19 \\ 3 \\ 31 \end{bmatrix}.$$

Compute the iteration matrix $T = M^{-1}N$ using the fact that $M = D$ and $N = -(L+U)$ for the Jacobi method. Determine whether $\rho(T) < 1$ *without* computing the eigenvalues of T explicitly.

5.1.3 The Gauss-Seidel Method

By modifying the Jacobi Method to use the most up-to-date information available, we obtain the **Gauss-Seidel Method**

$$x_i^{(k+1)} = \frac{1}{a_{ii}} \left(b_i - \sum_{j<i} a_{ij} x_j^{(k+1)} - \sum_{j>i} a_{ij} x_j^{(k)} \right). \tag{5.3}$$

This is equivalent to using the splitting $A = M - N$ where $M = D + L$ and $N = -U$; that is,

$$\mathbf{x}^{(k+1)} = (D + L)^{-1}[-U\mathbf{x}^{(k)} + \mathbf{b}].$$

Typically, this iteration converges more rapidly than the Jacobi method, but the Jacobi Method retains one significant advantage: because each component of $\mathbf{x}^{(k+1)}$ is computed independently of the others, the Jacobi method can be parallelized, whereas the Gauss-Seidel Method cannot, because the computation of $x_i^{(k+1)}$ depends on $x_j^{(k+1)}$ for $j < i$.

Exploration 5.1.5 It is easy to see from (5.2) and (5.3) that the Jacobi Method and the Gauss-Seidel Method perform the same number of floating-point operations during each iteration. However, when both methods are written in matrix-vector form, we see that the Jacobi Method requires solving a diagonal system of equations, which involves n operations, while the Gauss-Seidel Method requires solving a lower triangular system, which involves n^2 operations. Why is this not a contradiction?

Exploration 5.1.6 Modify your function `jacobi` from Exploration 5.1.2 to obtain a new function `[x,niter]=gaussseidel(A,b,tol)` that implements the Gauss-Seidel Method (5.3). Try your function on the system from Exploration 5.1.4. Compare its performance to that of your implementation of the Jacobi Method, in terms of execution time as measured by the `tic` and `toc` commands, and the number of iterations. Describe your observations.

5.1.4 Successive Overrelaxation

Both the Jacobi Method and Gauss-Seidel Method are guaranteed to converge if A is strictly row diagonally dominant. Furthermore, as will be discussed in an exercise, Gauss-Seidel is guaranteed to converge if A is symmetric positive definite. However, in certain important applications, both methods can converge quite slowly. To accelerate convergence, we first rewrite the Gauss-Seidel Method as follows:

$$x_i^{(k+1)} = x_i^{(k)} + \frac{1}{a_{ii}} \left[b_i - \sum_{j<i} a_{ij} x_j^{(k+1)} - \sum_{j>i} a_{ij} x_j^{(k)} - a_{ii} x_i^{(k)} \right], \quad i = 1, 2, \ldots, n.$$

The quantity in brackets indicates the direction of the step taken from $\mathbf{x}^{(k)}$ to $\mathbf{x}^{(k+1)}$. However, if the direction of this step corresponds closely to the step $\mathbf{x} - \mathbf{x}^{(k)}$ to the exact solution, it may be possible to accelerate convergence by increasing the length of this step. That is, we introduce a parameter ω so that

$$x_i^{(k+1)} = x_i^{(k)} + \frac{\omega}{a_{ii}} \left[b_i - \sum_{j<i} a_{ij} x_j^{(k+1)} - \sum_{j>i} a_{ij} x_j^{(k)} - a_{ii} x_i^{(k)} \right], \quad i = 1, 2, \ldots, n.$$

$$(5.4)$$

By choosing $\omega > 1$, which is called *over-relaxation*, we take a larger step in the direction of $[\mathbf{x}_{GS}^{(k+1)} - \mathbf{x}^{(k)}]$ than the Gauss-Seidel Method would call for, where $\mathbf{x}_{GS}^{(k+1)}$ is the iterate obtained from $\mathbf{x}^{(k)}$ by the Gauss-Seidel Method. Rewriting the above iteration as

$$a_{ii} x_i^{(k+1)} = (1 - \omega) a_{ii} x_i^{(k)} + \omega \left[b_i - \sum_{j<i} a_{ij} x_j^{(k+1)} - \sum_{j>i} a_{ij} x_j^{(k)} \right], \quad i = 1, 2, \ldots, n,$$

yields the **Method of Successive Over-Relaxation (SOR)**,

$$(D + \omega L)\mathbf{x}^{(k+1)} = [(1 - \omega)D - \omega U]\mathbf{x}^{(k)} + \omega \mathbf{b}. \tag{5.5}$$

Note that if $\omega = 1$, then SOR reduces to the Gauss-Seidel Method.

If we examine the iteration matrix T_ω for SOR, we have

$$T_\omega = (D + \omega L)^{-1}[(1 - \omega)D - \omega U].$$

Because the matrices $(D + \omega L)$ and $[(1 - \omega)D - \omega U]$ are both triangular, it follows that

$$\det(T_\omega) = \left(\prod_{i=1}^{n} a_{ii}^{-1}\right)\left(\prod_{i=1}^{n}(1 - \omega)a_{ii}\right) = (1 - \omega)^n.$$

Exploration 5.1.7 From the preceding discussion, find a lower and upper bound for the parameter ω.

Exploration 5.1.8 Compare (5.5) to (5.1) to obtain the matrices M and N for SOR. *Hint:* They are *not* the matrices $D + \omega L$ and $(1 - \omega)D - \omega U$!

Exploration 5.1.9 Modify your function `gaussseidel` from Exploration 5.1.6 to obtain a new function `[x,niter]=SOR(A,b,w,tol)` that implements the SOR method (5.4). The third input argument `w` represents the relaxation parameter ω. Compare its performance for various values of ω, in terms of both execution time as measured with the `tic` and `toc` commands, and the number of iterations.

In some cases, it is possible to analytically determine the optimal value of ω, which minimizes $\rho(T_\omega)$ and thus ensures the most rapid convergence possible. For example, if A is symmetric positive definite and tridiagonal, then the optimal value is

$$\omega = \frac{2}{1 + \sqrt{1 - [\rho(T_J)]^2}}, \tag{5.6}$$

where T_J is the iteration matrix $-D^{-1}(L + U)$ for the Jacobi method.

A natural criterion for stopping any iterative method is to check whether $\|\mathbf{x}^{(k)} - \mathbf{x}^{(k-1)}\|$ is less than some tolerance. However, if $\|T\| < 1$ in some natural matrix norm, then we have

$$\|\mathbf{x}^{(k)} - \mathbf{x}\| \leq \|T\|\|\mathbf{x}^{(k-1)} - \mathbf{x}\| \leq \|T\|\|\mathbf{x}^{(k-1)} - \mathbf{x}^{(k)}\| + \|T\|\|\mathbf{x}^{(k)} - \mathbf{x}\|,$$

which yields the estimate

$$\|\mathbf{x}^{(k)} - \mathbf{x}\| \leq \frac{\|T\|}{1 - \|T\|}\|\mathbf{x}^{(k)} - \mathbf{x}^{(k-1)}\|.$$

Therefore, the tolerance should be chosen with $\|T\|/(1 - \|T\|)$ in mind, as this can be quite large when $\|T\| \approx 1$.

5.1.5   Concept Check

1. Give some reasons why iterative methods for solving $A\mathbf{x} = \mathbf{b}$ are preferable to direct methods.
2. Define these terms: stationary iterative method, splitting, and iteration matrix.
3. When does a stationary iterative method converge?
4. Describe the three stationary iterative methods covered in this section. How do they relate to one another?
5. For what classes of matrices are these three stationary iterative methods guaranteed to converge?

5.2   Gradient Descent Methods

We have learned about *stationary* iterative methods for solving $A\mathbf{x} = \mathbf{b}$, that have the form of a fixed-point iteration. Now, we will consider an alternative approach to developing iterative methods, that leads to **non-stationary** iterative methods, in which **search directions** are used to progress from each iterate to the next. That is,

$$\mathbf{x}^{(k+1)} = \mathbf{x}^{(k)} + \alpha_k \mathbf{p}_k$$

where $\mathbf{p}_k$ is a search direction that is chosen so as to be approximately parallel to the error $\mathbf{e}_k = \mathbf{x} - \mathbf{x}^{(k)}$, and α_k is a constant that determines how far along that direction to proceed so that $\mathbf{x}^{(k+1)}$ will be, in some sense, as close to $\mathbf{x}$ as possible.

5.2.1   Steepest Descent

We assume that A is symmetric positive definite, and consider the problem of minimizing the function

$$\phi(\mathbf{x}) = \frac{1}{2}\mathbf{x}^T A \mathbf{x} - \mathbf{b}^T \mathbf{x}. \tag{5.7}$$

From multivariable calculus, it can be seen that solving this minimization problem is equivalent to solving $A\mathbf{x} = \mathbf{b}$.

Exploration 5.2.1 Prove that if $A \in \mathbb{R}^{n \times n}$ is symmetric positive definite, then the function $\phi(\mathbf{x})$ defined in (5.7) has a unique minimizer at the solution $\mathbf{x}$ of $A\mathbf{x} = \mathbf{b}$.

From any vector $\mathbf{x}_0 \in \mathbb{R}^n$, the **direction of steepest descent** of $\phi(\mathbf{x})$ from $\mathbf{x}_0$ is given by

$$-\nabla\phi(\mathbf{x}_0) = \mathbf{b} - A\mathbf{x}_0 = \mathbf{r}_0,$$

the **residual** vector. This suggests a simple non-stationary iterative method, which is called the **Method of Steepest Descent**. The basic idea is to choose the search direction $\mathbf{p}_k$ to be $\mathbf{r}_k = \mathbf{b} - A\mathbf{x}^{(k)}$, and then to choose α_k so as to minimize

$\phi(\mathbf{x}^{(k+1)}) = \phi(\mathbf{x}^{(k)} + \alpha_k \mathbf{r}_k)$. This entails solving a single-variable minimization problem to obtain α_k.

Exploration 5.2.2 Let the function $g(\alpha)$ be defined by

$$g(\alpha) = \phi(\mathbf{x}^{(k)} + \alpha \mathbf{r}_k).$$

Show that $g(\alpha)$ has a unique minimizer at

$$\alpha_k = \frac{\mathbf{r}_k^T \mathbf{r}_k}{\mathbf{r}_k^T A \mathbf{r}_k}.$$

We can now present the algorithm for the Method of Steepest Descent.

Algorithm 5.2.1 (Method of Steepest Descent) Let $A \in \mathbb{R}^{n \times n}$ be symmetric positive definite, and let $\mathbf{b} \in \mathbb{R}^n$. The following algorithm computes the solution $\mathbf{x}$ of the system of linear equations $A\mathbf{x} = \mathbf{b}$.

Choose an initial guess $\mathbf{x}^{(0)} \in \mathbb{R}^n$
for $k = 0, 1, 2, \ldots$ **do** until convergence
$\quad \mathbf{r}_k = \mathbf{b} - A\mathbf{x}^{(k)}$
$\quad \alpha_k = \frac{\mathbf{r}_k^T \mathbf{r}_k}{\mathbf{r}_k^T A \mathbf{r}_k}$
$\quad \mathbf{x}^{(k+1)} = \mathbf{x}^{(k)} + \alpha_k \mathbf{r}_k$
end for

The Method of Steepest Descent is effective when A is well-conditioned, but when A is ill-conditioned, convergence is very slow, because the level curves of ϕ become long, thin hyperellipsoids in which the direction of steepest descent does not yield much progress toward the minimum. Another problem with this method is that while it can be shown that $\mathbf{r}_{k+1}$ is orthogonal to $\mathbf{r}_k$, so that each direction is completely independent of the previous one, $\mathbf{r}_{k+1}$ is not necessarily independent of *previous* search directions.

Exploration 5.2.3 Show that each search direction $\mathbf{r}_k$ is orthogonal to the previous search direction $\mathbf{r}_{k-1}$.

In fact, even in the 2×2 case, where only two independent search directions are available, the method of steepest descent exhibits a "zig-zag" effect because it continually alternates between two orthogonal search directions, and the more ill-conditioned A is, the smaller each step tends to be.

> **Exploration 5.2.4** Write a MATLAB function
>
> ```
> x=steepestdescent(A,b,x0,tol)
> ```
>
> that implements Algorithm 5.2.1. The input argument x0 is the initial guess $\mathbf{x}^{(0)}$. Have the iteration terminate when the ℓ_∞-norm of the absolute difference between consecutive iterates is less than tol. Try your function on the matrices
>
> $$A = \begin{bmatrix} 2 & 1 \\ 1 & 2 \end{bmatrix}, \quad A = \begin{bmatrix} 4 & 2 \\ 2 & 1.00001 \end{bmatrix}$$
>
> with a randomly chosen right-hand side vector $\mathbf{b}$, and have your function plot the iterates in the xy-plane. Describe and explain the behavior of the algorithm in these cases.

5.2.2 Lanczos Iteration*

A more efficient iteration can be obtained if different search directions can be chosen so as to ensure that each residual is orthogonal to *all* previous residuals. While it would be preferable to require that all *search directions* are orthogonal, this goal is unrealistic, so we settle for orthogonality of the residuals instead. This orthogonality can be realized if we prescribe that each iterate $\mathbf{x}^{(k)}$ has the form

$$\mathbf{x}^{(k)} = \mathbf{x}^{(0)} + Q_k \mathbf{y}_k,$$

where $\mathbf{x}^{(0)}$ is an initial guess, Q_k is an $n \times k$ matrix with *orthonormal* columns, meaning that $Q_k^T Q_k = I$, and $\mathbf{y}_k$ is a k-vector of coefficients.

If we prescribe that the first column of Q_k is parallel to the initial residual $\mathbf{r}_1 = \mathbf{b} - A\mathbf{x}^{(0)}$, then, to ensure that each residual is orthogonal to all previous residuals, we first note that

$$\mathbf{b} - A\mathbf{x}^{(k)} = \beta_0 Q_k \mathbf{e}_1 - A Q_k \mathbf{y}_k,$$

where $\beta_0 = \|\mathbf{r}_1\|_2$. That is, the residual lies in the span of the space spanned by the columns of Q_k, and the vectors obtained by multiplying A by those columns. Therefore, if we choose the columns of Q_k so that they form an orthonormal basis for the k-dimensional **Krylov subspace**

$$\mathcal{K}(\mathbf{r}_1, A, k) = \text{span}\{\mathbf{r}_1, A\mathbf{r}_1, A^2\mathbf{r}_1, \ldots, A^{k-1}\mathbf{r}_1\},$$

then we ensure that the residual $\mathbf{r}_{k+1} = \mathbf{b} - A\mathbf{x}^{(k)}$ is in the span of the columns of Q_{k+1}. Next, we can guarantee that $\mathbf{r}_{k+1}$ is orthogonal to the columns of Q_k, which span the space that contains all previous residuals, by requiring

$$Q_k^T \mathbf{r}_{k+1} = Q_k^T(\mathbf{b} - A\mathbf{x}^{(k)}) = Q_k^T(\mathbf{b} - A\mathbf{x}^{(0)}) - Q_k^T A Q_k \mathbf{y}_k = \beta_0 \mathbf{e}_1 - T_k \mathbf{y}_k = \mathbf{0},$$

where

$$T_k = Q_k^T A Q_k.$$

It is easily seen that T_k is symmetric positive definite, since A is.

The columns of each matrix Q_k, denoted by $\mathbf{q}_1, \mathbf{q}_2, \ldots, \mathbf{q}_k$, are defined to be

$$\mathbf{q}_k = p_{k-1}(A)\mathbf{b},$$

where $p_0, p_1, \ldots$ define a sequence of **orthogonal polynomials** (see Section 8.2) with respect to the inner product

$$\langle p, q \rangle = \mathbf{b}^T p(A) q(A) \mathbf{b}.$$

That is, these polynomials satisfy

$$\langle p_i, p_j \rangle = \mathbf{b}^T p_i(A) p_j(A) \mathbf{b} = \begin{cases} 1 & i = j \\ 0 & i \neq j \end{cases}.$$

Like any sequence of orthogonal polynomials, they satisfy a **three-term recurrence relation**

$$\beta_j p_j(\lambda) = (\lambda - \alpha_j) p_{j-1}(\lambda) - \beta_{j-1} p_{j-2}(\lambda), \quad j \geq 1, \quad p_0(\lambda) \equiv \beta_0^{-1}, \quad p_{-1}(\lambda) \equiv 0,$$

which is obtained by applying Gram-Schmidt Orthogonalization to the monomial basis.

It follows from this three-term recurrence relation that

$$A\mathbf{q}_j = \beta_{j-1}\mathbf{q}_{j-1} + \alpha_j \mathbf{q}_j + \beta_j \mathbf{q}_{j+1}, \tag{5.8}$$

and therefore T_k is tridiagonal. In fact, by applying (5.8) for $j = 1, 2, \ldots, k$, we have

$$AQ_k = Q_k T_k + \beta_k \mathbf{q}_{k+1} \mathbf{e}_k^T,$$

where

$$T_k = \begin{bmatrix} \alpha_1 & \beta_1 & & & \\ \beta_1 & \alpha_2 & \beta_2 & & \\ & \ddots & \ddots & \ddots & \\ & & \beta_{k-2} & \alpha_{k-1} & \beta_{k-1} \\ & & & \beta_{k-1} & \alpha_k \end{bmatrix}. \tag{5.9}$$

Furthermore, the residual is given by

$$\mathbf{b} - A\mathbf{x}^{(k)} = \beta_0 Q_k \mathbf{e}_1 - AQ_k \mathbf{y}_k = \beta_0 Q_k \mathbf{e}_1 - Q_k T_k \mathbf{y}_k - \beta_k \mathbf{q}_{k+1} \mathbf{e}_k^T \mathbf{y}_k = -\beta_k \mathbf{q}_{k+1} y_k.$$

We now have the following algorithm for generating a sequence of approximations to the solution of $A\mathbf{x} = \mathbf{b}$, for which each residual is orthogonal to all previous residuals.

Algorithm 5.2.2 (Lanczos Iteration) Let $A \in \mathbb{R}^{n \times n}$ be symmetric positive definite and let $\mathbf{b} \in \mathbb{R}^n$. The following algorithm computes the solution $\mathbf{x}$ to the system of linear equations $A\mathbf{x} = \mathbf{b}$.

Choose an initial guess $\mathbf{x}^{(0)} \in \mathbb{R}^n$
$k = 0$, $\mathbf{v}_0 = \mathbf{b} - A\mathbf{x}^{(0)}$, $\mathbf{q}_0 = \mathbf{0}$
while not converged **do**
  $\beta_k = \|\mathbf{v}_k\|_2$
  $\mathbf{q}_{k+1} = \mathbf{v}_k / \beta_k$
  $k = k + 1$
  $\mathbf{v}_k = A\mathbf{q}_k$
  $\alpha_k = \mathbf{q}_k^T \mathbf{v}_k$
  $\mathbf{x}^{(k)} = \mathbf{x}^{(0)} + \beta_0 Q_k T_k^{-1} \mathbf{e}_1$
  $\mathbf{v}_k = \mathbf{v}_k - \alpha_k \mathbf{q}_k - \beta_{k-1} \mathbf{q}_{k-1}$
end while

Lanczos Iteration is not only used for solving linear systems; the matrix T_k is also useful for approximating extremal eigenvalues of A, and for approximating quadratic or bilinear forms involving functions of A, such as the inverse or exponential [Golub and Meurant (2009)].

Exploration 5.2.5 Write a MATLAB function

```
[x,niter]=lanczos(A,b,x0,tol,maxit)
```

that implements Algorithm 5.2.2. The third input argument x0 is the initial guess $\mathbf{x}^{(0)}$. Have the iteration terminate when the ℓ_∞-norm of the absolute difference between consecutive iterates is less than tol, or when the maximum number of iterations maxit is reached. The second output argument niter is the iteration count.

5.2.3 The Conjugate Gradient Method*

To improve the efficiency of Lanczos Iteration, the tridiagonal structure of T_k can be exploited so that the vector $\mathbf{y}_k$ can easily be obtained from $\mathbf{y}_{k-1}$ by a few simple recurrence relations. However, Lanczos iteration is not normally used directly to solve $A\mathbf{x} = \mathbf{b}$ for a symmetric positive definite matrix A, because it does not provide a simple method of computing $\mathbf{x}^{(k+1)}$ directly from $\mathbf{x}^{(k)}$, as in a general non-stationary iterative method. To rectify this, we examine the process of solving $T_k \mathbf{y}_k = \beta_0 \mathbf{e}_1$ more closely. If we write

$$T_{k+1} = \begin{bmatrix} T_k & \beta_k \mathbf{e}_k \\ \beta_k \mathbf{e}_k^T & \alpha_{k+1} \end{bmatrix}, \quad \mathbf{y}_{k+1} = \begin{bmatrix} \tilde{\mathbf{y}}_k \\ \psi_{k+1} \end{bmatrix} \tag{5.10}$$

then we have

$$T_k \tilde{\mathbf{y}}_k + \beta_k \psi_{k+1} \mathbf{e}_k = \beta_0 \mathbf{e}_1, \quad \beta_k \mathbf{e}_k^T \tilde{\mathbf{y}}_k + \alpha_{k+1} \psi_{k+1} = 0$$

which can be condensed to

$$\tilde{T}_k \tilde{\mathbf{y}}_k = \beta_0 \mathbf{e}_1, \quad \psi_{k+1} = -\alpha_{k+1}^{-1} \beta_k \mathbf{e}_k^T \tilde{\mathbf{y}}_k$$

where $\tilde{T}_k = T_k - \alpha_{k+1}^{-1} \beta_k^2 \mathbf{e}_k \mathbf{e}_k^T$. That is, $\tilde{T}_k$ is a **rank-one update** of T_k (see Section B.7.3).

Because T_k is tridiagonal, and symmetric positive definite, it has an LDL^T factorization

$$T_k = L_k D_k L_k^T,$$

where

$$L_k = \begin{bmatrix} 1 & 0 & \cdots & 0 \\ l_1 & 1 & 0 & 0 \\ \vdots & \ddots & \ddots & \vdots \\ 0 & \cdots & l_{k-1} & 1 \end{bmatrix}, \quad D_k = \begin{bmatrix} d_1 & 0 & \cdots & 0 \\ 0 & d_2 & & \vdots \\ \vdots & & \ddots & 0 \\ 0 & \cdots & 0 & d_k \end{bmatrix},$$

and the diagonal entries of D_k are positive. We can see that L_k has lower bandwidth 1. Furthermore, because $\tilde{T}_k$ and T_k differ only in the (k,k) entry, $\tilde{T}_{k-1}$ has the LDL^T factorization $\tilde{T}_k = L_k \tilde{D}_k L_k^T$, where $\tilde{D}_k$ also differs from D_k only in the (k,k) entry. It follows that if we define $\mathbf{w}_k = L_k^T \mathbf{y}_k$, and $\hat{\mathbf{w}}_k = L_k^T \tilde{\mathbf{y}}_k$, then we have

$$D_k \mathbf{w}_k = \tilde{D}_k \hat{\mathbf{w}}_k = L_k^{-1} \mathbf{e}_1$$

from which we can see that $\mathbf{w}_k$ and $\hat{\mathbf{w}}_k$ differ only in their last entries.

We therefore rewrite

$$\mathbf{x}^{(k)} - \mathbf{x}^{(0)} = Q_k \mathbf{y}_k = Q_k L_k^{-T} L_k^T \mathbf{y}_k = Q_k L_k^{-T} \mathbf{w}_k$$

where $L_k D_k \mathbf{w}_k = \beta_0 \mathbf{e}_1$. If we break the system $L_{k+1} D_{k+1} \mathbf{w}_{k+1} = \beta_0 \mathbf{e}_1$ down in the same way as in (5.10), we have

$$L_{k+1} D_{k+1} = \begin{bmatrix} L_k D_k & \mathbf{0} \\ l_k d_k \mathbf{e}_k^T & d_{k+1} \end{bmatrix}, \quad \mathbf{w}_{k+1} = \begin{bmatrix} \tilde{\mathbf{w}}_k \\ w_{k+1} \end{bmatrix},$$

and therefore

$$L_k D_k \tilde{\mathbf{w}}_k = \beta_0 \mathbf{e}_1, \quad l_k d_k \mathbf{e}_k^T \tilde{\mathbf{w}}_k + d_{k+1} w_{k+1} = 0.$$

That is, $\tilde{\mathbf{w}}_k = \mathbf{w}_k$. In other words, the first k elements of $\mathbf{w}_{k+1}$ are the elements of $\mathbf{w}_k$, which yields

$$\begin{aligned}
\mathbf{x}^{(k+1)} &= \mathbf{x}^{(0)} + Q_{k+1} L_{k+1}^{-T} \mathbf{w}_{k+1} \\
&= \mathbf{x}^{(0)} + Q_{k+1} L_{k+1}^{-T} \left(\begin{bmatrix} \mathbf{w}_k \\ 0 \end{bmatrix} + w_{k+1} \mathbf{e}_{k+1} \right) \\
&= \mathbf{x}^{(0)} + \begin{bmatrix} Q_k & \mathbf{q}_{k+1} \end{bmatrix} \begin{bmatrix} L_k^{-T} & \mathbf{z}_k \\ \mathbf{0}^T & 1 \end{bmatrix} \begin{bmatrix} \mathbf{w}_k \\ 0 \end{bmatrix} + w_{k+1} Q_{k+1} L_{k+1}^{-T} \mathbf{e}_{k+1} \\
&= \mathbf{x}^{(0)} + Q_k L_k^{-T} \mathbf{w}_k + w_{k+1} Q_{k+1} L_{k+1}^{-T} \mathbf{e}_{k+1} \\
&= \mathbf{x}^{(k)} + w_{k+1} Q_{k+1} L_{k+1}^{-T} \mathbf{e}_{k+1}.
\end{aligned}$$

For convenience, we define

$$\tilde{P}_k = Q_k L_k^{-T}.$$

Then, from

$$\begin{bmatrix} \tilde{\mathbf{p}}_1 & \cdots & \tilde{\mathbf{p}}_k \end{bmatrix} \begin{bmatrix} 1 & l_1 & \cdots & & 0 \\ 0 & 1 & 0 & & 0 \\ \vdots & & \ddots & \ddots & l_{k-1} \\ 0 & \cdots & & 0 & 1 \end{bmatrix}^T = \begin{bmatrix} \mathbf{q}_1 & \cdots & \mathbf{q}_k \end{bmatrix},$$

we see that the columns of $\tilde{P}_k$ satisfy

$$\tilde{\mathbf{p}}_1 = \mathbf{q}_1, \quad l_{j-1}\tilde{\mathbf{p}}_{j-1} + \tilde{\mathbf{p}}_j = \mathbf{q}_j, \quad j = 2, \ldots, k.$$

It follows that for $k = 0, 1, \ldots$, we can write

$$\mathbf{x}^{(k)} = \mathbf{x}^{(k-1)} + \nu_k \mathbf{p}_k, \quad k = 1, 2, \ldots \tag{5.11}$$

where ν_k is a scalar and, for $k > 1$, the **search direction** $\mathbf{p}_k$ is a linear combination of $\mathbf{q}_k$ and the previous search direction $\mathbf{p}_{k-1}$. For $k = 1$, we have $\mathbf{p}_k$ as a multiple of $\mathbf{q}_k$, which is itself a multiple of the initial $\mathbf{r}_1 = \mathbf{b} - A\mathbf{x}^{(0)}$.

Because we are free to scale the search directions in any way that is convenient, and the vectors $\mathbf{q}_1, \mathbf{q}_2, \ldots$ are the normalized residuals $\mathbf{r}_k = \mathbf{b} - A\mathbf{x}^{(k-1)}$, $k = 1, 2, \ldots$, we set $\mathbf{p}_1 = \mathbf{r}_1$ and prescribe

$$\mathbf{p}_k = \mathbf{r}_k + \mu_k \mathbf{p}_{k-1}, \quad k = 2, 3, \ldots. \tag{5.12}$$

It remains to choose the scalars μ_k and ν_k. To that end, we note that

$$(Q_k L_k^{-T})^T A Q_k L_k^{-T} = L_k^{-1} Q_k^T A Q_k L_k^{-T} = L_k^{-1} L_k D_k L_k^T L_k^{-T} = D_k,$$

which means that the columns of $Q_k L_k^{-T}$, which are parallel to the search directions, are not orthogonal but rather A-**orthogonal**, or A-**conjugate**. That is,

$$\mathbf{p}_j^T A \mathbf{p}_k = 0, \quad j \neq k. \tag{5.13}$$

Because the search directions $\mathbf{p}_k$, $k = 1, 2, \ldots$, are closely related to the residuals, which are gradients, and they are A-conjugate, this approach of computing $\mathbf{x}^{(k)}$ from these search directions is called the **Conjugate Gradient (CG) Method**.

We now use the requirement that the search directions are A-orthogonal, and the requirement that the residuals themselves are orthogonal, to determine μ_k and ν_k. If we multiply both sides of (5.11) by A and subtract from $\mathbf{b}$, we obtain

$$\mathbf{r}_{k+1} = \mathbf{r}_k - \nu_k A \mathbf{p}_k, \quad k = 1, 2, \ldots. \tag{5.14}$$

We then take the inner product of both sides with $\mathbf{r}_k$ to obtain

$$\nu_k = \frac{\mathbf{r}_k^T \mathbf{r}_k}{\mathbf{r}_k^T A \mathbf{p}_k}, \quad k = 1, 2, \ldots.$$

Using (5.12) to eliminate $\mathbf{r}_k$ in the denominator, and using (5.13), we obtain

$$\nu_k = \frac{\mathbf{r}_k^T \mathbf{r}_k}{\mathbf{p}_k^T A \mathbf{p}_k}, \quad k = 1, 2, \ldots.$$

Finally, by multiplying both sides of (5.12) by A, and then taking the inner product of both sides with $\mathbf{p}_{k-1}$, we obtain

$$\mu_k = -\frac{\mathbf{p}_{k-1}^T A \mathbf{r}_k}{\mathbf{p}_{k-1}^T A \mathbf{p}_{k-1}}, \quad k = 2, 3, \ldots.$$

To obtain a simpler expression for μ_k, and not have to store $A\mathbf{p}_{k-1}$ as well as $A\mathbf{p}_k$, we use (5.14) to eliminate $A\mathbf{p}_{k-1}$ and (5.12) to eliminate $\mathbf{p}_{k-1}$, which yields

$$
\begin{aligned}
\mu_k &= -\frac{\nu_{k-1}^{-1}(\mathbf{r}_{k-1} - \mathbf{r}_k)^T \mathbf{r}_k}{(\mathbf{r}_{k-1} + \mu_{k-1}\mathbf{p}_{k-2})^T A \mathbf{p}_{k-1}} \\
&= \frac{\nu_{k-1}^{-1}\mathbf{r}_k^T \mathbf{r}_k}{\mathbf{r}_{k-1}^T A \mathbf{p}_{k-1}} \\
&= \frac{\nu_{k-1}^{-1}\mathbf{r}_k^T \mathbf{r}_k}{\nu_{k-1}^{-1}\mathbf{r}_{k-1}^T(\mathbf{r}_{k-1} - \mathbf{r}_k)} \\
&= \frac{\mathbf{r}_k^T \mathbf{r}_k}{\mathbf{r}_{k-1}^T \mathbf{r}_{k-1}}, \quad k = 2, 3, \ldots,
\end{aligned}
$$

where, for convenience, we have defined $\mathbf{p}_0 = 0$.

From these formulas for $\mathbf{x}^{(k)}$, $\mathbf{p}_k$ and $\mathbf{r}_k$, we obtain the following algorithm.

Algorithm 5.2.3 (Conjugate Gradient Method) Given $A \in \mathbb{R}^{n \times n}$ that is symmetric positive definite, and $\mathbf{b} \in \mathbb{R}^n$, the following algorithm computes the solution of $A\mathbf{x} = \mathbf{b}$.

Choose an initial guess $\mathbf{x}^{(0)} \in \mathbb{R}^n$
$k = 1$, $\mathbf{r}_1 = \mathbf{b} - A\mathbf{x}^{(0)}$
for $k = 1, 2, \ldots$ **do** until convergence
    if $k > 1$ **then**
        $\mu_k = \mathbf{r}_k^T \mathbf{r}_k / \mathbf{r}_{k-1}^T \mathbf{r}_{k-1}$
        $\mathbf{p}_k = \mathbf{r}_k + \mu_k \mathbf{p}_{k-1}$
    else
        $\mathbf{p}_1 = \mathbf{r}_1$
    end if
    $\mathbf{v}_k = A\mathbf{p}_k$
    $\nu_k = \mathbf{r}_k^T \mathbf{r}_k / \mathbf{p}_k^T \mathbf{v}_k$
    $\mathbf{x}^{(k)} = \mathbf{x}^{(k-1)} + \nu_k \mathbf{p}_k$
    $\mathbf{r}_{k+1} = \mathbf{r}_k - \nu_k \mathbf{v}_k$
end for

An appropriate stopping criterion is that the norm of the residual $\mathbf{r}_{k+1}$ is smaller than some tolerance. It is also important to impose a maximum number of iterations. Note that only one matrix-vector multiplication per iteration is required.

> **Exploration 5.2.6** Write a MATLAB function
>
> $$[\texttt{x,flag}]=\texttt{cg(A,b,x0,tol,maxit)}$$
>
> that implements Algorithm 5.2.3, taking into account the suggestions for implementation given in the preceding paragraph. The third input argument x0 is the initial guess $\mathbf{x}^{(0)}$. Have the iteration terminate when
>
> $$\|\mathbf{r}_{k+1}\|_\infty < \texttt{tol}\|\mathbf{b}\|_\infty,$$
>
> and limit the number of iterations to maxit. The second output argument flag should be assigned the value 0 if the iteration converged to the specified tolerance, and 1 if it did not.

> **Exploration 5.2.7** Let $\mathbf{x}$ be the exact solution of $A\mathbf{x} = \mathbf{b}$, and let $\mathbf{e}_k = \mathbf{x} - \mathbf{x}^{(k)}$, where $\mathbf{x}^{(k)}$ is as defined in (5.11). Prove that $\mathbf{e}_k$ minimizes
>
> $$\|\mathbf{e}\|_A = \sqrt{\mathbf{e}^T A \mathbf{e}}$$
>
> over all $\mathbf{e} \in \mathbb{R}^n$ such that $\mathbf{e} = \mathbf{x} - \mathbf{x}^{(0)} + \mathbf{y}$, where $\mathbf{y} \in \mathcal{K}(\mathbf{r}_1, A, k)$.

Because of the A-orthogonality, and therefore linear independence, of the search directions, the Conjugate Gradient Method, in exact arithmetic, is guaranteed to converge to the exact solution of $A\mathbf{x} = \mathbf{b}$ in at most n iterations. In fact, it was for this reason that the Conjugate Gradient Method was originally viewed as a *direct* method when introduced in [Hestenes and Stiefel (1952)]. However, due to roundoff error, it was numerically unstable as a direct method and therefore set aside, until its promise as an iterative method was discovered [Reid (1971)].

5.2.4 Preconditioning

The Conjugate Gradient Method is far more effective than the Method of Steepest Descent, but it can also suffer from slow convergence when A is ill-conditioned. Therefore, the Conjugate Gradient Method is often paired with a **preconditioner** that transforms the problem $A\mathbf{x} = \mathbf{b}$ into an equivalent problem in which the matrix is close to I. The basic idea is to solve the problem

$$\tilde{A}\tilde{\mathbf{x}} = \tilde{\mathbf{b}},$$

where

$$\tilde{A} = C^{-1}AC^{-1}, \quad \tilde{\mathbf{x}} = C\mathbf{x}, \quad \tilde{\mathbf{b}} = C^{-1}\mathbf{b},$$

and C is symmetric positive definite. Modifying the Conjugate Gradient Method to solve this problem, we obtain the algorithm

$k = 1, C^{-1}\mathbf{r}_1 = C^{-1}(\mathbf{b} - A\mathbf{x}^{(0)})$
for $k = 1, 2, \ldots$ do until convergence
    if $k > 1$ then

$$\mu_k = \mathbf{r}_k^T C^{-2} \mathbf{r}_k / \mathbf{r}_{k-1}^T C^{-2} \mathbf{r}_{k-1}$$
$$C\mathbf{p}_k = C^{-1}\mathbf{r}_k + \mu_k C\mathbf{p}_{k-1}$$

else

$$C\mathbf{p}_1 = C^{-1}\mathbf{r}_1$$

end if

$$C^{-1}\mathbf{v}_k = C^{-1}A\mathbf{p}_k$$
$$\nu_k = \mathbf{r}_k^T C^{-2}\mathbf{r}_k / (C\mathbf{p}_k)^T (C^{-1}\mathbf{v}_k)$$
$$C\mathbf{x}^{(k)} = C\mathbf{x}^{(k-1)} + \nu_k C\mathbf{p}_k$$
$$C^{-1}\mathbf{r}_{k+1} = C^{-1}\mathbf{r}_k - \nu_k C^{-1}\mathbf{v}_k$$

end for

which, upon defining $M = C^2$, simplifies to the following.

Algorithm 5.2.4 (Preconditioned Conjugate Gradient) Given $A \in \mathbb{R}^{n \times n}$ and $M \in \mathbb{R}^{n \times n}$ that are symmetric positive definite, and $\mathbf{b} \in \mathbb{R}^n$, the following algorithm computes the solution of $A\mathbf{x} = \mathbf{b}$ using the Conjugate Gradient Method with M as a preconditioner.

Choose an initial guess $\mathbf{x}^{(0)} \in \mathbb{R}^n$
$k = 1$, $\mathbf{r}_1 = \mathbf{b} - A\mathbf{x}^{(0)}$
for $k = 1, 2, \ldots$ **do** until convergence
　　Solve $M\mathbf{z}_k = \mathbf{r}_k$
　　if $k > 1$ **then**
　　　　$\mu_k = \mathbf{r}_k^T \mathbf{z}_k / \mathbf{r}_{k-1}^T \mathbf{z}_{k-1}$
　　　　$\mathbf{p}_k = \mathbf{z}_k + \mu_k \mathbf{p}_{k-1}$
　　else
　　　　$\mathbf{p}_1 = \mathbf{z}_1$
　　end if
　　$\mathbf{v}_k = A\mathbf{p}_k$
　　$\nu_k = \mathbf{r}_k^T \mathbf{z}_k / \mathbf{p}_k^T \mathbf{v}_k$
　　$\mathbf{x}^{(k)} = \mathbf{x}^{(k-1)} + \nu_k \mathbf{p}_k$
　　$\mathbf{r}_{k+1} = \mathbf{r}_k - \nu_k \mathbf{v}_k$
end for

We see that the action of the transformation is only felt through the **preconditioner** $M = C^2$. Because a system involving M is solved during each iteration, it is essential that such a system is easily solved. One example of such a preconditioner is to define $M = HH^T$, where H is an "incomplete Cholesky factor" of A, which is a sparse matrix that approximates the true Cholesky factor.

Exploration 5.2.8 Modify your function cg from Exploration 5.2.6 to implement Algorithm 5.2.4. Your function must accept an additional optional input argument M that is a symmetric positive definite matrix of the same size as A, to be used as a preconditioner.

5.2.5 Concept Check

1. What is the general form of a non-stationary iterative method? What is meant by a search direction?
2. Explain the main idea behind the Method of Steepest Descent. To what kind of matrices can it be applied?
3. What is a drawback of the Method of Steepest Descent, and how does Lanczos Iteration address it?
4. How are the search directions chosen for the Conjugate Gradient Method?
5. What is preconditioning, and why is it necessary?

5.3 Other Krylov Subspace Methods*

So far, we have only seen Krylov subspace methods for symmetric positive definite matrices. We now generalize these methods to arbitrary square invertible matrices.

5.3.1 Minimum Residual Methods

In the derivation of the Conjugate Gradient Method, we made use of the **Krylov subspace**

$$\mathcal{K}(\mathbf{r}_1, A, k) = \text{span}\{\mathbf{r}_1, A\mathbf{r}_1, A^2\mathbf{r}_1, \dots, A^{k-1}\mathbf{r}_1\}$$

to obtain a solution of $A\mathbf{x} = \mathbf{b}$, where A was a symmetric positive definite matrix and $\mathbf{r}_1 = \mathbf{b} - A\mathbf{x}^{(0)}$ was the residual associated with the initial guess $\mathbf{x}^{(0)}$. Specifically, the Conjugate Gradient Method generated a sequence of approximate solutions $\mathbf{x}^{(k)}$, $k = 1, 2, \dots$, where each iterate had the form

$$\mathbf{x}^{(k)} = \mathbf{x}^{(0)} + Q_k \mathbf{y}_k = \mathbf{x}^{(0)} + \sum_{j=1}^{k} \mathbf{q}_j y_{jk}, \quad k = 1, 2, \dots, \tag{5.15}$$

and the columns $\mathbf{q}_1, \mathbf{q}_2, \dots, \mathbf{q}_k$ of Q_k formed an orthonormal basis of the Krylov subspace $\mathcal{K}(\mathbf{b}, A, k)$.

In this section, we examine whether a similar approach can be used for solving $A\mathbf{x} = \mathbf{b}$ even when the matrix A is *not* symmetric positive definite. That is, we seek a solution $\mathbf{x}^{(k)}$ of the form (5.15), where again the columns of Q_k form an orthonormal basis of the Krylov subspace $\mathcal{K}(\mathbf{b}, A, k)$. To generate this basis, we cannot simply generate a sequence of orthogonal polynomials using a three-term recurrence relation, as before, because the bilinear form

$$\langle f, g \rangle = \mathbf{b}^T f(A) g(A) \mathbf{b}$$

does not satisfy the requirements for a valid inner product when A is not symmetric positive definite. Therefore, we must use a different approach.

To that end, we make use of **Gram-Schmidt Orthogonalization**, introduced in Section 4.2.3. Suppose that we have already generated a sequence of orthonormal vectors $\mathbf{q}_1, \mathbf{q}_2, \dots, \mathbf{q}_k$ that span $\mathcal{K}(\mathbf{b}, A, k)$. To obtain a vector $\mathbf{q}_{k+1}$ such that

$\mathbf{q}_1, \mathbf{q}_2, \ldots, \mathbf{q}_{k+1}$ is an orthonormal basis for $\mathcal{K}(\mathbf{b}, A, k+1)$, we first compute $A\mathbf{q}_k$, and then *orthogonalize* this vector against $\mathbf{q}_1, \mathbf{q}_2, \ldots, \mathbf{q}_k$. This is accomplished as follows:

$$\mathbf{v}_k = A\mathbf{q}_k - \sum_{j=1}^{k} \mathbf{q}_j(\mathbf{q}_j^T A\mathbf{q}_k),$$

$$\mathbf{q}_{k+1} = \frac{\mathbf{v}_k}{\|\mathbf{v}_k\|_2}.$$

It can be verified directly that

$$\mathbf{q}_i^T \mathbf{v}_k = \mathbf{q}_i^T A\mathbf{q}_k - \sum_{j=1}^{k} \mathbf{q}_i^T \mathbf{q}_j(\mathbf{q}_j^T A\mathbf{q}_k) = \mathbf{q}_i^T A\mathbf{q}_k - \mathbf{q}_i^T \mathbf{q}_i(\mathbf{q}_i^T A\mathbf{q}_k) = 0,$$

so the vectors $\mathbf{q}_1, \mathbf{q}_2, \ldots, \mathbf{q}_{k+1}$ do indeed form an orthonormal set. To begin the sequence, we define $\mathbf{v}_0 = \mathbf{b} - A\mathbf{x}^{(0)}$ and let $\mathbf{q}_1 = \mathbf{v}_0/\|\mathbf{v}_0\|$.

If we define

$$h_{ij} = \mathbf{q}_i^T A\mathbf{q}_j, \quad i = 1, 2, \ldots, j, \quad h_{j+1,j} = \|\mathbf{v}_j\|_2,$$

it follows from the orthogonalization process that

$$A\mathbf{q}_j = \sum_{i=1}^{j} h_{ij}\mathbf{q}_i + h_{j+1,j}\mathbf{q}_{j+1}, \quad j = 1, 2, \ldots, k,$$

or, in matrix form,

$$AQ_k = Q_k H_k + h_{k+1,k}\mathbf{q}_{k+1}\mathbf{e}_k^T = Q_{k+1}\tilde{H}_k, \tag{5.16}$$

where H_k, as well as $\tilde{H}_k$, is an **upper Hessenberg matrix**

$$H_k = \begin{bmatrix} h_{11} & h_{12} & \cdots & & \cdots & h_{1j} \\ h_{21} & h_{22} & \cdots & & \cdots & h_{2j} \\ 0 & h_{32} & \ddots & & & \vdots \\ \vdots & 0 & \ddots & \ddots & & \vdots \\ 0 & \cdots & 0 & h_{k,k-1} & h_{kk} \end{bmatrix}, \quad \tilde{H}_k = \begin{bmatrix} H_k \\ h_{k+1,k}\mathbf{e}_k^T \end{bmatrix}.$$

That is, a matrix is upper Hessenberg if all entries below the subdiagonal are zero.

To find the vector $\mathbf{y}_k$ such that $\mathbf{x}^{(k)} = \mathbf{x}^{(0)} + Q_k\mathbf{y}_k$ is an approximate solution to $A\mathbf{x} = \mathbf{b}$, we aim to minimize the norm of the residual,

$$\begin{aligned} \|\mathbf{r}_{k+1}\|_2 &= \|\mathbf{b} - A\mathbf{x}^{(k)}\|_2 \\ &= \|\mathbf{v}_0 - AQ_k\mathbf{y}_k\|_2 \\ &= \|\|\mathbf{v}_0\|_2\mathbf{q}_1 - Q_{k+1}\tilde{H}_k\mathbf{y}_k\|_2 \\ &= \|\beta_0\mathbf{e}_1 - \tilde{H}_k\mathbf{y}_k\|_2, \end{aligned}$$

where $\beta_0 = \|\mathbf{v}_0\|_2$. This is an example of a **full-rank least-squares problem**, which can be solved using various approaches that are discussed in Chapter 4.

We now present the algorithm for solving $A\mathbf{x} = \mathbf{b}$ by minimizing the norm of the residual $\mathbf{r}_{k+1} = \mathbf{b} - A\mathbf{x}^{(k)}$. This method is the **Generalized Minimum Residual (GMRES) Method**.

Algorithm 5.3.1 (Generalized Minimum Residual Method) Let $A \in \mathbb{R}^{n \times n}$ be nonsingular, and let $\mathbf{b} \in \mathbb{R}^n$. The following algorithm computes the solution $\mathbf{x}$ of the system of linear equations $A\mathbf{x} = \mathbf{b}$.

Choose an initial guess $\mathbf{x}^{(0)} \in \mathbb{R}^n$
$\mathbf{v}_0 = \mathbf{b} - A\mathbf{x}^{(0)}$
$\beta_0 = \|\mathbf{v}_0\|_2$
$j = 0$
while $h_{j+1,j} > TOL$
$\quad \mathbf{q}_{j+1} = \mathbf{v}_j / h_{j+1,j}$
$\quad j = j + 1$
$\quad \mathbf{v}_j = A\mathbf{q}_j$
$\quad$ for $i = 1, 2, \ldots, j$
$\qquad h_{ij} = \mathbf{q}_i^T \mathbf{v}_j$
$\qquad \mathbf{v}_j = \mathbf{v}_j - h_{ij}\mathbf{q}_i$
$\quad$ end for
$\quad h_{j+1,j} = \|\mathbf{v}_j\|_2$
$\quad \mathbf{x}^{(j)} = \mathbf{x}^{(0)} + Q_j \mathbf{y}_j$ where $\|\beta_0 \mathbf{e}_1 - \tilde{H}_j \mathbf{y}_j\|_2 = $ minimum
end while

The iteration that produces the orthonormal basis $\mathbf{q}_1$, $\mathbf{q}_2$, ..., for the Krylov subspace $\mathcal{K}(A, \mathbf{b}, j)$ is known as **Arnoldi Iteration**.

Exploration 5.3.1 Write a MATLAB function

$$[\text{x,flag}] = \text{mygmres(A,b,x0,tol,maxit)}$$

to implement GMRES as described in Algorithm 5.3.1. The parameters `flag`, `x0`, `tol` and `maxit` have the same meaning as in Exploration 5.2.6.

Exploration 5.3.2 Explain how the exact solution of $A\mathbf{x} = \mathbf{b}$ can be obtained if the loop condition in Algorithm 5.3.1, $h_{j+1,j} > 0$, no longer holds. *Hint:* Examine (5.16).

Now, suppose that A is symmetric. From Arnoldi Iteration, we have

$$AQ_k = Q_k H_k + \beta_k \mathbf{q}_{k+1} \mathbf{e}_k^T, \quad \beta_k = h_{k+1,k}.$$

It follows from the orthonormality of the columns of Q_k that $H_k = Q_k^T A Q_k$, and therefore $H_k^T = Q_k^T A^T Q_k$. Therefore, if A is symmetric, so is H_k, but since H_k is also upper Hessenberg, this means that H_k is a *tridiagonal* matrix, which we refer to as T_k.

In fact, T_k is the very tridiagonal matrix obtained by applying the **Lanczos**

Iteration to A with initial vector $\mathbf{b}$. This can be seen from the fact that both algorithms generate an orthonormal basis for the Krylov subspace $\mathcal{K}(A, \mathbf{v}_0, k)$. Therefore, to solve $A\mathbf{x} = \mathbf{b}$ when A is symmetric, we can perform Lanczos Iteration to compute T_k, for $k = 1, 2, \ldots$, and at each iteration, solve the least squares problem of minimizing $\|\mathbf{b} - A\mathbf{x}^{(k)}\|_2 = \|\beta_0 \mathbf{e}_1 - \tilde{T}_k \mathbf{y}_k\|_2 = \text{minimum}$, where $\beta_0 = \|\mathbf{b} - A\mathbf{x}^{(0)}\|_2$ and $\tilde{T}_k$ is defined in the same way as $\tilde{H}_k$. This algorithm is known as the **MINRES (Minimum Residual)** method.

Exploration 5.3.3 Modify your function `mygmres` from Exploration 5.3.1 to obtain a new function `myminres` that implements MINRES for a symmetric matrix `A`.

One key difference between MINRES and GMRES is that in GMRES, the computation of $\mathbf{q}_{k+1}$ depends on $\mathbf{q}_1, \mathbf{q}_2, \ldots, \mathbf{q}_k$, whereas in MINRES, it only depends on $\mathbf{q}_{k-1}$ and $\mathbf{q}_k$, due to the three-term recurrence relation satisfied by the Lanczos vectors. It follows that the solution of the least squares problem during each iteration becomes substantially more expensive in GMRES, whereas this is not the case for MINRES; the approximate solution $\mathbf{x}^{(k)}$ can be updated from one iteration to the next. The following exploration guides the reader through the derivation of this efficient update, first described in [Paige and Saunders (1975)].

Exploration 5.3.4 Follow the approach used to derive equation (5.11), the main step of the Conjugate Gradient Method presented in Section 5.2.3, to obtain an efficient implementation of MINRES from Algorithm 5.3.1 for the case where A is symmetric. *Hint:* express $\tilde{H}_k^T \tilde{H}_k$ in terms of a rank-one update.

The consequence of this added expense for GMRES is that it must occasionally be *restarted* in order to contain the expense of solving these least squares problems.

Exploration 5.3.5 Modify your function `mygmres` from Exploration 5.3.1 to accept an additional optional input argument `restart` that indicates the number of iterations after which GMRES restarts with the latest iterate used as the new initial guess. Compare the performance of your function, both with and without restarting.

5.3.2 The Biconjugate Gradient Method

As we have seen, our first attempt to generalize the Conjugate Gradient Method to general matrices, GMRES, has the drawback that the computational expense of each iteration increases. To work around this problem, we consider whether it is possible to generalize Lanczos Iteration to unsymmetric matrices in such a way as to generate a Krylov subspace basis using a short recurrence relation, such as the three-term recurrence relation that is used in Lanczos Iteration.

This is possible, if we give up orthogonality. First, we let $\mathbf{r}_1 = \mathbf{b} - A\mathbf{x}^{(0)}$ and

$\hat{\mathbf{r}}_1 = \mathbf{b} - A^T \mathbf{x}^{(0)}$, where $\mathbf{x}^{(0)}$ is our initial guess. Then, we build bases for *two* Krylov subspaces,

$$\mathcal{K}(\mathbf{r}_1, A, k) = \text{span}\{\mathbf{r}_1, A\mathbf{r}_1, A^2\mathbf{r}_1, \ldots, A^{k-1}\mathbf{r}_1\},$$
$$\mathcal{K}(\hat{\mathbf{r}}_1, A^T, k) = \text{span}\{\hat{\mathbf{r}}_1, A^T\hat{\mathbf{r}}_1, (A^T)^2\hat{\mathbf{r}}_1, \ldots, (A^T)^{k-1}\hat{\mathbf{r}}_1\},$$

that are **biorthogonal**, as opposed to a single basis being orthogonal. That is, we construct a sequence $\mathbf{q}_1, \mathbf{q}_2, \ldots$ that spans a Krylov subspace generated by A, and a sequence $\hat{\mathbf{q}}_1, \hat{\mathbf{q}}_2, \ldots$ that spans a Krylov subspace generated by A^T, such that $\hat{\mathbf{q}}_i^T \mathbf{q}_j = \delta_{ij}$. In other words, if we define

$$Q_k = \begin{bmatrix} \mathbf{q}_1 & \mathbf{q}_2 & \cdots & \mathbf{q}_k \end{bmatrix}, \quad \hat{Q}_k = \begin{bmatrix} \hat{\mathbf{q}}_1 & \hat{\mathbf{q}}_2 & \cdots & \hat{\mathbf{q}}_k \end{bmatrix},$$

then $\hat{Q}_k^T Q_k = I$.

Since these bases span Krylov subspaces, it follows that their vectors have the form

$$\mathbf{q}_j = p_{j-1}(A)\mathbf{r}_1, \quad \hat{\mathbf{q}}_j = \hat{p}_{j-1}(A^T)\hat{\mathbf{r}}_1,$$

where p_i and $\hat{p}_i$ are polynomials of degree i. Then, using the framework of Gram-Schmidt Orthogonalization, we have

$$\beta_j \mathbf{q}_{j+1} = A\mathbf{q}_j - \sum_{i=1}^{j} h_{ij}\mathbf{q}_i, \quad h_{ij} = \hat{\mathbf{q}}_i^T A\mathbf{q}_j.$$

To ensure biorthogonality, we require that p_i be **formally orthogonal** to *all* polynomials of lesser degree. That is, $\langle p_i, g \rangle = 0$ for $g \in \mathcal{P}_{i-1}$, where $\mathcal{P}_n$ is the space of polynomials of degree at most n, and

$$\langle f, g \rangle = \hat{\mathbf{r}}_1^T f(A)g(A)\mathbf{r}_1 \tag{5.17}$$

is our bilinear form with which we define *formal* orthogonality. We say formal orthogonality because $\langle \cdot, \cdot \rangle$ does not actually define a valid inner product, due to A not being symmetric positive definite.

Exploration 5.3.6 Explain why the bilinear form $\langle \cdot, \cdot \rangle$ defined in (5.17) does not define a valid inner product, in the sense of the definition of an inner product given in Section B.15.

Suppose that $i < j - 1$. Then

$$h_{ij} = \hat{\mathbf{q}}_i^T A\mathbf{q}_j = \hat{\mathbf{r}}_1^T [\hat{p}_{i-1}(A^T)]^T Ap_{j-1}(A)\mathbf{r}_1 = \langle \lambda\hat{p}_{i-1}(\lambda), p_{j-1}(\lambda) \rangle = 0.$$

It follows that the $\mathbf{q}_j$ and $\hat{\mathbf{q}}_j$ can be obtained using three-term recurrence relations, as in Lanczos Iteration. These recurrence relations are

$$\beta_j \mathbf{q}_{j+1} = \mathbf{v}_j = (A - \alpha_j I)\mathbf{q}_j - \gamma_{j-1}\mathbf{q}_{j-1},$$
$$\gamma_j \hat{\mathbf{q}}_{j+1} = \hat{\mathbf{v}}_j = (A^T - \alpha_j I)\hat{\mathbf{q}}_j - \beta_{j-1}\hat{\mathbf{q}}_{j-1},$$

for $j = 1, 2, \ldots$. From these recurrence relations, we obtain

$$AQ_k = Q_k T_k + \beta_k \mathbf{q}_{k+1}\mathbf{e}_k^T, \quad A^T\hat{Q}_k = \hat{Q}_k T_k^T + \gamma_k \hat{\mathbf{q}}_{k+1}\mathbf{e}_k^T,$$

where

$$T_k = \hat{Q}_k^T A Q_k = \begin{bmatrix} \alpha_1 & \gamma_1 & & & \\ \beta_1 & \alpha_2 & \gamma_2 & & \\ & \beta_2 & \ddots & \ddots & \\ & & \ddots & \ddots & \gamma_{k-1} \\ & & & \beta_{k-1} & \alpha_k \end{bmatrix}.$$

This process of generating the bases Q_k and $\hat{Q}_k$, along with the tridiagonal matrix T_k, is known as the **Unsymmetric Lanczos Iteration**. The full algorithm is as follows:

$\mathbf{v}_0 = \mathbf{r}$
$\hat{\mathbf{v}}_0 = \hat{\mathbf{r}}_1$
$\mathbf{q}_0 = \mathbf{0}$
$\hat{\mathbf{q}}_0 = \mathbf{0}$
$\gamma_0 \beta_0 = \hat{\mathbf{v}}_0^T \mathbf{v}_0$
$k = 0$
while $\hat{\mathbf{v}}_k^T \mathbf{v}_k \neq 0$ **do**
    $\mathbf{q}_{k+1} = \mathbf{v}_k / \beta_k$
    $\hat{q}_{k+1} = \hat{\mathbf{v}}_k / \gamma_k$
    $k = k + 1$
    $\alpha_k = \hat{\mathbf{q}}_k^T A \mathbf{q}_k$
    $\mathbf{v}_k = (A - \alpha_k I)\mathbf{q}_k - \gamma_{k-1}\mathbf{q}_{k-1}$
    $\hat{\mathbf{v}}_k = (A^T - \alpha_k I)\hat{\mathbf{q}}_k - \beta_{k-1}\hat{\mathbf{q}}_{k-1}$
    $\gamma_k \beta_k = \hat{\mathbf{v}}_k^T \mathbf{v}_k$
end while

It should be noted that this algorithm gives flexibility in how to compute the off-diagonal entries γ_k and β_k of T_k. A typical approach is to compute

$$\beta_k = \sqrt{|\hat{\mathbf{v}}_k^T \mathbf{v}_k|}, \quad \gamma_k = \hat{\mathbf{v}}_k^T \mathbf{v}_k / \beta_k.$$

If $\mathbf{v}_k, \hat{\mathbf{v}}_k \neq \mathbf{0}$ but $\hat{\mathbf{v}}_k^T \mathbf{v}_k = 0$, then the iteration suffers what is known as **serious breakdown**. The iteration cannot continue due to division by zero, and no viable approximate solution is obtained.

> **Exploration 5.3.7** Explain why serious breakdown is not a concern for symmetric Lanczos Iteration.

If we then seek a solution to $A\mathbf{x} = \mathbf{b}$ of the form $\mathbf{x}^{(k)} = \mathbf{x}^{(0)} + Q_k \mathbf{y}_k$, then we can ensure that for $k \geq 1$, the residual $\mathbf{r}_{k+1} = \mathbf{b} - A\mathbf{x}^{(k)}$ is orthogonal to all columns of $\hat{Q}_k$ as follows:

$$0 = \hat{Q}_k^T \mathbf{r}_{k+1} = \hat{Q}_k^T (\mathbf{b} - A\mathbf{x}^{(k)}) = \hat{Q}_k^T (\beta_0 Q_k \mathbf{e}_1 - A Q_k \mathbf{y}_k) = \beta_0 \mathbf{e}_1 - T_k \mathbf{y}_k,$$

where for convenience we define $\beta_0 = \|\mathbf{r}_1\|_2$. Therefore, during each iteration we can solve the tridiagonal system of equations $T_k \mathbf{y}_k = \beta_0 \mathbf{e}_1$, as in Lanczos Iteration for solving $A\mathbf{x} = \mathbf{b}$ in the case where A is symmetric positive definite. However, as before, it would be preferable if we could obtain each approximate solution $\mathbf{x}^{(k)}$ through a simple update of $\mathbf{x}^{(k-1)}$ using an appropriate search direction.

Using similar manipulations to those used to derive the Conjugate Gradient Method from the (symmetric) Lanczos Iteration, we can obtain the **Biconjugate Gradient Method (BiCG)**:

Algorithm 5.3.2 (Biconjugate Gradient Method) Let $A \in \mathbb{R}^{n \times n}$ be nonsingular, and let $\mathbf{b} \in \mathbb{R}^n$. The following algorithm computes the solution $\mathbf{x}$ of the system of linear equations $A\mathbf{x} = \mathbf{b}$.

Choose an initial guess $\mathbf{x}^{(0)} \in \mathbb{R}^n$
$\mathbf{r}_1 = \mathbf{b} - A\mathbf{x}^{(0)}$
$\hat{\mathbf{r}}_1 = \mathbf{b} - A^T \mathbf{x}^{(0)}$
for $k = 1, 2, \ldots$ **do** until convergence
    if $k > 1$ **then**
        $\mu_k = \hat{\mathbf{r}}_k^T \mathbf{r}_k / \hat{\mathbf{r}}_{k-1}^T \mathbf{r}_{k-1}$
        $\mathbf{p}_k = \mathbf{r}_k + \mu_k \mathbf{p}_{k-1}$
        $\hat{\mathbf{p}}_k = \hat{\mathbf{r}}_k + \mu_k \hat{\mathbf{p}}_{k-1}$
    else
        $\mathbf{p}_1 = \mathbf{r}_1$
        $\hat{\mathbf{p}}_1 = \hat{\mathbf{r}}_1$
    end if
    $\mathbf{v}_k = A\mathbf{p}_k$
    $\nu_k = \hat{\mathbf{r}}_k^T \mathbf{r}_k / \hat{\mathbf{p}}_k^T \mathbf{v}_k$
    $\mathbf{x}^{(k)} = \mathbf{x}^{(k-1)} + \nu_k \mathbf{p}_k$
    $\mathbf{r}_{k+1} = \mathbf{r}_k - \nu_k \mathbf{v}_k$
    $\hat{\mathbf{r}}_{k+1} = \hat{\mathbf{r}}_k - \nu_k A^T \hat{\mathbf{p}}_k$
end for

Unfortunately, the biconjugate gradient method, as described above, is numerically unstable, even when used in conjunction with preconditioning. A variation of BiCG, known as BiCGSTAB, overcomes this instability [van der Vorst (2003)].

Exploration 5.3.8 Write a MATLAB function

```
[x,flag]=mybicg(A,b,x0,tol,maxit)
```

to implement the Biconjugate Gradient Method as described in Algorithm 5.3.2. The parameters `flag`, `x0`, `tol` and `maxit` have the same meaning as in Exploration 5.2.6.

An excellent online resource for testing iterative methods is the Matrix Market, which, as of this writing, can be found at `http://math.nist.gov/MatrixMarket`. It features approximately 500 sparse matrices from various applications, along with code written in C, FORTRAN and MATLAB for reading and writing matrices using its storage format.

Exploration 5.3.9 Visit the Matrix Market web site and download the MATLAB function `mmread`, along with the matrix named "ORSIRR 1" (this matrix can be found using the site's feature for browsing matrices by name). This matrix is 1030×1030 and unsymmetric. Test your functions `mygmres` and `mybicg` from Explorations 5.3.5 and 5.3.8, respectively, by solving a system involving this matrix and a randomly generated right-hand side. How does the performance of the two methods compare to one another, in terms of iteration count and execution time? Use `mygmres` with various values of the input argument `restart`.

5.3.3   Concept Check

1. What is a Krylov subspace?
2. What are the two minimum residual methods covered in this chapter, and when are they useful?
3. Why is one of the minimum residual methods substantially more costly than the other? How is this disadvantage addressed?
4. What does it mean for sets of vectors to be biorthogonal?
5. Briefly describe the Biconjugate Gradient Method. How does it relate to the Conjugate Gradient Method?

5.4   Additional Resources

Numerical linear algebra textbooks such as [Demmel (1997); Golub and van Loan (2012); Trefethen and Bau (1997)] offer in-depth coverage of iterative methods for linear systems. The texts [Greenbaum (1997); Saad (2003)] are devoted exclusively to the subject. The development of SOR is described in [Young (1950)]. The theory behind Krylov subspace methods, along with analysis of computational expense and roundoff error, is presented in [Liesen and Strakoš (2015)]. Several other Krylov subspace methods are not covered in this text, such as SYMMLQ [Paige and Saunders (1975)], QMR [Freund and Nachtigal (1991)] and LSQR [Paige and Saunders (1982)]. Beyond the scope of this text are multigrid methods [Trottenberg (2001)], for which a helpful resource for beginners is [Briggs, et al. (2000)]. A particularly active area of research is the development of effective preconditioners for various problems; for an example see [Elman, et al. (2005)].

5.5   Exercises

1. Consider the system $A\mathbf{x} = \mathbf{b}$, where
$$A = \begin{bmatrix} a & c & c \\ c & a & c \\ c & c & a \end{bmatrix}, \quad \mathbf{b} = \begin{bmatrix} a + 2c \\ a + 2c \\ a + 2c \end{bmatrix}.$$

It is easy to see that the exact solution is $\mathbf{x} = \begin{bmatrix} 1 & 1 & 1 \end{bmatrix}^T$. Try to solve this system using the Jacobi Method and Gauss-Seidel Method, in these cases:

   (a) $a = 10$, $c = 1$
   (b) $c = 10$, $a = 1$

What happens in these cases? Can you explain why? For either case, if both methods converge, which appears to be converging more rapidly?

2. Let A be a 2×2 symmetric positive definite matrix, and let T_J and T_{GS} be the iteration matrices for the Jacobi Method and Gauss-Seidel Method, respectively. Prove that $\rho(T_{GS}) = (\rho(T_J))^2$.

3. Let A be an $n \times n$ symmetric positive definite matrix. We consider a variation of the steepest descent method, in which the step size is a fixed constant α. Specifically,
$$\mathbf{x}^{(k+1)} = \mathbf{x}^{(k)} + \alpha(\mathbf{b} - A\mathbf{x}^{(k)}).$$

   (a) What is the iteration matrix for this method?
   (b) Derive a condition on α so that this iteration is guaranteed to converge. How should α relate to $\rho(A)$?
   (c) Find the optimal value of α that minimizes $\rho(T)$.
   (d) Express the minimum value of $\rho(T)$ in terms of $\kappa_2(A)$.

4. Let A be a symmetric positive definite matrix. Show that the A-**norm**
$$\|\mathbf{x}\|_A = \sqrt{\mathbf{x}^T A \mathbf{x}}$$

is indeed a norm, according to the definition of a vector norm from Section B.13.

5. Download the symmetric positive definite matrix BCSSTK16 from the Matrix Market web site at

$$\texttt{http://math.nist.gov/MatrixMarket/}$$

Then, apply SOR to the system $A\mathbf{x} = \mathbf{b}$ with this matrix, which is 4884×4884, and a random $\mathbf{b}$. Try SOR with five values of $\omega \in (0,2)$ and find an approximate "optimal" value $\hat{\omega}$. You may want to use polyfit to fit a polynomial to the obtained number of iterations as function of ω, and find its minimum. How does the choice of ω affect the rate of convergence?

6. We consider a modification of SOR for symmetric systems, called **Symmetric Successive Over-Relaxation (SSOR)**. The method proceeds as follows, for $k = 0, 1, 2, \ldots$:

1. Compute $\mathbf{y}^{(k+1)}$ from $\mathbf{x}^{(k)}$ using SOR. The components $y_i^{(k+1)}$ are computed in order: $i = 1, 2, \ldots, n$.
2. Compute $\mathbf{x}^{(k+1)}$ from $\mathbf{y}^{(k+1)}$, again using SOR, except that the components $x_i^{(k+1)}$ are computed in reverse order: $i = n, n-1, \ldots, 1$.

   (a) Write formulas for $y_i^{(k+1)}$ and $x_i^{(k+1)}$ analogous to (5.4).
   (b) Assume the symmetric matrix A is partitioned as $A = L + D + L^T$, where D is diagonal and L is lower triangular. Write SSOR in matrix form.
   (c) Construct the operator $\mathcal{S}_\omega = M_\omega^{-1} N_\omega$ such that
   $$\mathbf{x}^{(k+1)} = \mathcal{S}_\omega \mathbf{x}^{(k)} + M_\omega^{-1} \mathbf{b}.$$
   (d) Show that $\mathcal{S}_\omega$ is similar to a symmetric positive definite matrix. That is, show there exists an invertible matrix P such that $P^{-1} \mathcal{S}_\omega P$ is symmetric positive definite.

7. Let A be an $n \times n$ nonsingular and nonsymmetric matrix. We consider two iterative methods for solving $A\mathbf{x} = \mathbf{b}$:

   - GMRES, either with or without restarting, and
   - Conjugate Gradient, applied to the **normal equations** $A^T A\mathbf{x} = A^T \mathbf{b}$, since $A^T A$ is symmetric positive definite.

   Try these two approaches on a randomly generated matrix and right-hand side, with $n = 5000$. Compare the performance of these methods, in terms of iteration count and execution time.

8. (a) Let $A = M - N$ be a splitting of A, where M is symmetric positive definite and N is symmetric. Show that if $\lambda_{\min}(M) > \rho(N)$, then the iteration (5.1) converges.
   (b) Consider the following boundary value problem for the **Helmholtz equation** on a square R,
   $$\Delta u + \sigma(x, y)u = f(x, y), \quad (x, y) \in R,$$
   $$u = 0, \quad (x, y) \in \partial R.$$
   Discretization by finite differences (see Section 14.2) yields the system of linear equations
   $$(A + \Sigma)\mathbf{u} = \mathbf{f}$$
   where A is a discretization of the Laplacian, Σ is a diagonal matrix consisting of the values of $\sigma(x, y)$ at grid points in R, and $\mathbf{f}$ is a vector consisting of values of $f(x, y)$ at grid points in R.
   Suppose that we solve this system using the iteration (5.1) with $M = A$, $N = -\Sigma$. Use the result of part 8a to show that if
   $$\max_{(x,y) \in R} |\sigma(x, y)| < 2\pi^2,$$

then not only does this iteration converge, but the rate of convergence $\rho(M^{-1}N)$ is essentially independent of the grid spacing h used in the spatial discretization of the Laplacian. *Hint:* $\lambda_{\min}(A) = \frac{1}{h^2}[4 - 4\cos(\pi h)]$.

9. (a) Modify Algorithm 5.3.2 for the Biconjugate Gradient Method to include preconditioning, as was done for the Conjugate Gradient Method in Section 5.2.4. Specifically, apply BiCG to the system $\tilde{A}\tilde{\mathbf{x}} = \tilde{\mathbf{b}}$, where

$$\tilde{A} = L^{-1}AU^{-1}, \quad \tilde{\mathbf{x}} = U\mathbf{x}, \quad \tilde{\mathbf{b}} = L^{-1}\mathbf{b}.$$

Your final algorithm should not include L and U but instead the preconditioner $M = LU$.

(b) Modify your function `mybicg` from Exploration 5.3.8 to include preconditioning, as in Exploration 5.2.8. Allow the preconditioner to be specified using either one additional optional input argument that represents M, or two arguments that represent M implicitly as the product $M = M_1M_2$. Use the predefined variable `nargin` to distinguish between the two usages. What is the advantage of the latter approach to specifying the preconditioner?

(c) Try your modified function on a system $A\mathbf{x} = \mathbf{b}$, both with and without preconditioning. To obtain a preconditioner, use the `ilu` function. Compare the performance in terms of execution time and number of iterations.

10. In this exercise, we will prove that SOR applied to $A\mathbf{x} = \mathbf{b}$ converges if A is symmetric positive definite.

(a) Let $A = M - N$ be a splitting of A, and let $Q = M + M^T - A$. Prove that if both A and Q are symmetric positive definite, then the general stationary iteration (5.1) converges. *Hint:* Let $\mathbf{u}$ be an eigenvalue of $T = M^{-1}N$ with eigenvalue λ, and show that $|\lambda| < 1$ by obtaining an expression for $\mathbf{u}^T(M + M^T)\mathbf{u}$.

(b) Apply part 10a to the splitting used for SOR to show that SOR converges for $0 < \omega < 2$. Use the matrices M and N from Exploration 5.1.8.

(c) Suppose we wish to solve $A\mathbf{x} = \mathbf{b}$, for a general $n \times n$ matrix A. Show that if there exists a diagonal matrix D such that $B = DAD^{-1}$ is symmetric positive definite, then SOR converges when applied directly to $A\mathbf{x} = \mathbf{b}$.

Chapter 6

Eigenvalue Problems

One of the most fundamental problems in numerical linear algebra is the **eigenvalue problem**: finding a nonzero vector $\mathbf{x}$ and scalar λ such that

$$A\mathbf{x} = \lambda\mathbf{x},$$

where A is an $n \times n$ matrix. In this chapter, we will explore the numerical solution of this problem. We will see that methods for computing eigenvalues will be useful for solving least squares problems and differential equations, as well as fundamental problems such as computing roots of polynomials.

Section 6.1 provides an overview of properties of eigenvalues and eigenvectors, and related decompositions. In Section 6.2, we present the Power Method for computing a single eigenvalue-eigenvector pair, and show how it can be extended to compute multiple eigenpairs. This sets the stage for the **QR Algorithm** for computing all eigenvalues, which is developed in Section 6.3. Section 6.4 shows how the techniques presented in earlier sections can be adapted to the case in which the matrix is symmetric. In Section 6.5, these techniques are adapted even further to compute the **Singular Value Decomposition (SVD)**, which was introduced in Section 4.4 due to its importance for least squares problems.

6.1 Eigenvalues and Eigenvectors

In this section, we describe the eigenvalue problem, and establish preliminaries to facilitate discussion of numerical solution methods. It is recommended that the reader be familiar with concepts from linear algebra from Appendix B. In particular, Section B.12, that covers basic properties of eigenvalues, should be reviewed.

6.1.1 Definitions and Properties

Let $A \in \mathbb{C}^{n \times n}$. A *nonzero* vector $\mathbf{x}$ is called an **eigenvector** of A if there exists a scalar λ such that

$$A\mathbf{x} = \lambda\mathbf{x}.$$

197

The scalar λ is called an **eigenvalue** of A, and we say that $\mathbf{x}$ is an eigenvector of A *corresponding* to λ. We see that an eigenvector of A is a vector for which matrix-vector multiplication with A is equivalent to scalar multiplication by λ. This is important because the latter operation is substantially simpler and more efficient than the former–requiring only n floating-point operations rather than $O(n^2)$.

We say that a nonzero vector $\mathbf{y}$ is a **left eigenvector** of A if there exists a scalar λ such that

$$\lambda \mathbf{y}^H = \mathbf{y}^H A.$$

The superscript H refers to the **Hermitian transpose** (see Section B.7.2), which includes transposition and complex conjugation. That is, for any matrix A, $A^H = \overline{A^T}$, where the bar denotes complex conjugation. An eigenvector of A, as defined above, is sometimes called a **right eigenvector** of A, to distinguish from a left eigenvector.

Exploration 6.1.1 Show that $\mathbf{y}$ is a left eigenvector of A with eigenvalue λ if and only if $\mathbf{y}$ is a right eigenvector of A^H with eigenvalue $\overline{\lambda}$.

Because $\mathbf{x}$ is nonzero, it follows that if $\mathbf{x}$ is an eigenvector of A, then the matrix $A - \lambda I$ is **singular**, where λ is the corresponding eigenvalue. That is, λ satisfies the equation

$$\det(A - \lambda I) = 0.$$

The expression $\det(A - \lambda I)$ is a polynomial of degree n in λ, and therefore is called the **characteristic polynomial** of A (eigenvalues are sometimes called **characteristic values**). It follows from the fact that the eigenvalues of A are the roots of the characteristic polynomial that A has n eigenvalues, which can repeat, and can also be complex, even if A is real. However, if A is real, any complex eigenvalues must occur in complex-conjugate pairs, because in this case the characteristic polynomial has real coefficients.

The set of eigenvalues of A is called the **spectrum** of A, and denoted by $\lambda(A)$. This terminology explains why the magnitude of the largest eigenvalue of A is called the **spectral radius** of A, denoted by $\rho(A)$. The **trace** of A, denoted by $\mathrm{tr}(A)$, is defined to be the sum of the diagonal entries of A, but is also equal to the sum of the eigenvalues of A. Furthermore, $\det(A)$ is equal to the *product* of the eigenvalues of A.

Example 6.1.1 A 2×2 matrix

$$A = \begin{bmatrix} a & b \\ c & d \end{bmatrix}$$

has trace $\mathrm{tr}(A) = a + d$ and determinant $\det(A) = ad - bc$. Its characteristic polynomial is

$$\det(A - \lambda I) = \begin{vmatrix} a - \lambda & b \\ c & d - \lambda \end{vmatrix}$$

$$= (a - \lambda)(d - \lambda) - bc = \lambda^2 - (a + d)\lambda + (ad - bc)$$

$$= \lambda^2 - \operatorname{tr}(A)\lambda + \det(A).$$

From the quadratic formula, the eigenvalues are

$$\lambda_1 = \frac{a + d}{2} + \frac{\sqrt{(a - d)^2 + 4bc}}{2}, \quad \lambda_2 = \frac{a + d}{2} - \frac{\sqrt{(a - d)^2 + 4bc}}{2}.$$

It can be verified directly that the sum of these eigenvalues is equal to $\operatorname{tr}(A)$, and that their product is equal to $\det(A)$. $\square$

As shown in this example, the eigenvalues of a matrix can be obtained by computing the characteristic polynomial and then computing its roots, using algorithms such as those that will be presented in Chapter 10. In fact, this is how students are taught to compute eigenvalues of small matrices in their first linear algebra course. However, such an approach is not practical due to the lack of robustness of such algorithms, as well as the substantial computational expense and roundoff error that can be incurred from the computation of the characteristic polynomial. We will therefore explore other approaches in this chapter.

The MATLAB function `eig` computes and returns the eigenvalues of a given square matrix A. The simplest usage, `e=eig(A)`, returns the eigenvalues of the $n \times n$ matrix A in the column n-vector e. We will discuss other uses of `eig` later in this chapter. For now, though, we will use it to demonstrate that eigenvalues, and therefore polynomial roots, can be difficult to compute, even in MATLAB. It turns out that when it comes to practical computation, the relationship between eigenvalues and polynomial roots is the opposite of what it seems from an introductory linear algebra course.

The MATLAB function `compan` forms the **companion matrix** of a polynomial $p(\lambda) = a_0 + a_1\lambda + a_2\lambda^2 + \cdots + a_{n-1}\lambda^{n-1} + \lambda^n$, which is an $n \times n$ matrix C of the form

$$C = \begin{bmatrix} -a_{n-1} & -a_{n-2} & \cdots & -a_2 & -a_1 & -a_0 \\ 1 & & & & & \\ & 1 & & & & \\ & & \ddots & & & \\ & & & 1 & & \\ & & & & 1 & \end{bmatrix}.$$

The companion matrix C of $p(\lambda)$ has the property that its characteristic polynomial is equal to $p(\lambda)$, so the roots of $p(\lambda)$ are also the eigenvalues of C. In fact, this is how the MATLAB function `roots` computes the roots of a given polynomial: by forming its companion matrix C, and then calling `eig` to compute the eigenvalues of C, using the approach that will be presented in this chapter. The following exploration demonstrates the difficulty of the problem of computing polynomial roots, or, equivalently, eigenvalues.

Exploration 6.1.2 Recall from Section 1.2.18 that the function `poly` computes the coefficients of the monic polynomial that has the given roots. It therefore follows that `roots` and `poly` are, in a sense, inverses of one another. Try the statement `roots(poly(1:n))` for various values of n. What happens? The case n = 20 is known as **Wilkinson's example** [Wilkinson (1963)].

As in our treatment of systems of linear equations in Chapter 3, our development of effective algorithms for computing eigenvalues will be guided by the following questions:

- For which square matrices can the eigenvalues easily be obtained?
- How can the problem of computing the eigenvalues of a general square matrix be reduced to such a simple case?

Both of these questions will be addressed in the following discussion.

6.1.2 The Schur Decomposition

A subspace W of $\mathbb{C}^n$ is called an **invariant subspace** of A if, for any vector $\mathbf{x} \in W$, $A\mathbf{x} \in W$. Suppose that $\dim(W) = k$, and let X be an $n \times k$ matrix such that $\text{range}(X) = W$. That is, the columns of X form a **basis** for W (see Section B.4). Then, because each column of X is a vector in W, each column of AX is also a vector in W, and therefore is a linear combination of the columns of X. It follows that $AX = XB$, where B is a $k \times k$ matrix. The following result is easily proved.

Exploration 6.1.3 Let A, X and B be defined as above. Show $\lambda(B) \subseteq \lambda(A)$. If $\mathbf{y}$ is an eigenvector of B with corresponding eigenvalue λ, what is the eigenvector of A corresponding to λ?

If $k = n$, then X is an $n \times n$ invertible matrix, and it follows that A and B have the same eigenvalues. Furthermore, from $AX = XB$, we now have $B = X^{-1}AX$. We say that A and B are **similar matrices**, and that B is a **similarity transformation** of A.

Exploration 6.1.4 Let $A, B \in \mathbb{C}^{n \times n}$, and suppose that at least one of A and B is nonsingular. Prove that AB is similar to BA.

Similarity transformations are essential tools in algorithms for computing the eigenvalues of a matrix A, since the basic idea is to apply a sequence of similarity transformations to A, each of which preserves its eigenvalues, in order to obtain a new matrix B whose eigenvalues are easily obtained. For example, suppose that B

has a 2×2 *block* structure

$$B = \begin{bmatrix} B_{11} & B_{12} \\ 0 & B_{22} \end{bmatrix}, \tag{6.1}$$

where B_{11} is $p \times p$ and B_{22} is $q \times q$. We then have the following key result.

Theorem 6.1.2 Let B be defined as in (6.1). Then $\lambda(B) = \lambda(B_{11}) \cup \lambda(B_{22})$.

Exploration 6.1.5 Prove Theorem 6.1.2. *Hint:* Let $\mathbf{x} = \begin{bmatrix} \mathbf{x}_1^T & \mathbf{x}_2^T \end{bmatrix}^T$ be an eigenvector of B, where $\mathbf{x}_1 \in \mathbb{C}^p$ and $\mathbf{x}_2 \in \mathbb{C}^q$.

It follows that if we can use similarity transformations to reduce A to such a block structure, the problem of computing the eigenvalues of A *decouples* into two smaller problems of computing the eigenvalues of B_{ii} for $i = 1, 2$. Using an inductive argument, it can be shown that if A is block upper triangular, then the eigenvalues of A are equal to the union of the eigenvalues of the diagonal blocks. If each diagonal block is 1×1, then it follows that the eigenvalues of any upper triangular matrix are the diagonal entries. The same is true of any lower triangular matrix..

Example 6.1.3 The matrix

$$A = \begin{bmatrix} 1 & -2 & 3 & -3 & 4 \\ 0 & 4 & -5 & 6 & -5 \\ 0 & 0 & 6 & -7 & 8 \\ 0 & 0 & 0 & 7 & 0 \\ 0 & 0 & 0 & -8 & 9 \end{bmatrix}$$

has eigenvalues 1, 4, 6, 7, and 9. This is because A has a block upper triangular structure

$$A = \begin{bmatrix} A_{11} & A_{12} \\ 0 & A_{22} \end{bmatrix}, \quad A_{11} = \begin{bmatrix} 1 & -2 & 3 \\ 0 & 4 & -5 \\ 0 & 0 & 6 \end{bmatrix}, \quad A_{22} = \begin{bmatrix} 7 & 0 \\ -8 & 9 \end{bmatrix}.$$

Because both of these blocks are themselves triangular, their eigenvalues are equal to their diagonal entries, and the spectrum of A is the union of the spectra of these blocks. $\square$

Now that we have answered the first of the two questions posed earlier, by identifying a class of matrices for which eigenvalues are easily obtained, we move to the second question: how can a general square matrix A be reduced to (block) triangular form, in such a way that its eigenvalues are preserved? That is, we need to find a similarity transformation that can bring A closer to triangular form, in some sense.

With this goal in mind, suppose that $\mathbf{x} \in \mathbb{C}^n$ is a *normalized* eigenvector of $A \in \mathbb{C}^{n \times n}$ (that is, $\|\mathbf{x}\|_2 = 1$), with eigenvalue λ. Furthermore, suppose that P is a

Householder reflection (see Section 4.2.2) such that $P\mathbf{x} = \mathbf{e}_1$. Because P is complex, it is Hermitian, meaning that $P^H = P$, and **unitary**, meaning that $P^H P = I$. This the generalization of a real Householder reflection being symmetric and orthogonal. Therefore, P is its own inverse, so $P\mathbf{e}_1 = \mathbf{x}$. It follows that the matrix $P^H A P$, which is a similarity transformation of A, satisfies

$$P^H A P \mathbf{e}_1 = P^H A \mathbf{x} = \lambda P^H \mathbf{x} = \lambda P \mathbf{x} = \lambda \mathbf{e}_1.$$

That is, $\mathbf{e}_1$ is an eigenvector of $P^H A P$ with eigenvalue λ, and therefore $P^H A P$ has the block structure

$$P^H A P = \begin{bmatrix} \lambda & \mathbf{v}^H \\ \mathbf{0} & B \end{bmatrix}. \tag{6.2}$$

Therefore, $\lambda(A) = \{\lambda\} \cup \lambda(B)$, which means that we can now focus on the $(n-1) \times (n-1)$ matrix B to find the rest of the eigenvalues of A. This process of reducing the eigenvalue problem for A to that of B is called **deflation**.

Exploration 6.1.6 The procedure described in Section 4.2.2 for finding a Householder reflection P such that $P\mathbf{x} = \pm\|\mathbf{x}\|_2 \mathbf{e}_1$ assumed that $\mathbf{x}$ was real. How should this procedure be modified for the case in which $\mathbf{x}$ is complex?

Exploration 6.1.7 Write a MATLAB function

$$[P,e,B]=\texttt{deflate(A,x)}$$

that accepts as input an $n \times n$ matrix A and a vector x, that is assumed to be an eigenvector, and returns as output a Householder reflection P, eigenvalue e corresponding to x, and $(n-1) \times (n-1)$ matrix B such that $P' * A * P$ has the structure shown in (6.2).

Repeating this process of deflation, we obtain the **Schur Decomposition**

$$A = QTQ^H \tag{6.3}$$

where $T \in \mathbb{C}^{n \times n}$ is an upper triangular matrix whose diagonal entries are the eigenvalues of A, and $Q \in \mathbb{C}^{n \times n}$ is a unitary matrix. Every square complex matrix has a Schur Decomposition.

The MATLAB function `schur` computes the Schur Decomposition of a given matrix A. The statement

```
>> [Q,T]=schur(A)
```

computes matrices Q and T such that $A = QTQ^H$. If A is complex, then Q is unitary and T is upper triangular, with the eigenvalues of A along its diagonal. If A is real, then Q is a real orthogonal matrix and T is real and *block* upper triangular, with 1×1 diagonal blocks corresponding to real eigenvalues, and 2×2 diagonal blocks corresponding to complex-conjugate pairs of eigenvalues. This form of the Schur Decomposition is called the **Real Schur form**. If the complex Schur Decomposition is desired, one can use the statement [Q,T]=schur(A,'complex'). If only one output is specified in the call to schur, then only the matrix T is returned.

The columns of Q are called **Schur vectors**, and the first Schur vector $\mathbf{q}_1$ is an eigenvector as well. However, for a general matrix A, there is no simple relationship between the other Schur vectors of A and eigenvectors of A, as each Schur vector $\mathbf{q}_j$ satisfies $A\mathbf{q}_j = AQ\mathbf{e}_j = QT\mathbf{e}_j$. That is, $A\mathbf{q}_j$ is a linear combination of $\mathbf{q}_1, \ldots, \mathbf{q}_j$. It follows that for $j = 1, 2, \ldots, n$, the first j Schur vectors $\mathbf{q}_1, \mathbf{q}_2, \ldots, \mathbf{q}_j$ span an invariant subspace of A.

Exploration 6.1.8 Use the Schur Decomposition to prove that

$$\operatorname{tr}(A) \equiv \sum_{i=1}^{n} a_{ii} = \sum_{i=1}^{n} \lambda_i(A).$$

The Schur vectors *are* eigenvectors of A when A is a **normal** matrix, which means that $A^H A = A A^H$. Any real symmetric or skew-symmetric matrix, for example, is normal. It can be shown that in this case, the normalized eigenvectors of A form an orthonormal basis for $\mathbb{C}^n$.

Exploration 6.1.9 Let T be a normal, upper triangular matrix. Prove that T is diagonal.

Exploration 6.1.10 Use the result of Exploration 6.1.9 to show that if $A \in \mathbb{C}^{n \times n}$ is normal, then it has the Schur Decomposition

$$A = QDQ^H$$

where D is a diagonal matrix. Conclude that the eigenvectors of A form an orthonormal basis for $\mathbb{C}^n$.

It follows that if $\lambda_1, \lambda_2, \ldots, \lambda_n$ are the eigenvalues of A, with corresponding (orthonormal) eigenvectors $\mathbf{q}_1, \mathbf{q}_2, \ldots, \mathbf{q}_n$, then we have

$$AQ = QD, \quad Q = \begin{bmatrix} \mathbf{q}_1 & \cdots & \mathbf{q}_n \end{bmatrix}, \quad D = \operatorname{diag}(\lambda_1, \ldots, \lambda_n).$$

Because Q is a unitary matrix, it follows that

$$Q^H A Q = Q^H Q D = D,$$

and A is similar to a diagonal matrix. We therefore say that A is **diagonalizable**. Furthermore, because D can be obtained from A by a similarity transformation involving a unitary matrix, we say that A is **unitarily diagonalizable**.

6.1.3 Nonunitary Decompositions

Even if A is *not* a normal matrix, it may still be diagonalizable, meaning that there exists an invertible matrix P such that $P^{-1}AP = D$, where D is a diagonal matrix.

If this is the case, then, because $AP = PD$, the columns of P are eigenvectors of A, and the rows of P^{-1} are eigenvectors of A^T (as well as the left eigenvectors of A, if P is real). Recall that the MATLAB function eig, when used with only one output argument, returns a column vector containing the eigenvalues of a given matrix. The usage [X,D]=eig(A) returns a matrix X, the columns of which contain the eigenvectors of A, and a diagonal matrix D containing the eigenvalues of A. These matrices satisfy the relationship A*X = X*D.

Which kinds of non-normal matrices are diagonalizable? To help answer this question, suppose that λ_1 and λ_2 are *distinct* eigenvalues, with corresponding eigenvectors $\mathbf{x}_1$ and $\mathbf{x}_2$, respectively. Furthermore, suppose that $\mathbf{x}_1$ and $\mathbf{x}_2$ are linearly *dependent*. This means that they must be parallel; that is, there exists a nonzero constant c such that $\mathbf{x}_2 = c\mathbf{x}_1$. This implies that $A\mathbf{x}_2 = \lambda_2\mathbf{x}_2$ and $A\mathbf{x}_2 = cA\mathbf{x}_1 = c\lambda_1\mathbf{x}_1 = \lambda_1\mathbf{x}_2$. However, because $\lambda_1 \neq \lambda_2$, this is a contradiction. Therefore, $\mathbf{x}_1$ and $\mathbf{x}_2$ must be linearly *independent*.

Exploration 6.1.11 Use an inductive argument to show that if $\mathbf{x}_1, \mathbf{x}_2, \ldots, \mathbf{x}_k$ are eigenvectors of A with distinct corresponding eigenvalues $\lambda_1, \lambda_2, \ldots, \lambda_k$, then $\mathbf{x}_1, \mathbf{x}_2, \ldots, \mathbf{x}_k$ are linearly independent.

It follows that if A has n distinct eigenvalues, then it has a set of n linearly independent eigenvectors. If X is an $n \times n$ matrix whose columns are these eigenvectors, then $AX = XD$, where D is a diagonal matrix of the eigenvectors. Because the columns of X are linearly independent, X is invertible, and therefore $X^{-1}AX = D$, and A is diagonalizable.

Now, suppose that the eigenvalues of A are *not* distinct; that is, the characteristic polynomial has repeated roots. Then an eigenvalue with multiplicity m does not necessarily correspond to m linearly independent eigenvectors. To facilitate discussion of repeated eigenvalues, we introduce some terminology.

Definition 6.1.4 Let $A \in \mathbb{C}^{n \times n}$, and let $\lambda \in \mathbb{C}$ be an eigenvalue of A. The subspace of $\mathbb{C}^n$ spanned by all eigenvectors of A corresponding to λ is called an **eigenspace**.

By definition, an eigenvalue of $A \in \mathbb{C}^{n \times n}$ corresponds to at least one eigenvector. Because any nonzero scalar multiple of an eigenvector is also an eigenvector, an eigenvector defines a one-dimensional invariant subspace of $\mathbb{C}^n$. A repeated eigenvalue may have more than one linearly independent eigenvector; the set of all such eigenvectors is a basis of its eigenspace. Any invariant subspace of a diagonalizable matrix A is a union of eigenspaces.

Definition 6.1.5 The **algebraic multiplicity** of an eigenvalue λ is the number of times that λ occurs as a root of the characteristic polynomial. The **geometric multiplicity** of λ is the dimension of the eigenspace corresponding to λ, which is equal to the maximal size of a set of linearly independent eigenvectors corresponding to λ.

The geometric multiplicity of an eigenvalue λ is always less than or equal to its algebraic multiplicity. When it is strictly less, then we say that the eigenvalue is **defective**. The matrix A is said to be defective if any of its eigenvalues are defective. When both multiplicities are equal to one, then we say that the eigenvalue is **simple**.

Suppose that $A \in \mathbb{C}^{n \times n}$ is defective, which means its eigenvectors do not form a basis for $\mathbb{C}^n$. Since there is no similarity transformation that would reduce A to diagonal form, is there at least a similarity transformation that would reduce A to *block* diagonal form? The following explorations guide an investigation of this question.

Exploration 6.1.12 Let $\lambda \in \mathbb{C}$, and let $J \in \mathbb{C}^{n \times n}$ be defined by

$$J = \begin{bmatrix} \lambda & 1 & & & \\ & \lambda & 1 & & \\ & & \ddots & \ddots & \\ & & & \lambda & 1 \\ & & & & \lambda \end{bmatrix}.$$

Show that J has only one distinct eigenvalue of multiplicity n, and that $\mathbf{e}_1$ is the only linearly independent eigenvector.

Exploration 6.1.13 Let J be defined as in Exploration 6.1.12. Show that $(J - \lambda I)^n = 0$, and that $(J - \lambda I)^p \neq 0$ for $0 \leq p < n$. We say that $J - \lambda I$ is an example of a **nilpotent** matrix. *Hint:* Create such a matrix in MATLAB and compute powers of it. What do these powers look like?

Exploration 6.1.14 Show that $\mathbf{e}_j$, for $j = 2, 3, \ldots, n$, satisfy

$$(J - \lambda I)^j \mathbf{e}_j = \mathbf{0}, \quad (J - \lambda I)^k \mathbf{e}_j \neq \mathbf{0}, \quad k < j.$$

We say that these vectors are **generalized eigenvectors** of J.

Exploration 6.1.15 Let $A \in \mathbb{C}^{n \times n}$, and let $X \in \mathbb{C}^{n \times n}$ be an invertible matrix such that $AX = XJ$, where J is defined as in Exploration 6.1.12. Show that if $X = \begin{bmatrix} \mathbf{x}_1 & \mathbf{x}_2 & \cdots & \mathbf{x}_n \end{bmatrix}$, then $\mathbf{x}_1$ is the only linearly independent eigenvector of A, with eigenvalue λ, and that $\mathbf{x}_2, \ldots, \mathbf{x}_n$ are all generalized eigenvectors of A.

Generalizing from the preceding explorations, the **Jordan Canonical Form** of an $n \times n$ matrix A is a decomposition that yields information about the eigenspaces of A, as well as the multiplicities of the eigenvalues. It has the form

$$A = XJX^{-1}$$

where J has the block diagonal structure

$$J = \begin{bmatrix} J_1 & 0 & \cdots & 0 \\ 0 & J_2 & \ddots & \vdots \\ \vdots & \ddots & \ddots & 0 \\ 0 & \cdots & 0 & J_p \end{bmatrix}.$$

Each diagonal block J_p is a **Jordan block** that has the form

$$J_i = \begin{bmatrix} \lambda_i & 1 & & \\ & \lambda_i & \ddots & \\ & & \lambda_i & 1 \\ & & & \lambda_i \end{bmatrix}, \quad i = 1, 2, \ldots, p.$$

The number of Jordan blocks, p, is equal to the number of linearly independent eigenvectors of A. The diagonal entry of J_i, λ_i, is an eigenvalue of A. The number of Jordan blocks associated with λ_i is equal to the *geometric* multiplicity of λ_i. The sum of the sizes of these blocks is equal to the *algebraic* multiplicity of λ_i. If A is diagonalizable, then each Jordan block is 1×1.

Example 6.1.6 Consider a matrix with Jordan Canonical Form

$$J = \begin{bmatrix} 2 & 1 & 0 & & & \\ 0 & 2 & 1 & & & \\ 0 & 0 & 2 & & & \\ & & & 3 & 1 & \\ & & & 0 & 3 & \\ & & & & & 2 \end{bmatrix}.$$

The eigenvalues of this matrix are 2, with algebraic multiplicity 4, and 3, with algebraic multiplicity 2. The geometric multiplicity of the eigenvalue 2 is 2, because it is associated with two Jordan blocks. The geometric multiplicity of the eigenvalue 3 is 1, because it is associated with only one Jordan block. Therefore, there are a total of three linearly independent eigenvectors, and the matrix is not diagonalizable. □

The Jordan Canonical Form, while very informative about the eigensystem of A, is not practical to compute using floating-point arithmetic. This is due to the fact that while the eigenvalues of a matrix are continuous functions of its entries, the Jordan Canonical Form is not. If two computed eigenvalues are nearly equal, and their computed corresponding eigenvectors are nearly parallel, we do not know if they represent two distinct eigenvalues with linearly independent eigenvectors, or a multiple eigenvalue that could be defective. Nonetheless, the Jordan Canonical

Form is still useful in numerical analysis, as the discussion later in this section will demonstrate.

Exploration 6.1.16 As previously mentioned, the MATLAB function `eig`, when used with two output arguments, returns a matrix X and diagonal matrix D such that $AX = XD$, where the diagonal entries of D are the eigenvalues of A, and the columns of X are the corresponding eigenvectors. How does this function behave if A is *not* diagonalizable? More precisely, what properties does X have? Try `eig` on some of the non-diagonalizable matrices in this section, for which the eigenvectors and generalized eigenvectors can readily be obtained.

6.1.4 Perturbation Theory*

Just as the problem of solving a system of linear equations $Ax = b$ can be sensitive to perturbations in the data, the problem of computing the eigenvalues of a matrix can also be sensitive to perturbations in the matrix. We will now obtain some results concerning the extent of this sensitivity.

Theorem 6.1.7 (Gershgorin Circle Theorem) Let $A \in \mathbb{C}^{n \times n}$, and let $\lambda \in \lambda(A)$. Then λ lies within the union of the **Gershgorin Disks**

$$|\lambda - a_{ii}| \le \sum_{j=1, j \ne i}^{n} |a_{ij}|, \quad i = 1, 2, \dots, n.$$

Exploration 6.1.17 Prove Theorem 6.1.7 using the formula for a matrix-vector product from Section B.5.2, applied to the relationship $Ax = \lambda x$, where x is an eigenvector.

Example 6.1.8 The eigenvalues of the matrix

$$A = \begin{bmatrix} -5 & -1 & 1 \\ -2 & 2 & -1 \\ 1 & -3 & 7 \end{bmatrix}$$

are

$$\lambda(A) = \{6.4971, 2.7930, -5.2902\}.$$

The Gershgorin Disks are

$$D_1 = \{z \in \mathbb{C} \,|\, |z - 7| \le 4\},$$
$$D_2 = \{z \in \mathbb{C} \,|\, |z - 2| \le 3\},$$
$$D_3 = \{z \in \mathbb{C} \,|\, |z + 5| \le 2\}.$$

We see that each disk contains one eigenvalue. $\square$

It is important to note that while there are n eigenvalues and n Gershgorin Disks, it is not necessarily true that each disk contains an eigenvalue. The Gershgorin Circle Theorem only states that all of the eigenvalues are contained within the *union* of the disks.

Exploration 6.1.18 Write a utility MATLAB function
`drawcircle(x0,y0,r)` that plots a circle of radius `r` centered at $(x0, y0)$.

Exploration 6.1.19 Write a MATLAB function

$$\texttt{gershgorin(A)}$$

that plots the Gershgorin circles (that is, the boundaries of the Gershgorin Disks) for the given matrix A, using the function `drawcircle` from Exploration 6.1.18. Test your function on a sequence of matrices of the form $D + \epsilon E$, that are perturbations of a fixed diagonal matrix D. How do the Gershgorin Disks change as ϵ increases?

It would be desirable to have a concrete measure of the sensitivity of an eigenvalue, just as we have the condition number, introduced in Section 3.4.1, that measures the sensitivity of a system of linear equations. To that end, we assume that λ is a *simple* eigenvalue of $A \in \mathbb{C}^{n \times n}$, with eigenvector $\mathbf{x}$ and left eigenvector $\mathbf{y}$. From the fact that λ is a simple eigenvalue, we have $\text{rank}(A - \lambda I) = n - 1$. Because $\mathbf{y}$ is an eigenvector of A^H with eigenvalue $\bar{\lambda}$, $\mathbf{y}$ is orthogonal to $\text{range}(A - \lambda I)$. Therefore, $\mathbf{y}$ cannot also be orthogonal to $\mathbf{x}$, which spans $\text{null}(A - \lambda I)$, without being the zero vector, which contradicts the assumption that $\mathbf{y}$ is a left eigenvector. It follows that $\mathbf{y}^H \mathbf{x} \neq 0$, so we can therefore assume $\mathbf{x}$ and $\mathbf{y}$ are normalized so that $\mathbf{y}^H \mathbf{x} = 1$.

We now let A, λ, and $\mathbf{x}$ be functions of a parameter ϵ that satisfy

$$A(\epsilon)\mathbf{x}(\epsilon) = \lambda(\epsilon)\mathbf{x}(\epsilon), \quad A(\epsilon) = A + \epsilon F, \quad \|F\|_2 = 1.$$

Differentiating with respect to ϵ, and evaluating at $\epsilon = 0$, yields

$$F\mathbf{x} + A\mathbf{x}'(0) = \lambda\mathbf{x}'(0) + \lambda'(0)\mathbf{x}.$$

Taking the inner product (see Section B.7.3) of both sides with $\mathbf{y}^H$ and simplifying yields

$$\mathbf{y}^H F\mathbf{x} + \mathbf{y}^H A\mathbf{x}'(0) = \lambda\mathbf{y}^H\mathbf{x}'(0) + \lambda'(0)\mathbf{y}^H\mathbf{x}.$$

Because $\mathbf{y}$ is a left eigenvector corresponding to λ, and $\mathbf{y}^H\mathbf{x} = 1$, we have

$$\mathbf{y}^H F\mathbf{x} + \lambda\mathbf{y}^H\mathbf{x}'(0) = \lambda\mathbf{y}^H\mathbf{x}'(0) + \lambda'(0).$$

We conclude that

$$|\lambda'(0)| = |\mathbf{y}^H F\mathbf{x}| \leq \|\mathbf{y}\|_2\|F\|_2\|\mathbf{x}\|_2 \leq \|\mathbf{y}\|_2\|\mathbf{x}\|_2.$$

Because θ, the angle between $\mathbf{x}$ and $\mathbf{y}$, is given by

$$\cos\theta = \frac{\mathbf{y}^H\mathbf{x}}{\|\mathbf{y}\|_2\|\mathbf{x}\|_2} = \frac{1}{\|\mathbf{y}\|_2\|\mathbf{x}\|_2},$$

it follows that

$$|\lambda'(0)| \leq \frac{1}{|\cos\theta|}.$$

We define $1/|\cos\theta|$ to be the **condition number** of the simple eigenvalue λ. We require λ to be simple because otherwise, the angle between the left and right eigenvectors is not unique, due to the eigenvectors themselves not being unique.

The interpretation of the condition number is that an $O(\epsilon)$ perturbation in A can cause an $O(\epsilon/|\cos\theta|)$ perturbation in the eigenvalue λ. Therefore, if $\mathbf{x}$ and $\mathbf{y}$ are nearly orthogonal, a large change in the eigenvalue can occur. Furthermore, if the condition number is large, then A is close to a matrix with a multiple eigenvalue.

Example 6.1.9 The matrix

$$A = \begin{bmatrix} 3.1482 & -0.2017 & -0.5363 \\ 0.4196 & 0.5171 & 1.0888 \\ 0.3658 & -1.7169 & 3.3361 \end{bmatrix}$$

has a simple eigenvalue $\lambda = 1.9833$ with left and right eigenvectors

$$\mathbf{x} = \begin{bmatrix} 0.4150 \\ 0.6160 \\ 0.6696 \end{bmatrix}, \quad \mathbf{y} = \begin{bmatrix} -7.9435 \\ 83.0701 \\ -70.0066 \end{bmatrix},$$

such that $\mathbf{y}^H\mathbf{x} = 1$. It follows that the condition number of this eigenvalue is $\|\mathbf{x}\|_2\|\mathbf{y}\|_2 = 108.925$. In fact, the nearby matrix

$$B = \begin{bmatrix} 3.1477 & -0.2023 & -0.5366 \\ 0.4187 & 0.5169 & 1.0883 \\ 0.3654 & -1.7176 & 3.3354 \end{bmatrix}$$

has a double eigenvalue that is equal to 2. $\square$

We now consider the sensitivity of *repeated* eigenvalues. First, it is important to note that while the eigenvalues of a matrix A are continuous functions of the entries of A, they are not necessarily differentiable. To see this, we consider the matrix

$$A = \begin{bmatrix} 1 & a \\ \epsilon & 1 \end{bmatrix},$$

where $a > 0$. Computing its characteristic polynomial

$$\det(A - \lambda I) = \lambda^2 - 2\lambda + 1 - a\epsilon$$

and computing its roots yields the eigenvalues $\lambda = 1 \pm \sqrt{a\epsilon}$. Differentiating these eigenvalues with respect to ϵ yields

$$\frac{d\lambda}{d\epsilon} = \pm\frac{1}{2}\sqrt{\frac{a}{\epsilon}},$$

which is undefined at $\epsilon = 0$. In general, an $O(\epsilon)$ perturbation in A causes an $O(\epsilon^{1/p})$ perturbation in an eigenvalue associated with a $p \times p$ Jordan block [Wilkinson

(1965)], meaning that the "more defective" an eigenvalue is, the more sensitive it is.

Exploration 6.1.20 Find a $O(\epsilon)$ perturbation of the matrix J from Exploration 6.1.12 that has eigenvalues $\lambda_1, \lambda_2, \ldots, \lambda_n$ that all satisfy $|\lambda_i - \lambda| = O(\epsilon^{1/n})$. Check your perturbation in MATLAB.

We now consider the sensitivity of eigenvectors, or, more generally, invariant subspaces of a matrix A, such as a subspace spanned by the first k Schur vectors, which are the first k columns in a matrix Q such that $Q^H A Q$ is upper triangular. Suppose that an $n \times n$ matrix A has the Schur Decomposition

$$ A = QTQ^H, \quad Q = \begin{bmatrix} Q_1 \, Q_2 \end{bmatrix}, \quad T = \begin{bmatrix} T_{11} & T_{12} \\ 0 & T_{22} \end{bmatrix}, $$

where Q_1 is $n \times r$ and T_{11} is $r \times r$. We define the **separation** between the matrices T_{11} and T_{22} by

$$ \mathrm{sep}(T_{11}, T_{22}) = \min_{X \neq 0} \frac{\|T_{11}X - XT_{22}\|_F}{\|X\|_F}. $$

It can be shown that an $O(\epsilon)$ perturbation in A causes a $O(\epsilon/\mathrm{sep}(T_{11}, T_{22}))$ perturbation in the invariant subspace Q_1 [Stewart (1973)].

We now consider the case where $r = 1$, meaning that Q_1 is actually a vector $\mathbf{q}_1$, that is also an eigenvector, and T_{11} is its corresponding eigenvalue, λ. Then, we have

$$
\begin{aligned}
\mathrm{sep}(\lambda, T_{22}) &= \min_{X \neq 0} \frac{\|\lambda X - X T_{22}\|_F}{\|X\|_F} \\
&= \min_{\|\mathbf{y}\|_2 = 1} \|\mathbf{y}^H (T_{22} - \lambda I)\|_2 \\
&= \min_{\|\mathbf{y}\|_2 = 1} \|(T_{22} - \lambda I)^H \mathbf{y}\|_2 \\
&= \sigma_{\min}(T_{22} - \lambda I),
\end{aligned}
$$

since the Frobenius norm of a vector is equal to its vector 2-norm. Because the smallest singular value indicates the distance to a singular matrix, $\mathrm{sep}(\lambda, T_{22})$ provides a measure of the separation of λ from the other eigenvalues of A. It follows that eigenvectors are more sensitive to perturbation if the corresponding eigenvalues are clustered near one another.

Exploration 6.1.21 Create a random $n \times n$ matrix X in MATLAB; make sure it is invertible. Normalize each column of X so that it is a unit vector in the 2-norm. Then, create two diagonal $n \times n$ matrices D_1 and D_2, such that two of the diagonal entries of D_2 are very close to one another while those of D_1 are reasonably spread out. Then, compute $A_i = X D_i X_i^{-1}$ for $i = 1, 2$. Next, compute $\tilde{A}_i = A_i + \epsilon E$, where $\|E\|_2 = 1$. How do the eigenvectors of A_i compare to those of $\tilde{A}_i$?

It should be emphasized that there is no direct relationship between the sensitivity of an eigenvalue and the sensitivity of its corresponding invariant subspace. The sensitivity of a simple eigenvalue depends on the angle between its left and right eigenvectors, while the sensitivity of an invariant subspace depends on the clustering of the eigenvalues. Therefore, a sensitive eigenvalue, that is nearly defective, can be associated with an insensitive invariant subspace, if it is distant from other eigenvalues, while an insensitive eigenvalue can have a sensitive invariant subspace if it is very close to other eigenvalues.

6.1.5 Concept Check

1. How does MATLAB use methods for computing eigenvalues to compute the roots of a given polynomial?
2. What is an invariant subspace?
3. What is a similarity transformation, and why is it helpful for computing eigenvalues?
4. For what classes of square matrices are eigenvalues easily obtained?
5. What is deflation?
6. What is the Schur Decomposition?
7. What does it mean for a matrix to be diagonalizable?
8. What are the two types of multiplicities of an eigenvalue? How do they relate to one another?
9. What does it mean for an eigenvalue to be simple? Defective?
10. What is the Jordan Canonical Form, and why is it not practical to compute numerically?
11. State the Gershgorin Circle Theorem.
12. What factors influence the sensitivity of an eigenvalue? An eigenvector?

6.2 Power Iterations

Now that we have an understanding of the eigenvalue problem, in terms of existence, uniqueness and sensitivity of solutions, we can now turn our attention to the development of algorithms for its numerical solution for a given square matrix A. We have seen that if we can compute a single eigenvalue-eigenvector pair of A, then we can use deflation to bring A closer to triangular form. We therefore seek to compute such a pair, known as an **eigenpair**. For the remainder of this chapter, we assume that A is real.

6.2.1 The Power Method

For simplicity, we assume that A has eigenvalues $\lambda_1, \ldots, \lambda_n$ such that

$$|\lambda_1| > |\lambda_2| \geq |\lambda_3| \geq \cdots \geq |\lambda_n|.$$

We also assume that A is diagonalizable, meaning that it has n linearly independent eigenvectors $\mathbf{x}_1, \ldots, \mathbf{x}_n$ such that $A\mathbf{x}_i = \lambda_i \mathbf{x}_i$ for $i = 1, \ldots, n$.

Suppose that we continually multiply a given nonzero vector $\mathbf{x}^{(0)}$ by A, generating a sequence of vectors $\mathbf{x}^{(1)}, \mathbf{x}^{(2)}, \ldots$ defined by

$$\mathbf{x}^{(k)} = A\mathbf{x}^{(k-1)} = A^k \mathbf{x}^{(0)}, \quad k = 1, 2, \ldots.$$

Because A is diagonalizable, any vector in $\mathbb{R}^n$ is a linear combination of the eigenvectors, and therefore we can write

$$\mathbf{x}^{(0)} = c_1 \mathbf{x}_1 + c_2 \mathbf{x}_2 + \cdots + c_n \mathbf{x}_n.$$

We then have

$$\begin{aligned}
\mathbf{x}^{(k)} &= A^k \mathbf{x}^{(0)} \\
&= \sum_{i=1}^n c_i A^k \mathbf{x}_i \\
&= \sum_{i=1}^n c_i \lambda_i^k \mathbf{x}_i \\
&= \lambda_1^k \left[c_1 \mathbf{x}_1 + \sum_{i=2}^n c_i \left(\frac{\lambda_i}{\lambda_1} \right)^k \mathbf{x}_i \right].
\end{aligned}$$

Because $|\lambda_1| > |\lambda_i|$ for $i = 2, \ldots, n$, it follows that if we then normalize each $\mathbf{x}^{(k)}$, then the coefficients of $\mathbf{x}_i$, for $i = 2, \ldots, n$, converge to zero as $k \to \infty$. Therefore, the direction of $\mathbf{x}^{(k)}$ converges to that of $\mathbf{x}_1$. This leads to the most basic method of computing an eigenvalue and eigenvector, called the **Power Method**:

Algorithm 6.2.1 (Power Method) Given an $n \times n$ matrix A, with $|\lambda_1| > |\lambda_2|$, the following algorithm computes an eigenvector of A corresponding to λ_1.

Choose an initial vector $\mathbf{q}_0$ such that $\|\mathbf{q}_0\|_2 = 1$
for $k = 1, 2, \ldots$ **do** until convergence
    $\mathbf{z}_k = A\mathbf{q}_{k-1}$
    $\mathbf{q}_k = \mathbf{z}_k / \|\mathbf{z}_k\|_2$
end for

This algorithm continues until $\mathbf{q}_k$ converges to within some tolerance. If it converges, the limit is a unit vector that is a scalar multiple of $\mathbf{x}_1$, an eigenvector corresponding to the largest eigenvalue, λ_1. The rate of convergence is $|\lambda_1/\lambda_2|$, meaning that the distance between the subspaces span$\{\mathbf{q}_k\}$ and span$\{\mathbf{x}_1\}$ decreases by roughly this factor from iteration to iteration.

It follows that convergence can be slow if λ_2 is almost as large as λ_1, and in fact, the Power Method fails to converge if $|\lambda_2| = |\lambda_1|$, but $\lambda_2 \neq \lambda_1$ (for example, if they have opposite signs or are complex conjugates). It is worth noting the implementation detail that if λ_1 is negative, for example, it may appear that $\mathbf{q}_k$

"flip-flops" between two vectors. This is remedied by normalizing $\mathbf{q}_k$ so that it is not only a unit vector, but also has, for example, a positive first component.

> **Exploration 6.2.1** Write a MATLAB function
>
> $$[\text{q,e}]=\text{powermethod(A)}$$
>
> that implements Algorithm 6.2.1, noting the above implementation detail for the case of a negative λ_1. The output q is the eigenvector corresponding to the largest eigenvalue of A, and e is the corresponding eigenvalue. Use an approach similar to that of Exploration 6.1.21 to create a matrix for which the Power Method would fail. How does your implementation behave?

Once an eigenvector $\mathbf{x}_1$ is found, the corresponding eigenvalue λ_1 can be computed using a **Rayleigh quotient**

$$r(\mathbf{x}_1, A) \equiv \frac{\mathbf{x}_1^T A \mathbf{x}_1}{\mathbf{x}_1^T \mathbf{x}_1}. \tag{6.4}$$

Then, **deflation** can be carried out by constructing a Householder reflection P_1 so that $P_1 \mathbf{x}_1$ is a multiple of $\mathbf{e}_1$, as discussed in Section 6.1.2, and then $P_1 A P_1$ is a matrix with block upper triangular structure. This decouples the problem of computing the eigenvalues of A into the (solved) problem of computing λ_1, and that of computing the eigenvalues of the lower right $(n-1) \times (n-1)$ submatrix.

> **Exploration 6.2.2** Write a MATLAB function
>
> $$[\text{Q,T}]=\text{powerschur(A)}$$
>
> that accepts as input an $n \times n$ matrix A, that is assumed to have distinct, real eigenvalues, and returns as output matrices Q and T such that A = Q * T * Q' is the Schur Decomposition of A. Use your functions powermethod and deflate from Explorations 6.2.1 and 6.1.7, respectively.

The Power Method is particularly useful when applied to a **right stochastic matrix**, which is an $n \times n$ matrix P in which each entry p_{ij} is the probability of a transition from state i to state j. Because the entries are probabilities, it follows that

$$\sum_{j=1}^{n} p_{ij} = 1. \tag{6.5}$$

That is, all of the row sums must equal one. Equivalently, $P\mathbf{e} = \mathbf{e}$, where $\mathbf{e}$ is a vector with all entries equal to one. It follows that $\mathbf{e}$ is an eigenvector of P, with corresponding eigenvalue one. From (6.5), we also have $\|P\|_\infty = 1$. As shown in Section B.12, this implies that $\rho(P) \leq 1$, so the largest eigenvalue of P is equal to one.

The left eigenvector of P with eigenvalue one, that is also an eigenvector of P^T, is a vector in which the jth entry is the long-term probability of transitioning to state j. We can try to compute this vector using the Power Method.

Exploration 6.2.3 Use the `rand` function to create a random stochastic matrix P and use your `powermethod` function from Exploration 6.2.1 to compute the eigenvector of P^T corresponding to the largest eigenvalue of one.

Exploration 6.2.4 Use the `sprand` and `blkdiag` functions to create a block diagonal random stochastic matrix P with 10% of the entries nonzero, and use your `powermethod` function from Exploration 6.2.1 to compute the eigenvector of P^T corresponding to the largest eigenvalue of one. What happens?

Exploration 6.2.5 Repeat Exploration 6.2.4, but with matrix

$$\tilde{P} = \alpha P + (1 - \alpha)\mathbf{e}\mathbf{v}^T,$$

where α is a *damping factor* in $[0, 1)$, $\mathbf{e}$ is a vector of all ones, and $\mathbf{v}$ is a random vector of probabilities, with $\sum_{i=1}^{n} v_i = 1$. What happens this time?

Explorations 6.2.3-6.2.5 illustrate the main idea behind *PageRank*, an algorithm used by Google to sort search results.

6.2.2   Orthogonal Iteration

Computing all of the eigenvalues of A by using the Power Method to compute the largest one, and then deflating, can be impractical, even if A has distinct eigenvalues. This is due to the contamination of smaller eigenvalues by roundoff error from computing the larger ones and then deflating. An alternative is to compute several eigenvalues "at once" by using a *block* version of the Power Method, called **Orthogonal Iteration**. In this method, A is multiplied by an $n \times r$ matrix, with $r > 1$, and then the normalization of the vector computed by the Power Method is generalized to the *orthogonalization* of the block, through the "thin" QR Factorization discussed in Section 4.2.3. The method is as follows:

Algorithm 6.2.2 (Orthogonal Iteration) Given $A \in \mathbb{R}^{n \times n}$ and a positive integer r, the following algorithm computes an invariant subspace of A of dimension r.

Choose an $n \times r$ matrix U_0 such that $U_0^T U_0 = I_r$
for $k = 1, 2, \ldots$ **do** until convergence
    $Z_k = AU_{k-1}$
    $Z_k = U_k R_k$ ("thin" QR Factorization)
end for

Generally, this method computes a convergent sequence $\{U_k\}$, as long as U_0 is not deficient in the directions of certain eigenvectors of A^T, and $|\lambda_r| > |\lambda_{r+1}|$. From the relationship

$$R_k = U_k^T Z_k = U_k^T A U_{k-1},$$

we see that if U_k converges to a matrix U, then $U^T A U = R$ is upper triangular, and because $AU = UR$, the columns of U span an invariant subspace. Let $U^\perp \in \mathbb{R}^{n \times (n-r)}$ be a matrix whose columns span $(\text{range}(U))^\perp$, the **orthogonal complement** (see Definition 4.2.4) of $\text{range}(U)$. Then we have

$$\begin{bmatrix} U^T \\ (U^\perp)^T \end{bmatrix} A \begin{bmatrix} U & U^\perp \end{bmatrix} = \begin{bmatrix} U^T A U & U^T A U^\perp \\ (U^\perp)^T A U & (U^\perp)^T A U^\perp \end{bmatrix}$$

$$= \begin{bmatrix} R & U^T A U^\perp \\ 0 & (U^\perp)^T A U^\perp \end{bmatrix}.$$

That is, $\lambda(A) = \lambda(R) \cup \lambda((U^\perp)^T A U^\perp)$, and because R is upper triangular, the eigenvalues of R are its diagonal entries. We conclude that Orthogonal Iteration, when it converges, yields the largest r eigenvalues of A.

If we let $r = n$, and if the eigenvalues of A are separated in magnitude, then generally Orthogonal Iteration converges, yielding the Schur Decomposition of A, $A = UTU^T$. However, this convergence is generally quite slow. Before determining how convergence can be accelerated, we examine this instance of Orthogonal Iteration more closely. For each integer k, we define $T_k = U_k^T A U_k$. Then, from Algorithm 6.2.2, we have

$$T_{k-1} = U_{k-1}^T A U_{k-1} = U_{k-1}^T Z_k = (U_{k-1}^T U_k) R_k = Q_k R_k,$$

where $Q_k = U_{k-1}^T U_k$. Furthermore,

$$\begin{aligned} T_k &= U_k^T A U_k \\ &= U_k^T A U_{k-1} U_{k-1}^T U_k \\ &= U_k^T Z_k U_{k-1}^T U_k \\ &= U_k^T U_k R_k (U_{k-1}^T U_k) \\ &= R_k Q_k. \end{aligned}$$

That is, T_k is obtained from T_{k-1} by computing the QR Factorization of T_{k-1}, and then multiplying the factors in reverse order. Equivalently, T_k is obtained by applying an orthogonal similarity transformation to T_{k-1}, as

$$T_k = R_k Q_k = Q_k^T T_{k-1} Q_k.$$

If Orthogonal Iteration converges, then T_k converges to an upper triangular matrix $T = U^T A U$ whose diagonal entries are the eigenvalues of A. This simple process of repeatedly computing the QR Factorization, and multiplying the factors in reverse order, is called the **QR Iteration**, which proceeds as follows:

Algorithm 6.2.3 (QR Iteration) Given $A \in \mathbb{R}^{n \times n}$, the following algorithm computes the Schur Decomposition $A = QTQ^T$ of A.

Choose Q_0 so that $Q_0^T Q_0 = I_n$
$T_0 = Q_0^T A Q_0$
for $k = 1, 2, \ldots$ **do** until convergence
    $T_{k-1} = Q_k R_k$ (QR Factorization)
    $T_k = R_k Q_k$
end for

This version of Orthogonal Iteration serves as the cornerstone of an efficient algorithm for computing all of the eigenvalues of a matrix. Unfortunately, as described, QR Iteration is prohibitively expensive, because $O(n^3)$ operations are required in *each* iteration to compute the QR Factorization of T_{k-1}, and typically, many iterations are needed to obtain convergence. However, we will see that with a judicious choice of Q_0, the amount of computational effort can be drastically reduced.

Exploration 6.2.6 Write a MATLAB function `[Q,T]=qriter(A)` that implements Algorithm 6.2.3. What happens when it is applied to a random matrix? Try different kinds of matrices: symmetric or unsymmetric, matrices with real or complex eigenvalues, etc.

If A is a real matrix with complex eigenvalues, then QR Iteration will *not* converge, due to distinct eigenvalues having equal magnitude. However, the *structure* of the matrix T_k in QR Iteration generally will converge to "quasi-upper triangular" form, with 1×1 or 2×2 diagonal blocks corresponding to real eigenvalues or complex-conjugate pairs of eigenvalues, respectively. It is this type of convergence that we will seek in our continued development of the QR Iteration.

6.2.3 Concept Check

1. What is the Power Method? What does it compute?
2. How rapidly does the Power Method converge? When can it fail?
3. What is Orthogonal Iteration?
4. What is QR Iteration and how does it relate to the Power Method?
5. In what sense does QR Iteration converge when applied to a real matrix that has complex eigenvalues?

6.3 The QR Algorithm*

The QR Iteration from Algorithm 6.2.3 is our first attempt at an algorithm that computes all of the eigenvalues of a general square matrix. While it will accomplish this task in most cases, it is far from a practical algorithm. In this section, we use the QR Iteration as a starting point for a much more efficient algorithm, known as the **QR Algorithm**. This algorithm, developed by Francis [Francis (1961)], was named one of the "Top Ten Algorithms of the [20th] Century" [Cipra (2000)].

6.3.1 The Real Schur Form

Let $A \in \mathbb{R}^{n \times n}$. A may have complex eigenvalues, which must occur in complex-conjugate pairs. It is preferable that complex arithmetic be avoided when using QR Iteration to obtain the Schur Decomposition of A. However, in the algorithm for QR Iteration, if the matrix Q_0 used to compute $T_0 = Q_0^T A Q_0$ is real, then every matrix T_k generated by the iteration will also be real, so it will not be possible to obtain the Schur Decomposition.

We compromise by instead seeking the **Real Schur Decomposition** $A = QTQ^T$, introduced in the Section 6.1.2, where Q is a real, orthogonal matrix and T is a real, *quasi-upper triangular* matrix that has a *block* upper triangular structure

$$
T = \begin{bmatrix} T_{11} & T_{12} & \cdots & T_{1p} \\ 0 & T_{22} & \ddots & \vdots \\ 0 & & \ddots & \ddots & \vdots \\ 0 & 0 & 0 & T_{pp} \end{bmatrix},
$$

where each diagonal block T_{ii} is 1×1, corresponding to a real eigenvalue, or a 2×2 block, corresponding to a pair of complex eigenvalues that are conjugates of one another. If QR Iteration is applied to such a matrix, then the sequence $\{T_k\}$ will not converge to upper triangular form, but in most cases a block upper triangular structure will be obtained, which can then be used to compute all of the eigenvalues. Therefore, the iteration can be terminated when appropriate entries below the diagonal have been made sufficiently small.

6.3.2 Hessenberg Reduction

One significant drawback to QR Iteration is that each iteration is too expensive, as it requires $O(n^3)$ operations to compute the QR Factorization, and to multiply the factors in reverse order. Therefore, it is desirable to first use a similarity transformation $H = U^T A U$ to reduce A to a form for which the QR Factorization and matrix multiplication can be performed more efficiently. Certainly, a matrix H that is already "almost" upper triangular, in some sense, is desirable.

> **Exploration 6.3.1** Suppose we use a sequence of Householder reflections $P_1, P_2, \ldots, P_{n-1}$ to compute the QR Factorization $A = QR$ of A, and then let $H = Q^T A Q = RQ$. Why is this not a good choice for H? Explain in terms of the effect of these Householder reflections on the structure of A, particularly which rows or columns are affected by each transformation.

Now, suppose that instead of using the matrix Q that reduces A to upper triangular form, the first transformation P_1 is designed to zero all entries of the first column except the first *two*.

> **Exploration 6.3.2** Construct a random matrix $n \times n$ A, and a Householder reflection P_1 that zeros the last $n - 2$ entries of $A\mathbf{e}_1$. What is the structure of $P_1^T A P_1$? Explain why this transformation results in a much more favorable structure than a Householder reflection used in the QR Factorization of A, in terms of the rows and columns that are affected.

Continuing this reasoning with subsequent columns of A, we see that a sequence of orthogonal transformations can be used to reduce A to an **upper Hessenberg** matrix H, in which $h_{ij} = 0$ whenever $i > j + 1$. That is, all entries below the first *subdiagonal* are equal to zero.

Reduction to Hessenberg form can be accomplished by performing a sequence of Householder reflections $U = P_1 P_2 \cdots P_{n-2}$ on the columns of A, as in the following algorithm.

> **Algorithm 6.3.1 (Hessenberg Reduction)** Given an $n \times n$ real matrix A, the following algorithm computes an orthogonal matrix U and Hessenberg matrix H such that $H = U^T A U$.
>
> $U = I$, $H = A$
> **for** $j = 1, 2, \ldots, n - 2$ **do**
> $\quad \mathbf{v} = \text{house}(H(j + 1 : n, j))$, $c = 2/\mathbf{v}^T\mathbf{v}$
> $\quad H(j + 1 : n, j : n) = H(j + 1 : n, j : n) - c\mathbf{v}\mathbf{v}^T H(j + 1 : n, j : n)$
> $\quad H(1 : n, j + 1 : n) = H(1 : n, j + 1 : n) - cH(1 : n, j + 1 : n)\mathbf{v}\mathbf{v}^T$
> $\quad U(1 : n, j + 1 : n) = U(1 : n, j + 1 : n) - cU(1 : n, j + 1 : n)\mathbf{v}\mathbf{v}^T$
> **end for**

The function $\text{house}(\mathbf{x})$, described in Section 4.2.2, computes a vector $\mathbf{v}$ such that $P\mathbf{x} = I - c\mathbf{v}\mathbf{v}^T\mathbf{x} = \alpha\mathbf{e}_1$, where $c = 2/\mathbf{v}^T\mathbf{v}$ and $\alpha = \pm\|\mathbf{x}\|_2$. The algorithm for the Hessenberg reduction requires $O(n^3)$ operations, but it is performed only once, before the QR Iteration begins, so it still leads to a substantial reduction in the total number of operations that must be performed to compute the Real Schur Decomposition.

> **Exploration 6.3.3** Write a MATLAB function [U,H]=hessreduce(A) that implements Algorithm 6.3.1. If only one output argument is specified (as determined by checking the value of **nargout**), only return the Hessenberg matrix H, and do not accumulate the Householder reflections to obtain U. Comment on the difference in execution time between the cases in which U is computed, or not.

6.3.3 Hessenberg QR

It is particularly efficient to compute the QR Factorization of an upper Hessenberg, or simply Hessenberg, matrix, because it is only necessary to zero one entry in each column. Therefore, it can be accomplished with a sequence of $n-1$ Givens row rotations (see Section 4.2.1), which requires only $O(n^2)$ operations. Then, these same Givens rotations can be applied, in the same order, to the columns in order to complete the similarity transformation, or, equivalently, accomplish the task of multiplying the factors of the QR Factorization in reverse order.

More precisely, given a Hessenberg matrix H, we apply Givens row rotations $G_1^T, G_2^T, \ldots, G_{n-1}^T$ to H, where G_i^T rotates rows i and $i+1$, to obtain

$$G_{n-1}^T \cdots G_2^T G_1^T H = (G_1 G_2 \cdots G_{n-1})^T H = Q^T H = R,$$

where R is upper triangular. Then, we compute

$$\tilde{H} = Q^T H Q = RQ = R G_1 G_2 \cdots G_{n-1}$$

by applying column rotations to R, to obtain a new matrix $\tilde{H}$.

By considering which rows or columns the Givens rotations affect, it can be shown that Q is Hessenberg, and therefore $\tilde{H} = RQ$ is Hessenberg as well. The process of applying an orthogonal similarity transformation to a Hessenberg matrix to obtain a new Hessenberg matrix with the same eigenvalues that, hopefully, is closer to quasi-upper triangular form is called a **Hessenberg QR step**. The following algorithm performs a Hessenberg QR step, and also computes Q as a product of Givens column rotations, which is only necessary if the entire Real Schur Decomposition of A is required, as opposed to only the eigenvalues.

Algorithm 6.3.2 (Hessenberg QR Step) Given an $n \times n$ real Hessenberg matrix A, this algorithm overwrites H with $Q^T H Q$, where $H = QR$ is the QR Factorization of H.

for $j = 1, 2, \ldots, n-1$ **do**
$\quad [c,s] = \texttt{givens}(h_{jj}, h_{j+1,j})$
$\quad G_j = \begin{bmatrix} c & -s \\ s & c \end{bmatrix}$
$\quad H(j:j+1, j:n) = G_j^T H(j:j+1, :)$
end for
$Q = I$
for $j = 1, 2, \ldots, n-1$ **do**
$\quad H(1:j+1, j:j+1) = H(1:j+1, j:j+1) G_j$
$\quad Q(1:j+1, j:j+1) = Q(1:j+1, j:j+1) G_j$
end for

The function $\texttt{givens}(a, b)$, first described in Section 4.2.1, returns c and s such that

$$\begin{bmatrix} c & -s \\ s & c \end{bmatrix}^T \begin{bmatrix} a \\ b \end{bmatrix} = \begin{bmatrix} r \\ 0 \end{bmatrix}, \quad r = \sqrt{a^2 + b^2}.$$

Note that when performing row rotations, it is only necessary to update certain columns, and when performing column rotations, it is only necessary to update certain rows, because of the structure of the matrix at the time the rotation is performed; for example, after the first loop, H is upper triangular.

Exploration 6.3.4 Write a MATLAB function

$$[Q,H]=\text{hessqrstep(H)}$$

that implements Algorithm 6.3.2. If only one output argument is specified (as determined by checking the value of `nargout`), only return the updated Hessenberg matrix H, and do not accumulate the Givens rotations to compute Q. Comment on the difference in execution time between the cases in which Q is computed, or not.

Exploration 6.3.5 Update your function `qriter` from Exploration 6.2.6 so that it first calls your function `hessreduce` from Exploration 6.3.3 to reduce the input matrix A to a Hessenberg matrix H, and then calls your function `hessqrstep` from Exploration 6.3.4 to compute the QR Factorization of H and multiply the factors in reverse order to update H. Compare the performance of this version of `qriter`, in terms of both execution time and number of iterations, to the previous version from Exploration 6.2.6.

Exploration 6.3.6 Explain why, in Algorithm 6.3.2, it is necessary to first apply all Givens row rotations $G_1, G_2, \ldots, G_{n-1}$, and then apply all Givens column rotations, instead of applying G_1 on the left and right, and then G_2 on the left and right, and so on.

If a subdiagonal entry $h_{j+1,j}$ of a Hessenberg matrix H is equal to zero, then the problem of computing the eigenvalues of H *decouples* into two smaller problems of computing the eigenvalues of H_{11} and H_{22}, where

$$H = \begin{bmatrix} H_{11} & H_{12} \\ 0 & H_{22} \end{bmatrix}$$

and H_{11} is $j \times j$. Therefore, an efficient implementation of the QR Iteration on a Hessenberg matrix H focuses on a submatrix of H that is **unreduced**, meaning that all of its subdiagonal entries are nonzero. It is also important to monitor the subdiagonal entries after each iteration, to determine if any of them have become nearly zero, thus allowing further decoupling. Once no further decoupling is possible, H has been reduced to quasi-upper triangular form and the QR Iteration can terminate.

It is essential to choose an *maximal* unreduced diagonal block of H for applying a Hessenberg QR step. That is, the step must be applied to a submatrix H_{22} of H

that has the structure

$$H = \begin{bmatrix} H_{11} & H_{12} & H_{13} \\ 0 & H_{22} & H_{23} \\ 0 & 0 & H_{33} \end{bmatrix}$$

where H_{22} is unreduced. This condition ensures that the eigenvalues of H_{22} are also eigenvalues of H, as $\lambda(H) = \lambda(H_{11}) \cup \lambda(H_{22}) \cup \lambda(H_{33})$ when H is structured as above. Note that the size of either H_{11} or H_{33} may be 0×0, for example if H itself is unreduced.

The following property of unreduced Hessenberg matrices is useful for improving the efficiency of a Hessenberg QR step.

> **Theorem 6.3.3 (Implicit Q Theorem)** Let $A \in \mathbb{R}^{n \times n}$, and let Q and V be $n \times n$ orthogonal matrices such that $Q^T A Q = H$ and $V^T A V = G$ are both upper Hessenberg, and H is unreduced. If $Q = \begin{bmatrix} \mathbf{q}_1 \cdots \mathbf{q}_n \end{bmatrix}$ and $V = \begin{bmatrix} \mathbf{v}_1 \cdots \mathbf{v}_n \end{bmatrix}$, and if $\mathbf{q}_1 = \mathbf{v}_1$, then $\mathbf{q}_i = \pm \mathbf{v}_i$ for $i = 2, \ldots, n$, and $|h_{ij}| = |g_{ij}|$ for $i, j = 1, 2, \ldots, n$.

That is, if two orthogonal similarity transformations that reduce A to Hessenberg form have the same first column, then they are "essentially equal", as are the Hessenberg matrices.

> **Exploration 6.3.7** Prove the Theorem 6.3.3, the Implicit Q Theorem, by showing that the matrix $W = V^T Q$ is both orthogonal and upper triangular. How is W related to both G and H?

> **Exploration 6.3.8** Explain why the assumption that H is unreduced is essential for the proof of the Implicit Q Theorem.

The following explorations highlight another useful property of unreduced Hessenberg matrices, that will prove to be essential for us in making the QR Iteration more efficient.

> **Exploration 6.3.9** Let $H \in \mathbb{R}^{n \times n}$ be an unreduced Hessenberg matrix. Prove by induction that the first $n - 1$ columns of H are linearly independent.

In Exercise 2 of Chapter 4, it was shown that a matrix with full column rank has a unique QR Factorization. A similar approach can be used to show uniqueness in the case of an unreduced Hessenberg matrix, whether it has full column rank or not.

> **Exploration 6.3.10** Use the result of Exploration 6.3.9 to prove that the QR Factorization of an unreduced Hessenberg matrix is unique.

It follows that during QR Iteration, as long as we work with unreduced Hessenberg matrices, there is no need to perform QR with column pivoting, as in Section 4.3.1– simply carrying out Givens QR, as described in Algorithm 6.3.2, is sufficient.

6.3.4 Shifted QR Iteration

The efficiency of the QR Iteration for computing the eigenvalues of an $n \times n$ matrix A is significantly improved by first reducing A to a Hessenberg matrix H, so that only $O(n^2)$ operations per iteration are required, instead of $O(n^3)$. However, the iteration can still converges very slowly, so additional modifications are needed to make the QR Iteration a practical algorithm for computing the eigenvalues of a general matrix.

6.3.4.1 *Single Shift Strategy*

In general, the pth subdiagonal entry of H converges to zero at the rate

$$\left| \frac{\lambda_{p+1}}{\lambda_p} \right|,$$

where λ_p is the pth largest eigenvalue of A in magnitude. It follows that convergence can be particularly slow if eigenvalues are very close to one another in magnitude. Suppose that we *shift* H by a scalar μ, meaning that we compute the QR Factorization of $H - \mu I$ instead of H, and then update H to obtain a new Hessenberg $\tilde{H}$ by multiplying the QR factors in reverse order as before, but then adding μI. The algorithm is as follows.

Algorithm 6.3.4 (Shifted QR Iteration) Given a $A \in \mathbb{R}^{n \times n}$, the following algorithm computes the Real Schur Decomposition $A = QTQ^T$ of A.

Use Algorithm 6.3.1 to compute Q_0 so that $H_1 = Q_0^T A Q_0$ is upper
    Hessenberg
for $k = 1, 2, \ldots$ **do** until convergent
    Choose a shift μ_k
    $H_k - \mu_k I = Q_k R_k$ (QR Factorization)
    $H_{k+1} = R_k Q_k + \mu_k I$
end for

During each iteration, we have

$$\begin{aligned}
H_{k+1} &= R_k Q_k + \mu_k I \\
&= Q_k^T (H_k - \mu_k I) Q_k + \mu_k I \\
&= Q_k^T H_k Q_k - \mu_k Q_k^T Q_k + \mu_k I \\
&= Q_k^T H_k Q_k - \mu_k I + \mu_k I \\
&= Q_k^T H_k Q_k.
\end{aligned} \tag{6.6}$$

So, we are still performing an orthogonal similarity transformation of H_k, but with a different Q_k. It follows that the convergence rate becomes $|\lambda_{p+1} - \mu_k|/|\lambda_p - \mu_k|$. Therefore, if μ_k is close to an eigenvalue, convergence of a particular subdiagonal entry will be much more rapid.

> **Exploration 6.3.11** Suppose that the shifted Hessenberg QR step (6.6) is performed, with μ_k an exact eigenvalue of H_k. Prove that if H_k is unreduced, then $[H_{k+1}]_{n,n-1} = 0$; that is, decoupling occurs in just one iteration.

If μ_k is not an eigenvalue of H_k, but is still close to an eigenvalue, then $H_k - \mu_k I$ is nearly singular, which means that its columns are nearly linearly dependent. It follows that $[R_k]_{nn}$ is small, and it can be shown that $[H_{k+1}]_{n,n-1}$ is also small, and $[H_{k+1}]_{nn} \approx \mu_k$. Therefore, the problem is nearly decoupled, and μ_k is revealed by the structure of H_{k+1} as an approximate eigenvalue of H_k, and therefore of A as well.

This suggests the **single shift strategy**: using $[H_k]_{nn}$ as the shift μ_k during each iteration, because if $[H_k]_{n,n-1}$ is small compared to $[H_k]_{nn}$, then this choice of shift will drive $[H_k]_{n,n-1}$ toward zero. In fact, it can be shown that this strategy generally causes $[H_k]_{n,n-1}$ to converge to zero *quadratically*, meaning that only a few similarity transformations are needed to achieve decoupling. This improvement over the linear convergence rate reported earlier is due to the changing of the shift during each step.

Example 6.3.5 This example is adapted from one found in [Stewart (1973)]. Consider the 2×2 matrix

$$H_k = \begin{bmatrix} a & b \\ \epsilon & 0 \end{bmatrix}, \quad \epsilon > 0,$$

that arises naturally when using $[H_k]_{nn}$ as a shift. To compute the QR Factorization of H_k, we perform a single Givens rotation to obtain $H_k = G_k R_k$, where

$$G_k = \begin{bmatrix} c & -s \\ s & c \end{bmatrix}, \quad c = \frac{a}{\sqrt{a^2 + \epsilon^2}}, \quad s = \frac{\epsilon}{\sqrt{a^2 + \epsilon^2}}.$$

Performing the similarity transformation $H_{k+1} = G_k^T H_k G_k$ yields

$$
\begin{aligned}
H_{k+1} &= \begin{bmatrix} c & s \\ -s & c \end{bmatrix} \begin{bmatrix} a & b \\ \epsilon & 0 \end{bmatrix} \begin{bmatrix} c & -s \\ s & c \end{bmatrix} \\
&= \begin{bmatrix} ac^2 + bcs + \epsilon cs & bc^2 - acs - \epsilon s^2 \\ -acs - bs^2 + \epsilon c^2 & -bcs + as^2 - \epsilon cs \end{bmatrix} \\
&= \begin{bmatrix} a + bcs & bc^2 - \epsilon \\ -bs^2 & -bcs \end{bmatrix}.
\end{aligned}
$$

We see that the one subdiagonal entry is

$$-bs^2 = -b\frac{\epsilon^2}{\epsilon^2 + a^2},$$

compared to the original entry ϵ. It follows that if ϵ is small compared to a and b, then subsequent QR steps will cause the subdiagonal entry to converge to zero quadratically. For example, if

$$H = \begin{bmatrix} 0.6324 & 0.2785 \\ 0.0975 & 0.5469 \end{bmatrix},$$

then the value of h_{21} after each of the first four QR steps is 0.1574, -0.0038, 2.1072×10^{-5}, and -6.931×10^{-10}. $\square$

Exploration 6.3.12 Update your function `qriter` from Exploration 6.3.5 to incorporate the single shift strategy by carrying out the shifted Hessenberg QR step (6.6) with shift $\mu_k = [H_k]_{nn}$. Compare the performance of your updated function with the version from Exploration 6.3.5, in terms of number of iterations and execution time. Can you observe the quadratic convergence of subdiagonal entries to zero?

Exploration 6.3.13 How should the single shift strategy be adjusted if H_k is *not* an unreduced Hessenberg matrix? Modify your function `qriter` from Exploration 6.3.12 to incorporate this adjustment, and compare the performance of the two versions in terms of number of iterations and execution time.

The following alternative perspective is helpful for understanding why the single-shift strategy yields rapid convergence. We define

$$\hat{Q}_k = Q_1 Q_2 \cdots Q_k, \quad \hat{R}_k = R_k R_{k-1} \cdots R_1.$$

Then it can be shown that

$$\hat{Q}_k \hat{R}_k = (H - \mu_k I)(H - \mu_{k-1}I) \cdots (H - \mu_1 I). \tag{6.7}$$

Exploration 6.3.14 Prove (6.7) by induction.

If QR Iteration converges, the columns of $\hat{Q}_k$ converge to the Schur vectors of H. Furthermore, from (6.7) we obtain

$$\prod_{i=1}^{k}(H - \mu_i I)^T \hat{Q}_k = \hat{R}_k^T.$$

That is, the columns of $\hat{Q}_k$ are the result of k iterations of **Inverse Iteration** applied to H^T, with shifts $\mu_1, \mu_2, \ldots, \mu_k$.

For a general matrix A, Inverse Iteration is the Power Method applied to $(A - \mu I)^{-1}$, which converges to the eigenvector corresponding to the eigenvalue closest to μ. Therefore, the kth QR iteration is one iteration of the Power Method applied to $(H - \mu_k I)^{-T}$. As discussed in Section 6.4.3, Inverse Iteration is very rapidly convergent when used with a shift that is an approximate eigenvalue. It is also worth noting that

$$\mathbf{e}_n^T \hat{Q}_k^T H \hat{Q}_k \mathbf{e}_n = [H_{k+1}]_{nn}.$$

That is, the Rayleigh Quotient of A and the last column of $\hat{Q}_k$ is the shift for the next iteration, which indicates that this last column is converging to the *left* eigenvector corresponding to the eigenvalue to which the shifts are converging.

6.3.4.2 *Double Shift Strategy*

Unfortunately, the single shift strategy is not very effective if H has complex eigenvalues. An alternative is the **double shift strategy**, which is used if the two eigenvalues, μ_1 and μ_2, of the lower-right 2×2 block of H are complex. Then, these two eigenvalues are used as shifts in consecutive iterations to achieve quadratic convergence in the complex case as well. That is, we compute

$$H - \mu_1 I = Q_1 R_1$$

$$H_1 = R_1 Q_1 + \mu_1 I$$

$$H_1 - \mu_2 I = Q_2 R_2$$

$$H_2 = R_2 Q_2 + \mu_2 I. \tag{6.8}$$

Exploration 6.3.15 Update your function `qriter` from Exploration 6.3.12 to incorporate the double shift strategy by carrying out the shifted Hessenberg QR step (6.8) with shifts μ_1, μ_2 equal to the eigenvalues of the lower right 2×2 block of H if they are complex, and still using the single shift strategy (6.6) with $\mu = h_{nn}$ if they are real. Compare the performance of your updated function with the version from Exploration 6.3.12, in terms of number of iterations and execution time. Can you observe the quadratic convergence of subdiagonal entries to zero?

Exploration 6.3.16 Give two reasons why it is highly desirable to implement the double shift strategy in such a way as to avoid complex arithmetic, if possible.

To avoid complex arithmetic when using complex shifts, the **double implicit shift strategy** is used. We first note that because $\mu_1 = a + bi$ and $\mu_2 = a - bi$ are a complex-conjugate pair, it follows that $\mu_1 + \mu_2 = ab$ and $\mu_1 \mu_2 = a^2 + b^2$ are real. Therefore, the matrix

$$M = (H - \mu_2 I)(H - \mu_1 I) = H^2 - (\mu_1 + \mu_2)H + \mu_1 \mu_2 I$$

is real. From (6.7), we then have

$$M = (Q_1 Q_2)(R_2 R_1).$$

That is, $(Q_1 Q_2)(R_2 R_1)$ represents the QR Factorization of a real matrix, even though the individual matrices involved are complex. Furthermore,

$$
\begin{aligned}
H_2 &= R_2 Q_2 + \mu_2 I \\
&= Q_2^T Q_2 R_2 Q_2 + \mu_2 Q_2^T Q_2 \\
&= Q_2^T (Q_2 R_2 + \mu_2 I) Q_2 \\
&= Q_2^T H_1 Q_2 \\
&= Q_2^T (R_1 Q_1 + \mu_1 I) Q_2 \\
&= Q_2^T (Q_1^T Q_1 R_1 Q_1 + \mu_1 Q_1^T Q_1) Q_2 \\
&= Q_2^T Q_1^T (Q_1 R_1 + \mu_1 I) Q_1 Q_2 \\
&= Q_2^T Q_1^T H Q_1 Q_2.
\end{aligned}
$$

That is, Q_1Q_2 is the orthogonal matrix that implements the similarity transformation of H to obtain H_2. Therefore, we could use exclusively real arithmetic by forming M, computing its QR Factorization to obtain $M = ZR$, and then computing $H_2 = Z^T H Z$, since $Z = Q_1Q_2$, in view of the uniqueness of the QR Factorization. However, M is computed by squaring H, which requires $O(n^3)$ operations. Thus the efficiency gained by first reducing A to Hessenberg form has been lost.

We can work around this difficulty using the Implicit Q Theorem. Instead of forming M in its entirety, we only form its first column, which, being a second-degree polynomial of a Hessenberg matrix, has only three nonzero entries.

Exploration 6.3.17 Compute $M\mathbf{e}_1$, the first column of M.

We compute a Householder reflection P_0 that makes $M\mathbf{e}_1$ a multiple of $\mathbf{e}_1$. Then, we compute $P_0^T H P_0$, which is no longer Hessenberg, because P_0 operates on the first three rows and columns of H. Finally, we apply a series of Householder reflections $P_1, P_2, \ldots, P_{n-2}$ that restore Hessenberg form.

Because the reflections $P_1, P_2, \ldots, P_{n-2}$ do not affect the first row (when applied on the left) or column (when applied on the right), it follows that if we define $\tilde{Z} = P_0 P_1 P_2 \cdots P_{n-2}$, then Z and $\tilde{Z}$ have the same first column. Since both matrices implement similarity transformations that preserve the Hessenberg form of H, it follows from the Implicit Q Theorem that Z and $\tilde{Z}$ are essentially equal, and that they essentially produce the same updated matrix H_2. This variation of a Hessenberg QR step is called a **Francis QR Step** [Francis (1961)].

Exploration 6.3.18 In general, computing PA, where $A, P \in \mathbb{R}^{n \times n}$ and P is a Householder reflection, requires $O(n^2)$ floating-point operations. Explain why a Francis QR Step requires only $O(n^2)$ operations in total, even though it uses $n - 1$ Householder reflections.

Exploration 6.3.19 Describe, using pseudocode, how a Francis QR Step could be implemented using Givens rotations instead of Householder reflections.

Exploration 6.3.20 Update your function `qriter` from Exploration 6.3.15 so that during each iteration, instead of the previous implementation of the double shift strategy from (6.8), a Francis QR Step is performed as described in the preceding paragraph. Compare the performance of this version with the previous one from Exploration 6.3.15, in terms of execution time and number of iterations.

A Francis QR step requires about $10n^2$ operations, with an additional $10n^2$ operations if orthogonal transformations are being accumulated to obtain the entire Real Schur Decomposition. Generally, the entire QR Algorithm, including the initial reduction to Hessenberg form, requires about $10n^3$ operations, with an additional

$15n^3$ operations to compute the orthogonal matrix Q such that $A = QTQ^T$ is the Real Schur Decomposition of A.

After each QR iteration, it is necessary to check whether a subdiagonal entry of H is sufficiently small so that it can be "declared" to be zero, and therefore the eigenvalue problem can be decoupled. This allows the iteration to continue with smaller matrices, which substantially reduces computational expense because each iteration requires $O(p^2)$ floating-point operations when applied to a $p \times p$ matrix.

> **Exploration 6.3.21** Update your function `qriter` from Exploration 6.3.20 so that during each iteration, opportunities for decoupling are detected, so that the next iteration is applied to a maximal *unreduced* diagonal block of H. Compare the performance of this version with the previous one from Exploration 6.3.20, in terms of execution time and number of iterations.

6.3.5 Concept Check

1. What are two significant drawbacks of QR Iteration, as originally described?
2. What is the benefit of reducing a matrix to Hessenberg form before performing QR Iteration?
3. What is the single shift strategy, and how does it improve QR Iteration?
4. What is the double shift strategy, and why is it needed?
5. What is the Implicit Q Theorem, and how does it help with the efficient computation of eigenvalues?
6. Describe the main idea behind a Francis QR Step.

6.4 The Symmetric Eigenvalue Problem*

The eigenvalue problem for a real, symmetric matrix A, for which $A = A^T$, or a complex, *Hermitian* matrix A, for which $A = A^H$, is a considerable simplification of the eigenvalue problem for a general matrix. In this section, we investigate the ramifications of this simplification for numerical algorithms.

6.4.1 Properties

Consider the Schur Decomposition $A = QTQ^H$, where T is upper triangular. Then, if A is Hermitian, it follows that $T = T^H$. But because T is upper triangular, it follows that T must be diagonal. That is, any symmetric real matrix, or Hermitian complex matrix, is unitarily diagonalizable, as stated previously because A is normal. Furthermore, because the Hermitian transpose includes complex conjugation, T must equal its complex conjugate, which implies that the eigenvalues of A are real, even if A itself is complex.

Because the eigenvalues are real, we can order them. By convention, we prescribe that if A is an $n \times n$ Hermitian matrix, then it has eigenvalues

$$\lambda_1 \geq \lambda_2 \geq \cdots \geq \lambda_n.$$

Furthermore, by the **Courant-Fischer Minimax Theorem**, each of these eigenvalues has the following characterization:

$$\lambda_k = \max_{\dim(S)=k} \min_{\mathbf{y} \in S, \mathbf{y} \neq \mathbf{0}} \frac{\mathbf{y}^H A \mathbf{y}}{\mathbf{y}^H \mathbf{y}}. \tag{6.9}$$

That is, the kth largest eigenvalue of A is equal to the maximum, over all k-dimensional subspaces of $\mathbb{C}^n$, of the minimum value of the **Rayleigh quotient** defined in (6.4),

$$r(\mathbf{y}, A) = \frac{\mathbf{y}^H A \mathbf{y}}{\mathbf{y}^H \mathbf{y}}, \quad \mathbf{y} \neq \mathbf{0},$$

on each subspace. It follows that λ_1, the largest eigenvalue, is the absolute maximum value of the Rayleigh quotient on all of $\mathbb{C}^n$, while λ_n, the smallest eigenvalue, is the absolute minimum value.

Exploration 6.4.1 Let A be a Hermitian matrix. Prove that $\mathbf{y} \in \mathbb{C}^n$ is an eigenvector of A with corresponding eigenvalue λ if and only if $\mathbf{y}$ is a critical point of $r(\mathbf{y}, A)$, with $r(\mathbf{y}, A) = \lambda$.

Exploration 6.4.2 Let A be an $n \times n$ Hermitian matrix with Schur Decomposition $A = QDQ^H$. Use the Schur vectors to prove the Courant-Fischer Minimax Theorem (6.9). *Hint:* Use the fact that any subspace of dimension k must intersect the subspace span$\{\mathbf{q}_k, \ldots, \mathbf{q}_n\}$, which has dimension $n - k + 1$.

For the remainder of this section, we assume that A is a real symmetric matrix.

6.4.2   Perturbation Theory

In the symmetric case, the Gershgorin Disks become "Gershgorin intervals", because the eigenvalues of a symmetric matrix are real.

Example 6.4.1 The eigenvalues of the 3×3 symmetric matrix

$$A = \begin{bmatrix} -10 & -3 & 2 \\ -3 & 4 & -2 \\ 2 & -2 & 14 \end{bmatrix}$$

are

$$\lambda(A) = \{14.6515, 4.0638, -10.7153\}.$$

The Gershgorin intervals are

$$D_1 = \{x \in \mathbb{R} \mid |x - 14| \leq 4\},$$
$$D_2 = \{x \in \mathbb{R} \mid |x - 4| \leq 5\},$$
$$D_3 = \{x \in \mathbb{R} \mid |x + 10| \leq 5\}.$$

We see that each intervals contains one eigenvalue. In general, though, we can only be certain that the eigenvalues are contained in the *union* of these intervals. □

The characterization of the eigenvalues of a symmetric matrix as constrained maxima of the Rayleigh quotient leads to the following results about the eigenvalues of a perturbed symmetric matrix. As the eigenvalues are real, and therefore can be ordered, we denote by $\lambda_i(A)$ the ith largest eigenvalue of A.

Theorem 6.4.2 (Wielandt-Hoffman) If A and $A + E$ are $n \times n$ symmetric matrices, then

$$\sum_{i=1}^{n} (\lambda_i(A + E) - \lambda_i(A))^2 \leq \|E\|_F^2.$$

Exploration 6.4.3 Let $t \in \mathbb{R}$ and define $A(t) = A + tE$, where A and E are both symmetric. Furthermore, let $A(t) = Q(t)D(t)Q(t)^T$ be the Schur Decomposition of $A(t)$, for each $t \in \mathbb{R}$. Show that

$$\sum_{i=1}^{n} (\lambda_i(A + E) - \lambda_i(A))^2 = \|D(1) - D(0)\|_F^2.$$

Exploration 6.4.4 Assume that the matrices $Q(t)$ and $D(t)$ from Exploration 6.4.3 are continuously differentiable functions of t. Show that $D'(t) = \text{diag}(Q(t)^T E Q(t))$. *Hint:* Use the orthogonality of $Q(t)$ in conjunction with differentiation rules.

Exploration 6.4.5 Use the results of Explorations 6.4.3 and 6.4.4, in conjunction with the Fundamental Theorem of Calculus, to prove the Wielandt-Hoffman Theorem 6.4.2.

It is also possible to bound the distance between individual eigenvalues of A and $A + E$.

Theorem 6.4.3 If A and $A + E$ are $n \times n$ symmetric matrices, then

$$\lambda_n(E) \leq \lambda_k(A + E) - \lambda_k(A) \leq \lambda_1(E).$$

Furthermore,

$$|\lambda_k(A + E) - \lambda_k(A)| \leq \|E\|_2.$$

The second inequality in the above theorem follows directly from the first, as the 2-norm of the symmetric matrix E, being equal to its spectral radius, must be equal to the larger of the absolute value of $\lambda_1(E)$ or $\lambda_n(E)$. We leave the proof of the first inequality to the reader as an exercise.

Exploration 6.4.6 Use the Courant-Fischer Minimax Theorem (6.9) to prove the first inequality in Theorem 6.4.3.

For a symmetric matrix, or even a more general normal matrix, the left eigenvectors and right eigenvectors are the same, from which it follows that every simple eigenvalue is "perfectly conditioned"; that is, the condition number $1/|\cos\theta|$, introduced in Section 6.1.4, is equal to 1 because $\theta = 0$ in this case. However, the same results concerning the sensitivity of invariant subspaces from the nonsymmetric case apply in the symmetric case as well: such sensitivity increases as the eigenvalues become more clustered, even though there is no chance of a defective eigenvalue. This is because for a nondefective, repeated eigenvalue, there are infinitely many possible bases of the corresponding invariant subspace. Therefore, as the eigenvalues approach one another, the eigenvectors become more sensitive to small perturbations, for any matrix.

Let Q_1 be an $n \times r$ matrix with orthonormal columns, meaning that $Q_1^T Q_1 = I_r$. If these columns span an invariant subspace of an $n \times n$ symmetric matrix A, then $AQ_1 = Q_1 S$, where $S = Q_1^T A Q_1$. On the other hand, if range(Q_1) is *not* an invariant subspace, but the matrix

$$AQ_1 - Q_1 S = E_1$$

is small for any given $r \times r$ symmetric matrix S, then the columns of Q_1 define an *approximate* invariant subspace.

Exploration 6.4.7 Let A be an $n \times n$ real symmetric matrix, and let Q_1 be an $n \times r$ matrix with orthonormal columns. Use the invariance of the Frobenius norm under orthogonal transformations to prove that $\|AQ_1 - Q_1 S\|_F$ is minimized by choosing $S = Q_1^T A Q_1$.

Exploration 6.4.8 Let A and Q_1 be defined as in Exploration 6.4.7. Use your proof of the result in Exploration 6.4.7 to show that if $S = Q_1^T A Q_1$ has the Schur Decomposition $S = U\Theta U^T$, where $\Theta = \text{diag}(\theta_1, \ldots, \theta_r)$, and if $Q_1 U = \begin{bmatrix} \mathbf{y}_1 & \cdots & \mathbf{y}_r \end{bmatrix}$, then

$$\|A\mathbf{y}_k - \theta_k \mathbf{y}_k\|_2 \leq \|(I - Q_1 Q_1^T)AQ_1\|_2, \quad k = 1, 2, \ldots, r.$$

From Exploration 6.4.8, we see that if $\|E_1\|_2$ is small, then r eigenvalues of A are close to the eigenvalues of S, which are known as **Ritz values**, while the corresponding eigenvectors are called **Ritz vectors**. If $(\theta_k, \mathbf{u}_k)$ is an eigenvalue-eigenvector pair, or an **eigenpair** of S, then, because S is defined by $S = Q_1^T A Q_1$, $(\theta_k, \mathbf{y}_k)$, where $\mathbf{y}_k = Q_1 \mathbf{u}_k$, is also known as a **Ritz pair**.

An approximate invariant subspace of particular interest is when the columns of Q_1 form an orthonormal basis for a **Krylov subspace**

$$\mathcal{K}(\mathbf{q}_1, A, k) = \text{span}\{\mathbf{q}_1, A\mathbf{q}_1, A^2\mathbf{q}_1, \ldots, A^{k-1}\mathbf{q}_1\},$$

previously introduced in Section 5.2. The columns of Q_1 can be generated by the **Lanczos Iteration** presented in that section. We then have

$$AQ_1 = Q_1 T_k + \beta_k \mathbf{q}_{k+1} \mathbf{e}_k^T,$$

where T_k is as defined in (5.9). It follows from the proof of the result of Exploration 6.4.8 that if $T_k = U \Theta U^T$ is the Schur Decomposition of T_k, then the Ritz pairs satisfy

$$\|A\mathbf{y}_j - \theta_j \mathbf{y}_j\|_2 \leq |\beta_k||u_{kj}|, \quad j = 1, 2, \ldots, k.$$

We conclude that we can obtain an approximate invariant subspace by performing Lanczos Iteration on A with initial vector $\mathbf{q}_1$ until β_k is sufficiently small.

The extremal Ritz values produced by Lanczos Iteration tend to converge rapidly to the extremal eigenvalues of A, compared to the approximate eigenvalues obtained from the Power Method. An overview of the convergence behavior of Lanczos Iteration can be found in [Kaniel (1966); Paige (1971)].

6.4.3 Rayleigh Quotient Iteration

The Power Method, when applied to a symmetric matrix to obtain its largest eigenvalue, is more effective than for a general matrix: the rate of convergence of the approximations of λ_1 is $|\lambda_2/\lambda_1|^2$, meaning convergence is generally much more rapid than for an unsymmetric matrix with the same eigenvalues. A proof of this can be found in [Golub and van Loan (2012)].

Let A be an $n \times n$ symmetric matrix. Even more rapid convergence can be obtained if we consider a variation of the Power Method. **Inverse Iteration** is the Power Method applied to $(A - \mu I)^{-1}$. Let A have eigenvalues $\lambda_1, \ldots, \lambda_n$. Then, the eigenvalues of the matrix $(A - \mu I)^{-1}$ are $1/(\lambda_i - \mu)$, for $i = 1, 2, \ldots, n$. Therefore, this method finds the eigenvalue that is closest to μ, if that eigenvalue is unique. The algorithm is as follows:

Algorithm 6.4.4 (Inverse Iteration) Given an $n \times n$ symmetric matrix A and shift $\mu \in \mathbb{R}$, the following algorithm computes the eigenvector corresponding to the eigenvalue of A that is closest to μ.

Choose $\mathbf{x}_0$ so that $\|\mathbf{x}_0\|_2 = 1$
for $k = 0, 1, 2, \ldots$ **do** until convergence
    Solve $(A - \mu I)\mathbf{z}_k = \mathbf{x}_k$ for $\mathbf{z}_k$
    $\mathbf{x}_{k+1} = \mathbf{z}_k / \|\mathbf{z}_k\|_2$
end for

Now, suppose that we vary μ from iteration to iteration, by setting it equal to the **Rayleigh quotient**

$$r(\mathbf{x}, A) = \frac{\mathbf{x}^T A \mathbf{x}}{\mathbf{x}^T \mathbf{x}},$$

of which the eigenvalues of A are constrained extrema. We then obtain **Rayleigh Quotient Iteration:**

> **Algorithm 6.4.5 (Rayleigh Quotient Iteration)** Given an $n \times n$ symmetric matrix A, the following algorithm computes an eigenvalue-eigenvector pair of A.
>
> Choose a vector $\mathbf{x}_0$, $\|\mathbf{x}_0\|_2 = 1$
> **for** $k = 0, 1, 2, \ldots$ **do** until convergence
> $\mu_k = r(\mathbf{x}_k, A) = \mathbf{x}_k^T A \mathbf{x}_k$
>   Solve $(A - \mu_k I)\mathbf{z}_k = \mathbf{x}_k$ for $\mathbf{z}_k$
> $\mathbf{x}_{k+1} = \mathbf{z}_k / \|\mathbf{z}_k\|_2$
> **end for**

When this method converges, it does so *cubically* to an eigenvalue-eigenvector pair. The following explorations provide an outline of a proof.

> **Exploration 6.4.9** Let $A \in \mathbb{R}^{n \times n}$ be symmetric, and let $A = QDQ^T$ be the Schur Decomposition of A, where D is diagonal. Show that $r(\mathbf{x}, A) = r(\mathbf{y}, D)$ where $\mathbf{y} = Q^T \mathbf{x}$.

Therefore, it is sufficient to show cubic convergence of Rayleigh Quotient Iteration for a diagonal matrix D. Now, suppose that the iteration is converging to a particular eigenvector of D, which is simply a standard basis vector (see Section B.4).

> **Exploration 6.4.10** Assume that $D \in \mathbb{R}^{n \times n}$ is diagonal, and that $\mathbf{x}_k$, the kth iterate in Algorithm 6.4.5 applied to D, has the form $\mathbf{x}_k = \mathbf{e}_1 + \epsilon \mathbf{p}_k$ for some vector $\mathbf{p}_k$. Note that by Algorithm 6.4.5, $\|\mathbf{x}_k\|_2 = 1$. Show that $\mathbf{x}_{k+1} = \mathbf{e}_1 + \epsilon^3 \mathbf{p}_{k+1}$ for some vector $\mathbf{p}_{k+1}$.

> **Exploration 6.4.11** Use a similarity transformation $U^T D U$, for some orthogonal matrix U, to explain why, in Exploration 6.4.10, there is no loss of generality in assuming that $\mathbf{x}_k$ is converging to $\mathbf{e}_1$, as opposed to some other standard basis vector.

It should be noted that Inverse Iteration is also useful for a general (unsymmetric) matrix A, for finding *selected* eigenvectors after computing the Schur Decomposition $A = QTQ^T$, which reveals the eigenvalues of A, but not the eigenvectors. Then, a computed eigenvalue can be used as the shift μ, causing rapid convergence to a corresponding eigenvector. In fact, in practice a single iteration is sufficient [Peters and Wilkinson (1979)]. However, when no such information about eigenvalues is available, Inverse Iteration is far more practical for a symmetric matrix than an unsymmetric matrix, due to the superior convergence of the Power Method in the symmetric case.

> **Exploration 6.4.12** Explain why Rayleigh Quotient Iteration is not appropriate for an unsymmetric matrix, even though Inverse Iteration is.

6.4.4 The Symmetric QR Algorithm

A symmetric Hessenberg matrix is tridiagonal. Therefore, the same kind of Householder reflections that can be used to reduce a general matrix to Hessenberg form can be used to reduce a symmetric matrix A to a tridiagonal matrix T. However, the symmetry of A can be exploited to reduce the number of operations needed to apply each Householder reflection on the left and right of A.

Exploration 6.4.13 Adapt your function `hessreduce` from Exploration 6.3.3 to obtain a new function `[U,T]=trireduce(A)` that uses Householder reflections to reduce a symmetric real matrix `A` to tridiagonal form $T = U' * A * U$, where `U` is an orthogonal matrix. Make sure to take advantage of the structure of `A`, as it is being reduced, to eliminate unnecessary computations.

It can be verified by examining the structure of the matrices involved, and the rows and columns affected by Givens rotations, that if T is a symmetric tridiagonal matrix, and $T = QR$ is its QR Factorization, then Q is upper Hessenberg, and R is *upper-bidiagonal* (meaning that it is upper triangular, with upper bandwidth 1, so that all entries below the main diagonal and above the superdiagonal are zero). Furthermore, $\tilde{T} = RQ$ is also tridiagonal.

Because each Givens rotation only affects $O(1)$ nonzero entries of a tridiagonal matrix T, it follows that it only takes $O(n)$ floating-point operations to compute the QR Factorization of a tridiagonal matrix, and to multiply the factors in reverse order. However, to compute the eigenvectors of A as well as the eigenvalues, it is necessary to compute the product of all of the Givens rotations, which still takes $O(n^2)$ operations.

Exploration 6.4.14 Modify your function `hessqrstep` from Exploration 6.3.4 to obtain a new function `[Q,T]=symqrstep(T)` that uses Givens rotations to compute the QR Factorization $T = Q * R$ of a symmetric tridiagonal matrix `T`, and then overwrites `T` with RQ. As in Exploration 6.4.13, make sure to use the symmetry of `T` to eliminate computations that were necessary when applied to a Hessenberg matrix, but are unnecessary in this case.

The Implicit Q Theorem applies to symmetric matrices as well, meaning that if two orthogonal similarity transformations reduce a matrix A to *unreduced* tridiagonal form, and they have the same first column, then they are essentially equal, as are the tridiagonal matrices that they produce.

In the symmetric case, there is no need for a double-shift strategy, because the eigenvalues are real. However, the Implicit Q Theorem can be used for a different purpose: computing the similarity transformation to be used during each iteration without explicitly computing $T - \mu I$, where T is the tridiagonal matrix that is to be reduced to diagonal form. Instead, the first column of $T - \mu I$ can be computed, and then a Givens rotation can be applied to make it a multiple of $\mathbf{e}_1$. This can then be applied directly to T, followed by a series of Givens rotations to restore tridiagonal form. By the Implicit Q Theorem, this accomplishes the same effect as computing the QR Factorization $QR = T - \mu I$ and then computing $\tilde{T} = RQ + \mu I$.

Exploration 6.4.15 Let T be an unreduced tridiagonal matrix, and let μ be a chosen shift. Describe in detail how the structure of T is affected by the sequence of Givens rotations that are applied to compute $Q^T T Q$, where $T - \mu I = QR$, in which the first rotation is chosen to make the first column of $T - \mu I$ a multiple of $\mathbf{e}_1$. Where are "unwanted" nonzero entries after each rotation is applied, and which rotations should be used to zero them?

Exploration 6.4.16 Based on the result of Exploration 6.4.15, modify your code for a Francis QR step from Exploration 6.3.20 to work with a symmetric tridiagonal matrix T, using the Implicit Q Theorem to compute $\tilde{T} = RQ + \mu I$ where $T - \mu I = QR$ is the QR Factorization of $T - \mu I$, with μ a given shift. That is, instead of constructing a Householder transformation to apply to the first column of $M = H^2 - (\mu_1 + \mu_2)H + \mu_1 \mu_2 I$, your code will construct a Givens rotation to apply to the first column of $T - \mu I$. Then, use Givens rotations as in Exploration 6.3.19 to restore tridiagonal form, but taking symmetry into account.

While the shift $\mu = t_{nn}$ can always be used, it is actually more effective to use the **Wilkinson shift**, which is given by

$$\mu = t_{nn} + d - \text{sign}(d)\sqrt{d^2 + t_{n,n-1}^2}, \quad d = \frac{t_{n-1,n-1} - t_{nn}}{2}. \tag{6.10}$$

This expression yields the eigenvalue of the lower 2×2 block of T that is closer to t_{nn}. It can be shown that this choice of shift leads to *cubic* convergence of $t_{n,n-1}$ to zero [Wilkinson (1968)].

Exploration 6.4.17 Derive the formula (6.10).

The symmetric QR Algorithm is much faster than the unsymmetric QR Algorithm. A single QR step requires about $30n$ operations, because it operates on a tridiagonal matrix rather than a Hessenberg matrix, with an additional $6n^2$ operations for accumulating orthogonal transformations. The overall symmetric QR Algorithm requires $4n^3/3$ operations to compute only the eigenvalues, and approximately $8n^3$ additional operations to accumulate transformations. Because a symmetric matrix is unitarily diagonalizable, then the columns of the orthogonal matrix Q, such that $Q^T A Q$ is diagonal, contains the eigenvectors of A.

Exploration 6.4.18 Adapt your code from Exploration 6.3.21, combined with your code from Explorations 6.4.13 and 6.4.16, to obtain a function [Q,D]=symeig(A) that computes the Schur Decomposition $A = Q*D*Q'$ of a given symmetric matrix A.

6.4.5 Concept Check

1. What makes the symmetric eigenvalue problem so much simpler than the unsymmetric problem?
2. What is Rayleigh Quotient Iteration?
3. Why is a symmetric QR step (computing the QR Factorization, and then multiplying the factors in reverse order) so much more efficient than a QR step in the unsymmetric case?
4. What is the Wilkinson shift and how is it useful?
5. How are the eigenvectors of a symmetric matrix obtained?

6.5 The SVD Algorithm*

Let $A \in \mathbb{R}^{m \times n}$. Recall from Section 4.4 that we presented the **Singular Value Decomposition (SVD)** of A,

$$A = U \Sigma V^T,$$

where U is $m \times m$ and orthogonal, V is $n \times n$ and orthogonal, and Σ is an $m \times n$ diagonal matrix with nonnegative diagonal entries

$$\sigma_1 \geq \sigma_2 \geq \cdots \geq \sigma_p, \quad p = \min\{m, n\},$$

known as the **singular values** of A. As discussed in Section 4.4, the SVD is an extremely useful decomposition that yields much information about A, including its range, null space, rank, and 2-norm condition number. We now discuss a practical algorithm for computing the SVD of A, due to Golub and Kahan [Golub and Kahan (1965)].

Let U and V have column partitions

$$U = \begin{bmatrix} \mathbf{u}_1 \cdots \mathbf{u}_m \end{bmatrix}, \quad V = \begin{bmatrix} \mathbf{v}_1 \cdots \mathbf{v}_n \end{bmatrix}.$$

From the relations

$$A\mathbf{v}_j = \sigma_j \mathbf{u}_j, \quad A^T \mathbf{u}_j = \sigma_j \mathbf{v}_j, \quad j = 1, \ldots, p,$$

it follows that

$$A^T A \mathbf{v}_j = \sigma_j^2 \mathbf{v}_j.$$

That is, the squares of the singular values are the eigenvalues of $A^T A$, which is a symmetric matrix.

It follows that one approach to computing the SVD of A is to apply the symmetric QR Algorithm to $A^T A$ to obtain a decomposition $A^T A = V \Sigma^T \Sigma V^T$. Then, the relations $A\mathbf{v}_j = \sigma_j \mathbf{u}_j$, $j = 1, \ldots, p$, can be used in conjunction with the QR factorization with column pivoting to obtain U. However, this approach is not the most practical, because of the expense and loss of information incurred from computing $A^T A$ using floating-point arithmetic.

> **Exploration 6.5.1** Compute the singular values of a randomly generated matrix A using two different approaches: (1) using svd, and (2) using eig with $A^T A$, and then taking the square root. To what extent do these sets of singular values agree?

Instead, we can *implicitly* apply the symmetric QR Algorithm to $A^T A$ by working only on A. As the first step of the symmetric QR Algorithm is to use Householder reflections to reduce the matrix to tridiagonal form, we can use Householder reflections to instead reduce A to **upper bidiagonal form**

$$U_1^T A V_1 = B = \begin{bmatrix} d_1 & f_1 & & & \\ & d_2 & f_2 & & \\ & & \ddots & \ddots & \\ & & & d_{n-1} & f_{n-1} \\ & & & & d_n \end{bmatrix}.$$

It follows that $T = B^T B$ is symmetric and tridiagonal.

> **Exploration 6.5.2** Express the entries of T in terms of the entries of B.

> **Exploration 6.5.3** Adapt your function `trireduce` from Exploration 6.4.13 to reduce a given $m \times n$ matrix A to upper bidiagonal form using Householder reflections. The resulting function should have the interface `[U1,B,V1]=bireduce(A)`, that returns an orthogonal $m \times m$ matrix U1, an upper bidiagonal $m \times n$ matrix B, and an orthogonal $n \times n$ matrix V1 such that `U1' * A * V1 = B`. If only one output argument is specified, then your function must not accumulate Householder reflections to compute U1 and V1.

We could then apply the symmetric QR Algorithm directly to T, but, again, to avoid the loss of information from computing T explicitly, we implicitly apply the QR Algorithm to T by performing the following steps during each iteration:

1. Determine the first Givens row rotation G_1^T that *would* be applied to $T - \mu I$, where μ is the Wilkinson shift (6.10) from the symmetric QR Algorithm. This requires only computing the first column of T, which has only two nonzero entries.

> **Exploration 6.5.4** Use the result of Exploration 6.5.2 to express the Wilkinson shift (6.10) in terms of entries of B.

2. Apply G_1 as a *column* rotation to columns 1 and 2 of B to obtain $B_1 = BG_1$. This introduces an unwanted nonzero in the $(2,1)$ entry.

3. Apply a Givens row rotation H_1 to rows 1 and 2 to zero the $(2,1)$ entry of B_1, which yields $B_2 = H_1^T B G_1$. Then, B_2 has an unwanted nonzero in the $(1,3)$ entry.

4. Apply a Givens column rotation G_2 to columns 2 and 3 of B_2, which yields $B_3 = H_1^T B G_1 G_2$. This introduces an unwanted zero in the $(3,2)$ entry.

5. Continue applying alternating row and column rotations to "chase" the unwanted nonzero entry down the diagonals of B, until finally B is restored to upper bidiagonal form.

By the Implicit Q Theorem, since G_1 is the Givens rotation that would be applied to the first column of T, the column rotations that help restore upper bidiagonal form are essentially equal to those that would be applied to T if the symmetric QR Algorithm was being applied to T directly. Therefore, the symmetric QR Algorithm is being correctly applied, implicitly, to T by working only on B.

> **Exploration 6.5.5** Modify your code from Exploration 6.4.16 to implicitly apply a step of the symmetric QR Algorithm to an upper bidiagonal $m \times n$ matrix B with Wilkinson shift μ obtained from the appropriate entries of $T = B^T B$, where T is unreduced.

To detect decoupling, we note that if any superdiagonal entry f_i is small enough to be "declared" equal to zero, then decoupling has been achieved, because the ith subdiagonal entry of T is a multiple of f_i, and therefore such a subdiagonal entry must be zero as well. If a diagonal entry d_i becomes zero, then decoupling is also achieved, because row or column rotations can be used to zero an entire row or column of B. In summary, if any diagonal or superdiagonal entry of B becomes zero, then the tridiagonal matrix $T = B^T B$ is no longer unreduced.

> **Exploration 6.5.6** Suppose that after an implicit QR step, $d_k = 0$ for $k < n$. Devise a sequence of Givens row rotations that achieve decoupling in T.

> **Exploration 6.5.7** Suppose that after an implicit QR step, $d_n = 0$. Devise a sequence of Givens column rotations that will achieve decoupling in T.

Eventually, sufficient decoupling is achieved so that B is reduced to a diagonal matrix Σ. All Householder reflections that have pre-multiplied A, and all row rotations that have been applied to B, can be accumulated to obtain U, and all Householder reflections that have post-multiplied A, and all column rotations that have been applied to B, can be accumulated to obtain V.

> **Exploration 6.5.8** Adapt your code from Exploration 6.4.18, combined
> with your code from Explorations 6.5.3 and 6.5.5, to obtain a function
> [U,S,V]=mysvd(A) that computes the Singular Value Decomposition $A =$
> $U * S * V'$ of a given $m \times n$ real matrix A.

The relationship between the SVD and the symmetric eigenvalue problem can
be used to obtain theoretical results pertaining to the sensitivity of singular values.

> **Exploration 6.5.9** Let A be a real $m \times n$ matrix, with $m \geq n$. Use the
> SVD to obtain the Schur Decomposition of the $(m+n) \times (m+n)$ symmetric
> matrix
> $$\begin{bmatrix} 0 & A^T \\ A & 0 \end{bmatrix}.$$

> **Exploration 6.5.10** Use the result of Exploration 6.5.9 to prove analogues
> of Theorems 6.4.2 and 6.4.3 for singular values.

We also have this analogue of the Courant-Fischer Minimax Theorem (6.9) for
singular values:

> **Theorem 6.5.1** Let $A \in \mathbb{R}^{m \times n}$. Then, for $k = 1, 2, \ldots, \min\{m, n\}$,
> $$\sigma_k(A) = \max_{\dim(S)=k, \dim(T)=k} \min_{\mathbf{x} \in S, \mathbf{y} \in T} \frac{\mathbf{y}^T A \mathbf{x}}{\|\mathbf{y}\|_2 \|\mathbf{x}\|_2} = \max_{\dim(S)=k} \min_{\mathbf{x} \in S} \frac{\|A\mathbf{x}\|_2}{\|\mathbf{x}\|_2}.$$

> **Exploration 6.5.11** Prove Theorem 6.5.1 by adapting your proof of (6.9).

6.5.1 Concept Check

1. How is the SVD related to an eigenvalue problem?
2. What is the drawback to computing the singular values of A using the
   symmetric QR Algorithm?
3. What reduction is performed on A at the beginning of the SVD algorithm,
   using Householder reflections?
4. Once A is reduced using Householder reflections, how are Givens rotations
   used to reduce A to diagonal form?
5. How can sensitivity results for the symmetric eigenvalue problem be
   adapted to obtain sensitivity results for singular values?

6.6 Additional Resources

There are several textbooks on numerical linear algebra that offer extensive coverage
of techniques for computing eigenvalues and singular values, including [Demmel
(1997); Golub and van Loan (2012); Trefethen and Bau (1997); Watkins (2002)].
There are also books that are devoted entirely to eigenvalue problems, such as

[Stewart (2001)] and the classic text [Wilkinson (1965)]. For more information on the role of eigenvalue problem solvers in search engines, the reader is referred to [Berry and Browne (2005); Langville and Meyer (2006)].

Several topics related to eigenvalue problems are beyond the scope of this book: computing all the eigenvectors of an unsymmetric matrix [Bavely and Stewart (1979)], divide-and-conquer algorithms for symmetric tridiagonal matrices [Dongarra and Sorensen (1987)], and generalized eigenvalue problems and the QZ Method [Moler and Stewart (1973)]. Jacobi methods for computing eigenvalues of symmetric matrices [Jacobi (1846)] or singular values [Kogbetliantz (1955); Nash (1975)] are highlighted in Exercise 13; the reader is also referred to [Golub and van Loan (2012)] and the references therein.

The handbook [Wilkinson and Reinsch (1971)] set the stage for modern software for solving eigenvalue problems, such as EISPACK [Garbow, et al. (1972); Smith, et al. (1970)], since superseded by LAPACK [Anderson, et al. (1999)]. Methods for sparse matrices are available in ARPACK [Lehoucq, et al. (1998)].

6.7 Exercises

1. Let $A \in \mathbb{C}^{n \times n}$. Prove that A is similar to A^T. Is A similar to A^H? Prove or give a counterexample.

2. (a) Prove that if $\lambda = a + bi$ is an eigenvalue of an $n \times n$ *real* matrix A with eigenvector $\mathbf{x} \in \mathbb{C}^n$, then $\overline{\lambda} = a - bi$ is also an eigenvalue of A with eigenvector $\overline{\mathbf{x}}$.

   (b) A $n \times n$ matrix A is **skew-symmetric** if $A = -A^T$. Prove that if λ is an eigenvalue of A, then $\text{Re}(\lambda) = 0$. *Hint:* Consider the expression $\mathbf{x}^H A \mathbf{x}$, where $\mathbf{x}$ is an eigenvector of A.

   (c) Let $A \in \mathbb{R}^{n \times n}$ be skew-symmetric. Prove that if n is odd, then A must be singular.

3. (a) Prove by induction that if λ is an eigenvalue of A, then λ^k is an eigenvalue of A^k, where k is a positive integer.

   (b) A square matrix A is said to be **nilpotent** if $A^k = 0$ for some positive integer k. Prove that the eigenvalues of a nilpotent matrix are all equal to zero.

   (c) Prove that if A is both nilpotent and normal, then $A = 0$.

   (d) A square matrix A is said to be **idempotent** if $A^2 = A$. What are the possible eigenvalues of A?

   (e) Show that if A is idempotent, then so is $I - A$.

4. Compute the eigenvalues of the 5×5 matrix

$$A = \begin{bmatrix} 1 & 4 & 0 & 0 & 0 \\ 1 & 1 & 0 & 0 & 0 \\ 0 & 0 & 2 & 0 & 1 \\ 0 & 0 & 0 & 2 & 1 \\ 0 & 0 & 1 & 1 & 1 \end{bmatrix}$$

without computing a characteristic polynomial of degree 5. What are the algebraic and geometric multiplicities of the eigenvalues?

5. The scenario in this exercise arose in the first author's research [Palchak, et al. (2015)]. Let $A \in \mathbb{R}^{n \times n}$, where n is even, and suppose that $a_{ij} = 0$ whenever $i + j$ is odd. Describe a more efficient approach to computing the eigenvalues of A than simply applying the QR Algorithm to A directly.

6. Consider the matrix

$$A = \begin{bmatrix} 5 & 1 & 2 \\ 1 & 5 & 2 \\ 2 & 2 & 4 \end{bmatrix}.$$

   (a) What are the eigenvalues and eigenvectors of this matrix?

   (b) Apply the Power Method to A with initial vector $\mathbf{x}_0 = \begin{bmatrix} 1 & 0 & -1 \end{bmatrix}^T$. After each iteration, compute the Rayleigh Quotient $\mathbf{x}_k^T A \mathbf{x}_k / \mathbf{x}_k^T \mathbf{x}_k$. What happens, and can you explain the behavior? Be sure to perform at least 70 iterations.

7. Find the scalar μ that minimizes $\|A\mathbf{x} - \mu\mathbf{x}\|_2$ where $A \in \mathbb{R}^{n \times n}$ and $\mathbf{x} \in \mathbb{R}^n$, by treating it like any other least squares problem.

8. A **right stochastic matrix** is an $n \times n$ matrix P that describes a *Markov chain* over a finite *state space* with n states. As discussed in Section 6.2, each entry p_{ij} is the probability of moving from state i to state j in one step. Recall that because all of its row sums are equal to one, P has an eigenvalue of one, with corresponding eigenvector being a vector with all components equal.

   (a) Show that a product of right stochastic matrices is also a right stochastic matrix.

   (b) Given the interpretation of the entries of P, what is the interpretation of each entry of P^2? Of P^k, for any positive integer P?

   (c) Let $\mathbf{y} \in \mathbb{R}^n$, with y_i being the probability of being in state i, for $i = 1, 2, \ldots, n$. What is the interpretation of $\mathbf{y}^T P$?

   (d) Create a right stochastic matrix P and apply the Power Method to P^T to obtain the left eigenvector of P corresponding to $\lambda = 1$. Does the iteration converge?

   (e) Assume that $\lambda_1 = 1$ is a simple eigenvalue of P. Express $\lim_{k \to \infty} P^k$ as an outer product.

9. Let $A \in \mathbb{R}^{n \times n}$ be symmetric. What is the optimal shift μ so that the Power Method applied to $A - \mu I$ will converge most rapidly to the eigenvector corresponding to λ_1? To λ_n?

10. Compute the eigenvalues of the matrices

$$A_1 = \begin{bmatrix} 1 & 1 \\ 0 & 1 \end{bmatrix}, \quad A_2 = \begin{bmatrix} 1 & 1 \\ 10^{-4} & 1 \end{bmatrix}.$$

How does the change in the eigenvalues conform to theoretical expectations? Repeat this exercise with the $(1, 2)$ entry of both matrices being equal to 10^4. In both scenarios, what are the condition numbers of the eigenvalues of A_2?

11. Consider the matrix

$$A = \begin{bmatrix} -198 & -200 \\ -199 & 201 \end{bmatrix}.$$

(a) Compute the eigenvalues of A.

(b) Add 0.01 to a_{11} and compute the eigenvalues again.

(c) Compute the condition numbers of the eigenvalues of A. Does this explain the behavior that you have observed?

12. Let T be a symmetric tridiagonal matrix. Apply the following simplified versions of the QR Algorithm to T:

repeat until convergence
    Let $\mu = T_{kk}$, where $|T_{i+1,i}| < TOL$ for $i = k + 1, \ldots, n - 1$
    $T(1 : k, 1 : k) - \mu I = QR$
    $T(1 : k, 1 : k) = RQ + \mu I$
end repeat

and

repeat until convergence
    Let $\mu = \texttt{wilkinsonshift}(T(1 : k, 1 : k))$, where $|T_{i+1,i}| < TOL$
    for $i = k + 1, \ldots, n - 1$
    $T(1 : k, 1 : k) - \mu I = QR$
    $T(1 : k, 1 : k) = RQ + \mu I$
end repeat

where the function `wilkinsonshift` returns the result of (6.10) applied to the matrix passed as an input argument. Compare the performance of these two algorithms, in terms of both execution time as measured with the commands `tic` and `toc` and total number of iterations.

13. In this exercise we examine the idea behind **Jacobi methods** [Jacobi (1846)] for computing eigenvalues or singular values.

(a) Let $A \in \mathbb{R}^{2 \times 2}$ be symmetric. Derive formulas for the cosine-sine pair (c, s) such that

$$\begin{bmatrix} c & s \\ -s & c \end{bmatrix}^T \begin{bmatrix} a_{11} & a_{12} \\ a_{21} & a_{22} \end{bmatrix} \begin{bmatrix} c & s \\ -s & c \end{bmatrix} = \begin{bmatrix} \lambda_1 & 0 \\ 0 & \lambda_2 \end{bmatrix}.$$

That is, find the Givens rotation G such that $A = GDG^T$ is the Schur Decomposition of A.

(b) Let $A \in \mathbb{R}^{n \times n}$ be symmetric, and define

$$\text{off}(A) = \|A\|_F^2 - \sum_{i=1}^{n} a_{ii}^2.$$

That is, off(A) is the sum of squares of the off-diagonal entries of A. Let $B = Q^T A Q$ where $Q = G(i, j, c, s)$ is a Givens rotation of columns i and j, $i < j$, so that $b_{ij} = b_{ji} = 0$, using the 2×2 Schur Decomposition derived in part 13a applied to

$$A_{ij} = \begin{bmatrix} a_{ii} & a_{ij} \\ a_{ji} & a_{jj} \end{bmatrix}.$$

Prove that off$(B) <$ off(A). *Hint:* Use the fact that the Frobenius norm is invariant under orthogonal transformations, as proved in Exploration 4.4.6.

(c) Write a MATLAB function `[Q,D]=jacobieig(A,tol)` that repeatedly applies Givens row and column rotations to `A`, as described in part 13a, to zero the largest off-diagonal entry of `A` during each iteration. The iteration continues until off$(A) <$ `tol`. The output arguments are an orthogonal matrix `Q` that is the accumulation of all of the Givens rotations applied to `A`, and a diagonal matrix `D` that contains the eigenvalues of `A`.

(d) Modify the procedure outlined in this exercise to compute the SVD of an $m \times n$ real matrix A, using different Givens rotations on each side to compute the SVD of a 2×2 matrix A_{ij}. It is worth noting that the standard approach is to use a row rotation to symmetrize A_{ij} and then perform the symmetric 2×2 Schur Decomposition. Can you figure out how to compute the SVD of A_{ij} using only one rotation on each side?

PART III
Data Fitting and Function Approximation

Chapter 7

Polynomial Interpolation

Calculus provides many tools that can be used to understand the behavior of functions, but it is necessary for these functions to be continuous or differentiable. This presents a problem in many "real" applications, in which functions are used to model relationships between quantities, but our only knowledge of these functions consists of a set of discrete data points, where the data is obtained from measurements. Therefore, we need to be able to construct continuous functions based on discrete data.

The problem of constructing such a continuous function is called **data fitting**. In this chapter, we discuss a special case of data fitting known as **interpolation**, in which the goal is to find a function of a specified form to fit a set of data that imposes constraints, thus guaranteeing a unique solution that fits the data exactly, rather than approximately. The broader term "constraints" is used, rather than a simple requirement of fitting data points, since the description of the data may include additional information such as rates of change, or we may impose requirements that the fitting function have a certain number of continuous derivatives.

When it comes to the study of functions using calculus, polynomials are particularly simple to work with. Therefore, in this chapter we will focus on the problem of constructing a polynomial that fits given data. We first discuss some algorithms for computing the unique polynomial $p_n(x)$ of degree n that satisfies $p_n(x_i) = y_i$, $i = 0, \ldots, n$, where the points (x_i, y_i) are given. The $n + 1$ distinct points $x_0, x_1, \ldots, x_n$ are called **interpolation points**. The polynomial $p_n(x)$ is called the **interpolating polynomial** of the data (x_0, y_0), (x_1, y_1), ..., (x_n, y_n).

We begin in Section 7.1 with a discussion of the well-posedness of the problem of polynomial interpolation, and why the most straightforward approach of computing the interpolating polynomial is not the most practical one. In Sections 7.2 and 7.3, we examine two more effective approaches, which are **Lagrange interpolation** and **Newton interpolation**. Section 7.4 discusses the accuracy of polynomial interpolation in fitting a known function $f(x)$, and how it can be improved using **Chebyshev polynomials**. In Section 7.5 we relax the assumption that the interpolation points are distinct, to allow taking into account other information about the data such as rates of change. This leads to **osculatory interpolation**, including **Hermite interpolation**. Finally, in Section 7.6 we explore **piecewise polynomial interpolation**, or **spline interpolation**, in which a different polynomial is used to fit the data on each subinterval between interpolation points. This approach can be shown to improve accuracy compared to polynomial interpolation when the number of data points is large.

7.1   Existence and Uniqueness

Suppose that we express the interpolating polynomial $p_n(x)$ in the familiar **power form**

$$p_n(x) = a_0 + a_1 x + a_2 x^2 + \cdots + a_n x^n. \tag{7.1}$$

Then, a straightforward method of computing the interpolating polynomial is to solve the equations $p_n(x_i) = y_i$, for $i = 0, 1, \ldots, n$, for the unknown coefficients a_j, $j = 0, 1, \ldots, n$.

Exploration 7.1.1 Solve the system of equations

$$a_0 + a_1 x_0 = y_0,$$
$$a_0 + a_1 x_1 = y_1$$

for the coefficients of the linear function $p_1(x) = a_0 + a_1 x$ that interpolates the data (x_0, y_0), (x_1, y_1). What is the system of equations that must be solved to compute the coefficients a_0, a_1 and a_2 of the quadratic function $p_2(x) = a_0 + a_1 x + a_2 x^2$ that interpolates the data (x_0, y_0), (x_1, y_1), (x_2, y_2)? Express both systems of equations in matrix-vector form.

For general n, computing the coefficients $a_0, a_1, \ldots, a_n$ of $p_n(x)$ requires solving the system of linear equations $V_n \mathbf{a} = \mathbf{y}$, where the entries of V_n are defined by $[V_n]_{ij} = x_i^j$, $i, j = 0, \ldots, n$. The basis $\{1, x, \ldots, x^n\}$ of the space of polynomials of degree n is called the **monomial basis**, and the corresponding matrix V_n is called the **Vandermonde matrix** for the points $x_0, x_1, \ldots, x_n$. Because the x-values $x_0, x_1, \ldots, x_n$ are distinct, V_n is invertible (see Exploration 7.1.2), which ensures that the system $V_n \mathbf{a} = \mathbf{y}$ has a unique solution.

The MATLAB function p=polyfit(x,y,n) performs polynomial interpolation. The input arguments x and y are vectors consisting of the x- and y-coordinates of the data points to be fit. The input argument n is the degree of the interpolating polynomial; for the kind of interpolation described here, it should be set equal to one less than the length of x. The output p is a row vector of coefficients of the interpolating polynomial, ordered from highest degree to lowest, as used by other polynomial functions such as those introduced in Section 1.2.18. These coefficients are computed using the approach described in the preceding paragraph.

Unfortunately, this approach to computing $p_n(x)$ is not practical. Solving this system of equations requires $O(n^3)$ floating-point operations; we will see that computing $p_n(x)$ in $O(n^2)$ operations is possible. Furthermore, the Vandermonde matrix can be ill-conditioned (see Section 3.4.1), especially when the interpolation points $x_0, x_1, \ldots, x_n$ are close together. In the following sections, we will build our own versions of `polyfit` that employ more efficient approaches to interpolation.

Exploration 7.1.2 Write down the Vandermonde matrix V_n for the points $x_0, x_1, \ldots, x_n$. Show that

$$\det V_n = \prod_{i=0}^{n} \prod_{j=0}^{i-1} (x_i - x_j).$$

Conclude that the system of equations $V_n \mathbf{a} = \mathbf{y}$ has a unique solution.

Exploration 7.1.3 In this exploration, we consider another approach to proving the uniqueness of the interpolating polynomial. Let $p_n(x)$ and $q_n(x)$ be polynomials of degree n such that $p_n(x_i) = q_n(x_i) = y_i$ for $i = 0, 1, 2, \ldots, n$. Prove that $p_n(x) \equiv q_n(x)$ for all x.

The following exploration offers a preview of the more efficient approach to polynomial interpolation that will be covered in the next section.

Exploration 7.1.4 Suppose we express the interpolating polynomial of degree one in the form

$$p_1(x) = a_0(x - x_1) + a_1(x - x_0).$$

What is the matrix of the system of equations $p_1(x_i) = y_i$, for $i = 0, 1$? How should the form of the interpolating polynomial of degree two, $p_2(x)$, be chosen to obtain an equally simple system of equations to solve for the coefficients a_0, a_1, and a_2?

We conclude with a brief comment about sensitivity. Polynomial interpolation is an ill-conditioned problem if any of the interpolation points are too close together. This is related to the fact that a Vandermonde matrix becomes ill-conditioned under these circumstances. The following MATLAB session illustrates this sensitivity.

```
>> x=sort(rand(5,1))
x =
    0.1068
    0.4942
    0.6538
    0.7150
    0.7791
>> y=rand(5,1);
>> p=polyfit(x,y,4);
>> norm(y-polyval(p,x))/norm(y)
ans =
```

```
   3.7413e-13
>> x(5)=x(4)+1e-8
x =
   0.1068
   0.4942
   0.6538
   0.7150
   0.7150
>> p=polyfit(x,y,4);
Warning: Polynomial is badly conditioned. Add points with distinct X
values, reduce the degree of the polynomial, or try centering and
scaling as described in HELP POLYFIT.
> In polyfit (line 75)
>> norm(y-polyval(p,x))/norm(y)
ans =
   1.4508e-06
```

We see that when the problem of fitting the given data is ill-conditioned, the polynomial cannot even be counted on to actually do so with high accuracy.

7.1.1 Concept Check

1. What is data fitting and why is it necessary?
2. Why is it desirable to approximate functions using polynomials, as opposed to other types of functions?
3. Explain why the polynomial of degree n that interpolates the data (x_0, y_0), (x_1, y_1), ..., (x_n, y_n) is unique.
4. Why is it not practical to use a Vandermonde matrix to obtain an interpolating polynomial?

7.2 Lagrange Interpolation

As we have seen, expressing the interpolating polynomial $p_n(x)$ in power form (7.1) leads to an impractical approach to obtaining its coefficients. Therefore, we will construct $p_n(x)$ using a representation other than the monomial basis. That is, we will represent $p_n(x)$ as

$$p_n(x) = \sum_{i=0}^{n} c_i \varphi_i(x),$$

for some other choice of polynomials $\varphi_0(x), \varphi_1(x), \ldots, \varphi_n(x)$. This is equivalent to solving the linear system $P\mathbf{c} = \mathbf{y}$, where the matrix P has entries $p_{ij} = \varphi_j(x_i)$. By choosing the *basis functions* $\{\varphi_i(x)\}_{i=0}^{n}$ judiciously, we can obtain a simpler system of linear equations to solve.

7.2.1 Lagrange Polynomials

In **Lagrange interpolation**, the matrix P is simply the identity matrix, by virtue of the fact that the interpolating polynomial is written in the form

$$p_n(x) = \sum_{j=0}^{n} c_j \mathcal{L}_{n,j}(x),$$

where the polynomials $\{\mathcal{L}_{n,j}\}_{j=0}^{n}$ have the property that

$$\mathcal{L}_{n,j}(x_i) = \begin{cases} 1 \text{ if } i = j \\ 0 \text{ if } i \neq j \end{cases}.$$

The polynomials $\{\mathcal{L}_{n,j}\}$, $j = 0, \ldots, n$, are called the **Lagrange polynomials** for the interpolation points x_0, x_1, ..., x_n.

To obtain a formula for the Lagrange polynomials, we note that the above definition specifies the roots of $\mathcal{L}_{n,j}(x)$: x_i, for $i \neq j$. It follows that $\mathcal{L}_{n,j}(x)$ has the form

$$\mathcal{L}_{n,j}(x) = \beta_j \prod_{i=0, i \neq j}^{n} (x - x_i) \tag{7.2}$$

for some constant β_j. Substituting $x = x_j$ and requiring $\mathcal{L}_{n,j}(x_j) = 1$ yields $\beta_j = \prod_{i=0, i \neq j}^{n} \frac{1}{(x_j - x_i)}$. We conclude that

$$\mathcal{L}_{n,j}(x) = \prod_{i=0, i \neq j}^{n} \frac{x - x_i}{x_j - x_i}. \tag{7.3}$$

As the following result indicates, the problem of polynomial interpolation can easily be solved using Lagrange polynomials.

Theorem 7.2.1 Let $(x_0, y_0), (x_1, y_1), \ldots, (x_n, y_n)$ be $n + 1$ points in the xy-plane, with $x_0, x_1, \ldots, x_n$ distinct. Then the polynomial defined by

$$p_n(x) = \sum_{j=0}^{n} y_j \mathcal{L}_{n,j}(x) \tag{7.4}$$

is the unique polynomial of degree n that satisfies

$$p_n(x_i) = y_i, \quad i = 0, 1, \ldots, n.$$

Exploration 7.2.1 Prove Theorem 7.2.1.

We can see that unlike the approach to interpolation presented in Section 7.1, the task of obtaining the coefficients of the linear combination of Lagrange polynomials is trivial–they are simply the y-coordinates of the data points.

Example 7.2.2 We will use Lagrange interpolation to find the unique polynomial $p_3(x)$, of degree 3 or less, that agrees with the following data:

i	x_i	y_i
0	-1	3
1	0	-4
2	1	5
3	2	-6

In other words, we must have $p_3(-1) = 3$, $p_3(0) = -4$, $p_3(1) = 5$, and $p_3(2) = -6$.
First, we construct the Lagrange polynomials $\{\mathcal{L}_{3,j}(x)\}_{j=0}^3$, using the formula

$$\mathcal{L}_{n,j}(x) = \prod_{i=0, i \neq j}^{3} \frac{(x - x_i)}{(x_j - x_i)}.$$

This yields

$$\begin{aligned}
\mathcal{L}_{3,0}(x) &= \frac{(x - x_1)(x - x_2)(x - x_3)}{(x_0 - x_1)(x_0 - x_2)(x_0 - x_3)} \\
&= \frac{(x - 0)(x - 1)(x - 2)}{(-1 - 0)(-1 - 1)(-1 - 2)} \\
&= -\frac{1}{6}(x^3 - 3x^2 + 2x),
\end{aligned}$$

$$\begin{aligned}
\mathcal{L}_{3,1}(x) &= \frac{(x - x_0)(x - x_2)(x - x_3)}{(x_1 - x_0)(x_1 - x_2)(x_1 - x_3)} \\
&= \frac{(x + 1)(x - 1)(x - 2)}{(0 + 1)(0 - 1)(0 - 2)} \\
&= \frac{1}{2}(x^3 - 2x^2 - x + 2),
\end{aligned}$$

$$\begin{aligned}
\mathcal{L}_{3,2}(x) &= \frac{(x - x_0)(x - x_1)(x - x_3)}{(x_2 - x_0)(x_2 - x_1)(x_2 - x_3)} \\
&= \frac{(x + 1)(x - 0)(x - 2)}{(1 + 1)(1 - 0)(1 - 2)} \\
&= -\frac{1}{2}(x^3 - x^2 - 2x),
\end{aligned}$$

$$\begin{aligned}
\mathcal{L}_{3,3}(x) &= \frac{(x - x_0)(x - x_1)(x - x_2)}{(x_3 - x_0)(x_3 - x_1)(x_3 - x_2)} \\
&= \frac{(x + 1)(x - 0)(x - 1)}{(2 + 1)(2 - 0)(2 - 1)} \\
&= \frac{1}{6}(x^3 - x).
\end{aligned}$$

By substituting x_i for x in each Lagrange polynomial $\mathcal{L}_{3,j}(x)$, for $j = 0, 1, 2, 3$,
it can be verified that

$$\mathcal{L}_{3,j}(x_i) = \begin{cases} 1 \text{ if } i = j \\ 0 \text{ if } i \neq j \end{cases}.$$

It follows that the Lagrange interpolating polynomial $p_3(x)$ is given by

$$p_3(x) = \sum_{j=0}^{3} y_j \mathcal{L}_{3,j}(x)$$

$$= y_0 \mathcal{L}_{3,0}(x) + y_1 \mathcal{L}_{3,1}(x) + y_2 \mathcal{L}_{3,2}(x) + y_3 \mathcal{L}_{3,3}(x)$$

$$= (3)\left(-\frac{1}{6}\right)(x^3 - 3x^2 + 2x) + (-4)\frac{1}{2}(x^3 - 2x^2 - x + 2) +$$

$$(5)\left(-\frac{1}{2}\right)(x^3 - x^2 - 2x) + (-6)\frac{1}{6}(x^3 - x)$$

$$= -6x^3 + 8x^2 + 7x - 4.$$

Substituting each x_i, for $i = 0, 1, 2, 3$, into $p_3(x)$, we can verify that we obtain $p_3(x_i) = y_i$ in each case. $\square$

Exploration 7.2.2 Write a MATLAB function L=makelagrange(x) that accepts as input a vector x of length $n + 1$ consisting of the points $x_0, x_1, \ldots, x_n$, which must be distinct, and returns a $(n + 1) \times (n + 1)$ matrix L, each row of which consists of the coefficients of the Lagrange polynomial $\mathcal{L}_{n,j}$, $j = 0, 1, 2, \ldots, n$, with highest-degree coefficients in the first column. Use the poly function to construct polynomials that are designed to have certain roots, and use polyval to ensure that $\mathcal{L}_{n,j}(x_j) = 1$ for $j = 0, 1, \ldots, n$.

Exploration 7.2.3 Write a MATLAB function p=lagrangefit(x,y) that accepts as input vectors x and y of length $n + 1$ consisting of the x- and y-coordinates, respectively, of points $(x_0, y_0), (x_1, y_1), \ldots, (x_n, y_n)$, where the x-values must all be distinct, and returns a $(n+1)$-vector p consisting of the coefficients of the Lagrange interpolating polynomial $p_n(x)$, with highest-degree coefficient in the first position. Use your makelagrange function from Exploration 7.2.2. Test your function by comparing your output to that of the built-in function polyfit.

7.2.2 Barycentric Interpolation

The Lagrange interpolating polynomial can be inefficient to evaluate, as written, because it involves $O(n^2)$ subtractions for a polynomial of degree n. We can evaluate the interpolating polynomial more efficiently using a technique called **barycentric interpolation** [Berrut and Trefethen (2004)]. Starting with the Lagrange form of an interpolating polynomial $p_n(x)$ given by (7.4), we define the **barycentric weights** w_j, $j = 0, 1, \ldots, n$, by

$$w_j = \prod_{i=0, i \neq j}^{n} \frac{1}{x_j - x_i}. \tag{7.5}$$

Note that w_j is equal to the scaling constant β_j from (7.2). Next, we define

$$\pi_n(x) = (x - x_0)(x - x_1)\cdots(x - x_n).$$

Then, each Lagrange polynomial can be rewritten as

$$\mathcal{L}_{n,j}(x) = \frac{\pi_n(x)w_j}{x - x_j}, \quad x \neq x_j,$$

and the interpolant itself can be written as

$$p_n(x) = \pi_n(x) \sum_{j=0}^{n} \frac{y_j w_j}{x - x_j}. \tag{7.6}$$

However, if we let $y_j = 1$ for $j = 0, 1, \ldots, n$, we have

$$1 = \pi_n(x) \sum_{j=0}^{n} \frac{w_j}{x - x_j}. \tag{7.7}$$

Dividing (7.6) by (7.7) yields

$$p_n(x) = \frac{\displaystyle\sum_{j=0}^{n} \frac{y_j w_j}{x - x_j}}{\displaystyle\sum_{j=0}^{n} \frac{w_j}{x - x_j}}. \tag{7.8}$$

Although $O(n^2)$ products are needed to compute the barycentric weights, they need only be computed once, and then re-used for each x, which is not the case with the Lagrange form.

Exploration 7.2.4 Write a MATLAB function

$$w=\text{baryweights}(x)$$

that accepts as input a vector x of length $n + 1$, consisting of the distinct interpolation points $x_0, x_1, \ldots, x_n$, and returns a vector w of length $n + 1$ consisting of the barycentric weights w_j as defined in (7.5).

Exploration 7.2.5 Write a MATLAB function yy=lagrangeval(x,y,xx) that accepts as input vectors x and y of length $n + 1$ containing the data to be fit, and a vector of x-values xx. This function uses the barycentric weights (7.5), as computed by the function baryweights from Exploration 7.2.4, to compute the value of the Lagrange interpolating polynomial for the given data points at each x-value in xx using (7.8). The corresponding y-values must be returned in the vector yy, which must have the same dimensions as xx. *Note:* The formula (7.8) is not defined if $x = x_i$ for $i = 0, 1, \ldots, n$. Make sure your function handles this situation properly!

7.2.3 Concept Check

1. What properties characterize Lagrange polynomials?
2. What is the advantage of using Lagrange polynomials to express the interpolating polynomial, compared to using a Vandermonde matrix?
3. What is a drawback of using Lagrange polynomials?
4. What is the benefit of using barycentric interpolation?

7.3 Newton Interpolation

While the Lagrange polynomials are easy to compute, they are cumbersome to work with, and do not lend themselves to efficient evaluation of the interpolating polynomial $p_n(x)$. Furthermore, if new interpolation points are added, all of the Lagrange polynomials must be recomputed. Unfortunately, it is not uncommon, in practice, to add to an existing set of interpolation points. It may be determined after computing the kth-degree interpolating polynomial $p_k(x)$ of a function $f(x)$ that $p_k(x)$ is not a sufficiently accurate approximation of $f(x)$ on some domain. Therefore, an interpolating polynomial of higher degree must be computed, which requires additional interpolation points.

To address these issues, we consider the problem of computing the interpolating polynomial *recursively*. More precisely, let $k > 0$, and let $p_k(x)$ be the polynomial of degree k that interpolates the function $f(x)$ at the points $x_0, x_1, \ldots, x_k$. Ideally, we would like to be able to obtain $p_k(x)$ from polynomials of degree $k - 1$ that interpolate $f(x)$ at points chosen from among $x_0, x_1, \ldots, x_k$. The following result shows that this is possible.

Theorem 7.3.1 Let n be a positive integer, and let $f(x)$ be a function defined on a domain containing the $n + 1$ distinct points $x_0, x_1, \ldots, x_n$, and let $p_n(x)$ be the polynomial of degree n that interpolates $f(x)$ at the points $x_0, x_1, \ldots, x_n$. For each $i = 0, 1, \ldots, n$, we define $p_{n-1,i}(x)$ to be the polynomial of degree $n - 1$ that interpolates $f(x)$ at the points $x_0, x_1, \ldots, x_{i-1}, x_{i+1}, \ldots, x_n$. If i and j are distinct integers such that $0 \le i, j \le n$, then

$$p_n(x) = \frac{(x - x_j)p_{n-1,j}(x) - (x - x_i)p_{n-1,i}(x)}{x_i - x_j}.$$

Exploration 7.3.1 Prove Theorem 7.3.1.

This theorem can be used to compute the polynomial $p_n(x)$ itself, rather than its value at a given point. This yields an alternative method of constructing the interpolating polynomial, called **Newton interpolation**, that is more suitable for tasks such as inclusion of additional interpolation points. The basic idea is to represent interpolating polynomials using **Newton form**, which uses linear factors involving the interpolation points, instead of monomials of the form x^j. However, unlike Lagrange polynomials, which also include such factors, Newton form uses polynomials of increasing degree.

7.3.1   Divided Differences

It has been stated that if the interpolation points $x_0, \ldots, x_n$ are distinct, then the process of finding a polynomial that passes through the points (x_i, y_i), $i = 0, \ldots, n$, is equivalent to solving a system of linear equations $P\mathbf{c} = \mathbf{y}$ that has a unique solution. The matrix P is determined by the choice of basis for the space of polynomials of degree at most n. In Section 7.1, the monomial basis was used, which is a convenient choice, but P was a Vandermonde matrix, which made the coefficients difficult to obtain. By contrast, in Section 7.2, Lagrange polynomials were used, which are tedious to construct, but P was the identity matrix, so obtaining the coefficients was trivial.

In **Newton interpolation**, the basis functions are $\{\mathcal{N}_j(x)\}_{j=0}^n$, where

$$\mathcal{N}_0(x) = 1, \quad \mathcal{N}_j(x) = \prod_{k=0}^{j-1}(x - x_k), \quad j = 1, \ldots, n.$$

The numbers $x_0, x_1, \ldots, x_{n-1}$ that are subtracted from x in the linear factors that constitute these basis functions are referred to as the **centers** of the Newton form. Because each polynomial $\mathcal{N}_j(x)$ vanishes at $x_0, x_1, \ldots, x_{j-1}$, the matrix P is lower triangular (see Section B.10), so the system $P\mathbf{c} = \mathbf{y}$ is still easily solved, using **forward substitution** presented in Section 3.2.2. The advantage of Newton interpolation over Lagrange interpolation is that the interpolating polynomial is easily updated as interpolation points are added, since the basis functions $\{\mathcal{N}_j(x)\}$, $j = 0, \ldots, n$, do not change from the addition of the new points. While this is also true of the monomial basis, the relatively sparse structure of the matrix P used in Newton interpolation allows computation of the coefficients in only $O(n^2)$ floating-point operations, rather than $O(n^3)$ in the case of P being a Vandermonde matrix. In summary, Newton interpolation is an effective compromise between Vandermonde interpolation and Lagrange interpolation.

For convenience, we assume that the data consist of the values of a function $f(x)$ at $x_0, x_1, \ldots, x_n$; that is, $y_i = f(x_i)$ for $i = 0, 1, \ldots, n$. Using Theorem 7.3.1, it can be shown that the coefficients c_j of the Newton interpolating polynomial

$$p_n(x) = \sum_{j=0}^n c_j \mathcal{N}_j(x) \tag{7.9}$$

are given by

$$c_j = f[x_0, \ldots, x_j] \tag{7.10}$$

where $f[x_0, \ldots, x_j]$ denotes the **divided difference** of $x_0, \ldots, x_j$. The divided difference is defined as follows:

$$f[x_i] = y_i, \tag{7.11}$$

$$f[x_i, x_{i+1}] = \frac{f[x_{i+1}] - f[x_i]}{x_{i+1} - x_i}, \tag{7.12}$$

$$f[x_i, x_{i+1}, \ldots, x_{i+k}] = \frac{f[x_{i+1}, \ldots, x_{i+k}] - f[x_i, \ldots, x_{i+k-1}]}{x_{i+k} - x_i}. \tag{7.13}$$

This definition implies that for each nonnegative integer j, the divided difference $f[x_0, x_1, \ldots, x_j]$ only depends on the interpolation points $x_0, x_1, \ldots, x_j$ and the value of $f(x)$ at these points. It follows that the addition of new interpolation points does not change the coefficients $c_0, \ldots, c_n$. Specifically, we have

$$p_{n+1}(x) = p_n(x) + f[x_0, x_1, \ldots, x_n, x_{n+1}] \mathcal{N}_{n+1}(x).$$

This ease of updating makes Newton interpolation a very practical method of obtaining the interpolating polynomial.

> **Exploration 7.3.2** Use Theorem 7.3.1 and the definition of divided differences given by (7.11)-(7.13) to prove, by induction, that the coefficients $c_0, c_1, \ldots, c_n$ of $p_n(x)$ in the Newton form (7.9) are given by (7.10). *Hint:* In the induction step, use $i = 0$ and $j = n$ in Theorem 7.3.1.

The following result shows how the Newton interpolating polynomial bears a resemblance to a Taylor polynomial (see Section A.6).

> **Theorem 7.3.2** Let f be n times continuously differentiable on $[a, b]$, and let $x_0, x_1, \ldots, x_n$ be distinct points in $[a, b]$. Then there exists a number $\xi \in (a, b)$ such that $f[x_0, x_1, \ldots, x_n] = \frac{f^{(n)}(\xi)}{n!}$.

We now prove Theorem 7.3.2, beginning with the simplest cases.

> **Exploration 7.3.3** Prove Theorem 7.3.2 for the cases of $n = 1$ and $n = 2$, using the definition of divided differences, the Mean Value Theorem (Theorem A.5.3), and Taylor's Theorem (Theorem A.6.1).

The approach used in Exploration 7.3.3 is difficult to generalize to larger values of n, so we now consider a different approach.

> **Exploration 7.3.4** Let $p_n(x)$ be the interpolating polynomial for $f(x)$ at points $x_0, x_1, \ldots, x_n \in [a, b]$, and assume that f is n times differentiable. Use Rolle's Theorem (Theorem A.5.1) to prove that $p_n^{(n)}(\xi) = f^{(n)}(\xi)$ for some point $\xi \in (a, b)$.

> **Exploration 7.3.5** Use Exploration 7.3.4 to prove Theorem 7.3.2. *Hint:* Express $p_n(x)$ as the sum of $p_{n-1}(x)$ and an additional term, where $p_{n-1}(x)$ interpolates $f(x)$ at any n of the points $x_0, x_1, \ldots, x_n$.

7.3.2 Computing the Newton Interpolating Polynomial

We now describe how to compute the divided differences $c_j = f[x_0, x_1, \ldots, x_j]$ of the Newton interpolating polynomial $p_n(x)$, and how to evaluate $p_n(x)$ efficiently. The computation proceeds by filling in the entries of a **divided difference table**. This is a triangular table consisting of $n + 1$ columns, where n is the degree of the interpolating polynomial to be computed. For $j = 0, 1, \ldots, n$, the jth column contains $n + 1 - j$ entries, which are the divided differences $f[x_k, x_{k+1}, \ldots, x_{k+j}]$, for $k = 0, 1, \ldots, n - j$. A divided difference table for $n = 3$ has the form

$$
\begin{array}{llll}
f[x_i] & f[x_i, x_{i+1}] & f[x_i, \ldots, x_{i+2}] & f[x_i, \ldots, x_{i+3}] \\
\hline
f(x_0) & & & \\
 & f[x_0, x_1] & & \\
f(x_1) & & f[x_0, x_1, x_2] & \\
 & f[x_1, x_2] & & f[x_0, x_1, x_2, x_3] \\
f(x_2) & & f[x_1, x_2, x_3] & \\
 & f[x_2, x_3] & & \\
f(x_3) & & &
\end{array}
$$

We construct this table by filling in the $n + 1$ entries in column 0, which are the trivial divided differences $f[x_j] = f(x_j)$, for $j = 0, 1, \ldots, n$. Then, we use the recursive definition of the divided differences from (7.13) to fill in the entries of subsequent columns. Once the construction of the table is complete, we can obtain the coefficients of the Newton interpolating polynomial from the first entry in each column, which is $f[x_0, x_1, \ldots, x_j]$, for $j = 0, 1, \ldots, n$. The algorithm for constructing the divided difference table follows.

Algorithm 7.3.3 (Divided Difference Table) Given $n + 1$ distinct interpolation points $x_0, x_1, \ldots, x_n$, and the values of a function $f(x)$ at these points, the following algorithm computes the coefficients $c_j = f[x_0, x_1, \ldots, x_j]$ of the Newton interpolating polynomial $p_n(x)$.

for $i = 0, 1, \ldots, n$ **do**
    $d_{i,0} = f(x_i)$
end for
for $j = 1, 2, \ldots, n$ **do**
    for $i = n, n - 1, \ldots, j$ **do**
        $d_{i,j} = (d_{i,j-1} - d_{i-1,j-1})/(x_i - x_{i-j})$
    end for
end for
for $j = 0, 1, \ldots, n$ **do**
    $c_j = d_{j,j}$
end for

In a practical implementation of this algorithm, we do not need to store the entire table, because we only need the first entry in each column. Because each column has one fewer entry than the previous column, we can overwrite all of the other entries that we do not need. Algorithm 7.3.3 does *not* implement this idea, but it can be modified to do so.

Exploration 7.3.6 Modify Algorithm 7.3.3 so that it uses at most $O(n)$ storage; that is, it does not need to store a matrix $\{d_{i,j}\}_{i,j=0}^n$ of divided differences.

Example 7.3.4 We will use Newton interpolation to construct the third-degree polynomial $p_3(x)$ that fits the data from Example 7.2.2,

i	x_i	$f(x_i)$
0	-1	3
1	0	-4
2	1	5
3	2	-6

In other words, we must have $p_3(-1) = 3$, $p_3(0) = -4$, $p_3(1) = 5$, and $p_3(2) = -6$.

First, we construct the divided difference table from this data. The divided differences in the table are computed as follows:

$$f[x_0] = f(x_0) = 3, \quad f[x_1] = f(x_1) = -4,$$

$$f[x_2] = f(x_2) = 5, \quad f[x_3] = f(x_3) = -6,$$

$$f[x_0, x_1] = \frac{f[x_1] - f[x_0]}{x_1 - x_0} = \frac{-4 - 3}{0 - (-1)} = -7,$$

$$f[x_1, x_2] = \frac{f[x_2] - f[x_1]}{x_2 - x_1} = \frac{5 - (-4)}{1 - 0} = 9,$$

$$f[x_2, x_3] = \frac{f[x_3] - f[x_2]}{x_3 - x_2} = \frac{-6 - 5}{2 - 1} = -11,$$

$$f[x_0, x_1, x_2] = \frac{f[x_1, x_2] - f[x_0, x_1]}{x_2 - x_0} = \frac{9 - (-7)}{1 - (-1)} = 8,$$

$$f[x_1, x_2, x_3] = \frac{f[x_2, x_3] - f[x_1, x_2]}{x_3 - x_1} = \frac{-11 - 9}{2 - 0} = -10,$$

$$f[x_0, x_1, x_2, x_3] = \frac{f[x_1, x_2, x_3] - f[x_0, x_1, x_2]}{x_3 - x_0} = \frac{-10 - 8}{2 - (-1)} = -6.$$

The resulting divided difference table is

$f[x_0] = 3$			
	$f[x_0, x_1] = -7$		
$f[x_1] = -4$		$f[x_0, x_1, x_2] = 8$	
	$f[x_1, x_2] = 9$		$f[x_0, x_1, x_2, x_3] = -6$
$f[x_2] = 5$		$f[x_1, x_2, x_3] = -10$	
	$f[x_2, x_3] = -11$		
$f[x_3] = -6$			

It follows that the interpolating polynomial $p_3(x)$ can be expressed in Newton form as follows:

$$p_3(x) = \sum_{j=0}^{3} f[x_0, \ldots, x_j] \prod_{i=0}^{j-1} (x - x_i)$$

$$= f[x_0] + f[x_0, x_1](x - x_0) + f[x_0, x_1, x_2](x - x_0)(x - x_1) +$$
$$f[x_0, x_1, x_2, x_3](x - x_0)(x - x_1)(x - x_2)$$
$$= 3 - 7(x + 1) + 8(x + 1)x - 6(x + 1)x(x - 1).$$

We see that Newton interpolation produces an interpolating polynomial that is in the Newton form, with centers $x_0 = -1$, $x_1 = 0$, and $x_2 = 1$. $\square$

Exploration 7.3.7 Write a MATLAB function

$$c=\texttt{divdiffs(x,y)},$$

based on Algorithm 7.3.3, that computes the divided difference table from the given data stored in the input vectors x and y, and returns a vector c consisting of the divided differences $f[x_0, \ldots, x_j]$, $j = 0, 1, 2, \ldots, n$, where $n + 1$ is the length of both x and y.

Once the coefficients $c_0, c_1, \ldots, c_n$ of the Newton form (7.9) have been computed, we can use **nested multiplication** to evaluate the resulting interpolating polynomial, which is represented using the Newton form

$$p_n(x) = \sum_{j=0}^{n} c_j \mathcal{N}_j(x)$$

$$= \sum_{j=0}^{n} f[x_0, x_1, \ldots, x_j] \prod_{i=0}^{j-1} (x - x_i)$$

$$= f[x_0] + f[x_0, x_1](x - x_0) + f[x_0, x_1, x_2](x - x_0)(x - x_1) +$$
$$f[x_0, x_1, \ldots, x_n](x - x_0)(x - x_1) \cdots (x - x_{n-1}). \tag{7.14}$$

By writing $p_n(x)$ in **nested form**

$$p_n(x) = f[x_0] + (x - x_0)\left(f[x_0, x_1] + (x - x_1)\left(f[x_0, x_1, x_2] + \cdots +\right.\right.$$
$$f[x_0, x_1, \ldots, x_{n-1}] + (x - x_{n-1})f[x_0, x_1, \ldots, x_n]\right)\right)$$
$$= c_0 + (x - x_0)(c_1 + (x - x_1)(c_2 + \cdots + c_{n-1} + (x - x_{n-1})c_n)),$$

we obtain the following efficient algorithm for evaluating $p_n(x)$.

Algorithm 7.3.5 (Nested Multiplication) Given $n + 1$ distinct interpolation points $x_0, x_1, \ldots, x_n$ and the coefficients $c_j = f[x_0, x_1, \ldots, x_j]$, $j = 0, 1, \ldots, n$, of the Newton interpolating polynomial $p_n(x)$, the following algorithm computes $y = p_n(x)$ for a given real number x.

$b_n = c_n$
for $i = n - 1, n - 2, \ldots, 0$ **do**
$\quad b_i = c_i + (x - x_i)b_{i+1}$
end for
$y = b_0$

Exploration 7.3.8 Write a MATLAB function

$$yy = \texttt{newtonval(x,y,xx)}$$

that accepts as input vectors x and y, both of length $n + 1$, representing the data points $(x_0, y_0), (x_1, y_1), \ldots, (x_n, y_n)$, and a vector of x-values xx. This function uses Algorithm 7.3.5, as well as the function `divdiffs` from Exploration 7.3.7, to compute the value of the Newton interpolating polynomial for the given data points at each x-value in xx. The corresponding y-values must be returned in the vector yy, which must have the same dimensions as xx.

Nested multiplication has another useful application: the conversion of a polynomial from Newton form to **power form** (7.1). This is illustrated in the following example.

Example 7.3.6 Consider the interpolating polynomial obtained in the previous example,
$$p_3(x) = 3 - 7(x + 1) + 8(x + 1)x - 6(x + 1)x(x - 1).$$
We will use nested multiplication to write this polynomial in power form
$$p_3(x) = b_3 x^3 + b_2 x^2 + b_1 x + b_0.$$
This requires repeatedly applying nested multiplication to a polynomial of the form
$$p(x) = c_0 + c_1(x - x_0) + c_2(x - x_0)(x - x_1) + c_3(x - x_0)(x - x_1)(x - x_2),$$
and for each application it will perform the following steps,
$$b_3 = c_3$$
$$b_2 = c_2 + (x - x_2)b_3$$
$$b_1 = c_1 + (x - x_1)b_2$$
$$b_0 = c_0 + (x - x_0)b_1,$$
where, in this example, we will set $x = 0$ each time. The numbers b_0, b_1, b_2, and b_3 computed by the algorithm are the coefficients of $p(x)$ in the Newton form, with the centers x_0, x_1, and x_2 changed to 0, x_0, and x_1; that is,
$$p(x) = b_0 + b_1(x - 0) + b_2(x - 0)(x - x_0) + b_3(x - 0)(x - x_0)(x - x_1).$$

Note that $b_0 = p(0)$, as the main purpose of nested multiplication is to evaluate a polynomial in Newton form at a given point x.

We now carry out this process on our specific polynomial. Initially, we have

$$p(x) = 3 - 7(x + 1) + 8(x + 1)x - 6(x + 1)x(x - 1),$$

so the coefficients of $p(x)$ in this Newton form are

$$c_0 = 3, \quad c_1 = -7, \quad c_2 = 8, \quad c_3 = -6,$$

with the centers

$$x_0 = -1, \quad x_1 = 0, \quad x_2 = 1.$$

Applying nested multiplication to these coefficients and centers, with $z = 0$, yields

$$b_3 = -6$$
$$b_2 = 8 + (0 - 1)(-6) = 14$$
$$b_1 = -7 + (0 - 0)(14) = -7$$
$$b_0 = 3 + (0 - (-1))(-7) = -4.$$

It follows that

$$p(x) = -4 + (-7)(x - 0) + 14(x - 0)(x - (-1)) + (-6)(x - 0)(x - (-1))(x - 0)$$
$$= -4 - 7x + 14x(x + 1) - 6x^2(x + 1),$$

and the centers are now 0, -1, and 0.

For the second application of nested multiplication, we have

$$p(x) = -4 - 7x + 14x(x + 1) - 6x^2(x + 1),$$

so the coefficients of $p(x)$ in this Newton form are

$$c_0 = -4, \quad c_1 = -7, \quad c_2 = 14, \quad c_3 = -6,$$

with the centers

$$x_0 = 0, \quad x_1 = -1, \quad x_2 = 0.$$

Applying nested multiplication to these coefficients and centers, with $z = 0$, yields

$$b_3 = -6$$
$$b_2 = 14 + (0 - 0)(-6) = 14$$
$$b_1 = -7 + (0 - (-1))(14) = 7$$
$$b_0 = -4 + (0 - 0)(7) = -4.$$

It follows that

$$p(x) = -4 + 7(x - 0) + 14(x - 0)(x - 0) + (-6)(x - 0)(x - 0)(x - (-1))$$
$$= -4 + 7x + 14x^2 - 6x^2(x + 1),$$

and the centers are now 0, 0, and -1.

For the third and final application of nested multiplication, we have

$$p(x) = -4 + 7x + 14x^2 - 6x^2(x+1),$$

so the coefficients of $p(x)$ in this Newton form are

$$c_0 = -4, \quad c_1 = 7, \quad c_2 = 14, \quad c_3 = -6,$$

with the centers

$$x_0 = 0, \quad x_1 = 0, \quad x_2 = -1.$$

Applying nested multiplication to these coefficients and centers, with $z = 0$, yields

$$b_3 = -6$$
$$b_2 = 14 + (0 - (-1))(-6) = 8$$
$$b_1 = 7 + (0 - 0)(8) = 7$$
$$b_0 = -4 + (0 - 0)(7) = -4.$$

It follows that

$$p(x) = -4 + 7(x - 0) + 8(x - 0)(x - 0) + (-6)(x - 0)(x - 0)(x - 0)$$
$$= -4 + 7x + 8x^2 - 6x^3,$$

and the centers are now 0, 0, and 0. Since all of the centers are equal to zero, the polynomial is now in power form. $\square$

It should be noted that nested multiplication, as applied in the preceding example, is not the most efficient way to convert the Newton form of $p(x)$ to the power form. To see this, we observe that after one application of nested multiplication, we have

$$p(x) = b_0 + b_1(x - 0) + b_2(x - 0)(x - x_0) + b_3(x - 0)(x - x_0)(x - x_1)$$
$$= b_0 + (x - 0)[b_1 + b_2(x - x_0) + b_3(x - x_0)(x - x_1)].$$

Therefore, we can apply nested multiplication to the second-degree polynomial

$$q(x) = b_1 + b_2(x - x_0) + b_3(x - x_0)(x - x_1),$$

which is the quotient obtained by dividing $p(x)$ by $(x - 0)$. Because

$$p(x) = b_0 + (x - 0)q(x),$$

it follows that once we have changed all of the centers of $q(x)$ to be equal to 0, then all of the centers of $p(x)$ will be equal to 0 as well. In summary, we can convert a polynomial of degree n from Newton form to power form by applying nested multiplication n times, where the jth application is to a polynomial of degree $n - j + 1$, for $j = 1, 2, \ldots, n$.

Since the coefficients of the appropriate Newton form of each of these polynomials of successively lower degree are computed by the nested multiplication algorithm, it follows that we can implement this more efficient procedure simply by proceeding exactly as before, except that during the jth application of nested multiplication, we do not compute the coefficients $b_0, b_1, \ldots, b_{j-2}$, because they will not change anyway, as can be seen from the previous computations. For example, in the second application, we did not need to compute b_0, and in the third, we did not need to compute b_0 and b_1.

Exploration 7.3.9 Write a MATLAB function

$$p=powerform(x,c)$$

that accepts as input vectors x and c, both of length $n+1$, consisting of the interpolation points x_j and divided differences $f[x_0, x_1, \ldots, x_j]$, respectively, $j = 0, 1, \ldots, n$. The output is a $(n+1)$-vector consisting of the coefficients of the interpolating polynomial $p_n(x)$ in power form, ordered from highest degree to lowest.

Exploration 7.3.10 Write a MATLAB function

$$p=newtonfit(x,y)$$

that accepts as input vectors x and y of length $n+1$ consisting of the x- and y-coordinates, respectively, of points $(x_0, y_0), (x_1, y_1), \ldots, (x_n, y_n)$, where the x-values must all be distinct, and returns a $(n+1)$-vector p consisting of the coefficients of the Newton interpolating polynomial $p_n(x)$, in power form, with highest-degree coefficient in the first position. Use your **divdiffs** function from Exploration 7.3.7 and your **powerform** function from Exploration 7.3.9. Test your function by comparing your output to that of the built-in function **polyfit**.

When a polynomial is in power form, nested multiplication simplifies, as the centers are all equal to zero. The resulting algorithm is known as **Horner's Method**.

Algorithm 7.3.7 (Horner's Method) Given the coefficients c_j, $j = 0, 1, 2, \ldots, n$, of $p_n(x)$ in power form, the following algorithm computes $y = p_n(x)$ for a given real number x.

$y = c_n$
for $i = n-1, n-2, \ldots, 0$ **do**
    $y = c_i + xy$
end for

Using Horner's Method is equivalent to writing the polynomial $p_n(x)$ in the form

$$p_n(x) = a_0 + x(a_1 + x(a_2 + x(a_3 + \cdots + (a_{n-1} + xa_n)))),$$

which helps reduce the number of floating-point operations needed for evaluation.

7.3.3 Equally Spaced Points*

Suppose that the interpolation points $x_0, x_1, \ldots, x_n$ are equally spaced; that is, $x_i = x_0 + ih$ for some positive number h. In this case, the Newton interpolating polynomial can be simplified, since the denominators of all of the divided differences can be expressed in terms of the spacing h. If we define the **forward difference operator** Δ by

$$\Delta x_k = x_{k+1} - x_k,$$

where $\{x_k\}$ is any sequence, then the divided differences $f[x_0, x_1, \ldots, x_k]$ are given by

$$f[x_0, x_1, \ldots, x_k] = \frac{1}{k!h^k} \Delta^k f(x_0). \tag{7.15}$$

where Δ^k denotes k successive applications of the forward difference operator.

The interpolating polynomial can then be described by the **Newton forward-difference formula**

$$p_n(x) = f[x_0] + \sum_{k=1}^{n} \binom{s}{k} \Delta^k f(x_0), \tag{7.16}$$

where the new variable s is related to x by

$$s = \frac{x - x_0}{h},$$

and the **extended binomial coefficient** $\binom{s}{k}$ is defined by

$$\binom{s}{k} = \frac{s(s-1)(s-2)\cdots(s-k+1)}{k!},$$

where k is a nonnegative integer. It can be seen that this definition is consistent with that of binomial coefficients, for the case in which s is an integer greater than or equal to k.

Exploration 7.3.11 Use induction to prove (7.15). Then show that the Newton interpolating polynomial (7.14) reduces to (7.16) in the case of equally spaced interpolation points.

Example 7.3.8 We will use the Newton forward-difference formula

$$p_n(x) = f[x_0] + \sum_{k=1}^{n} \binom{s}{k} \Delta^k f(x_0)$$

to compute the interpolating polynomial $p_3(x)$ that fits the data

i	x_i	$f(x_i)$
0	-1	3
1	0	-4
2	1	5
3	2	-6

In other words, we must have $p_3(-1) = 3$, $p_3(0) = -4$, $p_3(1) = 5$, and $p_3(2) = -6$. Note that the interpolation points $x_0 = -1$, $x_1 = 0$, $x_2 = 1$, and $x_3 = 2$ are equally spaced, with spacing $h = 1$.

To apply the forward-difference formula, we define $s = (x - x_0)/h = x + 1$ and compute the extended binomial coefficients

$$\binom{s}{1} = s = x + 1, \quad \binom{s}{2} = \frac{s(s-1)}{2} = \frac{x(x+1)}{2},$$

$$\binom{s}{3} = \frac{s(s-1)(s-2)}{6} = \frac{(x+1)x(x-1)}{6},$$

and then the coefficients

$$
\begin{aligned}
f[x_0] &= f(x_0) = 3, \\
\Delta f(x_0) &= f(x_1) - f(x_0) = -7, \\
\Delta^2 f(x_0) &= \Delta(\Delta f(x_0)) \\
&= \Delta[f(x_1) - f(x_0)] \\
&= [f(x_2) - f(x_1)] - [f(x_1) - f(x_0)] \\
&= f(x_2) - 2f(x_1) + f(x_0) = 16, \\
\Delta^3 f(x_0) &= \Delta(\Delta^2 f(x_0)) \\
&= \Delta[f(x_2) - 2f(x_1) + f(x_0)] \\
&= [f(x_3) - f(x_2)] - 2[f(x_2) - f(x_1)] + [f(x_1) - f(x_0)] \\
&= f(x_3) - 3f(x_2) + 3f(x_1) - f(x_0) = -36.
\end{aligned}
$$

It follows that

$$
\begin{aligned}
p_3(x) &= f[x_0] + \sum_{k=1}^{3} \binom{s}{k} \Delta^k f(x_0) \\
&= 3 + \binom{s}{1} \Delta f(x_0) + \binom{s}{1} \Delta^2 f(x_0) + \binom{s}{2} \Delta^3 f(x_0) \\
&= 3 + (x+1)(-7) + \frac{x(x+1)}{2} 16 + \frac{(x+1)x(x-1)}{6}(-36) \\
&= 3 - 7(x+1) + 8(x+1)x - 6(x+1)x(x-1).
\end{aligned}
$$

Note that the forward-difference formula computes the same form of the interpolating polynomial as the Newton divided difference formula. □

Exploration 7.3.12 Define the **backward difference operator** ∇ by

$$\nabla x_k = x_k - x_{k-1},$$

for any sequence $\{x_k\}$. Then derive the **Newton backward-difference formula**

$$p_n(x) = f[x_n] + \sum_{k=1}^{n} (-1)^k \begin{pmatrix} -s \\ k \end{pmatrix} \nabla^k f(x_n),$$

where $s = (x - x_n)/h$, and the preceding definition of the extended binomial coefficient applies.

The quantities $\Delta^k f(x_0)$ can be computed efficiently using a process like that of constructing a divided difference table, except without denominators, as they are incorporated into the extended binomial coefficients $\begin{pmatrix} s \\ k \end{pmatrix}$, $k = 1, 2, \ldots, n$. These coefficients can be computed efficiently using the relationship

$$\begin{pmatrix} s \\ k \end{pmatrix} = \begin{pmatrix} s \\ k-1 \end{pmatrix} \frac{(s-k+1)}{k}, \quad k \geq 1.$$

These implementation details can be used in the following exploration.

Exploration 7.3.13 Look up the documentation for the MATLAB function `diff`. Then write functions `yy=newtonforwdiff(x,y,xx)` and `yy=newtonbackdiff(x,y,xx)` that use `diff` to implement the Newton forward-difference and Newton backward-difference formulas, respectively, and evaluate the interpolating polynomial $p_n(x)$, where $n = $ `length(x)` $- 1$, at the elements of `xx`. The resulting values must be returned in `yy`.

7.3.4 Concept Check

1. What advantage does Newton interpolation have over Lagrange interpolation?
2. What are divided differences, and what role do they play in Newton interpolation?
3. Describe the relationship between divided differences $f[x_0, x_1, \ldots, x_k]$ and higher-order derivatives of $f(x)$.
4. What is a divided difference table, and what role does it play in Newton interpolation?
5. Once the divided difference table is completed, how is the Newton interpolating polynomial obtained?
6. What is Horner's method?
7. What is nested multiplication, and how does it relate to Horner's method?
8. How can the interpolating polynomial be converted from Newton form to power form?

9. How does Newton interpolation simplify in the case of equally spaced points?

7.4   Error Analysis

In some applications, the interpolating polynomial $p_n(x)$ is used to fit a function $f(x)$ at the interpolation points $x_0, \ldots, x_n$, usually because $f(x)$ is not suitable for tasks such as differentiation or integration that are easy for polynomials, or because it is not practical to evaluate $f(x)$ at points other than the interpolation points because an explicit formula for $f(x)$ is not available. In such an application, it is desirable to determine how well $p_n(x)$ approximates $f(x)$ away from the interpolation points.

7.4.1   Error Estimation

From Theorem 7.3.2, we can obtain the following result.

Theorem 7.4.1 (Interpolation error) If f is $n + 1$ times continuously differentiable on $[a, b]$, and $p_n(x)$ is the unique polynomial of degree n that interpolates $f(x)$ at the $n + 1$ distinct points $x_0, x_1, \ldots, x_n$ in $[a, b]$, then for each $x \in [a, b]$,

$$f(x) - p_n(x) = \frac{f^{(n+1)}(\xi(x))}{(n+1)!} \prod_{j=0}^{n}(x - x_j),$$

where $\xi(x) \in (a, b)$.

It is interesting to note that the error closely resembles the Taylor remainder $R_n(x)$ from Theorem A.6.1.

Exploration 7.4.1 Prove Theorem 7.4.1. *Hint:* Work with the Newton interpolating polynomial for the points $x_0, x_1, \ldots, x_n, x$, and use the result of Theorem 7.3.2.

We now illustrate the application of Theorem 7.4.1.

Example 7.4.2 Let $f(x) = \sin x$, and let $p_1(x)$ be the first-degree polynomial that interpolates $f(x)$ at $x_0 = 0$ and $x_1 = 1$. Then, from Theorem 7.4.1, we have that for $x \in [0, 1]$,

$$f(x) - p_1(x) = -\frac{\sin \xi(x)}{2!} x(x - 1),$$

where $\xi(x) \in (0, 1)$. To obtain an upper bound on the error, we note that $\sin \theta$ is nonnegative and increasing for $\theta \in [0, 1]$, so $|\sin \theta| \leq \sin(1)$. To bound $|x(x-1)|$ on $[0, 1]$, we note that the function $g(x) = x(x - 1)$ vanishes at the endpoints of $[0, 1]$, and it has one critical point at $x = 1/2$. Therefore $|x(x - 1)| \leq 1/4$ on $[0, 1]$, and we have

$$|\sin x - p_1(x)| \le \frac{\sin(1)}{8} \approx 0.1052.$$

Using calculus, we find that on $[0,1]$, the function $h(x) = |\sin x - p_1(x)|$ has a maximum of approximately 0.06, which is consistent with our upper bound. $\square$

Exploration 7.4.2 Determine a bound on the error $|f(x) - p_2(x)|$ for x in $[0,1]$, where $f(x) = e^x$, and $p_2(x)$ is the interpolating polynomial of $f(x)$ at $x_0 = 0$, $x_1 = 0.5$, and $x_2 = 1$.

If the number of data points is large, then polynomial interpolation becomes problematic since high-degree interpolation yields oscillatory polynomials, when the data may fit a smooth function.

Example 7.4.3 Suppose that we wish to approximate the function $f(x) = 1/(1 + x^2)$ on the interval $[-5,5]$ with a tenth-degree interpolating polynomial that agrees with $f(x)$ at 11 equally-spaced points $x_0, x_1, \ldots, x_{10}$ in $[-5,5]$, where $x_j = -5 + j$, for $j = 0, 1, \ldots, 10$. Figure 7.1 shows that the resulting polynomial is not a good approximation of $f(x)$ on this interval, even though it agrees with $f(x)$ at the interpolation points. The following MATLAB session shows how the plot in the figure can be created.

```
>> % create vector of 11 equally spaced points in [-5,5]
>> x=linspace(-5,5,11);
>> % compute corresponding y-values
>> y=1./(1+x.^2);
>> % compute 10th-degree interpolating polynomial
>> p=polyfit(x,y,10);
>> % for plotting, create vector of 100 equally spaced points
>> xx=linspace(-5,5);
>> % compute corresponding y-values to plot function
>> yy=1./(1+xx.^2);
>> % plot function
>> plot(xx,yy)
>> % tell MATLAB that next plot should be superimposed on
>> % current one
>> hold on
>> % plot polynomial, using polyval to compute values
>> % and a red dashed curve
>> plot(xx,polyval(p,xx),'r--')
>> % indicate interpolation points on plot using circles
>> plot(x,y,'o')
>> % label axes
```

```
>> xlabel('x')
>> ylabel('y')
>> % set caption
>> title('Runge''s example')
```

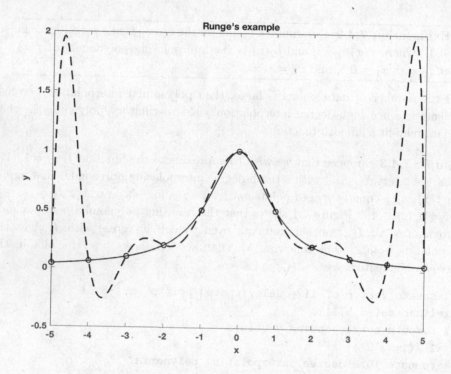

Fig. 7.1 The function $f(x) = 1/(1+x^2)$ (solid curve) cannot be interpolated accurately on $[-5, 5]$ using a tenth-degree polynomial (dashed curve) with equally-spaced interpolation points.

The example shown in Figure 7.1 is a well-known example of the difficulty of high-degree polynomial interpolation using equally-spaced points, and it is known as **Runge's Example** [Runge (1901)]. □

7.4.2 Chebyshev Interpolation

In general, it is not wise to use a high-degree interpolating polynomial and equally-spaced interpolation points to approximate a function on an interval $[a, b]$. Is it possible to choose the interpolation points so that the error is minimized? To answer this question, we introduce the **Chebyshev polynomials**

$$T_k(x) = \cos(k \cos^{-1}(x)), \quad |x| \le 1, \quad k = 0, 1, 2, \ldots. \tag{7.17}$$

Using (7.17) and the sum and difference formulas for cosine,

$$\cos(A + B) = \cos A \cos B - \sin A \sin B, \tag{7.18}$$
$$\cos(A - B) = \cos A \cos B + \sin A \sin B, \tag{7.19}$$

it can be shown that the Chebyshev polynomials satisfy the **three-term recurrence relation**

$$T_{k+1}(x) = 2x T_k(x) - T_{k-1}(x), \quad k \geq 1. \tag{7.20}$$

It can easily be seen from this relation, and the first two Chebyshev polynomials, that $T_k(x)$ is in fact a polynomial for all integers $k \geq 0$.

The Chebyshev polynomials have the following properties of interest:

1. The leading coefficient of $T_k(x)$ is 2^{k-1}.
2. $T_k(x)$ is an **even function** (that is, $T_k(-x) = T_k(x)$) if k is even, and an **odd function** (that is, $T_k(-x) = -T_k(x)$) if k is odd.
3. The zeros of $T_k(x)$, for $k \geq 1$, are

$$x_j = \cos \frac{(2j+1)\pi}{2k}, \quad j = 0, 1, 2, \ldots, k-1. \tag{7.21}$$

These roots are known as the **Chebyshev points**. Because these points are cosines of equally spaced angles in $(0, \pi)$, they are distributed non-uniformly within $(-1, 1)$. In particular, the points become closely clustered near the endpoints $x = \pm 1$. From property 1 above, we also have, for $k \geq 1$,

$$T_k(x) = 2^{k-1}(x - x_0)(x - x_1) \cdots (x - x_{k-1}). \tag{7.22}$$

This factorization of $T_k(x)$ will be useful to us later.

4. The extrema of $T_k(x)$ on $[-1, 1]$ are

$$\tilde{x}_j = \cos \frac{j\pi}{k}, \quad j = 0, 1, \ldots, k,$$

and the corresponding extremal values are ± 1.
5. $|T_k(x)| \leq 1$ on $[-1, 1]$ for all $k \geq 0$.

Exploration 7.4.3 Use (7.17), (7.18), and (7.19) to prove (7.20).

Exploration 7.4.4 Use (7.20) and induction to show that the leading coefficient of $T_k(x)$ is 2^{k-1}, for $k \geq 1$.

Exploration 7.4.5 Use the roots of cosine to compute the roots of $T_k(x)$. Show that they are real, distinct, and lie within $(-1, 1)$.

Let $f(x)$ be a function that is $(n+1)$ times continuously differentiable on $[a, b]$, and let $\tau_1, \tau_2, \ldots, \tau_{n+1}$ be the $n+1$ roots of the Chebyshev polynomial $T_{n+1}(t)$, given by (7.21) with $k = n+1$. If we approximate $f(x)$ by a nth-degree polynomial

$p_n(x)$ that interpolates $f(x)$ at the $n+1$ points $\xi_1, \xi_2, \ldots, \xi_{n+1}$ obtained by mapping these roots from $[-1, 1]$ to $[a, b]$, which are

$$\xi_j = \frac{1}{2}(b-a)\tau_j + \frac{1}{2}(a+b), \quad j = 1, 2, \ldots, n+1,$$

then the error in this approximation is

$$f(x) - p_n(x) = \frac{f^{(n+1)}(\xi)}{(n+1)!} \left(\frac{b-a}{2}\right)^{n+1} 2^{-n} T_{n+1}(t(x)),$$

where

$$t(x) = -1 + \frac{2}{b-a}(x-a)$$

is the linear map from $[a, b]$ to $[-1, 1]$. This is because

$$\prod_{j=0}^{n}(x - \xi_j) = \left(\frac{b-a}{2}\right)^{n+1} \prod_{j=0}^{n}(t(x) - \tau_j) = \left(\frac{b-a}{2}\right)^{n+1} 2^{-n} T_{n+1}(t(x)),$$

where we have used (7.22) with $k = n + 1$. From $|T_{n+1}(t)| \le 1$, we obtain

$$|f(x) - p_n(x)| \le \frac{(b-a)^{n+1}}{2^{2n+1}(n+1)!} \max_{\xi \in [a,b]} |f^{(n+1)}(\xi)|.$$

It can be shown that using Chebyshev points leads to much less error in the function $f(x) = 1/(1 + x^2)$ from Runge's Example [Quateroni and Saleri (2003)].

7.4.3 Concept Check

1. What is the error in the interpolating polynomial $p_n(x)$ of degree n that is used to interpolate $f(x)$ at the points $x_0, x_1, \ldots, x_n$?
2. What is Runge's Example and what is its significance?
3. What is a Chebyshev polynomial?
4. What is the three-term recurrence relation satisfied by the Chebyshev polynomials?
5. What are Chebyshev points? How are they distributed?
6. What is the benefit of using Chebyshev points for polynomial interpolation?

7.5 Osculatory Interpolation

Suppose that the interpolation points $x_0, x_1, \ldots, x_n$ are perturbed so that two neighboring points x_i and x_{i+1}, $0 \le i < n$, approach each other. What happens to the interpolating polynomial? In the limit, as $x_{i+1} \to x_i$, the interpolating polynomial $p_n(x)$ not only satisfies $p_n(x_i) = y_i$, but also the condition

$$p_n'(x_i) = \lim_{x_{i+1} \to x_i} \frac{y_{i+1} - y_i}{x_{i+1} - x_i},$$

assuming this limit exists. It follows that in order to ensure uniqueness, the data must specify the value of the derivative of the interpolating polynomial at x_i. In general, the inclusion of an interpolation point x_i k times within the set $x_0, \ldots, x_n$ must be accompanied by specification of $p_n^{(j)}(x_i)$, $j = 0, \ldots, k - 1$, in order to ensure a unique solution. These values, divided by $j!$, are used in place of divided differences of identical interpolation points in Newton interpolation, by Theorem 7.3.2.

Interpolation with repeated interpolation points is called **osculatory interpolation**, since it can be viewed as the limit of distinct interpolation points approaching one another, and the term "osculatory" is derived from the Latin word for "kiss".

Exploration 7.5.1 Use a divided difference table to compute the interpolating polynomial $p_2(x)$ for the function $f(x) = \cos x$ on $[0, \pi]$, with interpolation points $x_0 = x_1 = 0$ and $x_2 = \pi$. That is, $p_2(x)$ must satisfy $p_2(0) = f(0)$, $p_2'(0) = f'(0)$ and $p_2(\pi) = f(\pi)$. Then, extend $p_2(x)$ to a cubic polynomial $p_3(x)$ by also requiring that $p_3'(\pi) = f'(\pi)$, updating the divided difference table accordingly.

Exploration 7.5.2 Suppose that osculatory interpolation is used to construct the polynomial $p_n(x)$ that interpolates $f(x)$ at only one x-value, x_0, and satisfies $p_n(x_0) = f(x_0)$, $p_n'(x_0) = f'(x_0)$, $p_n''(x_0) = f''(x_0)$, and so on, up to $p_n^{(n)}(x_0) = f^{(n)}(x_0)$. What polynomial approximation of $f(x)$ is obtained?

7.5.1 Hermite Interpolation

In the case where each of the *distinct* interpolation points $x_0, x_1, \ldots, x_n$ is repeated exactly once, for a total of $2n + 2$ points including duplicates, the interpolating polynomial for a differentiable function $f(x)$ is called the **Hermite interpolating polynomial** of $f(x)$, and is denoted by $H_{2n+1}(x)$, since this polynomial must have degree $2n + 1$ in order to satisfy the $2n + 2$ constraints

$$H_{2n+1}(x_i) = f(x_i), \quad H_{2n+1}'(x_i) = f'(x_i), \quad i = 0, 1, \ldots, n. \tag{7.23}$$

To satisfy these constraints, we define, for $i = 0, 1, \ldots, n$,

$$\mathcal{H}_{2n+1,i}(x) = [\mathcal{L}_{n,i}(x)]^2 (1 - 2\mathcal{L}_{n,i}'(x_i)(x - x_i)), \tag{7.24}$$

$$\mathcal{K}_{2n+1,i}(x) = [\mathcal{L}_{n,i}(x)]^2 (x - x_i), \tag{7.25}$$

where, as before, $\mathcal{L}_{n,i}(x)$ is the ith Lagrange polynomial (7.3) for the interpolation points $x_0, x_1, \ldots, x_n$.

It can be verified directly that these polynomials satisfy, for $i, j = 0, 1, \ldots, n$,

$$\mathcal{H}_{2n+1,i}(x_j) = \delta_{ij}, \quad \mathcal{H}_{2n+1,i}'(x_j) = 0,$$

$$\mathcal{K}_{2n+1,i}(x_j) = 0, \quad \mathcal{K}_{2n+1,i}'(x_j) = \delta_{ij},$$

where δ_{ij} is the *Kronecker delta*

$$\delta_{ij} = \begin{cases} 1 & i = j \\ 0 & i \neq j \end{cases}.$$

It follows that

$$H_{2n+1}(x) = \sum_{i=0}^{n} [f(x_i)\mathcal{H}_{2n+1,i}(x) + f'(x_i)\mathcal{K}_{2n+1,i}(x)] \qquad (7.26)$$

is a polynomial of degree $2n + 1$ that satisfies the above constraints.

Exploration 7.5.3 Derive the formulas (7.24), (7.25) for $\mathcal{H}_{2n+1,i}(x)$ and $\mathcal{K}_{2n+1,i}(x)$, respectively, using the specified constraints for these polynomials. *Hint:* Use an approach similar to that used to derive the formula for Lagrange polynomials.

Exploration 7.5.4 Prove that the Hermite interpolating polynomial $H_{2n+1}(x)$ from (7.26) is the unique polynomial of degree $2n + 1$ that satisfies the conditions (7.23). *Hint:* Use the same approach as in Exploration 7.1.3, applied to the derivative. Rolle's Theorem (Theorem A.5.1) will be helpful.

Using a similar approach as for the Lagrange interpolating polynomial, the following result can be proved.

Theorem 7.5.1 Let f be $2n + 2$ times continuously differentiable on $[a, b]$, and let H_{2n+1} denote the Hermite interpolating polynomial of f with interpolation points $x_0, x_1, \ldots, x_n$ in $[a, b]$. Then there exists a point $\xi(x) \in (a, b)$ such that

$$f(x) - H_{2n+1}(x) = \frac{f^{(2n+2)}(\xi(x))}{(2n + 2)!}(x - x_0)^2(x - x_1)^2 \cdots (x - x_n)^2.$$

The proof will be left as an exercise.

7.5.2 Divided Differences

The Hermite interpolating polynomial can be described using Lagrange polynomials and their derivatives, but this representation is not practical because of the difficulty of differentiating and evaluating these polynomials. Instead, one can construct the Hermite interpolating polynomial using a divided difference table, as in Section 7.3.2, in which each entry corresponding to two identical interpolation points is filled with the value of $f'(x)$ at the common point. Then, the Hermite interpolating polynomial can be represented using the Newton form.

Example 7.5.2 We will use Hermite interpolation to construct the third-degree polynomial $H_3(x)$ that fits $f(x)$ and $f'(x)$ at $x_0 = 0$ and $x_1 = 1$. For convenience, we define new interpolation points z_i that list each (distinct) x-value twice. That is,

$$z_{2i} = z_{2i+1} = x_i, \quad i = 0, 1, \ldots, n,$$

where, in this example, $n = 1$. Our data is as follows:

i	z_i	$f(z_i)$	$f'(z_i)$
0,1	0	0	1
2,3	1	0	1

In other words, we must have $H_3(0) = 0$, $H_3'(0) = 1$, $H_3(1) = 0$, and $H_3'(1) = 1$. To include the values of $f'(x)$ at the two distinct interpolation points, we repeat each point once, so that the number of interpolation points, including repetitions, is equal to the number of constraints described by the data.

First, we construct the divided difference table from this data. The divided differences in the table are computed as follows:

$$f[z_0] = f(z_0) = 0, \quad f[z_1] = f(z_1) = 0, \quad f[z_2] = f(z_2) = 0, \quad f[z_3] = f(z_3) = 0,$$

$$f[z_0, z_1] = \frac{f[z_1] - f[z_0]}{z_1 - z_0} = f'(z_0) = 1,$$

$$f[z_1, z_2] = \frac{f[z_2] - f[z_1]}{z_2 - z_1} = \frac{0 - 0}{1 - 0} = 0,$$

$$f[z_2, z_3] = \frac{f[z_3] - f[z_2]}{z_3 - z_2} = f'(z_2) = 1,$$

$$f[z_0, z_1, z_2] = \frac{f[z_1, z_2] - f[z_0, z_1]}{z_2 - z_0} = \frac{0 - 1}{1 - 0} = -1,$$

$$f[z_1, z_2, z_3] = \frac{f[z_2, z_3] - f[z_1, z_2]}{z_3 - z_1} = \frac{1 - 0}{1 - 0} = 1,$$

$$f[z_0, z_1, z_2, z_3] = \frac{f[z_1, z_2, z_3] - f[z_0, z_1, z_2]}{z_3 - z_0} = \frac{1 - (-1)}{1 - 0} = 2.$$

Note that the values of the derivative are used whenever a divided difference of the form $f[z_i, z_{i+1}]$ is to be computed, where $z_i = z_{i+1}$. This makes sense because

$$\lim_{z_{i+1} \to z_i} f[z_i, z_{i+1}] = \lim_{z_{i+1} \to z_i} \frac{f(z_{i+1}) - f(z_i)}{z_{i+1} - z_i} = f'(z_i).$$

The resulting divided difference table is

$f[z_0] = 0$			
	$f[z_0, z_1] = 1$		
$f[z_1] = 0$		$f[z_0, z_1, z_2] = -1$	
	$f[z_1, z_2] = 0$		$f[z_0, z_1, z_2, z_3] = 2$
$f[z_2] = 0$		$f[z_1, z_2, z_3] = 1$	
	$f[z_2, z_3] = 1$		
$f[z_3] = 0$			

It follows that the interpolating polynomial $p_3(x)$ can be obtained using the Newton form as follows:

$$H_3(x) = \sum_{j=0}^{3} f[z_0, \ldots, z_j] \prod_{i=0}^{j-1} (x - z_i)$$

$$= f[z_0] + f[z_0, z_1](x - z_0) + f[z_0, z_1, z_2](x - z_0)(x - z_1) +$$
$$f[z_0, z_1, z_2, z_3](x - z_0)(x - z_1)(x - z_2)$$
$$= 0 + (x - 0) + (-1)(x - 0)(x - 0) + 2(x - 0)(x - 0)(x - 1)$$
$$= x - x^2 + 2x^2(x - 1).$$

We see that Hermite interpolation, using divided differences, produces an interpolating polynomial that is in Newton form, with centers $z_0 = 0$, $z_1 = 0$, and $z_2 = 1$.
□

Exploration 7.5.5 Use the Newton form of the Hermite interpolating polynomial to prove Theorem 7.5.1.

Exploration 7.5.6 Modify your function `divdiffs` from Exploration 7.3.7 to obtain a new function `c=hermdivdiffs(x,y,yp)` that computes the divided difference table from the given data stored in the input vectors x and y, where the elements of x are distinct, as well as the derivative values stored in yp. The function must return a vector c consisting of the divided differences $f[z_0, \ldots, z_j]$, $j = 0, 1, 2, \ldots, 2n + 1$, where $n + 1$ is the length of both x and y. *Note:* keep in mind that for Hermite interpolation, each distinct x-value must be listed twice in the divided difference table!

Exploration 7.5.7 Write a MATLAB function

$$H=hermpolyfit(x,y,yp)$$

that is similar to the built-in function `polyfit`, in that it returns a vector of coefficients, in power form, for the interpolating polynomial corresponding to the given data, except that Hermite interpolation is used instead of Lagrange interpolation. Use the function `hermdivdiffs` from Exploration 7.5.6 as well as the function `powerform` from Exploration 7.3.9.

7.5.3   Concept Check

1. What is osculatory interpolation?
2. When interpolation points are repeated, how is uniqueness of the interpolating polynomial ensured?
3. What is Hermite interpolation?
4. What is the error in the Hermite interpolating polynomial $H_{2n+1}(x)$ that interpolates $f(x)$ at the points $x_0, x_1, \ldots, x_n$?
5. How is a divided difference table used to compute a Hermite interpolating polynomial?

7.6 Piecewise Polynomial Interpolation

We have seen that high-degree polynomial interpolation can be problematic. However, if the fitting function is only required to have a few continuous derivatives, then one can instead construct a **piecewise polynomial** to fit the data. We now precisely define what we mean by a piecewise polynomial.

Definition 7.6.1 (Piecewise polynomial) Let $[a, b]$ be an interval that is divided into subintervals $[x_i, x_{i+1}]$, where $i = 0, \ldots, n-1$, $x_0 = a$ and $x_n = b$. A **piecewise polynomial** is a function $p(x)$ defined on $[a, b]$ by

$$p(x) = p_i(x), \quad x_{i-1} \le x \le x_i, \quad i = 1, 2, \ldots, n,$$

where, for $i = 1, 2, \ldots, n$, each function $p_i(x)$ is a polynomial defined on $[x_{i-1}, x_i]$. The **degree** of $p(x)$ is the maximum degree of each polynomial $p_i(x)$, for $i = 1, 2, \ldots, n$.

It is essential to note that by this definition, a piecewise polynomial defined on $[a, b]$ is equal to some polynomial on each subinterval $[x_{i-1}, x_i]$ of $[a, b]$, for $i = 1, 2, \ldots, n$, but a different polynomial may be used for each subinterval.

To study the accuracy of piecewise polynomials, we need to work with various function spaces, including **Sobolev spaces**; these function spaces are defined in Section B.13.3.

7.6.1 Piecewise Linear Approximation

We first consider one of the simplest types of piecewise polynomials, a piecewise linear polynomial. Let $f \in C[a, b]$. Given the points $x_0, x_1, \ldots, x_n$ defined as above, the **linear spline** $s(x)$ that interpolates f at these points is defined by

$$s(x) = f(x_{i-1}) \frac{x - x_i}{x_{i-1} - x_i} + f(x_i) \frac{x - x_{i-1}}{x_i - x_{i-1}}, \quad x \in [x_{i-1}, x_i], \quad i = 1, 2, \ldots, n.$$
(7.27)

The points $x_0, x_1, \ldots, x_n$ are called the **knots** of the spline.

Exploration 7.6.1 Given that $s(x)$ must satisfy $s(x_i) = f(x_i)$ for $i = 0, 1, 2, \ldots, n$, explain how the formula (7.27) can be derived.

Exploration 7.6.2 Given the points $(x_0, f(x_0))$, $(x_1, f(x_1))$, $\ldots$, $(x_n, f(x_n))$, explain how $s(x)$ can easily be graphed. How can the graph be produced in a single line of MATLAB code, given vectors x and y containing the x- and y- coordinates of these points, respectively?

If $f \in C^2[a, b]$, then by the error in Lagrange interpolation (Theorem 7.4.1), on each subinterval $[x_{i-1}, x_i]$, for $i = 1, 2, \ldots, n$, we have

$$f(x) - s(x) = \frac{f''(\xi)}{2} (x - x_{i-1})(x - x_i).$$

This leads to the following theorem.

Theorem 7.6.2 Let $f \in C^2[a, b]$, and let s be the piecewise linear spline defined by (7.27). For $i = 1, 2, \ldots, n$, let $h_i = x_i - x_{i-1}$, and define $h = \max_{1 \leq i \leq n} h_i$. Then

$$\|f - s\|_\infty \leq \frac{M}{8} h^2,$$

where $\|\cdot\|_\infty$ is the ∞-norm on $[a, b]$ (see Section B.13.3) and $\|f''(x)\|_\infty = M$.

Exploration 7.6.3 Prove Theorem 7.6.2.

In Section 7.4, it was observed in **Runge's Example** that even when $f(x)$ is smooth, an interpolating polynomial of $f(x)$ can be highly oscillatory, depending on the number and placement of interpolation points. By contrast, one of the most welcome properties of the linear spline $s(x)$ is that among all functions in $H^1(a, b)$ that interpolate $f(x)$ at the knots $x_0, x_1, \ldots, x_n$, it is the "flattest". That is, for any function $v \in H^1(a, b)$ that interpolates f at the knots,

$$\|s'\|_2 \leq \|v'\|_2. \tag{7.28}$$

See Section B.13.3 for the definition of the L^2-norm $\|\cdot\|_2$.

To prove this, we first write

$$\|v'\|_2^2 = \|v' - s'\|_2^2 + 2\langle v' - s', s' \rangle + \|s'\|_2^2,$$

where $\langle \cdot, \cdot \rangle$ is the standard **inner product** of real-valued functions on (a, b), defined in (B.8). We then note that on each subinterval $[x_{i-1}, x_i]$, since s is a linear function, s' is a constant function, which we denote by

$$s'(x) \equiv m_i = \frac{f(x_i) - f(x_{i-1})}{x_i - x_{i-1}}, \quad i = 1, 2, \ldots, n.$$

Exploration 7.6.4 Complete the proof of (7.28) by showing that

$$\langle v' - s', s' \rangle = \int_a^b [v'(x) - s'(x)]s'(x) \, dx = 0.$$

7.6.2   Cubic Spline Interpolation

A major drawback of the linear spline is that it does not have any continuous derivatives. This is significant when the function to be approximated, $f(x)$, is a smooth function. Therefore, it is desirable that a piecewise polynomial approximation possess a certain number of continuous derivatives. This requirement imposes additional constraints on the piecewise polynomial, and therefore the degree of the polynomials used on each subinterval must be chosen sufficiently high to ensure that these constraints can be satisfied.

We therefore define a **spline** of degree k to be a piecewise polynomial of degree k that has $k-1$ continuous derivatives. The most commonly used spline is a **cubic spline**, which we now define.

Definition 7.6.3 (Cubic Spline) Let $f(x)$ be function defined on an interval $[a, b]$, and let $x_0, x_1, \ldots, x_n$ be $n+1$ distinct points in $[a, b]$, where $a = x_0 < x_1 < \cdots < x_n = b$. A **cubic spline**, or **cubic spline interpolant**, is a piecewise polynomial $s(x)$ that satisfies the following conditions:

1. On each interval $[x_{i-1}, x_i]$, for $i = 1, \ldots, n$, $s(x) = s_i(x)$, where $s_i(x)$ is a cubic polynomial.
2. $s(x_i) = f(x_i)$ for $i = 0, 1, \ldots, n$.
3. $s(x)$ is twice continuously differentiable on (a, b).
4. Either of the following boundary conditions are satisfied:

   (a) $s''(a) = s''(b) = 0$, which is called **free** or **natural boundary conditions**, and
   (b) $s'(a) = f'(a)$ and $s'(b) = f'(b)$, which is called **clamped boundary conditions**.

If $s(x)$ satisfies free boundary conditions, we say that $s(x)$ is a **natural spline**. The points $x_0, x_1, \ldots, x_n$ are called the **knots** of $s(x)$.

Clamped boundary conditions are often preferable because they use more information about $f(x)$, which yields a spline that better approximates $f(x)$ on $[a, b]$. However, if information about $f'(x)$ is not available, then different boundary conditions, such as natural boundary conditions, must be used instead.

7.6.2.1 *Constructing Cubic Splines*

Suppose that we wish to construct a cubic spline interpolant $s(x)$ that fits the given data $(x_0, y_0), (x_1, y_1), \ldots, (x_n, y_n)$, where $a = x_0 < x_1 < \cdots < x_n = b$, and $y_i = f(x_i)$, for some known function $f(x)$ defined on $[a, b]$. From the preceding discussion, this spline is a piecewise polynomial of the form

$$s(x) = s_i(x) = d_i(x - x_{i-1})^3 + c_i(x - x_{i-1})^2 + b_i(x - x_{i-1}) + a_i, \quad x_{i-1} \le x \le x_i,$$
$$(7.29)$$

for $i = 1, 2, \ldots, n$. That is, the value of $s(x)$ is obtained by evaluating a different cubic polynomial for each subinterval $[x_{i-1}, x_i]$, for $i = 1, 2, \ldots, n$.

We now use the definition of a cubic spline to construct a system of equations that must be satisfied by the coefficients a_i, b_i, c_i and d_i for $i = 1, 2, \ldots, n$. We can then compute these coefficients by solving the system. Because $s(x)$ must fit the given data, we have

$$s(x_{i-1}) = a_i = y_{i-1} = f(x_{i-1}), \quad i = 1, 2, \ldots, n. \tag{7.30}$$

If we define $h_i = x_i - x_{i-1}$, for $i = 1, 2, \ldots, n$, and define $a_{n+1} = y_n$, then the requirement that $s(x)$ is continuous at the interior knots implies that we must have $s_i(x_i) = s_{i+1}(x_i)$ for $i = 1, 2, \ldots, n - 1$. Furthermore, because $s(x)$ must fit the given data, we must also have $s(x_n) = s_n(x_n) = y_n$. These conditions lead to the constraints

$$s_i(x_i) = d_i h_i^3 + c_i h_i^2 + b_i h_i + a_i = a_{i+1} = s_{i+1}(x_i), \quad i = 1, 2, \ldots, n. \qquad (7.31)$$

To ensure that $s(x)$ has a continuous first derivative at the interior knots, we require that $s_i'(x_i) = s_{i+1}'(x_i)$ for $i = 1, 2 \ldots, n - 1$, which imposes the constraints

$$s_i'(x_i) = 3d_i h_i^2 + 2c_i h_i + b_i = b_{i+1} = s_{i+1}'(x_i), \quad i = 1, 2, \ldots, n - 1. \qquad (7.32)$$

Similarly, to enforce continuity of the second derivative at the interior knots, we require that $s_i''(x_i) = s_{i+1}''(x_i)$ for $i = 1, 2, \ldots, n - 1$, which leads to the constraints

$$s_i''(x_i) = 6d_i h_i + 2c_i = 2c_{i+1} = s_{i+1}''(x_i), \quad i = 1, 2, \ldots, n - 1. \qquad (7.33)$$

There are $4n$ coefficients to determine, since there are n cubic polynomials, with 4 coefficients each. However, we have only prescribed $4n - 2$ constraints, so we must specify 2 more in order to determine a unique solution. If we use natural boundary conditions, then these constraints are

$$s_1''(x_0) = 2c_1 = 0, \qquad (7.34)$$
$$s_n''(x_n) = 3d_n h_n + c_n = 0. \qquad (7.35)$$

On the other hand, if we use clamped boundary conditions, then our additional constraints are

$$s_1'(x_0) = b_1 = z_0, \qquad (7.36)$$
$$s_n'(x_n) = 3d_n h_n^2 + 2c_n h_n + b_n = z_n, \qquad (7.37)$$

where $z_i = f'(x_i)$ for $i = 0, 1, \ldots, n$.

Having determined our constraints that must be satisfied by $s(x)$, we can set up a system of $4n$ linear equations based on these constraints, and then solve this system to determine the coefficients a_i, b_i, c_i, d_i for $i = 1, 2 \ldots, n$. However, it is not necessary to construct the matrix for such a system, because it is possible to instead solve a smaller system of only $O(n)$ equations obtained from the continuity conditions (7.32) and the boundary conditions (7.34), (7.35) or (7.36), (7.37), depending on whether natural or clamped boundary conditions, respectively, are imposed. This reduced system is accomplished by using equations (7.30), (7.31) and (7.33) to eliminate the a_i, b_i and d_i, respectively.

Exploration 7.6.5 Show that under natural boundary conditions, the coefficients $c_2, \ldots, c_n$ of the cubic spline (7.29) satisfy the system of equations $A\mathbf{c} = \mathbf{b}$, where

$$
A = \begin{bmatrix}
2(h_1 + h_2) & h_2 & 0 & \cdots & & 0 \\
h_2 & 2(h_2 + h_3) & h_3 & \ddots & & \vdots \\
0 & \ddots & \ddots & \ddots & & 0 \\
\vdots & \ddots & \ddots & \ddots & & h_{n-1} \\
0 & & \cdots & & 0 & h_{n-1} \; 2(h_{n-1} + h_n)
\end{bmatrix},
$$

$$
\mathbf{c} = \begin{bmatrix} c_2 \\ \vdots \\ c_n \end{bmatrix}, \quad
\mathbf{b} = \begin{bmatrix}
\frac{3}{h_2}(a_3 - a_2) - \frac{3}{h_1}(a_2 - a_1) \\
\vdots \\
\frac{3}{h_n}(a_{n+1} - a_n) - \frac{3}{h_{n-1}}(a_n - a_{n-1})
\end{bmatrix}.
$$

Exploration 7.6.6 Show that under clamped boundary conditions, the coefficients $c_1, \ldots, c_{n+1}$ of the cubic spline (7.29) satisfy the system of equations $A\mathbf{c} = \mathbf{b}$, where

$$
A = \begin{bmatrix}
2h_1 & h_1 & 0 & \cdots & & \cdots & 0 \\
h_1 & 2(h_1 + h_2) & h_2 & \ddots & & & \vdots \\
0 & h_2 & 2(h_2 + h_3) & h_3 & & & \vdots \\
\vdots & \ddots & \ddots & \ddots & \ddots & & 0 \\
\vdots & & & \ddots & h_{n-1} & 2(h_{n-1} + h_n) & h_n \\
0 & & \cdots & & 0 & h_n & 2h_n
\end{bmatrix},
$$

$$
\mathbf{c} = \begin{bmatrix} c_1 \\ c_2 \\ \vdots \\ c_{n+1} \end{bmatrix}, \quad
\mathbf{b} = \begin{bmatrix}
\frac{3}{h_1}(a_2 - a_1) - 3z_0 \\
\frac{3}{h_2}(a_3 - a_2) - \frac{3}{h_1}(a_2 - a_1) \\
\vdots \\
\frac{3}{h_n}(a_{n+1} - a_n) - \frac{3}{h_{n-1}}(a_n - a_{n-1}) \\
3z_n - \frac{3}{h_n}(a_{n+1} - a_n)
\end{bmatrix},
$$

and $c_{n+1} = s_n''(x_n)$.

Example 7.6.4 We will construct a cubic spline for the following data:

j	0	1	2	3	4
x_j	0	1/2	1	3/2	2
y_j	3	-4	5	-6	7

The spline, $s(x)$, will consist of four pieces $\{s_j(x)\}_{j=1}^4$, each of which is a cubic polynomial of the form

$$s_j(x) = a_j + b_j(x - x_{j-1}) + c_j(x - x_{j-1})^2 + d_j(x - x_{j-1})^3, \quad j = 1, 2, 3, 4.$$

We will impose natural boundary conditions on this spline, so it will satisfy the conditions $s''(0) = s''(2) = 0$, in addition to the "essential" conditions imposed on a spline: it must fit the given data and have continuous first and second derivatives on the interval $[0, 2]$.

These conditions lead to a system of linear equations that must be solved for the coefficients c_2, c_3, and c_4. We define $h = (2 - 0)/4 = 1/2$ to be the spacing between the interpolation points. From Exploration 7.6.5, we have

$$\frac{h}{3}(4c_2 + c_3) = \frac{y_2 - 2y_1 + y_0}{h},$$

$$\frac{h}{3}(c_2 + 4c_3 + c_4) = \frac{y_3 - 2y_2 + y_1}{h},$$

$$\frac{h}{3}(c_3 + 4c_4) = \frac{y_4 - 2y_3 + y_2}{h},$$

and $c_1 = 0$ from (7.34). Substituting $h = 1/2$ and the values of y_j, we obtain

$$\frac{1}{6}(4c_2 + c_3) = 32,$$

$$\frac{1}{6}(c_2 + 4c_3 + c_4) = -40,$$

$$\frac{1}{6}(c_3 + 4c_4) = 48.$$

This system has the solutions

$$c_2 = 516/7, \quad c_3 = -720/7, \quad c_4 = 684/7.$$

Using (7.30), (7.31), (7.33) and (7.35), we obtain

$$a_1 = 3, \quad a_2 = -4, \quad a_3 = 5, \quad a_4 = -6,$$

$$b_1 = -184/7, \quad b_2 = 74/7, \quad b_3 = -4, \quad b_4 = -46/7,$$

and

$$d_1 = 344/7, \quad d_2 = -824/7, \quad d_3 = 936/7, \quad d_4 = -456/7.$$

We conclude that the spline $s(x)$ that fits the given data, has two continuous derivatives on $[0, 2]$, and satisfies natural boundary conditions is

$$s(x) = \begin{cases} \frac{344}{7}x^3 - \frac{184}{7}x^2 + 3 & \text{if } x \in [0, 0.5] \\ -\frac{824}{7}(x - 1/2)^3 + \frac{516}{7}(x - 1/2)^2 + \frac{74}{7}(x - 1/2) - 4 & \text{if } x \in [0.5, 1] \\ \frac{936}{7}(x - 1)^3 - \frac{720}{7}(x - 1)^2 - 4(x - 1) + 5 & \text{if } x \in [1, 1.5] \\ -\frac{456}{7}(x - 3/2)^3 + \frac{684}{7}(x - 3/2)^2 - \frac{46}{7}(x - 3/2) - 6 & \text{if } x \in [1.5, 2] \end{cases}$$

The graph of the spline is shown in Figure 7.2. Note that the graph is a smooth curve, even though different cubic polynomials are used between each pair of knots.
□

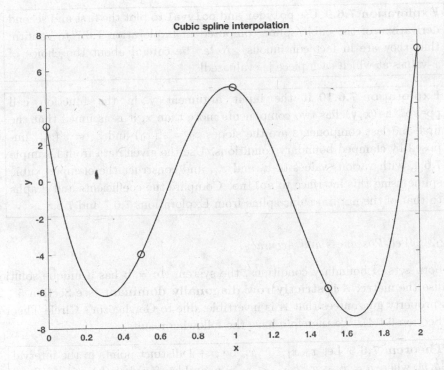

Fig. 7.2   From Example 7.6.4: Cubic spline that passes through the points $(0, 3)$, $(1/2, -4)$, $(1, 5)$, $(2, -6)$, and $(3, 7)$.

The MATLAB function `spline` can be used to construct cubic splines satisfying both natural (also known as *"not-a-knot"*) and clamped boundary conditions. The following explorations require reading the documentation for this function. They also require working with *structures* in MATLAB, which is a type of variable that groups related data using named *fields*.

Exploration 7.6.7 Use `spline` to construct a cubic spline for the data from Example 7.6.4. First, use the interface `pp=spline(x,y)`, where x and y are vectors consisting of the x- and y-coordinates, respectively, of the given data points, and pp is a MATLAB structure that represents the cubic spline $s(x)$. Examine the members of p and determine how to interpret them. Where do you see the coefficients computed in Example 7.6.4?

Exploration 7.6.8 The interface `yy=spline(x,y,xx)`, where xx is a vector of x-values at which the spline constructed from x and y should be evaluated, produces a vector yy of corresponding y-values. Use this interface on the data from Example 7.6.4 to reproduce Figure 7.2.

> **Exploration 7.6.9** Use `polyder` and `polyval` to plot the first and second derivatives of the cubic spline computed in Exploration 7.6.7 to confirm that they are in fact continuous. *Note:* be careful about the choice of x-values at which each piece is evaluated!

> **Exploration 7.6.10** If the input argument `y` in the function call `pp=spline(x,y)` has two components more than `x`, it is assumed that the first and last components are the slopes $z_0 = s'(x_0)$ and $z_n = s'(x_n)$ imposed by clamped boundary conditions. Use the given data from Example 7.6.4, with various values of z_0 and z_n, and construct the clamped cubic spline using this interface to `spline`. Compare the coefficients and graphs to that of the natural cubic spline from Explorations 7.6.7 and 7.6.8.

7.6.2.2 *Well-Posedness and Accuracy*

For both sets of boundary conditions, the system $Ac = b$ has a unique solution, because the matrix A is **strictly row diagonally dominant** (see Section 5.1.2). This property guarantees that A is invertible, due to Gershgorin's Circle Theorem (see Section 6.1.4). We therefore have the following results.

> **Theorem 7.6.5** Let $x_0, x_1, \ldots, x_n$ be $n + 1$ distinct points in the interval $[a, b]$, where $a = x_0 < x_1 < \cdots < x_n = b$, and let $f(x)$ be a function defined on $[a, b]$. Then f has a unique cubic spline interpolant $s(x)$ that is defined on the knots $x_0, x_1, \ldots, x_n$ that satisfies the natural boundary conditions $s''(a) = s''(b) = 0$.

> **Theorem 7.6.6** Let $x_0, x_1, \ldots, x_n$ be $n + 1$ distinct points in the interval $[a, b]$, where $a = x_0 < x_1 < \cdots < x_n = b$, and let $f(x)$ be a function defined on $[a, b]$ that is differentiable at a and b. Then f has a unique cubic spline interpolant $s(x)$ that is defined on the knots $x_0, x_1, \ldots, x_n$ that satisfies the clamped boundary conditions $s'(a) = f'(a)$ and $s'(b) = f'(b)$.

Just as the linear spline is the "flattest" interpolant, in an average sense, the natural cubic spline has the least "average curvature". Specifically, if $s(x)$ is the natural cubic spline for $f \in C[a, b]$ on $[a, b]$ with knots $a = x_0 < x_1 < \cdots < x_n = b$, and $v \in H^2(a, b)$ is any interpolant of f with these knots, then

$$\|s''\|_2 \leq \|v''\|_2.$$

This can be proved in the same way as the corresponding result for the linear spline. It is this property of the natural cubic spline, called the **smoothest interpolation property**, from which splines were named.

The following result, proved in [Schultz (1973)], provides insight into the accuracy with which a cubic spline interpolant $s(x)$ approximates a function $f(x)$.

Theorem 7.6.7 Let f be four times continuously differentiable on $[a, b]$, and assume that $\|f^{(4)}\|_\infty = M$, where $\|\cdot\|_\infty$ is the ∞-norm on $[a, b]$. Let $s(x)$ be the unique clamped cubic spline interpolant of $f(x)$ on the knots $x_0, x_1, \ldots, x_n$, where $a = x_0 < x_1 < \cdots < x_n < b$. Then for $x \in [a, b]$,

$$\|f(x) - s(x)\|_\infty \le \frac{5M}{384} \max_{1 \le i \le n} h_i^4,$$

where $h_i = x_i - x_{i-1}$.

A similar result applies in the case of natural boundary conditions [Birkhoff and De Boor (1964)].

7.6.2.3 *Hermite Cubic Splines*

We have seen that it is possible to construct a piecewise cubic polynomial that interpolates a function $f(x)$ at knots $a = x_0 < x_1 < \cdots < x_n = b$, that belongs to $C^2[a, b]$. Now, suppose that we also know the values of $f'(x)$ at the knots. We wish to construct a piecewise cubic polynomial $s(x)$ that agrees with $f(x)$, and whose derivative agrees with $f'(x)$ at the knots. This piecewise polynomial is called a **Hermite cubic spline**.

Because $s(x)$ is cubic on each subinterval $[x_{i-1}, x_i]$ for $i = 1, 2, \ldots, n$, there are $4n$ coefficients, and therefore $4n$ degrees of freedom, that can be used to satisfy any criteria that are imposed on $s(x)$. Requiring that $s(x)$ interpolates $f(x)$ at the knots, and that $s'(x)$ interpolates $f'(x)$ at the knots, imposes $2n + 2$ constraints on the coefficients. We can then use the remaining $2n - 2$ degrees of freedom to require that $s(x)$ belong to $C^1[a, b]$; that is, it is continuously differentiable on $[a, b]$. Note that unlike the cubic spline interpolant, the Hermite cubic spline does *not* have a continuous second derivative.

The following result provides insight into the accuracy with which a Hermite cubic spline interpolant $s(x)$ approximates a function $f(x)$.

Theorem 7.6.8 Let f be four times continuously differentiable on $[a, b]$, and assume that $\|f^{(4)}\|_\infty = M$, where $\|\cdot\|_\infty$ denotes the ∞-norm on $[a, b]$. Let $s(x)$ be the unique Hermite cubic spline interpolant of $f(x)$ on the knots $x_0, x_1, \ldots, x_n$, where $a = x_0 < x_1 < \cdots < x_n < b$. Then

$$\|f(x) - s(x)\|_\infty \le \frac{M}{384} \max_{1 \le i \le n} h_i^4,$$

where $h_i = x_i - x_{i-1}$.

This can be proved in the same way as the error bound for the linear spline, except that the error formula for Hermite interpolation is used instead of the error formula for Lagrange interpolation.

Exploration 7.6.11 Prove Theorem 7.6.8.

An advantage of Hermite cubic splines over cubic spline interpolants is that they are *local* approximations rather than *global*; that is, if the values of $f(x)$ and $f'(x)$ change at some knot x_i, only the polynomials defined on the pieces containing x_i need to be changed. In cubic spline interpolation, all pieces are coupled, so a change at one point changes the polynomials for all pieces. To see this, we represent the Hermite cubic spline using the same form as in the cubic spline interpolant,

$$s_i(x) = a_i + b_i(x - x_{i-1}) + c_i(x - x_{i-1})^2 + d_i(x - x_{i-1})^3, \quad x \in [x_{i-1}, x_i], \quad (7.38)$$

for $i = 1, 2, \ldots, n$. Then, the coefficients a_i, b_i, c_i, d_i can be determined explicitly in terms of only $f(x_{i-1})$, $f'(x_{i-1})$, $f(x_i)$ and $f'(x_i)$.

Exploration 7.6.12 Use the conditions

$$s_i(x_{i-1}) = f(x_{i-1}), \quad s_i(x_i) = f(x_i), \quad s_i'(x_{i-1}) = f'(x_{i-1}), \quad s_i'(x_i) = f'(x_i)$$

to obtain the values of the coefficients a_i, b_i, c_i, d_i in (7.38).

Exploration 7.6.13 Write a MATLAB function

```
hp=hermitespline(x,y)
```

that constructs a Hermite cubic spline for the data given in the vectors x and y. The output hp should be a structure that contains enough information so that the spline can be evaluated at any x-value without having to specify any additional parameters. Write a second function y=hsplineval(hp,x) that performs this evaluation.

7.6.3   Concept Check

1. What is piecewise polynomial interpolation?
2. What shortcoming of polynomial interpolation does piecewise polynomial interpolation address?
3. Describe one advantage and one disadvantage of piecewise linear interpolation.
4. What does it mean to say that a linear spline is the "flattest" interpolant?
5. What are the properties of a cubic spline?
6. Why are boundary conditions required for a cubic spline?
7. Describe two types of boundary conditions that can be imposed on a cubic spline.
8. How does cubic spline interpolation compare to Lagrange interpolation in terms of efficiency?
9. What is the "smoothest interpolation property" of cubic splines?
10. How are cubic splines computed in MATLAB?
11. List an advantage and disadvantage of using Hermite cubic splines, compared to conventional cubic splines.

7.7 Additional Resources

Polynomial interpolation is one of the most essential topics in numerical analysis, as it serves as the foundation of methods for numerical differentiation, numerical integration, and certain time-stepping methods for ODEs, as we will see later in this book. As such, polynomial interpolation is covered quite thoroughly in other textbooks; in particular we call the reader's attention [Ascher and Greif (2011); Burden and Faires (2004); Cheney and Kincaid (2008); Conte and de Boor (1972)]. Lagrange interpolation is covered in depth in [Ascher (2008); Ascher and Petzold (1998); Davis and Rabinowitz (1985)]. Also covered in [Ascher (2008)] is the use of Newton interpolation to efficiently update the interpolating polynomial for ENO ("essentially non-oscillatory") schemes for differential equations. A textbook that serves as a comprehensive reference on interpolation is [Davis (1975)].

The books [Fox and Parker (1968); Rivlin (1990)] are devoted to Chebyshev polynomials. Trefethen, et al. have developed `chebfun`, an open-source MATLAB toolbox for performing operations on functions, such as differentiation, integration, and solution of differential equations, in which functions are represented internally by their Chebyshev interpolants. The methods of `chebfun` take advantage of the fact that accurate high-degree polynomial interpolation is feasible when Chebyshev points are used as interpolation points. The `chebfun` toolbox is available at http://www.chebfun.org, and documentation is provided in [Driscoll, et al. (2014)].

The theory of splines is covered in [Ahlberg et al. (1967); Knott (2000); Schumaker (1981)]. Texts that focus on the computational aspects, including software, are [de Boor (1984); Dierckx (1993); Shikin and Plis (1995); Spath (1995)]. The application of splines to computer graphics and geometric modeling is covered in [Bartels et al. (1987); Farin (1990); Yamaguchi (1988)]. Beyond the scope of this text is B-splines, which are basis functions used to describe cubic splines. Coverage of this topic can be found in [Ascher and Greif (2011); de Boor (1984); Süli and Mayers (2003)]. Also presented in [Ascher and Greif (2011)] is multi-dimensional interpolation, including with radial basis functions (RBFs).

7.8 Exercises

1. Compute, by hand, the Lagrange and Newton forms of the interpolating polynomial $p_3(x)$ of degree three for $f(x) = x^4$ with interpolation points $x_i = i$ for $i = 0, 1, 2, 3$.

2. Compute, by hand, the Hermite interpolating polynomial $h_3(x)$ of degree three for $f(x) = x^4$, with $x_0 = 0$ and $x_1 = 3$. Plot the graphs of $f(x)$, $p_3(x)$ from Exercise 1, and $h_3(x)$ on the same plot. Compare the accuracy of these two approximations.

3. Use Theorems 7.4.1 and 7.5.1 to obtain an upper bound for the error in the interpolating polynomials computed in Exercises 1 and 2 on $[0, 3]$. What

step in this error estimation process is unnecessary if Chebyshev points are used as interpolation points? Explain why.

4. Describe, using pseudocode, an algorithm that evaluates a polynomial

$$p_n(x) = a_0 + a_1 x + a_2 x^2 + \cdots + a_n x^n$$

at a given x-value by computing the terms of $p_n(x)$, as written above, from left to right, instead of using Horner's Method. How many multiplications and additions are required? How does its efficiency compare to that of Horner's Method?

5. Given a polynomial $p_n(x)$ that interpolates the data (x_0, y_0), (x_1, y_1), ..., (x_n, y_n), determine how many additions and multiplications are required to evaluate $p_n(x)$ when it is in (a) Lagrange form, and (b) Newton form.

6. Show that the Lagrange polynomial $\mathcal{L}_{n,j}(x)$ defined in (7.3) can also be written as

$$\mathcal{L}_{n,j}(x) = \frac{\pi_n(x)}{(x - x_j)\pi_n'(x)},$$

where $\pi_n(x) = (x - x_0)(x - x_1)(x - x_2) \cdots (x - x_n)$.

7. Consider a *piecewise quadratic interpolant* $s(x)$ that interpolates a given function $f(x)$ at $x_0, x_1, \ldots, x_n$, defined by

$$s(x) = s_i(x) = a_i + b_i(x - x_{i-1}) + c_i(x - x_{i-1})^2, \quad x \in [x_{i-1}, x_i],$$

for $i = 1, 2, \ldots, n$. How many continuous derivatives can $s(x)$ be guaranteed to have? How many boundary conditions can be imposed? Describe a system of equations for obtaining the coefficients a_i, b_i, c_i for $i = 1, 2, \ldots, n$. Use boundary conditions analogous to clamped boundary conditions for cubic splines to ensure a unique solution.

8. Write your own MATLAB functions that implement the built-in functions `polyval`, `polyder` and `polyint`. Use Horner's Method to implement your version of `polyval`.

9. Write a MATLAB function

```
pp=clampedspline(x,y,spa,spb)
```

that uses the system of equations described in Exploration 7.6.6 to construct a cubic spline that interpolates the data described by the vectors `x` and `y` and also satisfies clamped boundary conditions $s'(a) = $ spa and $s'(b) = $ spb. The output `p` must be a structure similar to that returned by the built-in `spline` function that describes the computed spline.

10. We revisit the function from **Runge's Example**

$$f(x) = \frac{1}{1 + x^2}, \quad -5 \le x \le 5.$$

Plot the graph of this function on the interval $[-5, 5]$. On the same plot, graph the following approximations:

(a) The Lagrange interpolating polynomial of $f(x)$ of degree 10, with equally spaced interpolation points $x_i = -5 + i$, $i = 0, 1, 2, \ldots, 10$.

(b) The Lagrange interpolating polynomial of $f(x)$ of degree 10, with the Chebyshev points $x_0, x_1, \ldots, x_{10}$ equal to the roots of $T_{10}(x)$ mapped to the interval $[-5, 5]$

(c) A cubic spline that interpolates $f(x)$ at the equally spaced points from part 10a.

Comment on the accuracy of these approximations.

11. In this problem, we consider an alternative approach to that used in Exploration 7.4.1 to prove Theorem 7.4.1.

(a) Let $p_n(x)$ be the interpolating polynomial of $f(x)$ at $x_0, x_1, \ldots, x_n \in [a, b]$, and let $\pi_n(x) = (x - x_0)(x - x_1) \cdots (x - x_n)$. Define

$$\varphi(t) = f(t) - p_n(t) - \frac{f(x) - p_n(x)}{\pi_n(x)} \pi_n(t).$$

Prove Theorem 7.4.1 by showing that there exists $\xi \in (a, b)$ such that $\varphi^{(n+1)}(\xi) = 0$.

(b) Adapt the approach used in part 11a to prove Theorem 7.5.1.

Chapter 8

Approximation of Functions

In Chapter 7 we examined the problem of polynomial *interpolation*, in which a function $f(x)$ is approximated by a polynomial $p_n(x)$ that agrees with $f(x)$ at $n+1$ distinct points, based on the assumption that $p_n(x)$ will be, in some sense, a good approximation of $f(x)$ at other points. As we have seen, however, this assumption is not always valid, and in fact, such an approximation can be quite poor, as demonstrated by Runge's Example in Section 7.4.

Therefore, we consider an alternative approach to approximation of a function $f(x)$ on an interval $[a, b]$ by a polynomial, in which the polynomial is not required to agree with f at any specific points, but rather approximate f well in an "overall" sense, by not deviating much from f at *any* point in $[a, b]$. This requires that we define an appropriate notion of "distance" between functions that is, intuitively, consistent with our understanding of distance between numbers or points in space. To that end, we can use *vector norms*, as defined in Section B.13, where the vectors in question consist of the values of functions at selected points. In this case, the problem can be reduced to a **least squares problem**, as discussed in Chapter 4. This is discussed in Section 8.1.

Still, finding an approximation of $f(x)$ that is accurate with respect to any discrete, finite subset of the domain cannot guarantee that it accurately approximates $f(x)$ on the entire domain. Therefore, in Section 8.2 we generalize least squares approximations to a *continuous* setting by working with with norms on **function spaces**, which are vector spaces in which the vectors are functions. Such function spaces and norms are reviewed in Section B.13.3.

In the remainder of the chapter, we consider approximating $f(x)$ by functions other than polynomials. Section 8.3 presents an approach to approximating $f(x)$ with a *rational* function, to overcome the limitations of polynomial approximation, while Section 8.4 explores approximation through **trigonometric polynomials**, or sines and cosines, to capture the frequency content of $f(x)$. This will prove useful in Chapter 14, where we will solve partial differential equations.

8.1   Discrete Least Squares Approximation

As stated previously, one of the most fundamental problems in science and engineering is **data fitting**–constructing a function that, in some sense, conforms to given data points. In Chapter 7, we discussed two data-fitting techniques, polynomial interpolation and piecewise polynomial interpolation. Interpolation techniques, of any kind, construct functions that agree *exactly* with the data. That is, given points (x_1, y_1), (x_2, y_2), ..., (x_m, y_m), interpolation yields a function $f(x)$ such that $f(x_i) = y_i$ for $i = 1, 2, \ldots, m$, while also possibly satisfying other conditions.

However, fitting the data exactly may not be the best approach to describing the data with a function. We have seen that high-degree polynomial interpolation can yield oscillatory functions that behave very differently than a smooth function from which the data is obtained. Also, it may be pointless to try to fit data exactly, for if it is obtained by previous measurements or other computations, it may be erroneous. Therefore, we consider revising our notion of what constitutes a "best fit" of given data by a function.

Let $x_1, x_2, \ldots, x_m$ be distinct, and let

$$\mathbf{y} = \begin{bmatrix} y_1 \ y_2 \ \cdots \ y_m \end{bmatrix}^T, \quad \mathbf{f} = \begin{bmatrix} f(x_1) \ f(x_2) \ \cdots \ f(x_m) \end{bmatrix}^T$$

be m-vectors that store the *observed* values and *predicted* values, respectively. That is, $f(x)$ is viewed as predicting how y depends on x, based on the m observations. Furthermore, suppose that $f(x)$ is required to have the form

$$f(x) = \sum_{j=1}^{n} c_j \varphi_j(x) \tag{8.1}$$

for given linearly independent functions $\varphi_1(x), \varphi_2(x), \ldots, \varphi_n(x)$. Then, we can find $f(x)$ that is, in some sense, optimal by finding the coefficients $c_1, c_2, \ldots, c_n$ such that $\|\mathbf{f} - \mathbf{y}\|$ is minimized, for some choice of **vector norm** $\| \cdot \|$ (see Section B.13.1). As discussed in Section 4.1, a sensible choice is the ℓ_2-norm, as it is a differentiable function, and therefore this minimization problem lends itself to techniques from calculus. We therefore consider the problem of finding $f(x)$ of the form (8.1) for which

$$\|\mathbf{f} - \mathbf{y}\|_2^2 = \sum_{i=1}^{m} [f(x_i) - y_i]^2 = \sum_{i=1}^{m} \left[\sum_{j=1}^{n} c_j \varphi_j(x_i) - y_i \right]^2$$

is minimized. This is a **least squares problem**, a more abstract version of which was the subject of Chapter 4.

8.1.1   Linear Regression

We will first show how this problem is solved for the case in which $f(x)$ is a *linear* function of the form $f(x) = a_1 x + a_0$, and then generalize this solution to other types of functions. When $f(x)$ is linear, the least squares problem, known as **linear**

regression in this case, is the problem of finding constants a_0 and a_1 that minimize the sum of the squared *residuals*, which are the deviations of $f(x_i)$ from each y_i for $i = 1, 2, \ldots, m$. That is, we seek to minimize

$$R = \sum_{i=1}^{m} (a_1 x_i + a_0 - y_i)^2.$$

To minimize this expression, which is a function of a_0 and a_1, we must compute its partial derivatives with respect to a_0 and a_1. This yields

$$\frac{\partial R}{\partial a_0} = \sum_{i=1}^{m} 2(a_1 x_i + a_0 - y_i), \quad \frac{\partial R}{\partial a_1} = \sum_{i=1}^{m} 2(a_1 x_i + a_0 - y_i)x_i.$$

At a minimum, both of these partial derivatives must be equal to zero. This yields the system of linear equations

$$m a_0 + \left(\sum_{i=1}^{m} x_i \right) a_1 = \sum_{i=1}^{m} y_i,$$

$$\left(\sum_{i=1}^{m} x_i \right) a_0 + \left(\sum_{i=1}^{m} x_i^2 \right) a_1 = \sum_{i=1}^{m} x_i y_i.$$

These equations are called the **normal equations**.

> **Exploration 8.1.1** From multivariable calculus, a function of several variables has a local minimum at a point if its first partial derivatives are zero there, and if its matrix of second partial derivatives, the Hessian, is positive definite (see Section 3.3.3). Find the Hessian of R and use the Cauchy-Schwarz inequality (see Section B.13.1) to prove that it is positive definite.

Using the formula for the inverse of a 2×2 matrix,

$$\begin{bmatrix} a & b \\ c & d \end{bmatrix}^{-1} = \frac{1}{ad - bc} \begin{bmatrix} d & -b \\ -c & a \end{bmatrix},$$

we obtain the solutions

$$a_0 = \frac{\left(\sum_{i=1}^{m} x_i^2 \right)\left(\sum_{i=1}^{m} y_i \right) - \left(\sum_{i=1}^{m} x_i \right)\left(\sum_{i=1}^{m} x_i y_i \right)}{m \sum_{i=1}^{m} x_i^2 - \left(\sum_{i=1}^{m} x_i \right)^2},$$

$$a_1 = \frac{m \sum_{i=1}^{m} x_i y_i - \left(\sum_{i=1}^{m} x_i \right)\left(\sum_{i=1}^{m} y_i \right)}{m \sum_{i=1}^{m} x_i^2 - \left(\sum_{i=1}^{m} x_i \right)^2}.$$

> **Exploration 8.1.2** Prove that if the x-values $x_1, x_2, \ldots, x_m$ are not all the same, then the normal equations are guaranteed to have a unique solution.

Example 8.1.1 We wish to find the linear function $y = a_1 x + a_0$ that best approximates the data shown in the table below, in the least-squares sense.

i	x_i	y_i
1	2.0774	3.3123
2	2.3049	3.8982
3	3.0125	4.6500
4	4.7092	6.5576
5	5.5016	7.5173
6	5.8704	7.0415
7	6.2248	7.7497
8	8.4431	11.0451
9	8.7594	9.8179
10	9.3900	12.2477

Using the summations

$$\sum_{i=1}^{m} x_i = 56.2933, \quad \sum_{i=1}^{m} x_i^2 = 380.5426, \quad \sum_{i=1}^{m} y_i = 73.8373, \quad \sum_{i=1}^{m} x_i y_i = 485.9487,$$

we obtain

$$a_0 = \frac{380.5426 \cdot 73.8373 - 56.2933 \cdot 485.9487}{10 \cdot 380.5426 - 56.2933^2} = \frac{742.5703}{636.4906} = 1.1667,$$

$$a_1 = \frac{10 \cdot 485.9487 - 56.2933 \cdot 73.8373}{10 \cdot 380.5426 - 56.2933^2} = \frac{702.9438}{636.4906} = 1.1044.$$

We conclude that the linear function that best fits this data in the least-squares sense is

$$y = 1.1044x + 1.1667.$$

The data, and this function, are shown in Figure 8.1. We note that the line does not actually pass through any of the data points, but is near all of them. In fact, the average deviation of these points from the line is zero. □

Exploration 8.1.3 Write a MATLAB function

$$[s,b]=\texttt{leastsqline(x,y)}$$

that computes the slope $a_1 = $ s and y-intercept $a_0 = $ b of the line $y = a_1 x + a_0$ that best fits the data (x_i, y_i), $i = 1, 2, \ldots, m$ where $m = \texttt{length(x)}$, in the least-squares sense.

Exploration 8.1.4 Generalize the above derivation of the coefficients a_0 and a_1 of the least-squares line to obtain formulas for the coefficients a, b and c of the quadratic function $y = ax^2 + bx + c$ that best fits the data (x_i, y_i), $i = 1, 2, \ldots, m$, in the least-squares sense. Then generalize your function `leastsqline` from Exploration 8.1.3 to obtain a new function `leastsqquad` that computes these coefficients.

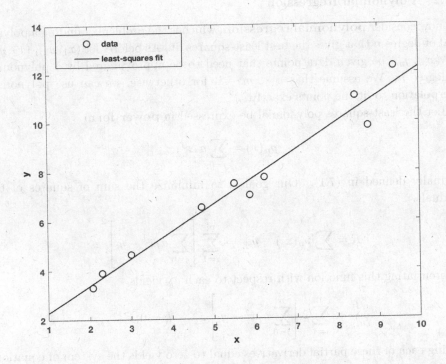

Fig. 8.1 From Example 8.1.1: Data points (x_i, y_i) (circles) and least-squares line (solid line)

It is interesting to note that if we define the $m \times 2$ matrix A, the 2-vector $\mathbf{a}$, and the m-vector $\mathbf{y}$ by

$$A = \begin{bmatrix} 1 & x_1 \\ 1 & x_2 \\ \vdots & \vdots \\ 1 & x_m \end{bmatrix}, \quad \mathbf{a} = \begin{bmatrix} a_0 \\ a_1 \end{bmatrix}, \quad \mathbf{y} = \begin{bmatrix} y_1 \\ y_2 \\ \vdots \\ y_m \end{bmatrix},$$

then $\mathbf{a}$ is the solution to the system of equations

$$A^T A \mathbf{a} = A^T \mathbf{y}.$$

These equations are the **normal equations** defined in Chapter 4. They arise from the problem of finding the vector $\mathbf{a}$ such that $\|A\mathbf{a} - \mathbf{y}\|_2$ is minimized. This norm is equivalent to the square root of the one we originally intended to minimize,

$$\sum_{i=1}^{m} (a_1 x_i + a_0 - y_i)^2,$$

but the normal equations also characterize the solution $\mathbf{a}$, an n-vector, to the more general linear least squares problem of minimizing $\|A\mathbf{a} - \mathbf{y}\|_2$ for any matrix A that is $m \times n$, where $m \geq n$, whose columns are linearly independent.

8.1.2   Polynomial Regression

We now consider **polynomial regression**, which is the problem of finding a polynomial of degree n that gives the best least-squares fit. As before, let (x_1, y_1), (x_2, y_2), $\ldots$, (x_m, y_m) be given data points that need to be approximated by a polynomial of degree n. We assume that $n < m - 1$, for otherwise, we can use polynomial interpolation to fit the points exactly.

Let the least-squares polynomial be expressed in **power form**

$$p_n(x) = \sum_{j=0}^{n} a_j x^j,$$

originally defined in (7.1). Our goal is to minimize the sum of squares of the residuals,

$$R = \sum_{i=1}^{m} [p_n(x_i) - y_i]^2 = \sum_{i=1}^{m} \left[\sum_{j=0}^{n} a_j x_i^j - y_i \right]^2.$$

Differentiating this function with respect to each a_k yields

$$\frac{\partial R}{\partial a_k} = \sum_{i=1}^{m} 2 \left[\sum_{j=0}^{n} a_j x_i^j - y_i \right] x_i^k, \quad k = 0, 1, \ldots, n.$$

Setting each of these partial derivatives equal to zero yields the system of equations

$$\sum_{j=0}^{n} \left(\sum_{i=1}^{m} x_i^{j+k} \right) a_j = \sum_{i=1}^{m} x_i^k y_i, \quad k = 0, 1, \ldots, n.$$

These are the **normal equations**. They are a generalization of the normal equations previously defined for the linear case, where $n = 1$. Solving this system yields the coefficients $\{a_j\}_{j=0}^{n}$ of the least-squares polynomial $p_n(x)$.

As in the linear case, the normal equations can be written in matrix-vector form

$$A^T A a = A^T \mathbf{y},$$

where

$$A = \begin{bmatrix} 1 & x_1 & x_1^2 & \cdots & x_1^n \\ 1 & x_2 & x_2^2 & \cdots & x_2^n \\ \vdots & \vdots & \vdots & \ddots & \vdots \\ 1 & x_m & x_m^2 & \cdots & x_m^n \end{bmatrix}, \quad \mathbf{a} = \begin{bmatrix} a_0 \\ a_1 \\ \vdots \\ a_n \end{bmatrix}, \quad \mathbf{y} = \begin{bmatrix} y_1 \\ y_2 \\ \vdots \\ y_m \end{bmatrix}. \tag{8.2}$$

The matrix A is called a **Vandermonde matrix** for the points $x_0, x_1, \ldots, x_m$. This matrix was previously seen in Section 7.1, in the case $m = n + 1$.

The normal equations equations can be used to compute the coefficients of *any* linear combination of functions $\{\varphi_j(x)\}_{j=0}^{n}$ that best fits data in the least-squares sense, provided that the values of these functions on $\{x_i\}_{i=1}^{m}$, as vectors, are linearly independent in $\mathbb{R}^m$ (see Section B.4). In this general case, the entries of the matrix A are given by $a_{ij} = \varphi_i(x_j)$, for $i = 1, 2, \ldots, m$ and $j = 0, 1, \ldots, n$.

Example 8.1.2 We wish to find the quadratic function $y = a_2x^2 + a_1x + a_0$ that best approximates the data shown in the following table, in the least-squares sense.

i	x_i	y_i
1	2.0774	2.7212
2	2.3049	3.7798
3	3.0125	4.8774
4	4.7092	6.6596
5	5.5016	10.5966
6	5.8704	9.8786
7	6.2248	10.5232
8	8.4431	23.3574
9	8.7594	24.0510
10	9.3900	27.4827

By defining

$$A = \begin{bmatrix} 1 & x_1 & x_1^2 \\ 1 & x_2 & x_2^2 \\ \vdots & \vdots & \vdots \\ 1 & x_{10} & x_{10}^2 \end{bmatrix}, \quad \mathbf{a} = \begin{bmatrix} a_0 \\ a_1 \\ a_2 \end{bmatrix}, \quad \mathbf{y} = \begin{bmatrix} y_1 \\ y_2 \\ \vdots \\ y_{10} \end{bmatrix},$$

and solving the normal equations

$$A^T A \mathbf{a} = A^T \mathbf{y},$$

we obtain the coefficients

$$c_0 = 4.7681, \quad c_1 = -1.5193, \quad c_2 = 0.4251,$$

and conclude that the quadratic function that best fits this data in the least-squares sense is

$$y = 0.4251x^2 - 1.5193x + 4.7681.$$

The data, and this function, are shown in Figure 8.2. As in Example 8.1.1, this quadratic does not pass through any of the data points exactly, but is near all of them, with the average deviation being zero. □

Exploration 8.1.5 Write a MATLAB function

```
a=leastsqpoly(x,y,n)
```

that computes the coefficients a_j, $j = 0, 1, \ldots, n$ of the polynomial of degree n that best fits the data (x_i, y_i) in the least-squares sense. Use the MATLAB function **vander** to easily construct the Vandermonde matrix A used in the normal equations. Make sure you solve the normal equations *without* explicitly computing $A^T A$. Test your function on the data from Example 8.1.2, with different values of n, but with $n < 10$. How does the residual $\|A\mathbf{a} - \mathbf{y}\|_2$ behave as n increases?

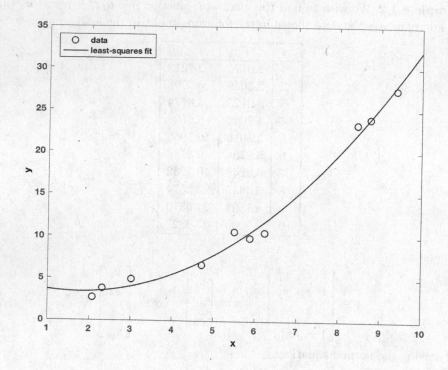

Fig. 8.2   From Example 8.1.2: Data points (x_i, y_i) (circles) and quadratic least-squares fit (solid curve)

Exploration 8.1.6 Test your function `leastsqpoly` from Exploration 8.1.5 to approximate the function $y = e^{-cx}$ on the interval $[0, 1]$ where c is a chosen positive constant. Experiment with different values of c, as well as m and n, the number of data points and degree of the approximating polynomial, respectively. What combination yields the smallest relative residual $\|A\mathbf{a} - \mathbf{y}\|_2 / \|\mathbf{y}\|_2$?

8.1.3   Nonlinear Least Squares

Least-squares fitting can also be used to fit data with functions that are not linear combinations of functions such as polynomials. Suppose we believe that given data points can best be matched to an exponential function of the form $y = be^{ax}$, where the constants a and b are unknown. Taking the natural logarithm of both sides of this equation yields

$$\ln y = \ln b + ax.$$

If we define $z = \ln y$ and $c = \ln b$, then the problem of fitting the original data points $\{(x_i, y_i)\}_{i=1}^{m}$ with an exponential function is transformed into the problem of fitting the data points $\{(x_i, z_i)\}_{i=1}^m$ with a linear function of the form $c + ax$, for unknown constants a and c.

Similarly, suppose the given data is believed to approximately conform to a function of the form $y = bx^a$, where the constants a and b are unknown. Taking the natural logarithm of both sides of this equation yields

$$\ln y = \ln b + a \ln x.$$

If we define $z = \ln y$, $c = \ln b$ and $w = \ln x$, then the problem of fitting the original data points $\{(x_i, y_i)\}_{i=1}^m$ with a constant times a power of x is transformed into the problem of fitting the data points $\{(w_i, z_i)\}_{i=1}^m$ with a linear function of the form $c + aw$, for unknown constants a and c.

Example 8.1.3 We wish to find the exponential function $y = be^{ax}$ that best approximates the data shown in the following table, in the least-squares sense.

i	x_i	y_i
1	2.0774	1.4509
2	2.3049	2.8462
3	3.0125	2.1536
4	4.7092	4.7438
5	5.5016	7.7260

By defining

$$A = \begin{bmatrix} 1 & x_1 \\ 1 & x_2 \\ \vdots & \vdots \\ 1 & x_5 \end{bmatrix}, \quad \mathbf{c} = \begin{bmatrix} c \\ a \end{bmatrix}, \quad \mathbf{z} = \begin{bmatrix} z_1 \\ z_2 \\ \vdots \\ z_5 \end{bmatrix},$$

where $c = \ln b$ and $z_i = \ln y_i$ for $i = 1, 2, \ldots, 5$, and solving the normal equations

$$A^T A \mathbf{c} = A^T \mathbf{z},$$

we obtain the coefficients

$$a = 0.4040, \quad b = e^c = e^{-0.2652} = 0.7670,$$

and conclude that the exponential function that best fits this data in the least-squares sense is

$$y = 0.7670 e^{0.4040x}.$$

The data, and this function, are shown in Figure 8.3. □

Exploration 8.1.7 Write a MATLAB function

$$[a,b] = \texttt{leastsqexp(x,y)}$$

that computes the coefficients a and b of a function $y = be^{ax}$ that fits the given data (x_i, y_i), $i = 1, 2, \ldots, m$ where $m = \texttt{length(x)}$, in the least squares sense.

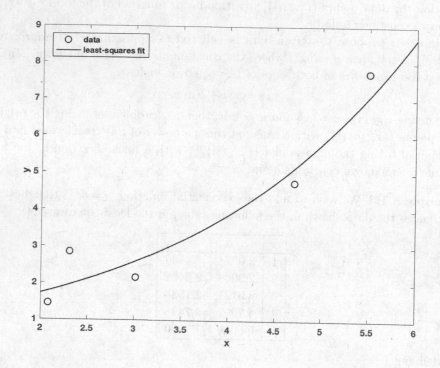

Fig. 8.3   From Example 8.1.3: Data points (x_i, y_i) (circles) and exponential least-squares fit (solid curve)

Exploration 8.1.8 Write a MATLAB function

$$[\texttt{a,b}]=\texttt{leastsqpower(x,y)}$$

that computes the coefficients a and b of a function $y = bx^a$ that fits the given data (x_i, y_i), $i = 1, 2, \ldots, m$ where $m = \texttt{length(x)}$, in the least squares sense.

8.1.4   Concept Check

1. When fitting a function to given data, why not simply use polynomial interpolation?
2. When approximating a function, what is the advantage of minimizing the ℓ_2-norm of the error, as opposed to some other norm?
3. What is meant by linear regression?
4. What is the system of linear equations that is solved to obtain a least squares approximation of a function?
5. How can techniques for computing least squares approximations be used to fit data with a function of the form be^{ax} or bx^a, that has two unknown

coefficients but is not a linear combination of two functions?

8.2 Continuous Least Squares Approximation

Now, suppose we have a *continuous* set of data. That is, we have a real-valued function $f(x)$ defined on an interval $[a, b]$, and we wish to approximate it as closely as possible, in some sense, on all of $[a, b]$ by a function $f_n(x)$ that is a linear combination of given real-valued functions $\{\varphi_j(x)\}_{j=0}^n$. If we choose m equally spaced points $\{x_i\}_{i=1}^m$ in $[a, b]$, and let $m \to \infty$, we obtain the **continuous least squares problem** of finding the function

$$f_n(x) = \sum_{j=0}^n c_j \varphi_j(x)$$

that minimizes

$$R = \|f_n - f\|_2^2 = \int_a^b [f_n(x) - f(x)]^2 \, dx = \int_a^b \left[\sum_{j=0}^n c_j \varphi_j(x) - f(x) \right]^2 \, dx,$$

where

$$\|f_n - f\|_2 = \left(\int_a^b [f_n(x) - f(x)]^2 \, dx \right)^{1/2}.$$

We refer to f_n as the **best approximation in** $\operatorname{span}\{\varphi_0, \varphi_1, \ldots, \varphi_n\}$ **to** f **in the** L^2**-norm on** (a, b). It is suggested that the reader review Section B.13.3 in which the L^2-norm is defined.

This minimization can be performed for $f \in C[a, b]$, the space of functions that are continuous on $[a, b]$, but it is not necessary for a function $f(x)$ to be continuous for $\|f\|_2$ to be defined. Rather, we consider the space $L^2(a, b)$, the space of real-valued functions such that $|f(x)|^2$ is *integrable* over (a, b). Both of these spaces, in addition to being **normed** spaces, are also **inner product spaces**, as they are equipped with an inner product

$$\langle f, g \rangle = \int_a^b f(x) g(x) \, dx. \qquad (8.3)$$

Such spaces are reviewed in Section B.15.

To obtain the coefficients $\{c_j\}_{j=0}^n$, we can proceed as in the discrete case. We compute the partial derivatives of R with respect to each c_k and obtain

$$\frac{\partial R}{\partial c_k} = \int_a^b \varphi_k(x) \left[\sum_{j=0}^n c_j \varphi_j(x) - f(x) \right] dx, \quad k = 0, 1, \ldots, n,$$

and requiring that each partial derivative be equal to zero yields the **normal equations**

$$\sum_{j=0}^n \left[\int_a^b \varphi_k(x) \varphi_j(x) \, dx \right] c_j = \int_a^b \varphi_k(x) f(x) \, dx, \quad k = 0, 1, \ldots, n.$$

We can then solve this system of equations to obtain the coefficients $\{c_j\}_{j=0}^n$. This system can be solved as long as the functions $\{\varphi_j(x)\}_{j=0}^n$ are *linearly independent* in $C[a,b]$. That is, the condition

$$\sum_{j=0}^n c_j \varphi_j(x) \equiv 0, \quad x \in [a,b],$$

is only true if $c_0 = c_1 = \cdots = c_n = 0$.

Exploration 8.2.1 Prove that the functions $\{\varphi_j(x)\}_{j=0}^n$ are linearly independent in $C[a,b]$ if, for $j = 0, 1, \ldots, n$, $\varphi_j(x)$ is a polynomial of degree j.

Exploration 8.2.2 Let A be the $(n+1) \times (n+1)$ matrix defined by

$$a_{ij} = \int_a^b \varphi_i(x)\varphi_j(x)\, dx,$$

where the functions $\{\varphi_j(x)\}_{j=0}^n$ are real-valued functions that are linearly independent in $C[a,b]$. Prove that A is symmetric positive definite. How does this guarantee that the solution of the normal equations yields a minimum rather than a maximum or saddle point? Why is the assumption of linear independence essential?

Example 8.2.1 We approximate $f(x) = e^x$ on the interval $[0, 5]$ by a fourth-degree polynomial

$$f_4(x) = c_0 + c_1 x + c_2 x^2 + c_3 x^3 + c_4 x^4.$$

The normal equations have the form

$$\sum_{j=0}^n a_{ij} c_j = b_i, \quad i = 0, 1, \ldots, 4,$$

or, in matrix-vector form, $A\mathbf{c} = \mathbf{b}$, where

$$a_{ij} = \int_0^5 x^i x^j \, dx = \int_0^5 x^{i+j} \, dx = \frac{5^{i+j+1}}{i+j+1}, \quad i,j = 0, 1, \ldots, 4,$$

$$b_i = \int_0^5 x^i e^x \, dx, \quad i = 0, 1, \ldots, 4.$$

Integration by parts yields the relation

$$b_i = 5^i e^5 - i b_{i-1}, \quad b_0 = e^5 - 1.$$

Solving this system of equations yields the polynomial

$$f_4(x) = 2.3002 - 6.226x + 9.5487x^2 - 3.86x^3 + 0.6704x^4.$$

As Figure 8.4 shows, this polynomial is barely distinguishable from e^x on $[0, 5]$.

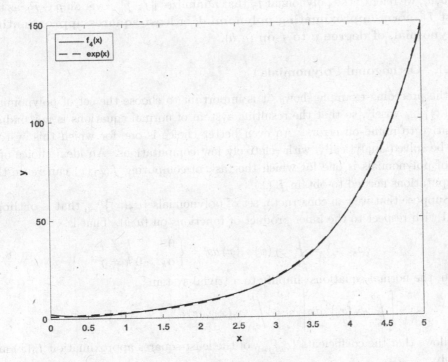

Fig. 8.4 From Example 8.2.1: Graphs of $f(x) = e^x$ (dashed curve) and the 4th-degree continuous least-squares polynomial approximation $f_4(x)$ on $[0, 5]$ (solid curve)

However, it should be noted that the matrix A is closely related to the $n \times n$ **Hilbert matrix** H_n, which has entries

$$[H_n]_{ij} = \frac{1}{i + j - 1}, \quad 1 \leq i, j \leq n.$$

This matrix is famous for being highly **ill-conditioned** (See Section 3.4.1), meaning that solutions to systems of linear equations involving this matrix that are computed using floating-point arithmetic are highly sensitive to roundoff error. In fact, the matrix A in this example has a **condition number** of 1.56×10^7, which means that a $O(\epsilon)$ perturbation in the right-hand side vector $\mathbf{b}$, with entries b_i, can cause a $O(\epsilon \times 10^7)$ change in the solution $\mathbf{c}$. $\square$

Exploration 8.2.3 Repeat Example 8.2.1 with $f(x) = x^7$. What happens to the coefficients $\{c_j\}_{j=0}^4$ if the right-hand side vector $\mathbf{b}$ is perturbed?

For the remainder of this section, we restrict ourselves to the case where the functions $\{\varphi_j(x)\}_{j=0}^n$ are polynomials. These polynomials form a **basis** (see Section B.4) of $\mathcal{P}_n$, the vector space of polynomials of degree at most n. Then, for $f \in$

$L^2(a, b)$, we refer to the polynomial f_n that minimizes $\|f-p\|_2$ over all $p \in \mathcal{P}_n$ as the **best L^2-norm approximating polynomial**, or **least-squares approximating polynomial**, of degree n to f on (a, b).

8.2.1   Orthogonal Polynomials

As the preceding example shows, it is important to choose the set of polynomials $\{\varphi_j(x)\}_{j=0}^n$ wisely, so that the resulting system of normal equations is not unduly sensitive to round-off errors. An even better choice is one for which this system can be solved analytically, with relatively few computations. An ideal choice of a set of polynomials is one for which the task of computing $f_{n+1}(x)$ can reuse the computations needed to obtain $f_n(x)$.

Suppose that we can construct a set of polynomials $\{\varphi_j(x)\}_{j=0}^n$ that is **orthogonal** with respect to the inner product of functions on (a, b). That is,

$$\langle \varphi_k, \varphi_j \rangle = \int_a^b \varphi_k(x) \varphi_j(x)\, dx = \begin{cases} 0 & k \neq j \\ \alpha_k > 0 & k = j \end{cases}. \tag{8.4}$$

Then, the normal equations simplify to a trivial system

$$\left[\int_a^b [\varphi_k(x)]^2\, dx \right] c_k = \int_a^b \varphi_k(x) f(x)\, dx, \quad k = 0, 1, \ldots, n.$$

It follows that the coefficients $\{c_j\}_{j=0}^n$ of the least-squares approximation $f_n(x)$ are simply

$$c_k = \frac{\langle \varphi_k, f \rangle}{\|\varphi_k\|_2^2}, \quad k = 0, 1, \ldots, n.$$

If the constants $\{\alpha_k\}_{k=0}^n$ in (8.4) satisfy $\alpha_k = 1$ for $k = 0, 1, \ldots, n$, then we say that the orthogonal set of functions $\{\varphi_j(x)\}_{j=0}^n$ is **orthonormal**. In that case, the solution to the continuous least-squares problem is simply given by

$$c_k = \langle \varphi_k, f \rangle, \quad k = 0, 1, \ldots, n. \tag{8.5}$$

Next, we will learn how sets of orthogonal polynomials can be constructed.

8.2.2   Construction of Orthogonal Polynomials

Recall from Section 4.2.3 the process known as **Gram-Schmidt Orthogonalization** for obtaining a set of orthogonal vectors $\mathbf{p}_1, \mathbf{p}_2, \ldots, \mathbf{p}_n$ from a set of linearly independent vectors $\mathbf{a}_1, \mathbf{a}_2, \ldots, \mathbf{a}_n$:

$$\mathbf{p}_1 = \mathbf{a}_1,$$

$$\mathbf{p}_2 = \mathbf{a}_2 - \frac{\mathbf{p}_1^T \mathbf{a}_2}{\mathbf{p}_1^T \mathbf{p}_1} \mathbf{p}_1,$$

$$\vdots$$

$$\mathbf{p}_n = \mathbf{a}_n - \sum_{j=0}^{n-1} \frac{\mathbf{p}_j^T \mathbf{a}_n}{\mathbf{p}_j^T \mathbf{p}_j} \mathbf{p}_j.$$

By normalizing each vector $\mathbf{p}_j$, we obtain a unit vector

$$\mathbf{q}_j = \frac{1}{\|\mathbf{p}_j\|_2} \mathbf{p}_j,$$

and a set of orthonormal vectors $\{\mathbf{q}_j\}_{j=1}^n$, in that they are orthogonal ($\mathbf{q}_k^T \mathbf{q}_j = 0$ for $k \neq j$), and unit vectors ($\mathbf{q}_j^T \mathbf{q}_j = 1$).

We can use a similar process to compute a set of **orthogonal polynomials** $\{p_j(x)\}_{j=0}^n$. For convenience, we will require that all polynomials in the set be **monic**; that is, their leading (highest-degree) coefficient must be equal to 1. We first define $p_0(x) = 1$. Then, because $p_1(x)$ is supposed to be of degree 1, it must have the form $p_1(x) = x - \alpha_1$ for some constant α_1. To ensure that $p_1(x)$ is orthogonal to $p_0(x)$, we compute their inner product, and obtain

$$0 = \langle p_0, p_1 \rangle = \langle 1, x - \alpha_1 \rangle,$$

so we must have

$$\alpha_1 = \frac{\langle 1, x \rangle}{\langle 1, 1 \rangle}.$$

For $j > 1$, we start by setting $p_j(x) = x p_{j-1}(x)$, since p_j should be of degree one greater than that of p_{j-1}, and this satisfies the requirement that p_j be monic. Then, we need to subtract polynomials of lower degree to ensure that p_j is orthogonal to p_i, for $i < j$. To that end, we apply Gram-Schmidt Orthogonalization and obtain

$$p_j(x) = x p_{j-1}(x) - \sum_{i=0}^{j-1} \frac{\langle p_i, x p_{j-1} \rangle}{\langle p_i, p_i \rangle} p_i(x).$$

However, by the definition of the inner product, $\langle p_i, x p_{j-1} \rangle = \langle x p_i, p_{j-1} \rangle$. Furthermore, because $x p_i$ is of degree $i + 1$, and p_{j-1} is orthogonal to *all* polynomials of degree less than $j - 1$, it follows that $\langle p_i, x p_{j-1} \rangle = 0$ whenever $i < j - 2$.

We have shown that sequences of orthogonal polynomials satisfy a **three-term recurrence relation**

$$p_j(x) = (x - \alpha_j) p_{j-1}(x) - \beta_{j-1}^2 p_{j-2}(x), \quad j > 1, \tag{8.6}$$

where the **recursion coefficients** α_j and β_{j-1}^2 are defined to be

$$\alpha_j = \frac{\langle p_{j-1}, x p_{j-1} \rangle}{\langle p_{j-1}, p_{j-1} \rangle}, \quad j > 1,$$

$$\beta_j^2 = \frac{\langle p_{j-1}, x p_j \rangle}{\langle p_{j-1}, p_{j-1} \rangle} = \frac{\langle x p_{j-1}, p_j \rangle}{\langle p_{j-1}, p_{j-1} \rangle} = \frac{\langle p_j, p_j \rangle}{\langle p_{j-1}, p_{j-1} \rangle} = \frac{\|p_j\|_2^2}{\|p_{j-1}\|_2^2}, \quad j \geq 1.$$

Note that $\langle x p_{j-1}, p_j \rangle = \langle p_j, p_j \rangle$ because both p_j and $x p_{j-1}$ are monic, and differ from one another by a polynomial of degree at most $j - 1$, which is orthogonal to p_j. The recurrence relation (8.6) is also valid for $j = 1$, provided that we define $p_{j-1}(x) \equiv 0$, and α_1 is defined as above. That is,

$$p_1(x) = (x - \alpha_1) p_0(x), \quad \alpha_1 = \frac{\langle p_0, x p_0 \rangle}{\langle p_0, p_0 \rangle}.$$

If we also define the recursion coefficient β_0 by

$$\beta_0^2 = \langle p_0, p_0 \rangle,$$

and then define

$$q_j(x) = \frac{p_j(x)}{\beta_0 \beta_1 \cdots \beta_j},$$

then the polynomials $q_0, q_1, \ldots, q_n$ are also orthogonal, and

$$\langle q_j, q_j \rangle = \frac{\langle p_j, p_j \rangle}{\beta_0^2 \beta_1^2 \cdots \beta_j^2} = \langle p_j, p_j \rangle \frac{\langle p_{j-1}, p_{j-1} \rangle}{\langle p_j, p_j \rangle} \cdots \frac{\langle p_0, p_0 \rangle}{\langle p_1, p_1 \rangle} \frac{1}{\langle p_0, p_0 \rangle} = 1.$$

That is, these polynomials are *orthonormal*.

Exploration 8.2.4 Compute the first three monic orthogonal polynomials with respect to the inner product

$$\langle f, g \rangle = \int_0^1 f(x) g(x) \, dx.$$

Exploration 8.2.5 Write a MATLAB function

$$\texttt{P=orthpoly(a,b,n)}$$

that computes the coefficients of monic orthogonal polynomials on the interval (a, b), up to and including degree n, and stores their coefficients in the rows of the $(\mathrm{n}+1) \times (\mathrm{n}+1)$ matrix P. Use MATLAB's polynomial functions from Section 1.2.18 to evaluate the required inner products.

8.2.3   Legendre Polynomials

If we consider the inner product

$$\langle f, g \rangle = \int_{-1}^1 f(x) g(x) \, dx,$$

then, by Gram-Schmidt Orthogonalization, a sequence of orthogonal polynomials, with respect to this inner product, can be defined as follows:

$$L_0(x) = 1, \tag{8.7}$$

$$L_1(x) = x, \tag{8.8}$$

$$L_{j+1}(x) = \frac{2j+1}{j+1} x L_j(x) - \frac{j}{j+1} L_{j-1}(x), \quad j = 1, 2, \ldots \tag{8.9}$$

These are known as the **Legendre polynomials** [Le Gendre (1785)]. One of their most important applications is in the construction of Gauss quadrature rules (see Section 9.7). Specifically, the roots of $L_n(x)$, for $n \geq 1$, are the nodes of a Gauss quadrature rule for the interval $(-1, 1)$. However, they can also be used to easily compute continuous least-squares polynomial approximations, as the following example shows.

Example 8.2.2 We will use Legendre polynomials to approximate $f(x) = \cos x$ on $[-\pi/2, \pi/2]$ by a quadratic polynomial. First, we note that the first three Legendre polynomials, which are the ones of degree 0, 1 and 2, are

$$L_0(x) = 1, \quad L_1(x) = x, \quad L_2(x) = \frac{1}{2}(3x^2 - 1).$$

However, it is not practical to use these polynomials directly to approximate $f(x)$, because they are orthogonal with respect to the inner product defined on the interval $(-1, 1)$, and we wish to approximate $f(x)$ on $(-\pi/2, \pi/2)$.

To obtain orthogonal polynomials on $(-\pi/2, \pi/2)$, we replace x by $2t/\pi$, where t belongs to $[-\pi/2, \pi/2]$, in the Legendre polynomials, which yields

$$\tilde{L}_0(t) = 1, \quad \tilde{L}_1(t) = \frac{2t}{\pi}, \quad \tilde{L}_2(t) = \frac{1}{2}\left(\frac{12}{\pi^2}t^2 - 1\right).$$

Then, we can express our quadratic approximation $f_2(x)$ of $f(x)$ as the linear combination

$$f_2(x) = c_0 \tilde{L}_0(x) + c_1 \tilde{L}_1(x) + c_2 \tilde{L}_2(x),$$

where

$$c_j = \frac{\langle f, \tilde{L}_j \rangle}{\langle \tilde{L}_j, \tilde{L}_j \rangle}, \quad j = 0, 1, 2.$$

Computing these inner products yields

$$\langle f, \tilde{L}_0 \rangle = \int_{-\pi/2}^{\pi/2} \cos t \, dt = 2,$$

$$\langle f, \tilde{L}_1 \rangle = \int_{-\pi/2}^{\pi/2} \frac{2t}{\pi} \cos t \, dt = 0,$$

$$\langle f, \tilde{L}_2 \rangle = \int_{-\pi/2}^{\pi/2} \frac{1}{2}\left(\frac{12}{\pi^2}t^2 - 1\right) \cos t \, dt = \frac{2}{\pi^2}(\pi^2 - 12),$$

$$\langle \tilde{L}_0, \tilde{L}_0 \rangle = \int_{-\pi/2}^{\pi/2} 1 \, dt = \pi,$$

$$\langle \tilde{L}_1, \tilde{L}_1 \rangle = \int_{-\pi/2}^{\pi/2} \left(\frac{2t}{\pi}\right)^2 dt = \frac{8\pi}{3},$$

$$\langle \tilde{L}_2, \tilde{L}_2 \rangle = \int_{-\pi/2}^{\pi/2} \left[\frac{1}{2}\left(\frac{12}{\pi^2}t^2 - 1\right)\right]^2 dt = \frac{\pi}{5}.$$

It follows that

$$c_0 = \frac{2}{\pi}, \quad c_1 = 0, \quad c_2 = \frac{2}{\pi^2}\frac{5}{\pi}(\pi^2 - 12) = \frac{10}{\pi^3}(\pi^2 - 12),$$

and therefore

$$f_2(x) = \frac{2}{\pi} + \frac{5}{\pi^3}(\pi^2 - 12)\left(\frac{12}{\pi^2}x^2 - 1\right) \approx 0.98016 - 0.4177x^2.$$

This approximation is shown in Figure 8.5. $\square$

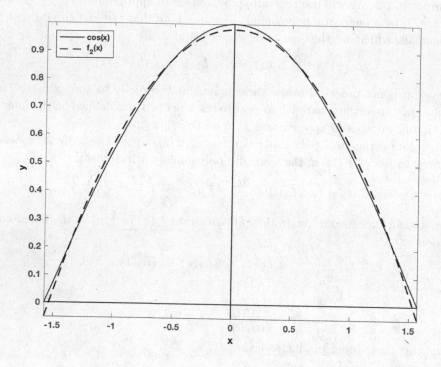

Fig. 8.5   Graph of $\cos x$ (blue curve) and its continuous least-squares quadratic approximation (dashed curve) on $(-\pi/2, \pi/2)$

Exploration 8.2.6 Write a MATLAB script that computes the coefficients of the Legendre polynomials up to a given degree n, using the recurrence relation (8.9) and the function `conv` for multiplying polynomials. Then, plot the graphs of these polynomials on the interval $(-1, 1)$. What properties can you observe in these graphs? Is there any symmetry to them?

Exploration 8.2.7 Prove by induction that the Legendre polynomial $L_j(x)$ is an odd function (that is, $L_j(-x) = -L_j(x)$) if j is odd, and an even function (that is, $L_j(-x) = L_j(x)$) if j is even.

Exploration 8.2.8 Let A be the Vandermonde matrix from (8.2), where the points $x_1, x_2, \ldots, x_m$ are equally spaced points in the interval $(-1, 1)$. Construct this matrix in MATLAB for a small chosen value of n and a large value of m, and then compute the **QR factorization** of A (see Section 4.2). How do the columns of Q relate to the Legendre polynomials?

8.2.4 Chebyshev Polynomials

It is possible to compute sequences of orthogonal polynomials with respect to other inner products, that differ beyond just a change in the interval of integration. A generalization of the inner product (8.3) that we have been using is defined by

$$\langle f, g \rangle = \int_a^b f(x)g(x)w(x)\,dx, \tag{8.10}$$

where $w(x)$ is a **weight function**. To be a weight function, it is required that $w(x) \geq 0$ on (a, b), and that $w(x) \neq 0$ on any subinterval of (a, b). So far, we have only considered the case of $w(x) \equiv 1$.

Exploration 8.2.9 Prove that the discussion of Section 8.2.2 also applies when using the inner product

$$\langle f, g \rangle = \int_a^b f(x)g(x)w(x)\,dx,$$

where $w(x)$ is a weight function. That is, polynomials orthogonal with respect to this inner product also satisfy a three-term recurrence relation, with analogous definitions of the recursion coefficients α_j and β_j.

Another weight function of interest is

$$w(x) = \frac{1}{\sqrt{1-x^2}}, \quad -1 < x < 1.$$

A sequence of polynomials that is orthogonal with respect to this weight function, and the associated inner product

$$\langle f, g \rangle = \int_{-1}^1 f(x)g(x)\frac{1}{\sqrt{1-x^2}}\,dx$$

is the sequence of **Chebyshev polynomials**, previously introduced in Section 7.4.2:

$$T_0(x) = 1,$$
$$T_1(x) = x,$$
$$T_{j+1}(x) = 2xT_j(x) - T_{j-1}(x), \quad j = 1, 2, \ldots$$

which can also be defined by

$$T_j(x) = \cos(j\cos^{-1}x), \quad -1 \leq x \leq 1.$$

It is interesting to note that if we let $x = \cos\theta$, then

$$\langle f, T_j \rangle = \int_{-1}^1 f(x)\cos(j\cos^{-1}x)\frac{1}{\sqrt{1-x^2}}\,dx$$

$$= \int_0^\pi f(\cos\theta)\cos j\theta\,d\theta.$$

In Section 8.4, we will investigate continuous and discrete least-squares approximation of functions by linear combinations of **trigonometric polynomials** such as $\cos j\theta$ or $\sin j\theta$, which will reveal how these coefficients $\langle f, T_j \rangle$ can be computed very rapidly.

Exploration 8.2.10 Write a MATLAB function

$$\texttt{fn=best2normapprox(f,a,b,n,w)}$$

that computes a vector **fn** consisting of the coefficients of $f_n(x)$, a polynomial of degree **n** that is the best L^2-norm approximation on $(\texttt{a}, \texttt{b})$ of the given function $f(x)$ represented by the function handle **f**. "Best L^2-norm approximation" is defined in terms of the inner product

$$\langle f, g \rangle = \int_a^b f(x)g(x)w(x)\,dx,$$

where **w** is a function handle for the weight function $w(x)$. Use the built-in MATLAB function `integral` to evaluate the required inner products. Make the fifth argument **w** an optional argument, using $w(x) \equiv 1$ as a default.

Exploration 8.2.11 Compute the best L^2-norm approximating polynomial of degree 3 to the functions $f(x) = e^x$ and $g(x) = \sin \pi x$ on $(-1, 1)$, using both Legendre and Chebyshev polynomials. Comment on the accuracy of these approximations.

8.2.5 Error Analysis

Let $p \in \mathcal{P}_n$, where $\mathcal{P}_n$ is the space of polynomials of degree at most n, and let f_n be the best L^2-norm approximating polynomial of degree n to $f \in L^2(a, b)$. As before, we assume the polynomials $q_0(x), q_1(x), \ldots, q_n(x)$ are *orthonormal*, in the sense that

$$\langle q_j, q_k \rangle = \int_a^b q_j(x)q_k(x)w(x)\,dx = \delta_{jk}, \quad j, k = 0, 1, \ldots, n.$$

Then, from (8.5) we have

$$f_n(x) = \sum_{j=0}^n \langle q_j, f \rangle q_j(x). \tag{8.11}$$

This form of $f_n(x)$ can be used to prove the following result.

Theorem 8.2.3 The polynomial $f_n \in \mathcal{P}_n$ is the best L^2-norm approximating polynomial of degree n to $f \in L^2(a, b)$ if and only if

$$\langle f - f_n, p \rangle = 0$$

for all $p \in \mathcal{P}_n$.

Exploration 8.2.12 Use (8.11) to prove one part of Theorem 8.2.3: assume f_n is the best L^2-norm approximating polynomial to $f \in L^2(a, b)$, and show that $\langle f - f_n, p \rangle = 0$ for any $p \in \mathcal{P}_n$.

Exploration 8.2.13 Use the Cauchy-Schwarz inequality (see Section B.13.1) to prove the converse of Exploration 8.2.12: that if $f \in L^2(a,b)$ and $\langle f - f_n, p \rangle = 0$ for an arbitrary $p \in \mathcal{P}_n$, then f_n is the best L^2-norm approximating polynomial of degree n to f; that is,

$$\|f - f_n\|_2 \leq \|f - p\|_2.$$

Hint: By the assumptions, $f - f_n$ is orthogonal to *any* polynomial in $\mathcal{P}_n$.

8.2.6 Roots of Orthogonal Polynomials*

Finally, we prove one property of orthogonal polynomials that will prove useful in our upcoming discussion in Chapter 9 of the role of orthogonal polynomials in numerical integration. Let $\varphi_j(x)$ be a polynomial of degree $j \geq 1$ that is orthogonal to all polynomials of lower degree, with respect to the inner product

$$\langle f, g \rangle = \int_a^b f(x)g(x)w(x)\,dx.$$

We will prove, by contradiction, that the roots of φ_j are real, distinct, and lie in (a,b). First, we note that φ_j has real coefficients, as a consequence of Gram-Schmidt orthogonalization.

- *The roots are real:* If $j = 1$, this is trivially true. Suppose $j \geq 2$ and that φ_j has a complex root $c + di$. Because φ_j has real coefficients, $c - di$ is also a root, and therefore φ_j has the factorization $\varphi_j(x) = ((x-c)^2 + d^2)q_{j-2}(x)$, where q_{j-2} has degree $j - 2$. We then have

$$\langle \varphi_j, q_{j-2} \rangle = \int_a^b ((x-c)^2 + d^2)[q_{j-2}(x)]^2 w(x)\,dx > 0,$$

because the integrand is nonnegative on (a,b), but because φ_j is orthogonal to all polynomials of lesser degree, $\langle \varphi_j, q_{j-2} \rangle = 0$, which is a contradiction.

- *The roots are distinct:* Again, this is trivially true if $j = 1$. Suppose $j \geq 2$ and that φ_j has a root c of multiplicity at least 2. Then φ_j has the factorization $\varphi_j(x) = (x-c)^2 q_{j-2}(x)$, where q_{j-2} has degree $j - 2$. We then have

$$\langle \varphi_j, q_{j-2} \rangle = \int_a^b (x-c)^2 [q_{j-2}(x)]^2 w(x)\,dx > 0,$$

because the integrand is nonnegative on (a,b), but because φ_j is orthogonal to all polynomials of lesser degree, $\langle \varphi_j, q_{j-2} \rangle = 0$, which is a contradiction.

- *The roots are in (a,b):* Suppose φ_j has a root c that is outside (a,b). Then φ_j has the factorization $\varphi_j(x) = (x-c)q_{j-1}(x)$, where q_{j-1} has degree $j-1$. We then have

$$\langle \varphi_j, q_{j-1} \rangle = \int_a^b (x-c)[q_{j-1}(x)]^2 w(x)\,dx \neq 0,$$

because the integrand does not change sign on (a, b), but because φ_j is orthogonal to all polynomials of lesser degree, $\langle \varphi_j, q_{j-1} \rangle = 0$, which is a contradiction.

Exploration 8.2.14 Use your function `orthpoly` from Exploration 8.2.5 to generate orthogonal polynomials of a fixed degree n for various weight functions. How does the distribution of the roots of $p_n(x)$ vary based on where the weight function has smaller or larger values? *Hint:* Consider the distribution of the roots of Chebyshev polynomials, and their weight function $w(x) = (1 - x^2)^{-1/2}$.

8.2.7   Concept Check

1. What is the continuous least squares problem? Contrast with the discrete least squares problem from Section 8.1.
2. When approximating a function by a polynomial, why not use the monomial basis?
3. What is the process used to construct a sequence of orthogonal polynomials?
4. Name two families of orthogonal polynomials. On which intervals are they orthogonal, and with respect to which weight functions?
5. What properties do the roots of orthogonal polynomials have?

8.3   Rational Approximation*

In some cases, it is not practical to approximate a given function $f(x)$ by a polynomial, because it simply cannot capture the behavior of $f(x)$ unless it has an impractically high degree. This is because higher-degree polynomials tend to be oscillatory, so if $f(x)$ is mostly smooth, the degree n of an approximating polynomial $f_n(x)$ must be unreasonably high. Therefore, in this section we consider an alternative to polynomial approximation.

8.3.1   Padé Approximants

Specifically, we seek a *rational function* of the form

$$r_{m,n}(x) = \frac{p_m(x)}{q_n(x)} = \frac{a_0 + a_1 x + a_2 x^2 + \cdots + a_m x^m}{b_0 + b_1 x + b_2 x^2 + \cdots + b_n x^n},$$

where $p_m(x)$ and $q_n(x)$ are polynomials of degree m and n, respectively. For convenience, we impose $b_0 = 1$, since otherwise the other coefficients can simply be scaled.

To construct $p_m(x)$ and $q_n(x)$, we generalize approximation of $f(x)$ by a Taylor polynomial (see Theorem A.6.1) of degree n. Consider the error

$$E(x) = f(x) - r_{m,n}(x) = \frac{f(x)q_n(x) - p_m(x)}{q_n(x)}. \tag{8.12}$$

As in Taylor polynomial approximation, our goal is to choose the coefficients of p_m and q_n so that

$$E(0) = E'(0) = E''(0) = \cdots = E^{(m+n)}(0) = 0.$$

That is, 0 is a root of multiplicity $m+n+1$. It follows that x^{m+n+1} is included in the factorization of the numerator of $E(x)$.

For convenience, we express p and q as polynomials of degree $m+n$, by padding them with coefficients that are zero: $a_{m+1} = a_{m+2} = \cdots = a_{m+n} = 0$ and $b_{n+1} = b_{n+2} = \cdots = b_{n+m} = 0$. Taking a Maclaurin expansion of $f(x)$ (that is, a Taylor expansion around $x_0 = 0$),

$$f(x) = \sum_{i=0}^{\infty} c_i x^i, \quad c_i = \frac{f^{(i)}(0)}{i!},$$

we obtain the following expression for the numerator of (8.12):

$$f(x)q_n(x) - p_m(x) = \sum_{i=0}^{\infty} c_i x^i \sum_{j=0}^{m+n} b_j x^j - \sum_{i=0}^{m+n} a_i x^i \qquad (8.13)$$

$$= \sum_{i=0}^{\infty} \sum_{j=0}^{m+n} c_i b_j x^{i+j} - \sum_{i=0}^{m+n} a_i x^i$$

$$= \sum_{i=0}^{\infty} \sum_{j=0}^{\min(m+n,i)} b_j c_{i-j} x^i - \sum_{i=0}^{m+n} a_i x^i$$

$$= \sum_{i=0}^{m+n} \left[\sum_{j=0}^{i} b_j c_{i-j} - a_i \right] x_i + \sum_{i=m+n+1}^{\infty} \sum_{j=0}^{m+n} b_j c_{i-j} x^i.$$

We can then ensure that 0 is a root of multiplicity $m+n+1$ if the numerator has no terms of degree $m+n$ or less. That is, each coefficient of x^i in the first summation must equal zero.

This entails solving the system of $m+n+1$ equations

$$c_0 = a_0,$$

$$c_1 + b_1 c_0 = a_1,$$

$$c_2 + b_1 c_1 + b_2 c_0 = a_2,$$

$$\vdots \qquad (8.14)$$

$$c_n + b_1 c_{n-1} + \cdots + b_n c_0 = a_n,$$

$$c_{n+1} + b_1 c_n + \cdots + b_n c_1 = a_{n+1},$$

$$\vdots$$

$$c_{n+m} + b_1 c_{n-1+m} + \cdots + b_n c_m = a_{m+n}.$$

This is a system of $m+n+1$ linear equations in the $m+n+1$ unknowns $b_1, b_2, \ldots, b_n$, $a_0, a_1, \ldots, a_m$. The resulting rational function $r_{m,n}(x)$ is called a **Padé approximant** of $f(x)$ [Padé (1892)]. We now illustrate the computation of these coefficients.

Example 8.3.1 We approximate $f(x) = e^{-x}$ by a rational function of the form

$$r_{2,3}(x) = \frac{a_0 + a_1 x + a_2 x^2}{1 + b_1 x + b_2 x^2 + b_3 x^3}.$$

The Maclaurin series for $f(x)$ has coefficients $c_j = (-1)^j/j!$. The system of equations (8.14) becomes

$$c_0 = a_0,$$
$$c_1 + b_1 c_0 = a_1,$$
$$c_2 + b_1 c_1 + b_2 c_0 = a_2,$$
$$c_3 + b_1 c_2 + b_2 c_1 + b_3 c_0 = 0,$$
$$c_4 + b_1 c_3 + b_2 c_2 + b_3 c_1 = 0,$$
$$c_5 + b_1 c_4 + b_2 c_3 + b_3 c_2 = 0.$$

This can be written as $A\mathbf{x} = \mathbf{b}$, where

$$A = \begin{bmatrix} -1 & & & & & \\ & -1 & & c_0 & & \\ & & -1 & c_1 & c_0 & \\ & & & c_2 & c_1 & c_0 \\ & & & c_3 & c_2 & c_1 \\ & & & c_4 & c_3 & c_2 \end{bmatrix}, \quad \mathbf{x} = \begin{bmatrix} a_0 \\ a_1 \\ a_2 \\ b_1 \\ b_2 \\ b_3 \end{bmatrix}, \quad \mathbf{b} = \begin{bmatrix} -c_0 \\ -c_1 \\ -c_2 \\ -c_3 \\ -c_4 \\ -c_5 \end{bmatrix}.$$

We see that Gaussian Elimination (see Section 3.1) can be carried out by eliminating at most $n-1$ entries in columns $m+2, \ldots, m+n$. After that, the matrix will be reduced to upper triangular form so that back substitution can be carried out. If pivoting is required, it need be carried out only on the last n rows, because due to the block upper triangular structure of A, it follows that A is nonsingular if and only if the lower right $n \times n$ block is.

After carrying out Gaussian Elimination for this example, with Maclaurin series coefficients $c_j = (-1)^j/j!$, we obtain the rational approximation

$$e^{-x} \approx r_{2,3}(x) = \frac{p_2(x)}{q_3(x)} = \frac{\frac{1}{20}x^2 - \frac{2}{5}x + 1}{\frac{1}{60}x^3 + \frac{3}{20}x^2 + \frac{3}{5}x + 1}.$$

Plotting the error in this approximation on the interval $[0, 1]$, we see that the error is maximum at $x = 1$, at roughly 4.5×10^{-5}. □

Exploration 8.3.1 Write a MATLAB function `[p,q]=padeapprox(c,m,n)` that computes vectors `p` and `q` consisting of the coefficients of the polynomials $p_m(x)$ and $q_n(x)$, respectively, such that $r_{m,n}(x) = p_m(x)/q_n(x)$ is the Padé approximant of degree m, n for the function $f(x)$ with Maclaurin series coefficients $c_0, c_1, \ldots, c_{m+n}$ stored in the vector `c`.

8.3.2 Continued Fraction Form

We now examine the process of efficiently evaluating $r_{m,n}(x)$. A natural approach is to apply **nested multiplication** (see Section 7.3) to $p_m(x)$ and $q_n(x)$.

Example 8.3.2 If we apply nested multiplication to $p_2(x)$ and $q_3(x)$ from Example 8.3.1, we obtain

$$p_2(x) = 1 + x\left(-\frac{2}{5} + \frac{1}{20}x\right), \quad q_3(x) = 1 + x\left(\frac{3}{5} + x\left(\frac{3}{20} + \frac{1}{60}x\right)\right).$$

It follows that evaluating $r_{2,3}(x)$ requires 5 multiplications, 5 additions, and one division. An alternative approach is to represent $r_{2,3}(x)$ as a **continued fraction** [Ralston and Rabinowitz (1978)]. We have

$$r_{2,3}(x) = \frac{p_2(x)}{q_3(x)} = \frac{\dfrac{1}{20}x^2 - \dfrac{2}{5}x + 1}{\dfrac{1}{60}x^3 + \dfrac{3}{20}x^2 + \dfrac{3}{5}x + 1}$$

$$= \frac{3}{\dfrac{x^3 + 9x^2 + 36x + 60}{x^2 - 8x + 20}}$$

$$= \frac{3}{x + 17 + \dfrac{152x - 280}{x^2 - 8x + 20}}$$

$$= \frac{3}{x + 17 + \dfrac{152}{\dfrac{x^2 - 8x + 20}{x - 35/19}}}$$

$$= \frac{3}{x + 17 + \dfrac{152}{x - \dfrac{117}{19} + \dfrac{3125/361}{x - 35/19}}}$$

In this form, evaluation of $r_{2,3}(x)$ requires three divisions, no multiplications, and five additions, resulting in significantly more efficiency than using nested multiplication on $p_2(x)$ and $q_3(x)$. □

It is important to note that the efficiency of this approach comes from the ability to make the polynomial in each denominator monic–that is, having a leading coefficient of one–to remove the need for a multiplication.

Exploration 8.3.2 Write a MATLAB function

$$y=contfrac(p,q,x)$$

that takes as input polynomials $p(x)$ and $q(x)$, represented as vectors of coefficients p and q, respectively, and outputs $y = p(x)/q(x)$ by evaluating $p(x)/q(x)$ as a continued fraction. *Hint:* Use the MATLAB function deconv to divide polynomials.

Exploration 8.3.3 Write a MATLAB function

```
cf=contfracform(p,q)
```

that takes as input polynomials $p(x)$ and $q(x)$, represented as vectors of coefficients p and q, respectively, and outputs a MATLAB structure cf that contains sufficient information to evaluate $p(x)/q(x)$ using continued fraction form. What information should cf contain, to use as little storage as possible?

8.3.3 Chebyshev Rational Approximation

One drawback of the Padé approximant is that while it is highly accurate near $x = 0$, it loses accuracy as x moves away from zero. Certainly it is straightforward to perform Taylor expansion around a different center x_0, which ensures similar accuracy near x_0, but it would be preferable to instead compute an approximation that is accurate on an entire interval $[a, b]$.

To that end, we can employ the **Chebyshev polynomials**, previously discussed in Section 7.4.2. Just as they can help reduce the error in polynomial interpolation over an interval, they can also improve the accuracy of a rational approximation over an interval. For simplicity, we consider the interval $(-1, 1)$, on which each Chebyshev polynomial $T_k(x)$ satisfies $|T_k(x)| \le 1$, but the approach described here can readily be applied to an arbitrary interval through shifting and scaling as needed.

The main idea is to use $T_k(x)$ in place of x^k in constructing our rational approximation. That is, our rational approximation now has the form

$$r_{m,n}(x) = \frac{p_m(x)}{q_n(x)} = \frac{\sum_{k=0}^{m} a_k T_k(x)}{\sum_{k=0}^{n} b_k T_k(x)}.$$

If we also expand $f(x)$ in a series of Chebyshev polynomials,

$$f(x) = \sum_{k=0}^{\infty} c_k T_k(x), \tag{8.15}$$

then the error in our approximation is, by analogy with (8.13),

$$E(x) = f(x) - r_{m,n}(x)$$
$$= \frac{f(x)q_n(x) - p_m(x)}{q_n(x)}$$
$$= \frac{1}{q_n(x)} \left[\sum_{i=0}^{\infty} \sum_{j=0}^{n} c_i b_j T_i(x) T_j(x) - \sum_{i=0}^{m} a_i T_i(x) \right].$$

By applying the identity

$$T_i(x) T_j(x) = \frac{1}{2}[T_{i+j}(x) + T_{|i-j|}(x)], \tag{8.16}$$

we obtain the error

$$E(x) = \frac{1}{q_n(x)} \left[\frac{1}{2} \sum_{i=0}^{\infty} \sum_{j=0}^{n} c_i b_j [T_{i+j}(x) + T_{|i-j|}(x)] - \sum_{i=0}^{m} a_i T_i(x) \right]$$

$$= \frac{1}{q_n(x)} \left\{ \sum_{i=0}^{\infty} c_i T_i(x) - \sum_{i=0}^{m} a_i T_i(x) + \right.$$

$$\left. \frac{1}{2} \sum_{j=1}^{n} b_j \left[\sum_{i=j}^{\infty} c_{i-j} T_i(x) + \sum_{i=1}^{j} c_{j-i} T_i(x) + \sum_{i=0}^{\infty} c_{i+j} T_i(x) \right] \right\}. \quad (8.17)$$

The coefficients $\{a_j\}_{i=0}^{m}$, $\{b_j\}_{j=1}^{n}$ are then determined by requiring that the coefficient of $T_i(x)$ in $E(x)$ vanishes, for $i = 0, 1, 2, \ldots, m+n$.

To obtain the coefficients $\{c_j\}_{j=0}^{\infty}$ in the series expansion of $f(x)$ from (8.15), we use the fact that the Chebyshev polynomials are orthogonal on $(-1, 1)$ with respect to the weight function $w(x) = (1 - x^2)^{-1/2}$. By taking the inner product of both sides of (8.15), formulas for c_j can be obtained.

> **Exploration 8.3.4** Derive a formula for the coefficients c_j, $j = 0, 1, 2, \ldots$, of the expansion of $f(x)$ in a series of Chebyshev polynomials in (8.15).

> **Exploration 8.3.5** Prove (8.16).

Example 8.3.3 We consider the approximation of $f(x) = e^{-x}$ by a rational function of the form

$$r_{2,3}(x) = \frac{a_0 T_0(x) + a_1 T_1(x) + a_2 T_2(x)}{1 + b_1 T_1(x) + b_2 T_2(x) + b_3 T_3(x)}.$$

The Chebyshev series (8.15) for $f(x)$ has coefficients c_j that can be obtained using the result of Exploration 8.3.4. The system of equations implied by (8.17) becomes

$$c_0 + \frac{1}{2}(b_1 c_1 + b_2 c_2 + b_3 c_3) = a_0,$$

$$c_1 + b_1 c_0 + \frac{1}{2}(b_1 c_2 + b_2 c_1 + b_2 c_3 + b_3 c_2 + b_3 c_4) = a_1,$$

$$c_2 + b_2 c_0 + \frac{1}{2}(b_1 c_1 + b_1 c_3 + b_2 c_4 + b_3 c_1 + b_3 c_5) = a_2,$$

$$c_3 + b_3 c_0 + \frac{1}{2}(b_1 c_2 + b_1 c_4 + b_2 c_1 + b_2 c_5 + b_3 c_6) = 0,$$

$$c_4 + \frac{1}{2}(b_1 c_3 + b_1 c_5 + b_2 c_2 + b_2 c_6 + b_3 c_1 + b_3 c_7) = 0,$$

$$c_5 + \frac{1}{2}(b_1 c_4 + b_1 c_6 + b_2 c_3 + b_2 c_7 + b_3 c_2 + b_3 c_8) = 0.$$

This can be written as $A\mathbf{x} = \mathbf{b}$, where

$$A = \begin{bmatrix} -1 & & & \\ & -1 & & \\ & & -1 & \\ & & & \\ & & & \end{bmatrix} + \frac{1}{2}\begin{bmatrix} c_0 & & \\ & c_0 & \\ & & c_0 \\ & & \\ & & \end{bmatrix} +$$

$$\frac{1}{2}\begin{bmatrix} & c_0 & c_1 & c_2 \\ & c_1 & c_0 & c_1 \\ & c_2 & c_1 & c_0 \\ & c_3 & c_2 & c_1 \\ & c_4 & c_3 & c_2 \end{bmatrix} + \frac{1}{2}\begin{bmatrix} c_1 & c_2 & c_3 \\ c_2 & c_3 & c_4 \\ c_3 & c_4 & c_5 \\ c_4 & c_5 & c_6 \\ c_5 & c_6 & c_7 \\ c_6 & c_7 & c_8 \end{bmatrix}, \tag{8.18}$$

$$\mathbf{x} = \begin{bmatrix} a_0 \\ a_1 \\ a_2 \\ b_1 \\ b_2 \\ b_3 \end{bmatrix}, \quad \mathbf{b} = \begin{bmatrix} -c_0 \\ -c_1 \\ -c_2 \\ -c_3 \\ -c_4 \\ -c_5 \end{bmatrix}. \tag{8.19}$$

After carrying out Gaussian Elimination for this example, we obtain the rational approximation

$$e^{-x} \approx r_{2,3}(x) = \frac{p_2(x)}{q_3(x)} \approx \frac{0.0231x^2 - 0.3722x + 0.9535}{0.0038x^3 + 0.0696x^2 + 0.5696x + 1}.$$

Plotting the error in this approximation on the interval $(-1, 1)$, we see that the error is maximum at $x = -1$, at roughly 1.1×10^{-5}, which is less than one-fourth of the error in the Padé approximant on $[0, 1]$. In fact, on $[0, 1]$, the error is maximum at $x = 0$ and is only 4.1×10^{-6}. $\square$

The system of equations $A\mathbf{x} = \mathbf{b}$ is most easily set up using (8.18), (8.19) as a guide. For general m and n, there are $m + n + 1$ equations and unknowns, and the coefficients $c_0, c_1, \ldots, c_{2n+m}$ are needed to specify p_m and q_n completely.

The last n columns of the third matrix in the decomposition of A in (8.18) is a **Toeplitz matrix**, a matrix in which all entries along any diagonal are equal. A general $m \times n$ Toeplitz T matrix has the form

$$T = \begin{bmatrix} a_0 & a_1 & a_2 & \cdots & a_{n-1} \\ a_{-1} & a_0 & a_1 & \ddots & a_{n-2} \\ a_{-2} & \ddots & \ddots & \ddots & \vdots \\ \vdots & \ddots & & a_{-1} & a_0 & a_1 \\ a_{-m+1} & a_{-m+2} & \cdots & & a_{-1} & a_0 \end{bmatrix}.$$

In MATLAB, such a matrix can be constructed using the `toeplitz` function. Similarly, the last n columns of the fourth matrix in (8.18) is a **Hankel matrix**, a symmetric matrix in which all entries along any *skew-diagonal* are equal. A general $m \times n$ Hankel matrix H has the form

$$H = \begin{bmatrix} a_0 & a_1 & a_2 & \cdots & a_{n-1} \\ a_1 & a_2 & & & a_n \\ a_2 & & & & a_{n+1} \\ \vdots & & & & \vdots \\ a_{m-1} & a_m & a_{m+1} & \cdots & a_{m+n-2} \end{bmatrix}.$$

The `hankel` function in MATLAB can be used to create a Hankel matrix.

Exploration 8.3.6 Write a MATLAB function

$$[\text{p},\text{q}]=\text{chebyrat}(\text{c},\text{m},\text{n})$$

that accepts as inputs a vector c consisting of the coefficients $c_0, c_1, \ldots, c_{m+n}$ in the expansion of a given function $f(x)$ in a series of Chebyshev polynomials as in (8.15), along with the degrees m and n of the numerator and denominator, respectively, of a rational Chebyshev interpolant $r_{m,n}(x)$ of $f(x)$. The output must be row vectors p and q containing the coefficients of the polynomials $p_m(x)$ and $q_n(x)$ for the numerator and denominator, respectively, of $r_{m,n}(x)$.

8.3.4 Concept Check

1. What is the need for rational approximation of functions?
2. What is the Padé approximant?
3. What is continued fraction form and how is it beneficial?
4. What is an advantage of Chebyshev rational approximation over Padé approximants?

8.4 Trigonometric Interpolation

In many application areas, such as differential equations and signal processing, it is more useful to express a given function $u(x)$ as a linear combination of sines and cosines, rather than polynomials. In differential equations, this form of approximation is beneficial due to the simplicity of the derivatives of sines and cosines, and in signal processing, one can readily analyze the frequency content of $u(x)$. In this section, we develop efficient algorithms for approximation of functions by linear combinations of such trigonometric functions.

8.4.1 Fourier Series

Suppose that a function $u(x)$ defined on the interval $[0, L]$ is intended to satisfy **periodic boundary conditions** $u(0) = u(L)$. Then, since $\sin \omega x$ and $\cos \omega x$ are both 2π-periodic when ω is an integer, $u(x)$ can be expressed in terms of both sines and cosines, as follows:

$$u(x) = \frac{a_0}{2} + \sum_{\omega=1}^{\infty} a_\omega \cos \frac{2\pi\omega x}{L} + b_\omega \sin \frac{2\pi\omega x}{L}, \tag{8.20}$$

where, for $\omega = 0, 1, 2, \ldots$, the coefficients a_ω and b_ω are defined by

$$a_\omega = \frac{2}{L} \int_0^L u(x) \cos \frac{2\pi\omega x}{L} \, dx, \quad b_\omega = \frac{2}{L} \int_0^L u(x) \sin \frac{2\pi\omega x}{L} \, dx. \tag{8.21}$$

This series representation (8.20) of $u(x)$ is called the **Fourier series** of $u(x)$.

The formulas for the coefficients $\{a_\omega\}$, $\{b_\omega\}$ in (8.21) are obtained using the fact that the functions $\{\cos(2\pi\omega x/L)\}_{\omega=0}^{\infty}$, $\{\sin(2\pi\omega x/L)\}_{\omega=1}^{\infty}$ are *orthogonal* with respect to the inner product

$$\langle f, g \rangle = \int_0^L \overline{f(x)} g(x) \, dx, \tag{8.22}$$

which can be established using trigonometric identities. The complex conjugation of $f(x)$ in (8.22) is necessary to ensure that even if u is complex, the L^2-**norm** $\| \cdot \|_2$ defined by

$$\|u\|_2 = \sqrt{\langle u, u \rangle} \tag{8.23}$$

satisfies one of the essential properties of norms, that the norm of a function must be nonnegative. Previously, in our inner product defined by (8.3), we did not use complex conjugation because it was defined for a space of real-valued functions.

Exploration 8.4.1 Prove that if m, n are integers, then

$$\left\langle \cos \frac{2\pi m x}{L}, \cos \frac{2\pi n x}{L} \right\rangle = \begin{cases} 0 & m \neq n, \\ L/2 & m = n, n \neq 0, \\ L & m = n = 0 \end{cases}$$

$$\left\langle \sin \frac{2\pi m x}{L}, \sin \frac{2\pi n x}{L} \right\rangle = \begin{cases} 0 & m \neq n, \\ L/2 & m = n \end{cases}$$

$$\left\langle \cos \frac{2\pi m x}{L}, \sin \frac{2\pi n x}{L} \right\rangle = 0$$

where the inner product $\langle f, g \rangle$ is as defined in (8.22).

Alternatively, we can use the relation $e^{i\theta} = \cos \theta + i \sin \theta$ to express the solution in terms of complex exponentials,

$$u(x) = \frac{1}{\sqrt{L}} \sum_{\omega=-\infty}^{\infty} \hat{u}(\omega) e^{2\pi i \omega x/L}, \tag{8.24}$$

where

$$\hat{u}(\omega) = \frac{1}{\sqrt{L}} \int_0^L e^{-2\pi i \omega x/L} u(x)\,dx. \tag{8.25}$$

Like the sines and cosines in (8.20), the functions $e^{2\pi i \omega x/L}$ are orthogonal with respect to the inner product (8.22). Specifically, we have

$$\left\langle e^{2\pi i \omega x/L}, e^{2\pi i \eta x/L} \right\rangle = \begin{cases} L & \omega = \eta \\ 0 & \omega \neq \eta \end{cases}. \tag{8.26}$$

This explains the presence of the scaling constant $1/\sqrt{L}$ in (8.24). It normalizes the functions $e^{2\pi i \omega x/L}$ so that they form an orthonormal set, meaning that they are orthogonal to one another, and have unit L^2-norm.

> **Exploration 8.4.2** Prove (8.26).

If $f \in L^2(0, L)$ (see Section B.13.3) is piecewise continuous, the following identity, known as **Parseval's identity**, is satisfied:

$$\sum_{\omega=-\infty}^{\infty} |\hat{f}(\omega)|^2 = \|f\|_2^2, \tag{8.27}$$

where the norm $\|\cdot\|_2$ is as defined in (8.23).

> **Exploration 8.4.3** Prove (8.27).

8.4.2 The Discrete Fourier Transform

Suppose that we define a grid on the interval $[0, L]$, consisting of the N equally-spaced points $x_j = j\Delta x$, where $\Delta x = L/N$, for $j = 0, \ldots, N-1$. For convenience, we assume N is even. Given an L-periodic function $f(x)$, we would like to compute an approximation to its Fourier series of the form

$$f_N(x) = \frac{1}{\sqrt{L}} \sum_{\omega=-N/2+1}^{N/2} e^{2\pi i \omega x/L} \tilde{f}(\omega), \tag{8.28}$$

where each coefficient $\tilde{f}(\omega)$ approximates the corresponding coefficient $\hat{f}(\omega)$ of the true Fourier series. Ideally, this approximate series should satisfy

$$f_N(x_j) = f(x_j), \quad j = 0, 1, \ldots, N-1. \tag{8.29}$$

That is, $f_N(x)$ should be an *interpolant* of $f(x)$, with the N points x_j, $j = 0, 1, \ldots, N-1$, as the interpolation points.

8.4.2.1   *Fourier Interpolation*

The problem of finding this interpolant, called the **Fourier interpolant** of f, has a unique solution that can easily be computed. The coefficients $\tilde{f}(\omega)$ are obtained by approximating the integrals that defined the coefficients of the Fourier series:

$$\tilde{f}(\omega) = \frac{1}{\sqrt{L}} \sum_{j=0}^{N-1} e^{-2\pi i \omega x_j / L} f(x_j) \Delta x, \quad \omega = -N/2 + 1, \dots, N/2. \tag{8.30}$$

Because the functions $\{e^{2\pi i \omega x / L}\}_{\omega = -N/2+1}^{N/2}$ are orthogonal with respect to the discrete inner product

$$\langle u, v \rangle_N = \Delta x \sum_{j=0}^{N-1} \overline{u(x_j)} v(x_j), \tag{8.31}$$

it is straightforward to verify that $f_N(x)$ does in fact satisfy the conditions (8.29). Note that the discrete inner product (8.31) is an approximation of the continuous inner product (8.22) by a Riemann sum. It is also equal to the standard inner product of vectors in $\mathbb{C}^N$, scaled by Δx.

From (8.28) and (8.29), we have

$$f(x_j) = \frac{1}{\sqrt{L}} \sum_{\eta = -N/2+1}^{N/2} e^{2\pi i \eta x_j / L} \tilde{f}(\eta). \tag{8.32}$$

Multiplying both sides by $\Delta x e^{-2\pi i \omega x_j / L}$, and summing from $j = 0$ to $j = N - 1$ yields

$$\Delta x \sum_{j=0}^{N-1} e^{-2\pi i \omega x_j / L} f(x_j) = \Delta x \frac{1}{\sqrt{L}} \sum_{j=0}^{N-1} \sum_{\eta = -N/2+1}^{N/2} e^{-2\pi i \omega x_j / L} e^{2\pi i \eta x_j / L} \tilde{f}(\eta), \tag{8.33}$$

or

$$\Delta x \sum_{j=0}^{N-1} e^{-2\pi i \omega x_j / L} f(x_j) = \frac{1}{\sqrt{L}} \sum_{\eta = -N/2+1}^{N/2} \tilde{f}(\eta) \left[\Delta x \sum_{j=0}^{N-1} e^{-2\pi i \omega x_j / L} e^{2\pi i \eta x_j / L} \right]. \tag{8.34}$$

Because

$$\left\langle e^{2\pi i \omega x / L}, e^{2\pi i \eta x / L} \right\rangle_N = \begin{cases} L & \omega = \eta \,(\mathrm{mod}\ N) \\ 0 & \omega \neq \eta \,(\mathrm{mod}\ N) \end{cases}, \tag{8.35}$$

all terms in the outer sum on the right side of (8.34) vanish except for $\eta = \omega$, and we obtain the formula (8.30). It should be noted that the algebraic operations performed on (8.32) are equivalent to taking the discrete inner product of both sides of (8.32) with $e^{2\pi i \omega x / L}$.

Exploration 8.4.4 Prove (8.35). *Hint:* Use formulas associated with geometric series.

The process of obtaining the approximate Fourier coefficients as in (8.30) is called the **Discrete Fourier Transform (DFT)** of $f(x)$. The **Discrete Inverse Fourier Transform** is given by (8.32). As at the beginning of this section, we can also work with the real form of the Fourier interpolant,

$$f_N(x) = \frac{\tilde{a}_0}{2} + \sum_{\omega=1}^{N/2-1} \tilde{a}_\omega \cos \frac{2\pi\omega x}{L} + \tilde{b}_\omega \sin \frac{2\pi\omega x}{L} + \tilde{a}_{N/2} \cos \frac{\pi N x}{L}, \qquad (8.36)$$

where the coefficients $\tilde{a}_\omega$, $\tilde{b}_\omega$ approximate a_ω, b_ω from (8.21).

Exploration 8.4.5 Express the coefficients $\tilde{a}_\omega$, $\tilde{b}_\omega$ of the real form of the Fourier interpolant (8.36) in terms of the coefficients $\tilde{f}(\omega)$ from the complex exponential form (8.28).

Exploration 8.4.6 Why is there no need for a coefficient $\tilde{b}_{N/2}$ in (8.36)?

Exploration 8.4.7 Use the result of Exploration 8.4.4 to prove the following discrete orthogonality relations:

$$\left\langle \cos \frac{2\pi m x}{L}, \cos \frac{2\pi n x}{L} \right\rangle_N = \begin{cases} 0 & m \neq n, \\ L/2 & m = n, n \neq 0, \\ L & m = n = 0 \end{cases}$$

$$\left\langle \sin \frac{2\pi m x}{L}, \sin \frac{2\pi n x}{L} \right\rangle_N = \begin{cases} 0 & m \neq n, \\ L/2 & m = n \end{cases}$$

$$\left\langle \cos \frac{2\pi m x}{L}, \sin \frac{2\pi n x}{L} \right\rangle_N = 0$$

where m and n are integers, and the discrete inner product $\langle f, g \rangle_N$ is as defined in (8.31).

Exploration 8.4.8 Prove the following discrete analogue of Parseval's identity (8.27),

$$\sum_{\omega=-N/2+1}^{N/2} |\tilde{f}(\omega)|^2 = \sum_{j=0}^{N-1} |f(x_j)|^2.$$

8.4.2.2 *Aliasing*

Suppose we have $N = 128$ data points sampled from the following function over $[0, 2\pi]$:

$$f(x) = \sin(10x) + \text{noise}. \qquad (8.37)$$

The function $f(x)$, shown in Figure 8.6(a), has small high-frequency oscillations. However, by taking the Discrete Fourier Transform (Figure 8.6(b)), we can extract the original sine wave quite easily. The DFT shows two distinct spikes, corresponding to frequencies of $\omega = \pm 10$, that is, the frequencies of the original sine wave. The first $N/2 + 1$ values of the DFT correspond to frequencies of $0 \leq \omega \leq \omega_{max}$, where $\omega_{max} = N/2$. The remaining $N/2 - 1$ values of the DFT correspond to the frequencies $-\omega_{max} < \omega < 0$. The main idea behind **denoising** is to remove components of the Fourier series that have high frequency and small amplitude that is characteristic of noise.

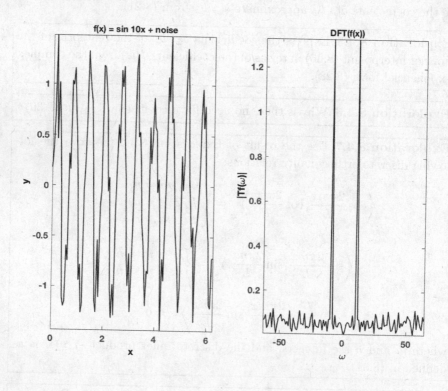

Fig. 8.6   (a) Left plot: noisy signal (b) Right plot: Discrete Fourier Transform

The DFT only includes a finite range of frequencies. If there are frequencies beyond this range present in the Fourier series, an effect known as **aliasing** occurs. The effect of aliasing is shown in Figure 8.7: it "folds" these frequencies back into the computed DFT. Specifically,

$$\tilde{f}(\omega) = \sum_{\ell=-\infty}^{\infty} \hat{f}(\omega + \ell N), \quad -N/2 + 1 \leq \omega \leq N/2. \tag{8.38}$$

Aliasing can have disastrous consequences. If, for example, functions with high-frequency content are multiplied pointwise, then the DFT of the product will be "contaminated" by the highest-frequency Fourier series coefficients, which are outside the range $[-N/2+1, N/2]$, that alias to lower frequencies that are in range. This can cause differentiation performed via the DFT, which is used in spectral methods for PDEs as discussed in Section 14.6, to be highly inaccurate. Such adverse consequences of aliasing can be avoided by *filtering* the DFT coefficients, which means that coefficients corresponding to high frequencies are set equal to zero. Details on various strategies for dealing with aliasing can be found in [Boyd (2001)].

Exploration 8.4.9 Use (8.25) and (8.30) to prove (8.38).

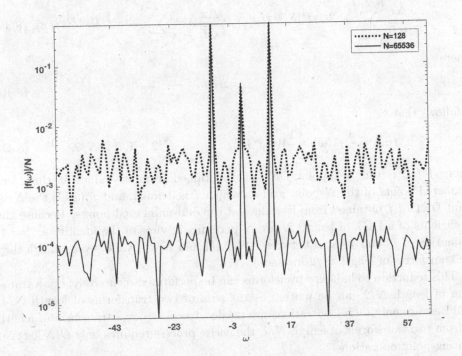

Fig. 8.7 Aliasing effect on noisy signal: the solid curve represents the 65536-point DFT of a function, while the dotted curve depicts the 128-point DFT of the same function. The coefficients $\hat{f}(\omega)$, for ω outside $[-63, 64]$, are added to coefficients inside this interval to obtain the DFT coefficients for the 128-point case.

8.4.3 The Fast Fourier Transform

The Discrete Fourier Transform, as it was presented earlier in this section, requires $O(N^2)$ operations to compute. In fact, the DFT can be computed much more efficiently by using the **Fast Fourier Transform (FFT)**. The FFT arises by noting that when N is even, a DFT of length N can be written as the sum of two Fourier transforms, each of length $N/2$. One of these transforms is formed from the even-numbered points of the original set of N, and the other transform is formed from the odd-numbered points.

We have

$$\tilde{f}(\omega) = \frac{\Delta x}{\sqrt{L}} \sum_{j=0}^{N-1} e^{-2\pi i j \omega/N} f(x_j)$$

$$= \frac{\Delta x}{\sqrt{L}} \sum_{j=0}^{N/2-1} e^{-2\pi i \omega(2j)/N} f(x_{2j}) + \frac{\Delta x}{\sqrt{L}} \sum_{j=0}^{N/2-1} e^{-2\pi i \omega(2j+1)/N} f(x_{2j+1})$$

$$= \frac{\Delta x}{\sqrt{L}} \sum_{j=0}^{N/2-1} e^{-2\pi i \omega j/(N/2)} f(x_{2j}) + \frac{\Delta x}{\sqrt{L}} W^\omega \sum_{j=0}^{N/2-1} e^{-2\pi i \omega j/(N/2)} f(x_{2j+1}) \quad (8.39)$$

where

$$W = e^{-2\pi i/N}. \tag{8.40}$$

It follows that

$$\tilde{f}(\omega) = \frac{1}{2}\tilde{f}^e(\omega) + \frac{1}{2}W^\omega \tilde{f}^o(\omega), \quad \omega = -N/2+1, \ldots, N/2, \tag{8.41}$$

where $\tilde{f}^e(\omega)$ is the $N/2$-point DFT of f obtained from its values at the even-numbered points of the N-point grid on which f is defined, and $\tilde{f}^o(\omega)$ is the $N/2$-point DFT of f obtained from its values at the odd-numbered points. Because the coefficients of a DFT of length N are N-periodic, in view of the identity $e^{2\pi i} = 1$, evaluation of $\tilde{f}^e$ and $\tilde{f}^o$ at ω between $-N/2+1$ and $N/2$ is valid, even though they are transforms of length $N/2$ instead of N.

This reduction to half-size transforms can be performed recursively; i.e. a transform of length $N/2$ can be written as the sum of two transforms of length $N/4$, etc. Because only $O(N)$ operations are needed to construct a transform of length N from two transforms of length $N/2$, the entire process requires only $O(N \log_2 N)$ floating-point operations.

Exploration 8.4.10 Write two functions to compute the DFT of a function $f(x)$ defined on $[0, L]$, represented by a N-vector f that contains its values at $x_j = j\Delta x$, $j = 0, 1, 2, \ldots, N-1$, where $j = L/N$. For the first function, use the formula (8.28), and for the second, use recursion and the formula (8.41) for the FFT. Compare the efficiency of your functions for different values of N. How does the execution time increase as N increases?

8.4.4   Convergence and Gibbs' Phenomenon*

The Fourier series for an L-periodic function $f(x)$ will converge to $f(x)$ at any point in $[0, L]$ at which f is continuously differentiable. If f has a jump discontinuity at a point c, then the series will converge to $\frac{1}{2}[f(c^+) + f(c^-)]$, where

$$f(c^+) = \lim_{x \to c^+} f(x), \quad f(c^-) = \lim_{x \to c^-} f(x). \tag{8.42}$$

If $f(x)$ is *not* L-periodic, then there is a jump discontinuity in the L-periodic exten-sion of $f(x)$ beyond $[0, L]$, and the Fourier series will again converge to the average of the values of $f(x)$ on either side of this discontinuity.

Such discontinuities pose severe difficulties for trigonometric interpolation, be-cause the basis functions $e^{i\omega x}$ grow more oscillatory as $|\omega|$ increases. In particular, the truncated Fourier series of a function $f(x)$ with a jump discontinuity at $x = c$ exhibits what is known as **Gibbs' phenomenon**, first discussed in [Wilbraham (1848)], in which oscillations appear on either side of $x = c$, even if $f(x)$ itself is smooth there.

Convergence of the Fourier series of f is more rapid when f is smooth. In particular, if f is p-times differentiable and its pth derivative is at least *piecewise* continuous (that is, continuous except possibly for jump discontinuities), then the coefficients of the complex exponential form of the Fourier series satisfy

$$|\hat{f}(\omega)| \leq \frac{C}{|\omega|^{p+1} + 1} \tag{8.43}$$

for some constant C that is independent of ω [Gustafsson et al. (1995)].

Exploration 8.4.11 Generate a random vector of DFT coefficients that satisfy the decay rate (8.43), for some value of p. Then, perform an inverse FFT to obtain the truncated Fourier series (8.28), and plot the resulting function $f_N(x)$. How does the behavior of the function change as p de-creases?

Exploration 8.4.12 Demonstrate Gibbs' phenomenon by plotting trun-cated Fourier series of the function $f(x) = x$ on $[0, 2\pi]$. Use the formula (8.28), evaluated on a finer grid (that is, using $\tilde{N}$ equally spaced points in $[0, 2\pi]$, where $\tilde{N} \gg N$). What happens as N increases?

8.4.5 Concept Check

1. What is a Fourier series?
2. What are some applications of Fourier series?
3. What property of trigonometric polynomials is very helpful for computing Fourier series coefficients? item What is Parseval's Identity?
4. What is the Discrete Fourier Transform (DFT)? How does it relate to Fourier interpolation?
5. How does the Fourier interpolant of $f(x)$ relate to its Fourier series?
6. What is aliasing and what problem does it pose for Fourier interpolation?
7. What is the Fast Fourier Transform (FFT)? How fast is it?
8. What is Gibbs' phenomenon?
9. How does the smoothness of a function relate to the behavior of its Fourier series coefficients?

8.5   Additional Resources

Thorough treatments of orthogonal polynomials, in the context of analytical or numerical methods for the solution of differential equations, can be found in [Arfken, et al. (2012); Shen, et al. (2011)]. Beyond the scope of this book is the construction of **minimax polynomials**, that minimize the L^∞-norm of the error on an interval. The reader is referred to [Atkinson (1989); Süli and Mayers (2003)].

Approximation via rational functions, including Padé approximants and continued fractions, is covered extensively by [Ralston and Rabinowitz (1978)]; see also [Burden and Faires (2004); Powell (1981)]. The use of Padé approximants for matrix function approximation is treated in [Golub and van Loan (2012)].

An in-depth treatment of Fourier interpolation, for the purpose of solving time-dependent PDEs, is featured in [Gustafsson et al. (1995)]. Texts on spectral methods for PDEs such as [Boyd (2001); Shen, et al. (2011)] are also valuable resources. Implementation details, such as handling non-uniformly spaced data, or any number of data points besides powers of 2, can be found in [Press, et al. (2007)]. The FFTW library [Frigo and Johnson (2005)] includes several DFT subroutines, and is the foundation of the DFT functions in MATLAB.

The *wavelet transform*, unlike the Fourier transform, provides a decomposition of a function into a linear combination of basis functions that are not only orthogonal but also localized in space. For background on wavelets, the reader is referred to [Daubechies (1992); Frazier (1999); Hubbard (1998); Mallat (2009)].

8.6   Exercises

1. Let $f(x) = e^{-x^2}$. Compute the following approximations of $f(x)$ on $[-1, 1]$:

    (a) the polynomial $p_4(x)$ of degree at most four that minimizes

    $$\|p_4 - f\|_2 = \left(\int_{-1}^{1} |p_4(x) - f(x)|^2 \, dx \right)^{1/2}.$$

    (b) the Padé approximant $r_{2,4}(x)$ of $f(x)$.
    (c) the Fourier interpolant $\tilde{f}_N(x)$ on $[-1, 1]$ with $N = 32$.

    For each approximation, compute the error using the L^2-norm from part 1a. Use the MATLAB function `integral` to evaluate any required integrals. Compare the accuracy of the three approximations.

2. We approximate $f(x) = e^{-10x}$ on the interval $[0, 1]$ so that the approximation $\tilde{f}(x)$ satisfies $\|f - \tilde{f}\|_2 \leq 10^{-3}$. Try the following forms for $\tilde{f}(x)$:

    (a) a polynomial of degree n, and
    (b) a rational function $r_{m,n}(x) = p_m(x)/q_n(x)$.

    How large do these degrees have to be to ensure sufficient accuracy?

3. Compute the Fourier interpolants of the following functions on $[0, 1]$, with $N = 64$ grid points:

(a) $f_1(x) = -4x^3 + 6x^2$

(b) $f_2(x) = 12x^4 - 24x^3 + 11x^2 + x$

(c) $f_3(x) = 16x^4 - 32x^3 + 16x^2$

In each case, plot $|\tilde{f}(\omega)|$ versus ω. What do you observe about the decay rate of the Fourier coefficients? Can you explain your observations?

4. Suppose we wish to compute a polynomial approximation to a given function $f(x)$ on $[-1,1]$ of the form

$$f_N(x) = \sum_{k=0}^{N} c_k T_k(x),$$

where $T_k(x)$ is the Chebyshev polynomial of degree k, and the coefficients c_k, $k = 0, 1, \ldots, N$, are chosen so as to minimize

$$\|f - f_N\|_2 = \left(\int_{-1}^{1} \frac{1}{\sqrt{1-x^2}} |f(x) - f_N(x)|^2 \, dx \right)^{1/2}.$$

Derive an algorithm for approximating $\{c_k\}_{k=0}^{N}$ using the FFT.

5. Write a MATLAB script that uses (8.9) to compute the coefficients of the Legendre polynomials up to degree 10. Compare the computed coefficients to the exact coefficients of the Legendre polynomials. Can you explain why significant discrepancies occur? How can (8.9) be used to obtain the values of Legendre polynomials at select points while avoiding this inaccuracy?

6. Let $p_0, p_1, \ldots, p_N$ be a sequence of polynomials orthogonal with respect to a given inner product (8.10), and, for $k = 0, 1, \ldots, N$, let f_k be the polynomial of degree k that minimizes $\|f - f_k\|_2$, where $\|f\|_2$ is the norm induced by this inner product. Show that for $k \geq 1$, $\|f - f_k\|_2 \leq \|f - f_{k-1}\|_2$. Under what circumstance is there equality?

7. Write a MATLAB function v=dftpoly(p,a,b,N) that computes the Fourier series coefficients $\hat{p}(\omega)$, $\omega = -N/2 + 1, \ldots, N/2$, on the interval $[a, b]$, of a polynomial represented by the row vector of coefficients p. For each ω, the integral (8.25) is to be computed *exactly* using repeated integration by parts, with the help of MATLAB's polynomial functions.

8. A matrix $C \in \mathbb{R}^{n \times n}$ is said to be a **circulant matrix** if it is of the form

$$C = \begin{bmatrix} c_0 & c_1 & c_2 & \cdots & c_{n-1} \\ c_{n-1} & c_0 & c_1 & & c_{n-2} \\ \vdots & c_{n-1} & c_0 & \ddots & \vdots \\ c_2 & & \ddots & \ddots & c_1 \\ c_1 & c_2 & & c_{n-1} & c_0 \end{bmatrix}.$$

(a) Let $h = 2\pi/n$. How does the n-point DFT of $f(x)$ relate to the DFT of $f(x + kh)$, where k is an integer?

(b) Let ω be an integer. Compute the matrix-vector product $C e_\omega$, where

$$e_\omega = \begin{bmatrix} 1 & e^{i\omega h} & e^{i\omega 2h} & \cdots & e^{i\omega(n-1)h} \end{bmatrix}^T.$$

Use the result of part 8a to help.

(c) Use the result of part 8b to write down formulas for the eigenvalues and eigenvectors of C.

9. Derive an analogue of (8.41) for the case where N is a multiple of 3.

10. Let $\mathbf{x} \in \mathbb{R}^N$ be a vector of N equally spaced points in $[0, 1]$, including both endpoints. Suppose we wish to fit a polynomial of degree $n < N$ to given data $(x_1, y_1), \ldots, (x_N, y_N)$, in the least squares sense. For this problem, we will let $n = 4$.

(a) As in Section 8.1.2, set up the normal equations, using the monomial basis $\{1, x, x^2, x^3, x^4\}$ evaluated at the elements of $\mathbf{x}$ to obtain the matrix A. What is the condition number of A?

(b) Repeat part 10a, except using orthogonal polynomials on $[0, 1]$ obtained by composing the Legendre polynomials with the change of variable from $[0, 1]$ to $[-1, 1]$. Can you explain the difference in the condition number?

(c) Compute $A^T A$, where A is the matrix computed in part 10b. Using the sense in which these polynomials are orthogonal, explain why the off-diagonal entries are relatively small. Can you explain why some are almost exactly zero while others are not?

(d) Change the vector $\mathbf{x}$ so that it consists of the *midpoints* of N subintervals of width h. Form the matrix A from part 10b again, and compute $A^T A$. Compare to $A^T A$ computed in part 10c. Can you explain your observations?

Chapter 9

Differentiation and Integration

The solution of many mathematical models requires performing the basic operations of calculus: differentiation and integration. In this chapter, we will learn several techniques for approximating a derivative of a function $f(x)$ at a point x_0, and a definite integral of a function $f(x)$ over an interval $[a, b]$. As we will see, our discussion of polynomial interpolation from Chapter 7 will play an essential role, as polynomials are the easiest functions on which to perform these operations.

Section 9.1 presents techniques for approximating derivatives based on **finite differences**, as opposed to the "infinitely small" differences used to define the derivative. A recently-developed technique for approximating **weak derivatives** is also presented. Section 9.2 begins our exploration of numerical integration by introducing the concept of a **quadrature rule**. A quadrature rule is similar to a Riemann sum that is used to define the definite integral, in that it is a sum of values of the integrand $f(x)$ at selected points, called **nodes**, that are multiplied by scaling factors called **weights**. A class of quadrature rules of particular interest are **interpolatory quadrature rules**, which compute the exact integrals of polynomial interpolants of the integrand.

Subsequent sections cover specific types of quadrature rules, that are characterized by the placement of their nodes and determination of their weights. Section 9.3 covers Newton-Cotes rules, which are interpolatory quadrature rules in which the nodes are equally spaced. Because these rules are not effective on large intervals or with a large number of nodes, **composite quadrature** is introduced in Section 9.4 as an alternative approach.

Because the accuracy of both finite differences and composite quadrature rules are dependent on a spacing h between points substituted into $f(x)$, Section 9.5 presents **extrapolation** as a means of obtaining greater accuracy from previously computed approximations. Section 9.6 describes **adaptive quadrature**, in which error estimation is used to reduce the number of nodes needed to obtain a desired level of accuracy. **Gauss quadrature**, a highly accurate and robust integration technique based on orthogonal polynomials from Section 8.2, is presented in Section 9.7. The chapter concludes in Section 9.8 with generalization of these integration techniques to higher-dimensional integrals.

9.1   Numerical Differentiation

We first discuss how Taylor expansion (defined in Theorem A.6.1) and polynomial interpolation can be applied to help solve a fundamental problem from calculus, which is the problem of computing the derivative of a given function $f(x)$ at a given point $x = x_0$. The techniques presented in this section will be essential ingredients in methods that we will see later in this book for solving problems that arise in countless applications: nonlinear equations in Chapter 10, optimization problems in Chapter 11, and differential equations in Part V. The basics of derivatives are reviewed in Section A.2.

9.1.1   Derivation Using Taylor Expansion

Recall from Definition A.2.1 that the derivative of $f(x)$ at a point x_0, denoted by $f'(x_0)$, is defined by

$$f'(x_0) = \lim_{h \to 0} \frac{f(x_0 + h) - f(x_0)}{h}.$$

This definition suggests a method for approximating $f'(x_0)$. If we choose h to be a small positive constant, then

$$f'(x_0) \approx \frac{f(x_0 + h) - f(x_0)}{h}.$$

This approximation is called the **forward difference formula**.

To estimate the accuracy of this approximation, we note that if $f''(x)$ exists on $[x_0, x_0 + h]$, then, by Taylor's Theorem, $f(x_0 + h) = f(x_0) + f'(x_0)h + f''(\xi)h^2/2$, where $\xi \in (x_0, x_0 + h)$. Solving for $f'(x_0)$, we obtain

$$f'(x_0) = \frac{f(x_0 + h) - f(x_0)}{h} - \frac{f''(\xi)}{2}h, \tag{9.1}$$

so the error in the forward difference formula is $O(h)$. We say that this formula is **first-order accurate**.

The forward-difference formula is called a **finite difference approximation** to $f'(x_0)$, because it approximates $f'(x)$ using values of $f(x)$ at points that have a small, but finite, distance between them, as opposed to the definition of the derivative, that takes a limit and therefore computes the derivative using an "infinitely small" value of h. The forward-difference formula, however, is just one example of a finite difference approximation. If we replace h by $-h$ in the forward-difference formula, where h is still positive, we obtain the **backward-difference formula**

$$f'(x_0) = \frac{f(x_0) - f(x_0 - h)}{h} + \frac{f''(\xi)}{2}h, \quad \xi \in (x_0 - h, x_0). \tag{9.2}$$

Like the forward-difference formula, the backward difference formula is first-order accurate.

If we average these two approximations, we obtain the **centered difference formula**

$$f'(x_0) \approx \frac{f(x_0 + h) - f(x_0 - h)}{2h}.$$

To determine the accuracy of this approximation, we assume that $f'''(x)$ exists on the interval $[x_0 - h, x_0 + h]$, and then apply Taylor's Theorem again to obtain the **Taylor expansions**

$$f(x_0 + h) = f(x_0) + f'(x_0)h + \frac{f''(x_0)}{2}h^2 + \frac{f'''(\xi_+)}{6}h^3,$$

$$f(x_0 - h) = f(x_0) - f'(x_0)h + \frac{f''(x_0)}{2}h^2 - \frac{f'''(\xi_-)}{6}h^3,$$

where $\xi_+ \in (x_0, x_0 + h)$ and $\xi_- \in (x_0 - h, x_0)$. Subtracting the second equation from the first and solving for $f'(x_0)$ yields

$$f'(x_0) = \frac{f(x_0 + h) - f(x_0 - h)}{2h} - \frac{f'''(\xi_+) + f'''(\xi_-)}{12}h^2.$$

Suppose that f''' is continuous on $[x_0 - h, x_0 + h]$. By the Intermediate Value Theorem (Theorem A.1.8), $f'''(x)$ must assume every value between $f'''(\xi_-)$ and $f'''(\xi_+)$ on the interval (ξ_-, ξ_+), including the average of these two values. Therefore, we can simplify this equation to

$$f'(x_0) = \frac{f(x_0 + h) - f(x_0 - h)}{2h} - \frac{f'''(\xi)}{6}h^2, \tag{9.3}$$

where $\xi \in (x_0 - h, x_0 + h)$. Because the error is $O(h^2)$, we say that the centered-difference formula is **second-order accurate**. This is due to the cancellation of the $O(h)$ terms involving $f''(x_0)$.

Example 9.1.1 Consider the function

$$f(x) = \frac{\sin^2\left(\frac{\sqrt{x^2+x}}{\cos x - x}\right)}{\sin\left(\frac{\sqrt{x-1}}{\sqrt{x^2+1}}\right)}.$$

Our goal is to compute $f'(0.25)$. Using the Quotient Rule and the Chain Rule, we obtain

$$f'(x) = \frac{2\sin\left(\frac{\sqrt{x^2+x}}{\cos x - x}\right)\cos\left(\frac{\sqrt{x^2+x}}{\cos x - x}\right)\left[\frac{2x+1}{2\sqrt{x^2+1}(\cos x - x)} + \frac{\sqrt{x^2+1}(\sin x+1)}{(\cos x - x)^2}\right]}{\sin\left(\frac{\sqrt{x-1}}{\sqrt{x^2+1}}\right)} -$$

$$\frac{\sin\left(\frac{\sqrt{x^2+x}}{\cos x - x}\right)\cos\left(\frac{\sqrt{x-1}}{\sqrt{x^2+1}}\right)\left[\frac{1}{2\sqrt{x}\sqrt{x^2+1}} - \frac{x(\sqrt{x-1})}{(x^2+1)^{3/2}}\right]}{\sin^2\left(\frac{\sqrt{x-1}}{\sqrt{x^2+1}}\right)}.$$

Evaluating this monstrous function at $x = 0.25$ yields $f'(0.25) = -9.066698770$.

Alternatively, we can use the centered difference formula (9.3) with $x = 0.25$ and $h = 0.005$. This yields the approximation

$$f'(0.25) \approx \frac{f(0.255) - f(0.245)}{0.01} = -9.067464295,$$

which has absolute error 7.7×10^{-4}. While this complicated function must be evaluated twice to obtain this approximation, that is still much less work than using differentiation rules to compute $f'(x)$, and then evaluating $f'(x)$, which is much more complicated than $f(x)$. $\square$

Example 9.1.2 We will use Taylor expansion to derive an approximation of $f'(x_0)$ of the form

$$f'(x_0) = \frac{1}{h}[c_0 f(x_0) + c_1 f(x_0 + h) + c_2 f(x_0 + 2h)] + \text{error}. \qquad (9.4)$$

Such a formula is useful when there is no information available about $f(x)$ for $x < x_0$. The leading factor of $1/h$ is not essential, but it is used so that the coefficients c_0, c_1, c_2 will be independent of h. The Taylor expansions of $f(x_0 + h)$ and $f(x_0 + 2h)$ are

$$f(x_0 + h) = f(x_0) + f'(x_0)h + \frac{1}{2}f''(x_0)h^2 + \frac{1}{6}f'''(\xi_1)h^3,$$

$$f(x_0 + 2h) = f(x_0) + f'(x_0)2h + \frac{1}{2}f''(x_0)4h^2 + \frac{1}{6}f'''(\xi_2)8h^3,$$

where $\xi_1 \in (x_0, x_0 + h)$ and $\xi_2 \in (x_0, x_0 + 2h)$. Substituting these expansions into (9.4), we obtain

$$f'(x_0) = \frac{1}{h}\left\{ c_0 f(x_0) + c_1 \left[f(x_0) + f'(x_0)h + \frac{1}{2}f''(x_0)h^2 + \frac{1}{6}f'''(\xi_1)h^3 \right] + \right.$$

$$\left. c_2 \left[f(x_0) + f'(x_0)2h + \frac{1}{2}f''(x_0)4h^2 + \frac{1}{6}f'''(\xi_2)8h^3 \right] \right\} + \text{error}.$$

$$= \frac{1}{h}\left\{ (c_0 + c_1 + c_2)f(x_0) + h(c_1 + 2c_2)f'(x_0) + h^2 \left(\frac{1}{2}c_1 + 2c_2 \right) f''(x_0) + \right.$$

$$\left. c_1 \frac{1}{6}f'''(\xi_1)h^3 + c_2 \frac{1}{6}f'''(\xi_2)8h^3 \right\} + \text{error}.$$

Because the right side must equal $f'(x_0)$, we can see that the coefficients c_0, c_1 and c_2 satisfy the equations

$$c_0 + c_1 + c_2 = 0,$$

$$c_1 + 2c_2 = 1,$$

$$c_1 + 4c_2 = 0.$$

This system of linear equations has the unique solution

$$c_0 = -\frac{3}{2}, \quad c_1 = 2, \quad c_2 = -\frac{1}{2},$$

which yields the approximation

$$f'(x_0) \approx \frac{-3f(x_0) + 4f(x_0 + h) - f(x_0 + 2h)}{2h}.$$

To determine the error, we use the Taylor remainder terms in the expansions of $f(x_0 + h)$ and $f(x_0 + 2h)$ to obtain

$$f'(x_0) = \frac{-3f(x_0) + 4f(x_0 + h) - f(x_0 + 2h)}{2h} - \frac{1}{3}h^2[f'''(\xi_1) - 2f'''(\xi_2)].$$

We see that this formula is second-order accurate. Unfortunately, we cannot use the Intermediate Value Theorem, as before, to combine the two values of $f'''(x)$ into a single value. This is because the coefficients of these values are not the same sign, so they cannot be expressed as an average. Later in this section, we will learn how to derive more concise error formulas in such cases. $\square$

A similar approach can be used to obtain finite difference approximations of $f'(x_0)$ involving any points of our choosing, and with an arbitrarily high order of accuracy, provided that sufficiently many points are used. In general, a finite difference approximation of the form

$$f'(x_0) \approx \frac{1}{h} \sum_{i=-j}^{k} c_i f(x_0 + ih) \tag{9.5}$$

is nth-order accurate, where $n = j + k$. That is, the error is $O(h^n)$. In some cases, such as the centered difference formula, a coefficient c_i may equal zero, thus improving efficiency. In Example 9.1.2, we used $j = 0$ and $k = 2$, whereas for the centered difference, $j = 1$ and $k = 1$.

To obtain the coefficients c_i, $i = -j, \ldots, k$, we set up and solve a system of $n+1$ linear equations in $n+1$ unknowns, as in Example 9.1.2. The right-side values are all 0, except for the second one that is equal to 1, which corresponds to the first derivative (since the first derivative is included in the second term of a Taylor expansion). On the left side, for $i = -j, \ldots, k$, the coefficient of c_i in row ℓ, for $\ell = 0, 1, \ldots, n$, is i^ℓ.

Exploration 9.1.1 Following the approach used in Example 9.1.2, use Taylor expansions of $f(x_0 \pm jh)$, for $j = 1, 2$, to derive a finite difference approximation of $f'(x_0)$ that is fourth-order accurate.

Exploration 9.1.2 Generalizing the process carried out by hand in Example 9.1.2 and Exploration 9.1.1, write a MATLAB function c=makediffrule(p) that takes as input a row vector of integers p and returns in a vector c the coefficients of a finite-difference approximation of $f'(x_0)$ that has the form (9.5), where the indices $-j, \ldots, k$ to be used are stored in the vector p.

As we have seen in examples, the error term includes values of $f^{(n+1)}(x)$, where n is the order of accuracy. It follows that if $f(x)$ is a polynomial of degree at most n, then the finite-difference approximation actually yields the exact value of $f'(x_0)$. In other words, for a general function $f(x)$, the finite difference approximation of $f'(x_0)$ is the exact derivative of the polynomial $p_n(x)$ that interpolates $f(x)$ at the $n + 1$ points $x_0 + ih$, $i = -j, \ldots, k$. We will now use this fact to improve our approach to approximating derivatives and analyzing error.

9.1.2 Derivation Using Lagrange Interpolation

While Taylor's Theorem can be used to derive formulas with higher-order accuracy simply by evaluating $f(x)$ at more points, this process can be tedious due to the need to solve a system of linear equations, and the derivation of a concise error formula, involving just one value of a higher-order derivative of f, can be difficult.

An alternative approach is to compute the derivative of the interpolating polynomial that fits $f(x)$ at these points. Specifically, suppose we want to compute the derivative at a point x_0 using the data

$$(x_{-j}, y_{-j}), \ldots, (x_{-1}, y_{-1}), (x_0, y_0), (x_1, y_1), \ldots, (x_k, y_k),$$

where j and k are known nonnegative integers, $x_{-j} < x_{-j+1} < \cdots < x_{k-1} < x_k$, and $y_i = f(x_i)$ for $i = -j, \ldots, k$. Then, a finite difference formula for $f'(x_0)$ can be obtained by analytically computing the derivatives of the **Lagrange polynomials** $\{\mathcal{L}_{n,i}(x)\}_{i=-j}^{k}$ (see Section 7.2) for these points, where $n = j + k$, and the values of these derivatives at x_0 are the proper weights for the function values $y_{-j}, \ldots, y_k$. If $f(x)$ is $n + 1$ times continuously differentiable on $[x_{-j}, x_k]$, then we obtain an approximation of the form

$$f'(x_0) = \sum_{i=-j}^{k} f(x_i)\mathcal{L}_{n,i}'(x_0) + \frac{f^{(n+1)}(\xi)}{(n+1)!} \prod_{i=-j, i\neq 0}^{k} (x_0 - x_i), \qquad (9.6)$$

where $\xi \in (x_{-j}, x_k)$.

Exploration 9.1.3 Prove (9.6) by differentiating the error formula for Lagrange interpolation from Theorem 7.4.1 and evaluating the result at x_0. *Hint:* Use the fact that the unknown point ξ in the error formula depends on x.

Among the best-known finite difference formulas that can be derived using this approach is the second-order-accurate three-point formula

$$f'(x_0) = \frac{-3f(x_0) + 4f(x_0 + h) - f(x_0 + 2h)}{2h} + \frac{f'''(\xi)}{3}h^2, \quad \xi \in (x_0, x_0+2h). \ (9.7)$$

This formula was derived in Example 9.1.2, but the error term was not; it can be obtained using (9.6). As noted in the example, this formula is useful when there is no information available about $f(x)$ for $x < x_0$. If there is no information available about $f(x)$ for $x > x_0$, then we can replace h by $-h$ in the above formula to obtain a second-order-accurate three-point formula that uses the values of $f(x)$ at x_0, $x_0 - h$ and $x_0 - 2h$.

Another formula is the five-point formula

$$f'(x_0) = \frac{f(x_0 - 2h) - 8f(x_0 - h) + 8f(x_0 + h) - f(x_0 + 2h)}{12h} + \frac{f^{(5)}(\xi)}{30}h^4, \quad ,$$

where $\xi \in (x_0 - 2h, x_0 + 2h)$. This formula is fourth-order accurate. The reason it is called a five-point formula, even though it uses the value of $f(x)$ at four points, is that it is derived from the Lagrange polynomials for the five points $x_0 - 2h, x_0 - h, x_0, x_0 + h$, and $x_0 + 2h$. However, $f(x_0)$ is not used in the formula because $\mathcal{L}_{4,0}'(x_0) = 0$, where $\mathcal{L}_{4,0}(x)$ is the Lagrange polynomial that is equal to one at x_0 and zero at the other four points.

If we do not have any information about $f(x)$ for $x < x_0$, then we can use the following five-point formula that actually uses the values of $f(x)$ at five points,

$$f'(x_0) = \frac{-25f(x_0) + 48f(x_0 + h) - 36f(x_0 + 2h) + 16f(x_0 + 3h) - 3f(x_0 + 4h)}{12h}$$

$$+ \frac{f^{(5)}(\xi)}{5} h^4,$$

where $\xi \in (x_0, x_0 + 4h)$. As before, we can replace h by $-h$ to obtain a similar formula that approximates $f'(x_0)$ using the values of $f(x)$ at $x_0, x_0 - h, x_0 - 2h, x_0 - 3h$, and $x_0 - 4h$.

Exploration 9.1.4 Use (9.6) to derive a general error formula for the approximation of $f'(x_0)$ in the case where $x_i = x_0 + ih$, for $i = -j, \ldots, k$. Use the preceding examples to check the correctness of your error formula.

Example 9.1.3 We will demonstrate how the three-point formula (9.7) for approximating $f'(x_0)$ can be constructed by interpolating $f(x)$ at the points x_0, $x_0 + h$, and $x_0 + 2h$ using a second-degree polynomial $p_2(x)$, and then approximating $f'(x_0)$ by $p_2'(x_0)$. Using Lagrange interpolation, we obtain

$$p_2(x) = f(x_0)\mathcal{L}_{2,0}(x) + f(x_0 + h)\mathcal{L}_{2,1}(x) + f(x_0 + 2h)\mathcal{L}_{2,2}(x),$$

where $\{\mathcal{L}_{2,j}(x)\}_{j=0}^2$ are the Lagrange polynomials for the points x_0, $x_1 = x_0 + h$ and $x_2 = x_0 + 2h$. Recall that these polynomials satisfy

$$\mathcal{L}_{2,j}(x_k) = \delta_{jk} = \begin{cases} 1 & \text{if } j = k \\ 0 & \text{otherwise} \end{cases}.$$

Using the formula for the Lagrange polynomials,

$$\mathcal{L}_{2,j}(x) = \prod_{i=0, i \neq j}^{2} \frac{(x - x_i)}{(x_j - x_i)},$$

we obtain

$$\mathcal{L}_{2,0}(x) = \frac{(x - (x_0 + h))(x - (x_0 + 2h))}{(x_0 - (x_0 + h))(x_0 - (x_0 + 2h))}$$

$$= \frac{x^2 - (2x_0 + 3h)x + (x_0 + h)(x_0 + 2h)}{2h^2},$$

$$\mathcal{L}_{2,1}(x) = \frac{(x - x_0)(x - (x_0 + 2h))}{(x_0 + h - x_0)(x_0 + h - (x_0 + 2h))}$$

$$= \frac{x^2 - (2x_0 + 2h)x + x_0(x_0 + 2h)}{-h^2},$$

$$\mathcal{L}_{2,2}(x) = \frac{(x - x_0)(x - (x_0 + h))}{(x_0 + 2h - x_0)(x_0 + 2h - (x_0 + h))}$$

$$= \frac{x^2 - (2x_0 + h)x + x_0(x_0 + h)}{2h^2}.$$

It follows that

$$\mathcal{L}'_{2,0}(x) = \frac{2x - (2x_0 + 3h)}{2h^2},$$

$$\mathcal{L}'_{2,1}(x) = -\frac{2x - (2x_0 + 2h)}{h^2},$$

$$\mathcal{L}'_{2,2}(x) = \frac{2x - (2x_0 + h)}{2h^2}.$$

We conclude that $f'(x_0) \approx p'_2(x_0)$, where

$$p'_2(x_0) = f(x_0)\mathcal{L}'_{2,0}(x_0) + f(x_0 + h)\mathcal{L}'_{2,1}(x_0) + f(x_0 + 2h)\mathcal{L}'_{2,2}(x_0)$$

$$\approx f(x_0)\frac{-3}{2h} + f(x_0 + h)\frac{2}{h} + f(x_0 + 2h)\frac{-1}{2h}$$

$$\approx \frac{-3f(x_0) + 4f(x_0 + h) - f(x_0 + 2h)}{2h}.$$

From (9.6), it can be shown (see Exploration 9.1.4) that the error in this approximation is $O(h^2)$, and that this formula is exact when $f(x)$ is a polynomial of degree 2 or less. The error formula is given in (9.7). Note that this finite difference formula is the same as the one obtained in Example 9.1.2 using Taylor expansion. □

Because the points x_i, $i = -j, \ldots, k$, are equally spaced, the Lagrange interpolation can be simplified by using the change of variable $s = (x - x_0)/h$, which transforms the interpolation points $x_{-j}, \ldots, x_k$ to the integers $-j, \ldots, k$.

Exploration 9.1.5 Create a new version of `makediffrule` (see Exploration 9.1.2) so that it uses Lagrange interpolation rather than Taylor expansion. Based on your solution to Exploration 9.1.4, make your function return a second output `err` which is the constant C such that the error in (9.6) is of the form $Ch^n f^{(n+1)}(\xi)$, where $n = j + k$.

9.1.3 Higher-Order Derivatives

The approaches of combining Taylor expansions or differentiating Lagrange polynomials can be used to approximate higher-order derivatives. In Example 2.1.20, the second derivative was approximated using a centered difference formula,

$$f''(x_0) = \frac{f(x_0 + h) - 2f(x_0) + f(x_0 - h)}{h^2} - \frac{h^2}{12}f^{(4)}(\xi), \tag{9.8}$$

where $\xi \in (x_0 - h, x_0 + h)$. This formula is second-order accurate.

Exploration 9.1.6 Use Taylor expansions (Example 9.1.2) and Lagrange polynomials (Example 9.1.3) to derive the finite difference formula

$$f''(x_0) \approx \frac{f(x_0) - 2f(x_0 + h) + f(x_0 + 2h)}{h^2}. \tag{9.9}$$

What is the order of accuracy? What is the error formula?

Error formulas for finite-difference approximations of higher-order derivatives are more difficult to obtain than for approximations of first derivatives. A straightforward generalization of the approach used to obtain (9.6) is not feasible because it would entail computing derivatives of divided differences. The following exploration describes an alternative approach.

Exploration 9.1.7 To obtain an error formula for the approximation (9.9) derived in Exploration 9.1.6, we define

$$\varphi(x) = f(x) - p_2(x) - \frac{f''(x_0) - p_2''(x_0)}{\pi''(x_0)}\pi(x),$$

where $p_2(x)$ interpolates $f(x)$ at $x_0, x_0 + h$ and $x_0 + 2h$, and $\pi(x) = (x - x_0)(x - x_0 - h)(x - x_0 - 2h)$. Use Rolle's Theorem (Theorem A.5.1) repeatedly to show that there exists a point $\xi \in (x_0, x_0 + 2h)$ such that $\varphi'''(\xi) = 0$. Use this result to obtain an error formula for (9.9).

This approach would not work for deriving the error formula for (9.8), because in this case, $\pi''(x_0) = 0$. Instead, we make the observation that a divided difference $f[x_0, x_1, \ldots, x_k]$ is independent of the order in which the points $x_0, x_1, \ldots, x_k$ are listed, as divided differences are coefficients of an interpolating polynomial in Newton form, which also does not depend on the ordering of these points.

Exploration 9.1.8 Show that

$$\frac{d}{dx}f[x_0, x_1, \ldots, x_k, x] = f[x_0, x_1, \ldots, x_k, x, x].$$

Exploration 9.1.9 Use the result of Exploration 9.1.8 to derive the error formula from (9.8) by taking the second derivative of the interpolation error

$$f(x) - p_2(x) = f[x_0 - h, x_0, x_0 + h, x](x - x_0 + h)(x - x_0)(x - x_0 - h).$$

Exploration 9.1.10 Generalize your function `makediffrule` from Exploration 9.1.5 so that it computes the coefficients of a finite difference approximation to a derivative of a given order, which is specified as an input argument. The output argument `err` from is not required in this case.

9.1.4 Sensitivity

Based on the error formula for each of these finite difference approximations, one would expect that it is possible to obtain an approximation that is accurate to within machine precision simply by choosing h sufficiently small. In the following exploration, we can put this expectation to the test.

> **Exploration 9.1.11** Use the centered difference formula (9.3) to approximate $f'(x_0)$ for $f(x) = \sin x$, $x_0 = 1.2$, and $h = 10^{-d}$ for $d = 1, 2, \ldots, 15$. Compare the error in each approximation with the error formula given in (9.3). How does the actual error compare to theoretical expectations?

The reason for the discrepancy observed in Exploration 9.1.11 is that the error formula in (9.3), or any other finite difference approximation, only accounts for **discretization error** (see Section 2.1.1), not **roundoff error** (see Section 2.2.2).

In a practical implementation of finite difference formulas, it is essential to note that roundoff error in evaluating $f(x)$ is bounded independently of the spacing h. It follows that the roundoff error in the approximation of $f'(x)$ actually *increases* as h decreases, because the errors incurred by evaluating $f(x)$ are divided by h. Therefore, one must choose h sufficiently small so that the finite difference formula can produce an approximation with small discretization error, and sufficiently large so that this approximation is not too contaminated by roundoff error.

The following exploration considers the conditioning of the problem of computing a derivative. Because finite difference approximations are based on polynomial interpolation, the approximate derivative of $f(x)$ is the exact derivative of a polynomial that approximates $f(x)$. It follows that the error in polynomial interpolation is **backward error** from Section 2.1.4, which is related to **forward error** (in this case, the error in numerical differentiation) by the condition number.

> **Exploration 9.1.12** For the finite difference approximations considered in this section, use the accompanying error formulas, along with the error in polynomial interpolation from Theorem 7.4.1, to estimate the condition number of numerical differentiation. For each finite difference approximation considered, be sure to use the polynomial interpolant on which the approximate derivative is based (that is, the polynomial that interpolates $f(x)$ at the same points), and obtain an upper bound on the interpolation error on $[x_0 - jh, x_0 + kh]$, where j, k are as defined in (9.5). For centered difference formulas, include x_0 as an interpolation point.

9.1.5 Differentiation Matrices*

It is often necessary to compute derivatives of a function $f(x)$ at a set of points within a given domain. Suppose that both $f(x)$ and $f'(x)$ are represented by vectors $\mathbf{f}$ and $\mathbf{g}$, respectively, whose elements are the values of f and f' at N selected points. Then, in view of the linearity of differentiation, $\mathbf{f}$ and $\mathbf{g}$ should be related by a linear transformation. That is, $\mathbf{g} \approx D\mathbf{f}$, where D is an $N \times N$ matrix that uses finite difference formulas to approximate $f'(x)$ at each point with linear combinations of the elements of $\mathbf{f}$. In this context, D is called a **differentiation matrix**.

Example 9.1.4 We construct a differentiation matrix for functions defined on $[0, 1]$, and satisfying the *boundary conditions* $f(0) = f(1) = 0$. Let $x_1, x_2, \ldots, x_n$ be

n equally spaced points in $(0,1)$, defined by $x_i = ih$, where $h = 1/(n+1)$. If we use the forward difference approximation, we then have

$$f'(x_1) \approx \frac{f(x_2) - f(x_1)}{h},$$

$$f'(x_2) \approx \frac{f(x_3) - f(x_2)}{h},$$

$$\vdots$$

$$f'(x_{n-1}) \approx \frac{f(x_n) - f(x_{n-1})}{h},$$

$$f'(x_n) \approx \frac{0 - f(x_n)}{h}.$$

Writing these equations in matrix-vector form, we obtain a relation of the form $\mathbf{g} \approx D\mathbf{f}$, where

$$\mathbf{g} = \begin{bmatrix} f'(x_1) \\ f'(x_2) \\ \vdots \\ f'(x_n) \end{bmatrix}, \quad \mathbf{f} = \begin{bmatrix} f(x_1) \\ f(x_2) \\ \vdots \\ f(x_n) \end{bmatrix}, \quad D = \frac{1}{h} \begin{bmatrix} -1 & 1 & & & \\ & -1 & 1 & & \\ & & \ddots & \ddots & \\ & & & -1 & 1 \\ & & & & -1 \end{bmatrix}.$$

The entries of D can be determined from the coefficients of each value $f(x_j)$ used to approximate $f'(x_i)$, for $i = 1, 2, \ldots, n$. From the structure of this upper bidiagonal matrix, it follows that we can approximate $f'(x)$ at these grid points by a matrix-vector multiplication which costs only $O(n)$ floating-point operations.

Now, suppose that we instead impose **periodic boundary conditions** $f(0) = f(1)$. In this case, we again use n equally spaced points, but including the left boundary: $x_i = ih$, $i = 0, 1, \ldots, n-1$, where $h = 1/n$. Using forward differencing again, we have the same approximations as before, except

$$f'(x_{n-1}) \approx \frac{f(1) - f(x_{n-1})}{h} = \frac{f(0) - f(x_{n-1})}{h} = \frac{f(x_0) - f(x_{n-1})}{h}.$$

It follows that the differentiation matrix is

$$D = \frac{1}{h} \begin{bmatrix} -1 & 1 & & & \\ & -1 & 1 & & \\ & & \ddots & \ddots & \\ & & & -1 & 1 \\ 1 & & & & -1 \end{bmatrix}.$$

Note the "wrap-around" effect in which the superdiagonal appears to continue past the last column into the first column. For this reason, D is an example of what is called a **circulant matrix**, which was introduced in Exercise 8 of Chapter 8. □

Exploration 9.1.13 What are the differentiation matrices based on the backward difference (9.2) and centered difference (9.3) for functions defined on $[0, 1]$, for (a) boundary conditions $f(0) = f(1) = 0$, and (b) periodic boundary conditions $f(0) = f(1)$?

Exploration 9.1.14 What are the differentiation matrices corresponding to (9.8) for functions defined on $[0, 1]$, for (a) boundary conditions $f(0) = f(1) = 0$, and (b) periodic boundary conditions $f(0) = f(1)$?

Other types of differentiation matrices will be covered in Chapter 14.

9.1.6   Weak Derivatives*

As can be seen from their error formulas, the accuracy of finite difference formulas presented in this section relies on the function being sufficiently smooth, with possibly disastrous results if this is not the case. This makes these formulas useless in applications such as differential equations that do not have *classical solutions*, which possess the required number of derivatives. Instead, we have to settle for *weak solutions*. This leads to the concept of a **weak derivative**. A function $g(x)$ that is locally integrable on $[a, b]$ is the weak derivative of $f(x)$, also locally integrable on $[a, b]$, if

$$\int_a^b g(x)\varphi(x)\,dx = f(x)\varphi(x)\big|_a^b - \int_a^b f(x)\varphi'(x)\,dx$$

for any infinitely differentiable function $v(x)$ that has compact support on (a, b); that is, $v(a) = v(b) = 0$.

In [Feng, et al. (2016)], **discrete derivatives** are defined as follows. Let $I_j = [a, b]$ be divided into subintervals $I = [x_{j-1}, x_j]$, $j = 1, 2, \ldots, n$, where $a = x_0 < x_1 < \cdots < x_{n-1} < x_n = b$. Then, for $j = 1, 2, \ldots, n$, let $\mathbb{P}_r(I_j)$ be the space of polynomials on I_j of degree at most r, and let $V_r^h = \prod_{i=1}^n \mathbb{P}_r(I_j)$. That is, functions in V_r^h are piecewise polynomials, with the pieces being the subintervals $\{I_j\}_{j=1}^n$, but unlike splines from Section 7.6, they are not required to be continuous. Next, we define the **jump** of a function across subintervals as follows:

$$[u]\big|_{x_j} \equiv \begin{cases} u(x_j^-) - u(x_j^+) & j = 1, 2, \ldots, n-1, \\ u(x_j) & j = 0, n \end{cases}$$

where $u(a^-) = \lim_{x \to a^-} u(x)$, $u(a^+) = \lim_{x \to a^+} u(x)$.

Then, the **average** of u between subintervals is

$$\{u\}\big|_{x_j} \equiv \begin{cases} \frac{1}{2}(u(x_j^-) + u(x_j^+)) & j = 1, 2, \ldots, n-1, \\ u(x_j) & j = 0, n \end{cases}$$

Next, we define the **trace** operators

$$Q^{\pm}(u) = \{u\} \pm \frac{1}{2}[u]$$

on $\{x_1, \ldots, x_{n-1}\}$, and $Q^{\pm}(u) = u$ on $\{x_0, x_n\}$. Finally, the discrete derivatives of u, denoted by $\partial_h^{\pm} u$, are defined by

$$\langle \partial_h^{\pm} u, \varphi_h \rangle_{I_j} = Q^{\pm}(u)(x)[\varphi_h(x)]\big|_{x_{j-1}}^{x_j} - \langle u, \varphi_h' \rangle_{I_j}, \qquad \varphi_h \in V_r^h \qquad (9.10)$$

$$\partial_h u = \frac{1}{2}(\partial_h^+ u + \partial_h^- u), \qquad (9.11)$$

where

$$\langle f, g \rangle_{I_j} = \int_{x_{j-1}}^{x_j} f(x)g(x)\, dx$$

is the standard inner product of real-valued functions on I_j.

To compute $\partial_h^{\pm} u$ on each subinterval I_j, we prescribe the form

$$\partial_h^{\pm} u = \sum_{i=0}^{r} \alpha_i \varphi_i,$$

where $\{\varphi_i\}_{i=0}^r$ is a basis for $\mathbb{P}_r$. Then, from (9.10) we obtain

$$\sum_{i=0}^{r} \alpha_i \langle \varphi_i, \varphi_k \rangle_{I_j} = Q^{\pm}(u)(x)[\varphi_k(x)]\big|_{x_{j-1}}^{x_j} - \langle u, \varphi_k' \rangle_{I_j}, \qquad k = 0, 1, \ldots, r.$$

This yields a system of linear equations for the coefficients $\{\alpha_i\}_{i=0}^r$. This system is particularly simple to solve if we use a basis of **orthogonal polynomials**, such as the **Legendre polynomials** featured in Section 8.2.3.

Example 9.1.5 Let

$$u(x) = \begin{cases} \sin x & x < 0 \\ \sin 2x & x \geq 0 \end{cases}.$$

Figure 9.1 shows the graphs of $u(x)$ and $\partial_h u(x)$ as defined in (9.11). Note that $u(x)$ is continuous, but not differentiable at $x = 0$. However, $\partial_h u(x)$ is an accurate approximation of the left-hand derivative for $x < 0$, and the right-hand derivative for $x > 0$. The weak derivative was computed using MATLAB code that, as of this writing, is posted at `http://www.math.utk.edu/~schnake/research/`. □

9.1.7 Concept Check

1. Write down the formulas for approximation of the first derivative at x_0 using a (a) forward difference, (b) backward difference, and (c) centered difference, with spacing h, along with their error formulas.
2. What does it mean to say a finite difference approximation has a given order of accuracy?
3. How can Lagrange interpolation be used to obtain finite difference approximations of derivatives?
4. Explain why roundoff error is a significant problem for numerical differentiation.
5. What is a differentiation matrix?
6. What is a weak derivative and what is its significance?

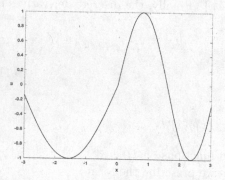 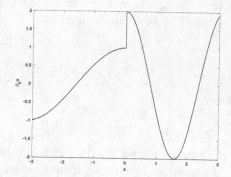

Fig. 9.1   From Example 9.1.5. Left plot: the function $u(x) = \sin x$ for $x < 0$, $u(x) = \sin 2x$ for $x \geq 0$. Right plot: weak derivative $\partial_h u(x)$

9.2   Numerical Integration

Numerous applications call for the computation of the integral of some function $f : \mathbb{R} \to \mathbb{R}$ over an interval $[a, b]$,

$$I[f] = \int_a^b f(x)\, dx.$$

In some cases, $I[f]$ can be computed by applying the Fundamental Theorem of Calculus and computing

$$I[f] = F(b) - F(a),$$

where $F(x)$ is an *antiderivative* of f, meaning that $F'(x) = f(x)$. Unfortunately, this is not practical if an antiderivative of f is not available. In such cases, numerical techniques must be employed instead.

For example, the arc length of a curve $y = f(x)$ for $a \leq x \leq b$, given by

$$L = \int_a^b \sqrt{1 + [f'(x)]^2}\, dx,$$

generally does not lend itself to anti-differentiation. In statistics, probabilities are often represented by integrals that cannot be evaluated exactly, such as those involving a normal distribution. For example, the probability that a normally distributed random variable X is between x_1 and x_2 is given by

$$P(x_1 \leq X \leq x_2) = \frac{1}{\sqrt{2\pi\sigma^2}} \int_{x_1}^{x_2} e^{-\frac{(x-\mu)^2}{2\sigma^2}}\, dx,$$

where μ and σ are the mean and standard deviation, respectively. This integral cannot be evaluated using anti-differentiation techniques taught in calculus, so numerical methods are again required.

There are also instances in which $I[f]$ needs to be computed even though a formula for the integrand $f(x)$ is not available. For example, if $v(t)$ represents the velocity of an object at time t, then $I[v] = \int_{t_1}^{t_2} v(t)\,dt$ represents the (net) displacement of the object between times t_1 and t_2. If $v(t)$ is represented only by its values at various times in $[t_1, t_2]$, then $I[v]$ must be approximated using a numerical method. This kind of situation, in which the integrand is only known at discrete points, also arises in methods for solving differential equations, as will be seen in Part V. Still, the basics of integration play a vital role in the development and analysis of numerical techniques, so they are reviewed in Section A.4.

9.2.1 Quadrature Rules

Clearly, if f is a Riemann integrable function and $\{R_n\}_{n=1}^{\infty}$ is a sequence of Riemann sums that converges to $I[f]$, then any particular Riemann sum R_n can be viewed as an approximation of $I[f]$. However, such an approximation is usually not practical since a large value of n may be necessary to achieve sufficient accuracy.

> **Exploration 9.2.1** Write a MATLAB script that computes the Riemann sum R_n for
> $$\int_0^1 x^2\,dx = \frac{1}{3},$$
> where the *left* endpoint of each subinterval is used to obtain the height of the corresponding rectangle. How large must n, the number of subintervals, be to obtain an approximate answer that is accurate to within 10^{-5}?

Instead, we use a **quadrature rule** to approximate $I[f]$. A quadrature rule $Q_n[f]$ is a sum of the form

$$Q_n[f] = \sum_{i=1}^{n} f(x_i)w_i, \tag{9.12}$$

where the points x_i, $i = 1, \ldots, n$, are called the **nodes** of the quadrature rule, and the numbers w_i, $i = 1, \ldots, n$, are the **weights**.

The objective in designing quadrature rules is to achieve sufficient accuracy in approximating $I[f]$, for any Riemann integrable function f, while using as few nodes as possible to maximize efficiency. To determine suitable nodes and weights, we consider the following questions, as part of a recurring theme in this book:

- For what functions f is $I[f]$ easy to compute?
- Given a general Riemann integrable function f, can $I[f]$ be approximated by the integral of a function g for which $I[g]$ is easy to compute?

9.2.2 Interpolatory Quadrature

One class of functions for which integrals are easily evaluated is the class of polynomial functions. If we choose n nodes $x_1, \ldots, x_n$, then any polynomial $p_{n-1}(x)$ of degree $n - 1$ can be written in the form

$$p_{n-1}(x) = \sum_{i=1}^{n} p_{n-1}(x_i)\mathcal{L}_{n-1,i}(x),$$

where $\mathcal{L}_{n-1,i}(x)$ is the ith Lagrange polynomial for the points $x_1, \ldots, x_n$. It follows from the linearity of the integral that

$$
\begin{aligned}
I[p_{n-1}] &= \int_a^b p_{n-1}(x)\,dx \\
&= \sum_{i=1}^{n} p_{n-1}(x_i)\left(\int_a^b \mathcal{L}_{n-1,i}(x)\,dx\right) \\
&= \sum_{i=1}^{n} p_{n-1}(x_i)w_i \\
&= Q_n[p_{n-1}]
\end{aligned}
$$

where

$$w_i = \int_a^b \mathcal{L}_{n-1,i}(x)\,dx, \quad i = 1,\ldots,n, \tag{9.13}$$

are the weights of a quadrature rule with nodes $x_1, \ldots, x_n$.

Therefore, *any* n-point quadrature rule with weights chosen as in (9.13) computes $I[f]$ *exactly* when f is a polynomial of degree less than n. For a more general function f, we can use this quadrature rule to approximate $I[f]$ by $I[p_{n-1}]$, where p_{n-1} is the polynomial that interpolates f at the nodes $x_1, \ldots, x_n$. Quadrature rules that use the weights defined above for given nodes $x_1, \ldots, x_n$ are called **interpolatory** quadrature rules. We say that an interpolatory quadrature rule has **degree of accuracy** n if it integrates polynomials of degree n exactly, but is not exact for polynomials of degree $n + 1$.

Exploration 9.2.2 Use MATLAB's polynomial functions to write a function I=polydefint(p,a,b) that computes and returns the definite integral of a polynomial with coefficients stored in the vector p over the interval [a, b].

Exploration 9.2.3 Use your function polydefint from Exploration 9.2.2, in conjunction with your function makelagrange from Exploration 7.2.2, to write a function w=interpweights(x,a,b) that returns a vector of weights w for an interpolatory quadrature rule for the interval [a, b] with nodes stored in the vector x.

> **Exploration 9.2.4** Use your function `interpweights` from Exploration 9.2.3 to write a function `I=interpquad(f,a,b,x)` that approximates $I[f]$ over $[a, b]$ using an interpolatory quadrature rule with nodes stored in the vector `x`. The input argument `f` must be a function handle. Test your function by using it to evaluate the integrals of polynomials of various degrees, comparing the results to the exact integrals returned by your function `polydefint` from Exploration 9.2.2.

9.2.3 Sensitivity

To examine the sensitivity of $I[f]$, we use the ∞-norm of a function $f(x)$ (see Section B.13.3), defined by

$$\|f\|_\infty = \max_{x \in [a,b]} |f(x)|$$

and let $\hat{f}$ be a perturbation of f that is also Riemann integrable. Then the **absolute condition number** (see Section 2.1.4) of the problem of computing $I[f]$ can be approximated by

$$\frac{|I[\hat{f}] - I[f]|}{\|\hat{f} - f\|_\infty} = \frac{|I[\hat{f} - f]|}{\|\hat{f} - f\|_\infty}$$
$$\leq \frac{I[|\hat{f} - f|]}{\|\hat{f} - f\|_\infty}$$
$$\leq \frac{(b - a)\|\hat{f} - f\|_\infty}{\|\hat{f} - f\|_\infty}$$
$$\leq (b - a),$$

from which it follows that the problem is fairly well-conditioned. Similarly, perturbations of the endpoints a and b do not lead to large perturbations in $I[f]$, in most cases.

> **Exploration 9.2.5** What is the relative condition number of the problem of computing $I[f]$?

If the weights w_i, $i = 1, \ldots, n$, are nonnegative, then the quadrature rule is stable, as its absolute condition number can be bounded by $(b - a)$, which is the same absolute condition number as the underlying integration problem. However, if any of the weights are negative, then the condition number can be arbitrarily large.

> **Exploration 9.2.6** Find the absolute condition number of the problem of computing $Q_n[f]$ for a general quadrature n-node rule of the form (9.12).

9.2.4 Concept Check

1. What is a quadrature rule? How does it relate to the definition of a definite integral in terms of Riemann sums?

2. What is an interpolatory quadrature rule? How are the weights defined for such a rule?

3. What does it mean to say an interpolatory quadrature rule has degree of accuracy p?

4. Is numerical integration a well-conditioned or ill-conditioned problem? Explain.

5. Is the problem of computing $Q_n[f]$, for a general n-node quadrature rule, well-conditioned or ill-conditioned? Discuss.

9.3 Newton-Cotes Rules

The family of **Newton-Cotes** quadrature rules consists of interpolatory quadrature rules in which the nodes are equally spaced points within the interval $[a, b]$. There are two types of Newton-Cotes rules: **open**, which do *not* include the endpoints $x = a$ and $x = b$ among the nodes, and **closed** rules, which do include the endpoints. The most commonly used Newton-Cotes rules are:

- The **Midpoint Rule**, which is an open rule with one node, is defined by

$$\int_a^b f(x)\,dx \approx (b-a)f\left(\frac{a+b}{2}\right). \tag{9.14}$$

  It is of degree one, and it is based on the principle that the area under $f(x)$ can be approximated by the area of a rectangle with width $b-a$ and height $f(m)$, where $m = (a+b)/2$ is the midpoint of the interval $[a, b]$.

- The **Trapezoidal Rule**, which is a closed rule with two nodes, is defined by

$$\int_a^b f(x)\,dx \approx \frac{b-a}{2}[f(a) + f(b)]. \tag{9.15}$$

  It is of degree one, and it is based on the principle that the area under $f(x)$ from $x = a$ to $x = b$ can be approximated by the area of a trapezoid with heights $f(a)$ and $f(b)$ and width $b-a$.

- **Simpson's Rule**, which is a closed rule with three nodes, is defined by

$$\int_a^b f(x)\,dx \approx \frac{b-a}{6}\left[f(a) + 4f\left(\frac{a+b}{2}\right) + f(b)\right]. \tag{9.16}$$

  It is of degree three, and it is derived by computing the integral of the quadratic polynomial that interpolates $f(x)$ at the points a, $(a+b)/2$, and b.

Example 9.3.1 Let $f(x) = x^3$, $a = 0$ and $b = 1$. We have

$$\int_a^b f(x)\,dx = \int_0^1 x^3\,dx = \left.\frac{x^4}{4}\right|_0^1 = \frac{1}{4}.$$

Approximating this integral with the Midpoint Rule yields

$$\int_0^1 x^3\, dx \approx (1-0)\left(\frac{0+1}{2}\right)^3 = \frac{1}{8}.$$

Using the Trapezoidal Rule, we obtain

$$\int_0^1 x^3\, dx \approx \frac{1-0}{2}[0^3 + 1^3] = \frac{1}{2}.$$

Finally, Simpson's Rule yields

$$\int_0^1 x^3\, dx \approx \frac{1-0}{6}\left[0^3 + 4\left(\frac{0+1}{2}\right)^3 + 1^3\right] = \frac{1}{6}\left[0 + 4\frac{1}{8} + 1\right] = \frac{1}{4}.$$

That is, the approximation of the integral by Simpson's Rule is actually exact, which is expected because Simpson's Rule is of degree three. On the other hand, if we approximate the integral of $f(x) = x^4$ from 0 to 1, Simpson's Rule yields $5/24$, while the exact value is $1/5$. Still, this is a better approximation than those obtained using the Midpoint Rule $(1/16)$ or the Trapezoidal Rule $(1/2)$. $\square$

Exploration 9.3.1 Write MATLAB functions I=quadmidpoint(f,a,b), I=quadtrapezoidal(f,a,b) and I=quadsimpsons(f,a,b) that implement the Midpoint Rule, Trapezoidal Rule and Simpson's Rule, respectively, to approximate the integral of $f(x)$, represented by the function handle f, over the interval $[a, b]$.

Exploration 9.3.2 Use your code from Exploration 9.2.4 to write a function I=quadnewtoncotes(f,a,b,n) to integrate $f(x)$, represented by the function handle f, over $[a, b]$ using an n-node *closed* Newton-Cotes rule.

9.3.1 Error Analysis

The error in any interpolatory quadrature rule defined on an interval $[a, b]$, such as a Newton-Cotes rule, can be obtained by computing the integral from a to b of the error in the polynomial interpolant on which the rule is based, using the error formula from Theorem 7.4.1. Recall from Section 2.1.3 that in **backward error analysis**, the approximate solution is the exact solution of a nearby problem. In the context of interpolatory quadrature, the approximate integral of $f(x)$ is the exact integral of a polynomial that interpolates $f(x)$ at the nodes. Therefore, we can view the interpolation error as the **backward error** in interpolatory quadrature.

For the Trapezoidal Rule, which is obtained by integrating a linear polynomial that interpolates the integrand $f(x)$ at $x = a$ and $x = b$, this approach to error analysis yields

$$\int_a^b f(x)\, dx - \frac{b-a}{2}[f(a) + f(b)] = \int_a^b \frac{f''(\xi(x))}{2}(x-a)(x-b)\, dx,$$

where $\xi(x)$ lies in (a,b) for $a < x < b$. The function $(x-a)(x-b)$ does not change sign on $[a,b]$, which allows us to apply the **Weighted Mean Value Theorem for Integrals** (Theorem A.5.5) and obtain a more useful expression for the error,

$$\int_a^b f(x)\,dx - \frac{b-a}{2}[f(a)+f(b)] = \frac{f''(\eta)}{2}\int_a^b (x-a)(x-b)\,dx = -\frac{f''(\eta)}{12}(b-a)^3,$$

(9.17)

where $a < \eta < b$. Because the error depends on the second derivative, it follows that the Trapezoidal Rule is exact for any linear function.

A similar approach can be used to obtain expressions for the error in the Midpoint Rule and Simpson's Rule, although the process is somewhat more complicated. This is because the functions $(x-m)$, for the Midpoint Rule, and $(x-a)(x-m)(x-b)$, for Simpson's Rule, where in both cases $m = (a+b)/2$, change sign on $[a,b]$, thus making the Weighted Mean Value Theorem for Integrals impossible to apply in the same straightforward manner as it was for the Trapezoidal Rule. We instead use the following approach, illustrated for the Midpoint Rule and adapted from a similar proof for Simpson's Rule from [Süli and Mayers (2003)]. We assume that f is twice continuously differentiable on $[a,b]$. First, we make a change of variable

$$x = \frac{a+b}{2} + \frac{b-a}{2}t, \quad t \in [-1,1],$$

to map the interval $[-1,1]$ to $[a,b]$, and then define $F(t) = f(x(t))$. The error in the Midpoint Rule is then given by

$$\int_a^b f(x)\,dx - (b-a)f\left(\frac{a+b}{2}\right) = \frac{b-a}{2}\left[\int_{-1}^1 F(\tau)\,d\tau - 2F(0)\right].$$

We now define

$$G(t) = \int_{-t}^t F(\tau)\,d\tau - 2tF(0).$$

It is easily seen that the error in the Midpoint Rule is $\frac{1}{2}(b-a)G(1)$. We then define

$$H(t) = G(t) - t^3 G(1).$$

Because $H(0) = H(1) = 0$, it follows from Rolle's Theorem that there exists a point $\xi_1 \in (0,1)$ such that $H'(\xi_1) = 0$. However, from

$$H'(0) = G'(0) = [F(t) + F(-t)]|_{t=0} - 2F(0) = 0,$$

it follows from Rolle's Theorem that there exists a point $\xi_2 \in (0,\xi_1)$ such that $H''(\xi_2) = 0$.

From

$$H''(t) = G''(t) - 6tG(1) = F'(t) - F'(-t) - 6tG(1),$$

and the Mean Value Theorem, we obtain, for some $\xi_3 \in (-1,1)$,

$$0 = H''(\xi_2) = 2\xi_2 F''(\xi_3) - 6\xi_2 G(1),$$

or

$$G(1) = \frac{1}{3}F''(\xi_3) = \frac{1}{3}\left(\frac{b-a}{2}\right)^2 f''(x(\xi_3)).$$

Multiplying by $(b-a)/2$ yields the error in the Midpoint Rule. That is,

$$\int_a^b f(x)\,dx - (b-a)f\left(\frac{a+b}{2}\right) = \frac{f''(\eta)}{24}(b-a)^3, \qquad (9.18)$$

and for Simpson's Rule,

$$\int_a^b f(x)\,dx - \frac{b-a}{6}\left[f(a) + 4f\left(\frac{a+b}{2}\right) + f(b)\right] = -\frac{f^{(4)}(\eta)}{90}\left(\frac{b-a}{2}\right)^5, \qquad (9.19)$$

where, in both cases, η is some point in (a, b).

It follows that the Midpoint Rule is exact for any linear function, just like the Trapezoidal Rule, even though it uses one less interpolation point, because of the cancellation that results from choosing the midpoint of $[a, b]$ as the interpolation point. Similar cancellation causes Simpson's Rule to be exact for polynomials of degree three or less, even though it is obtained by integrating a quadratic interpolant over $[a, b]$.

> **Exploration 9.3.3** Adapt the approach in the preceding derivation of the error formula (9.18) for the Midpoint Rule to obtain the error formula (9.19) for Simpson's Rule.

In general, the degree of accuracy of Newton-Cotes rules can be determined by performing a Taylor expansion of the integrand $f(x)$ around the midpoint of $[a, b]$, $m = (a+b)/2$. This technique can be used to show that n-point Newton-Cotes rules with an odd number of nodes have degree n, which is surprising since, in general, interpolatory n-point quadrature rules have degree $n - 1$. This extra degree of accuracy is due to the symmetry of the leading error terms.

> **Exploration 9.3.4** Prove that an n-node Newton-Cotes rule has degree of accuracy n if n is odd, and $n - 1$ if n is even.

9.3.2 Higher-Order Rules

Unfortunately, Newton-Cotes rules are not practical when the number of nodes is large, due to the inaccuracy of high-degree polynomial interpolation using equally spaced points. Furthermore, for $n \geq 11$, n-point Newton-Cotes rules have at least one negative weight, and therefore such rules can be ill-conditioned. This can be seen by revisiting **Runge's Example** from Section 7.4, and attempting to approximate

$$\int_{-5}^5 \frac{1}{1+x^2}\,dx \qquad (9.20)$$

using a Newton-Cotes rule. As n increases, the approximate integral does not converge to the exact result.

> **Exploration 9.3.5** What is the smallest value of n for which a n-node Newton-Cotes rule has a negative weight?

> **Exploration 9.3.6** Use your code from Exploration 9.3.2 to evaluate the integral from (9.20) for increasing values of **n**, the number of nodes, and describe the behavior of the error as **n** increases.

9.3.3 Concept Check

1. What is a Newton-Cotes rule?
2. What does it mean to say a Newton-Cotes rule is open or closed?
3. Write down the following Newton-Cotes rules, along with their error formulas: (a) Trapezoidal Rule, (b) Midpoint Rule, and (c) Simpson's Rule.
4. What is the degree of accuracy of a general Newton-Cotes rule?
5. What is problematic about higher-order Newton-Cotes rules?

9.4 Composite Rules

When using a quadrature rule to approximate $I[f]$ on some interval $[a, b]$, the error is proportional to h^r, where $h = b - a$ and r is some positive integer. Therefore, if the interval $[a, b]$ is large, the error can be quite large as well. It is thus advisable to divide $[a, b]$ into smaller intervals, use a quadrature rule to compute the integral of f on each subinterval, and add the results to approximate $I[f]$. Such a scheme is called a **composite quadrature rule**.

It can be shown that the approximate integral obtained using a composite rule that divides $[a, b]$ into n subintervals will converge to $I[f]$ as $n \to \infty$, provided that the maximum width of the n subintervals approaches zero, and the quadrature rule used on each subinterval has a degree of at least zero and positive weights. It should be noted that using *closed* quadrature rules on each subinterval improves efficiency, because the nodes on the endpoints of each subinterval, except for a and b, are shared by two quadrature rules. As a result, fewer function evaluations are necessary, compared to a composite rule that uses *open* rules with the same number of nodes.

9.4.1 Composite Newton-Cotes Rules

We will now state some of the most well-known composite quadrature rules. In the following discussion, we assume that the interval $[a, b]$ is divided into n subintervals of equal width $h = (b - a)/n$, and that these subintervals have endpoints $[x_{i-1}, x_i]$, where $x_i = a + ih$, for $i = 0, 1, 2, \ldots, n$. Given such a partition of $[a, b]$, we can approximate $I[f]$ using

- the **Composite Midpoint Rule**

$$\int_a^b f(x)\, dx \approx 2h[f(x_1) + f(x_3) + \cdots + f(x_{n-1})], \quad n \text{ is even}, \qquad (9.21)$$

- the **Composite Trapezoidal Rule**

$$\int_a^b f(x)\, dx \approx \frac{h}{2}[f(x_0) + 2f(x_1) + 2f(x_2) + \cdots + 2f(x_{n-1}) + f(x_n)], \quad (9.22)$$

- or the **Composite Simpson's Rule**

$$\int_a^b f(x)\, dx \approx \frac{h}{3}[f(x_0) + 4f(x_1) + 2f(x_2) + 4f(x_3) + \cdots +$$
$$2f(x_{n-2}) + 4f(x_{n-1}) + f(x_n)], \qquad (9.23)$$

for which n is required to be even, as in the Composite Midpoint Rule.

Exploration 9.4.1 Write MATLAB functions

```
I=quadcompmidpt(f,a,b,n)
I=quadcomptrap(f,a,b,n)
I=quadcompsimp(f,a,b,n)
```

that implement the Composite Midpoint rule (9.21), Composite Trapezoidal Rule (9.22), and Composite Simpson's Rule (9.23), respectively, to approximate the integral of $f(x)$, represented by the function handle f, over [a, b] with n + 1 nodes $x_0, x_1, \ldots, x_n$.

9.4.2   Error Analysis

To obtain the error in each of these composite rules, we can sum the errors in the corresponding basic rules over the n subintervals. For the Composite Trapezoidal Rule (9.22), we use (9.17) on each subinterval to obtain

$$E_{\text{trap}} = \int_a^b f(x)\, dx - \frac{h}{2}\left[f(x_0) + 2\sum_{i=1}^{n-1} f(x_i) + f(x_n)\right]$$

$$= -\sum_{i=1}^n \frac{f''(\eta_i)}{12}(x_i - x_{i-1})^3$$

$$= -\frac{h^3}{12}\sum_{i=1}^n f''(\eta_i)$$

$$= -\frac{h^3}{12}nf''(\eta)$$

$$= -\frac{f''(\eta)}{12}(b-a)h^2, \qquad (9.24)$$

where, for $i = 1, \ldots, n$, η_i belongs to (x_{i-1}, x_i), and $a < \eta < b$. The replacement of $\sum_{i=1}^h f''(\eta_i)$ by $nf''(\eta)$ is justified by the Intermediate Value Theorem, as in

Section 9.1.1, provided that $f''(x)$ is continuous on $[a, b]$. We see that the Composite Trapezoidal Rule is **second-order accurate**, as the error is $O(h^2)$. Furthermore, its *degree of accuracy*, which is the highest degree of polynomial that is guaranteed to be integrated exactly, is the same as for the basic Trapezoidal Rule, which is one.

Similarly, for the Composite Midpoint Rule (9.21), we apply (9.18) on $n/2$ subintervals, each of width $2h$, to obtain

$$E_{\text{mid}} = \int_a^b f(x)\,dx - 2h \sum_{i=1}^{n/2} f(x_{2i-1}) = \sum_{i=1}^{n/2} \frac{f''(\eta_i)}{24}(2h)^3 = \frac{f''(\eta)}{6}(b-a)h^2.$$

Although it appears that the Composite Midpoint Rule is less accurate than the Composite Trapezoidal Rule, it should be noted that it uses about half as many function evaluations. This is because the Basic Midpoint Rule is applied $n/2$ times on subintervals of width $2h$, whereas the Composite Trapezoidal Rule uses n subintervals. Rewriting the Composite Midpoint Rule in such a way that it uses n function evaluations, each on a subinterval of width h, we obtain

$$\int_a^b f(x)\,dx = h \sum_{i=1}^{n} f\left(x_{i-1} + \frac{h}{2}\right) + \frac{f''(\eta)}{24}(b-a)h^2, \qquad (9.25)$$

which reveals that the Composite Midpoint Rule is generally more accurate.

Finally, for the Composite Simpson's Rule (9.23), we apply (9.19) on $n/2$ subintervals, each of width $2h$, to obtain

$$E_{\text{simp}} = -\sum_{i=1}^{n/2} \frac{f^{(4)}(\eta_i)}{90} h^5 = -\frac{f^{(4)}(\eta)}{180}(b-a)h^4. \qquad (9.26)$$

The error is $O(h^4)$, so we say that the Simpson's Rule is **fourth-order accurate**.

Exploration 9.4.2 Derive the error formula (9.26) for the Composite Simpson's Rule (9.23).

Example 9.4.1 We wish to approximate

$$\int_0^1 e^x\,dx$$

using composite quadrature, to 3 decimal places. That is, the error must be less than 10^{-3}. This requires choosing the number of subintervals, n, sufficiently large so that an upper bound on the error is less than 10^{-3}.

For the Composite Trapezoidal Rule, the error is

$$E_{\text{trap}} = -\frac{f''(\eta)}{12}(b-a)h^2 = -\frac{e^\eta}{12n^2},$$

since $f(x) = e^x$, $a = 0$ and $b = 1$, which yields $h = (b-a)/n = 1/n$. Since $0 \le \eta \le 1$, and e^x is increasing, the factor e^η is bounded above by $e^1 = e$. It follows that $|E_{\text{trap}}| < 10^{-3}$ if

$$\frac{e}{12n^2} < 10^{-3} \quad \Longrightarrow \quad \frac{1000e}{12} < n^2 \quad \Longrightarrow \quad n > 15.0507.$$

Therefore, the error will be sufficiently small provided that we choose $n \geq 16$.

On the other hand, if we use the Composite Simpson's Rule, the error is

$$E_{\text{simp}} = -\frac{f^{(4)}(\eta)}{180}(b-a)h^4 = -\frac{e^\eta}{180n^4}$$

for some η in $[0, 1]$, which is less than 10^{-3} in absolute value if

$$n > \left(\frac{1000e}{180}\right)^{1/4} \approx 1.9713,$$

so $n = 2$ is sufficient. That is, we can approximate the integral to 3 decimal places by setting $h = (b-a)/n = (1-0)/2 = 1/2$ and computing

$$\int_0^1 e^x \, dx \approx \frac{h}{3}[e^{x_0} + 4e^{x_1} + e^{x_2}] = \frac{1/2}{3}[e^0 + 4e^{1/2} + e^1] \approx 1.71886,$$

whereas the exact value is approximately 1.71828. $\square$

Exploration 9.4.3 Apply your functions from Exploration 9.4.1 to approximate the integrals

$$\int_0^1 \sqrt{x} \, dx, \quad \int_1^2 \sqrt{x} \, dx.$$

Use different values of n, the number of subintervals. How does the accuracy increase as n increases? Explain any discrepancy between the observed behavior and theoretical expectations.

9.4.3   Concept Check

1. What is a composite quadrature rule?
2. What is the advantage of using a composite quadrature rule instead of an interpolatory rule, given the same number of nodes for both?
3. What is the advantage of building a composite quadrature rule from a closed quadrature rule instead of an open rule?
4. Write down the formulas for the following composite quadrature rules, along with their error formulas: (a) Composite Trapezoidal Rule, (b) Composite Midpoint Rule, and (c) Composite Simpson's Rule.
5. What does it mean to say that a composite quadrature rule has order of accuracy p?

9.5   Extrapolation to the Limit

We have seen that the accuracy of methods for computing integrals or derivatives of a function $f(x)$ depends on the spacing h between points at which f is evaluated, and that the approximation tends to the exact value as this spacing tends to 0.

Suppose that a uniform spacing h is used. We denote by $F(h)$ the approximation computed using the spacing h, from which it follows that the exact value is given by $F(0)$. Let p be the order of accuracy in our approximation; that is,

$$F(h) = a_0 + a_1 h^p + O(h^r), \quad r > p, \tag{9.27}$$

where a_0 is the exact value $F(0)$. Then, if we choose a value for h and compute $F(h)$ and $F(h/q)$ for some positive integer q, then we can neglect the $O(h^r)$ terms and solve a system of two equations for the unknowns a_0 and a_1, thus obtaining an approximation that is rth order accurate. If we can describe the error in this approximation in the same way that we can describe the error in our original approximation $F(h)$, we can repeat this process to obtain an approximation that is even *more* accurate.

9.5.1 Richardson Extrapolation

This process of extrapolating from $F(h)$ and $F(h/q)$ to approximate $F(0)$ with a higher order of accuracy is called **Richardson extrapolation** [Richardson (1911)]. We now illustrate how beneficial it can be for improving the accuracy of such approximations.

Example 9.5.1 Consider the function

$$f(x) = \frac{\sin^2\left(\frac{\sqrt{x^2+x}}{\cos x - x}\right)}{\sin\left(\frac{\sqrt{x}-1}{\sqrt{x^2+1}}\right)}.$$

Our goal is to compute $f'(0.25)$ as accurately and efficiently as possible. Using a centered difference approximation,

$$f'(x) = \frac{f(x+h) - f(x-h)}{2h} + O(h^2),$$

with $x = 0.25$ and $h = 0.01$, we obtain the approximation

$$f'(0.25) \approx \frac{f(0.26) - f(0.24)}{0.02} = -9.06975297890147,$$

which has absolute error 3.0×10^{-3}, and if we use $h = 0.005$, we obtain the approximation

$$f'(0.25) \approx \frac{f(0.255) - f(0.245)}{0.01} = -9.06746429492149,$$

which has absolute error 7.7×10^{-4}. As expected, the error decreases by a factor of approximately 4 when we reduce the step size h by a factor of 2, because the error in the centered difference formula is of $O(h^2)$.

We can obtain a more accurate approximation by applying *Richardson Extrapolation* to these approximations. We define the function $D_1(h)$ to be the centered difference approximation to $f'(0.25)$ obtained using the step size h. Then, with $h = 0.01$, we have

$$D_1(h) = -9.06975297890147, \quad D_1(h/2) = -9.06746429492149,$$

and the exact value is given by $D_1(0) = -9.06669877124279$. Because the error in the centered difference approximation satisfies

$$D_1(h) = D_1(0) + K_1 h^2 + K_2 h^4 + K_3 h^6 + O(h^8), \qquad (9.28)$$

where the constants K_1, K_2 and K_3 depend on the derivatives of $f(x)$ at $x = 0.25$, it follows that

$$D_1(h/2) = D_1(0) + K_1 \frac{h^2}{2^2} + K_2 \frac{h^4}{2^4} + K_3 \frac{h^6}{2^6} + O(h^8). \qquad (9.29)$$

We neglect the terms that are $O(h^4)$ in (9.28) and (9.29). Solving (9.29) for $K_1 h^2$ and substituting into the truncated (9.28) yields

$$D_1(h) = D_1(0) + 2^2(D_1(h/2) - D_1(0)).$$

Solving for the exact value $D_1(0)$, we obtain the new approximation

$$D_2(h) = D_1(h/2) + \frac{D_1(h/2) - D_1(h)}{2^2 - 1} = -9.06670140026149,$$

which has fourth-order accuracy. Specifically, if we denote the exact value by $D_2(0)$, we have

$$D_2(h) = D_2(0) + \tilde{K}_2 h^4 + \tilde{K}_3 h^6 + O(h^8),$$

where the constants $\tilde{K}_2$ and $\tilde{K}_3$ are independent of h.

Now, suppose that we compute

$$D_1(h/4) = \frac{f(x + h/4) - f(x - h/4)}{2(h/4)}$$

$$= \frac{f(0.2525) - f(0.2475)}{0.005} = -9.06689027527046,$$

which has an absolute error of 1.9×10^{-4}, we can use extrapolation again to obtain a second fourth-order accurate approximation,

$$D_2(h/2) = D_1(h/4) + \frac{D_1(h/4) - D_1(h/2)}{3} = -9.06669893538678,$$

which has absolute error of 1.7×10^{-7}. It follows from the form of the error in $D_2(h)$ that we can use extrapolation on $D_2(h)$ and $D_2(h/2)$ to obtain a *sixth*-order accurate approximation,

$$D_3(h) = D_2(h/2) + \frac{D_2(h/2) - D_2(h)}{2^4 - 1} = -9.06669877106180,$$

which has an absolute error of 1.8×10^{-10}. $\square$

Exploration 9.5.1 Use Taylor expansion to prove (9.28); that is, the error in the centered difference approximation can be expressed as a sum of terms involving only even powers of h.

Exploration 9.5.2 Based on the preceding example, give a general formula for Richardson extrapolation, applied to the approximation $F(h)$ from (9.27), that uses $F(h)$ and $F(h/q)$, for some integer $q > 1$, to obtain an approximation of $F(0)$ that is of order r.

9.5.2   The Euler-Maclaurin Expansion*

In the previous example, it was stated that the error in the centered difference approximation could be expressed as a sum of terms involving even powers of the spacing h (with the proof assigned to the reader). We would like to use Richardson Extrapolation to enhance the accuracy of approximate integrals computed using the Composite Trapezoidal Rule, but first we must determine the form of the error in these approximations. We have established that the Composite Trapezoidal Rule is second-order accurate, but if Richardson Extrapolation is used once to eliminate the $O(h^2)$ portion of the error, we do not know the order of the error that remains.

Suppose that $F(t)$ is differentiable on $(0,1)$. From integration by parts, and a judicious choice of constant of integration of $-1/2$, we obtain

$$\int_0^1 F(t)\,dt = \frac{1}{2}[F(0) + F(1)] - \int_0^1 \left(t - \frac{1}{2}\right) F'(t)\,dt.$$

The first term on the right side is the basic Trapezoidal Rule approximation of the integral on the left side. The second term on the right side is the error in this approximation. If F is $2k$-times differentiable on $(0,1)$, and we apply integration by parts $2k - 1$ times, we obtain

$$\int_0^1 F(t)\,dt = \frac{1}{2}[F(0) + F(1)] +$$

$$\left[-q_2(t)F'(t) + q_3(t)F''(t) - \cdots - q_{2k}(t)F^{(2k-1)}(t)\right]\Big|_0^1 +$$

$$\int_0^1 q_{2k}(t)F^{(2k)}(t)\,dt,$$

where the sequence of polynomials $q_1(t), \ldots, q_{2k}(t)$ satisfy

$$q_1(t) = t - \frac{1}{2}, \quad q'_{r+1}(t) = q_r(t), \quad r = 1, 2, \ldots, 2k - 1.$$

Our goal is to make this representation of the error as simple as possible. To that end, we need to determine which choice of the constant of integration for each $q_j(t)$, $j = 2, 3, \ldots, 2k$, is most advantageous.

We begin with

$$q_2(t) = \frac{1}{2}t^2 - \frac{1}{2}t + C_2.$$

We see that $q_2(0) = q_2(1) = C_2$, so regardless of the choice of C_2, we can factor $q_2(t)$ out of the boundary term $q_2(t)F'(t)|_0^1 = q_2(1)F'(t)|_0^1$. However, by integration, we obtain

$$q_3(t) = \frac{1}{6}t^3 - \frac{1}{4}t^2 + C_2 t + C_3.$$

We then have $q_3(0) = C_3$ and $q_3(1) = -1/12 + C_2 + C_3$. If we try to make $q_2(t)$ vanish at $t = 0, 1$ by setting $C_2 = 0$, then the boundary term involving $q_3(t)$ cannot be simplified in a similar manner. Therefore, we set $C_2 = 1/12$, so that $q_3(0) = q_3(1)$.

Next, we integrate again and obtain

$$q_4(t) = \frac{1}{24}t^4 - \frac{1}{12}t^3 + \frac{1}{24}t^2 + C_3 t + C_4,$$

which yields $q_4(0) = C_4$ and $q_4(1) = C_3 + C_4$. It follows that the boundary term involving $q_4(t)$ can be simplified like the term involving $q_2(t)$ only if $C_3 = 0$. Continuing in this fashion, we obtain

$$q_j(t) = \frac{1}{j!}B_j(t), \quad j = 0, 1, 2, \ldots,$$

where $B_j(t)$ is the **Bernoulli polynomial** of degree j. These polynomials have the property that $B_j(0) = B_j(1)$ for j even, and $B_j(0) = -B_j(1)$ for j odd. Furthermore, if j is odd and $j > 1$, then $B_j(0) = B_j(1) = 0$. It follows that after $2k$ applications of integration by parts, we obtain

$$\int_0^1 F(t)\,dt = \frac{1}{2}[F(0) + F(1)] - \sum_{r=1}^{k} q_{2r}(1)[F^{(2r-1)}(1) - F^{(2r-1)}(0)] +$$

$$\int_0^1 q_{2k}(t)F^{(2k)}(t)\,dt.$$

Using this expression for the error in the context of the Composite Trapezoidal Rule, applied to the integral of a $2k$-times differentiable function $f(x)$ on a general interval $[a, b]$, yields the **Euler-Maclaurin Expansion**

$$\int_a^b f(x)\,dx = \frac{h}{2}\left[f(a) + 2\sum_{i=1}^{n-1} f(x_i) + f(b)\right] - \sum_{r=1}^{k} c_r h^{2r}[f^{(2r-1)}(b) - f^{(2r-1)}(a)] +$$

$$h^{2k}\sum_{i=1}^{n}\int_{x_{i-1}}^{x_i} q_{2k}(t)f^{(2k)}(x)\,dx, \tag{9.30}$$

where, for each $i = 1, 2, \ldots, n$, $t = \frac{1}{h}(x - x_{i-1})$, and the constants

$$c_r = q_{2r}(1) = \frac{B_{2r}}{(2r)!}, \quad r = 1, 2, \ldots, k$$

are closely related to the **Bernoulli numbers** B_r.

It can be seen from this expansion that the error E_{trap} in the Composite Trapezoidal Rule, like the error in the centered difference approximation of the derivative, has the form

$$E_{\text{trap}} = K_1 h^2 + K_2 h^4 + K_3 h^6 + \cdots + O(h^{2k}), \tag{9.31}$$

where the constants K_i are independent of h, provided that the integrand is at least $2k$ times continuously differentiable. This knowledge of the error provides guidance on how Richardson Extrapolation can be repeatedly applied to approximations obtained using the Composite Trapezoidal Rule at different spacings in order to obtain higher-order accurate approximations.

It can also be seen from the Euler-Maclaurin Expansion that the Composite Trapezoidal Rule is particularly accurate when the integrand is a periodic function, of period $b - a$, as this causes the terms involving the derivatives of the integrand at a and b to vanish. Specifically, if $f(x)$ is periodic with period $b - a$, and is at least $2k$ times continuously differentiable, then the error in the Composite Trapezoidal Rule approximation to $\int_a^b f(x) \, dx$, with spacing h, is $O(h^{2k})$, rather than $O(h^2)$. It follows that if $f(x)$ is infinitely differentiable, such as a finite linear combination of sines or cosines, then the Composite Trapezoidal Rule has an *exponential* order of accuracy, meaning that as $h \to 0$, the error converges to zero more rapidly than *any* power of h.

Exploration 9.5.3 Obtain explicit formulas for the polynomials $q_r(t)$ for $r = 4, 5, 6$. Look up the Bernoulli polynomials and confirm that the relationship $q_r(t) = \frac{1}{r!} B_r(t)$ continues to hold.

Exploration 9.5.4 Use the Composite Trapezoidal Rule to integrate $f(x) = \sin k\pi x$, where k is an integer, over $[0, 1]$. How does the error behave?

9.5.3 Romberg Integration

Richardson extrapolation, presented earlier in this section, is not only used to compute more accurate approximations of derivatives, but is also used as the foundation of a numerical integration scheme called **Romberg integration** [Romberg (1955)]. In this scheme, the integral

$$I[f] = \int_a^b f(x) \, dx$$

is approximated using the Composite Trapezoidal Rule with step sizes $h_k = (b - a)2^{-k}$, $k = 0, 1, 2, \ldots$. Then, for each k, Richardson extrapolation is used $k - 1$ times with previously computed approximations to improve the order of accuracy as much as possible.

More precisely, suppose that we compute approximations $T_{1,1}$ and $T_{2,1}$ to the integral, using the Composite Trapezoidal Rule with one and two subintervals, respectively. That is,

$$T_{1,1} = \frac{b - a}{2} [f(a) + f(b)]$$

$$T_{2,1} = \frac{b - a}{4} \left[f(a) + 2f\left(\frac{a + b}{2}\right) + f(b) \right].$$

Suppose that f has continuous derivatives of all orders on $[a, b]$. Then, by the Euler-Maclaurin Expansion (9.30), the Composite Trapezoidal Rule, for a general number of subintervals n, satisfies

$$\int_a^b f(x) \, dx = \frac{h}{2} \left[f(a) + 2 \sum_{j=1}^{n-1} f(x_j) + f(b) \right] + \sum_{i=1}^{\infty} K_i h^{2i},$$

where $h = (b - a)/n$, $x_j = a + jh$, and the constants $\{K_i\}_{i=1}^{\infty}$ depend only on the derivatives of f. It follows that we can use Richardson Extrapolation to compute an approximation with a higher order of accuracy. If we denote the exact value of the integral by $I[f]$ then we have

$$T_{1,1} = I[f] + K_1 h^2 + O(h^4),$$
$$T_{2,1} = I[f] + K_1(h/2)^2 + O(h^4).$$

Neglecting the $O(h^4)$ terms, we have a system of equations that we can solve for K_1 and $I[f]$. The value of $I[f]$, which we denote by $T_{2,2}$, is an improved approximation given by

$$T_{2,2} = T_{2,1} + \frac{T_{2,1} - T_{1,1}}{3}.$$

It follows from the representation of the error in the Composite Trapezoidal Rule that $I[f] = T_{2,2} + O(h^4)$.

Suppose that we compute another approximation $T_{3,1}$ using the Composite Trapezoidal Rule with 4 subintervals. Then, as before, we can use Richardson Extrapolation with $T_{2,1}$ and $T_{3,1}$ to obtain a new approximation $T_{3,2}$ that is fourth-order accurate. Now, however, we have two approximations, $T_{2,2}$ and $T_{3,2}$, that satisfy

$$T_{2,2} = I[f] + \tilde{K}_2 h^4 + O(h^6),$$
$$T_{3,2} = I[f] + \tilde{K}_2(h/2)^4 + O(h^6),$$

for some constant $\tilde{K}_2$. It follows that we can apply Richardson Extrapolation to these approximations to obtain a new approximation $T_{3,3}$ that is *sixth-order* accurate. We can continue this process to obtain as high an order of accuracy as we wish. We now describe the entire algorithm.

Algorithm 9.5.2 (Romberg Integration) Given a positive integer J, an interval $[a, b]$ and a function $f(x)$, the following algorithm computes an approximation to $I[f] = \int_a^b f(x)\, dx$ that is accurate to order $2J$.

$h = b - a$
for $j = 1, 2, \ldots, J$ **do**
    $T_{j,1} = \frac{h}{2}\left[f(a) + 2\sum_{i=1}^{2^{j-1}-1} f(a + ih) + f(b)\right]$
    for $k = 2, 3, \ldots, j$ **do**
        $T_{j,k} = T_{j,k-1} + \frac{T_{j,k-1} - T_{j-1,k-1}}{4^{k-1} - 1}$
    end for
    $h = h/2$
end for

It should be noted that in a practical implementation, $T_{j,1}$ can be computed more efficiently by using $T_{j-1,1}$, because $T_{j-1,1}$ already includes more than half of the function values used to compute $T_{j,1}$, and they are weighted correctly relative to one another. It follows that for $j > 1$, if we split the summation in the algorithm into two summations containing odd- and even-numbered terms, respectively, we obtain

$$T_{j,1} = \frac{h}{2}\left[f(a) + 2\sum_{i=1}^{2^{j-2}} f(a + (2i-1)h) + 2\sum_{i=1}^{2^{j-2}-1} f(a + 2ih) + f(b)\right]$$

$$= \frac{h}{2}\left[f(a) + 2\sum_{i=1}^{2^{j-2}-1} f(a + 2ih) + f(b)\right] + \frac{h}{2}\left[2\sum_{i=1}^{2^{j-2}} f(a + (2i-1)h)\right]$$

$$= \frac{1}{2}T_{j-1,1} + h\sum_{i=1}^{2^{j-2}} f(a + (2i-1)h).$$

In summary, the number of functions evaluations required to compute $T_{j,1}$, for $j = 1, 2, \ldots, J$, is equal to the number of evaluations required to compute just $T_{J,1}$.

Example 9.5.3 We will use Romberg integration to obtain a sixth-order accurate approximation to

$$\int_0^1 e^{-x^2}\,dx,$$

an integral that *cannot* be evaluated using the Fundamental Theorem of Calculus. We begin by using the Trapezoidal Rule, or, equivalently, the Composite Trapezoidal Rule

$$\int_a^b f(x)\,dx \approx \frac{h}{2}\left[f(a) + \sum_{j=1}^{n-1} f(x_j) + f(b)\right], \quad h = \frac{b-a}{n}, \quad x_j = a + jh,$$

with $n = 1$ subintervals. Since $h = (b-a)/n = (1-0)/1 = 1$, we have

$$T_{1,1} = \frac{1}{2}[f(0) + f(1)] = 0.68393972058572,$$

which has an absolute error of 6.3×10^{-2}.

If we bisect the interval $[0, 1]$ into two subintervals of equal width, and approximate the area under e^{-x^2} using two trapezoids, then we are applying the Composite Trapezoidal Rule with $n = 2$ and $h = (1-0)/2 = 1/2$, which yields

$$T_{2,1} = \frac{0.5}{2}[f(0) + 2f(0.5) + f(1)] = 0.73137025182856,$$

which has an absolute error of 1.5×10^{-2}. As expected, the error is reduced by a factor of roughly 4 when the step size is halved, since the error in the Composite Trapezoidal Rule is of $O(h^2)$. Now, we can use Richardson Extrapolation to obtain a more accurate approximation,

$$T_{2,2} = T_{2,1} + \frac{T_{2,1} - T_{1,1}}{3} = 0.74718042890951,$$

which has an absolute error of 3.6×10^{-4}. Because the error in the Composite Trapezoidal Rule satisfies

$$\int_a^b f(x)\,dx = \frac{h}{2}\left[f(a) + \sum_{j=1}^{n-1} f(x_j) + f(b)\right] + K_1 h^2 + K_2 h^4 + K_3 h^6 + O(h^8),$$

where the constants K_1, K_2 and K_3 depend on the derivatives of $f(x)$ on $[a, b]$ and are independent of h, we can conclude that $T_{2,1}$ has fourth-order accuracy.

We can obtain a second approximation with fourth-order accuracy by using the Composite Trapezoidal Rule with $n = 4$ to obtain a third approximation with second-order accuracy. We set $h = (1-0)/4 = 1/4$, and then compute

$$T_{3,1} = \frac{0.25}{2}\left[f(0) + 2[f(0.25) + f(0.5) + f(0.75)] + f(1)\right] = 0.74298409780038,$$

which has an absolute error of 3.8×10^{-3}. Now, we can apply Richardson Extrapolation to $T_{2,1}$ and $T_{3,1}$ to obtain

$$T_{3,2} = T_{3,1} + \frac{T_{3,1} - T_{2,1}}{3} = 0.74685537979099,$$

which has an absolute error of 3.1×10^{-5}. This significant decrease in error from $T_{2,2}$ is to be expected, since both $T_{2,2}$ and $T_{3,2}$ have fourth-order accuracy, and $T_{3,2}$ is computed using half the step size of $T_{2,2}$.

It follows from the error term in the Composite Trapezoidal Rule, and the formula for Richardson Extrapolation, that

$$T_{2,2} = \int_0^1 e^{-x^2}\,dx + \tilde{K}_2 h^4 + O(h^6), \quad T_{3,2} = \int_0^1 e^{-x^2}\,dx + \tilde{K}_2 \left(\frac{h}{2}\right)^4 + O(h^6).$$

Therefore, we can use Richardson Extrapolation with these two approximations to obtain a new approximation

$$T_{3,3} = T_{3,2} + \frac{T_{3,2} - T_{2,2}}{2^4 - 1} = 0.74683370984975,$$

which has an absolute error of 9.6×10^{-6}. Because $T_{3,3}$ is a linear combination of $T_{3,2}$ and $T_{2,2}$ in which the terms of order h^4 cancel, we can conclude that $T_{3,3}$ is of sixth-order accuracy. $\square$

Exploration 9.5.5 Write a MATLAB function

$$\texttt{I=quadromberg(f,a,b,J)}$$

that implements Algorithm 9.5.2 for Romberg integration described in this section. Apply it to approximate the integrals

$$\int_0^1 e^x\,dx, \quad \int_0^1 \frac{1}{1+x^2}\,dx, \quad \int_0^1 \sqrt{x}\,dx.$$

How does the accuracy of the approximations improve as the number of extrapolations, J, increases? Explain any discrepancy in the observed behavior.

9.5.4   Concept Check

1. What is Richardson extrapolation?
2. What is the Euler-Maclaurin Expansion, and what does it reveal about the Composite Trapezoidal Rule?
3. What is the advantage of using the Composite Trapezoidal Rule with integrands that are periodic functions?
4. What is Romberg integration? What two techniques does it combine?
5. What is the order of accuracy of the approximation $T_{J,J}$ of $I[f]$ obtained using Algorithm 9.5.2?

9.6   Adaptive Quadrature

Composite rules can be used to implement an **automatic** quadrature procedure, in which the all of the subintervals of $[a, b]$ are continually subdivided until sufficient accuracy is achieved. However, this approach is impractical since small subintervals are not necessary in regions where the integrand is smooth.

An alternative is **adaptive quadrature** [Rice (1975)]. Adaptive quadrature is a technique in which the interval $[a, b]$ is divided into n subintervals $[a_j, b_j]$, for $j = 0, 1, \ldots, n-1$, and a quadrature rule, such as the Trapezoidal Rule or Simpson's Rule, is used on each subinterval to compute

$$I_j[f] = \int_{a_j}^{b_j} f(x)\, dx,$$

as in any composite quadrature rule. However, in adaptive quadrature, a subinterval $[a_j, b_j]$ is subdivided only if it is determined that the quadrature rule has not computed $I_j[f]$ with sufficient accuracy.

9.6.1   Error Estimation

To make this determination, we use the quadrature rule on $[a_j, b_j]$ to obtain an approximation Q_1, and then use the corresponding composite rule on $[a_j, b_j]$, with two subintervals, to compute a second approximation Q_2. If Q_1 and Q_2 are sufficiently close, then it is reasonable to conclude that these two approximations are accurate, so there is no need to subdivide $[a_j, b_j]$. Otherwise, we divide $[a_j, b_j]$ into two subintervals, and repeat this process on these subintervals. We apply this technique to all subintervals, until we can determine that the integral of f over each one has been computed with sufficient accuracy. By subdividing only when it is necessary, we avoid unnecessary computation and obtain the same accuracy as with composite rules or automatic quadrature procedures, but with much less computational effort.

How do we determine whether Q_1 and Q_2 are sufficiently close? Suppose that the composite rule has order of accuracy p. Then, Q_1 and Q_2 should satisfy

$$I_j[f] - Q_2 \approx \frac{1}{2^p}(I_j[f] - Q_1),$$

where, as above, $I_j[f]$ is the exact value of the integral over $[a_j, b_j]$. We then have

$$I_j[f] - Q_2 \approx \frac{1}{2^p}(I_j[f] - Q_2 + Q_2 - Q_1)$$

which can be rearranged to obtain

$$I_j[f] - Q_2 \approx \frac{1}{2^p - 1}(Q_2 - Q_1).$$

Thus we have obtained an *error estimate* in terms of our two approximations.

9.6.2 An Adaptive Quadrature Algorithm

We now describe an algorithm for adaptive quadrature. This algorithm uses the Trapezoidal Rule to integrate over intervals, and intervals are subdivided as necessary into two subintervals of equal width. The algorithm uses a data structure called a **stack** in order to keep track of the subintervals over which f still needs to be integrated. A stack is essentially a list of elements, where the elements, in this case, are subintervals. An element is added to the stack using a **push** operation, and is removed using a **pop** operation. Working with a stack is often described using the phrase "last-in-first-out," because the most recent element to be pushed onto the stack is the first element to be popped. This corresponds to our intuitive notion of a stack of objects, in which objects are placed on top of the stack and are removed from the top as well.

Algorithm 9.6.1 (Adaptive Quadrature) Given a function $f(x)$ that is integrable on an interval $[a, b]$, the following algorithm computes an approximation I to $I[f] = \int_a^b f(x)\, dx$ that is accurate to within $(b - a)TOL$, where TOL is a given error tolerance.

S is an empty stack
$push(S, [a, b])$
$I = 0$
while S is not empty **do**
    $[a, b] = pop(S)$ (the interval $[a, b]$ on top of S is removed from S)
    $f_a = f(a), f_b = f(b)$
    $Q_1 = (b - a)/2[f_a + f_b]$ (Trapezoidal Rule)
    $m = (a + b)/2$
    $Q_2 = (b - a)/4[f_a + 2f(m) + f_b]$ (Composite Trapezoidal Rule)
    if $|Q_1 - Q_2| < 3(b - a)TOL$ **then**
        $I = I + Q_2$
    else
        $push(S, [a, m])$
        $push(S, [m, b])$
    end if
end while

Throughout the execution of the loop in the above algorithm, the stack S contains all intervals over which f still needs to be integrated to within the desired accuracy. Initially, the only such interval is the original interval $[a, b]$. As long as intervals remain on the stack, the interval on top of the stack is removed, and we attempt to integrate over it. If we obtain a sufficiently accurate result, then we are finished with the interval. Otherwise, the interval is bisected into two subintervals, both of which are pushed on the stack so that they can be processed later. Once the stack is empty, we know that we have accurately integrated f over a collection of intervals whose union is the original interval $[a, b]$, so the algorithm can terminate.

Example 9.6.2 We will use *adaptive quadrature* to compute the integral

$$\int_0^{\pi/4} e^{3x} \sin 2x \, dx$$

to within $(\pi/4)10^{-4}$. Let $f(x) = e^{3x} \sin 2x$ denote the integrand. First, we use Simpson's Rule, or, equivalently, the Composite Simpson's Rule with $n = 2$ subintervals, to obtain an approximation Q_1 to this integral. We have

$$Q_1 = \frac{\pi/4}{6}[f(0) + 4f(\pi/8) + f(\pi/4)] = 2.58369640324748.$$

Then, we divide the interval $[0, \pi/4]$ into two subintervals of equal width, $[0, \pi/8]$ and $[\pi/8, \pi/4]$, and integrate over each one using Simpson's Rule to obtain a second approximation Q_2. This is equivalent to using the Composite Simpson's Rule on $[0, \pi/4]$ with $n = 4$ subintervals. We obtain

$$Q_2 = \frac{\pi/8}{6}[f(0) + 4f(\pi/16) + f(\pi/8)] + \frac{\pi/8}{6}[f(\pi/8) + 4f(3\pi/16) + f(\pi/4)]$$

$$= \frac{\pi/16}{3}[f(0) + 4f(\pi/16) + 2f(\pi/8) + 4f(3\pi/16) + f(\pi/4)]$$

$$= 2.58770145345862.$$

Now, we need to determine whether the approximation Q_2 is sufficiently accurate. Because the error in the Composite Simpson's Rule is $O(h^4)$, where h is the width of each subinterval used in the rule, it follows that the actual error in Q_2 satisfies

$$|Q_2 - I[f]| \approx \frac{1}{15}|Q_2 - Q_1|,$$

where $I[f]$ is the exact value of the integral of f. We find that the relation

$$|Q_2 - I[f]| \approx \frac{1}{15}|Q_2 - Q_1| < \frac{\pi}{4}10^{-4}$$

is *not* satisfied, so we must divide the interval $[0, \pi/4]$ into two subintervals of equal width, $[0, \pi/8]$ and $[\pi/8, \pi/4]$, and use the Composite Simpson's Rule with these smaller intervals in order to achieve the desired accuracy.

First, we work with the interval $[0, \pi/8]$. Proceeding as before, we use the Composite Simpson's Rule with $n = 2$ and $n = 4$ subintervals to obtain approximations Q_1 and Q_2 to the integral of $f(x)$ over this interval. We have

$$Q_1 = \frac{\pi/8}{6}[f(0) + 4f(\pi/16) + f(\pi/8)] = 0.33088926959519.$$

and

$$Q_2 = \frac{\pi/16}{6}[f(0) + 4f(\pi/32) + f(\pi/16)] + \frac{\pi/16}{6}[f(\pi/16) + 4f(3\pi/32) + f(\pi/8)]$$

$$= \frac{\pi/32}{3}[f(0) + 4f(\pi/32) + 2f(\pi/16) + 4f(3\pi/32) + f(\pi/8)]$$

$$= 0.33054510467064.$$

Since these approximations satisfy the relation

$$|Q_2 - I[f]| \approx \frac{1}{15}|Q_2 - Q_1| < \frac{\pi}{8}10^{-4},$$

where $I[f]$ denotes the exact value of the integral of f over $[0, \pi/8]$, we have achieved sufficient accuracy on this interval and we do not need to subdivide it further. The more accurate approximation Q_2 can be included in our approximation to the integral over the original interval $[0, \pi/4]$.

Now, we need to achieve sufficient accuracy on the remaining subinterval, which is $[\pi/4, \pi/8]$. As before, we compute the approximations Q_1 and Q_2 of the integral of f over this interval and obtain

$$Q_1 = \frac{\pi/8}{6}[f(\pi/8) + 4f(3\pi/16) + f(\pi/4)] = 2.25681218386343.$$

and

$$Q_2 = \frac{\pi/16}{6}[f(\pi/8) + 4f(5\pi/32) + f(3\pi/16)] +$$

$$\frac{\pi/16}{6}[f(3\pi/16) + 4f(7\pi/32) + f(\pi/4)]$$

$$= \frac{\pi/32}{3}[f(\pi/8) + 4f(5\pi/32) + 2f(3\pi/16) + 4f(7\pi/32) + f(\pi/4)]$$

$$= 2.25801455892266.$$

Since these approximations do not satisfy the relation

$$|Q_2 - I[f]| \approx \frac{1}{15}|Q_2 - Q_1| < \frac{\pi}{8}10^{-4},$$

where $I[f]$ denotes the exact value of the integral of f over $[\pi/8, \pi/4]$, we have not achieved sufficient accuracy on this interval and we need to subdivide it into two subintervals of equal width, $[\pi/8, 3\pi/16]$ and $[3\pi/16, \pi/4]$, and use the Composite Simpson's Rule with these smaller intervals in order to achieve the desired accuracy. The discrepancy in these two approximations to the integral of f over $[\pi/8, \pi/4]$ is larger than the discrepancy in the two approximations of the integral over $[0, \pi/8]$ because even though these intervals have the same width, the derivatives of f are larger on $[\pi/8, \pi/4]$, and therefore the error in the Composite Simpson's Rule is larger.

We continue the process of adaptive quadrature on the interval $[\pi/8, 3\pi/16]$. As before, we compute the approximations Q_1 and Q_2 of the integral of f over this interval and obtain

$$Q_1 = \frac{\pi/16}{6}[f(\pi/8) + 4f(5\pi/32) + f(3\pi/16)] = 0.72676545197054.$$

and

$$Q_2 = \frac{\pi/32}{6}[f(\pi/8) + 4f(9\pi/64) + f(5\pi/32)] +$$

$$\frac{\pi/32}{6}[f(5\pi/32) + 4f(11\pi/64) + f(3\pi/16)]$$

$$= \frac{\pi/64}{3}[f(\pi/8) + 4f(9\pi/64) + 2f(5\pi/32) + 4f(11\pi/64) + f(3\pi/16)]$$

$$= 0.72677918153379.$$

Since these approximations satisfy the relation

$$|Q_2 - I[f]| \approx \frac{1}{15}|Q_2 - Q_1| < \frac{\pi}{16}10^{-4},$$

where $I[f]$ denotes the exact value of the integral of f over $[\pi/8, 3\pi/16]$, we have achieved sufficient accuracy on this interval and we do not need to subdivide it further. The more accurate approximation Q_2 can be included in our approximation to the integral over the original interval $[0, \pi/4]$.

Now, we work with the interval $[3\pi/16, \pi/4]$. Proceeding as before, we use the Composite Simpson's Rule with $n = 2$ and $n = 4$ subintervals to obtain approximations Q_1 and Q_2 to the integral of $f(x)$ over this interval. We have

$$Q_1 = \frac{\pi/16}{6}[f(3\pi/16) + 4f(7\pi/32) + f(\pi/4)] = 1.53124910695212.$$

and

$$Q_2 = \frac{\pi/32}{6}[f(3\pi/16) + 4f(13\pi/64) + f(7\pi/32)] +$$

$$\frac{\pi/32}{6}[f(7\pi/32) + 4f(15\pi/64) + f(\pi/4)]$$

$$= \frac{\pi/64}{3}[f(3\pi/16) + 4f(13\pi/64) + 2f(7\pi/32) + 4f(15\pi/64) + f(\pi/4)]$$

$$= 1.53131941583939.$$

Since these approximations satisfy the relation

$$|Q_2 - I[f]| \approx \frac{1}{15}|Q_2 - Q_1| < \frac{\pi}{16}10^{-4},$$

where $I[f]$ denotes the exact value of the integral of f over $[3\pi/16, \pi/4]$, we have achieved sufficient accuracy on this interval and we do not need to subdivide it further. The more accurate approximation Q_2 can be included in our approximation to the integral over the original interval $[0, \pi/4]$.

We conclude that the integral of $f(x)$ over $[0, \pi/4]$ can be approximated by the sum of our approximate integrals over $[0, \pi/8]$, $[\pi/8, 3\pi/16]$, and $[3\pi/16, \pi/4]$, which yields

$$\int_0^{\pi/4} e^{3x} \sin 2x \, dx \approx 0.33054510467064 + 0.72677918153379 + 1.53131941583939$$

$$\approx 2.58864370204382.$$

Since the exact value is 2.58862863250716, the absolute error is -1.507×10^{-5}, which is less in magnitude than our desired error bound of $(\pi/4)10^{-4} \approx 7.854 \times 10^{-5}$. This is because on each subinterval, we ensured that our approximation was accurate to within 10^{-4} times the width of the subinterval, so that when we added these approximations, the total error in the sum would be bounded by 10^{-4} times the width of the union of these subintervals, which is the original interval $[0, \pi/4]$. The graph of the integrand over the interval of integration is shown in Figure 9.2. Note that the nodes used during the overall process are not equally spaced, as more nodes were needed in the right half of the interval to achieve sufficient accuracy. $\square$

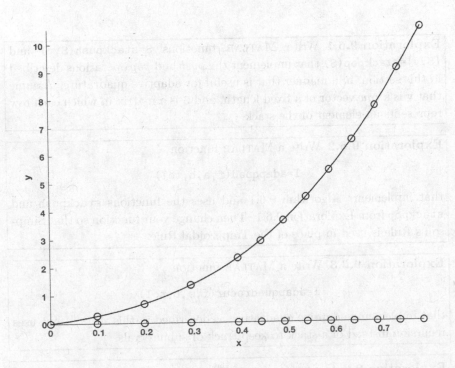

Fig. 9.2   Graph of $f(x) = e^{3x} \sin 2x$ on $[0, \pi/4]$, with quadrature nodes from Example 9.6.2 shown on the graph and on the x-axis.

9.6.3   Implementation Issues

Adaptive quadrature can be very effective, but it should be used cautiously, for the following reasons:

- The integrand is only evaluated at a few points within each subinterval. Such sampling can miss a portion of the integrand whose contribution to the integral can be misjudged.
- Regions in which the function is not smooth will still only make a small contribution to the integral if the region itself is small, so this should be taken into account to avoid unnecessary function evaluations.
- Adaptive quadrature can be very inefficient if the integrand has a discontinuity within a subinterval, since repeated subdivision will occur. This is unnecessary if the integrand is smooth on either side of the discontinuity, so subintervals should be chosen so that discontinuities occur between subintervals whenever possible.

Exploration 9.6.1 Write MATLAB functions `S=stackpush(S,v)` and `[S,v]=stackpop(S)` that implement the *push* and *pop* operations described in this section, in a manner that is useful for adaptive quadrature. Assume that `v` is a *row* vector of a fixed length, and `S` is a matrix in which each row represents an element of the stack.

Exploration 9.6.2 Write a MATLAB function

$$I=\text{adapquad}(f,a,b,\text{tol})$$

that implements Algorithm 9.6.1 and uses the functions `stackpush` and `stackpop` from Exploration 9.6.1. Then change your function so that Simpson's Rule is used in place of the Trapezoidal Rule.

Exploration 9.6.3 Write a MATLAB function

$$I=\text{adapquadrecur}(f,a,b,\text{tol})$$

that implements adaptive quadrature as described in this section, but uses recursion instead of a stack to keep track of subintervals.

Exploration 9.6.4 Let $f(x) = e^{-1000(x-c)^2}$, where c is a parameter. Use your function `adapquadrecur` from Exploration 9.6.3 to approximate $\int_0^1 f(x)\,dx$ for the cases $c = 1/8$ and $c = 1/4$. Explain the difference in performance between these two cases.

Exploration 9.6.5 Explain why a straightforward implementation of adaptive quadrature using recursion, as in your function `adapquadrecur` from Exploration 9.6.3, is not as efficient as it could be. What can be done to make it more efficient? Modify your implementation accordingly.

9.6.4 Concept Check

1. What is an automatic quadrature procedure?
2. What is the drawback of automatic quadrature?
3. What is adaptive quadrature?
4. How is the error estimated in adaptive quadrature?
5. What are the drawbacks of adaptive quadrature?

9.7 Gauss Quadrature

Previously, we learned that a Newton-Cotes quadrature rule with n nodes can have degree n, while a general n-node interpolatory quadrature rule has degree at least $n - 1$. Therefore, it is natural to ask whether it is possible to select the nodes and weights of an n-node interpolatory quadrature rule so that the rule has degree greater than n. **Gauss quadrature rules** [Abramowitz and Stegun (1972)] have the surprising property that they can be used to integrate polynomials of degree $2n - 1$ exactly using only n nodes.

9.7.1 Direct Construction

Gauss quadrature rules can be constructed using a technique known as **moment matching**, or **direct construction**. For any nonnegative integer k, the kth *moment* is defined to be

$$\mu_k = \int_a^b x^k \, dx.$$

For given n, our goal is to select weights and nodes so that the first $2n$ moments are computed exactly; i.e.,

$$\mu_k = \sum_{i=1}^n w_i x_i^k, \quad k = 0, 1, \ldots, 2n - 1. \tag{9.32}$$

Since we have $2n$ free parameters, it is reasonable to think that suitable nodes and weights can be found. Unfortunately, this system of equations is nonlinear, so it can be quite difficult to solve.

Exploration 9.7.1 Use direct construction to solve the equations (9.32) for the case of $n = 2$ on the interval $(a, b) = (-1, 1)$ for the nodes x_1, x_2 and weights w_1, w_2.

9.7.2 Orthogonal Polynomials

Suppose $g(x)$ is a polynomial of degree at most $2n-1$. For convenience, we will write $g \in \mathcal{P}_{2n-1}$, where, for any natural number k, $\mathcal{P}_k$ denotes the space of polynomials of degree at most k. We shall obtain nodes $\{x_i\}_{i=1}^n$ and weights $\{w_i\}_{i=1}^n$ such that

$$I[g] = \int_a^b g(x)\,dx = \sum_{i=1}^n \dot{w}_i g(x_i).$$

That is, the quadrature rule is *exact* for such integrands, because the error is zero.

To accomplish this, we shall construct a family of **orthogonal polynomials** $\{q_i(x)\}_{i=0}^n$, as in Section 8.2, so that

$$\langle q_r, q_s \rangle = \int_a^b q_r(x)q_s(x)\,dx = \begin{cases} 0 & r \neq s, \\ 1 & r = s. \end{cases}$$

We then choose the nodes $\{x_i\}_{i=1}^n$ to be the roots of the nth-degree polynomial $q_n(x)$ in this family, which are real, distinct and lie within (a,b), as proved in Section 8.2.6. Next, we construct the interpolant of degree $n-1$, denoted $p_{n-1}(x)$, of $g(x)$ through the nodes:

$$p_{n-1}(x) = \sum_{i=1}^n g(x_i)\mathcal{L}_{n-1,i}(x),$$

where, for $i = 1, \ldots, n$, $\mathcal{L}_{n-1,i}(x)$ is the ith Lagrange polynomial (7.3) for the points $x_1, \ldots, x_n$.

We shall now look at the interpolation error function

$$e(x) = g(x) - p_{n-1}(x).$$

Clearly, since $g \in \mathcal{P}_{2n-1}$, $e \in \mathcal{P}_{2n-1}$. Since $p_{n-1}(x)$ interpolates $g(x)$ at the nodes, which are the roots of $q_n(x)$, $e(x)$ has roots at each of these points. It follows that we can factor $e(x)$ so that

$$e(x) = q_n(x)r(x),$$

where $r \in \mathcal{P}_{n-1}$. It follows from the fact that $q_n(x)$ is orthogonal to *any* polynomial in $\mathcal{P}_{n-1}$ that the integral of g can then be written as

$$\begin{aligned}
I[g] &= \int_a^b p_{n-1}(x)\,dx + \int_a^b q_n(x)r(x)\,dx \\
&= \int_a^b p_{n-1}(x)\,dx + \langle q_n, r \rangle \\
&= \int_a^b \sum_{i=1}^n g(x_i)\mathcal{L}_{n-1,i}(x)\,dx + 0 \\
&= \sum_{i=1}^n g(x_i) \int_a^b \mathcal{L}_{n-1,i}(x)\,dx \\
&= \sum_{i=1}^n g(x_i)w_i
\end{aligned}$$

where

$$w_i = \int_a^b \mathcal{L}_{n-1,i}(x)\, dx, \quad i = 1, 2, \ldots, n. \tag{9.33}$$

We conclude that this Gauss quadrature rule is exact for all $g \in \mathcal{P}_{2n-1}$. For a more general function $f(x)$, the error in the Gauss quadrature approximation of $I[f]$ can be obtained from the expression for Hermite interpolation error presented in Section 7.5.1, as we will investigate later in this section.

While it is more difficult to obtain the nodes and weights for Gauss quadrature than for other quadrature rules we have seen, it is not necessary to go to the trouble of actually computing them, for small values of n. Tables of nodes and weights for generic intervals, such as $[-1, 1]$, can be looked up in several books or online. These nodes and weights can then be transformed for a given interval $[a, b]$, as illustrated in the following example.

Example 9.7.1 We will use Gauss quadrature to approximate the integral

$$\int_0^1 e^{-x^2}\, dx.$$

The particular Gauss quadrature rule that we will use consists of five nodes x_1, x_2, x_3, x_4 and x_5, and five weights w_1, w_2, w_3, w_4 and w_5. To determine the proper nodes and weights, we use the fact that the nodes and weights of a 5-point Gauss rule for integrating over the interval $[-1, 1]$ are given by

i	Nodes $r_{5,i}$	Weights $c_{5,i}$
1	0.9061798459	0.2369268850
2	0.5384693101	0.4786286705
3	0.0000000000	0.5688888889
4	−0.5384693101	0.4786286705
5	−0.9061798459	0.2369268850

To obtain the corresponding nodes and weights for integrating over $[0, 1]$, we can use the fact that in general,

$$\int_a^b f(x)\, dx = \int_{-1}^1 f\left(\frac{b-a}{2}t + \frac{a+b}{2}\right) \frac{b-a}{2}\, dt,$$

as can be shown using the change of variable $x = [(b-a)/2]t + (a+b)/2$ that maps $[a, b]$ into $[-1, 1]$. We then have

$$\int_a^b f(x)\, dx = \int_{-1}^1 f\left(\frac{b-a}{2}t + \frac{a+b}{2}\right) \frac{b-a}{2}\, dt$$

$$\approx \sum_{i=1}^5 f\left(\frac{b-a}{2}r_{5,i} + \frac{a+b}{2}\right) \frac{b-a}{2} c_{5,i}$$

$$\approx \sum_{i=1}^5 f(x_i) w_i,$$

where
$$x_i = \frac{b-a}{2} r_{5,i} + \frac{a+b}{2}, \quad w_i = \frac{b-a}{2} c_{5,i}, \quad i = 1, \ldots, 5.$$
In this example, $a = 0$ and $b = 1$, so the nodes and weights for a 5-point Gauss quadrature rule for integrating over $[0, 1]$ are given by
$$x_i = \frac{1}{2} r_{5,i} + \frac{1}{2}, \quad w_i = \frac{1}{2} c_{5,i}, \quad i = 1, \ldots, 5,$$
which yields

i	Nodes x_i	Weights w_i
1	0.95308992295	0.11846344250
2	0.76923465505	0.23931433525
3	0.50000000000	0.28444444444
4	0.23076534495	0.23931433525
5	0.04691007705	0.11846344250

It follows that
$$\int_0^1 e^{-x^2} \, dx \approx \sum_{i=1}^5 e^{-x_i^2} w_i$$

$$\approx 0.11846344250 e^{-0.95308992295^2} + 0.23931433525 e^{-0.76923465505^2} +$$
$$0.28444444444 e^{-0.5^2} + 0.23931433525 e^{-0.23076534495^2} +$$
$$0.11846344250 e^{-0.04691007705^2}$$

$$\approx 0.74682412673352.$$

Since the exact value is 0.74682413281243, the absolute error is -6.08×10^{-9}, which is remarkably accurate considering that only fives nodes are used. $\square$

The high degree of accuracy of Gauss quadrature rules make them the most commonly used rules in practice. However, they are not without their drawbacks:

- They are not *progressive*, which means that the nodes must be recomputed whenever additional degrees of accuracy are desired. An alternative is to use **Gauss-Kronrod rules** [Laurie (1997)]. A $(2n + 1)$-point Gauss-Kronrod rule uses the nodes of the n-point Gauss rule. For this reason, practical quadrature procedures use both the Gauss rule and the corresponding Gauss-Kronrod rule to estimate accuracy.

- Because the nodes are the roots of a polynomial, if they are not published in a table as in the preceding example, they must be computed using traditional root-finding methods (such as those in Chapter 10, or by computing the eigenvalues of the polynomial's **companion matrix** as described in Section 6.1.1), which are not always reliable. Errors in the computed nodes lead to lost degrees of accuracy in the approximate integral. In practice, however, this does not normally cause significant difficulty, as the error in a quadrature rule is generally a continuous function of the nodes and weights.

It is worth noting that the Gauss quadrature nodes on $(-1, 1)$ are the roots of the **Legendre polynomials**, which were introduced in Section 8.2.3. This will be used in the following exploration.

Exploration 9.7.2 Write a MATLAB function

$$\texttt{I=gaussquadrule(f,a,b,n)}$$

that approximate the integral of $f(x)$, implemented by the function handle $\texttt{f}$, over $[a, b]$ with a n-node Gauss quadrature rule. Use your function $\texttt{interpquad}$ from Exploration 9.2.4 as well as your function $\texttt{makelegendre}$ from Section 8.2. Test your function by comparing its output to that of the built-in function $\texttt{integral}$. How does its accuracy compare to that of your function $\texttt{quadnewtoncotes}$ from Exploration 9.3.2?

Exploration 9.7.3 A 5-node Gauss quadrature rule is exact for the integrand $f(x) = x^8$, while a 5-node Newton-Cotes rule is not. How important is it that the Gauss quadrature nodes be computed with high accuracy? Investigate this by approximating $\int_{-1}^{1} x^8 \, dx$ using a 5-node interpolatory quadrature rule with nodes

$$x_i = \theta \tilde{x}_i + (1 - \theta)\hat{x}_i,$$

where $\{\tilde{x}_i\}_{i=1}^5$ and $\{\hat{x}_i\}_{i=1}^5$ are the nodes for a 5-node Gauss and Newton-Cotes rule, respectively, and $\theta \in [0, 1]$. Use your function $\texttt{interpquad}$ from Exploration 9.2.4 and let θ vary from 0 to 1. How does the error behave as θ increases?

9.7.3 Error Analysis

It is easy to show that the weights w_i are positive. Since the interpolation basis functions $\mathcal{L}_{n-1,i}$ belong to $\mathcal{P}_{n-1}$, it follows that $\mathcal{L}_{n-1,i}^2 \in \mathcal{P}_{2n-2}$, and therefore

$$0 < \int_a^b \mathcal{L}_{n-1,i}^2(x) \, dx = \sum_{j=0}^{n-1} w_j \mathcal{L}_{n-1,i}^2(x_j) = w_i.$$

Note that we have thus obtained an alternative formula to (9.33) for the weights.

This formula also arises from an alternative approach to constructing Gauss quadrature rules, from which a representation of the error can easily be obtained. We construct the **Hermite interpolating polynomial** $H_{2n-1}(x)$ of our integrand $f(x)$, using the Gauss quadrature nodes as interpolation points, that satisfies the $2n$ conditions

$$H_{2n-1}(x_i) = f(x_i), \quad H'_{2n-1}(x_i) = f'(x_i), \quad i = 1, 2, \ldots, n.$$

We recall from Section 7.5.1 that this interpolant has the form

$$H_{2n-1}(x) = \sum_{i=1}^n f(x_i)\mathcal{H}_{2n-1,i}(x) + \sum_{i=1}^n f'(x_i)\mathcal{K}_{2n-1,i}(x),$$

where, as in our previous discussion of Hermite interpolation,

$$\mathcal{H}_{2n-1,i}(x_j) = \delta_{ij}, \quad \mathcal{H}'_{2n-1,i}(x_j) = 0,$$

$$\mathcal{K}_{2n-1,i}(x_j) = 0, \quad \mathcal{K}'_{2n-1,i}(x_j) = \delta_{ij},$$

for $i, j = 1, 2, \ldots, n$. Then, we have

$$\int_a^b H_{2n-1}(x)\, dx = \sum_{i=1}^n f(x_i) \int_a^b \mathcal{H}_{2n-1,i}(x)\, dx + \sum_{i=1}^n f'(x_i) \int_a^b \mathcal{K}_{2n-1,i}(x)\, dx.$$

We recall from Section 7.5.1 that

$$\mathcal{H}_{2n-1,i}(x) = \mathcal{L}_{n-1,i}(x)^2[1 - 2\mathcal{L}'_{n-1,i}(x_i)(x - x_i)],$$

$$\mathcal{K}_{2n-1,i}(x) = \mathcal{L}_{n-1,i}(x)^2(x - x_i), \quad i = 1, 2, \ldots, n,$$

and for convenience, we define

$$\pi_n(x) = (x - x_1)(x - x_2) \cdots (x - x_n), \tag{9.34}$$

and recall the result of Exercise 6 from Chapter 7,

$$\mathcal{L}_{n-1,i}(x) = \frac{\pi_n(x)}{(x - x_i)\pi'_n(x_i)}, \quad i = 1, 2, \ldots, n. \tag{9.35}$$

Exploration 9.7.4 Use (9.35) and the fact that $\pi_n(x)$ from (9.34) is orthogonal to all polynomials of lesser degree to show that

$$\int_a^b \mathcal{H}_{2n-1,i}(x)\, dx = \int_a^b \mathcal{L}_{n-1,i}(x)^2\, dx,$$

$$\int_a^b \mathcal{K}_{2n-1,i}(x)\, dx = 0.$$

We conclude that

$$\int_a^b H_{2n-1}(x)\, dx = \sum_{i=1}^n f(x_i)w_i,$$

where, as before,

$$w_i = \int_a^b \mathcal{L}_{n-1,i}(x)^2\, dx = \int_a^b \mathcal{L}_{n-1,i}(x)\, dx, \quad i = 1, 2, \ldots, n.$$

We can then use the error in the Hermite interpolating polynomial from Theorem 7.5.1 to obtain the following result.

Theorem 9.7.2 (Gauss Quadrature Error) Let f be $2n + 2$ times continuously differentiable on $[a, b]$, and let $x_1, x_2, \ldots, x_n$ and $w_1, w_2, \ldots, w_n$ be the nodes and weights, respectively, of a Gauss quadrature rule on $[a, b]$. Then

$$I[f] = \int_a^b f(x)\, dx = \sum_{i=1}^n f(x_i)w_i + \frac{f^{(2n)}(\xi)}{(2n)!} \int_a^b \prod_{i=1}^n (x - x_i)^2\, dx,$$

where $\xi \in (a, b)$.

Exploration 9.7.5 Prove Theorem 9.7.2. *Hint:* The Weighted Mean Value Theorem for Integrals (Theorem A.5.5) will be helpful.

In addition to this error formula, we can easily obtain qualitative bounds on the error. For instance, if we know that the even derivatives of f are positive, then we know that the quadrature rule yields a lower bound for $I[f]$. Similarly, if the even derivatives of f are negative, then the quadrature rule gives an upper bound.

Finally, it can be shown that as $n \to \infty$, the n-node Gauss quadrature approximation of $I[f]$ converges to $I[f]$. The key to the proof is the fact that the weights are guaranteed to be positive. The following classical theorem is also helpful.

Theorem 9.7.3 (Weierstrass Approximation Theorem) Let $f(x)$ be continuous on $[a, b]$. Then, for any $\epsilon > 0$, there exists a polynomial $p(x)$ such that $\max_{a \leq x \leq b} |f(x) - p(x)| < \epsilon$.

Exploration 9.7.6 Use the Weierstrass Approximation Theorem, to prove the following: let $f(x)$ be continuous on $[a, b]$, and let $Q_n[f]$ be the approximation of $I[f] = \int_a^b f(x)\, dx$ by a n-node Gauss quadrature rule. Then

$$\lim_{n \to \infty} Q_n[f] = I[f].$$

Hint: Use Definition A.1.4, in conjunction with Theorem 9.7.3 to decompose the error $|Q_n[f] - I[f]|$ into multiple terms that are easier to bound.

Such a result does not hold for Newton-Cotes rules, because the sum of the absolute values of the weights cannot be bounded, due to the presence of negative weights.

9.7.4 Other Weight Functions

In Section 8.2 we learned how to construct sequences of orthogonal polynomials for the inner product

$$\langle f, g \rangle = \int_a^b f(x)g(x)w(x)\, dx,$$

where f and g are real-valued functions on (a, b) and $w(x)$ is a **weight function** satisfying $w(x) \geq 0$ on (a, b), and $w(x) \neq 0$ on any subinterval of (a, b). These orthogonal polynomials can be used to construct Gauss quadrature rules for integrals with weight functions, in a similar manner to how they were constructed earlier in this section for the case $w(x) \equiv 1$.

Exploration 9.7.7 Let $w(x)$ be a weight function. Derive a Gauss quadrature rule of the form

$$\int_a^b f(x)w(x)\, dx = \sum_{i=1}^{n} f(x_i)w_i + \text{error}$$

that is exact for $f \in \mathcal{P}_{2n-1}$. What is the error formula?

Of particular interest is the interval $(-1, 1)$ with the weight function $w(x) = 1/\sqrt{1 - x^2}$, as the orthogonal polynomials for the corresponding inner product are the **Chebyshev polynomials**, introduced in Section 7.4.2. Unlike other Gauss quadrature rules, there are simple formulas for the nodes and weights in this case.

Exploration 9.7.8 Use trigonometric identities to prove that

$$\cos\theta + \cos 3\theta + \cos 5\theta + \cdots + \cos(2n - 1)\theta = \frac{\sin 2n\theta}{2\sin\theta},$$

when θ is not an integer multiple of π.

Exploration 9.7.9 Use the result of Exploration 9.7.8 and direct construction to derive the nodes and weights for an n-node Gauss quadrature rule of the form

$$\int_{-1}^{1} (1 - x^2)^{-1/2} f(x)\, dx = \sum_{i=1}^{n} f(x_i) w_i,$$

that is exact when $f \in \mathcal{P}_{2n-1}$. Specifically, require that this quadrature rule is exact for the Chebyshev polynomials up to degree $2n - 1$. That is,

$$\sum_{i=1}^{n} T_k(x_i) w_i = \int_{-1}^{1} (1 - x^2)^{-1/2} T_k(x)\, dx = \begin{cases} \pi & k = 0 \\ 0 & k \neq 0 \end{cases},$$

for $k = 0, 1, 2, \ldots, 2n - 1$. *Hint:* Use the fact that, from Section 8.2.4, the Chebyshev polynomials are orthogonal polynomials on $(-1, 1)$ with respect to the weight function $(1 - x^2)^{-1/2}$. What does this tell you about what the Gauss quadrature nodes should be?

It is worth noting that this weight function for Chebyshev polynomials is singular at the endpoints of the interval $(-1, 1)$. This suggests an approach for approximating certain *improper integrals* in which the integrand is singular at one or both endpoints of the interval of integration (a, b). A portion of the integrand that does not change sign on (a, b) can be treated as a weight function, so that Gauss quadrature rules can still be constructed, as long as integrals of polynomials with respect to this weight function exist.

9.7.5   Prescribing Nodes*

We have seen that for an integrand $f(x)$ that has even derivatives that do not change sign on (a, b), it can be determined that the Gauss quadrature approximation of $I[f]$ is either an upper bound or a lower bound. By prescribing either or both of the endpoints $x = a$ or $x = b$ as quadrature nodes, we can obtain additional bounds and *bracket* the exact value of $I[f]$. However, it is important to prescribe such nodes in a manner that, as much as possible, maintains the high degree of accuracy of the quadrature rule.

A Gauss quadrature rule with $n + 1$ nodes is exact for any integrand in $\mathcal{P}_{2n+1}$.

Our goal is to construct a quadrature rule with $n + 1$ nodes, one of which is at $x = a$, that computes

$$I[f] = \int_a^b f(x)w(x)\,dx,$$

for a given weight function $w(x)$, exactly when $f \in \mathcal{P}_{2n}$. That is, prescribing a node reduces the degree of accuracy by only one. Such a quadrature rule is called a **Gauss-Radau quadrature rule** [Abramowitz and Stegun (1972)].

We begin by dividing $f(x)$ by $(x - a)$, which yields

$$f(x) = (x - a)q_{2n-1}(x) + f(a).$$

We then construct a n-node Gauss quadrature rule

$$\int_a^b g(x)\tilde{w}(x)\,dx = \sum_{i=1}^n g(\tilde{x}_i)\tilde{w}_i + \text{error}$$

for the weight function $\tilde{w}(x) = (x - a)w(x)$. It is clear that $\tilde{w}(x) > 0$ on (a, b). Because this rule is exact for $g \in \mathcal{P}_{2n-1}$, we have

$$
\begin{aligned}
I[f] &= \int_a^b q_{2n-1}(x)\tilde{w}(x)\,dx + f(a)\int_a^b w(x)\,dx \\
&= \sum_{i=1}^n q_{2n-1}(\tilde{x}_i)\tilde{w}_i + f(a)\int_a^b w(x)\,dx \\
&= \sum_{i=1}^n \frac{f(\tilde{x}_i) - f(a)}{\tilde{x}_i - a}\tilde{w}_i + f(a)\int_a^b w(x)\,dx \\
&= \sum_{i=1}^n f(\tilde{x}_i)w_i + f(a)\left[\int_a^b w(x)\,dx - \sum_{i=1}^n w_i\right],
\end{aligned}
$$

where $w_i = \tilde{w}_i/(\tilde{x}_i - a)$ for $i = 1, 2, \ldots, n$.

By defining

$$w_0 = \int_a^b w(x)\,dx - \sum_{i=1}^n w_i,$$

and defining the nodes

$$x_0 = a, \quad x_i = \tilde{x}_i, \quad i = 1, 2, \ldots, n,$$

we obtain a quadrature rule

$$I[f] = \sum_{i=0}^n f(x_i)w_i + \text{error}$$

that is exact for $f \in \mathcal{P}_{2n}$. Clearly the weights $w_1, w_2, \ldots, w_n$ are positive; it can be shown that $w_0 > 0$ by noting that it is the error in the Gauss quadrature approximation of

$$\tilde{I}\left[\frac{1}{x - a}\right] = \int_a^b \frac{1}{x - a}\tilde{w}(x)\,dx.$$

It can also be shown that if the integrand $f(x)$ is sufficiently differentiable and satisfies $f^{(2n+1)} < 0$ on (a, b), then this Gauss-Radau rule yields an upper bound for $I[f]$.

Exploration 9.7.10 Following the discussion above, derive a Gauss-Radau quadrature rule in which a node is prescribed at $x = b$. Prove that the weights $w_1, w_2, \ldots, w_{n+1}$ are positive. Does this rule yield an upper bound or lower bound for the integrand $f(x) = 1/(x - c)$, where $c < a$?

Exploration 9.7.11 Derive formulas for the nodes and weights for a **Gauss-Lobatto quadrature rule** [Abramowitz and Stegun (1972)], in which nodes are prescribed at $x = a$ *and* $x = b$. Specifically, the rule must have $n + 1$ nodes $x_0 = a < x_1 < x_2 < \cdots < x_{n-1} < x_n = b$. Prove that the weights $w_0, w_1, \ldots, w_n$ are positive. What is the degree of accuracy of this rule?

Exploration 9.7.12 Explain why developing a Gauss-Radau rule by prescribing a node at $x = c$, where $c \in (a, b)$, is problematic.

9.7.6   Concept Check

1. What is a Gauss quadrature rule?
2. What is the degree of accuracy of a Gauss quadrature rule with n nodes?
3. How are the nodes of a Gauss quadrature rule obtained?
4. What is the error formula for a n-node Gauss quadrature rule?
5. How can Gauss quadrature nodes and weights on a general interval $[a, b]$ be obtained with minimal computational effort?
6. What are two drawbacks of Gauss quadrature rules?
7. What are Gauss-Radau and Gauss-Lobatto rules?
8. What is the benefit of using Gauss rules in combination with Gauss-Radau or Gauss-Lobatto rules?

9.8   Multiple Integrals

As many problems in scientific computing involve two- or three-dimensional domains, it is essential to be able to compute integrals over such domains. In this section, we explore the generalization of techniques for integrals of functions of one variable to such multivariable cases.

9.8.1   Double Integrals

Double integrals can be evaluated using the following strategies:

- If a two-dimensional domain Ω can be decomposed into rectangles, then the integral of a function $f(x, y)$ over Ω can be computed by evaluating integrals of the form

$$I[f] = \int_a^b \int_c^d f(x,y) \, dy \, dx. \tag{9.36}$$

Then, to evaluate $I[f]$, one can use a **Cartesian product rule**, whose nodes and weights are obtained by combining one-dimensional quadrature rules that are applied to each dimension. For example, if functions of x are integrated along the line between $x = a$ and $x = b$ using nodes x_i and weights w_i, for $i = 1, \ldots, n$, and if functions of y are integrated along the line between $y = c$ and $y = d$ using nodes y_i and weights z_i, for $i = 1, \ldots, m$, then the resulting Cartesian product rule

$$Q_{n,m}[f] = \sum_{i=1}^{n} \sum_{j=1}^{m} f(x_i, y_j) w_i z_j$$

has nodes (x_i, y_j) and corresponding weights $w_i z_j$ for $i = 1, \ldots, n$ and $j = 1, \ldots, m$.

- If the domain Ω can be described as the region between two curves $y_1(x)$ and $y_2(x)$ for $x \in [a, b]$, then we can write

$$I[f] = \iint_\Omega f(x,y) \, dA$$

as an **iterated integral**

$$I[f] = \int_a^b \int_{y_1(x)}^{y_2(x)} f(x,y) \, dy \, dx$$

which can be evaluated by applying a one-dimensional quadrature rule to compute the **outer integral**

$$I[f] = \int_a^b g(x) \, dx$$

where $g(x)$ is evaluated by using a one-dimensional quadrature rule to compute the **inner integral**

$$g(x) = \int_{y_1(x)}^{y_2(x)} f(x,y) \, dy.$$

- For various simple regions such as triangles, there exist **cubature rules** that are not combinations of one-dimensional quadrature rules. Cubature rules are more direct generalizations of quadrature rules, in that they evaluate the integrand at selected nodes and use weights determined by the geometry of the domain and the placement of the nodes.

It should be noted that all of these strategies apply to certain special cases. The first algorithm capable of integrating over a general two-dimensional domain was presented in [Lambers and Rice (1991)]. This algorithm combines the second and third strategies described above, decomposing the domain into subdomains that are either triangles or regions between two curves.

Example 9.8.1 We will use the Composite Trapezoidal Rule with $m = n = 2$ to evaluate the double integral

$$\int_0^{1/2} \int_0^{1/2} e^{y-x} \, dy \, dx.$$

The Composite Trapezoidal Rule with $n = 2$ subintervals is

$$\int_a^b f(x) \, dx \approx \frac{\Delta x}{2} \left[f(a) + 2f\left(\frac{a+b}{2}\right) + f(b) \right], \quad \Delta x = \frac{b-a}{n}.$$

If $a = 0$ and $b = 1/2$, then $\Delta x = (1/2 - 0)/2 = 1/4$ and this simplifies to

$$\int_0^{1/2} f(x) \, dx \approx \frac{1}{8}[f(0) + 2f(1/4) + f(1/2)].$$

We first use this rule to evaluate the "single" integral

$$\int_0^{1/2} g(x) \, dx$$

where

$$g(x) = \int_0^{1/2} e^{y-x} \, dy.$$

This yields

$$\int_0^{1/2} \int_0^{1/2} e^{y-x} \, dy \, dx = \int_0^{1/2} g(x) \, dx$$

$$\approx \frac{1}{8}[g(0) + 2g(1/4) + g(1/2)]$$

$$\approx \frac{1}{8}\left[\int_0^{1/2} e^{y-0} \, dy + 2\int_0^{1/2} e^{y-1/4} \, dy + \int_0^{1/2} e^{y-1/2} \, dy \right].$$

Now, to evaluate each of these integrals, we use the Composite Trapezoidal Rule in the y-direction with $m = 2$. If we let Δy denote the step size in the y-direction, we have $\Delta y = (1/2 - 0)/2 = 1/4$, and therefore we have

$$\int_0^{1/2} \int_0^{1/2} e^{y-x} \, dy \, dx \approx \frac{1}{8}\left[\int_0^{1/2} e^{y-0} \, dy + 2\int_0^{1/2} e^{y-1/4} \, dy + \int_0^{1/2} e^{y-1/2} \, dy \right]$$

$$\approx \frac{1}{8}\left[\frac{1}{8}\left[e^{0-0} + 2e^{1/4-0} + e^{1/2-0} \right] + \right.$$

$$2\frac{1}{8}\left[e^{0-1/4} + 2e^{1/4-1/4} + e^{1/2-1/4} \right] +$$

$$\left. \frac{1}{8}\left[e^{0-1/2} + 2e^{1/4-1/2} + e^{1/2-1/2} \right] \right]$$

$$\approx \frac{1}{64}\left[e^0 + 2e^{1/4} + e^{1/2} \right] +$$

$$\frac{1}{32}\left[e^{-1/4} + 2e^0 + e^{1/4} \right] +$$

$$\frac{1}{64}\left[e^{-1/2} + 2e^{-1/4} + e^0 \right]$$

$$\approx \frac{3}{32}e^0 + \frac{1}{16}e^{-1/4} + \frac{1}{64}e^{-1/2} + \frac{1}{16}e^{1/4} + \frac{1}{64}e^{1/2}$$

$$\approx 0.25791494889765.$$

The exact value, to 15 digits, is 0.255251930412762. The error is 2.66×10^{-3}, which is to be expected due to the use of few subintervals, and the fact that the Composite Trapezoidal Rule is only second-order-accurate. $\square$

Example 9.8.2 We will use the Composite Simpson's Rule with $n = 2$ and $m = 4$ to evaluate the double integral

$$\int_0^1 \int_x^{2x} x^2 + y^3 \, dy \, dx.$$

In this case, the domain of integration described by the limits is not a rectangle, but a triangle defined by the lines $y = x$, $y = 2x$, and $x = 1$. The Composite Simpson's Rule with $n = 2$ subintervals is

$$\int_a^b f(x) \, dx \approx \frac{\Delta x}{3} \left[f(a) + 4f\left(\frac{a+b}{2}\right) + f(b) \right], \quad \Delta x = \frac{b-a}{n}.$$

If $a = 0$ and $b = 1$, then $\Delta x = (1-0)/2 = 1/2$, and this simplifies to

$$\int_0^{1/2} f(x) \, dx \approx \frac{1}{6}[f(0) + 4f(1/2) + f(1)].$$

We first use this rule to evaluate the "single" integral

$$\int_0^1 g(x) \, dx$$

where

$$g(x) = \int_x^{2x} x^2 + y^3 \, dy.$$

This yields

$$\int_0^1 \int_x^{2x} x^2 + y^3 \, dy \, dx = \int_0^1 g(x) \, dx$$
$$\approx \frac{1}{6}[g(0) + 4g(1/2) + g(1)]$$
$$\approx \frac{1}{6} \left[\int_0^0 y^3 \, dy + 4 \int_{1/2}^1 \left(\frac{1}{2}\right)^2 + y^3 \, dy + \int_1^2 1^2 + y^3 \, dy \right].$$

The first integral will be zero, since the limits of integration are equal. To evaluate the second and third integrals, we use the Composite Simpson's Rule in the y-direction with $m = 4$. If we let Δy denote the step size in the y-direction, we have $\Delta y = (2x - x)/4 = x/4$, and therefore we have $\Delta y = 1/8$ for the second integral and $\Delta y = 1/4$ for the third. This yields

$$\int_0^1 \int_x^{2x} x^2 + y^3 \, dy \, dx \approx \frac{1}{6} \left[4 \int_{1/2}^1 \left(\frac{1}{2}\right)^2 + y^3 \, dy + \int_1^2 1^2 + y^3 \, dy \right]$$

$$\approx \frac{1}{6} \left\{ 4\frac{1}{24} \left[\left(\frac{1}{4} + \left(\frac{1}{2}\right)^3\right) + 4\left(\frac{1}{4} + \left(\frac{5}{8}\right)^3\right) + \right. \right.$$

$$2\left(\frac{1}{4} + \left(\frac{3}{4}\right)^3\right) + 4\left(\frac{1}{4} + \left(\frac{7}{8}\right)^3\right) + \left.\left(\frac{1}{4} + 1^3\right)\right] +$$

$$\frac{1}{12}\left[(1+1^3) + 4\left(1 + \left(\frac{5}{4}\right)^3\right) + \right.$$

$$\left. 2\left(1 + \left(\frac{3}{2}\right)^3\right) + 4\left(1 + \left(\frac{7}{4}\right)^3\right) + (1+2^3)\right] \right\}$$

$$\approx 1.03125.$$

The exact value is 1. The error 3.125×10^{-2} is rather large, which is to be expected due to the poor distribution of nodes through the triangular domain of integration. A better distribution is achieved if we use $n = 4$ and $m = 2$, which yields the much more accurate approximation of 1.001953125. □

Exploration 9.8.1 Write a MATLAB function

$$I=\texttt{dblintcomptrap(f,a,b,c,d,n,m)}$$

that approximates (9.36) using the Composite Trapezoidal Rule (9.22) with n subintervals in the x-direction and x subintervals in the y-direction.

Exploration 9.8.2 Generalize your function `dblintcomptrap` from Exploration 9.8.1 so that the arguments c and d can be either scalars or function handles. If they are function handles, then your function approximates the integral

$$I[f] = \int_a^b \int_{c(x)}^{d(x)} f(x,y) \, dy \, dx.$$

Hint: Use the MATLAB function `isnumeric` to determine whether c and d are numbers.

Exploration 9.8.3 Generalize your function `dblintcomptrap` from Exploration 9.8.2 so that the arguments a, b, c and d can be either scalars or function handles. If a and b are function handles, then your function approximates the integral

$$I[f] = \int_c^d \int_{a(y)}^{b(y)} f(x,y) \, dx \, dy.$$

Exploration 9.8.4 Use the error formula for the Composite Trapezoidal Rule to obtain an error formula for a Cartesian product rule such as the one implemented in Exploration 9.8.1. As in that exploration, assume that m subintervals are used in the x-direction and n subintervals in the y-direction. *Hint:* First, apply the single-variable error formula to the integral

$$\int_a^b g(x)\,dx, \quad g(x) = \int_c^d f(x,y)\,dy.$$

Exploration 9.8.5 Modify your function `dblintcomptrap` from Exploration 9.8.3 to obtain a new function `dblintcompsimp` that uses the Composite Simpson's Rule (9.23) instead of the Composite Trapezoidal Rule.

Exploration 9.8.6 Write a MATLAB function

$$\texttt{I=dblinttriangle(f,x,y,n,m)}$$

that approximates the integral of $f(x,y)$, represented by the function handle `f`, over the triangle T with vertices stored in the 3-vectors `x` and `y`. Use a translation that shifts $(\texttt{x(1)}, \texttt{y(1)})$ to the origin, and a Givens rotation (see Section 4.2.1) to transform T to a new triangle $\tilde{T}$, in which the edge of T from (x_1, y_1) to (x_2, y_2) is mapped to an edge of $\tilde{T}$ that begins at the origin and lies on the positive y-axis. Then, use your function `dblintcompsimp` from Exploration `exerdblintgensimp` to approximate the integral

$$\int\int_T f(x,y)\,dx\,dy = \int\int_{\tilde{T}} f(x_1 + c\tilde{x} - s\tilde{y}, y_1 + s\tilde{x} + c\tilde{y})\,d\tilde{x}\,d\tilde{y},$$

where $[c,s] = \texttt{givens}(x_2 - x_1, y_2 - y_1)$, with the function `givens` defined as in Exploration 4.2.2.

9.8.2 Higher Dimensions

The techniques presented in this section can be adapted for the purpose of approximating triple integrals. This adaptation is carried out in the following explorations.

Exploration 9.8.7 Write a MATLAB function

$$\texttt{I=tripintcomptrap(f,a,b,c,d,s,t,n,m,p)}$$

that approximates

$$\int_a^b \int_c^d \int_s^t f(x,y,z)\,dz\,dy\,dx$$

using the Composite Trapezoidal Rule with n subintervals in the x-direction, m subintervals in the y-direction, and p subintervals in the z-direction. *Hint:* Use your function `dblintcomptrap` from Exploration 9.8.1.

Exploration 9.8.8 Modify your function `tripintcomptrap` from Exploration 9.8.7 to obtain a function `tripintcompsimp` that uses the Composite Simpson's Rule in each direction. Then, use an approach similar to that used in Exploration 9.8.4 to obtain an error formula for the Cartesian product rule used in `tripintcompsimp`.

Exploration 9.8.9 Generalize your function `tripintcomptrap` from Exploration 9.8.7 so that any of the arguments a, b, c, d, s and t can be either scalars or function handles. For example, if a and b are scalars, c and d are function handles that have two input arguments, and s and t are function handles that have one input argument, then your function approximates the integral

$$I[f] = \int_a^b \int_{s(x)}^{t(x)} \int_{c(x,z)}^{d(x,z)} f(x, y, z)\, dy\, dz\, dx.$$

Hint: Use the MATLAB function `isnumeric` to determine whether any arguments are numbers, and the function `nargin` to determine how many arguments a function handle requires.

In more than three dimensions, generalizations of quadrature rules are not practical, since the number of function evaluations needed to attain sufficient accuracy grows very rapidly as the number of dimensions increases. An alternative is the **Monte Carlo Method**, which samples the integrand at n randomly selected points and attempts to compute the mean value of the integrand on the entire domain. The method converges rather slowly, but its convergence rate depends only on n, not the number of dimensions.

9.8.3 Concept Check

1. What is a Cartesian product rule?
2. How can a double integral be expressed as two single integrals?
3. What is a cubature rule? How does it differ from a Cartesian product rule?
4. What is problematic about generalization of quadrature rules to integrals over higher-dimensional domains?
5. What are Monte Carlo methods for integrals, and for what kinds of integrals are they used?

9.9 Additional Resources

Differentiation matrices are covered in greater detail in [Trefethen (2000)], in the context of *spectral methods*, which are introduced in Chapter 14. Beyond the scope of this text is **automatic differentiation**, also known as *algorithmic differentiation*. The main idea is to generate code for the derivative of a function f by repeatedly applying the Chain Rule to produce statements that represent the derivatives of the statements that implement $\mathtt{f}$. This topic is covered in [Griewank (2000); Nocedal and Wright (2006)].

The classic text [Davis and Rabinowitz (1985)] includes a wealth of knowledge about numerical integration. In addition to expansion on the material presented in this chapter, their text also covers topics beyond the scope of this book, such as Monte Carlo techniques and improper integrals, which were briefly alluded to in Section 9.7. Other general references on numerical integration include [Engels (1980); Evans (1993); Krommer and Ueberhuber (1998)].

Tables of nodes and weights for Gauss quadrature rules can be found in many resources, including [Abramowitz and Stegun (1972); Stroud and Secrest (1966); Zwillinger (1996)]. Extrapolation techniques are covered extensively in [Brezinski (1991); Joyce (19971); Wimp (1981)]. The paper [Gander and Gautschi (2000)] provides an overview of adaptive quadrature, including the choice of termination criteria and deficiencies in known adaptive quadrature routines. Further discussion of drawbacks of adaptive quadrature can be found in [Lyness (1983); Lyness and Kaganove (1976)]. An algorithm for approximating double integrals with 1-D adaptive quadrature routines is presented in [Fritsch et al. (1981)]. General references on the approximation of multiple integrals are [Cools (1997); Haber (1970); Lyness and Cools (1993); Sloan and Joe (1994); Stroud (1972)]. Monte Carlo methods in particular are covered in [Hastie et al. (2001)].

QUADPACK is a library of routines written in FORTRAN 77 for approximating one-dimensional integrals. As of this writing, the last stable release was thirty years ago, but its routines are still used in software packages such as Octave and R. QUADPACK is available at

```
http://nines.cs.kuleuven.be/software/QUADPACK
```

and its routines are documented in [Piessens et al. (1983)].

9.10 Exercises

1. Use Taylor expansion to derive a finite difference approximation to $f'''(x_0)$ that uses the points $x_0 \pm h$ and $x_0 \pm 2h$. What is the error formula for this approximation?

2. Consider the general two-point approximation of the second derivative,

$$f''(x_0) \approx 2\frac{h_1 f(x_0 - h_2) - (h_1 + h_2)f(x_0) + h_2 f(x_0 + h_1)}{h_1 h_2 (h_1 + h_2)}.$$

   (a) Prove that this approximation is equal to $2f[x_0 - h_2, x_0, x_0 + h_1]$, or 2 times the second-order divided difference (as in Section 7.3) of f over the points $x_0 - h_2, x_0,$ and $x_0 + h_1$.

   (b) Prove that this approximation is first-order accurate unless $h_1 = h_2$.

3. Let $f(x) = \sin x$ and $x_0 = 1.2$. Consider the centered-difference approximation of $f''(x_0)$,

$$f''(x_0) = \frac{f(x_{-1}) - 2f(x_0) + f(x_1)}{h^2} + O(h^2),$$

where $x_i = x_0 + ih$, and the error formula is derived in Exploration 9.1.6. In this exercise, we account for roundoff error in this formula by replacing each expression of the form $f(x_i)$ with $f(x_i)(1 + \epsilon_i)$, where we assume $|\epsilon_i| \leq \mathbf{u}$ with $\mathbf{u}$ being the machine precision from Definition 2.2.9.

(a) Show that the error in this approximation satisfies

$$\left| f''(x_0) - \frac{f(x_{-1}) - 2f(x_0) + f(x_1)}{h^2} \right| \leq \frac{Mh^2}{12} + \frac{4\mathbf{u}}{h^2},$$

where $|f^{(4)}(x)| \leq M$ on $[x_1, x_2]$.

(b) Find the value of h that minimizes the error bound in part 3a.

(c) Use this approximation to approximate $f''(x_0)$, with $h = 2^{-p}$ for $p = 0, 1, 2, \ldots, 16$. Plot the errors against these values of h on a logarithmic scale (that is, use `loglog`). Estimate the value of h at which the minimum error is achieved. How does this compare to the optimal h found in part 3b? Use the value of the MATLAB variable `eps` for $\mathbf{u}$.

4. The following tabular data is used to describe a function $f(x)$:

x	0	1.5	2	3	4.5	5	6
$f(x)$	1	1.5	3	4	4.5	7	7.5

Use three-point finite-difference approximations to approximate $f'(x)$ at each of the x-values listed in the table. *Note:* it will be easiest to use Lagrange interpolation, rather than Taylor expansion. You may use `polyfit` to construct interpolating polynomials that fit the given data, rather than deriving general finite-difference formulas. Do the results seem reasonable when compared to a graph of $f(x)$ based on this data?

5. Derive Simpson's Rule by explicitly computing the weights using (9.13).

6. Let $H_3(x)$ be the Hermite interpolating polynomial (see Section 7.5.1) of $f(x)$ at $x = a, b$. Use this polynomial to derive the **Corrected Trapezoidal Rule**

$$\int_a^b f(x)\,dx = \frac{b-a}{2}[f(a) + f(b)] + \frac{(b-a)^2}{12}[f'(a) - f'(b)] +$$

$$\frac{(b-a)^5}{720} f^{(5)}(\xi), \quad \xi \in (a, b).$$

Include the derivation of the error formula in your answer. How does the accuracy of this rule compare to Simpson's Rule, when applied to $I[f] = \int_0^1 e^x\,dx$? *Hint:* Use the Newton form of $H_3(x)$.

7. (a) Compute the weights $w_1, w_2, \ldots, w_n$ for closed Newton-Cotes rules with $n = 4$ and $n = 5$ nodes on $[-1, 1]$.

(b) Compare the weights from part 7a to the weights for the Trapezoidal Rule and Simpson's Rule on $[-1, 1]$, which are also closed Newton-Cotes rules. What similarity do you notice?

(c) Prove that this property of the weights applies to a general interpolatory quadrature rule on $[a, b]$.

8. Assume that $f''(x) \geq 0$ on $[a, b]$. Let $M_n[f]$ be the approximation of $I[f]$ computed using the n-node Composite Midpoint Rule on $[a, b]$, and similarly, let $T_n[f]$ be the approximation computed using the Composite Trapezoidal Rule. Prove that

$$M_n[f] \leq I[f] \leq T_n[f].$$

9. Is it generally more accurate to apply a n-node Newton-Cotes rule Q_n on each half of $[a, b]$, or to apply a $2n$-node Newton-Cotes rule Q_{2n} to all of $[a, b]$? Explain your reasoning.

10. The **gamma function** $\Gamma(z)$ is defined by

$$\Gamma(z) = \int_0^\infty x^{z-1} e^{-x} \, dx.$$

It is worth noting that if z is a positive integer, then $\Gamma(z) = (z - 1)!$ It can be evaluated in MATLAB using the **gamma** function. Use the following approaches to approximate $\Gamma(z)$ for various values of $z \in [1, 10]$:

(a) Apply adaptive quadrature on a finite interval $[0, b]$, where b is chosen so that $\int_b^\infty x^{z-1} e^{-x}$ is negligibly small. Explain your approach to choosing b.

(b) Use **Gauss-Laguerre quadrature**, which is Gauss quadrature with weight function $w(x) = e^{-x}$ on the interval $[0, \infty)$. The nodes and weights can be looked up.

How do these approaches compare in terms of accuracy and efficiency?

11. Let p_n denote the area of a n-sided regular polygon inscribed within the unit circle. Similarly, let q_n denote the area of a n-sided regular polygon that circumscribes the unit circle. Then

$$p_n = n \sin \frac{\pi}{n}, \quad q_n = n \tan \frac{\pi}{n}.$$

Since the area of the unit circle is π, p_n and q_n are both approximations of π that become more accurate as n increases.

(a) Let $h = 1/n$. Show that p_n and q_n satisfy

$$p_n = \pi + a_1 h^2 + a_2 h^4 + a_3 h^6 + \cdots$$
$$q_n = \pi + b_1 h^2 + b_2 h^4 + b_3 h^6 + \cdots$$

(b) Use Richardson Extrapolation to obtain $O(h^6)$-accurate approximations of π from the values p_4, p_8, p_{16} and q_4, q_8, q_{16}.

12. Use Richardson Extrapolation, applied to the centered difference formula for the second derivative, to compute a 6th-order accurate approximation of $f''(x_0)$, where $f(x) = \sqrt{x}$ and $x_0 = 1$. Explain why Richardson extrapolation for this problem is much more effective than it is when combined with the Composite Trapezoidal Rule to implement Romberg Integration for approximating $I[f] = \int_0^1 \sqrt{x}\,dx$.

13. (a) Write a MATLAB function y=piecewisef(x) that evaluates the piecewise function

$$f(x) = \begin{cases} e^x & x < 0 \\ e^{1-x} & x \geq 0 \end{cases}$$

   and also plots the point (x, y) with a circle.

   (b) Use your function adapquadrecur from Exploration 9.6.3 to approximate $\int_{-1}^2 f(x)\,dx$ via adaptive quadrature, with tolerance 10^{-3}. By using piecewisef, the points at which $f(x)$ is evaluated will be displayed, as long as the command hold on is used first. It is recommended that you plot the graph of $f(x)$ on $[-1, 2]$ first. What causes adaptive quadrature to be very inefficient in this case?

   (c) How can you modify your function to address this inefficiency?

14. Let $R = [-1, 1] \times [-1, 1]$, and let (x_0, y_0) be a point *outside* of R. If there is a uniform charge distribution within R, then the electrostatic potential at (x_0, y_0) given by

$$\Phi(x_0, y_0) = \iint_R \frac{dx\,dy}{\sqrt{(x - x_0)^2 + (y - y_0)^2}}.$$

Use your function quadcomptrap from Exploration 9.8.1 to approximate this integral for points (x_0, y_0) chosen to be equally spaced points within the square $R_0 = [2, 10] \times [2, 10]$. Then, use the MATLAB function surf to plot the graph of $\Phi(x, y)$ on R_0, which is a surface. *Note:* this will require storing the values of $\Phi(x_0, y_0)$ in a matrix.

PART IV
Nonlinear Equations and Optimization

Chapter 10

Zeros of Nonlinear Functions

To this point, we have only considered the solution of linear equations. We now explore the much more difficult problem of solving nonlinear equations of the form

$$f(x) = 0,$$

where $f(x) : D \subseteq \mathbb{R} \to \mathbb{R}$ can be any continuous function. A solution x^* of such a nonlinear equation is called a **root** or a **zero** of the function f. Each of the remaining chapters in this book will rely on the root-finding techniques presented in this chapter as essential ingredients in algorithms for solving optimization problems (Chapter 11) and differential equations (Part V).

The solution methods considered in this chapter will be *iterative* in nature, like some of the methods for systems of linear equations presented in Chapter 5. Therefore, in Section 10.1, we will discuss the types of convergence that iterative methods can exhibit, as well as the well-posedness of the problem of solving $f(x) = 0$. In Section 10.2 we will learn about the Method of Bisection, an inefficient yet reliable method for finding zeros. Section 10.3 will introduce Fixed-Point Iteration, which is not only a useful root-finding method in its own right, but also a framework for other methods. One such method is Newton's Method, a rapidly convergent iteration which will be presented in Section 10.4, along with a variation, the Secant Method. Finally, Section 10.5 will highlight how the previous methods can be modified or combined to improve efficiency or reliability.

10.1 Solution by Iteration

In general, nonlinear equations cannot be solved in a finite sequence of steps. While systems of linear equations can be solved using **direct methods** such as Gaussian Elimination, nonlinear equations usually require **iterative methods**. In iterative methods, an approximate solution is refined with each iteration until it is determined to be sufficiently accurate, at which time the iteration terminates. Since it is desirable for iterative methods to converge to the solution as rapidly as possible, it is necessary to be able to measure the rate with which an iterative method converges.

To that end, we assume that an iterative method generates a sequence of iterates $x^{(0)}, x^{(1)}, x^{(2)}, \ldots$ that converges to the exact solution x^*. Ideally, we would like the error in a given iterate $x^{(k+1)}$ to be much smaller than the error in the previous iterate $x^{(k)}$. For example, if the error is raised to a power greater than one from iteration to iteration, then once the error is less than one, it will approach zero very rapidly. This leads to the following definition.

Definition 10.1.1 (Order and Rate of Convergence) Let $\{x^{(k)}\}_{k=0}^{\infty}$ be a sequence in $\mathbb{R}$ that converges to $x^* \in \mathbb{R}$ and assume that $x^{(k)} \neq x^*$ for each k. We say that the **order of convergence** of $\{x^{(k)}\}$ to x^* is **order** r, with **asymptotic error constant** C, if

$$\lim_{k \to \infty} \frac{|x^{(k+1)} - x^*|}{|x^{(k)} - x^*|^r} = C,$$

where $r \geq 1$. If $r = 1$, then the number $\rho = -\log_{10} C$ is called the **asymptotic rate of convergence**.

If $r = 1$, and $0 < C < 1$, we say that convergence is **linear**. If $r = 1$ and $C = 0$, or if $1 < r < 2$ for any positive C, then we say that convergence is **superlinear**. If $r = 2$, then the method converges **quadratically**, and if $r = 3$, we say it converges **cubically**, and so on. Note that the value of C need only be less than one in the case of linear convergence.

When convergence is linear, the asymptotic rate of convergence ρ indicates the number of correct decimal digits obtained in a single iteration. In other words, $\lfloor 1/\rho \rfloor + 1$ iterations are required to obtain an additional correct decimal digit, where $\lfloor x \rfloor$ is the **floor** of x, which is the largest integer that is less than or equal to x.

10.1.1   Existence and Uniqueness

It is easy to think of examples in which the equation $f(x) = 0$, where f is continuous on its domain, can have a unique solution, infinitely many solutions, or no solution at all.

Exploration 10.1.1 List three functions $f(x)$, $g(x)$ and $h(x)$ such that the equation $f(x) = 0$ has a unique solution, the equation $g(x) = 0$ has infinitely many solutions, and the equation $h(x) = 0$ has no solution.

For a general equation $f(x) = 0$, it is not possible to characterize the conditions under which a solution exists or is unique. However, in some situations, it is possible to determine existence analytically. The **Intermediate Value Theorem** (Theorem A.1.8) implies that if a continuous function $f(x)$ satisfies $f(a) < 0$ and $f(b) > 0$ (or vice versa), where $a < b$, then $f(x) = 0$ for some $x \in (a, b)$.

One useful result from calculus that can be used to establish existence and, in some sense, uniqueness of a solution is the **Inverse Function Theorem**, which states that if $f'(x_0) \neq 0$, then, within some neighborhood of x_0, f is invertible and the equation $f(x) = y$ has a unique solution for all y near $f(x_0)$.

If $f(x_0) = 0$ and $f'(x_0) \neq 0$, then we say that x_0 is a **simple root** of $f(x)$. On the other hand, if $f'(x_0) = 0$, then $f(x)$ is said to be **degenerate** at x_0, and x_0 is a **double root** of $f(x)$. This term comes from the fact that $f(x)$ can be written as $f(x) = (x - x_0)^2 g(x)$, where $g(x)$ is defined at x_0. More generally, if $f^{(j)}(x_0) = 0$ for $j = 0, \ldots, m - 1$, for $m > 1$, then x_0 is a **multiple root**, or root of multiplicity m. We will see that degeneracy can cause difficulties when trying to solve nonlinear equations numerically.

10.1.2 Sensitivity of Solutions

Recall from Section 2.1.4 the **absolute condition number** of a function $f(x)$ is a measure of how a perturbation in x, denoted by $x + \epsilon$ for some small ϵ, is amplified by $f(x)$. Using the Mean Value Theorem (Theorem A.5.3), we have

$$|f(x + \epsilon) - f(x)| = |f'(c)||\epsilon|$$

where c is between x and $x + \epsilon$. With ϵ being small, the absolute condition number can be approximated by $|f'(x)|$, the factor by which the perturbation in x is amplified to measure the perturbation in $f(x)$.

In solving a nonlinear equation in one dimension, we are trying to solve an **inverse problem**; that is, instead of computing $y = f(x)$ (the **forward problem**), we are computing $x^* = f^{-1}(0)$, assuming that f is invertible near the root x^*. It follows from the differentiation rule

$$\frac{d}{dx}[f^{-1}(x)] = \frac{1}{f'(f^{-1}(x))}$$

that the absolute condition number for solving $f(x) = 0$ is approximately $1/|f'(x^*)|$. This discussion can be generalized to higher dimensions, where the condition number is obtained using the norm of the **Jacobian matrix** (see Section 11.1.1).

Using **backward error analysis** (see Section 2.1.3), we assume that the approximate solution $\hat{x} = \hat{f}^{-1}(0)$, obtained by evaluating an approximation of f^{-1} at the exact input $y = 0$, can also be viewed as evaluating the exact function f^{-1} at a nearby input $\hat{y} = \epsilon$. That is, the approximate solution $\hat{x} = f^{-1}(\epsilon)$ is the exact solution of a "nearby" equation.

From this viewpoint, it can be seen from a graph that if $|f'|$ is large near x^*, which means that the condition number of the problem $f(x) = 0$ is small (that is, the problem is **well-conditioned**), then even if ϵ is relatively large, $\hat{x} = f^{-1}(\epsilon)$ is close to x^*. On the other hand, if $|f'|$ is small near x^*, so that the problem is **ill-conditioned**, then even if ϵ is small, $\hat{x}$ can be far away from x^*. These contrasting situations are illustrated in Figure 10.1.

10.1.3 Concept Check

1. Define the order of convergence of a sequence.

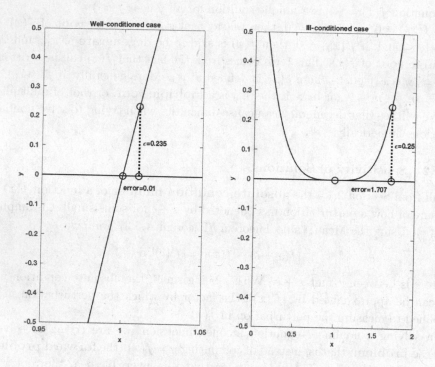

Fig. 10.1   Left plot: Well-conditioned problem of solving $f(x) = 0$. $f'(x^*) = 24$, and an approximate solution $\hat{x} = f^{-1}(\epsilon)$ has small error relative to ϵ. Right plot: Ill-conditioned problem of solving $f(x) = 0$. $f'(x^*) = 0$, and $\hat{x}$ has large error relative to ϵ.

2. What is the difference between the *order* of convergence of a sequence and its *rate* of convergence?

3. What does it mean when we say a sequence converges superlinearly?

4. When can we be certain that the equation $f(x) = 0$ has a solution? Is there any assurance of uniqueness?

5. When is the problem of solving $f(x) = 0$ ill-conditioned: when the graph of $f(x)$ is very steep near a root, or very "flat"?

10.2   The Bisection Method

Suppose that $f(x)$ is a continuous function that changes sign on the interval (a, b). Then, by the Intermediate Value Theorem, $f(x) = 0$ for some $x \in (a, b)$. How can we find the solution, knowing that it lies in this interval? To determine how to proceed, we consider some examples in which such a sign change occurs. We work with the functions

$$f(x) = x - \cos x, \quad g(x) = e^x \cos(x^2),$$

on the intervals $[0, \pi/2]$ and $[0, \pi]$, respectively. The graphs of these functions are shown in Figure 10.2.

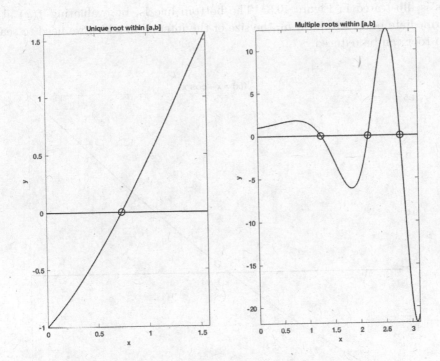

Fig. 10.2 Illustrations of the Intermediate Value Theorem. Left plot: $f(x) = x - \cos x$ has a unique root on $[0, \pi/2]$. Right plot: $g(x) = e^x \cos(x^2)$ has multiple roots on $[0, \pi]$.

It can be seen that $f(a)$ and $f(b)$ have different signs, and since both functions are continuous on $[a, b]$, the Intermediate Value Theorem guarantees the existence of a root in (a, b). However, both of these intervals are rather large, so we cannot obtain a useful approximation of a root from this information alone. At each root in these examples, $f(x)$ changes sign, so $f(x) > 0$ for x on one side of the root x^*, and $f(x) < 0$ on the other side. Therefore, if we can find two values a' and b' such that $f(a')$ and $f(b')$ have different signs, but a' and b' are very close to one another, then we can accurately approximate x^*.

10.2.1 The Bisection Idea

Consider the first example in Figure 10.2, that has a unique root. We have $f(0) < 0$ and $f(\pi/2) > 0$. From the graph, we see that if we evaluate f at *any* point x_0 in $(0, \pi/2)$, and we do not "get lucky" and happen to choose the root, then we have either of the following:

- $f(x_0) > 0$, in which case f has a root on $(0, x_0)$, because f changes sign on this interval.
- $f(x_0) < 0$, in which case f has a root on $(x_0, \pi/2)$, because of a sign change.

This is illustrated in Figure 10.3. The bottom line is, by evaluating $f(x)$ at an intermediate point within (a, b), the size of the interval in which we need to search for a root can be reduced.

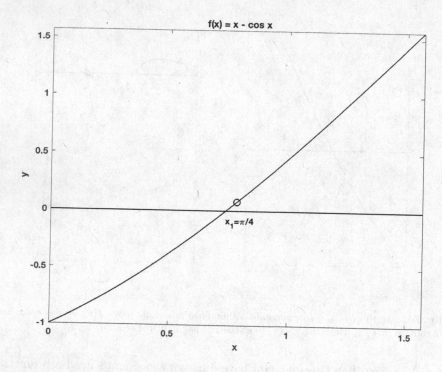

Fig. 10.3   Because $f(\pi/4) > 0$, $f(x)$ has a root in $(0, \pi/4)$.

The **Bisection Method** proceeds by reducing the size of the interval in which a solution is known to exist. Suppose that we evaluate $f(m)$, where $m = (a + b)/2$. If $f(m) = 0$, then we have found a root x^*. Otherwise, f must change sign on the interval (a, m) or (m, b), since $f(a)$ and $f(b)$ have different signs and therefore $f(m)$ must have a different sign from one of these values.

Let us try this approach on the function $f(x) = x - \cos x$, on $[a, b] = [0, \pi/2]$. This example can be set up in MATLAB as follows:

```
>> a=0;
>> b=pi/2;
>> f=@(x)(x-cos(x));
```

To help visualize the results of the iterative process that we will carry out to find an approximate solution x^*, we will also graph $f(x)$ on $[a, b]$:

```
>> fplot(f(x),a,b)
>> axis tight
```

```
>> xlabel('x')
>> ylabel('y')
>> title('f(x) = x - cos x')
>> hold on
>> plot([ 0 pi/2 ],[ 0 0 ],'k')
```

The last statement is used to plot the relevant portion of the x-axis, so that we can more easily visualize our progress toward computing a root.

Now, we can begin searching for an approximate root of $f(x)$. We can reduce the size of our search space by evaluating $f(x)$ at *any* point in (a, b), but for convenience, and to ensure as rapid a reduction as possible, we choose the midpoint:

```
>> m=(a+b)/2;
>> plot(m,f(m),'ro')
```

The `plot` statement is used to plot the point $(m, f(m))$ on the graph of $f(x)$, using a red circle. You have now reproduced Figure 10.3, except that your figure is in color.

Now, we examine the values of f at a, b and the midpoint m:

```
>> f(a)
ans =
    -1
>> f(m)
ans =
    0.078291382210901
>> f(b)
ans =
    1.570796326794897
```

We can see that $f(a)$ and $f(m)$ have different signs, so a root exists within (a, m). We can therefore update our search space $[a, b]$ accordingly:

```
>> b=m;
```

We then repeat the process, working with the midpoint of our new interval:

```
>> m=(a+b)/2;
>> plot(m,f(m),'ro')
```

Now, it does not matter at which endpoint of our interval $f(x)$ has a positive or negative value; we only need the signs of $f(a)$ and $f(b)$ to be different. Therefore, we can simply check whether the product of the values of f at the endpoints is negative. If $f(m)$ is nonzero, then because $f(a)f(b) < 0$, $f(m)$ must have a different sign from exactly one of $f(a)$ or $f(b)$.

```
>> f(a)*f(m)
```

```
ans =
   0.531180450812563
>> f(m)*f(b)
ans =
  -0.041586851697525
```

We see that the sign of f changes on (m, b), so we update a to reflect that this is our new interval to search:

```
>> a=m;
>> m=(a+b)/2;
>> plot(m,f(m),'ro')
```

The progress toward a root can be seen in Figure 10.4.

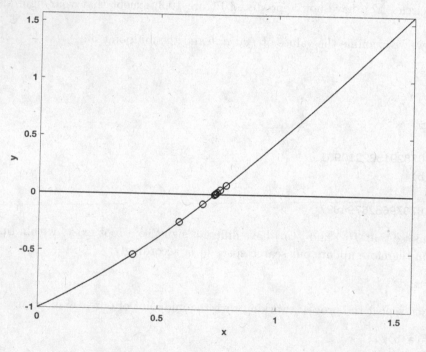

Fig. 10.4   Progress of the Bisection method toward finding a root of $f(x) = x - \cos x$ on $(0, \pi/2)$

Exploration 10.2.1 Repeat this process a few more times: check whether f changes sign on (a, m) or (m, b), update $[a, b]$ accordingly, and then compute a new midpoint m. After plotting each new midpoint, what behavior can you observe? Are the midpoints converging, and if so, are they converging to a root of f? Check by evaluating f at each midpoint.

10.2.2 Practical Implementation

We can continue this process until the interval $[a, b]$ is sufficiently small, in which case we are close to a solution. By including these steps in a loop, we obtain the following algorithm that implements the approach that we have been carrying out.

Algorithm 10.2.1 (Bisection Method) Let f be a continuous function on the interval $[a, b]$ that changes sign on (a, b). The following algorithm computes an approximate solution x^* of $f(x) = 0$, with absolute error tolerance TOL.

for $j = 1, 2, \ldots$ **do** until convergence
    $m = (a + b)/2$
    if $f(m) = 0$ **or** $b - a < 2 \cdot TOL$ **then**
        $x^* = m$
        return x^*
    end if
    if $f(a)f(m) < 0$ **then**
        $b = m$
    else
        $a = m$
    end if
end for

At the beginning, it is known that (a, b) contains a solution. During each iteration, this algorithm updates the interval (a, b) by checking whether f changes sign in the first half (a, m), or in the second half (m, b). Once the correct half is found, the interval (a, b) is set equal to that half. Therefore, at the beginning of *each* iteration, it is known that the current interval (a, b) contains a solution.

Exploration 10.2.2 Implement Algorithm 10.2.1 in a MATLAB function `[x,niter]=bisection(f,a,b,tol)` that accepts as input a function handle `f` for $f(x)$, the endpoints `a` and `b` of an interval $[a, b]$, and an absolute error tolerance `tol`. The output `niter` is the number of iterations; that is, the number of times that the midpoint m of $[a, b]$ is examined. Use the `error` function to ensure that if f does not change sign on $[a, b]$, the function immediately exits with an appropriate error message.

10.2.3 Convergence

In comparison to other methods, including some that we will discuss later in this chapter, the Bisection Method tends to converge rather slowly, but it is also guaranteed to converge. These qualities can be seen in the following result concerning the accuracy of the Bisection Method.

Theorem 10.2.2 Let f be continuous on $[a, b]$, and assume that $f(a)f(b) < 0$. For each $k = 1, 2, \ldots$, let $x^{(k)}$ be the kth iterate (that is, midpoint of (a, b)) that is produced by the Bisection Method. Then the sequence $\{x^{(k)}\}_{n=1}^{\infty}$ converges to a number x^* in (a, b) such that $f(x^*) = 0$, and each iterate $x^{(k)}$ satisfies

$$|x^{(k)} - x^*| \leq \frac{b-a}{2^k}.$$

Exploration 10.2.3 Prove Theorem 10.2.2.

Exploration 10.2.4 On a computer using the IEEE double-precision floating-point system, what is the largest number of iterations of Bisection that is practical to perform? Justify your answer.

It should be noted that because the kth iterate can lie anywhere within the interval (a, b) that is used during the kth iteration, it is possible that the error bound given by this theorem may be quite conservative.

Example 10.2.3 We seek a solution of the equation $f(x) = 0$, where

$$f(x) = x^2 - x - 1.$$

Because $f(1) = -1$ and $f(2) = 1$, and f is continuous, we can use the Intermediate Value Theorem to conclude that $f(x) = 0$ has a solution in the interval $(1, 2)$, since $f(x)$ must assume every value between -1 and 1 in this interval.

We use the Bisection Method to find a solution. First, we compute the midpoint of the interval, which is $(1 + 2)/2 = 1.5$. Since $f(1.5) = -0.25$, we see that $f(x)$ changes sign between $x = 1.5$ and $x = 2$, so we can apply the Intermediate Value Theorem again to conclude that $f(x) = 0$ has a solution in the interval $(1.5, 2)$.

Continuing this process, we compute the midpoint of the interval $(1.5, 2)$, which is $(1.5 + 2)/2 = 1.75$. Since $f(1.75) = 0.3125$, we see that $f(x)$ changes sign between $x = 1.5$ and $x = 1.75$, so we conclude that there is a solution in the interval $(1.5, 1.75)$. The table below shows the outcome of several more iterations of this procedure. Each row shows the current interval (a, b) in which we know that a solution exists, as well as the midpoint of the interval, given by $(a + b)/2$, and the value of f at the midpoint. Note that from iteration to iteration, only one of a or b changes, and the endpoint that changes is always set equal to the midpoint.

a	b	$m = (a+b)/2$	$f(m)$
1	2	1.5	−0.25
1.5	2	1.75	0.3125
1.5	1.75	1.625	0.015625
1.5	1.625	1.5625	−0.12109
1.5625	1.625	1.59375	−0.053711
1.59375	1.625	1.609375	−0.019287
1.609375	1.625	1.6171875	−0.0018921
1.6171875	1.625	1.62109325	0.0068512
1.6171875	1.62109325	1.619140625	0.0024757
1.6171875	1.619140625	1.6181640625	0.00029087

The correct solution, to ten decimal places, is 1.6180339887, which is the number known as the **golden ratio**. □

Exploration 10.2.5 The function $f(x) = x^2 - 2x - 2$ has two real roots, one positive and one negative. Find two disjoint intervals $[a_1, b_1]$ and $[a_2, b_2]$ that can be used with Bisection to find the negative and positive roots, respectively. Why is it not practical to use a much larger interval that contains both roots, so that Bisection can supposedly find one of them?

For the Bisection Method, it is easier to determine the order of convergence if we use a different measure of the error in each iterate (midpoint) $x^{(k)}$ than $|x^{(k)} - x^*|$. Since each iterate is contained within an interval $[a_k, b_k]$, where $b_k - a_k = 2^{1-k}(b-a)$ with $[a, b]$ being the original interval, it follows that we can bound the error $x^{(k)} - x^*$ by $e_k = (b_k - a_k)/2$. Using this measure, we can easily conclude that Bisection converges linearly, with asymptotic error constant $1/2$.

10.2.4 Concept Check

1. Describe the Bisection Method for finding a solution of $f(x) = 0$. What condition must $f(x)$ satisfy for this method to be effective?
2. What theorem from calculus is used to justify the Bisection Method?
3. What is the order of convergence of the Bisection Method? What is its asymptotic error constant?
4. Give an advantage and disadvantage of the Bisection Method compared to other root-finding methods.
5. Suppose that the Bisection Method is used to find a root of a function $f(x)$ on the interval $[0, 1]$. How many iterations are needed to ensure that a root is found with error at most 0.01?

10.3 Fixed-Point Iteration

A nonlinear equation of the form $f(x) = 0$ can be rewritten to obtain an equation of the form

$$x = g(x),$$

in which case the solution is a **fixed point**, or **stationary point**, of the function g. In this section, we present a simple yet powerful algorithm for the solution of an equation of this form.

10.3.1 Solution by Successive Substitution

The formulation of the original problem $f(x) = 0$ into one of the form $x = g(x)$ leads to a simple solution method known as **Fixed-Point Iteration**, or **simple iteration**, which we now describe.

Algorithm 10.3.1 (Fixed-Point Iteration) Let g be a continuous function defined on the interval $[a, b]$. The following algorithm computes a solution $x^* \in (a, b)$ of the equation $x = g(x)$.

Choose an initial guess $x^{(0)}$ in $[a, b]$.
for $k = 0, 1, 2, \ldots$ **do** until convergence
$\qquad x^{(k+1)} = g(x^{(k)})$
end for
$x^* = x^{(k)}$

A simple convergence test would be, for example, checking whether $|x^{(k+1)} - x^{(k)}|$ is sufficiently small. We will discuss stopping criteria later in this section.

When rewriting the equation $f(x) = 0$ in the form $x = g(x)$, it is essential to choose the function g wisely. One guideline is to choose

$$g(x) = x - \phi(x)f(x), \tag{10.1}$$

where the function $\phi(x)$ is, ideally, nonzero except possibly at a solution of $f(x) = 0$. This can be satisfied by choosing $\phi(x)$ to be constant, but this can fail, as the following example illustrates.

Example 10.3.2 Consider the equation

$$x + \ln x = 0.$$

By the Intermediate Value Theorem, this equation has a solution in the interval $[0.5, 0.6]$. Furthermore, this solution is unique. To see this, let $f(x) = x + \ln x$. Then $f'(x) = 1 + 1/x > 0$ on the domain of f, which means that f is increasing on its entire domain. Therefore, it is not possible for $f(x) = 0$ to have more than one solution.

We consider using Fixed-Point Iteration to solve the equivalent equation

$$x = g(x) = x - \phi(x)f(x) = x - (1)(x + \ln x) = -\ln x.$$

That is, we choose $\phi(x) \equiv 1$ in (10.1). Let us try applying Fixed-Point Iteration in MATLAB, with the initial guess being the midpoint of the interval $[0.5, 0.6]$:

```
>> g=@(x)(-log(x))
g =
  function_handle with value:
    @(x)(-log(x))
>> x=0.55;
>> x=g(x)
x =
   0.597837000755620
>> x=g(x)
x =
   0.514437136173803
>> x=g(x)
x =
   0.664681915480620
```

> **Exploration 10.3.1** Try this for a few more iterations. What happens?

Clearly, we need to use a different approach for converting our original equation $f(x) = 0$ to an equivalent equation of the form $x = g(x)$, since we are obtaining iterates outside of the interval within which we know the solution exists.

What went wrong? To help us answer this question, we examine the error $e_k = x^{(k)} - x^*$. Suppose that $x = g(x)$ has a solution x^* in (a, b), as it does in this example, and that g is also *continuously* differentiable on (a, b), as is the case in this example. We can use the Mean Value Theorem to obtain

$$e_{k+1} = x^{(k+1)} - x^* = g(x^{(k)}) - g(x^*) = g'(\xi_k)(x^{(k)} - x^*) = g'(\xi_k)e_k,$$

where ξ_k lies between $x^{(k)}$ and x^*.

We do not yet know the conditions under which Fixed-Point Iteration will converge, but if it *does* converge, then it follows from the continuity of g' at x^* that it does so linearly with asymptotic error constant $|g'(x^*)|$, since, by the definition of ξ_k and the continuity of g',

$$\lim_{k \to \infty} \frac{|e_{k+1}|}{|e_k|} = \lim_{k \to \infty} |g'(\xi_k)| = |g'(x^*)|.$$

Recall, though, that for linear convergence, the asymptotic error constant $C = |g'(x^*)|$ must satisfy $C < 1$. Unfortunately, with $g(x) = -\ln x$, we have $|g'(x)| = |-1/x| > 1$ on the interval $[0.5, 0.6]$, so it is not surprising that the iteration diverged.

What if we could convert the original equation $f(x) = 0$ into an equation of the form $x = g(x)$ so that g' satisfied $|g'(x)| < 1$ on an interval $[a, b]$ where a fixed point was known to exist? What we can do is take advantage of the differentiation rule

$$\frac{d}{dx}[f^{-1}(x)] = \frac{1}{f'(f^{-1}(x))}$$

and apply $g^{-1}(x) = e^{-x}$ to both sides of the equation $x = g(x)$ to obtain

$$g^{-1}(x) = g^{-1}(g(x)) = x,$$

which, in this case, simplifies to

$$x = e^{-x}.$$

The new function $g(x) = e^{-x}$ satisfies $|g'(x)| < 1$ on $[0.5, 0.6]$, as $g'(x) = -e^{-x}$, and $e^{-x} < 1$ when the argument x is positive. What happens if you try Fixed-Point Iteration with this choice of g?

```
>> g=@(x)(exp(-x))
g =

    function_handle with value:
    @(x)(exp(-x))
>> x=0.55;
>> x=g(x)
x =

   0.576949810380487
>> x=g(x)
x =

   0.561608769952327
>> x=g(x)
x =

   0.570290858658895
```

This is more promising.

> **Exploration 10.3.2** Continue this process to confirm that the iteration is in fact converging.

☐

Having seen what can go wrong if we are not careful in applying Fixed-Point Iteration, we should now address the questions of existence and uniqueness of a fixed point. The following result, first proved in [Brouwer (1911)], answers the first of these questions.

> **Theorem 10.3.3 (Brouwer Fixed Point Theorem)** Let g be continuous on $[a, b]$. If $g(x) \in [a, b]$ for each $x \in [a, b]$, then g has a fixed point in $[a, b]$.

> **Exploration 10.3.3** Use the Intermediate Value Theorem to prove The-
> orem 10.3.3.

Given a continuous function g that is known to have a fixed point in an interval $[a, b]$, we can try to find this fixed point by repeatedly evaluating g at points in $[a, b]$ until we find a point x for which $g(x) = x$, as in Algorithm 10.3.1. However, just because g has a fixed point does not mean that this iteration will necessarily converge. We will now investigate this further.

10.3.2 Convergence Analysis

Under what circumstances will Fixed-Point Iteration converge to a fixed point x^*? We say that a function g that is continuous on $[a, b]$ satisfies a **Lipschitz condition** on $[a, b]$ if there exists a positive constant L such that

$$|g(x) - g(y)| \le L|x - y|, \quad x, y \in [a, b].$$

The constant L is called a **Lipschitz constant**. If, in addition, $L < 1$, we say that g is a **contraction** on $[a, b]$.

If we denote the error in $x^{(k)}$ by $e_k = x^{(k)} - x^*$, we can see from the fact that $g(x^*) = x^*$ that if $x^{(k)} \in [a, b]$, then

$$|e_{k+1}| = |x^{(k+1)} - x^*| = |g(x^{(k)}) - g(x^*)| \le L|x^{(k)} - x^*| \le L|e_k| < |e_k|.$$

Therefore, if g satisfies the conditions of the Brouwer Fixed-Point Theorem, and g is a contraction on $[a, b]$, and $x^{(0)} \in [a, b]$, then Fixed-Point Iteration is *convergent* to x^*. Furthermore, the fixed point x^* must be *unique*, for if there exist two distinct fixed points x^* and y^* in $[a, b]$, then, by the Lipschitz condition, we have

$$0 < |x^* - y^*| = |g(x^*) - g(y^*)| \le L|x^* - y^*| < |x^* - y^*|,$$

which is a contradiction; thus, $x^* = y^*$. We summarize our findings with the statement of the following result, first established in [Banach (1922)].

> **Theorem 10.3.4 (Contraction Mapping Theorem)** Let g be a con-
> tinuous function on the interval $[a, b]$. If $g(x) \in [a, b]$ for each $x \in [a, b]$, and
> if there exists a constant $0 < L < 1$ such that
> $$|g(x) - g(y)| \le L|x - y|, \quad x, y \in [a, b],$$
> then g has a unique fixed point x^* in $[a, b]$, and the sequence of iterates
> $\{x^{(k)}\}_{k=0}^{\infty}$ produced by Fixed-Point Iteration converges to x^*, for any initial
> guess $x^{(0)} \in [a, b]$.

In general, when Fixed-Point Iteration converges, it does so at a rate that varies *inversely* with the Lipschitz constant L.

If g satisfies the conditions of the Contraction Mapping Theorem with Lipschitz constant L, then Fixed-Point Iteration achieves at least linear convergence, with an asymptotic error constant that is bounded above by L. This value can be used to estimate the total number of iterations needed for a specified degree of accuracy.

From the Lipschitz condition, we have, for $k \geq 1$,

$$|x^{(k)} - x^*| \leq L|x^{(k-1)} - x^*| \leq L^k|x^{(0)} - x^*|.$$

From

$$|x^{(0)} - x^*| \leq |x^{(0)} - x^{(1)}| + |x^{(1)} - x^*| \leq |x^{(0)} - x^{(1)}| + L|x^{(0)} - x^*|$$

we obtain

$$|x^{(k)} - x^*| \leq \frac{L^k}{1 - L}|x^{(1)} - x^{(0)}|. \tag{10.2}$$

We can bound the number of iterations after performing a single iteration, as long as the Lipschitz constant L is known.

Exploration 10.3.4 Use (10.2) to obtain a lower bound on the number of iterations required to ensure that $|x^{(k)} - x^*| \leq \epsilon$ for some error tolerance ϵ.

We can now develop a practical implementation of Fixed-Point Iteration.

Exploration 10.3.5 Write a MATLAB function

$$\texttt{[x,niter]=fixedpt(g,x0,tol)}$$

that implements Algorithm 10.3.1 to solve $x = g(x)$ with initial guess x0, except that instead of using the absolute difference between iterates to test for convergence, the error estimate (10.2) is compared to the specified tolerance $\texttt{tol}$, and a minimum number of iterations is determined based on the result of Exploration 10.3.4. Estimate L with the lower bound

$$L \geq \max_{1 \leq j \leq k} \left| \frac{g(x_j) - g(x_{j-1})}{x_j - x_{j-1}} \right|.$$

The output arguments $\texttt{x}$ and $\texttt{niter}$ are the computed solution x^* and number of iterations, respectively. Test your function on Example 10.3.2.

We know that Fixed-Point Iteration will converge to the unique fixed point in $[a, b]$ if g satisfies the conditions of the Contraction Mapping Theorem. However, if g is differentiable on $[a, b]$, its derivative can be used to obtain an alternative criterion for convergence that can be more practical than computing the Lipschitz constant L. We can see from the Mean Value Theorem and the fact that $g(x^*) = x^*$ that

$$x^{(k+1)} - x^* = g(x^{(k)}) - g(x^*) = g'(\xi_k)(x^{(k)} - x^*)$$

where ξ_k lies between $x^{(k)}$ and x^*. However, from

$$|g'(\xi_k)| = \left| \frac{g(x^{(k)}) - g(x^*)}{x^{(k)} - x^*} \right|$$

it follows that if $|g'(x)| \le L$ on (a, b), where $L < 1$, then the Contraction Mapping Theorem applies. This leads to the following result.

> **Theorem 10.3.5 (Fixed-Point Theorem)** Let g be a continuous function on the interval $[a, b]$, and let g be differentiable on (a, b). If $g(x) \in [a, b]$ for each $x \in [a, b]$, and if there exists a constant $L < 1$ such that
>
> $$|g'(x)| \le L, \quad x \in (a, b),$$
>
> then the sequence of iterates $\{x^{(k)}\}_{k=0}^{\infty}$ converges to the unique fixed point x^* of g in $[a, b]$, for any initial guess $x^{(0)} \in [a, b]$.

Using the Mean Value Theorem, it can also be shown that if g' is continuous at x^* and $|g'(x^*)| < 1$, then Fixed-Point Iteration is **locally convergent**; that is, it converges if $x^{(0)}$ is chosen sufficiently close to x^* [Süli and Mayers (2003)].

It can be seen from the preceding discussion why $g'(x)$ must be bounded away from 1 on (a, b), as opposed to the weaker condition $|g'(x)| < 1$ on (a, b). If $g'(x)$ is allowed to approach 1 as x approaches a point $c \in (a, b)$, then it is possible that the error e_k might not approach zero as k increases, in which case Fixed-Point Iteration would not converge.

> **Exploration 10.3.6** Find a function g and interval $[a, b]$ such that g continuous on $[a, b]$ and differentiable on (a, b), but does *not* satisfy a Lipschitz condition on $[a, b]$ for any Lipschitz constant L.

The derivative can also be used to indicate why Fixed-Point Iteration might *not* converge.

Example 10.3.6 The function $g(x) = x^2 + \frac{3}{16}$ has two fixed points, $x_1^* = 1/4$ and $x_2^* = 3/4$, as can be determined by solving the quadratic equation $x^2 + \frac{3}{16} = x$. If we consider the interval $[0, 3/8]$, then g satisfies the conditions of the Fixed-Point Theorem, as $g'(x) = 2x < 1$ on this interval, and therefore Fixed-Point Iteration will converge to x_1^* for any $x^{(0)} \in [0, 3/8]$.

On the other hand, $g'(3/4) = 3/2 > 1$. Therefore, it is not possible for g to satisfy the conditions of the Fixed-Point Theorem on any interval containing $x = 3/$. Furthermore, if $x^{(0)}$ is chosen so that $1/4 < x^{(0)} < 3/4$, then Fixed-Point Iteration will converge to $x_1^* = 1/4$, whereas if $x^{(0)} > 3/4$, then Fixed-Point Iteration *diverges*. $\square$

The fixed point $x_2^* = 3/4$ in the preceding example is an **unstable fixed point** of g, meaning that *no* choice of $x^{(0)}$ yields a sequence of iterates that converges to x_2^*. The fixed point $x_1^* = 1/4$ is a **stable fixed point** of g, meaning that *any* choice of $x^{(0)}$ that is sufficiently close to x_1^* yields a sequence of iterates that converges to x_1^*.

The preceding example shows that Fixed-Point Iteration applied to an equation of the form $x = g(x)$ can fail to converge to a fixed point x^* if $|g'(x^*)| > 1$. We wish to determine whether this condition indicates non-convergence in general. If $|g'(x^*)| > 1$, and g' is continuous in a neighborhood of x^*, then there exists an interval $|x - x^*| \leq \delta$ such that $|g'(x)| > 1$ on the interval. If $x^{(k)}$ lies within this interval, it follows from the Mean Value Theorem that

$$|x^{(k+1)} - x^*| = |g(x^{(k)}) - g(x^*)| = |g'(\eta)||x^{(k)} - x^*|,$$

where η lies between $x^{(k)}$ and x^*. Because η is also in this interval, we have

$$|x^{(k+1)} - x^*| > |x^{(k)} - x^*|.$$

In other words, the error in the iterates *increases* whenever they fall within a sufficiently small interval containing the fixed point. Because of this increase, the iterates must eventually fall outside of the interval. Therefore, it is not possible to find a k_0, for any given δ, such that $|x^{(k)} - x^*| \leq \delta$ for all $k \geq k_0$. We have thus proven the following result.

Theorem 10.3.7 Let g have a fixed point at x^*, and let g' be continuous in a neighborhood of x^*. If $|g'(x^*)| > 1$, then Fixed-Point Iteration does not converge to x^* for any initial guess $x^{(0)}$ except in a finite number of iterations.

Now, suppose that in addition to the conditions of the Fixed-Point Theorem, we assume that $g'(x^*) = 0$, and that g is twice continuously differentiable on $[a, b]$. Then, using Taylor's Theorem, we obtain

$$e_{k+1} = g(x^{(k)}) - g(x^*) = g'(x^*)(x^{(k)} - x^*) + \frac{1}{2}g''(\xi_k)(x^{(k)} - x^*)^2 = \frac{1}{2}g''(\xi_k)e_k^2,$$

where ξ_k lies between $x^{(k)}$ and x^*. It follows that for any initial iterate $x^{(0)} \in [a, b]$, Fixed-Point Iteration converges at least *quadratically*, with asymptotic error constant $|g''(x^*)/2|$. Soon, this will be exploited to obtain a quadratically convergent method for solving nonlinear equations of the form $f(x) = 0$.

10.3.3   Relaxation*

Now that we understand the convergence behavior of Fixed-Point Iteration, we consider the application of Fixed-Point Iteration to the solution of an equation of the form $f(x) = 0$.

Example 10.3.8 We use Fixed-Point Iteration to solve the equation $f(x) = 0$, where $f(x) = x - \cos x - 2$. It makes sense to work with the equation $x = g(x)$, where $g(x) = \cos x + 2$.

Where should we look for a solution to this equation? For example, consider the interval $[0, \pi/4]$. On this interval, $g'(x) = -\sin x$, which certainly satisfies the condition $|g'(x)| \leq \rho < 1$ where $\rho = \sqrt{2}/2$, but g does *not* map this interval into itself, as required by the Brouwer Fixed-Point Theorem. On the other hand, if we consider the interval $[1, 3]$, it can readily be confirmed that $g(x)$ maps this interval to itself, as $1 \leq 2 + \cos x \leq 3$ for all real x, so a fixed point exists in this interval.

First, let's set up a figure with a graph of $g(x)$:

```
>> g=@(x)(cos(x)+2);
>> fplot(g,1,3)
>> hold on
>> plot([ 1 3 ],[ 1 3 ],'k--')
>> xlabel('x')
>> ylabel('y')
>> title('g(x) = cos x + 2')
```

Exploration 10.3.7 Go ahead and try Fixed-Point Iteration on $g(x)$, with initial guess $x^{(0)} = 2$. What happens?

The behavior is quite interesting, as the iterates seem to bounce back and forth. This is illustrated in Figure 10.5. Continuing, we see that convergence is achieved, but it

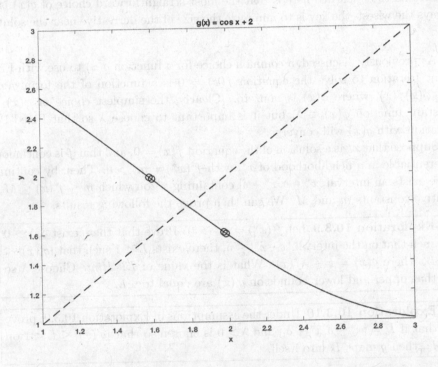

Fig. 10.5   Fixed-Point Iteration applied to $g(x) = \cos x + 2$, from Example 10.3.8.

is quite slow. An examination of the derivative explains why: $g'(x) = -\sin x$, and we have $|g'(\pi/2)| = |-\sin \pi/2| = 1$, so the conditions of the Fixed-Point Theorem are *not* satisfied—in fact, we could not be assured of convergence at all, though it does occur in this case.

An examination of the iterates shown in Figure 10.5, along with an indication of the solution, suggests how convergence can be accelerated. What if we used the *average* of two consecutive iterates, x and $g(x)$, at each iteration? That is, we solve $x = h(x)$, where

$$h(x) = \frac{1}{2}[x + g(x)] = \frac{1}{2}[x + \cos x + 2].$$

It can be verified that if $x = h(x)$, then $f(x) = 0$. However, we have

$$h'(x) = \frac{1}{2}[1 - \sin x],$$

and how large can this be on the interval $[1, 3]$? In this case, the Fixed-Point Theorem *does* apply.

Exploration 10.3.8 Try Fixed-Point Iteration with $h(x)$, and with initial guess $x^{(0)} = 2$. What behavior can you observe?

The lesson to be learned here is that the most straightforward choice of $g(x)$ is not always the wisest–the key is to minimize the size of the derivative near the solution. □

As previously discussed, a common choice for a function $g(x)$ to use with Fixed-Point Iteration to solve the equation $f(x) = 0$ is a function of the form $g(x) = x - \phi(x)f(x)$, where $\phi(x)$ is nonzero. Clearly, the simplest choice of $\phi(x)$ is a constant function $\phi(x) \equiv \lambda$, but it is important to choose λ so that Fixed-Point Iteration with $g(x)$ will converge.

Suppose that x^* is a solution of the equation $f(x) = 0$, and that f is continuously differentiable in a neighborhood of x^*, with $f'(x^*) = m^* > 0$. Then, by continuity, there exists an interval $[x^* - \delta, x^* + \delta]$ containing x^* on which $m \le f'(x) \le M$, for positive constants m and M. We can then prove the following results.

Exploration 10.3.9 Let $f'(x^*) = m^* > 0$. Prove that there exist $\delta, \lambda > 0$ such that on the interval $|x - x^*| \le \delta$, there exists $L < 1$ such that $|g'(x)| \le L$, where $g(x) = x - \lambda f(x)$. What is the value of L? *Hint:* Choose λ so that upper and lower bounds on $g'(x)$ are equal to $\pm L$.

Exploration 10.3.10 Under the assumptions of Exploration 10.3.9, prove that if $I_\delta = [x^* - \delta, x^* + \delta]$, and $\lambda > 0$ is chosen so that $|g'(x)| \le L < 1$ on I_δ, then g maps I_δ into itself.

We conclude from the preceding two explorations that the Fixed-Point Theorem (Theorem 10.3.5) applies, and therefore Fixed-Point Iteration converges linearly to x^* for any choice of $x^{(0)}$ in $[x^* - \delta, x^* + \delta]$, with asymptotic error constant $|1 - \lambda m^*| \le L$.

In summary, if f is continuously differentiable in a neighborhood of a root x^* of $f(x) = 0$, and $f(x^*)$ is nonzero, then there exists a constant λ such that Fixed-Point Iteration with $g(x) = x - \lambda f(x)$ converges to x^* for $x^{(0)}$ chosen sufficiently close to x^*. This approach to Fixed-Point Iteration, with a constant ϕ, is known as

relaxation. Convergence can be accelerated by allowing λ to vary from iteration to iteration. Intuitively, an effective choice is to try to minimize $|g'(x)|$ near x^* by setting $\lambda = 1/f'(x^{(k)})$, for each k, so that $g'(x^{(k)}) = 1 - \lambda f'(x^{(k)}) = 0$. This results in linear convergence with an asymptotic error constant of 0, which indicates faster than linear convergence. We will see that convergence is actually *quadratic*.

10.3.4   Concept Check

1. Describe the method of Fixed-Point Iteration for finding a solution of the equation $x = g(x)$.
2. Under what conditions does the equation $x = g(x)$ have a solution? Under what additional conditions is this solution unique?
3. How rapidly does Fixed-Point Iteration converge, and what is the asymptotic error constant?
4. How can Fixed-Point Iteration be applied to solve an equation of the form $f(x) = 0$?
5. Under what conditions does Fixed-Point Iteration fail to converge to a solution x^* of $x = g(x)$, even if the initial guess $x^{(0)}$ is chosen to be very near x^*?
6. What is meant by relaxation in the context of solving the equation $f(x) = 0$?

10.4   Newton's Method and the Secant Method

To develop a more effective method for solving this problem of computing a solution to $f(x) = 0$, we can address the following questions:

- Are there cases in which this problem easy to solve, and if so, how do we solve it in such cases?
- Is it possible to extend our method of solving the problem in these "easy" cases to more general cases?

As demonstrated (and stated) earlier, a recurring theme in this book is that these questions are useful for solving a variety of problems.

10.4.1   Newton's Method

For the problem at hand, we ask whether the equation $f(x) = 0$ is easy to solve for any particular choice of the function f. This is certainly the case if f is a linear function, which can be written in the form $f(x) = m(x - a) + b$, where m and b are constants and $m \neq 0$. Setting $f(x) = 0$ yields the equation

$$m(x - a) + b = 0,$$

which can easily be solved for x to obtain the unique solution

$$x = a - \frac{b}{m}.$$

We now consider the case where f is *not* a linear function. For example, recall when we solved the equation $f(x) = x - \cos x$ using the Bisection Method. In Figure 10.2, note that near the root, f is well-approximated by a linear function. How shall we exploit this? Using Taylor's Theorem (Theorem A.6.1), it is simple to construct a linear function that approximates $f(x)$ near a given point x_0. This function is simply the first Taylor polynomial of $f(x)$ with center x_0,

$$P_1(x) = f(x_0) + f'(x_0)(x - x_0).$$

This function has a useful geometric interpretation, as its graph is the tangent line of $f(x)$ at the point $(x_0, f(x_0))$.

We will illustrate this in MATLAB, for the example $f(x) = x - \cos x$, and initial guess $x_0 = 1$. The following code plots $f(x)$ and the tangent line at $(x_0, f(x_0))$.

```
>> f=@(x)(x-cos(x));
>> a=0.5;
>> b=1.5;
>> % plot f(x) on [a,b]
>> fplot(f,a,b)
>> hold on
>> % plot x-axis
>> plot([ a b ],[ 0 0 ],'k')
>> x0=1;
>> % plot initial guess on graph of f(x)
>> plot(x0,f(x0),'ro')
>> % f'(x) = 1 + sin(x)
>> % slope of tangent line: m = f'(x0)
>> m=1+sin(x0);
>> % plot tangent line using points x=a,b
>> plot([ a b ],[ f(x0) + m*([ a b ] - x0) ],'r')
>> xlabel('x')
>> ylabel('y')
```

> **Exploration 10.4.1** Rearrange the formula for the tangent line approximation $P_1(x)$ to obtain a formula for its x-intercept x_1. Compute this value in MATLAB and plot the point $(x_1, f(x_1))$ as a red '+'.

The plot that should be obtained from the preceding code and Exploration 10.4.1 is shown in Figure 10.6.

As can be seen in Figure 10.6, we can obtain an approximate solution to the equation $f(x) = 0$ by determining where the linear function $P_1(x)$ is equal to zero. If the resulting value, x_1, is not a solution, then we can repeat this process, approximating f by a linear function near x_1 and once again determining where this approximation is equal to zero.

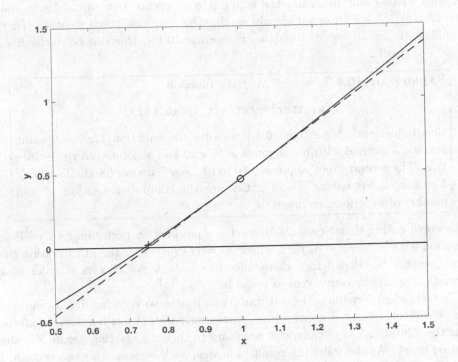

Fig. 10.6 Approximating a root of $f(x) = x - \cos x$ (solid curve) using the tangent line of $f(x)$ at $x_0 = 1$ (dashed line).

Exploration 10.4.2 Modify the above MATLAB statements to effectively "zoom in" on the graph of $f(x)$ near $x = x_1$, the zero of the tangent line approximation $P_1(x)$ above. Use the tangent line at $(x_1, f(x_1))$ to compute a second approximation x_2 of the root of $f(x)$. What do you observe?

The algorithm that results from repeating this process of approximating a root of $f(x)$ using tangent line approximations is known as **Newton's Method**, which we now describe in detail.

Algorithm 10.4.1 (Newton's Method) Let $f : D \subseteq \mathbb{R} \to \mathbb{R}$ be a differentiable function. The following algorithm computes a solution x^* of the equation $f(x) = 0$.

Choose an initial guess $x^{(0)}$
for $k = 0, 1, 2, \ldots$ **do** until convergence
$$x^{(k+1)} = x^{(k)} - \frac{f(x^{(k)})}{f'(x^{(k)})}$$
end for
$x^* = x^{(k)}$

As with Fixed-Point Iteration, the simplest convergence test would be to check whether $|x^{(k+1)} - x^{(k)}|$ is sufficiently small. One can also check whether $|f(x^{(k)})|$ is sufficiently small, but as discussed in Section 10.1.2, this can be misleading if $|f'(x^*)|$ is small.

Exploration 10.4.3 Write a MATLAB function

```
[x,niter]=newton(f,fp,x0,tol)
```

that implements Algorithm 10.4.1 to solve the equation $f(x) = 0$ using Newton's Method with initial guess $x^{(0)} = $ x0 and absolute error tolerance tol. The second input argument fp must be a function handle for $f'(x)$. The output arguments x and niter are the computed solution x^* and number of iterations, respectively.

It is worth noting that Newton's Method is equivalent to performing Fixed-Point Iteration with $g(x) = x - \lambda_k f(x)$, where $\lambda_k = 1/f'(x^{(k)})$ for the kth iteration that computes $x^{(k+1)}$. Recall that this choice of constant was chosen in order make relaxation as rapidly convergent as possible.

In fact, when Newton's Method converges, it does so very rapidly. However, it can be difficult to ensure convergence, particularly if $f(x)$ has horizontal tangents near the solution x^*. Typically, it is necessary to choose a starting iterate $x^{(0)}$ that is close to x^*. As the following result indicates, such a choice, if close enough, is indeed sufficient for convergence [Süli and Mayers (2003)].

Theorem 10.4.2 (Convergence of Newton's Method) Let f be twice continuously differentiable on the interval $[a, b]$, and suppose that $f(c) = 0$ and $f'(c) \neq 0$ for some $c \in [a, b]$. Then there exists a $\delta > 0$ such that Newton's Method applied to $f(x)$ converges to c for any initial guess $x^{(0)}$ in the interval $[c - \delta, c + \delta]$.

Example 10.4.3 We will use Newton's Method to compute a root of $f(x) = x - \cos x$. Since $f'(x) = 1 + \sin x$, it follows that in Newton's Method, we can obtain the next iterate $x^{(k+1)}$ from the previous iterate $x^{(k)}$ by

$$x^{(k+1)} = x^{(k)} - \frac{f(x^{(k)})}{f'(x^{(k)})} = x^{(k)} - \frac{x^{(k)} - \cos x^{(k)}}{1 + \sin x^{(k)}} = \frac{x^{(k)} \sin x^{(k)} + \cos x^{(k)}}{1 + \sin x^{(k)}}.$$

We choose our starting iterate $x^{(0)} = 1$, and compute the next several iterates as follows:

$$x^{(1)} = \frac{(1)\sin 1 + \cos 1}{1 + \sin 1} = 0.750363867840244,$$

$$x^{(2)} = \frac{x^{(1)} \sin x^{(1)} + \cos x^{(1)}}{1 + \sin x^{(1)}} = 0.739112890911362,$$

$$x^{(3)} = 0.739085133385284,$$

$$x^{(4)} = 0.739085133215161,$$

$$x^{(5)} = 0.739085133215161.$$

Since the fourth and fifth iterates agree to 15 decimal places, we assume that 0.739085133215161 is a correct solution to $f(x) = 0$, to at least 15 decimal places. □

Exploration 10.4.4 How many correct decimal places are obtained in each $x^{(k)}$ in Example 10.4.3? What does this suggest about the order of convergence of Newton's Method?

We can see from this example that Newton's Method converged to a root far more rapidly than the Bisection method did when applied to the same function. However, unlike Bisection, Newton's Method is not guaranteed to converge. Whereas Bisection is guaranteed to converge to a root in $[a, b]$ if $f(a)f(b) < 0$, we must be careful about our choice of the initial guess $x^{(0)}$ for Newton's Method, as the following example illustrates.

Example 10.4.4 Newton's Method can be used to compute the reciprocal of a number a *without performing any divisions*. The reciprocal, $1/a$, satisfies the equation $f(x) = 0$, where

$$f(x) = a - \frac{1}{x}.$$

Since

$$f'(x) = \frac{1}{x^2},$$

it follows that in Newton's Method, we can obtain the next iterate $x^{(k+1)}$ from the previous iterate $x^{(k)}$ by

$$x^{(k+1)} = x^{(k)} - \frac{a - 1/x^{(k)}}{1/[x^{(k)}]^2} = x^{(k)} - \frac{a}{1/[x^{(k)}]^2} + \frac{1/x^{(k)}}{1/[x^{(k)}]^2} = 2x^{(k)} - a[x^{(k)}]^2.$$

Note that no divisions are necessary to obtain $x^{(k+1)}$ from $x^{(k)}$. In the 1960's, this iteration was actually used on IBM System/360 computers to carry out division.

We use this iteration to compute the reciprocal of $a = 12$. Choosing our starting iterate to be $x^{(0)} = 0.1$, we compute the next several iterates as follows:

$$x^{(1)} = 2(0.1) - 12(0.1)^2 = 0.08,$$

$$x^{(2)} = 2(0.12) - 12(0.12)^2 = 0.0832,$$

$$x^{(3)} = 0.0833312,$$

$$x^{(4)} = 0.08333333333279,$$

$$x^{(5)} = 0.08333333333333.$$

We conclude that 0.08333333333333 is an accurate approximation to the correct solution.

Now, suppose we repeat this process, but with an initial iterate of $x^{(0)} = 1$. Then, we have

$$x^{(1)} = 2(1) - 12(1)^2 = -10,$$

$$x^{(2)} = 2(-10) - 12(-10)^2 = -1220,$$

$$x^{(3)} = 2(-1220) - 12(-1220)^2 = -17863240.$$

It is clear that this sequence of iterates is not going to converge to the correct solution. In general, for this iteration to converge to the reciprocal of a, the initial iterate $x^{(0)}$ must be chosen so that $0 < x^{(0)} < 2/a$. This condition guarantees that the next iterate $x^{(1)}$ will at least be positive. The contrast between the two choices of $x^{(0)}$ are illustrated in Figure 10.7 for $a = 8$. $\square$

10.4.2   Convergence Analysis

We now analyze the convergence of Newton's Method applied to the equation $f(x) = 0$, where we assume that f is twice continuously differentiable near the exact solution x^*. As before, we define $e_k = x^{(k)} - x^*$ to be the error after k iterations. Using a Taylor expansion around $x^{(k)}$, we obtain

$$
\begin{aligned}
e_{k+1} &= x^{(k+1)} - x^* \\
&= x^{(k)} - \frac{f(x^{(k)})}{f'(x^{(k)})} - x^* \\
&= e_k - \frac{f(x^{(k)})}{f'(x^{(k)})} \\
&= e_k - \frac{1}{f'(x^{(k)})}\left[f(x^*) - f'(x^{(k)})(x^* - x^{(k)}) - \frac{1}{2}f''(\xi_k)(x^{(k)} - x^*)^2\right] \\
&= \frac{f''(\xi_k)}{2f'(x^{(k)})}e_k^2
\end{aligned}
$$

where ξ_k is between $x^{(k)}$ and x^*.

Because, for each k, ξ_k lies between $x^{(k)}$ and x^*, ξ_k converges to x^* as well. By the continuity of f'', we conclude that Newton's Method converges quadratically to x^*, with asymptotic error constant

$$C = \left|\frac{f''(x^*)}{2f'(x^*)}\right|.$$

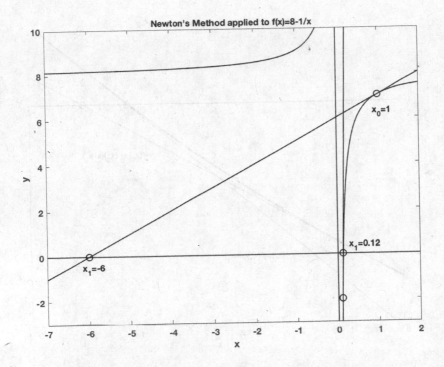

Fig. 10.7 Newton's Method used to compute the reciprocal of 8 by solving the equation $f(x) = 8 - 1/x = 0$. When $x^{(0)} = 0.1$, the tangent line of $f(x)$ at $(x^{(0)}, f(x^{(0)}))$ crosses the x-axis at $x^{(1)} = 0.12$, which is close to the exact solution. When $x^{(0)} = 1$, the tangent line crosses the x-axis at $x^{(1)} = -6$, which causes searching to continue on the wrong portion of the graph, so the sequence of iterates does not converge to the correct solution.

Example 10.4.5 Suppose that Newton's Method is used to find the solution of $f(x) = 0$, where $f(x) = x^2 - 2$. We examine the error $e_k = x^{(k)} - x^*$, where $x^* = \sqrt{2}$ is the exact solution. The first two iterations are illustrated in Figure 10.8. Continuing, we obtain

k	$x^{(k)}$	$|e_k|$
0	1	0.41421356237310
1	1.5	0.08578643762690
2	1.41666666666667	0.00245310429357
3	1.41421568627457	0.00000212390141
4	1.41421356237469	0.00000000000159

We can determine analytically that Newton's Method converges quadratically, and in this example, the asymptotic error constant is $|f''(\sqrt{2})/2f'(\sqrt{2})| \approx 0.35355$. Examining the numbers in the table above, we can see that the number of correct decimal places approximately doubles with each iteration, which is typical of

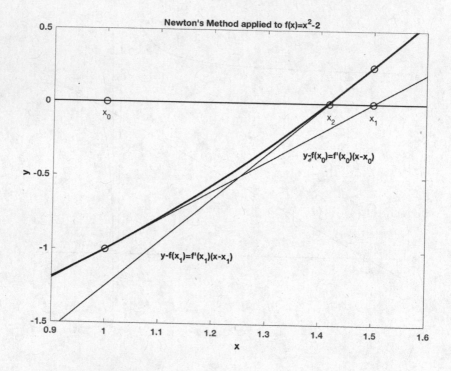

Fig. 10.8   Newton's Method applied to $f(x) = x^2 - 2$, from Example 10.4.5. The bold curve is the graph of f. The initial iterate $x^{(0)}$ is chosen to be 1. The tangent line of $f(x)$ at the point $(x^{(0)}, f(x^{(0)}))$ is used to approximate $f(x)$, and it crosses the x-axis at $x^{(1)} = 1.5$, which is much closer to the exact solution than $x^{(0)}$. Then, the tangent line at $(x^{(1)}, f(x^{(1)}))$ is used to approximate $f(x)$, and it crosses the x-axis at $x^{(2)} = 1.41\bar{6}$, which is already very close to the exact solution.

quadratic convergence. Furthermore, we have

$$\frac{|e_4|}{|e_3|^2} \approx 0.35352,$$

so the actual behavior of the error is consistent with the behavior that is predicted by theory. $\square$

It is easy to see from the above analysis, however, that if $f'(x^*)$ is very small, or zero, then convergence can be very slow, or may not even occur.

Example 10.4.6 The function

$$f(x) = (x - 1)^2 e^x$$

has a double root at $x^* = 1$, and therefore $f'(x^*) = 0$. Therefore, the previous convergence analysis does not apply. Instead, we obtain

$$e_{k+1} = x^{(k+1)} - 1$$

$$= x^{(k)} - \frac{f(x^{(k)})}{f'(x^{(k)})} - 1$$

$$= x^{(k)} - \frac{(x^{(k)} - 1)^2 e^{x^{(k)}}}{[2(x^{(k)} - 1) + (x^{(k)} - 1)^2]e^{x^{(k)}}} - 1$$

$$= e_k - \frac{e_k^2}{2e_k + e_k^2}$$

$$= \frac{x^{(k)}}{x^{(k)} + 1} e_k.$$

It follows that if we choose $x^{(0)} > 0$, then Newton's Method converges to $x^* = 1$ *linearly*, with asymptotic error constant $C = \frac{1}{2}$. $\square$

Exploration 10.4.5 Prove that if $f(x^*) = 0$, $f'(x^*) = 0$, and $f''(x^*) \neq 0$, then Newton's Method converges linearly with asymptotic error constant $C = \frac{1}{2}$. That is, the result obtained in Example 10.4.6 applies to a general twice-differentiable function with a double root.

Normally, convergence of Newton's Method is only assured if $x^{(0)}$ is chosen sufficiently close to x^*. However, in some cases, it is possible to prove that Newton's Method converges quadratically on an interval, under certain conditions on the sign of the derivatives of f on that interval. For example, suppose that on the interval $I_\delta = [x^*, x^* + \delta]$, $f'(x) > 0$ and $f''(x) > 0$, so that f is increasing and concave up on this interval.

Let $x^{(k)} \in I_\delta$. Then, from

$$x^{(k+1)} = x^{(k)} - \frac{f(x^{(k)})}{f'(x^{(k)})},$$

we have $x^{(k+1)} < x^{(k)}$, because f, being equal to zero at x^* and increasing on I_δ, must be positive at $x^{(k)}$. However, because

$$x^{(k+1)} - x^* = \frac{f''(\xi_k)}{2f'(x^{(k)})}(x^{(k)} - x^*)^2, \quad \xi_k \in (x^*, x^{(k)}),$$

and f' and f'' are both positive at $x^{(k)}$, we must also have $x^{(k+1)} > x^*$.

It follows that the sequence $\{x^{(k)}\}$ is monotonic and bounded, and therefore must be convergent to a limit $x_* \in I_\delta$. From the convergence of the sequence and the determination of $x^{(k+1)}$ from $x^{(k)}$, it follows that $f(x_*) = 0$. However, f is positive on $(x^*, x^* + \delta]$, which means that we must have $x_* = x^*$, so Newton's Method converges to x^*. Using the previous analysis, it can be shown that this convergence is quadratic.

Exploration 10.4.6 Prove that Newton's Method converges quadratically if $f'(x) > 0$ and $f''(x) < 0$ on an interval $[a, b]$ that contains x^* and $x^{(0)}$, where $x^{(0)} < x^*$. What goes wrong if $x^{(0)} > x^*$?

10.4.3 The Secant Method

One drawback of Newton's Method is that it is necessary to evaluate $f'(x)$ at each iterate $x^{(k)}$, which may not be practical for some choices of f. The **Secant Method** avoids this issue by using a finite difference (see Section 9.1) to approximate the derivative. As a result, $f(x)$ is approximated by a **secant line** through two points on the graph of f, rather than a tangent line through one point on the graph.

Since a secant line is defined using two points on the graph of $f(x)$, as opposed to a tangent line that requires information at only one point on the graph, it is necessary to choose two initial iterates $x^{(0)}$ and $x^{(1)}$. Then, as in Newton's Method, the next iterate $x^{(2)}$ is then obtained by computing the x-value at which the secant line passing through the points $(x^{(0)}, f(x^{(0)}))$ and $(x^{(1)}, f(x^{(1)}))$ has a y-coordinate of zero. This yields the equation

$$\frac{f(x^{(1)}) - f(x^{(0)})}{x^{(1)} - x^{(0)}}(x^{(2)} - x^{(1)}) + f(x^{(1)}) = 0$$

which has the solution

$$x^{(2)} = x^{(1)} - \frac{f(x^{(1)})(x^{(1)} - x^{(0)})}{f(x^{(1)}) - f(x^{(0)})}.$$

This leads to the following algorithm.

Algorithm 10.4.7 (Secant Method) Let $f : D \subseteq \mathbb{R} \to \mathbb{R}$ be a continuous function. The following algorithm computes a solution x^* of the equation $f(x) = 0$.

Choose two initial guesses $x^{(0)}$ and $x^{(1)}$
$f_0 = f(x^{(0)})$, $f_1 = f(x^{(1)})$
for $k = 1, 2, 3, \ldots$ **do**
    $x^{(k+1)} = x^{(k)} - f_1 \frac{x^{(k)} - x^{(k-1)}}{f_1 - f_0}$
    $f_0 = f_1$
    $f_1 = f(x^{(k+1)})$
end for
$x^* = x^{(k)}$

Exploration 10.4.7 Write a MATLAB function

$$[\texttt{x,niter}]=\texttt{secant(f,x0,x1,tol)}$$

that implements Algorithm 10.4.7 to solve the equation $\texttt{f}(x) = 0$ using the Secant Method with initial guesses $x^{(0)} = \texttt{x0}$ and $x^{(1)} = \texttt{x1}$, and absolute error tolerance $\texttt{tol}$. The output arguments $\texttt{x}$ and $\texttt{niter}$ are the computed solution x^* and number of iterations, respectively.

Like Newton's Method, it is necessary to choose the starting iterate $x^{(0)}$ to be reasonably close to the solution x^*. Convergence is not as rapid as that of Newton's

Method, since the secant-line approximation of f is not as accurate as the tangent-line approximation employed by Newton's Method.

Example 10.4.8 We will use the Secant Method to solve the equation $f(x) = 0$, where $f(x) = x^2 - 2$. This method requires that we choose two initial iterates $x^{(0)}$ and $x^{(1)}$, and then compute subsequent iterates using the formula

$$x^{(k+1)} = x^{(k)} - \frac{f(x^{(k)})(x^{(k)} - x^{(k-1)})}{f(x^{(k)}) - f(x^{(k-1)})}, \quad k = 1, 2, 3, \ldots.$$

We choose $x^{(0)} = 1$ and $x^{(1)} = 1.5$. Applying the above formula, we obtain

$$x^{(2)} = 1.4,$$
$$x^{(3)} = 1.413793103448276,$$
$$x^{(4)} = 1.414215686274510,$$
$$x^{(5)} = 1.414213562057320.$$

As we can see, the iterates produced by the Secant Method are converging to the exact solution $x^* = \sqrt{2}$, but not as rapidly as those produced by Newton's Method. $\square$

Exploration 10.4.8 How many correct decimal places are obtained in each $x^{(k)}$ in the preceding example? What does this suggest about the order of convergence of the Secant Method?

We now prove that the Secant Method converges if $x^{(0)}$ is chosen sufficiently close to a solution x^* of $f(x) = 0$, if f is continuously differentiable near x^* and $f'(x^*) = m^* \neq 0$. Without loss of generality, we assume $m^* > 0$. Then, by the continuity of f', there exists an interval $I_\delta = [x^* - \delta, x^* + \delta]$ and a constant α, $0 < \alpha < 1/3$, such that

$$(1 - \alpha)m^* \leq f'(x) \leq (1 + \alpha)m^*, \quad x \in I_\delta.$$

It follows from the Mean Value Theorem that

$$x^{(k+1)} - x^* = x^{(k)} - x^* - f(x^{(k)})\frac{x^{(k)} - x^{(k-1)}}{f(x^{(k)}) - f(x^{(k-1)})}$$

$$= x^{(k)} - x^* - \frac{f'(\xi_k)(x^{(k)} - x^*)}{f'(\eta_k)}$$

$$= \left[1 - \frac{f'(\xi_k)}{f'(\eta_k)}\right](x^{(k)} - x^*),$$

where ξ_k lies between $x^{(k)}$ and x^*, and η_k lies between $x^{(k)}$ and $x^{(k-1)}$. Therefore, if $x^{(k-1)}$ and $x^{(k)}$ are in I_δ, then so are ξ_k and η_k, and $x^{(k+1)}$ satisfies

$$|x^{(k+1)} - x^*| \leq \max\left\{\left|1 - \frac{1 + \alpha}{1 - \alpha}\right|, \left|1 - \frac{1 - \alpha}{1 + \alpha}\right|\right\} |x^{(k)} - x^*| \leq \frac{2\alpha}{1 - \alpha}|x^{(k)} - x^*|.$$

We conclude that if $x^{(0)}, x^{(1)} \in I_\delta$, then all subsequent iterates lie in I_δ, and the Secant Method converges at least linearly, with asymptotic error constant $C = 2\alpha/(1-\alpha) < 1$. Because $\alpha \in (0, 1/3)$ is arbitrary, the asymptotic error constant can be made arbitrarily small, thus implying *superlinear* convergence.

The order of convergence of the Secant Method can be determined using the following result. We assume that $\{x^{(k)}\}_{k=0}^{\infty}$ is the sequence of iterates produced by the Secant Method for solving $f(x) = 0$, and that this sequence converges to a simple root x^*.

Exploration 10.4.9 Compute

$$S = \lim_{x^{(k-1)}, x^{(k)} \to x^*} \frac{x^{(k+1)} - x^*}{(x^{(k)} - x^*)(x^{(k-1)} - x^*)}.$$

Hint: Take one limit at a time, and use Taylor expansion. Assume that x^* is not a multiple root.

It follows from the preceding exploration that for k sufficiently large,

$$|x^{(k+1)} - x^*| \approx S|x^{(k)} - x^*||x^{(k-1)} - x^*|.$$

We assume that $\{x^{(k)}\}$ converges to x^* of order r. Then, dividing both sides of the above relation by $|x^{(k)} - x^*|^r$, we obtain

$$\frac{|x^{(k+1)} - x^*|}{|x^{(k)} - x^*|^r} \approx S|x^{(k)} - x^*|^{1-r}|x^{(k-1)} - x^*|.$$

Because r is the order of convergence, the left side must converge to a positive constant C as $k \to \infty$. It follows that the right side must converge to a positive constant as well, as must its reciprocal. In other words, there must exist positive constants C_1 and C_2 such that

$$\frac{|x^{(k)} - x^*|}{|x^{(k-1)} - x^*|^r} \to C_1, \quad \frac{|x^{(k)} - x^*|^{r-1}}{|x^{(k-1)} - x^*|} \to C_2.$$

Therefore, the ratios of the exponents must be the same; that is,

$$\frac{1}{r} = \frac{r-1}{1}.$$

This yields the quadratic equation

$$r^2 - r - 1 = 0,$$

which has the solutions

$$r_1 = \frac{1 + \sqrt{5}}{2} \approx 1.618, \quad r_2 = \frac{1 - \sqrt{5}}{2} \approx -0.618.$$

Since we must have $r \geq 1$, the order of convergence is 1.618.

Exploration 10.4.10 Use the value of S from Exploration 10.4.9, as well as the preceding discussion, to obtain the asymptotic error constant for the Secant Method.

Exploration 10.4.11 Use both Newton's Method and the Secant Method to compute a root of the same polynomial. For both methods, count the number of floating-point operations required in each iteration, and the number of iterations required to achieve convergence with the same error tolerance. Which method requires fewer floating-point operations?

10.4.4 Concept Check

1. How does Newton's Method approximate $f(x)$ for the purpose of finding an approximate root?
2. How rapidly can Newton's Method converge, and what is its asymptotic error constant? Under what circumstances is convergence slowed?
3. List two drawbacks of Newton's Method.
4. In what sense does the Secant Method approximate Newton's Method?
5. What is the order of convergence of the Secant Method?

10.5 Improvements to Root-Finding Methods*

In this section, we examine how the methods presented in this chapter can be modified or combined to produce new methods that are more efficient or robust.

10.5.1 Convergence Acceleration

Suppose that a sequence $\{x^{(k)}\}_{k=0}^{\infty}$ converges linearly to a limit x^*, in such a way that if k is sufficiently large, then $x^{(k)} - x^*$ has the same sign; that is, $\{x^{(k)}\}$ converges **monotonically** to x^*. It follows from the linear convergence of $\{x^{(k)}\}$ that for sufficiently large k,

$$\frac{x^{(k+2)} - x^*}{x^{(k+1)} - x^*} \approx \frac{x^{(k+1)} - x^*}{x^{(k)} - x^*}. \qquad (10.3)$$

Solving for x^* yields

$$x^* \approx x^{(k)} - \frac{(x^{(k+1)} - x^{(k)})^2}{x^{(k+2)} - 2x^{(k+1)} + x^{(k)}}. \qquad (10.4)$$

Exploration 10.5.1 Solve (10.3) for x^* to obtain (10.4).

Based on this approximation, we can construct a sequence $\{\hat{x}^{(k)}\}_{k=0}^{\infty}$, where

$$\hat{x}^{(k)} = x^{(k)} - \frac{(x^{(k+1)} - x^{(k)})^2}{x^{(k+2)} - 2x^{(k+1)} + x^{(k)}},$$

that also converges to x^*. This sequence has the following desirable property.

Theorem 10.5.1 Suppose that the sequence $\{x^{(k)}\}_{k=0}^{\infty}$ converges linearly to a limit x^* and that for k sufficiently large, $(x^{(k+1)} - x^*)(x^{(k)} - x^*) > 0$. Then, if the sequence $\{\hat{x}^{(k)}\}_{k=0}^{\infty}$ is defined by

$$\hat{x}^{(k)} = x^{(k)} - \frac{(x^{(k+1)} - x^{(k)})^2}{x^{(k+2)} - 2x^{(k+1)} + x^{(k)}}, \quad k = 0, 1, 2, \ldots,$$

then

$$\lim_{k \to \infty} \frac{\hat{x}^{(k)} - x^*}{x^{(k)} - x^*} = 0.$$

In other words, the sequence $\{\hat{x}^{(k)}\}$ converges to x^* more rapidly than $\{x^{(k)}\}$ does.

Exploration 10.5.2 Prove Theorem 10.5.1. Assume that the sequence $\{x^{(k)}\}_{k=0}^{\infty}$ converges linearly with asymptotic error constant C, where $0 < C < 1$. *Hint:* Take one limit at a time, as in Exploration 10.4.9.

Recall from Section 7.3.3 the **forward difference operator** Δ, defined by

$$\Delta x^{(k)} = x^{(k+1)} - x^{(k)}.$$

We have

$$\begin{aligned}
\Delta^2 x^{(k)} &= \Delta(x^{(k+1)} - x^{(k)}) \\
&= (x^{(k+2)} - x^{(k+1)}) - (x^{(k+1)} - x^{(k)}) \\
&= x^{(k+2)} - 2x^{(k+1)} + x^{(k)},
\end{aligned}$$

and therefore $\hat{x}^{(k)}$ can be rewritten as

$$\hat{x}^{(k)} = x^{(k)} - \frac{(\Delta x^{(k)})^2}{\Delta^2 x^{(k)}}, \quad k = 0, 1, 2, \ldots$$

For this reason, the method of accelerating the convergence of $\{x^{(k)}\}$ by constructing $\{\hat{x}^{(k)}\}$ is called **Aitken's Δ^2 Method** [Aitken (1932)].

A slight variation of this method, called **Steffensen's Method**, can be used to accelerate the convergence of Fixed-Point Iteration, which, as previously discussed, is linearly convergent. The basic idea is as follows:

1. Choose an initial iterate $x^{(0)}$
2. Compute $x^{(1)}$ and $x^{(2)}$ using Fixed-Point Iteration
3. Use Aitken's Δ^2 Method to compute $\hat{x}^{(0)}$ from $x^{(0)}$, $x^{(1)}$, and $x^{(2)}$.
4. Repeat steps 2 and 3 with $x^{(0)} = \hat{x}^{(0)}$.

The principle behind Steffensen's Method is that $\hat{x}^{(0)}$ is thought to be a better approximation to the fixed point x^* than $x^{(2)}$, so it should be used as the next iterate for Fixed-Point Iteration. The algorithm follows.

> **Algorithm 10.5.2 (Steffensen's Method)** Given a function $g(x)$ with a fixed point x^*, the following algorithm computes a solution x^* of the equation $x = g(x)$.
>
> Choose an initial guess $x_0^{(0)}$
> **for** $k = 0, 1, 2, \ldots$ **do** until convergence
> $$x_1^{(k)} = g(x_0^{(k)})$$
> $$x_2^{(k)} = g(x_1^{(k)})$$
> $$x_0^{(k+1)} = x_0^{(k)} - \frac{(x_1^{(k)} - x_0^{(k)})^2}{x_2^{(k)} - 2x_1^{(k)} + x_0^{(k)}}$$
> **end while**
> $x^* = x^{(k)}$

Example 10.5.3 We wish to find the unique fixed point of the function $f(x) = \cos x$ on the interval $[0, 1]$. If we use Fixed-Point Iteration with $x^{(0)} = 0.5$, then we obtain the following iterates from the formula $x^{(k+1)} = g(x^{(k)}) = \cos(x^{(k)})$. All iterates are rounded to five decimal places.

$$x^{(1)} = 0.87758,$$
$$x^{(2)} = 0.63901,$$
$$x^{(3)} = 0.80269,$$
$$x^{(4)} = 0.69478,$$
$$x^{(5)} = 0.76820.$$

These iterates show little sign of converging, as they are oscillating around the fixed point. If, instead, we use Fixed-Point Iteration accelerated by Aitken's Δ^2 method, we obtain a new sequence of iterates $\{\hat{x}^{(k)}\}$, where

$$\hat{x}^{(k)} = x^{(k)} - \frac{(\Delta x^{(k)})^2}{\Delta^2 x^{(k)}}$$
$$= x^{(k)} - \frac{(x^{(k+1)} - x^{(k)})^2}{x^{(k+2)} - 2x^{(k+1)} + x^{(k)}},$$

for $k = 0, 1, 2, \ldots$. The first few iterates of this sequence are

$$\hat{x}^{(0)} = 0.73139,$$
$$\hat{x}^{(1)} = 0.73609,$$
$$\hat{x}^{(2)} = 0.73765,$$
$$\hat{x}^{(3)} = 0.73847,$$
$$\hat{x}^{(4)} = 0.73880.$$

Clearly, these iterates are converging much more rapidly than Fixed-Point Iteration, as they are not oscillating around the fixed point, but convergence is still linear.

Finally, we try Steffensen's Method. We begin with the first three iterates of Fixed-Point Iteration,

$$x_0^{(0)} = 0.5, \quad x_1^{(0)} = \cos(x_0^{(0)}) = 0.87758, \quad x_2^{(0)} = \cos(x_1^{(0)}) = 0.63901.$$

Then, we use the formula from Aitken's Δ^2 Method to compute

$$x_0^{(1)} = x_0^{(0)} - \frac{(x_1^{(0)} - x_0^{(0)})^2}{x_2^{(0)} - 2x_1^{(0)} + x_0^{(0)}} = 0.73139.$$

We use this value to restart Fixed-Point Iteration and compute two iterates, which are

$$x_1^{(1)} = \cos(x_0^{(1)}) = 0.74425, \quad x_2^{(1)} = \cos(x_1^{(1)}) = 0.73560.$$

Repeating this process, we apply the formula from Aitken's Δ^2 Method to the iterates $x_0^{(1)}$, $x_1^{(1)}$ and $x_2^{(1)}$ to obtain

$$x_0^{(2)} = x_0^{(1)} - \frac{(x_1^{(1)} - x_0^{(1)})^2}{x_2^{(1)} - 2x_1^{(1)} + x_0^{(1)}} = 0.739076.$$

Restarting Fixed-Point Iteration with $x_0^{(2)}$ as the initial iterate, we obtain

$$x_1^{(2)} = \cos(x_0^{(2)}) = 0.739091, \quad x_2^{(2)} = \cos(x_1^{(2)}) = 0.739081.$$

The most recent iterate $x_2^{(2)}$ is correct to five decimal places.

Using all three methods to compute the fixed point to ten decimal digits of accuracy, we find that Fixed-Point Iteration requires 57 iterations. Aitken's Δ^2 Method requires us to compute 25 iterates of the modified sequence $\{\hat{x}^{(k)}\}$, which in turn requires 27 iterates of the sequence $\{x^{(k)}\}$, where the first iterate x_0 is given. Steffensen's Method requires us to compute $x_2^{(3)}$, which means that only 11 iterates need to be computed, 8 of which require a function evaluation. $\square$

Exploration 10.5.3 Write a MATLAB function

$$[\texttt{x,niter}]=\texttt{steffensen(g,x0,tol)}$$

that modifies your function `fixedpt` from Exploration 10.3.5 to accelerate convergence using Steffensen's Method. Test your function on the equation from Example 10.3.2. How much more rapidly do the iterates converge?

As Example 10.5.3 demonstrated, the convergence of Steffensen's Method is quite accelerated indeed. In fact, we have the following result, proved in [Isaacson and Keller (1966)].

Theorem 10.5.4 Let $\{x_k^{(0)}\}_{k=0}^{\infty}$ be the sequence of iterates generated by Steffensen's Method (Algorithm 10.5.2) from Fixed-Point Iteration applied to $x = g(x)$. Assume Fixed-Point Iteration converges to a root x^* that has multiplicity m, with order of convergence r.

- If $r \geq 2$, then the order of convergence of $\{x_k^{(0)}\}$ to x^* is $2r - 1$.
- If $r = 1$ and $m = 1$, then $\{x_k^{(0)}\}$ converges quadratically to x^*.
- If $r = 1$ and $m > 1$, then $\{x_k^{(0)}\}$ converges linearly to x^*, with asymptotic error constant $1 - 1/m$.

Exploration 10.5.4 Prove the second part of Theorem 10.5.4 in the case $g'(x^*) \neq 1$ by proving that if we define

$$G(x) = x - \frac{(g(x) - x)^2}{g(g(x)) - 2g(x) + x},$$

then Steffensen's Method applied to $x = g(x)$ is equivalent to Fixed-Point Iteration with $x = G(x)$, and $G'(x^*) = 0$. Explain why this implies quadratic convergence.

10.5.2 Safeguarded Methods

It is natural to ask whether it is possible to combine the rapid convergence of methods such as Newton's Method with "safe" methods such as the Bisection Method that, while slow, are guaranteed to converge. This leads to the concept of **safeguarded methods**, which maintain an interval within which a solution is known to exist, as in the Bisection Method, but use a method such as Newton's Method to find a solution within that interval. If an iterate falls outside this interval, the safe procedure is used to refine the interval before trying the rapid method again.

An example of a safeguarded method is the **Method of Regula Falsi**, which is also known as the **Method of False Position**. It is a modification of the Secant Method in which the two initial iterates $x^{(0)}$ and $x^{(1)}$ are chosen so that $f(x^{(0)}) \cdot f(x^{(1)}) < 0$, thus guaranteeing that a solution lies between $x^{(0)}$ and $x^{(1)}$. This condition also guarantees that the next iterate $x^{(2)}$ will lie between $x^{(0)}$ and $x^{(1)}$, as can be seen by applying the Intermediate Value Theorem to the secant line passing through $(x^{(0)}, f(x^{(0)}))$ and $(x^{(1)}, f(x^{(1)}))$.

It follows that if $f(x^{(2)}) \neq 0$, then a solution must lie between $x^{(0)}$ and $x^{(2)}$, or between $x^{(1)}$ and $x^{(2)}$. In the first scenario, we use the secant line passing through $(x^{(0)}, f(x^{(0)}))$ and $(x^{(2)}, f(x^{(2)}))$ to compute the next iterate $x^{(3)}$. Otherwise, we use the secant line passing through $(x^{(1)}, f(x^{(1)}))$ and $(x^{(2)}, f(x^{(2)}))$. Continuing in this fashion, we obtain a sequence of smaller and smaller intervals that are guaranteed to contain a solution, as in the Bisection Method, but the interval is updated using a superlinearly convergent method, the Secant Method.

Algorithm 10.5.5 (Method of Regula Falsi) Let $f : D \subseteq \mathbb{R} \to \mathbb{R}$ be a continuous function that changes sign on the interval (a, b). The following algorithm computes a solution x^* of the equation $f(x) = 0$.

repeat until convergence
  $c = b - \frac{f(b)(b-a)}{f(b)-f(a)}$
  if $f(a)f(c) < 0$ **then**
    $b = c$
  else
    $a = c$
  end if
end repeat
$x^* = c$

Example 10.5.6 We use the Method of Regula Falsi to solve $f(x) = 0$ where $f(x) = x^2 - 2$. First, we must choose two initial guesses $x^{(0)}$ and $x^{(1)}$ such that $f(x)$ changes sign between $x^{(0)}$ and $x^{(1)}$. Choosing $x^{(0)} = 1$ and $x^{(1)} = 1.5$, we see that $f(x^{(0)}) = f(1) = -1$ and $f(x^{(1)}) = f(1.5) = 0.25$, so these choices are suitable.

Next, we use the Secant Method to compute the next iterate $x^{(2)}$ by determining the point at which the secant line passing through the points $(x^{(0)}, f(x^{(0)}))$ and $(x^{(1)}, f(x^{(1)}))$ intersects the line $y = 0$. We have

$$x^{(2)} = x^{(0)} - \frac{f(x^{(0)})(x^{(1)} - x^{(0)})}{f(x^{(1)}) - f(x^{(0)})} = 1 - \frac{(-1)(1.5 - 1)}{0.25 - (-1)} = 1.4.$$

Computing $f(x^{(2)})$, we obtain $f(1.4) = -0.04 < 0$. Since $f(x^{(2)}) < 0$ and $f(x^{(1)}) > 0$, we can use the Intermediate Value Theorem to conclude that a solution exists in the interval $(x^{(2)}, x^{(1)})$. Therefore, we compute $x^{(3)}$ by determining where the secant line through the points $(x^{(1)}, f(x^{(1)}))$ and $f(x^{(2)}, f(x^{(2)}))$ intersects the line $y = 0$. Using the formula for the Secant Method, we obtain $x^{(3)} = 1.41379$.

Since $f(x^{(3)}) < 0$ and $f(x^{(2)}) < 0$, we do not know that a solution exists in the interval $(x^{(2)}, x^{(3)})$. However, we do know that a solution exists in the interval $(x^{(3)}, x^{(1)})$, because $f(x^{(1)}) > 0$. Therefore, instead of proceeding as in the Secant Method and using the Secant line determined by $x^{(2)}$ and $x^{(3)}$ to compute $x^{(4)}$, we use the secant line determined by $x^{(1)}$ and $x^{(3)}$ to compute $x^{(4)}$.

By the Intermediate Value Theorem, and the fact that this secant line intersects the graph of $f(x)$ at $x = x^{(1)}$ and $x = x^{(3)}$, we have that $x^{(4)}$ must fall between $x^{(3)}$ and $x^{(1)}$, so then we can continue looking for a solution in the interval $(x^{(3)}, x^{(4)})$ or the interval $(x^{(4)}, x^{(1)})$, depending on whether $f(x^{(4)})$ is positive or negative. In this sense, the Method of Regula Falsi is very similar to Bisection. The only difference is in how we divide the interval (a, b) that is known to contain a solution. $\square$

In the previous example, the right endpoint of each interval, $x^{(1)} = 1.5$, was not changing, while the left endpoint progressed toward a root. This is a disturbing development, because it indicates that unlike Bisection, the length of the search interval does not converge to zero; rather, the interval converges to $(x^*, x^{(1)})$. This can actually occur in a situation that is easy to characterize.

Exploration 10.5.5 Suppose that $f(a)f(b) < 0$, and that f is twice continuously differentiable on (a, b). Prove that if f'' does not change sign on (a, b), then one of the endpoints a or b in Algorithm 10.5.5 will never change. Prove that Regula Falsi converges linearly in this case, and compute the asymptotic error constant.

This issue can be fixed with a small modification, resulting in the **Illinois Algorithm** [Dahlquist and Björck (1974)]. The idea is this: if $f(x^{(k+1)})$ and $f(x^{(k)})$ have the same sign, then this indicates that the endpoint of the interval opposite $x^{(k)}$, which we will denote by $x^{(K)}$, is the one that is not changing. Therefore, we compute the next secant line between the points $(x^{(k+1)}, f(x^{(k+1)}))$ and $(x^{(K)}, \frac{1}{2}f(x^{(K)}))$, to force $x^{(k+2)}$ to be on the same side of the graph of $f(x)$ as $x^{(K)}$.

To implement this idea, we make the following change from Algorithm 10.5.5: instead of maintaining endpoints of a search interval (a, b) in such a way that $a < b$, we prescribe that $a = x^{(k-1)}$ and $b = x^{(k)}$ for $k = 1, 2, \ldots$, so that b is always referring to the previous iterate, rather than the right endpoint of the search interval. This facilitates the determination of whether it is necessary to alter the secant line.

Algorithm 10.5.7 (Illinois Algorithm) Let $f : D \subseteq \mathbb{R} \to \mathbb{R}$ be a continuous function that changes sign on the interval (a, b). The following algorithm computes a solution x^* of the equation $f(x) = 0$.

$f_a = f(a)$, $f_b = f(b)$
repeat until convergence
    $c = b - \dfrac{f_b(b-a)}{f_b - f_a}$
    $f_c = f(c)$
    if $f_b f_c < 0$ **then**
        $a = b$, $f_a = f_b$
    else
        $f_a = f_a/2$
    end if
    $b = c$, $f_b = f_c$
end repeat
$x^* = c$

In the case where $f(b)$ and $f(c)$ have the same sign, we infer that the endpoint a is not changing from iteration to iteration, so we reduce its y-value to alter the secant line. Otherwise, we set $a = b$ so that the next iterate, $x^{(k+2)}$, will be computed from the previous two iterates, $x^{(k)} = b$ and $x^{(k+1)} = c$. If we apply this algorithm to the function from Example 10.5.6, then we see that convergence is superlinear.

Exploration 10.5.6 Write a MATLAB function

$$[x,niter]=illinois(f,a,b,tol)$$

that implements Algorithm 10.5.7. The fourth input argument `tol` is an absolute error tolerance used to check for convergence, and the second output argument `niter` is the iteration count. Compare the performance of this algorithm to Newton's Method and the Secant Method, in terms of both the iteration count and the execution time as measured using the MATLAB commands `tic` and `toc`.

Exploration 10.5.7 Write a MATLAB function

$$[x,niter]=regulaaitken(f,a,b,tol)$$

that implements Algorithm 10.5.5, with the modification that if one of the endpoints a or b does not change over three consecutive iterations, then Aitken's Δ^2 Method (10.4) is used to obtain the next iterate. How does the performance of this method compare to that of the Illinois Algorithm from Exploration 10.5.6?

10.5.3 Concept Check

1. Describe Aitken's Δ^2 Method. To what kind of sequences can it be applied?
2. What is Steffensen's Method, and how does it relate to Aitken's Δ^2 Method?
3. What is meant by a safeguarded root-finding method?
4. What is the Method of Regula Falsi, and on what other root-finding methods is it based?
5. What is a drawback of Regula Falsi, and how can it be addressed?

10.6 Additional Resources

Basic root-finding methods are covered in any numerical analysis textbook; see, for example [Burden and Faires (2004); Cheney and Kincaid (2008); Conte and de Boor (1972); Dahlquist and Björck (1974); Heath (2002); Isaacson and Keller (1966)]. More detailed discussion can be found in books dedicated to the subject, such as [Ostrowski (1966); Traub (1964)]. Some methods that were not covered in this chapter are **inverse interpolation** and **linear fractional interpolation**, which are presented in [Heath (2002)]. Linear fractional interpolation is extended to more general rational functions in [Larkin (1980)].

Müller's Method, covered in [Burden and Faires (2004)], is a generalization of the Secant Method that uses a quadratic approximation that interpolates the last three iterates, rather than a linear approximation passing through the last two. It is a particularly useful method for finding roots of polynomials. All of the roots of a polynomial can be obtained by computing the eigenvalues of its **companion matrix**, using the methods of Chapter 6; this approach is discussed in [Edelman and Murakami (1995)].

10.7 Exercises

1. Determine the number of iterations required for the Bisection Method to compute a root of the function $f(x) = x^2 - x - 1$ on the interval $[1, 2]$, with error at most 10^{-6}. Try the Bisection Method on this function with this interval, using this many iterations. How does the actual error compare to the tolerance of 10^{-6}?

2. Use Newton's Method to derive an iteration of the form $x^{(k+1)} = g(x^{(k)})$ for computing the cube root of a given number a. The function g should be simplified as much as possible, as in Example 10.4.3. Use this iteration to compute $\sqrt[3]{2}$, with initial guess $x^{(0)} = 1$. Examine the error after each iteration. Does Newton's Method appear to converge quadratically? Justify your answer.

3. Consider the function $f(x) = x^3 + x - 5$.

   (a) Use the Intermediate Value Theorem to prove that $f(x)$ has a root in $(0, 2)$. Then use the derivative to show that this root is unique.

   (b) Apply the following methods to find the root of $f(x)$ in $(0, 2)$:

       i. Bisection
       ii. Fixed-Point Iteration with initial guess $x^{(0)} = 1$. Verify that your choice of $g(x)$ satisfies the conditions for convergence in Theorem 10.3.5.
       iii. Newton's Method with initial guess $x^{(0)} = 1$.
       iv. The Secant Method with initial guesses $x^{(0)} = 0$, $x^{(1)} = 2$.

   Compare the performance of these methods. How many iterations does each require? Does the convergence match theoretical expectations?

4. Show that the Secant Method (Algorithm 10.4.7) is equivalent to the iteration

$$x^{(k+1)} = \frac{x^{(k-1)} f(x^{(k)}) - x^{(k)} f(x^{(k-1)})}{f(x^{(k)}) - f(x^{(k-1)})}, \quad k = 1, 2, 3, \ldots.$$

   Which of these two iterations is more susceptible to issues with floating-point arithmetic? Explain.

5. Suppose that Newton's Method is modified so as to avoid the expense of computing a derivative. The modified iteration is

$$x^{(k+1)} = x^{(k)} - \frac{f(x^{(k)})}{d},$$

   where d is a constant. Use Theorem 10.3.5 to derive a condition on d that ensures convergence of this iteration.

6. Consider Fixed-Point Iteration applied to the equation $x = g(x)$, where g has a unique fixed point x^*. Show that if $g'(x^*) = g''(x^*) = \cdots = g^{(r)}(x^*) = 0$, then the order of convergence is $r + 1$.

7. Consider the function $f(x) = x^3 - 3x + 2$, that has roots at $x = 1$ and $x = 2$.

   (a) Rewrite the equation $f(x) = 0$ in the form $x = g(x)$, in three different ways.

   (b) For each of these equations, and for each root, determine whether Fixed-Point Iteration will converge *without* actually performing any iterations.

8. Consider the function $f(x) = x^3 - 6x^2 + 12x - 8$.

   (a) Apply Newton's Method to compute a root of $f(x)$, with initial guess $x^{(0)} = 1$. How rapidly does the iteration converge? Use a graph to explain the convergence behavior.

   (b) Would Bisection be a viable choice of algorithm for finding the root instead? Explain why or why not.

9. Consider the function $f(x) = -3x^4 + 7x^2$.

   (a) Apply Newton's Method with initial guess $x^{(0)} = 1$, and also $x^{(0)} = -1$. What happens?

   (b) Try again with initial guesses $x^{(0)} = 1.5$ and $x^{(0)} = 0.1$. What happens now? Explain the behavior.

10. The **van der Waals equation of state**

$$\left(p + \frac{n^2 a}{V^2} \right) (V - nb) = nRT$$

relates the pressure p, volume V, and temperature T of a gas, in a more realistic way than the **ideal gas law**

$$p V = nRT.$$

In both equations, R is the universal gas constant, and n is the number of moles. The constants a and b depend on the gas. Suppose the gas is hydrogen, for which $a = 0.2476$ L^{2}bar/mol^2 and $b = 0.02661$ L/mol. Let $P = 1$ kPa, $V = 1$ L, $R = 0.083146$, and $T = 300°$ K. Use Newton's Method to solve the van der Waals equation for n, using the ideal gas law to obtain an initial guess.

11. In [Keller (1984)] Keller derived the formula

$$P = \frac{1+p}{2} \left(\frac{p}{1 - p + p^2} \right)^{21}$$

for the probability that a racquetball player will shut out their opponent, given the probability p that the player wins a specific rally. Use a root-finding method to determine the value of p at which $P = 0.5$; that is, the player will shut out their opponent in half of the matches they play. Use the fact that p represents a probability to choose any initial guesses.

12. Adapt the approach used to derive Newton's Method to obtain a new root-finding method that converges *cubically*. Specifically, for each iteration, instead of computing each new iterate $x^{(k+1)}$ by finding the x-intercept of the tangent line approximation of $f(x)$ at $x^{(k)}$, approximate $f(x)$ by a *quadratic* function: $P_2(x)$, the second-degree Taylor polynomial of $f(x)$ centered at $x^{(k)}$. Then, set $x^{(k+1)}$ equal to the root of $P_2(x)$ that is closest to $x^{(k)}$. Try this method on the function $f(x) = x^4 - 4x^3 + 2x^2 - 5x + 3$, which has real roots in the intervals $[0, 1]$ and $[3, 4]$. Does it converge cubically? What is a reason why this method should not be used, regardless of its convergence behavior?

13. Suppose $f(x)$ has a multiple root at $x = a$. Then a is a simple root of the function $g(x) = f(x)/f'(x)$. Apply Newton's Method to $g(x)$ to obtain a new iterative method for computing a multiple root of $f(x)$ (that is, derive a formula for $x^{(k+1)}$ in terms of f and its derivatives, not g). Implement this method in MATLAB to compute the double root of the function $f(x) = (x - 1)^2 e^x$ from Example 10.4.6. Perform a convergence analysis to determine the order of convergence, and compare to the behavior observed in your implementation.

14. Write a MATLAB function to compute *all* of the roots of a given function $f(x)$ on a given interval $[a, b]$. The function must proceed as follows:

    (a) Divide $[a, b]$ into 100 subintervals $[a_i, b_i]$, $i = 1, 2, \ldots, 100$, that are of equal width.

    (b) For each $i = 1, 2, \ldots, 100$, determine whether the sign of $f(x)$ changes on the subinterval $[a_i, b_i]$. If it does, then use Newton's Method with initial guess equal to the midpoint of this subinterval. Use an error tolerance of 10^{-10}.

    (c) When using Newton's Method, if any iterate should fall outside the interval $[a_i, b_i]$, reject the iterate and instead perform three iterations of Bisection of $f(x)$ on $[a_i, b_i]$ to obtain a new initial guess for Newton's Method. Update a_i and b_i to be the endpoints of the interval produced by Bisection.

    Use your function to compute all of the roots of the function

    $$f(x) = \begin{cases} \frac{\sin x}{x} & x \neq 0 \\ 1 & x = 0 \end{cases}.$$

    This function is known as the *sinc function*.

15. Let $a > 0$. Recall from Example 10.4.4 that Newton's Method, when applied to the function $f(x) = a - \frac{1}{x}$, yields the iteration

    $$x^{(k+1)} = 2x^{(k)} - a[x^{(k)}]^2,$$

    which converges to $1/a$, if it converges. That is, this iteration can compute the reciprocal of a without performing any divisions.

(a) In Example 10.4.4, it was stated that the condition $x^{(0)} < 2/a$ was *necessary* for convergence, for otherwise $x^{(1)} < 0$. Prove that if $0 < x^{(k)} < 2/a$, then $0 < x^{(k+1)} < 2/a$ as well. Explain why this implies that the condition $0 < x^{(0)} < 2/a$ is *sufficient* for convergence.

(b) Suppose that a has the binary floating-point representation (see Definition 2.2.1)

$$a = m2^{E},$$

where m is the mantissa and E is an integer exponent. We also assume that this representation is **normalized** (see Section 2.2.1.2), so that m has the form

$$m = 1 + f,$$

where $0 < f < 1$. How can this representation of a be used to easily obtain a similar floating-point representation of an initial guess $x^{(0)}$ that is guaranteed to satisfy $0 < x^{(0)} < 2/a$? *Hint:* Come up with a floating-point representation of $1/a$ in terms of the components of the representation of a.

(c) Implement a function in MATLAB to compute the reciprocal of a given positive number a using the above iteration and the initial guess from part (b). *Hint:* Use the result of Exploration 2.2.5.

Chapter 11

Optimization

It is intuitively clear that for countless endeavors in science, engineering or business, the objective can be expressed in terms of maximizing or minimizing some quantity. Examples include maximizing profit or minimizing cost in business, maximizing stiffness (or, equivalently, minimizing strain energy) in structural design, or, in geophysics, maximizing the quality of fit between a mathematical model of a physical process and data observed from that process. This chapter provides an introduction to numerical methods for finding a minimum of a function of one or several variables, called an **objective function**. We focus exclusively on minimization, since maximizing f is equivalent to minimizing $-f$.

In Section 11.1, we show how the zero-finding methods of Chapter 10 can be generalized to the problem of solving a system of nonlinear equations. These methods will be applied in Section 11.2 to find a critical point of a function of several variables, at which a local minimum may occur. Section 11.3 will present examples of a different class of methods for finding the minimum of a function of one variable, that do not require derivatives.

11.1 Systems of Nonlinear Equations

The techniques presented in Chapter 10 for solving a single nonlinear equation of the form $f(x) = 0$, or $x = g(x)$, can be generalized to solve a *system* of n nonlinear equations in n unknowns $x_1, x_2, \ldots, x_n$, using concepts and techniques from numerical linear algebra that were presented in Chapters 3 and 5.

11.1.1 Fixed-Point Iteration

In Section 10.3, we learned how to use **Fixed-Point Iteration** to solve a single nonlinear equation of the form

$$f(x) = 0 \tag{11.1}$$

by first transforming the equation into one of the form

$$x = g(x). \tag{11.2}$$

Then, after choosing an initial guess $x^{(0)}$, we compute a sequence of iterates by

$$x^{(k+1)} = g(x^{(k)}), \quad k = 0, 1, 2, \ldots,$$

that, hopefully, converges to a solution of the original equation.

Recall from Theorem 10.3.5 that if g is continuous and maps an interval I into itself, then g has a **fixed point** (also called a **stationary point**) x^* in I, which is a point that satisfies (11.2). That is, a solution to (11.1) exists within I. Furthermore, if there is a constant $\rho < 1$ such that

$$|g'(x)| \le \rho, \quad x \in I, \tag{11.3}$$

then this fixed point is *unique*.

We now generalize Fixed-Point Iteration to the problem of solving a system of n nonlinear equations in n unknowns,

$$f_1(x_1, x_2, \ldots, x_n) = 0,$$
$$f_2(x_1, x_2, \ldots, x_n) = 0,$$
$$\vdots$$
$$f_n(x_1, x_2, \ldots, x_n) = 0.$$

For simplicity, we express this system of equations in vector form,

$$\mathbf{F}(\mathbf{x}) = \mathbf{0}, \tag{11.4}$$

where $\mathbf{F} : D \subseteq \mathbb{R}^n \to \mathbb{R}^n$ is a *vector-valued* function of n variables represented by the vector $\mathbf{x} = (x_1, x_2, \ldots, x_n)$, and $f_1, f_2, \ldots, f_n$ are the **component functions**, or **coordinate functions**, of $\mathbf{F}$ (see Section A.1.5).

We now define Fixed-Point Iteration for solving (11.4). First, we transform this system of equations into an equivalent system of the form

$$\mathbf{x} = \mathbf{G}(\mathbf{x}).$$

One approach to doing this is to solve the ith equation in the original system for x_i. This is analogous to the derivation of the Jacobi Method from Section 5.1.2 for solving systems of linear equations. Next, we choose an initial guess $\mathbf{x}^{(0)}$. Then, we compute subsequent iterates by

$$\mathbf{x}^{(k+1)} = \mathbf{G}(\mathbf{x}^{(k)}), \quad k = 0, 1, 2, \ldots.$$

Exploration 11.1.1 Write a MATLAB function

`[x,niter]=fixedptsys(G,x0,tol)`

that performs Fixed-Point Iteration to solve the equation $\mathbf{x} = \mathbf{G}(\mathbf{x})$ with an initial guess `x0` and absolute error tolerance `tol`. Use Algorithm 10.3.1 as a starting point, with appropriate generalization to functions of several variables. The output argument `niter` is the number of iterations performed.

The existence and uniqueness of fixed points of vector-valued functions of several variables can be described in an analogous manner to how it is described in the single-variable case. The function $\mathbf{G}$ has a fixed point in a domain $D \subseteq \mathbb{R}^n$ if $\mathbf{G}$ is continuous on D and $\mathbf{G}$ maps D into D. Furthermore, by analogy with (11.3), if $\mathbf{G}$ has continuous first partial derivatives on D and there exists a constant $\rho < 1$ such that, in some natural matrix norm (see Section B.13.2),

$$\|J_{\mathbf{G}}(\mathbf{x})\| \leq \rho, \quad \mathbf{x} \in D, \tag{11.5}$$

where

$$J_{\mathbf{G}}(\mathbf{x}) = \begin{bmatrix} \frac{\partial g_1(\mathbf{x})}{\partial x_1} & \frac{\partial g_1(\mathbf{x})}{\partial x_2} & \cdots & \frac{\partial g_1(\mathbf{x})}{\partial x_n} \\ \frac{\partial g_2(\mathbf{x})}{\partial x_1} & \frac{\partial g_2(\mathbf{x})}{\partial x_2} & \cdots & \frac{\partial g_2(\mathbf{x})}{\partial x_n} \\ \vdots & \cdots & & \vdots \\ \frac{\partial g_n(\mathbf{x})}{\partial x_1} & \frac{\partial g_n(\mathbf{x})}{\partial x_2} & \cdots & \frac{\partial g_n(\mathbf{x})}{\partial x_n} \end{bmatrix}$$

is the **Jacobian matrix** of first partial derivatives of $\mathbf{G}$ evaluated at $\mathbf{x}$, then $\mathbf{G}$ has a *unique* fixed point $\mathbf{x}^*$ in D. Furthermore, Fixed-Point Iteration is guaranteed to converge linearly to $\mathbf{x}^*$ for any initial guess chosen in D. This can be seen by computing a multivariable Taylor expansion (see Theorem A.6.6) of the error $\mathbf{x}^{(k+1)} - \mathbf{x}^*$ around $\mathbf{x}^*$.

Exploration 11.1.2 Use a multivariable Taylor expansion to prove that if $\mathbf{G}$ satisfies the assumptions in the preceding discussion (that it is continuous, maps D into itself, has continuous first partial derivatives, and satisfies $\|J_{\mathbf{G}}(\mathbf{x})\| \leq \rho < 1$ for $\mathbf{x} \in D$ and any natural matrix norm $\| \cdot \|$), then $\mathbf{G}$ has a unique fixed point $\mathbf{x}^* \in D$ and Fixed-Point Iteration will converge to $\mathbf{x}^*$ for any initial guess $\mathbf{x}^{(0)} \in D$.

Exploration 11.1.3 Under the assumptions of Exploration 11.1.2, obtain a bound for the error after k iterations, $\|\mathbf{x}^{(k)} - \mathbf{x}^*\|$, in terms of the initial difference $\|\mathbf{x}^{(1)} - \mathbf{x}^{(0)}\|$. *Hint:* Refer to (10.2).

It is interesting to note that the convergence of Fixed-Point Iteration for functions of several variables can, in some cases, be accelerated by using an approach similar to how the Jacobi method for linear systems is modified to obtain the Gauss-Seidel method (see Section 5.1.3). That is, when computing $\mathbf{x}_i^{(k+1)}$ by evaluating $g_i(\mathbf{x}^{(k)})$, we replace $\mathbf{x}_j^{(k)}$, for $j < i$, by $\mathbf{x}_j^{(k+1)}$, since it has already been computed (assuming all components of $\mathbf{x}^{(k+1)}$ are computed in order). Therefore, as in Gauss-Seidel, we are using the most up-to-date information available when computing each iterate.

We have actually seen the condition (11.5) before, in Section 5.1.1 about stationary iterative methods, such as Jacobi and Gauss-Seidel, that have the form

$$M\mathbf{x}^{(k+1)} = N\mathbf{x}^{(k)} + \mathbf{b}. \tag{11.6}$$

This iteration can be written in the form $\mathbf{x}^{(k+1)} = \mathbf{g}(\mathbf{x}^{(k)})$, where
$$\mathbf{g}(\mathbf{x}) = (M^{-1}N)\mathbf{x} + M^{-1}\mathbf{b},$$
from which it follows that $T = M^{-1}N$ is the Jacobian matrix of $\mathbf{g}$. In Section 5.1, we learned that for such iterations, (11.5) is sufficient for convergence, but the condition $\rho(T) < 1$ is necessary *and* sufficient. We can observe similar convergence behavior in the nonlinear case as well.

Example 11.1.1 Consider the system of equations
$$x_2 = x_1^2,$$
$$x_1^2 + x_2^2 = 1.$$
The first equation describes a parabola, while the second describes the unit circle. By graphing both equations, it can easily be seen that this system has two solutions, one of which lies in the first quadrant $(x_1, x_2 > 0)$.

To solve this system using Fixed-Point Iteration, we solve the second equation for x_1, and obtain the equivalent system
$$x_1 = \sqrt{1 - x_2^2},$$
$$x_2 = x_1^2.$$
If we consider the rectangle
$$D = \{(x_1, x_2) \,|\, 0 \leq x_1 \leq 1, 0 \leq x_2 \leq 1\},$$
we see that the function
$$\mathbf{G}(x_1, x_2) = \left(\sqrt{1 - x_2^2}, x_1^2 \right)$$
maps D into itself.

Because $\mathbf{G}$ is also continuous on D, it follows that $\mathbf{G}$ has a fixed point in D. However, $\mathbf{G}$ has the Jacobian matrix
$$J_{\mathbf{G}}(\mathbf{x}) = \begin{bmatrix} 0 & -x_2/\sqrt{1 - x_2^2} \\ 2x_1 & 0 \end{bmatrix},$$
which cannot satisfy $\|J_{\mathbf{G}}\| < 1$ on D. Therefore, we cannot guarantee that Fixed-Point Iteration with this choice of $\mathbf{G}$ will converge, and, in fact, it can be shown that it does *not* converge. Instead, the iterates tend to approach the corners of D, at which they remain.

In an attempt to achieve convergence, we note that $\partial g_2/\partial x_1 = 2x_1 > 1$ near the fixed point. Therefore, we modify $\mathbf{G}$ as follows:
$$\mathbf{G}(x_1, x_2) = \left(\sqrt{x_2}, \sqrt{1 - x_1^2} \right).$$
For this choice of $\mathbf{G}$, $J_{\mathbf{G}}$ still has partial derivatives that are greater than 1 in magnitude near the fixed point. However, there is one crucial distinction: near the fixed point, $\rho(J_{\mathbf{G}}) < 1$, whereas with the original choice of $\mathbf{G}$, $\rho(J_{\mathbf{G}}) > 1$. Attempting Fixed-Point Iteration with the new $\mathbf{G}$, we see that convergence is actually achieved, although it is slow. $\square$

It can be seen from this example that the condition (11.5) for the convergence of Fixed-Point Iteration is sufficient, but not necessary.

11.1.2 Newton's Method

Suppose that Fixed-Point Iteration is being used to solve an equation of the form $\mathbf{F}(\mathbf{x}) = \mathbf{0}$, where $\mathbf{F}$ is a vector-valued function of n variables, by transforming it into an equation of the form $\mathbf{x} = \mathbf{G}(\mathbf{x})$. Furthermore, suppose that $\mathbf{G}$ is known to map a domain $D \subseteq \mathbb{R}^n$ into itself, so that a fixed point exists in D. We have learned that the number ρ from (11.5) provides an indication of the rate of convergence, in the sense that as the iterates converge to a fixed point $\mathbf{x}^*$, if they converge, the error satisfies

$$\|\mathbf{x}^{(k+1)} - \mathbf{x}^*\| \le \rho \|\mathbf{x}^{(k)} - \mathbf{x}^*\|.$$

Furthermore, as the iterates converge, a suitable value for ρ is given by $\rho(J_{\mathbf{G}}(\mathbf{x}^*))$, the spectral radius of the Jacobian matrix at the fixed point.

Therefore, it makes sense to ask: what if this spectral radius is equal to zero? In that case, if the first partial derivatives are continuous near $\mathbf{x}^*$, and the second partial derivatives are continuous and bounded at $\mathbf{x}^*$, then Fixed-Point Iteration converges *quadratically*. That is, there exists a constant M such that

$$\|\mathbf{x}^{(k+1)} - \mathbf{x}^*\| \le M \|\mathbf{x}^{(k)} - \mathbf{x}^*\|^2.$$

Exploration 11.1.4 Let $\mathbf{G} : D \subseteq \mathbb{R}^n \to \mathbb{R}^n$ be a twice continuously differentiable function that has a fixed point $\mathbf{x}^{(0)} \in D$. Furthermore, assume that $J_{\mathbf{G}}(\mathbf{x}^*) = 0$ and that the second partial derivatives of the component functions g_i, $i = 1, 2, \ldots, n$ of $\mathbf{G}$ are bounded on D. Use a multivariable Taylor expansion to prove that Fixed-Point Iteration applied to $\mathbf{G}$ converges quadratically for any initial guess $\mathbf{x}^{(0)} \in D$.

In Section 10.4 we learned that for a single nonlinear equation $f(x) = 0$, **Newton's Method** generally achieves quadratic convergence. Recall that this method computes iterates by

$$x^{(k+1)} = x^{(k)} - \frac{f(x^{(k)})}{f'(x^{(k)})}, \quad k = 0, 1, 2, \ldots,$$

where $x^{(0)}$ is an initial guess. We now wish to generalize this method to systems of nonlinear equations.

Consider the Fixed-Point Iteration function

$$\mathbf{G}(\mathbf{x}) = \mathbf{x} - [J_{\mathbf{F}}(\mathbf{x})]^{-1}\mathbf{F}(\mathbf{x}). \tag{11.7}$$

Then, it can be shown by direct differentiation that the Jacobian matrix $J_{\mathbf{G}}(\mathbf{x})$ of this function is equal to the zero matrix at $\mathbf{x} = \mathbf{x}^*$, a solution to $\mathbf{F}(\mathbf{x}) = \mathbf{0}$. If we define

$$\mathbf{x} = \begin{bmatrix} x_1 \\ x_2 \\ \vdots \\ x_n \end{bmatrix}, \quad \mathbf{F}(\mathbf{x}) = \begin{bmatrix} f_1(x_1, x_2, \ldots, x_n) \\ f_2(x_1, x_2, \ldots, x_n) \\ \vdots \\ f_n(x_1, x_2, \ldots, x_n) \end{bmatrix}, \quad \mathbf{G}(\mathbf{x}) = \begin{bmatrix} g_1(x_1, x_2, \ldots, x_n) \\ g_2(x_1, x_2, \ldots, x_n) \\ \vdots \\ g_n(x_1, x_2, \ldots, x_n) \end{bmatrix},$$

where f_i and g_i, $i = 1, 2, \ldots, n$ are the coordinate functions of $\mathbf{F}$ and $\mathbf{G}$, respectively, then we have

$$\frac{\partial}{\partial x^{(k)}} g_i(\mathbf{x}) = \frac{\partial}{\partial x^{(k)}} \left[x_i - \sum_{j=1}^{n} b_{ij}(\mathbf{x}) f_j(\mathbf{x}) \right]$$

$$= \delta_{ik} - \sum_{j=1}^{n} b_{ij}(\mathbf{x}) \frac{\partial}{\partial x^{(k)}} f_j(\mathbf{x}) - \sum_{j=1}^{n} \frac{\partial}{\partial x^{(k)}} b_{ij}(\mathbf{x}) f_j(\mathbf{x}),$$

where $b_{ij}(\mathbf{x})$ is the (i, j) element of $[J_{\mathbf{F}}(\mathbf{x})]^{-1}$.

Exploration 11.1.5 Prove that if $\mathbf{G}(\mathbf{x})$ is defined as in (11.7) and $\mathbf{F}(\mathbf{x}^*) = \mathbf{0}$, then $J_{\mathbf{G}}(\mathbf{x}^*) = 0$.

We see that this choice of Fixed-Point Iteration is a direct generalization of Newton's Method to systems of equations, in which the division by $f'(x^{(k)})$ is replaced by multiplication by the inverse of $J_{\mathbf{F}}(\mathbf{x}^{(k)})$, the total derivative of $\mathbf{F}(\mathbf{x})$.

The algorithm for Newton's Method for a system of equations follows.

Algorithm 11.1.2 (Newton's Method for Systems) Given a function $\mathbf{F} : D \subseteq \mathbb{R}^n \to \mathbb{R}^n$, the following algorithm computes a solution $\mathbf{x}^* \in \mathbb{R}^n$ of the equation $\mathbf{F}(\mathbf{x}) = \mathbf{0}$.

Choose an initial guess $\mathbf{x}^{(0)} \in \mathbb{R}^n$
for $k = 0, 1, 2, \ldots$ **do** until convergence
    $\mathbf{y}_k = \mathbf{F}(\mathbf{x}^{(k)})$
    Solve $[J_{\mathbf{F}}(\mathbf{x}^{(k)})]\mathbf{s}_k = -\mathbf{y}_k$ for $\mathbf{s}_k$
    $\mathbf{x}^{(k+1)} = \mathbf{x}^{(k)} + \mathbf{s}_k$
end while
$\mathbf{x}^* = \mathbf{x}^{(k)}$

The vector $\mathbf{s}_k$ can be obtained using any method for solving a system of linear equations, such as those covered in Chapter 3. Convergence can be determined by examining the magnitude of $\mathbf{s}_k$ or $\mathbf{y}_k$.

Example 11.1.3 Recall the system of equations from Example 11.1.1, rewritten as

$$x_2 - x_1^2 = 0,$$
$$x_1^2 + x_2^2 - 1 = 0.$$

Fixed-Point iteration, for a straightforward choice of $\mathbf{G}(\mathbf{x})$, converged rather slowly for this system, if it converged at all. Now, we apply Newton's Method to this system. We have

$$\mathbf{F}(x_1, x_2) = \begin{bmatrix} x_2 - x_1^2 \\ x_1^2 + x_2^2 - 1 \end{bmatrix}, \quad J_{\mathbf{F}}(x_1, x_2) = \begin{bmatrix} -2x_1 & 1 \\ 2x_1 & 2x_2 \end{bmatrix}.$$

Using the formula for the inverse of a 2×2 matrix, we obtain the iteration

$$\begin{bmatrix} x_1^{(k+1)} \\ x_2^{(k+1)} \end{bmatrix} = \begin{bmatrix} x_1^{(k)} \\ x_2^{(k)} \end{bmatrix} + \frac{1}{4x_1^{(k)} x_2^{(k)} + 2x_1^{(k)}} \begin{bmatrix} 2x_2^{(k)} & -1 \\ -2x_1^{(k)} & -2x_1^{(k)} \end{bmatrix} \begin{bmatrix} x_2^{(k)} - (x_1^{(k)})^2 \\ (x_1^{(k)})^2 + (x_2^{(k)})^2 - 1 \end{bmatrix}.$$

Implementing this iteration in MATLAB, we see that it converges quite rapidly, much more so than Fixed-Point Iteration. Note that in order for the iteration to not break down, we must have $x_1^{(k)} \neq 0$ and $x_2^{(k)} \neq -1/2$. $\square$

Example 11.1.4 We solve the system of equations $\mathbf{F}(\mathbf{x}) = \mathbf{0}$, where

$$\mathbf{F}(x, y, z) = \begin{bmatrix} x^2 + y^2 + z^2 - 3 \\ x^2 + y^2 - z - 1 \\ x + y + z - 3 \end{bmatrix}.$$

We will begin to use Newton's Method to solve this system of equations, with initial guess $\mathbf{x}^{(0)} = (x^{(0)}, y^{(0)}, z^{(0)}) = (1, 0, 1)$. We have

$$J_{\mathbf{F}}(x, y, z) = \begin{bmatrix} 2x & 2y & 2z \\ 2x & 2y & -1 \\ 1 & 1 & 1 \end{bmatrix}.$$

Therefore, each Newton iterate $\mathbf{x}^{(k+1)}$ is obtained by solving the system of equations

$$J_{\mathbf{F}}(x^{(k)}, y^{(k)}, z^{(k)})(\mathbf{x}^{(k+1)} - \mathbf{x}^{(k)}) = -\mathbf{F}(x^{(k)}, y^{(k)}, z^{(k)}),$$

or

$$\begin{bmatrix} 2x^{(k)} & 2y^{(k)} & 2z^{(k)} \\ 2x^{(k)} & 2y^{(k)} & -1 \\ 1 & 1 & 1 \end{bmatrix} \begin{bmatrix} x^{(k+1)} - x^{(k)} \\ y^{(k+1)} - y^{(k)} \\ z^{(k+1)} - z^{(k)} \end{bmatrix} = - \begin{bmatrix} (x^{(k)})^2 + (y^{(k)})^2 + (z^{(k)})^2 - 3 \\ (x^{(k)})^2 + (y^{(k)})^2 - z^{(k)} - 1 \\ x^{(k)} + y^{(k)} + z^{(k)} - 3 \end{bmatrix}.$$

Setting $k = 0$ and substituting $(x^{(0)}, y^{(0)}, z^{(0)}) = (1, 0, 1)$ yields the system

$$\begin{bmatrix} 2 & 0 & 2 \\ 2 & 0 & -1 \\ 1 & 1 & 1 \end{bmatrix} \begin{bmatrix} x^{(1)} - 1 \\ y^{(1)} \\ z^{(1)} - 1 \end{bmatrix} = \begin{bmatrix} 1 \\ 1 \\ 1 \end{bmatrix},$$

which has the solution $\mathbf{x}^{(1)} = (\frac{3}{2}, \frac{1}{2}, 1)$. Repeating this process with $k = 1$ yields the system

$$\begin{bmatrix} 3 & 1 & 2 \\ 3 & 1 & -1 \\ 1 & 1 & 1 \end{bmatrix} \begin{bmatrix} x^{(2)} - \frac{3}{2} \\ y^{(2)} - \frac{1}{2} \\ z^{(2)} - 1 \end{bmatrix} = \begin{bmatrix} -\frac{1}{2} \\ -\frac{1}{2} \\ 0 \end{bmatrix},$$

which has the solution $\mathbf{x}^{(2)} = (\frac{5}{4}, \frac{3}{4}, 1)$. A similar process yields the next iterate $\mathbf{x}^{(3)} = (\frac{9}{8}, \frac{7}{8}, 1)$.

It can be seen that these iterates are converging to $(1, 1, 1)$, which is the exact solution. However, if we instead use the initial guess $\mathbf{x}^{(0)} = (0, 0, 0)$, then we obtain

$$J_{\mathbf{F}}(0, 0, 0) = \begin{bmatrix} 0 & 0 & 0 \\ 0 & 0 & -1 \\ 1 & 1 & 1 \end{bmatrix},$$

which is not invertible. The system $J_{\mathbf{F}}(0, 0, 0)\mathbf{s}_0 = -\mathbf{F}(0, 0, 0)$ does not have a solution, and therefore Newton's Method fails. $\square$

Exploration 11.1.6 Write a MATLAB function

$$[\texttt{x,niter}]=\texttt{newtonsys(F,JF,x0,tol)}$$

that solves the system of equations $\mathbf{F}(\mathbf{x}) = \mathbf{0}$ using Newton's Method, where F is a function handle that represents $\mathbf{F}(\mathbf{x})$, and JF is a function handle that takes x as input and returns the matrix $J_{\mathbf{F}}(\mathbf{x})$. The parameters x0 and tol are the initial guess and absolute error tolerance, respectively. The output argument niter is the number of iterations performed. Test your function on the system from Example 11.1.3.

11.1.3   Broyden's Method

One of the drawbacks of using Newton's Method to solve a system of nonlinear equations $\mathbf{F}(\mathbf{x}) = \mathbf{0}$ is the computational expense that must be incurred during each iteration to evaluate the partial derivatives of $\mathbf{F}$ at $\mathbf{x}^{(k)}$, and then solve a system of linear equations involving the resulting Jacobian matrix. The algorithm does not facilitate the re-use of data from previous iterations, and in some cases evaluation of the partial derivatives can be unnecessarily costly.

An alternative is to modify Newton's Method so that *approximate* partial derivatives are used, as in the Secant Method for a single nonlinear equation, since the slightly slower convergence is offset by the improved efficiency of each iteration. Unfortunately, simply replacing the analytical Jacobian matrix of $\mathbf{F}$ with a matrix consisting of finite difference approximations of the partial derivatives does not do much to reduce the cost of each iteration, because the cost of solving the system of linear equations is unchanged.

However, because the Jacobian matrix consists of the partial derivatives evaluated at an element of a convergent sequence, intuitively Jacobian matrices from consecutive iterations are "near" one another in some sense, which suggests that it should be possible to cheaply update an approximate Jacobian matrix from iteration to iteration, in such a way that the *inverse* of the Jacobian matrix can be updated efficiently as well.

This is the case when a matrix has the form

$$B = A + \mathbf{u}\mathbf{v}^T,$$

where $\mathbf{u}$ and $\mathbf{v}$ are given vectors. This modification of A to obtain B is called a **rank-one update**, since $\mathbf{u}\mathbf{v}^T$, an **outer product**, has rank one (see Section B.7.3). To obtain B^{-1} from A^{-1}, we note that if

$$A\mathbf{x} = \mathbf{u},$$

then

$$B\mathbf{x} = (A + \mathbf{u}\mathbf{v}^T)\mathbf{x} = (1 + \mathbf{v}^T\mathbf{x})\mathbf{u},$$

which yields

$$B^{-1}\mathbf{u} = \frac{1}{1 + \mathbf{v}^T A^{-1}\mathbf{u}} A^{-1}\mathbf{u}.$$

On the other hand, if $\mathbf{x}$ is such that $\mathbf{v}^T A^{-1}\mathbf{x} = 0$, then

$$BA^{-1}\mathbf{x} = (A + \mathbf{u}\mathbf{v}^T)A^{-1}\mathbf{x} = \mathbf{x},$$

which yields

$$B^{-1}\mathbf{x} = A^{-1}\mathbf{x}.$$

This takes us to the following more general problem: given a matrix C, we wish to construct a matrix D such that the following conditions are satisfied:

- $D\mathbf{w} = \mathbf{z}$, for given vectors $\mathbf{w}$ and $\mathbf{z}$
- $D\mathbf{y} = C\mathbf{y}$, if $\mathbf{y}$ is orthogonal to a given vector $\mathbf{g}$.

In our application, $C = A^{-1}$, $D = B^{-1}$, $\mathbf{w} = \mathbf{u}$, $\mathbf{z} = 1/(1 + \mathbf{v}^T A^{-1}\mathbf{u})A^{-1}\mathbf{u}$, and $\mathbf{g} = A^{-T}\mathbf{v}$. To solve this problem, we set

$$D = C + \frac{(\mathbf{z} - C\mathbf{w})\mathbf{g}^T}{\mathbf{g}^T\mathbf{w}}. \tag{11.8}$$

It can be verified directly that D satisfies the above conditions.

Exploration 11.1.7 Prove that the matrix D defined in (11.8) satisfies $D\mathbf{w} = \mathbf{z}$ and $D\mathbf{y} = C\mathbf{y}$ for $\mathbf{g}^T\mathbf{y} = 0$.

Applying this definition of D, we obtain

$$
\begin{aligned}
B^{-1} &= A^{-1} + \frac{\left(\frac{1}{1+\mathbf{v}^T A^{-1}\mathbf{u}} A^{-1}\mathbf{u} - A^{-1}\mathbf{u}\right)\mathbf{v}^T A^{-1}}{\mathbf{v}^T A^{-1}\mathbf{u}} \\
&= A^{-1} - \frac{A^{-1}\mathbf{u}\mathbf{v}^T A^{-1}}{1 + \mathbf{v}^T A^{-1}\mathbf{u}}.
\end{aligned} \tag{11.9}
$$

This formula for the inverse of a rank-one update is known as the **Sherman-Morrison Formula**.

Exploration 11.1.8 Verify the final form of the Sherman-Morrison formula given in (11.9).

We now return to the problem of approximating the Jacobian of $\mathbf{F}$, and efficiently obtaining its inverse, at each iterate $\mathbf{x}^{(k)}$. We begin with an exact Jacobian, $A_0 = J_{\mathbf{F}}(\mathbf{x}^{(0)})$, and use A_0 to compute the first iterate, $\mathbf{x}^{(1)}$, using Newton's Method. Then, we recall that for the Secant Method, we use the approximation

$$f'(x^{(1)}) \approx \frac{f(x^{(1)}) - f(x^{(0)})}{x^{(1)} - x^{(0)}}.$$

Generalizing this approach to a system of equations, we seek an approximation A_1 to $J_{\mathbf{F}}(\mathbf{x}^{(1)})$ that has these properties:

- $A_1(\mathbf{x}^{(1)} - \mathbf{x}^{(0)}) = \mathbf{F}(\mathbf{x}^{(1)}) - \mathbf{F}(\mathbf{x}^{(0)})$. This condition is called the **Secant Condition**.
- If $\mathbf{z}^T(\mathbf{x}^{(1)} - \mathbf{x}^{(0)}) = 0$, then $A_1\mathbf{z} = J_{\mathbf{F}}(\mathbf{x}^{(0)})\mathbf{z} = A_0\mathbf{z}$.

It follows from previous discussion that

$$A_1 = A_0 + \frac{\mathbf{y}_1 - A_0\mathbf{s}_0}{\mathbf{s}_0^T \mathbf{s}_0}\mathbf{s}_0^T,$$

where $\mathbf{s}_0 = \mathbf{x}^{(1)} - \mathbf{x}^{(0)}$ and $\mathbf{y}_1 = F(\mathbf{x}^{(1)}) - \mathbf{F}(\mathbf{x}^{(0)})$. Furthermore, once we have computed A_0^{-1}, we have, by the Sherman-Morrison formula,

$$A_1^{-1} = A_0^{-1} - \frac{A_0^{-1}\left(\frac{\mathbf{y}_1 - A_0\mathbf{s}_0}{\mathbf{s}_0^T \mathbf{s}_0}\mathbf{s}_0^T\right)A_0^{-1}}{1 + \mathbf{s}_0^T A_0^{-1}\left(\frac{\mathbf{y}_1 - A_0\mathbf{s}_0}{\mathbf{s}_0^T \mathbf{s}_0}\right)} = A_0^{-1} + \frac{(\mathbf{s}_0 - A_0^{-1}\mathbf{y}_1)\mathbf{s}_0^T A_0^{-1}}{\mathbf{s}_0^T A_0^{-1}\mathbf{y}_1}.$$

Then, as A_1 is an approximation to $J_F(\mathbf{x}^{(1)})$, we can obtain our next iterate $\mathbf{x}^{(2)}$ as follows:

$$A_1\mathbf{s}_1 = -\mathbf{F}(\mathbf{x}^{(1)}), \quad \mathbf{x}^{(2)} = \mathbf{x}^{(1)} + \mathbf{s}_1.$$

Repeating this process, we obtain the following algorithm.

Algorithm 11.1.5 (Broyden's Method) Given a function $\mathbf{F} : D \subseteq \mathbb{R}^n \to \mathbb{R}^n$, the following algorithm computes a solution $\mathbf{x}^* \in \mathbb{R}^n$ of the equation $\mathbf{F}(\mathbf{x}) = \mathbf{0}$.

Choose an initial guess $\mathbf{x}^{(0)} \in \mathbb{R}^n$
$A_0 = J_\mathbf{F}(\mathbf{x}^{(0)})$
$\mathbf{s}_0 = -A_0^{-1}\mathbf{F}(\mathbf{x}^{(0)})$
$\mathbf{x}^{(1)} = \mathbf{x}^{(0)} + \mathbf{s}_0$
for $k = 1, 2, \ldots$, do until convergence
    $\mathbf{y}_k = \mathbf{F}(\mathbf{x}^{(k)}) - \mathbf{F}(\mathbf{x}^{(k-1)})$
    $\mathbf{w}_k = A_{k-1}^{-1}\mathbf{y}_k$
    $c = 1/\mathbf{s}_{k-1}^T\mathbf{w}_k$
    $A_k^{-1} = A_{k-1}^{-1} + c(\mathbf{s}_{k-1} - \mathbf{w}_k)\mathbf{s}_{k-1}^T A_{k-1}^{-1}$
    $\mathbf{s}_k = -A_k^{-1}\mathbf{F}(\mathbf{x}^{(k)})$
    $\mathbf{x}^{(k+1)} = \mathbf{x}^{(k)} + \mathbf{s}_k$
end while
$\mathbf{x}^* = \mathbf{x}^{(k)}$

Note that it is not necessary to compute A_k for $k \geq 1$; only A_k^{-1} is needed. It follows that no systems of linear equations need to be solved during an iteration; only matrix-vector multiplications are required, thus saving an order of magnitude of computational effort during each iteration compared to Newton's Method, in exchange for the expense of computing the inverse of the initial Jacobian matrix.

Exploration 11.1.9 Write a MATLAB function

$$[x,niter]=broyden(F,JF,x0,tol)$$

that solves the system of equations $\mathbf{F}(\mathbf{x}) = \mathbf{0}$ using Broyden's method, where F is a function handle that represents $\mathbf{F}(\mathbf{x})$ and JF is a function handle that takes x as input and returns the matrix $J_{\mathbf{F}}(\mathbf{x})$. The parameters x0 and tol are the initial guess and absolute error tolerance, respectively. The output argument niter is the number of iterations performed. Test your function on the system from Example 11.1.3 and compare the efficiency to that of Newton's Method as implemented in Exploration 11.1.6.

11.1.4 Concept Check

1. Given that a function $\mathbf{G} : \mathbb{R}^n \to \mathbb{R}^n$ has a fixed point in a domain D, what is a condition on $\mathbf{G}$ that ensures that the fixed point is unique, and that Fixed-Point Iteration will converge to it?
2. How can ideas from iterative methods for linear systems be used to accelerate the convergence of Fixed-Point Iteration?
3. How is Newton's Method extended to functions of several variables?
4. Under what circumstance can Newton's Method break down?
5. What drawback of Newton's Method is addressed by Broyden's Method?
6. What is the Sherman-Morrison Formula, and how is it beneficial?

11.2 Optimization Based on Newton's Method

Let $f : D \subseteq \mathbb{R}^n \to \mathbb{R}$ be a function we wish to minimize; that is, f is our **objective function**. When f is differentiable, finding the minimum, or **optimization**, of f can be accomplished by finding its **critical points**, at which $\nabla f(\mathbf{x}) = \mathbf{0}$. However, solving this system of equations can be quite difficult. Therefore, it is often necessary to use numerical methods that compute an *approximate* solution, such as the root-finding methods presented in the previous section. Before we begin our exploration of such methods, we introduce the following definitions that will be needed.

Definition 11.2.1 Let $C \subseteq \mathbb{R}^n$ be a convex set, and let $f : C \to \mathbb{R}$. Then $f(\mathbf{x})$ is **convex** on C if

$$f(\lambda \mathbf{x} + (1-\lambda)\mathbf{y}) \leq \lambda f(\mathbf{x}) + (1-\lambda)f(\mathbf{y})$$

for all $\mathbf{x}, \mathbf{y} \in C$ and $\lambda \in [0,1]$. If

$$f(\lambda \mathbf{x} + (1-\lambda)\mathbf{y}) < \lambda f(\mathbf{x}) + (1-\lambda)f(\mathbf{y})$$

for all $\mathbf{x}, \mathbf{y} \in C$, $\mathbf{x} \neq \mathbf{y}$, and $\lambda \in (0,1)$, then we say that $f(\mathbf{x})$ is **strictly convex**.

The main benefit of knowing whether a function is convex, as far as optimization is concerned, is provided by the following theorem.

Theorem 11.2.2 Let $C \subseteq \mathbb{R}^n$ be a convex set, and let $f : C \to \mathbb{R}$ be convex on C. Then any local minimizer of $f(\mathbf{x})$ is a global minimizer. Furthermore, if $f(\mathbf{x})$ is strictly convex on C, then any local minimizer of $f(\mathbf{x})$ is the unique strict global minimizer of $f(\mathbf{x})$ on C.

A comprehensive treatment of methods for optimizing nonlinear functions can be found in [Peressini et al. (1988)]. Here, we limit our discussion to methods that are closely related to root-finding methods presented in Chapter 10.

11.2.1   Newton's Method

Suppose that we wish to minimize a function $f(\mathbf{x})$, where $\mathbf{x} \in D \subseteq \mathbb{R}^n$, and therefore need to solve the system of equations $\nabla f(\mathbf{x}) = \mathbf{0}$. Then, we can apply a method for solving the system of equations $\mathbf{g}(\mathbf{x}) = \mathbf{0}$, where $\mathbf{g}(\mathbf{x}) = \nabla f(\mathbf{x})$, and $J_\mathbf{g}(\mathbf{x}) = H_f(\mathbf{x})$, which is the *Hessian* of f, defined by

$$H_f(\mathbf{x}) = \begin{bmatrix} \frac{\partial^2 f}{\partial x_1^2}(\mathbf{x}) & \frac{\partial^2 f}{\partial x_1 \partial x_2}(\mathbf{x}) & \cdots & \frac{\partial^2 f}{\partial x_1 \partial x_n}(\mathbf{x}) \\ \frac{\partial^2 f}{\partial x_2 \partial x_1}(\mathbf{x}) & \frac{\partial^2 f}{\partial x_2^2}(\mathbf{x}) & \cdots & \frac{\partial^2 f}{\partial x_2 \partial x_n}(\mathbf{x}) \\ \vdots & \vdots & \ddots & \vdots \\ \frac{\partial^2 f}{\partial x_n \partial x_1}(\mathbf{x}) & \frac{\partial^2 f}{\partial x_n \partial x_2}(\mathbf{x}) & \cdots & \frac{\partial^2 f}{\partial x_n^2}(\mathbf{x}) \end{bmatrix}.$$

The Newton's Method step then takes the form

$$\mathbf{x}^{(k+1)} = \mathbf{x}^{(k)} + \mathbf{s}_k, \quad \mathbf{s}_k = -[H_f(\mathbf{x}^{(k)})]^{-1}\nabla f(\mathbf{x}^{(k)}). \tag{11.10}$$

When used for minimization, Newton's Method approximates $f(\mathbf{x})$ by a quadratic function near $\mathbf{x}^{(k)}$, using multivariable Taylor expansion. This approximation is

$$f_k(\mathbf{x}) = f(\mathbf{x}^{(k)}) + \nabla f(\mathbf{x}^{(k)})^T(\mathbf{x} - \mathbf{x}^{(k)}) + \frac{1}{2}(\mathbf{x} - \mathbf{x}^{(k)})^T H_f(\mathbf{x}^{(k)})(\mathbf{x} - \mathbf{x}^{(k)}).$$

Then, we compute the unique critical point of $f_k(\mathbf{x})$, which is the unique solution of $\nabla f_k(\mathbf{x}) = \mathbf{0}$. Note that $\nabla f_k(\mathbf{x}) = \nabla f(\mathbf{x}^{(k)})$, and $H_{f_k}(\mathbf{x}) = H_f(\mathbf{x}^{(k)})$.

If $H_f(\mathbf{x}^{(k)})$ is positive definite, then this critical point is also guaranteed to be the unique strict global minimizer of $f_k(\mathbf{x})$. For quadratic functions in general, we have this useful result.

Theorem 11.2.3 Let A be an $n \times n$ symmetric positive definite matrix, let $\mathbf{b} \in \mathbb{R}^n$, and let $a \in \mathbb{R}$. Then the quadratic function

$$f(\mathbf{x}) = a + \mathbf{b}^T\mathbf{x} + \frac{1}{2}\mathbf{x}^T A\mathbf{x}$$

is strictly convex and has a unique strict global minimizer $\mathbf{x}^*$, where $A\mathbf{x}^* = -\mathbf{b}$. For any initial guess $\mathbf{x}^{(0)}$, Newton's Method applied to $f(\mathbf{x})$ converges to $\mathbf{x}^*$ in one step; that is, $\mathbf{x}^* = \mathbf{x}^{(1)}$.

> **Exploration 11.2.1** Prove Theorem 11.2.3. *Hint:* Use the result of Exploration 4.1.1.

If $f(\mathbf{x})$ is not a quadratic function, then Newton's Method will generally not compute a minimizer of $f(\mathbf{x})$ in one step, even if its Hessian $H_f(\mathbf{x})$ is positive definite. However, in this case, Newton's Method is guaranteed to make progress, as the following theorem indicates.

> **Theorem 11.2.4** Let $\{\mathbf{x}^{(k)}\}_{k=0}^{\infty}$ be the sequence of Newton iterates for the function $f(\mathbf{x})$. If $H_f(\mathbf{x}^{(k)})$ is positive definite and if $\nabla f(\mathbf{x}^{(k)}) \neq \mathbf{0}$, then the vector
> $$\mathbf{s}_k = -[H_f(\mathbf{x}^{(k)})]^{-1} \nabla f(\mathbf{x}^{(k)})$$
> from $\mathbf{x}^{(k)}$ to $\mathbf{x}^{(k+1)}$ is a descent direction for $f(\mathbf{x})$; that is,
> $$f(\mathbf{x}^{(k)} + t\mathbf{s}_k) < f(\mathbf{x}^{(k)})$$
> for t sufficiently small.

> **Exploration 11.2.2** Prove Theorem 11.2.4. *Hint:* Let $\varphi(t) = f(\mathbf{x}^{(k)} + t\mathbf{s}_k)$. How does $\varphi(t)$ behave for t near 0?

In the following example and other examples in this section, for convenience we will occasionally identify points (x, y) with vectors $\begin{bmatrix} x & y \end{bmatrix}^T$.

Example 11.2.5 Let
$$f(x, y) = x^4 + 2x^2y^2 + y^4 - 4x^3 - 4x^2y - 4xy^2 - 4y^3 + 8x^2 + 8xy + 8y^2 - 8x - 8y + 9.$$
Then we have
$$\nabla f(x, y) = \begin{bmatrix} 4x^3 + 4xy^2 - 12x^2 - 8xy - 4y^2 + 16x + 8y - 8 \\ 4x^2y + 4y^3 - 4x^2 - 8xy - 12y^2 + 8x + 16y - 8 \end{bmatrix}$$
and
$$H_f(x, y) = \begin{bmatrix} 12x^2 + 4y^2 - 24x - 8y + 16 & 8xy - 8x - 8y + 8 \\ 8xy - 8x - 8y + 8 & 4x^2 + 12y^2 - 8x - 24y + 16 \end{bmatrix}.$$
Let $\mathbf{x}^{(0)} = (0, 0)$. Then
$$\nabla f(\mathbf{x}^{(0)}) = \begin{bmatrix} -8 \\ -8 \end{bmatrix} = -8 \begin{bmatrix} 1 \\ 1 \end{bmatrix},$$
and
$$H_f(\mathbf{x}^{(0)}) = \begin{bmatrix} 16 & 8 \\ 8 & 16 \end{bmatrix} = 8 \begin{bmatrix} 2 & 1 \\ 1 & 2 \end{bmatrix}.$$
Then, using the formula for the inverse of a 2×2 matrix,
$$\begin{bmatrix} a & b \\ c & d \end{bmatrix}^{-1} = \frac{1}{ad - bc} \begin{bmatrix} d & -b \\ -c & a \end{bmatrix},$$

we obtain

$$\mathbf{s}_0 = -[H_f(\mathbf{x}^{(0)})]^{-1}\nabla f(\mathbf{x}^{(0)}) = \frac{1}{3}\begin{bmatrix} 2 & -1 \\ -1 & 2 \end{bmatrix}\begin{bmatrix} 1 \\ 1 \end{bmatrix} = \frac{1}{3}\begin{bmatrix} 1 \\ 1 \end{bmatrix}.$$

It follows that

$$\mathbf{x}^{(1)} = \mathbf{x}^{(0)} + \mathbf{s}_0 = (0,0) + \frac{1}{3}(1,1) = \left(\frac{1}{3}, \frac{1}{3}\right).$$

It can be seen that in general,

$$\mathbf{x}^{(k+1)} = \left(\frac{1}{3}, \frac{1}{3}\right) + \frac{2}{3}\mathbf{x}^{(k)},$$

and therefore the Newton iterates converge to $(1,1)$, which is the unique strict global minimizer of $f(x,y)$. □

Exploration 11.2.3 We have seen that Newton's Method, when it converges, generally does so quadratically. Did it converge quadratically in Example 11.2.5? Explain why or why not.

Exploration 11.2.4 Write a MATLAB function

$$[\texttt{x,niter]=fminnewton(f,gf,Hf,x0,tol)}$$

that uses Newton's Method (11.10) to find a local minimum of the function $f(\mathbf{x})$ represented by the function handle `f`. The input arguments `gf` and `Hf` are assumed to be function handles for functions that accept a vector `x` as an input argument and return a column vector for the gradient of f at `x` and a matrix for the Hessian of f at `x`, respectively. The input vector `x0` is the initial guess, and the input argument `tol` is the tolerance for the convergence check. Terminate the iteration when $\|\mathbf{s}_k\| < \texttt{tol}$. The outputs `x` and `niter` are the minimizer and iteration count, respectively.

11.2.2 The Method of Steepest Descent

When it is not possible to find the minimum of a function analytically, and therefore must use an iterative method for obtaining an approximate solution, Newton's Method can be an effective method, but it can also be unreliable, as we have seen when using it to find a root of a single equation. Therefore, we now consider another approach. Before we begin, we need the following definition.

Definition 11.2.6 A continuous function $f(\mathbf{x})$ that is defined on all of $\mathbb{R}^n$ is **coercive** if

$$\lim_{\|\mathbf{x}\|\to\infty} f(\mathbf{x}) = +\infty.$$

That is, for any constant $M > 0$ there exists a constant $R_M > 0$ such that $\|f(\mathbf{x})\| > M$ whenever $\|\mathbf{x}\| > R_M$.

From multivariable calculus, given a function $f : \mathbb{R}^n \to \mathbb{R}$ that is differentiable at $\mathbf{x}_0$, the *direction of steepest descent* is the vector $-\nabla f(\mathbf{x}_0)$. To see this, consider the function

$$\varphi(t) = f(\mathbf{x}_0 + t\mathbf{u}),$$

where $\mathbf{u}$ is a *unit* vector; that is, $\|\mathbf{u}\| = 1$. Then, by the Chain Rule,

$$\begin{aligned}
\varphi'(t) &= \frac{\partial f}{\partial x_1}\frac{\partial x_1}{\partial t} + \cdots + \frac{\partial f}{\partial x_n}\frac{\partial x_n}{\partial t} \\
&= \frac{\partial f}{\partial x_1}u_1 + \cdots + \frac{\partial f}{\partial x_n}u_n \\
&= \nabla f(\mathbf{x}_0 + t\mathbf{u})^T\mathbf{u},
\end{aligned}$$

and therefore

$$\varphi'(0) = \nabla f(\mathbf{x}_0)^T\mathbf{u} = \|\nabla f(\mathbf{x}_0)\|\cos\theta,$$

where θ is the angle between $\nabla f(\mathbf{x}_0)$ and $\mathbf{u}$. It follows that $\varphi'(0)$ is minimized when $\theta = \pi$, which yields

$$\mathbf{u} = -\frac{\nabla f(\mathbf{x}_0)}{\|\nabla f(\mathbf{x}_0)\|}, \quad \varphi'(0) = -\|\nabla f(\mathbf{x}_0)\|.$$

We can therefore reduce the problem of minimizing a function of several variables to a single-variable minimization problem, by finding the minimum of $\varphi(t)$ for this choice of $\mathbf{u}$. That is, we find the value of t, for $t > 0$, that minimizes

$$\varphi_0(t) = f(\mathbf{x}_0 - t\nabla f(\mathbf{x}_0)).$$

After finding the minimizer t_0, we can set

$$\mathbf{x}_1 = \mathbf{x}_0 - t_0\nabla f(\mathbf{x}_0)$$

and continue the process, by searching from $\mathbf{x}_1$ in the direction of $-\nabla f(\mathbf{x}_1)$ to obtain $\mathbf{x}_2$ by minimizing $\varphi_1(t) = f(\mathbf{x}_1 - t\nabla f(\mathbf{x}_1))$, and so on.

This is the **Method of Steepest Descent**: given an initial guess $\mathbf{x}^{(0)}$, the method computes a sequence of iterates $\{\mathbf{x}^{(k)}\}$, where

$$\mathbf{x}^{(k+1)} = \mathbf{x}^{(k)} - t_k\nabla f(\mathbf{x}^{(k)}), \quad k = 0, 1, 2, \ldots,$$

and $t_k > 0$ minimizes the function

$$\varphi_k(t) = f(\mathbf{x}^{(k)} - t\nabla f(\mathbf{x}^{(k)})).$$

Example 11.2.7 We apply the Method of Steepest Descent to the function

$$f(x_1, x_2) = 4x_1^2 - 4x_1x_2 + 2x_2^2 - 8x_1 + 4x_2 + 4$$

with initial point $\mathbf{x}^{(0)} = (1, 1)$. We first compute the steepest descent direction from

$$\nabla f(x_1, x_2) = \begin{bmatrix} 8x_1 - 4x_2 - 8 \\ 4x_2 - 4x_1 + 4 \end{bmatrix}$$

to obtain

$$\nabla f(\mathbf{x}^{(0)}) = \nabla f(1,1) = \begin{bmatrix} -4 \\ 4 \end{bmatrix}.$$

We then minimize the function

$$\varphi(t) = f(\mathbf{x}^{(0)} - t\nabla f(\mathbf{x}^{(0)})) = f(1 + 4t, 1 - 4t)$$

by computing

$$\varphi'(t) = -\nabla f(1 + 4t, 1 - 4t)^T \begin{bmatrix} -4 \\ 4 \end{bmatrix}$$

$$= -\left[8(1 + 4t) - 4(1 - 4t) - 8\ 4(1 - 4t) - 4(1 + 4t) + 4 \right] \begin{bmatrix} -4 \\ 4 \end{bmatrix}$$

$$= -\left[48t - 4\ -32t + 4 \right] \begin{bmatrix} -4 \\ 4 \end{bmatrix}$$

$$= 320t - 32.$$

This strictly convex function has a strict global minimum when $\varphi'(t) = 320t - 32 = 0$, or $t = 1/10$, as can be seen by noting that $\varphi''(t) = 320 > 0$. We therefore set

$$\mathbf{x}^{(1)} = \mathbf{x}^{(0)} - \frac{1}{10}\nabla f(\mathbf{x}^{(0)}) = \begin{bmatrix} 1 \\ 1 \end{bmatrix} - \frac{1}{10}\begin{bmatrix} -4 \\ 4 \end{bmatrix} = \begin{bmatrix} 7/5 \\ 3/5 \end{bmatrix}.$$

Continuing the process, we have

$$\nabla f(\mathbf{x}^{(1)}) = \nabla f(7/5, 3/5) = \begin{bmatrix} 4/5 \\ 4/5 \end{bmatrix},$$

which, it is interesting to note, is orthogonal to $\nabla f(\mathbf{x}^{(0)})$. By defining

$$\varphi(t) = f(\mathbf{x}^{(1)} - t\nabla f(\mathbf{x}^{(1)})) = f(7/5 - 4t/5, 3/5 - 4t/5)$$

we obtain

$$\varphi'(t) = -\left[\frac{8(7-4t)}{5} - \frac{4(3-4t)}{5} - 8\ \frac{4(3-4t)}{5} - \frac{4(7-4t)}{5} + 4 \right] \begin{bmatrix} 4/5 \\ 4/5 \end{bmatrix} = \frac{64t - 32}{25}.$$

We have $\varphi'(t) = 0$ when $t = 1/2$, and because $\varphi''(t) = 64/25$, this critical point is a strict global minimizer. We therefore set

$$\mathbf{x}^{(2)} = \mathbf{x}^{(1)} - \frac{1}{10}\nabla f(\mathbf{x}^{(1)}) = \begin{bmatrix} 7/5 \\ 3/5 \end{bmatrix} - \frac{1}{2}\begin{bmatrix} 4/5 \\ 4/5 \end{bmatrix} = \begin{bmatrix} 1 \\ 1/5 \end{bmatrix}.$$

Repeating this process yields $\nabla f(\mathbf{x}^{(2)}) = \left[-4/5\ 4/5 \right]^T$, which is, again, orthogonal to $\nabla f(\mathbf{x}^{(1)})$, and $\mathbf{x}^{(3)} = (27/25, 3/25)$. We can see that the Method of Steepest Descent produces a sequence of iterates $\mathbf{x}^{(k)}$ that is converging to the strict global minimizer of $f(x_1, x_2)$ at $\mathbf{x}^* = (1, 0)$. $\square$

The following theorems describe some important properties of the Method of Steepest Descent.

Theorem 11.2.8 Let $f : \mathbb{R}^n \to \mathbb{R}$ be continuously differentiable on $\mathbb{R}^n$, and let $\mathbf{x}^{(0)} \in \mathbb{R}^n$. Let $t^* > 0$ be the minimizer of the function

$$\varphi(t) = f(\mathbf{x}^{(k)} - t\nabla f(\mathbf{x}^{(k)})), \quad t \geq 0 \qquad (11.11)$$

and let $\mathbf{x}^{(k+1)} = \mathbf{x}^{(k)} - t^*\nabla f(\mathbf{x}^{(k)})$. Then

$$f(\mathbf{x}^{(k+1)}) < f(\mathbf{x}^{(k)}).$$

Exploration 11.2.5 Prove Theorem 11.2.8 using a similar approach to that used in Exploration 11.2.2. Conclude that the Method of Steepest Descent is guaranteed to make at least some progress toward a minimizer $\mathbf{x}^*$ during each iteration.

It was no accident that in the previous example, each steepest descent direction was orthogonal to the previous one.

Theorem 11.2.9 Let $f : \mathbb{R}^n \to \mathbb{R}$ be continuously differentiable on $\mathbb{R}^n$, and let $\mathbf{x}^{(k)}$ and $\mathbf{x}^{(k+1)}$, for $k \geq 0$, be two consecutive iterates produced by the Method of Steepest Descent. Then the steepest descent directions from $\mathbf{x}^{(k)}$ and $\mathbf{x}^{(k+1)}$ are orthogonal; that is,

$$\nabla f(\mathbf{x}^{(k)})^T \nabla f(\mathbf{x}^{(k+1)}) = 0.$$

Exploration 11.2.6 Prove Theorem 11.2.9 by finding a critical point of $\varphi(t)$ from (11.11).

That is, the Method of Steepest Descent pursues completely independent search directions from one iteration to the next. However, as we have seen with linear systems in Section 5.2, in some cases this causes the method to "zig-zag" from the initial iterate $\mathbf{x}_0$ to the minimizer $\mathbf{x}^*$.

We have seen that Newton's Method can fail to converge to a solution if the initial iterate is not chosen wisely. For certain functions, however, the Method of Steepest Descent can be shown to be much more reliable.

Theorem 11.2.10 Let $f : \mathbb{R}^n \to \mathbb{R}$ be a coercive function with continuous first partial derivatives on $\mathbb{R}^n$. Then, for any initial guess $\mathbf{x}^{(0)}$, the sequence of iterates produced by the Method of Steepest Descent from $\mathbf{x}^{(0)}$ contains a subsequence that converges to a critical point of f.

This result can be proved by applying the **Bolzano-Weierstrass Theorem**, which states that any bounded sequence contains a convergent subsequence. The sequence $\{f(\mathbf{x}^{(k)})\}_{k=0}^{\infty}$ is a decreasing sequence, as indicated by Theorem 11.2.8, and it is a bounded sequence, because $f(\mathbf{x})$ is continuous and coercive and therefore has a global minimum $f(\mathbf{x}^*)$. It follows that the sequence $\{\mathbf{x}^{(k)}\}_{k=0}^{\infty}$ is also bounded, for a coercive function cannot be bounded above on an unbounded set.

By the Bolzano-Weierstrass Theorem, $\{\mathbf{x}^{(k)}\}$ has a convergent subsequence $\{\mathbf{x}^{(k_p)}\}$, which can be shown to converge to a critical point of $f(\mathbf{x})$. Intuitively, as $\mathbf{x}^{(k+1)} = \mathbf{x}^{(k)} - t^*\nabla f(\mathbf{x}^{(k)})$ for some $t^* > 0$, convergence of $\{\mathbf{x}^{(k_p)}\}$ implies that

$$0 = \lim_{p\to\infty} \mathbf{x}^{(k_{p+1})} - \mathbf{x}^{(k_p)} = -\lim_{p\to\infty} \sum_{i=k_p}^{k_{p+1}-1} t_i^*\nabla f(\mathbf{x}^{(i)}), \quad t_i^* > 0,$$

which suggests the convergence of $\nabla f(\mathbf{x}^{(k_p)})$ to zero. A full proof can be found in [Peressini et al. (1988)].

11.2.3 Line Search

In the Method of Steepest Descent, as it has just been described, a one-dimensional minimization problem is solved to determine the scaling factor t_k for the step $\mathbf{x}^{(k+1)} = \mathbf{x}^{(k)} - t_k\nabla f(\mathbf{x}^{(k)})$. However, solving this minimization problem may require substantial computational effort, when a much less expensive approach may still yield a t_k that ensures that progress is made toward a minimum. Here, we consider a couple of **line search** strategies.

For the first one, we choose an initial guess for t_k (for example, one may simply choose $t_k = 1$), and check whether $f(\mathbf{x}^{(k+1)}) < f(\mathbf{x}^{(k)})$, where, as before, $\mathbf{x}^{(k+1)} = \mathbf{x}^{(k)} - t_k\nabla f(\mathbf{x}^{(k)})$. If not, then we divide t_k by 2, and try again. Because $-\nabla f(\mathbf{x}^{(k)})$ is a descent direction, it is guaranteed that there exists $t_k > 0$ such that $f(\mathbf{x}^{(k+1)}) < f(\mathbf{x}^{(k)})$, so at some point, this iteration will terminate with a viable choice of t_k. This approach is an example of a **weak line search**.

A second example of a weak line search involves **osculatory interpolation**, which was covered in Section 7.5. We again choose an initial value for t_k, such as $t_k = 1$, and then construct the second-degree interpolating polynomial $p_2(t)$ for $\varphi_k(t) = f(\mathbf{x}^{(k)} - t\nabla f(\mathbf{x}^{(k)}))$ that satisfies these conditions:

$$p_2(0) = \varphi_k(0) = f(\mathbf{x}^{(k)}), \quad p_2(t_k) = \varphi_k(t_k), \quad p_2'(0) = \varphi_k'(0) = -\|\nabla f(\mathbf{x}^{(k)})\|_2^2.$$

We then find the minimizer of $p_2(t)$ on $[0,1]$, denoted by t^*, and set $t_k = t^*$.

> **Exploration 11.2.7** Prove that $p_2(t^*) < p_2(0)$, where t^* is the minimizer of $p_2(t)$ on $[0, t_k]$.

These are just simple examples of line search strategies. More sophisticated strategies can be found in [Kelley (1999)].

11.2.4 Secant Methods

Let $\mathbf{g} : \mathbb{R}^n \to \mathbb{R}^n$ be continuously differentiable. Previously, we learned that Newton's Method for solving the system of nonlinear equations $\mathbf{g}(\mathbf{x}) = \mathbf{0}$,

$$\mathbf{x}^{(k+1)} = \mathbf{x}^{(k)} - [J_{\mathbf{g}}(\mathbf{x}^{(k)})]^{-1}\mathbf{g}(\mathbf{x}^{(k)}),$$

can be made more efficient by approximating the Jacobian matrix $J_{\mathbf{g}}(\mathbf{x}^{(k)})$ by a *rank-one update* of the previous approximate Jacobian matrix, as follows:

$$J_{k+1} = J_k + \frac{(\mathbf{y}_k - J_k\mathbf{s}_k)\mathbf{s}_k^T}{\mathbf{s}_k^T\mathbf{s}_k},$$

where $J_0 = J_{\mathbf{g}}(\mathbf{x}^{(0)})$, $\mathbf{s}_k = \mathbf{x}^{(k+1)} - \mathbf{x}^{(k)}$, and $\mathbf{y}_k = \mathbf{g}(\mathbf{x}^{(k+1)}) - \mathbf{g}(\mathbf{x}^{(k)})$. This is the essential idea behind **Broyden's Method** (Algorithm 11.1.5).

Furthermore, by applying the **Sherman-Morrison Formula** (11.9), we can use the following iteration to update the inverses of these approximate Jacobian matrices, instead of computing the matrices J_k themselves:

$$J_{k+1}^{-1} = J_k^{-1} + \frac{J_k^{-1}[(\mathbf{y}_k - J_k\mathbf{s}_k)\mathbf{s}_k^T]J_k^{-1}}{1 + \mathbf{s}_k^T J_k^{-1}(\mathbf{y}_k - J_k\mathbf{s}_k)}.$$

Now, we would like to solve the problem of minimizing a twice continuously differentiable function $f : \mathbb{R}^n \to \mathbb{R}$ by applying Broyden's Method, in either form, to solve the system of nonlinear equations $\nabla f(\mathbf{x}) = \mathbf{0}$.

Unfortunately, one problem with this approach is that it is not possible to ensure that the approximate Jacobian matrices J_k of ∇f, which are also approximate Hessians of f, are positive definite, because a rank-one update does not provide enough flexibility to ensure that each J_k has this property *and* also satisfies the **Secant Condition** $J_{k+1}\mathbf{s}_k = \mathbf{y}_k$. Therefore, we modify the idea behind Broyden's Method, and instead apply a *rank-two update* to J_k, or J_k^{-1}, during each iteration. We consider two such approaches: one that modifies J_k, and one that modifies J_k^{-1}.

11.2.4.1   The BFGS Method

In the first approach, we prescribe that J_{k+1} has the form

$$J_{k+1} = J_k + c_k(\mathbf{u}_k \mathbf{u}_k^T) + d_k(\mathbf{v}_k \mathbf{v}_k^T),$$

where the scalars c_k and d_k, and the vectors $\mathbf{u}_k$ and $\mathbf{v}_k$, are to be determined. The reason for this form is that if $\mathbf{x} \neq \mathbf{0}$, then

$$\mathbf{x}^T J_{k+1}\mathbf{x}_k = \mathbf{x}^T J_k \mathbf{x} + c_k \mathbf{x}^T(\mathbf{u}_k \mathbf{u}_k^T)\mathbf{x} + d_k \mathbf{x}^T(\mathbf{v}_k \mathbf{v}_k^T)\mathbf{x}$$
$$= \mathbf{x}^T J_k \mathbf{x} + c_k(\mathbf{x}^T \mathbf{u}_k)^2 + d_k(\mathbf{x}^T \mathbf{v}_k)^2.$$

This form gives us more control over the signs of the terms in the quadratic form $\mathbf{x}^T J_{k+1}\mathbf{x}$, which helps us ensure that J_{k+1} is positive definite, given that J_k is positive definite.

Because J_{k+1} must satisfy the Secant Condition $J_{k+1}\mathbf{s}_k = \mathbf{y}_k$, we have

$$\mathbf{y}_k - J_k\mathbf{s}_k = c_k(\mathbf{u}_k^T \mathbf{s}_k)\mathbf{u}_k + d_k(\mathbf{v}_k^T \mathbf{s}_k)\mathbf{v}_k.$$

Since the right side of the equation must be a linear combination of $\mathbf{y}_k$ and $J_k\mathbf{s}_k$, and it is already a linear combination of $\mathbf{u}_k$ and $\mathbf{v}_k$, we set

$$\mathbf{u}_k = \mathbf{y}_k, \quad \mathbf{v}_k = J_k\mathbf{s}_k,$$

which forces us to also set

$$c_k = \frac{1}{\mathbf{y}_k^T \mathbf{s}_k}, \quad d_k = -\frac{1}{\mathbf{s}_k^T J_k\mathbf{s}_k}.$$

In summary, we have

$$J_{k+1} = J_k + \frac{\mathbf{y}_k \mathbf{y}_k^T}{\mathbf{y}_k^T \mathbf{s}_k} - \frac{J_k\mathbf{s}_k(J_k\mathbf{s}_k)^T}{\mathbf{s}_k^T J_k\mathbf{s}_k}.$$

This is the main ingredient of what is called the **BFGS (Broyden-Fletcher-Goldfarb-Shanno) Method** [Fletcher (1987)].

Exploration 11.2.9 Use the Cauchy-Schwarz Inequality to show that if $\mathbf{s}_k^T \mathbf{y}_k > 0$, then J_{k+1} is positive definite.

Once J_k is computed, the BFGS Method computes $\mathbf{x}_{k+1}$ as follows:

$$\mathbf{x}^{(k+1)} = \mathbf{x}^{(k)} - t_k J_k^{-1}\nabla f(\mathbf{x}^{(k)}),$$

where the parameter $t_k > 0$ is chosen so as to ensure that $f(\mathbf{x}^{(k+1)}) < f(\mathbf{x}^{(k)})$. For example, t_k can be chosen by minimizing the function

$$\varphi_k(t) = f(\mathbf{x}^{(k)} + t_k\mathbf{p}_k),$$

where $\mathbf{p}_k = -J_k^{-1}\nabla f(\mathbf{x}^{(k)})$. This approach to choosing t_k, which is a **line search**, is similar to what is done in the Method of Steepest Descent. Alternatively, one of the **weak line search** strategies described in Section 11.2.3 can be used.

One last detail is the choice of the initial matrix J_0. While Broyden's Method uses $J_0 = J_{\nabla f}(\mathbf{x}^{(0)})$, an alternative is to choose J_0 to be an arbitrary positive definite matrix, to avoid the expense of computing an exact Jacobian. For example, choosing $J_0 = I$ makes $\mathbf{s}_0$ a steepest descent direction.

Example 11.2.11 We apply the BFGS Method to the function

$$f(x_1, x_2) = 4x_1^2 - 4x_1 x_2 + 2x_2^2 - 4x_2 + 4,$$

with initial guess $\mathbf{x}^{(0)} = (0,0)$. We begin by setting $J_0 = I$. From

$$\nabla f(x, y) = \begin{bmatrix} 8x_1 - 4x_2 \\ -4x_1 + 4x_2 - 4 \end{bmatrix},$$

we obtain

$$\mathbf{p}_0 = -J_0^{-1} \nabla f(\mathbf{x}_0) = -\nabla f(0,0) = \begin{bmatrix} 0 \\ 4 \end{bmatrix}.$$

To obtain the parameter t_0, we minimize the function

$$\varphi_0(t) = f((0,0) + t(0,4)) = f(0, 4t) = 32t^2 - 16t.$$

From $\varphi_0'(t) = 64t - 16$, we obtain the minimizer $t_0 = \frac{1}{4}$, which then yields

$$\mathbf{x}^{(1)} = \mathbf{x}^{(0)} + t_0 \mathbf{p}_0 = \begin{bmatrix} 0 \\ 0 \end{bmatrix} + \frac{1}{4} \begin{bmatrix} 0 \\ 4 \end{bmatrix} = \begin{bmatrix} 0 \\ 1 \end{bmatrix}.$$

For the second iteration, we update J_0 as follows:

$$J_1 = J_0 + \frac{\mathbf{y}_0 \mathbf{y}_0^T}{\mathbf{y}_0^T \mathbf{s}_0} - \frac{J_0 \mathbf{s}_0 (J_0 \mathbf{s}_0)^T}{\mathbf{s}_0^T J_0 \mathbf{s}_0}$$

where

$$\mathbf{s}_0 = \mathbf{x}^{(1)} - \mathbf{x}^{(0)} = \begin{bmatrix} 0 \\ 1 \end{bmatrix} - \begin{bmatrix} 0 \\ 0 \end{bmatrix} = \begin{bmatrix} 0 \\ 1 \end{bmatrix},$$

$$\mathbf{y}_0 = \nabla f(\mathbf{x}^{(1)}) - \nabla f(\mathbf{x}^{(0)})$$
$$= \nabla f(0,1) - \nabla f(0,0)$$
$$= \begin{bmatrix} -4 \\ 4 \end{bmatrix}.$$

This yields

$$J_1 = I + \frac{\mathbf{y}_0 \mathbf{y}_0^T}{\mathbf{y}_0^T \mathbf{s}_0} - \frac{\mathbf{s}_0 \mathbf{s}_0^T}{\mathbf{s}_0^T \mathbf{s}_0}$$

$$= \begin{bmatrix} 1 & 0 \\ 0 & 1 \end{bmatrix} + \frac{1}{4} \begin{bmatrix} 16 & -16 \\ -16 & 16 \end{bmatrix} - \begin{bmatrix} 0 & 0 \\ 0 & 1 \end{bmatrix}$$

$$= \begin{bmatrix} 5 & -4 \\ -4 & 4 \end{bmatrix}.$$

It can be verified by checking the minors that J_1 is positive definite.
    From

$$J_1^{-1} = \frac{1}{5(4) - (-4)(-4)} \begin{bmatrix} 4 & 4 \\ 4 & 5 \end{bmatrix} = \begin{bmatrix} 1 & 1 \\ 1 & 5/4 \end{bmatrix}$$

we obtain

$$\mathbf{p}_1 = -J_1^{-1}\nabla f(\mathbf{x}^{(1)}) = -\begin{bmatrix} 1 & 1 \\ 1 & 5/4 \end{bmatrix}\begin{bmatrix} -4 \\ 0 \end{bmatrix} = \begin{bmatrix} 4 \\ 4 \end{bmatrix}.$$

To obtain the parameter t_1, we minimize the function

$$\begin{aligned}
\varphi_1(t) &= f(\mathbf{x}^{(1)} + t\mathbf{p}_1) \\
&= f((0,1) + t(4,4)) \\
&= f(4t, 1 + 4t) \\
&= 4(4t)^2 - 4(1 + 4t)(4t) + 2(1 + 4t)^2 - 4(1 + 4t) + 4 \\
&= 2 + 32t^2 - 16t.
\end{aligned}$$

From $\varphi_1'(t) = 64t - 16$, we again obtain $t_1 = \frac{1}{4}$, which then yields

$$\mathbf{x}^{(2)} = \mathbf{x}^{(1)} + t_1\mathbf{p}_1 = \begin{bmatrix} 0 \\ 1 \end{bmatrix} + \frac{1}{4}\begin{bmatrix} 4 \\ 4 \end{bmatrix} = \begin{bmatrix} 1 \\ 2 \end{bmatrix}.$$

This happens to be the minimizer of $f(x, y)$, as

$$f(x_1, x_2) = (x_2 - 2x_1)^2 + (x_2 - 2)^2 \geq 0,$$

which equals zero at $(x_1, x_2) = (1, 2)$. $\square$

Exploration 11.2.10 Adapt your function `fminnewton` from Exploration 11.2.4 to obtain a new function `[x,niter]=fminbfgs(f,gf,Hf,x0,tol)` that uses BFGS to find a minimum of $f(\mathbf{x})$ that is implemented via the function handle `f`. The input arguments `gf` and `Hf` are function handles for ∇f and H_f, the fourth input argument `x0` is the initial guess, and the fifth input argument `tol` is a tolerance for the convergence check. Terminate the iteration when $\|\mathbf{s}_k\| < $ `tol`. The outputs `x` and `niter` are the minimizer and iteration count, respectively. Use $J_0 = I$. How does the speed of this function compare to that of `fminnewton` from Exploration 11.2.4?

11.2.4.2 *The DFP Method*

In the second approach, we develop an iteration that updates J_k^{-1} directly, rather than J_k. Let $\tilde{J}_k$ be an approximation to $[J_{\nabla f}(\mathbf{x}^{(k)})]^{-1}$. Then $\tilde{J}_k$ must satisfy the **Inverse Secant Condition**

$$\tilde{J}_{k+1}\mathbf{y}_k = \mathbf{s}_k.$$

As in the BFGS Method, we prescribe that $\tilde{J}_{k+1}$ so that it has the form

$$\tilde{J}_{k+1} = \tilde{J}_k + c_k(\mathbf{u}_k\mathbf{u}_k^T) + d_k(\mathbf{v}_k\mathbf{v}_k^T),$$

where the scalars c_k and d_k, and the vectors $\mathbf{u}_k$ and $\mathbf{v}_k$, are to be determined.

Because $\tilde{J}_{k+1}$ must satisfy the Inverse Secant Condition $\tilde{J}_{k+1}\mathbf{y}_k = \mathbf{s}_k$, we have

$$\mathbf{s}_k - \tilde{J}_k\mathbf{y}_k = c_k(\mathbf{u}_k^T\mathbf{y}_k)\mathbf{u}_k + d_k(\mathbf{v}_k^T\mathbf{y}_k)\mathbf{v}_k.$$

Since the right side of the equation must be a linear combination of $\mathbf{s}_k$ and $\tilde{J}_k\mathbf{y}_k$, and it is already a linear combination of $\mathbf{u}_k$ and $\mathbf{v}_k$, we set

$$\mathbf{u}_k = \mathbf{s}_k, \quad \mathbf{v}_k = \tilde{J}_k\mathbf{y}_k,$$

which forces us to also set

$$c_k = \frac{1}{\mathbf{s}_k^T\mathbf{y}_k}, \quad d_k = -\frac{1}{\mathbf{y}_k^T\tilde{J}_k\mathbf{y}_k}.$$

In summary, we have

$$\tilde{J}_{k+1} = \tilde{J}_k + \frac{\mathbf{s}_k\mathbf{s}_k^T}{\mathbf{y}_k^T\mathbf{s}_k} - \frac{\tilde{J}_k\mathbf{y}_k(\tilde{J}_k\mathbf{y}_k)^T}{\mathbf{y}_k^T\tilde{J}_k\mathbf{y}_k}.$$

This is the main ingredient of what is called the **DFP (Davidon-Fletcher-Powell) Method** [Fletcher (1987)]. As in BFGS, it can be shown that if $\mathbf{s}_k^T\mathbf{y}_k > 0$, then $\tilde{J}_{k+1}$ is positive definite.

Once $\tilde{J}_k$ is computed, the DFP Method computes $\mathbf{x}^{(k+1)}$ as follows:

$$\mathbf{x}^{(k+1)} = \mathbf{x}^{(k)} - t_k\tilde{J}_k\nabla f(\mathbf{x}^{(k)}),$$

where the parameter $t_k > 0$ is chosen so as to ensure that $f(\mathbf{x}^{(k+1)}) < f(\mathbf{x}^{(k)})$. For example, t_k can be chosen by minimizing the function

$$\varphi_k(t) = f(\mathbf{x}^{(k)} + t_k\mathbf{s}_k),$$

where $\mathbf{s}_k = -\tilde{J}_k\nabla f(\mathbf{x}^{(k)})$, as in the BFGS Method. Alternatively, a weak line search strategy can be used.

As in the BFGS Method, the DFP Method can benefit from the choice of an arbitrary positive definite matrix $\tilde{J}_0$, instead of using the Broyden's Method choice of $\tilde{J}_0 = [J_{\nabla f}(\mathbf{x}^{(0)})]^{-1}$, but in this case the benefit is greater because computing an inverse matrix is very expensive. By choosing $\tilde{J}_0 = I$, we simply use the steepest descent direction.

Exploration 11.2.11 Adapt your function `fminbfgs` from Exploration 11.2.10 to obtain a new function `[x,niter]=fmindfp(f,gf,Hf,x0,tol)` that uses DFP to find a minimum of $f(\mathbf{x})$ that is implemented via the function handle `f`. Use $J_0 = I$. How does the speed of this function compare to that of `fminbfgs`?

11.2.5 Nonlinear Least Squares*

Let $\mathbf{g} : D \subseteq \mathbb{R}^n \to \mathbb{R}^m$, with $m \geq n$, and let $\mathbf{b} \in \mathbb{R}^m$. We consider the problem of finding $\mathbf{x} \in D$ that minimizes

$$\varphi(\mathbf{x}) = \frac{1}{2}\|\mathbf{b} - \mathbf{g}(\mathbf{x})\|_2^2.$$

In Chapter 4, we considered the case of $\mathbf{g}(\mathbf{x}) = A\mathbf{x}$, where $A \in \mathbb{R}^{m\times n}$, which is the linear least squares problem. Now, however, we allow $\mathbf{g}$ to be a nonlinear function of $\mathbf{x}$, which gives us a **nonlinear least squares problem** to solve.

The gradient of $\varphi(\mathbf{x})$ is given by

$$\nabla\varphi(\mathbf{x}) = A^T(\mathbf{g}(\mathbf{x}) - \mathbf{b}), \tag{11.12}$$

where $A = J_{\mathbf{g}}(\mathbf{x})$ is the Jacobian matrix of $\mathbf{g}$, and its Hessian is

$$H_{\varphi}(\mathbf{x}) = A^T A + C, \tag{11.13}$$

where

$$c_{ij} = \sum_{k=1}^{m} \frac{\partial^2 g_k}{\partial x_i \partial x_j}(g_k - b_k), \quad i,j = 1,2,\ldots,n.$$

Exploration 11.2.12 Verify (11.12) and (11.13).

Suppose that we have an initial guess $\mathbf{x}^{(0)}$, and we seek a next iterate of the form $\mathbf{x}^{(1)} = \mathbf{x}^{(0)} + \mathbf{p}_0$, where $\mathbf{p}_0$ is a search direction. Replacing $\mathbf{g}(\mathbf{x})$ in $\nabla\varphi(\mathbf{x})$ with a linear approximation in the equation $\nabla\varphi(\mathbf{x}^{(1)}) = \mathbf{0}$, we obtain the system of equations

$$\mathbf{0} = \nabla\varphi(\mathbf{x}^{(1)}) = A^T(\mathbf{g}(\mathbf{x}^{(0)}) + A\mathbf{p}_0 - \mathbf{b})$$

which, upon rearranging, becomes

$$A^T A\mathbf{p}_0 = A^T(\mathbf{b} - \mathbf{g}(\mathbf{x}^{(0)})).$$

These are the normal equations for the linear least squares problem

$$\|\mathbf{g}(\mathbf{x}^{(0)}) + A\mathbf{p}_0 - \mathbf{b}\|_2 = \text{minimum}.$$

Continuing this process, we obtain the following algorithm.

Algorithm 11.2.12 (Gauss-Newton Method) Given $\mathbf{g} : D \subseteq \mathbb{R}^n \to \mathbb{R}^m$, $\mathbf{b} \in \mathbb{R}^m$, the following algorithm computes a minimizer $\mathbf{x}^* \in \mathbb{R}^n$ of $\|\mathbf{b} - \mathbf{g}(\mathbf{x})\|_2$.

Choose an initial guess $\mathbf{x}^{(0)} \in D$
for $k = 0, 1, 2, \ldots$ **do** until convergence
    $\mathbf{r}_k = \mathbf{b} - \mathbf{g}(\mathbf{x}^{(k)})$
    $A = J_{\varphi}(\mathbf{x}^{(k)})$
    Solve the linear least squares problem $\|A\mathbf{p}_k - \mathbf{r}_k\|_2 = \text{minimum}$
    $\mathbf{x}^{(k+1)} = \mathbf{x}^{(k)} + \mathbf{p}_k$
end while
$\mathbf{x}^* = \mathbf{x}^{(k)}$

An appropriate convergence test can be, for example, checking whether $\|\mathbf{p}_k\|$ or $\|A^T \mathbf{r}_k\|$ is less than some tolerance.

> **Exploration 11.2.13** Write a MATLAB function
>
> $$[\texttt{x,niter}]=\texttt{gaussnewton(g,b,tol)}$$
>
> that implements Algorithm 11.2.12. The output argument `niter` is the number of iterations performed, and the input argument `tol` is an error tolerance to be used in the convergence test of your choice.

It is interesting to note that if Newton's Method is used to minimize $\varphi(\mathbf{x})$, then the system of equations that it would solve at each iteration would feature the matrix $A^T A + C$, as it is the Hessian of φ, rather than $A^T A$. While this seems like a glaring omission in Gauss-Newton, there are two things to consider:

- As $\mathbf{x}^{(k)}$ converges, $\|\mathbf{r}_k\|$ decreases, and therefore $\|C\|$ decreases as well, because C is a linear combination of matrices, in which the coefficients are the components of $\mathbf{r}_k$.
- $A^T A$ is symmetric positive definite, whereas $A^T A + C$ may not be.

In particular, if the minimum residual is small, then Gauss-Newton converges more rapidly, because it more closely resembles Newton's Method.

11.2.6 Concept Check

1. What does it mean for a function to be convex? What can be said about a minimizer of a convex function?
2. How is Newton's Method for a system of nonlinear equations applied to find the minimum of a function of several variables?
3. What is the main idea behind the Method of Steepest Descent?
4. What advantage does the Method of Steepest Descent have over Newton's Method?
5. What does it mean for a function to be coercive? What can be said of the Method of Steepest Descent when applied to a coercive function?
6. What is the problem with applying Broyden's Method directly to the problem of finding a minimum?
7. Describe the main ideas behind the two modifications of Broyden's Method presented in this section. How do they "fix" Broyden's Method to be suitable for optimization? How do they differ from one another?
8. Describe the Gauss-Newton Method for solving a nonlinear least squares problem. Compare and contrast with Newton's Method.

11.3 Derivative-Free Optimization*

In some applications, it is necessary to find the minimum of a function without being able to evaluate its derivatives, as Newton's Method would require. In this section, we describe a couple of methods for **derivative-free optimization**, which use only values of the objective function $f(x)$. Unlike Newton's Method, the derivative-free methods presented in this chapter cannot be generalized to functions of several variables. References to derivative-free optimization techniques for the multivariable case are given in Section 11.4.

11.3.1   Golden Section Search

A continuous function $f(x)$ is said to be **unimodal** on an interval $[a, b]$ if it has a unique local minimum (or maximum) on $[a, b]$, and is *strictly* increasing or decreasing at all other points in $[a, b]$. We seek the unique minimum of a unimodal function $f(x)$ on $[0, 1]$. Suppose that we evaluate $f(x)$ at $x = a$ and $x = b$, where $0 < a < b < 1$. We consider two cases:

- If $f(a) \leq f(b)$, then the minimum must lie in $[0, b)$. We therefore continue searching on this interval, with a playing the role of b.
- If $f(b) < f(a)$, then the minimum must lie in $(a, 1]$. We therefore continue searching on this interval, with b playing the role of a.

In either case, we want the next search to be a scaled-down version of the current search, meaning that the ratios of successive interval widths should be the same. Otherwise, the size of the interval may decrease too slowly, leading to unnecessary iterations.

More precisely, if $f(a) \leq f(b)$, then the intervals $(0, 1)$ and $(0, b)$ are replaced by $(0, b)$ and $(0, a)$, respectively. The requirement that widths of corresponding intervals have equal ratios yields the equation

$$\frac{b}{1} = \frac{a}{b}.$$

Similarly, if $f(b) < f(a)$, then the intervals $(0, 1)$ and $(0, a)$ are replaced by $(a, 1)$ and (a, b), respectively, and the same requirement yields the equation

$$\frac{a}{1} = \frac{b - a}{1 - a}.$$

Substituting the first equation into the second yields the quadratic equation

$$b^2 + b - 1 = 0.$$

The only positive root is $b = (\sqrt{5} - 1)/2 \approx 0.618$, which then yields $a = b^2 = 1 - b \approx 0.382$. This leads to the following algorithm.

> **Algorithm 11.3.1 (Golden Section Search)** Given a function $f(x)$ and interval $[a, b]$, such that $f(x)$ is unimodal on $[a, b]$, the following algorithm computes an approximate minimizer x^* of $f(x)$ on $[a, b]$, with a given absolute error tolerance ϵ.
>
> $\rho = (\sqrt{5} - 1)/2$
> **while** $(b - a)/2 > \epsilon$ **do**
> $c = a + (1 - \rho)(b - a)$
> $d = a + \rho(b - a)$
>   **if** $f(c) \leq f(d)$ **then**
> $b = d$
>   **else**
> $a = c$
>   **end if**
> **end while**
> $x^* = (a + b)/2$

The name "Golden Section Search" comes from the fact that $1/\rho = (\sqrt{5} + 1)/2 \approx 1.618$, which is known as the **golden ratio**, due to its many applications and appearances in nature. This number appeared in Section 10.4, as the order of convergence of the Secant Method.

In the above algorithm, if $f(c) \leq f(d)$, then, in the next iteration, d is equal to the value of c from the previous iteration. Similarly, if $f(c) > f(d)$, then, in the next iteration, c is equal to the value of d from the previous iteration. It follows that there is only one *new* function evaluation needed in each iteration.

> **Exploration 11.3.1** Write a MATLAB function
>
> $$x=\text{mingss(f,a,b,tol)}$$
>
> that implements Algorithm 11.3.1, in such a way that only one evaluation of f is performed during each iteration. The input argument tol is the absolute error tolerance.

Like the Bisection Method presented in Section 10.2, Golden Section Search reduces the size of the interval $[a, b]$ by a fixed fraction. Therefore, the number of iterations needed to obtain a minimum to a specified tolerance can be determined in advance.

> **Exploration 11.3.2** Let x^* be the minimizer of $f(x)$ on $[a, b]$. Show that after k iterations, the midpoint $m = (a + b)/2$ of the interval $[a, b]$ satisfies
>
> $$|m - x^*| \leq \rho^k (b - a)/2.$$
>
> Use this error bound to obtain a lower bound on the number of iterations required to obtain an absolute error of at most ϵ.

> **Exploration 11.3.3** How does Golden Section Search behave if the unique minimum of $f(x)$ on $[a, b]$ occurs at one of the endpoints?

11.3.2 Successive Parabolic Interpolation

Golden Section Search is a very simple and robust algorithm for finding the unique minimum of a function on an interval, but as we have seen, its convergence is slow. Then again, this method only uses function values for comparison, which gives little guidance on how to narrow the search for a minimum. We therefore consider an alternative approach that makes better use of the information available.

Recall from Section 10.4 that the Secant Method finds a zero of a function $f(x)$ by finding the zero of a linear approximation, obtained by interpolating f at two distinct points. Similarly, we can find a minimum of a function $f(x)$ by finding the minimum of a quadratic approximation, obtained by interpolating f at three distinct points.

Specifically, given three distinct points $(a, f(a))$, $(b, f(b))$, and $(c, f(c))$, we compute the Newton interpolating polynomial (7.14) of $f(x)$:

$$p_2(x) = f(a) + f[a, b](x - a) + f[a, b, c](x - a)(x - b).$$

Then, we find this interpolant's unique extremum by finding the zero of its derivative:

$$p_2'(x) = f[a, b] + f[a, b, c](2x - a - b).$$

We therefore have

$$x = \frac{a + b}{2} - \frac{f[a, b]}{2f[a, b, c]}.$$

Substituting the definitions of the divided differences yields

$$x = \frac{a + b}{2} - \frac{(f(b) - f(a))(c - a)(c - b)}{2[(f(c) - f(b))(b - a) - (f(b) - f(a))(c - b)]}.$$

The overall algorithm is as follows.

> **Algorithm 11.3.2 (Successive Parabolic Interpolation)** Given a function $f(x)$, the following algorithm computes a minimizer x^* of $f(x)$.
>
> Choose three initial guesses $x^{(0)}, x^{(1)}, x^{(2)}$
> $d_1 = x^{(1)} - x^{(0)}, d_2 = x^{(2)} - x^{(1)}$
> $y_0 = f(x^{(0)}), y_1 = f(x^{(1)}), y_2 = f(x^{(2)})$
> $z_1 = y_1 - y_0, z_2 = y_2 - y_1$
> **for** $k = 2, 3, \ldots$ **do until convergence**
> $\quad x^{(k+1)} = \frac{x^{(k-2)} + x^{(k-1)}}{2} - \frac{z_{k-1}(x^{(k)} - x^{(k-2)})d_k}{2(z_k d_{k-1} - z_{k-1} d_k)}$
> $\quad d_{k+1} = x^{(k+1)} - x^{(k)}$
> $\quad y_{k+1} = f(x^{(k+1)})$
> $\quad z_{k+1} = y_{k+1} - y_k$
> **end while**
> $x^* = x^{(k)}$

Note that as with Golden Section Search, there is only one new function evaluation per iteration. A convergence test can include a check of $|x^{(k+1)} - x^{(k)}|$. One variation of the above algorithm is to have the three points used during each iteration to be the ones for which the function values are the smallest, rather than the last ones in the sequence.

Exploration 11.3.4 Write a MATLAB function

$$[x,niter]=minspi(f,x0,x1,x2,tol)$$

that implements Algorithm 11.3.2. The input argument f is assumed to be a function handle. The output argument `niter` is the number of iterations performed. Use the convergence test of your choice, based on the input argument `tol`.

Unlike Golden Section Search, Successive Parabolic Interpolation does not have any guarantee of convergence; like the Secant Method, it can fail to converge due to poorly chosen initial guesses.

Exploration 11.3.5 What is an easy way to check whether the new approximate minimum $x^{(k+1)}$ is actually a *maximum* of the quadratic approximation of $f(x)$, rather than a minimum? Incorporate this check into your function `minspi` from Exploration 11.3.4 to have the function abort and print an error message advising that the user choose different initial guesses.

11.3.3 Concept Check

1. How does Golden Section Search get its name?
2. Name one advantage and one disadvantage of Golden Section Search, compared to other optimization methods.
3. What is the basic idea behind Successive Parabolic Interpolation?
4. Name one advantage and one disadvantage of Successive Parabolic Interpolation, compared to Golden Section Search.
5. What is a significant limitation of both Golden Section Search and Successive Parabolic Interpolation?

11.4 Additional Resources

In this chapter, we have barely scratched the surface of the vast field of optimization. In particular, we have not covered **constrained optimization**, though examples of this in the context of linear least squares problems were covered in Section 4.5. Textbooks offering in-depth coverage of optimization, either unconstrained or constrained, are [Fletcher (1987); Gill, et al. (1981); Nocedal and Wright (2006); Peressini et al. (1988)]. As for software, MATLAB sells an optimization toolbox, and Stanford Business Software, Inc. offers a variety of libraries at its web site, `http://www.sbsi-sol-optimize.com`. The CPLEX package, developed by IBM and hosted at their web site, is free for academic use or for small-scale problems.

Regarding derivative-free optimization, an algorithm specifically for functions of several variables is the *Nelder-Mead Method*, which is covered in [Sauer (2012)]. A free software package for derivative-free optimization problems is HOPSPACK, developed by Sandia National Laboratories. For more information, the reader is referred to `https://software.sandia.gov/trac/hopspack/wiki`.

11.5 Exercises

1. Consider the system of equations

$$x_1 = 5x_1 + 2x_2,$$
$$x_2 = 2x_1 + 2x_2.$$

   (a) Try solving this system using Fixed-Point Iteration applied to the equation $\mathbf{x} = \mathbf{G}(\mathbf{x})$, where $\mathbf{G}(\mathbf{x})$ is defined to be the right-hand side of the system. Use the initial guess $\mathbf{x}^{(0)} = \begin{bmatrix} 1 & 1 \end{bmatrix}^T$. What happens?

   (b) Solve the equivalent system $\mathbf{x} = \mathbf{G}^{-1}(\mathbf{x})$, with the same initial guess. What happens?

   (c) Use eigenvalues to explain the behavior that you have observed.

2. Try using Newton's Method to solve the following systems of equations. In both cases, use the initial guess $\mathbf{x}^{(0)} = \begin{bmatrix} 1 & 1 \end{bmatrix}^T$.

   (a)

$$3x_1^2 + x_1 x_2 - 2x_2^2 = 0,$$
$$x_1^2 - x_1 - 2x_2 + 3 = 0.$$

   (b)

$$3x_1^2 + x_1 x_2 - 2x_2^2 = 0,$$
$$x_1^2 + x_1 - 2x_2 + 1 = 0.$$

   What happens in each case? Can you explain the behavior that you observe?

3. By Taylor's Theorem, if $f : \mathbb{R}^n \to \mathbb{R}$ has continuous second partial derivatives in a closed ball around $\mathbf{x}^{(0)} \in \mathbb{R}^n$, then

$$f(\mathbf{x}) = f(\mathbf{x}^{(0)}) + \nabla f(\mathbf{x}^{(0)})^T(\mathbf{x} - \mathbf{x}^{(0)}) + \frac{1}{2} \sum_{i,j=1}^{n} \frac{\partial^2 f}{\partial x_i \partial x_j}(\xi_{ij})(x_i - x_i^{(0)}),$$

   where, for $i, j = 1, \ldots, n$, ξ_{ij} is a point on the line segment between $\mathbf{x}^{(0)}$ and $\mathbf{x}$.

(a) Let $\mathbf{p} = \mathbf{x} - \mathbf{x}^{(0)}$. Show that

$$f(\mathbf{x}) = f(\mathbf{x}^{(0)}) + \nabla f(\mathbf{x}^{(0)})^T \mathbf{p} + \frac{1}{2} \mathbf{p}^T H \mathbf{p},$$

where H is symmetric.

(b) Show that if $\mathbf{F} : \mathbb{R}^n \to \mathbb{R}^n$ has component functions that have continuous second partial derivatives in a closed ball around $\mathbf{x}^{(0)}$, then for $\mathbf{x}$ in this ball, there exists a constant ρ such that

$$\|\mathbf{F}(\mathbf{x}) - \mathbf{F}(\mathbf{x}^{(0)}) - J_{\mathbf{F}}(\mathbf{x}^{(0)})(\mathbf{x} - \mathbf{x}^{(0)})\|_2 \leq \frac{1}{2} \rho \|\mathbf{x} - \mathbf{x}^{(0)}\|_2^2.$$

(c) Use the preceding result to show that if $\mathbf{F}(\mathbf{x}^*) = 0$ and $J_{\mathbf{F}}(\mathbf{x}^*)$ is nonsingular, then, for $\mathbf{x}^{(0)}$ sufficiently close to $\mathbf{x}^*$, Newton's Method converges quadratically.

4. Use the Method of Steepest Descent to minimize the function

$$f(\mathbf{x}) = \frac{1}{2} \mathbf{x}^T A \mathbf{x} - \mathbf{x}^T \mathbf{b},$$

where $\mathbf{b} = \begin{bmatrix} 2 & -8 \end{bmatrix}^T$ and A is given by

(a) $A = \begin{bmatrix} 3 & 2 \\ 2 & 6 \end{bmatrix}$

(b) $A = \begin{bmatrix} 1000 & -999 \\ -999 & 1001 \end{bmatrix}$.

How does the method perform for these two cases, in terms of the number of iterations required to find the minimum within a certain tolerance? Explain your observations.

5. Let $A \in \mathbb{R}^{n \times n}$ be invertible, and let $U, V \in \mathbb{R}^{n \times k}$. Use an approach analogous to that used in Section 11.1.3 to derive the **Sherman-Morrison-Woodbury formula**

$$(A + UV^T)^{-1} = A^{-1} - A^{-1}U(I + V^T A^{-1} U)V^T A^{-1}.$$

6. Use the Sherman-Morrison-Woodbury formula from Exercise 5 to show that inverse of the approximate Jacobian matrix J_{k+1} in BFGS is given by

$$J_{k+1}^{-1} = J_k^{-1} + \frac{(\mathbf{s}_k^T \mathbf{y}_k + \mathbf{y}_k^T J_k^{-1} \mathbf{y}_k)\mathbf{s}_k \mathbf{s}_k^T}{(\mathbf{s}_k^T \mathbf{y}_k)^2} - \frac{J_k^{-1} \mathbf{y}_k \mathbf{s}_k^T + \mathbf{s}_k \mathbf{y}_k^T J_k^{-1}}{\mathbf{s}_k^T \mathbf{y}_k}.$$

Then, modify your function `fminbfgs` from Exploration 11.2.10 to use this formula to update J_k^{-1} rather than J_k. How does this affect the efficiency of this function?

7. Define

$$\mathbf{g}(\mathbf{x}) = \begin{bmatrix} 4x_1^2 x_2^3 - 3x_1^2 x_2^3 \\ 2x_1^3 x_2^3 + 3x_1 x_2^3 \\ 2x_1^2 x_2 - x_1^2 \end{bmatrix}, \quad \mathbf{b} = \begin{bmatrix} 60 \\ 133 \\ 10 \end{bmatrix},$$

and

$$\varphi(\mathbf{x}) = \frac{1}{2} \|\mathbf{b} - \mathbf{g}(\mathbf{x})\|_2^2.$$

(a) Use Newton's Method to minimize $\varphi(\mathbf{x})$, using initial guess $\mathbf{x}^{(0)} = \begin{bmatrix} 1.5 & 1.5 \end{bmatrix}^T$, and then with initial guess $\mathbf{x}^{(0)} = \begin{bmatrix} 1.7 & 1.7 \end{bmatrix}^T$. Can you explain the behavior observed in both cases?

(b) Solve the same optimization problem using the Gauss-Newton Method, again with initial guess $\mathbf{x}^{(0)} = \begin{bmatrix} 1.5 & 1.5 \end{bmatrix}^T$. How does the behavior contrast with that of Newton's Method? Explain this contrast.

8. Consider the function

$$f(x) = 10x^4 - 81x^3 + 226x^2 - 249x + 98.$$

(a) Find the local minima and maxima of this function on the interval $[0, 4]$.

(b) Try your functions `mingss` (Golden Section Search) and `minspi` (Successive Parabolic Interpolation) from Explorations 11.3.1 and 11.3.4, respectively. Use various intervals (for Golden Section Search) and sets of initial guesses (from Successive Parabolic Interpolation) based on your knowledge of the local extrema. Is the behavior what you would expect? Discuss your observations. Which method is more efficient, in terms of number of iterations and execution time?

PART V
Differential Equations

Chapter 12

Initial Value Problems

In this chapter, we begin our exploration of the development of numerical methods for solving **differential equations**, which are equations that depend on derivatives of unknown quantities. Differential equations arise in mathematical models of a wide variety of phenomena, such as propagation of waves, diffusion of heat energy, population growth, or motion of fluids. Solutions of differential equations yield valuable insight about such phenomena, and therefore techniques for solving differential equations are among the most essential methods of applied mathematics.

We begin with examples of mathematical models based on differential equations. Newton's Second Law states

$$F = ma = m\frac{dv}{dt},$$

where F, m, a, and v represent force, mass, acceleration, and velocity, respectively. We use this law to develop a mathematical model for the velocity of a falling object that includes a differential equation. The forces on the falling object include gravity and air resistance, or drag; to simplify the discussion, we neglect any other forces.

The force due to gravity is equal to mg, where g is the acceleration due to gravity, and the drag force is equal to $-\gamma v$, where γ is the drag coefficient. We use downward orientation, so that gravity is acting in the positive (downward) direction and drag is acting in the negative (upward) direction. In summary, we have

$$F = mg - \gamma v.$$

Combining with Newton's Second Law yields the differential equation

$$m\frac{dv}{dt} = mg - \gamma v \qquad (12.1)$$

for the velocity v of the falling object.

Another example of a mathematical model is a differential equation for the population p of a species, which can have the form

$$\frac{dp}{dt} = rp - d, \qquad (12.2)$$

where the constant r is the rate of reproduction of the species. In general, r is called a *rate constant* or *growth rate*. The constant d indicates the number of specimens that die per unit of time, perhaps due to predation or other causes.

A differential equation such as this one does not have a unique solution, as it does not include enough information. Typically, the differential equation is paired with an **initial condition** of the form

$$y(t_0) = y_0,$$

where t_0 represents an **initial time** and y_0 is an **initial value**. The differential equation, together with the initial condition, is called an **initial value problem** (**IVP**). As discussed in the next section, under certain assumptions, it can be proven that an initial value problem has a unique solution. This chapter explores the numerical solution of initial value problems. Chapter 13 investigates the numerical solution of **boundary value problems (BVP)**, which are differential equations defined on a spatial domain, such as a bounded interval $[a, b]$, paired with **boundary conditions** that ensure a unique solution. In Chapter 14, we consider numerical methods for **partial differential equations**, which combine aspects of methods for both IVPs and BVPs.

Section 12.1 lists basic properties of differential equations that aid in classifying them, and discusses the well-posedness of IVPs. We then begin our exploration of **time-stepping** methods for IVPs in Section 12.2, which introduces **one-step methods**. These methods, including (Forward and Backward) **Euler's Method** and **Runge-Kutta methods**, use only information from a single time step to compute an approximate solution at the next time step. By contrast, Section 12.3 presents **multistep methods**, which use information further back in time to improve efficiency. In Section 12.4 we will analyze the accuracy and robustness of one-step and multistep methods, and learn how to easily determine whether they are **convergent**; that is, whether the computed solution converges to the exact solution as the length of the time step approaches zero. Section 12.5 describes how we can modify one-step and multistep methods to obtain **adaptive time-stepping** methods, that adjust the time step length in such a way as to preserve accuracy while improving efficiency. Finally, in Section 12.6 we generalize time-stepping methods to *systems* of ODEs, as well as ODEs with higher-order derivatives.

12.1 Basics of Differential Equations

In this section, we cover essential properties of differential equations that will be needed in our study of numerical methods for their solution.

12.1.1 Classification of Differential Equations

There are many types of differential equations, and a wide variety of solution techniques, even for equations of the same type, let alone different types. We now introduce some terminology that aids in classification of differential equations and, by extension, selection of solution techniques.

- An **ordinary differential equation**, or ODE, is an equation that depends on one or more derivatives of functions of a single variable. Differential equations given in the preceding examples are all ordinary differential equations, and we will consider these equations exclusively in this chapter and the next. The differential equations (12.1) and (12.2) presented at the beginning of this chapter are ODEs.

- A **partial differential equation**, or PDE, is an equation that depends on one or more *partial* derivatives of functions of several variables. In many cases, PDEs are solved by reduction to multiple ODEs. Numerical methods for PDEs will be presented in Chapter 14.

Example 12.1.1 The **heat equation**

$$\frac{\partial u}{\partial t} = k^2 \frac{\partial^2 u}{\partial x^2},$$

where k is a constant, is an example of a partial differential equation, as its solution $u(x, t)$ is a function of two independent variables, and the equation includes partial derivatives with respect to both variables. □

- The **order** of a differential equation is the order of the highest derivative of any unknown function in the equation.

Example 12.1.2 Equations (12.1) and (12.2) are examples of the differential equation

$$\frac{dy}{dt} = ay - b,$$

where a and b are constants. This equation is a *first-order* differential equation, as only the first derivative of the solution $y(t)$ appears in the equation. On the other hand, the ODE

$$y'' + 3y' + 2y = 0$$

is a *second-order* differential equation, whereas the PDE known as the **beam equation**

$$u_{tt} = -u_{xxxx}$$

is a *fourth-order* differential equation. □

- A differential equation is **linear** if it can be written in the form

$$\sum_{|\alpha| \leq m} a_\alpha(\mathbf{x}) D^\alpha u = 0, \tag{12.3}$$

where m is the order of the equation. Here, α is a **multi-index** that indicates the order of differentiation with respect to each independent variable, and $|\alpha|$ is the sum of the indices in α. The coefficients $\{a_\alpha\}_{|\alpha| \leq m}$ can only depend on the independent variables of the PDE.

Example 12.1.3 The **heat equation** in 2-D,

$$u_t - (u_{xx} + u_{yy}) = 0,$$

is a linear second-order partial differential equation. It can be written in the form (12.3) as follows:

$$a_{(0,0,1)} D^{(0,0,1)} u + a_{(2,0,0)} D^{(2,0,0)} u + a_{(0,2,0)} D^{(0,2,0)} u = 0,$$

where $a_{(0,0,1)} = 1$ and $a_{(2,0,0)} = a_{(0,2,0)} = -1$. All multi-indices use the ordering (x, y, t) of the independent variables. For example, the multi-index $(0, 2, 0)$ corresponds to the second partial derivative with respect to y. $\square$

A differential equation that is not linear is said to be **nonlinear**. Nonlinear equations are, in general, very difficult to solve, so in many cases one approximates a nonlinear equation by a linear equation, called a **linearization**, that is more readily solved.

Example 12.1.4 The ODEs

$$y' + 3t^2 y = e^t, \quad y'' + (\sin t) y' + ty = 0$$

are examples of linear differential equations. Note that coefficients of these equations may be functions of the independent variable t, but not of the dependent variable y. On the other hand, the PDE known as **Burgers' equation**,

$$u_t + u u_x = 0,$$

is a nonlinear differential equation. A linearization may be obtained by replacing the coefficient u of u_x with a constant or a function of only x or t, but not u. $\square$

For most of this chapter, we limit ourselves to numerical methods for the solution of first-order ODEs. In Section 12.6, we consider systems of first-order ODEs, which allows these numerical methods to be applied to higher-order ODEs.

12.1.2 Existence and Uniqueness of Solutions

Consider the general first-order **initial value problem**, or **IVP**, that has the form

$$y' = f(t, y), \quad t_0 < t \le T, \tag{12.4}$$

$$y(t_0) = y_0. \tag{12.5}$$

We would like to have an understanding of when this problem can be solved, and whether any solution that can be obtained is unique. The following notion of continuity, applied previously in Section 10.3 to establish convergence criteria for Fixed-Point Iteration, is helpful for this purpose.

> **Definition 12.1.5 (Lipschitz condition)** A function $f(t,y)$ satisfies a **Lipschitz condition** in y on $D \subset \mathbb{R}^2$ if
>
> $$|f(t, y_2) - f(t, y_1)| \leq L|y_2 - y_1|, \quad (t, y_1), (t, y_2) \in D, \qquad (12.6)$$
>
> for some constant $L > 0$, which is called a **Lipschitz constant** for f. We also say that f is **Lipschitz continuous** in y on D.

If $\partial f / \partial y$ exists on D, and $|\partial f / \partial y| \leq L$ on D, then by applying the Mean Value Theorem (Theorem A.5.3), we can conclude that f is Lipschitz continuous in y on D. The converse also holds.

When solving a problem numerically, it is not sufficient to know that a solution exists and is unique. As discussed in Section 2.1.4, if a small change in the problem data can cause a substantial change in the solution, then the problem is **ill-conditioned**, and a numerical solution is therefore unreliable, because it could be unduly influenced by roundoff error. The following definition characterizes problems involving differential equations for which numerical solution is feasible.

> **Definition 12.1.6 (Well-posed problem)** A differential equation of any type, in conjunction with any other information such as an initial condition, is said to describe a **well-posed problem** if it satisfies three conditions, known as **Hadamard's conditions** for well-posedness:
>
> - A solution of the problem exists.
> - A solution of the problem is unique.
> - The unique solution *depends continuously* on the problem data.
>
> If a problem is not well-posed, then it is said to be **ill-posed**.

The "problem data" in this definition may include, for example, initial values or coefficients of the differential equation. We recall that these conditions were stated in Section 2.1.1, in the context of error analysis for more general problems.

We are now ready to describe a class of initial-value problems that can be solved numerically.

> **Theorem 12.1.7 (Existence-Uniqueness, Well-Posedness)** Let $D = [t_0, T] \times \mathbb{R}$, and let $f(t, y)$ be continuous on D. If f satisfies a Lipschitz condition on D in y, then the initial value problem (12.4), (12.5) has a unique solution $y(t)$ on $[t_0, T]$. Furthermore, the problem is well-posed.

This theorem can be proved using Fixed-Point Iteration (see Section 10.3), in which the Lipschitz condition on f is used to prove that the iteration converges [Birkhoff and Rota (1989)].

Exploration 12.1.1 Consider the initial value problem

$$y' = 3y + 2t, \quad 0 < t \le 1, \quad y(0) = 1.$$

Show that this problem is well-posed.

Exploration 12.1.2 Show that there exists a $T > 0$ such that the IVP for the **doomsday equation**

$$y' = y^{1.1}, \quad 0 < t \le T, \quad y(0) = 1$$

is ill-posed.

12.1.3 Concept Check

1. For each of the following differential equations, state their order, and classify them as: ordinary or partial, and linear or nonlinear.

   (a) $u_t = u_{xx}$
   (b) $u_t = u_{xx} + u_{yy} + u - u^3$
   (c) $u_{xx} = f$
   (d) $y'' + xy' = 0$
   (e) $y'' + yy' = x$

2. What is an initial value problem?
3. What does it mean for a function to be Lipschitz continuous?
4. If a function is differentiable with respect to a particular independent variable and also Lipschitz continuous in that variable, what can we conclude about its derivative with respect to that variable?
5. What does it mean for a problem to be well-posed or ill-posed?
6. Under what conditions is an initial value problem guaranteed to have a unique solution?

12.2 One-Step Methods

Numerical methods for the initial-value problem (12.4), (12.5) can be developed using Taylor expansion (see Theorem A.6.1). We wish to approximate the solution at times t_n, $n = 1, 2, \ldots$, where

$$t_n = t_0 + nh,$$

with h being a chosen **time step**. Computing approximate solution values in this manner is called **time-stepping** or **time-marching**. Taking a Taylor expansion of the exact solution $y(t)$ at $t = t_{n+1}$ around the center $t = t_n$, we obtain

$$y(t_{n+1}) = y(t_n) + hy'(t_n) + \frac{h^2}{2}y''(\xi), \tag{12.7}$$

where $t_n < \xi < t_{n+1}$.

12.2.1 Euler's Method

Using the fact that $y' = f(t, y)$, we obtain a numerical scheme by truncating the Taylor expansion after the second term. The result is a **difference equation**

$$y_{n+1} = y_n + hf(t_n, y_n), \tag{12.8}$$

where each y_n, for $n = 1, 2, \ldots$, is an approximation of $y(t_n)$. This method is called **Euler's Method**, the simplest example of what is known as a **one-step method**.

We say that a time-stepping method **converges** if it produces a sequence of approximate solution values $y_1, y_2, \ldots$ that satisfies

$$\lim_{h \to 0} \max_{0 \leq n \leq (T-t_0)/h} |y(t_n) - y_n| = 0. \tag{12.9}$$

We now need to determine whether Euler's Method converges. To that end, we attempt to bound the error at time t_n. We begin with a comparison of the difference equation and the Taylor expansion of the exact solution,

$$y_{n+1} = y_n + hf(t_n, y_n), \tag{12.10}$$

$$y(t_{n+1}) = y(t_n) + hf(t_n, y(t_n)) + \frac{h^2}{2}y''(\xi). \tag{12.11}$$

It follows that if we define $e_n = y_n - y(t_n)$, then

$$e_{n+1} = e_n + h[f(t_n, y_n) - f(t_n, y(t_n))] - \frac{h^2}{2}y''(\xi).$$

We assume that $y \in C^2[t_0, T]$; that is, y is twice continuously differentiable on $[t_0, T]$ (see Section B.13.3). Using the assumption that f satisfies a Lipschitz condition (12.6) in y, we obtain

$$|e_{n+1}| \leq (1 + hL)|e_n| + \frac{h^2 M}{2}, \tag{12.12}$$

where

$$|y''(t)| \leq M, \quad t_0 \leq t \leq T,$$

and L is the Lipschitz constant for f in y on $[t_0, T] \times \mathbb{R}$.

Applying the relationship (12.12) repeatedly yields

$$|e_n| \leq (1 + hL)^n |e_0| + \frac{h^2 M}{2} \sum_{i=0}^{n-1} (1 + hL)^i$$

$$\leq \frac{h^2 M}{2} \frac{(1 + hL)^n - 1}{(1 + hL) - 1}$$

$$\leq \frac{h^2 M}{2} \frac{[e^{hL}]^n - 1}{hL}$$

$$\leq \frac{hM}{2L}[e^{L(t_n - t_0)} - 1].$$

Here, we have used the formula for the partial sum of a geometric series,

$$1 + r + r^2 + \cdots + r^{n-1} = \frac{r^n - 1}{r - 1}.$$

We conclude that for $t_0 \leq t_n \leq T$,

$$|y(t_n) - y_n| \leq \frac{hM}{2L}[e^{L(t_n - t_0)} - 1] \leq \frac{hM}{2L}[e^{L(T - t_0)} - 1]. \qquad (12.13)$$

That is, as $h \to 0$, the solution obtained using Euler's Method converges to the exact solution. Furthermore, in the sense of Definition 2.1.16, the rate of convergence is $O(h)$; that is, Euler's Method is **first-order accurate**.

This convergence analysis, however, assumes exact arithmetic. To properly account for roundoff error, we note that the approximate solution values $\tilde{y}_n$, $n = 0, 1, 2, \ldots$, satisfy the modified difference equation

$$\tilde{y}_{n+1} = \tilde{y}_n + hf(t_n, \tilde{y}_n) + \delta_{n+1}, \quad \tilde{y}_0 = y_0 + \delta_0, \qquad (12.14)$$

where, for $n = 0, 1, 2, \ldots$, $|\delta_n| \leq \delta$, which is $O(\mathbf{u})$, where $\mathbf{u}$ is the the machine precision (i.e., unit roundoff) introduced in Section 2.2.1. Note that even the initial value $\tilde{y}_0$ has an error term, which arises from representation of y_0 in the floating-point system.

Exploration 12.2.1 Repeat the convergence analysis for Euler's Method on (12.14) to obtain the error bound

$$|\tilde{y}_n - y(t_n)| \leq \frac{1}{L}\left(\frac{hM}{2} + \frac{\delta}{h}\right)[e^{L(t_n - t_0)} - 1] + \delta e^{L(t_n - t_0)}.$$

What happens to this error bound as $h \to 0$? What is an optimal choice of h so that the error bound is minimized?

We conclude our discussion of Euler's Method with an example of how the previous convergence analyses can be used to select a suitable time step h.

Example 12.2.1 Consider the IVP

$$y' = -y, \quad 0 < t \leq 10, \quad y(0) = 1.$$

We know that the exact solution is $y(t) = e^{-t}$. Euler's Method applied to this problem yields the difference equation

$$y_{n+1} = y_n - hy_n = (1 - h)y_n, \quad y_0 = 1.$$

We wish to select h so that the error at time $T = 10$ is less than 0.001. To that end, we use the error bound

$$|y(t_n) - y_n| \leq \frac{hM}{2L}[e^{L(t_n - t_0)} - 1],$$

with $M = 1$, since $y''(t) = e^{-t}$, which satisfies $0 < y''(t) \leq 1$ on $[0, 10]$, and $L = 1$, since $f(t, y) = -y$ satisfies $|\partial f/\partial y| = |-1| \equiv 1$. Substituting $t_n = 10$ and $t_0 = 0$ yields

$$|y(10) - y_n| \leq \frac{h}{2}[e^{10} - 1] \approx 1101.27h.$$

Ensuring that the error at this time is less than 10^{-3} requires choosing $h < 9.08 \times 10^{-8}$. However, the bound on the error at $t = 10$ is quite crude. Applying Euler's Method with this time step yields a solution whose error at $t = 10$ is approximately 2×10^{-11}. □

> **Exploration 12.2.2** As a follow-up to Example 12.2.1, include roundoff error in the error analysis and determine the optimal time step using the result of Exploration 12.2.1. Use $\delta = 2\mathbf{u}$, where $\mathbf{u} = 2^{-53}$, the unit roundoff for double-precision floating-point numbers. What is the actual error with this optimal choice of h?

12.2.2 Solving IVPs in MATLAB

MATLAB provides several ODE solvers [Shampine and Reichelt (1997)]. To solve an IVP of the form

$$y' = f(t,y), \quad t_0 < t \leq T, \quad y(t_0) = y_0,$$

one can use, for example, the command

```
>> [t,y]=ode23(f,[ t0 T ],y0);
```

where f is a function handle for $f(t,y)$. The first output t is a column vector consisting of times $t_0, t_1, \ldots, t_n = T$, where n is the number of time steps. The second output y is an $n \times m$ matrix, where m is the length of y0. The ith row of y consists of the values of $y(t_i)$, for $i = 1, 2, \ldots, n$. This is the simplest usage of one of the ODE solvers; additional interfaces are described in the documentation.

> **Exploration 12.2.3**
> (a) Write a MATLAB function
>
> $$[T,Y]=\text{eulersmethod}(f,tspan,y0,h)$$
>
> that solves a given IVP of the form (12.4), (12.5) using Euler's Method (12.8). Assume that tspan is a vector of the form $\begin{bmatrix} t_0 & T \end{bmatrix}$ that contains the initial and final times, as in the typical usage of MATLAB ODE solvers. The output T must be a column vector of time values, and the output Y must be a matrix, each row of which represents the computed solution at the corresponding time value in the same row of T.
> (b) Test your function on the IVP from Example 12.2.1 with h=0.1 and h=0.01, and compute the error at the final time t using the known exact solution. What happens to the error as h decreases? Is the behavior what you would expect based on theory?

> **Exploration 12.2.4** Explain why it is not a good idea to carry out time-stepping using a while loop that simply increments t by h as long as $t < T$, even if $(T - t_0)/h$ is an integer.

12.2.3 Runge-Kutta methods

We have seen that Euler's Method (12.8) is first-order accurate. Now, we will use Taylor expansion to design methods that have a higher order of accuracy, based on the criterion that the error incurred during a single time step, due to truncation of Taylor expansions, is higher-order in h.

A numerical method that uses higher-order Taylor expansion directly to approximate $y(t_{n+1})$ is not practical, because it requires partial derivatives of f with respect to t and y. However, such a Taylor expansion be useful in the *design* of a higher-order method, if not its implementation. Our approach will be to use evaluations of f at carefully chosen values of its arguments, t and y, in order to create an approximation that is just as accurate as a higher-order Taylor expansion of $y(t + h)$. To find the right values of t and y at which to evaluate f, we need to take a *multivariable* Taylor expansion of f evaluated at these (unknown) values, and then match the resulting numerical scheme to a Taylor expansion of $y(t + h)$ around t.

We now illustrate our proposed approach in order to obtain a method that is second-order accurate; that is, the leading neglected term in our Taylor expansion is $O(h^3)$, just as the leading neglected term in the Taylor expansion (12.7) that led to Euler's Method was $O(h^2)$. The proposed method has the form

$$y_{n+1} = y_n + ahf(t_n, y_n) + bhf(t_n + \alpha, y_n + \beta), \qquad (12.15)$$

where a, b, α, and β are to be determined. To ensure second-order accuracy, we must match the Taylor expansion of the exact solution,

$$y(t + h) = y(t) + hf(t, y(t)) + \frac{h^2}{2}\frac{d}{dt}[f(t, y(t))] + \frac{h^3}{6}\frac{d^2}{dt^2}[f(\xi, y(\xi))],$$

where $t < \xi < t + h$, to

$$y(t + h) = y(t) + ahf(t, y(t)) + bhf(t + \alpha, y(t) + \beta).$$

After simplifying by removing terms or factors that already match, we see that we only need to match

$$f(t, y(t)) + \frac{h}{2}\frac{d}{dt}[f(t, y(t))] + \frac{h^2}{6}\frac{d^{2'}}{dt^2}[f(t, y)]$$

with

$$af(t, y(t)) + bf(t + \alpha, y(t) + \beta),$$

at least up to and including terms of $O(h)$, so that overall, the expansions will match at least up to and including terms that are $O(h^2)$.

Applying the multivariable version of Taylor's Theorem to f (Theorem A.6.6), we obtain

$$af(t, y(t)) + bf(t + \alpha, y(t) + \beta) = af(t, y(t)) + bf(t, y(t)) + b\alpha\frac{\partial f}{\partial t}(t, y(t)) +$$

$$b\beta\frac{\partial f}{\partial y}(t, y(t)) + \cdots . \qquad (12.16)$$

Meanwhile, computing the full derivative with respect to t in the Taylor expansion of the solution yields

$$f(t, y(t)) + \frac{h}{2}\frac{d}{dt}[f(t, y(t))] = f(t, y(t)) + \frac{h}{2}\frac{\partial f}{\partial t}(t, y(t)) + \frac{h}{2}\frac{\partial f}{\partial y}(t, y(t))f(t, y(t)) + O(h^2).$$

$$(12.17)$$

Comparing terms of the right sides of (12.16) and (12.17) yields

$$a + b = 1, \quad b\alpha = \frac{h}{2}, \quad b\beta = \frac{h}{2} f(t, y(t)).$$

If we let $a = 0$, then we have $b = 1$, $\alpha = h/2$, and $\beta = hf(t, y(t))/2$. Substituting these values into the form (12.15) yields the numerical scheme

$$y_{n+1} = y_n + hf\left(t_n + \frac{h}{2}, y_n + \frac{h}{2} f(t_n, y_n)\right). \tag{12.18}$$

This scheme is known as the **Midpoint Method**, or the **Explicit Midpoint Method**. Note that it evaluates f at the midpoints of the intervals $[t_n, t_{n+1}]$ and $[y_n, y_{n+1}]$, where the midpoint in y is approximated using Euler's Method (12.8) with time step $h/2$. Because any terms in the Taylor expansion omitted from (12.16) are $O(h^2)$, the Midpoint Method is indeed second-order accurate.

The Midpoint Method is the simplest example of a **Runge-Kutta Method**, which is the name given to any of a class of time-stepping schemes that are derived by matching multivariable Taylor expansions of $f(t, y)$ with terms in a Taylor expansion of $y(t + h)$. Another often-used Runge-Kutta Method is the **Modified Euler Method**, which is obtained by setting $a = 1/2$ in (12.15). The resulting scheme is

$$y_{n+1} = y_n + \frac{h}{2}[f(t_n, y_n) + f(t_{n+1}, y_n + hf(t_n, y_n))], \tag{12.19}$$

also known as the **Explicit Trapezoidal Method**, as it resembles the Trapezoidal Rule (9.15) from numerical integration. This method is also second-order accurate.

Exploration 12.2.5 Use Taylor expansion to prove that the Explicit Trapezoidal Method (12.19) is second-order accurate.

However, the best-known Runge-Kutta method is the **Fourth-Order Runge-Kutta Method**, hereafter abbreviated as **RK4**, which uses *four* evaluations of f during each time step. The method proceeds as follows:

$$k_1 = hf(t_n, y_n),$$
$$k_2 = hf\left(t_n + \frac{h}{2}, y_n + \frac{1}{2}k_1\right),$$
$$k_3 = hf\left(t_n + \frac{h}{2}, y_n + \frac{1}{2}k_2\right),$$
$$k_4 = hf(t_{n+1}, y_n + k_3),$$
$$y_{n+1} = y_n + \frac{1}{6}(k_1 + 2k_2 + 2k_3 + k_4). \tag{12.20}$$

In a sense, this method is similar to Simpson's Rule (9.16) from numerical integration, which is also fourth-order accurate, as values of f at the midpoint in time, $t_n + h/2$, are given four times as much weight as values at the endpoints t_n and t_{n+1}.

The values $k_1, \ldots, k_4$ are referred to as **stages**; more precisely, a stage of a Runge-Kutta Method is an evaluation of $f(t, y)$, and the number of stages of a Runge-Kutta Method is the number of evaluations required per time step. We therefore say that (12.20) is a four-stage, fourth-order method, while the Midpoint Method (12.18) and Explicit Trapezoidal Method (12.19) are both two-stage, second-order methods. We will see in Section 12.5 that the number of stages does not always correspond to the order of accuracy.

Example 12.2.2 We compare Euler's Method with RK4 on the initial value problem

$$y' = -2ty, \quad 0 < t \le 1, \quad y(0) = 1,$$

which has the exact solution $y(t) = e^{-t^2}$. We use a time step of $h = 0.1$ for both

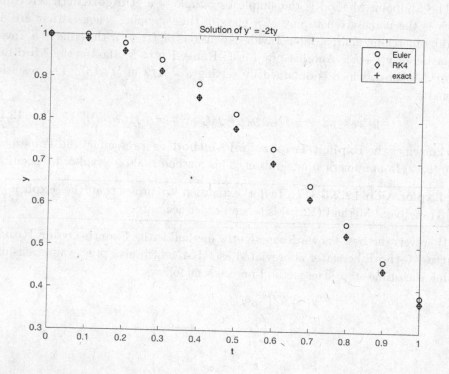

Fig. 12.1   From Example 12.2.2: Solutions of $y' = -2ty$, $y(0) = 1$ on $[0, 1]$, computed using Euler's Method (12.8) and RK4 (12.20)

methods. The computed solutions and exact solution are shown in Figure 12.1.

It can be seen that RK4 is far more accurate than Euler's Method, which is first-order accurate. In fact, the solution computed using RK4 is visually indistinguishable from the exact solution. At the final time $T = 1$, the relative error in the solution computed using Euler's Method is 0.038, while the relative error in the solution computing using RK4 is 4.4×10^{-6}.

The following table shows the absolute error, measured by $\max_{0 \le n \le 1/h} |y(t_n) - y_n|$, for both methods and two different values of h.

Method	$h = 0.1$	$h = 0.01$
Euler's Method	3.48×10^{-2}	3.28×10^{-3}
RK4	1.63×10^{-6}	1.64×10^{-10}

We see that as h decreases by a factor of 10, the error in Euler's Method also decreases by a factor of roughly 10, while the error in RK4 decreases by a factor of roughly 10^4. This is consistent with the error in Euler's Method being $O(h)$, while the error in RK4 is $O(h^4)$. $\square$

Exploration 12.2.6 Modify your function `eulersmethod` from Exploration 12.2.3 to obtain a new function `[T,Y]=rk4(f,tspan,y0,h)` that implements RK4 (12.20).

12.2.4 Implicit Methods

Suppose that we integrate the ODE $y' = f(t, y)$ from t_n to t_{n+1}, which yields

$$y(t_{n+1}) = y(t_n) + \int_{t_n}^{t_{n+1}} f(s, y(s)) \, ds. \tag{12.21}$$

If we approximate the integral using a one-node quadrature rule of the form (9.12) with node t_{n+1} and weight h, we obtain a one-step method

$$y_{n+1} = y_n + hf(t_{n+1}, y_{n+1}), \tag{12.22}$$

known as **Backward Euler's Method**, also simply known as **Backward Euler**. Like Euler's Method (12.8), Backward Euler's Method is first-order accurate. To avoid ambiguity, Euler's Method is sometimes referred to as **Forward Euler**.

Backward Euler's Method contrasts with Euler's Method because it is an **implicit** method, due to the evaluation of $f(t, y)$ at (t_{n+1}, y_{n+1}). It follows that it is generally necessary to solve a nonlinear equation to obtain y_{n+1} from y_n, using methods such as those presented in Chapter 10, or Section 11.1 for systems of ODEs. This additional computational effort is offset by the fact that implicit methods can generally be used with a larger time step h than **explicit** methods such as Euler's Method, which do not evaluate $f(t, y)$ at any point involving y_{n+1}. This will be discussed in more detail in Section 12.4.4.

If we apply the Trapezoidal Rule (9.15) to the integral in (12.21), we obtain

$$y_{n+1} = y_n + \frac{h}{2}[f(t_n, y_n) + f(t_{n+1}, y_{n+1})], \tag{12.23}$$

known as the **Trapezoidal Method**, also known as the **Implicit Trapezoidal Method** to distinguish from the Explicit Trapezoidal Method (12.19). Like the

Explicit Trapezoidal Method, the Implicit Trapezoidal Method is second-order accurate. The proof of this will be left as an exercise.

Exploration 12.2.7 Write a MATLAB function

$$[T,Y]=\texttt{backwardeuler(f,tspan,y0,h)}$$

that implements Backward Euler's Method (12.22). Use the Secant Method to solve for y_{n+1} at each time step. For initial guesses, use y_n and $y_n + hf(t_n, y_n)$, the approximation of y_{n+1} obtained using (Forward) Euler's Method (12.8).

Exploration 12.2.8 Suppose that Fixed-Point Iteration (see Section 10.3) is used to solve for y_{n+1} in Backward Euler's Method (12.22). What is the function g in the equation $y_{n+1} = g(y_{n+1})$? Assuming that g satisfies the condition for a fixed point to exist, how should h be chosen to help ensure convergence of Fixed-Point Iteration?

Exploration 12.2.9 Repeat Exploration 12.2.8 for the Implicit Trapezoidal Method (12.23).

12.2.5 Concept Check

1. What does it mean for a time-stepping method to converge?
2. What does it mean for a time-stepping method to be explicit or implicit?
3. List three MATLAB functions that can be used to solve ODEs.
4. For each of the following methods, provide the formula for the method. Also state its order of accuracy, and whether it is implicit or explicit, if this information is not provided in the name of the method.

   (a) Euler's Method
   (b) Modified Euler Method
   (c) Backward Euler's Method
   (d) Fourth-Order Runge-Kutta Method (RK4)
   (e) Implicit Trapezoidal Method

5. What is a stage of a Runge-Kutta Method?

12.3 Multistep Methods

All of the numerical methods that we have developed for solving initial value problems are classified as **one-step methods**, because they only use information about the solution at time t_n to approximate the solution at time t_{n+1}. As n increases, that means that there are additional values of the solution, at previous times, that could be helpful, but are unused. **Multistep methods** are time-stepping methods that do use this information. A general multistep method has the form

$$\sum_{i=0}^{m} \alpha_i y_{n+1-i} = h \sum_{i=0}^{m} \beta_i f(t_{n+1-i}, y_{n+1-i}),$$

where m is the number of steps in the method ($m = 1$ for a one-step method), and h is the time step size, as before.

An important distinction between multistep methods and one-step methods is that a multistep method requires only one *new* evaluation of $f(t, y)$ during each time step. By contrast, Runge-Kutta methods require multiple evaluations of $f(t, y)$ per time step, referred to as *stages*. It follows that multistep methods are substantially more efficient than multi-stage one-step methods, per time step.

By convention, $\alpha_0 = 1$, so that y_{n+1} can be conveniently expressed in terms of other y-values. If $\beta_0 = 0$, the multistep method is said to be *explicit*, because then y_{n+1} can be described using an explicit formula, whereas if $\beta_0 \neq 0$, the method is *implicit*, because then an equation, generally nonlinear, must be solved to compute y_{n+1}. For a general implicit multistep method, for which $\beta_0 \neq 0$, Newton's Method (Algorithm 10.4.1) can be applied to compute a root of the function

$$F(y) = \alpha_0 y + \sum_{i=1}^{m} \alpha_i y_{n+1-i} - h\beta_0 f(t_{n+1}, y) - h \sum_{i=1}^{s} \beta_i f_{n+1-i},$$

where, for convenience, we define $f_j = f(t_j, y_j)$. The resulting iteration is

$$
\begin{aligned}
y_{n+1}^{(k+1)} &= y_{n+1}^{(k)} - \frac{F(y_{n+1}^{(k)})}{F'(y_{n+1}^{(k)})} \\
&= y_{n+1}^{(k)} - \frac{\alpha_0 y_{n+1}^{(k)} + \sum_{i=1}^{m} \alpha_i y_{n+1-i} - h\beta_0 f(t_{n+1}, y_{n+1}^{(k)}) - h \sum_{i=1}^{m} \beta_i f_{n+1-i}}{\alpha_0 - h\beta_0 f_y(t_{n+1}, y_{n+1}^{(k)})},
\end{aligned}
$$

with $y_{n+1}^{(0)} = y_n$. If one does not wish to compute f_y, then the Secant Method (Algorithm 10.4.7) can be used instead.

12.3.1 Adams Methods

Like Backward Euler's Method (12.22) **Adams Methods** [Bashforth and Adams (1883)] are derived from the integral form of the ODE,

$$y(t_{n+1}) = y(t_n) + \int_{t_n}^{t_{n+1}} f(s, y(s)) \, ds.$$

The general idea behind Adams Methods is to approximate the above integral using polynomial interpolation of f at the points $t_{n+1-m}, t_{n+2-m}, \ldots, t_n$ if the method is explicit, and t_{n+1} as well if the method is implicit. In all Adams Methods, $\alpha_0 = 1$, $\alpha_1 = -1$, and $\alpha_i = 0$ for $i = 2, \ldots, m$.

Explicit Adams Methods are called **Adams-Bashforth Methods**. To derive an Adams-Bashforth Method, we interpolate f at the points $t_n, t_{n-1}, \ldots, t_{n-m+1}$

with a polynomial of degree $m - 1$. We then integrate this polynomial exactly. It follows that the constants β_i, $i = 1, \ldots, m$, are the integrals of the corresponding Lagrange polynomials (7.3) from t_n to t_{n+1}, divided by h. It follows from the error formula for polynomial interpolation (Theorem 7.4.1) that the error in an m-step Adams-Bashforth Method is $O(h^m)$. This will be discussed further in Section 12.4.

Example 12.3.1 We derive the three-step Adams-Bashforth Method,

$$y_{n+1} = y_n + h(\beta_1 f_n + \beta_2 f_{n-1} + \beta_3 f_{n-2}),$$

where, as before, we let $f_j = f(t_j, y_j)$. The constants β_i, $i = 1, 2, 3$, are obtained by evaluating the integral from t_n to t_{n+1} of a polynomial $p_2(t)$ that interpolates $f(t, y(t))$ at t_n, t_{n-1} and t_{n-2}. We write

$$p_2(t) = \sum_{i=0}^{2} f_{n-i} \mathcal{L}_{2,i}(t),$$

where $\mathcal{L}_{2,i}(t)$ is the ith Lagrange polynomial for the interpolation points t_n, t_{n-1} and t_{n-2}. Because our final method expresses y_{n+1} as a linear combination of y_n and values of f, it follows that the constants β_i, $i = 1, 2, 3$, are the integrals of these Lagrange polynomials from t_n to t_{n+1}, divided by h.

Using a change of variable $u = (t_{n+1} - s)/h$, we can instead interpolate at the points $u = 1, 2, 3$, thus simplifying the integration. If we define $\tilde{p}_2(u) = p_2(s) = p_2(t_{n+1} - hu)$ and $\tilde{\mathcal{L}}_{2,i}(u) = \mathcal{L}_{2,i}(t_{n+1} - hu)$, then we obtain

$$\int_{t_n}^{t_{n+1}} f(s, y(s)) \, ds = \int_{t_n}^{t_{n+1}} p_2(s) \, ds$$

$$= h \int_0^1 \tilde{p}_2(u) \, du$$

$$= h \int_0^1 f_n \tilde{\mathcal{L}}_{2,0}(u) + f_{n-1} \tilde{\mathcal{L}}_{2,1}(u) + f_{n-2} \tilde{\mathcal{L}}_{2,2}(u) \, du$$

$$= h \left[f_n \int_0^1 \frac{(u-2)(u-3)}{(1-2)(1-3)} \, du + f_{n-1} \int_0^1 \frac{(u-1)(u-3)}{(2-1)(2-3)} \, du + \right.$$
$$\left. f_{n-2} \int_0^1 \frac{(u-1)(u-2)}{(3-1)(3-2)} \, du \right]$$

$$= h \left(\frac{23}{12} f_n - \frac{4}{3} f_{n-1} + \frac{5}{12} f_{n-2} \right).$$

We conclude that the three-step Adams-Bashforth Method is

$$y_{n+1} = y_n + \frac{h}{12} (23 f_n - 16 f_{n-1} + 5 f_{n-2}). \tag{12.24}$$

It can be shown that this method is third-order accurate. $\square$

Exploration 12.3.1 Show that for an m-step Adams-Bashforth Method, the coefficients $\beta_1, \beta_2, \ldots, \beta_m$ are given by

$$\beta_i = \int_0^1 \prod_{j=1, j \neq i}^{m} \frac{u - j}{i - j} \, du.$$

The same approach can be used to derive an implicit Adams Method, which is known as an **Adams-Moulton Method**. The only difference is that because t_{n+1} is an interpolation point, after the change of variable to u, the interpolation points $0, 1, 2, \ldots, m$ are used. Because the resulting interpolating polynomial is of degree one greater than in the explicit case, the error in an m-step Adams-Moulton Method is $O(h^{m+1})$, as opposed to $O(h^m)$ for an m-step Adams-Bashforth Method.

Exploration 12.3.2 Derive the four-step Adams-Bashforth Method

$$y_{n+1} = y_n + \frac{h}{24}(55f_n - 59f_{n-1} + 37f_{n-2} - 9f_{n-3}) \qquad (12.25)$$

and the three-step Adams-Moulton Method

$$y_{n+1} = y_n + \frac{h}{24}(9f_{n+1} + 19f_n - 5f_{n-1} + f_{n-2}). \qquad (12.26)$$

What is the order of accuracy of each of these methods?

12.3.2 Predictor-Corrector Methods

An Adams-Moulton Method can be impractical because, being implicit, it requires an iterative method for solving nonlinear equations, such as Fixed-Point Iteration (see Section 10.3), and this method must be applied during every time step. An alternative is to pair an Adams-Bashforth Method with an Adams-Moulton Method to obtain an **Adams-Moulton Predictor-Corrector Method** [Moulton (1926)]. Such a method proceeds as follows:

- *Predict:* Use the Adams-Bashforth Method to compute a first approximation to y_{n+1}, which we denote by $\tilde{y}_{n+1}$.
- *Evaluate:* Evaluate f at this value, computing $f(t_{n+1}, \tilde{y}_{n+1})$.
- *Correct:* Use the Adams-Moulton Method to compute y_{n+1}, but instead of solving an equation, use $f(t_{n+1}, \tilde{y}_{n+1})$ in place of $f(t_{n+1}, y_{n+1})$ so that the Adams-Moulton Method can be used as if it was an *explicit* method.
- *Evaluate:* Evaluate f at the newly computed value of y_{n+1}, computing $f(t_{n+1}, y_{n+1})$, to use during the next time step.

Example 12.3.2 We illustrate the predictor-corrector approach with the two-step Adams-Bashforth Method

$$y_{n+1} = y_n + \frac{h}{2}(3f_n - f_{n-1})$$

and the two-step Adams-Moulton Method

$$y_{n+1} = y_n + \frac{h}{12}(5f_{n+1} + 8f_n - f_{n-1}).$$ (12.27)

First, we apply the Adams-Bashforth Method, and compute

$$\tilde{y}_{n+1} = y_n + \frac{h}{2}(3f_n - f_{n-1}).$$

Then, we compute $f(t_{n+1}, \tilde{y}_{n+1})$ and apply the Adams-Moulton Method, to compute

$$y_{n+1} = y_n + \frac{h}{12}[5f(t_{n+1}, \tilde{y}_{n+1}) + 8f_n - f_{n-1}].$$

This new value of y_{n+1} is used when evaluating $f_{n+1} = f(t_{n+1}, y_{n+1})$ during the next time step. $\square$

One drawback of multistep methods is that because they rely on values of the solution from previous time steps, they cannot be used during the first time steps, because not enough values are available. Therefore, it is necessary to use a one-step method, with at least the same order of accuracy, to compute enough **starting values** of the solution to be able to use the multistep method. For example, to use the three-step Adams-Bashforth Method, it is necessary to first use a one-step method such as RK4 (12.20) to compute y_1 and y_2, and then the Adams-Bashforth Method can be used to compute y_3 using y_2, y_1 and y_0.

Exploration 12.3.3 How many starting values are needed to use an m-step multistep method?

Exploration 12.3.4 Write a MATLAB function

      `[T,Y]=adamsbashforth4(f,tspan,y0,h)`

that implements the 4-step Adams-Bashforth Method (12.25) to solve the given IVP (12.4), (12.5). Use RK4 to generate the necessary starting values. Use different values of h on the IVP from Example 12.2.2 to confirm that your method is fourth-order accurate. Compare the performance of your method to that of RK4, for the same value of h, in terms of both accuracy and efficiency.

Exploration 12.3.5 Write a MATLAB function

      `[T,Y]=predictcorrect(f,tspan,y0,h)`

that implements an Adams-Moulton Predictor-Corrector Method using a four-step predictor (12.25) and three-step corrector (12.26). Use RK4 to generate the necessary starting values. Use different values of h on a sample IVP to confirm that your method is fourth-order accurate.

12.3.3 Backward Differentiation Formulas

Another class of multistep methods, known as **Backward Differentiation Formulas (BDF)** [Süli and Mayers (2003)], can be derived using polynomial interpolation as in Adams Methods, but for a different purpose–to approximate the *derivative* of y at t_{n+1}, rather than its integral from t_n to t_{n+1}. This approximation is then equated to $f(t_{n+1}, y_{n+1})$. It follows that all methods based on BDFs are implicit, and they all satisfy $\beta_0 = 1$, with $\beta_i = 0$ for $i = 1, 2, \ldots, m$.

More precisely, a BDF has the form

$$\sum_{i=0}^{m} \alpha_i y_{n+1-i} = h f_{n+1},$$

where

$$\alpha_i = \mathcal{L}'_{m,i}(t_{n+1}),$$

and $\mathcal{L}_{m,i}(t)$, for $i = 0, 1, \ldots, m$, are the Lagrange polynomials for the interpolation points $t_{n+1-m}, \ldots, t_n, t_{n+1}$. In view of the result of Exploration 9.1.4, the error in an m-step BDF is $O(h^m)$.

Exploration 12.3.6 Show that a 1-step BDF is simply Backward Euler's Method (12.22).

Exploration 12.3.7 Derive a 2-step BDF. How do the coefficients α_i, $i = 0, 1, 2$, relate to those of the 3-point second-order numerical differentiation formula (9.7)?

12.3.4 Concept Check

1. What is a multistep method, as opposed to a one-step method?
2. Give one advantage and one disadvantage of using a multistep method to solve an IVP instead of a one-step method.
3. What characterizes the coefficients of an Adams Method? A BDF? What is the order of accuracy of each type of method?
4. How are the coefficients of an Adams Method obtained?
5. What is a predictor-corrector method? Give an advantage of using a predictor-corrector method rather than an Adams-Moulton Method.
6. What is the basic idea behind a Backward Differentiation Formula?

12.4 Convergence Analysis

We have previously determined that when applying Euler's Method (12.8) to the initial value problem (12.4), (12.5), the error in the computed solution satisfies the error bound

$$|y_n - y(t_n)| \leq \frac{Mh}{2L}[e^{L(t_n - t_0)} - 1],$$

where L is the Lipschitz constant for f and M is an upper bound on $|y''(t)|$. This error bound indicates that the numerical solution **converges** to the exact solution at $h \to 0$; that is,

$$\lim_{h \to 0} \max_{0 \le n \le (T-t_0)/h} |y(t_n) - y_n| = 0.$$

It would be desirable to be able to prove that a numerical method converges without having to proceed through the same cumbersome error analysis that was carried out with Euler's Method, since other methods are more sophisticated.

To that end, we define two properties that a numerical method must have in order to be convergent.

Definition 12.4.1

- **Local truncation error:** a numerical method for the initial-value problem (12.4), (12.5) that has the form

$$y_{n+1} = G(t_n, y_{n+1-m}, \ldots, y_{n+1}, h)$$

has **local truncation error** $\tau_n(h)$ defined by

$$\tau_n(h) = \frac{1}{h}[y(t_{n+1}) - G(t_n, y(t_{n+1-m}), \ldots, y(t_{n+1}), h)],$$

where $y(t)$ is the exact solution of (12.4).

- **Consistency:** a numerical method is said to be **consistent** if

$$\lim_{h \to 0} \max_{0 \le n \le (T-t_0)/h} |\tau_n(h)| = 0,$$

where $\tau_n(h)$ is the local truncation error at time t_n.

- **Stability:** a numerical method is said to be **stable** if, for h sufficiently small, there exists a constant K independent of h such that for any two numerical solutions y_n and $\tilde{y}_n$,

$$|y_n - \tilde{y}_n| \le K|y_0 - \tilde{y}_0|, \quad 0 \le n \le (T - t_0)/h.$$

Informally, a consistent method converges to the differential equation as $h \to 0$, and the solution computed using a stable method is not overly sensitive to perturbations in the initial data. While the difference in solutions is allowed to grow over time, it is "controlled" growth, meaning that the rate of growth is independent of the step size h.

Recall from Section 12.1 that the exact solution of a well-posed IVP depends continuously on the problem data. The above definition of stability imposes a requirement that as $h \to 0$, the computed solution also behaves continuously on the data. That is, the solution produced by a stable numerical method is not unduly affected by data error or roundoff error.

12.4.1 Consistency

The definition of consistency in Definition 12.4.1 can be cumbersome to apply directly to a given method. Therefore, we consider one-step and multistep methods separately to obtain simple approaches for determining whether a given method is consistent, and if so, its order of accuracy.

12.4.1.1 One-Step Methods

We adapt our previous convergence analysis of Euler's Method from Section 12.2 to analyze the convergence of a general one-step method of the form

$$y_{n+1} = y_n + h\Phi(t_n, y_n, y_{n+1}, h), \tag{12.28}$$

for some continuous function $\Phi(t, y, z, h)$. We define the **local truncation error** of this one-step method by

$$\tau_n(h) = \frac{y(t_{n+1}) - y(t_n)}{h} - \Phi(t_n, y(t_n), y(t_n + h), h).$$

That is, the local truncation error is the result of substituting the exact solution into the approximation of the ODE by the numerical method.

> **Exploration 12.4.1** Find the local truncation error of the Modified Euler Method (12.19).

As $h \to 0$ and $n \to \infty$, in such a way that $t_0 + nh = t \in [t_0, T]$, we obtain

$$\tau_n(h) \to y'(t) - \Phi(t, y(t), y(t), 0).$$

It follows from Definition 12.4.1 that the one-step method is **consistent** if

$$\Phi(t, y, y, 0) = f(t, y).$$

That is, a consistent one-step method, rearranged algebraically so that $\Phi(t, y, z, h)$ is isolated on one side, converges to the ODE as $h \to 0$.

Example 12.4.2 Consider the Midpoint Method (12.18), for which

$$\Phi(t, y, z, h) = f\left(t + \frac{h}{2}, y + \frac{h}{2}f(t, y)\right).$$

We note that $\Phi(t, y, z, h)$ does not depend on z, as the Midpoint Method is explicit. To check consistency, we compute

$$\Phi(t, y, y, 0) = f\left(t + \frac{0}{2}, y + \frac{0}{2}f(t, y)\right) = f(t, y),$$

and conclude that it is consistent. To obtain the order of accuracy, we compute the local truncation error by performing a Taylor expansion of $y(t_{n+1})$ around t_n, and a multivariable Taylor expansion of $\Phi(t_n, y(t_n), y(t_n + h), h)$ around $(t_n, y(t_n))$. This yields

$$
\begin{aligned}
\tau_n(h) &= \frac{y(t_{n+1}) - y(t_n)}{h} - f\left(t_n + \frac{h}{2}, y(t_n) + \frac{h}{2}f(t_n, y(t_n))\right) \\
&= \frac{hf(t_n, y(t_n)) + \frac{1}{2}h^2 y''(t_n) + O(h^3)}{h} - \left[f(t_n, y(t_n)) + \frac{h}{2}\frac{\partial f}{\partial t}(t_n, y(t_n)) + \right.\\
&\quad \left. \frac{h}{2}\frac{\partial f}{\partial y}(t_n, y(t_n))f(t_n, y(t_n))\right] + O(h^2) \\
&= \frac{hf(t_n, y(t_n)) + \frac{1}{2}h^2\left[\frac{\partial f}{\partial t}(t_n, y(t_n)) + \frac{\partial f}{\partial y}(t_n, y(t_n))f(t_n, y(t_n))\right] + O(h^3)}{h} - \\
&\quad \left[f(t_n, y(t_n)) + \frac{h}{2}\frac{\partial f}{\partial t}(t_n, y(t_n)) + \frac{h}{2}\frac{\partial f}{\partial y}(t_n, y(t_n))f(t_n, y(t_n))\right] + O(h^2) \\
&= O(h^2).
\end{aligned}
$$

We conclude that the local truncation error in the Midpoint Method is $O(h^2)$. $\square$

Exploration 12.4.2 Verify that RK4 (12.20) is consistent.

12.4.1.2   *Multistep Methods*

For multistep methods, we must define consistency slightly differently, because we must account for the fact that a multistep method requires starting values that are computed using another method. Therefore, we say that a multistep method is consistent if its own local truncation error $\tau_n(h)$ approaches zero as $h \to 0$, *and* if the one-step method used to compute its starting values is also consistent.

To compute the local truncation error of Adams Methods, we integrate the error in the polynomial interpolation used to derive the method, from t_n to t_{n+1}. For the explicit m-step method, this yields

$$
\tau_n(h) = \frac{1}{h}\int_{t_n}^{t_{n+1}} \frac{f^{(m)}(\xi, y(\xi(t)))}{m!}(t - t_n)(t - t_{n-1})\cdots(t - t_{n-m+1})\,dt.
$$

Using the substitution $u = (t_{n+1} - t)/h$, and the Weighted Mean Value Theorem for Integrals (Theorem A.5.5), yields

$$
\tau_n(h) = \frac{f^{(m)}(\xi, y(\xi))}{m!}h^m(-1)^m \int_0^1 (u - 1)(u - 2)\cdots(u - m)\,du.
$$

Evaluating the integral yields the constant in the error term. We also use the fact that $y' = f(t, y)$ to replace $f^{(m)}(\xi, y(\xi))$ with $y^{(m+1)}(\xi)$. Obtaining the local truncation error for an implicit, Adams-Moulton Method can be accomplished in the same way, except that t_{n+1} is also used as an interpolation point.

For a general multistep method, we substitute the exact solution into the method, as in one-step methods, and obtain

$$
\tau_n(h) = \frac{\sum_{j=0}^{m} \alpha_j y(t_{n+1-j}) - h\sum_{j=0}^{m} \beta_j f(t_{n+1-j}, y(t_{n+1-j}))}{h\sum_{j=0}^{m} \beta_j},
$$

where the scaling by $h \sum_{j=0}^{m} \beta_j$ is designed to make this definition of local truncation error consistent with that of one-step methods.

By replacing each evaluation of $y(t)$ by a Taylor expansion around t_{n+1}, we obtain

$$\tau_n(h) = \frac{1}{h \sum_{j=0}^{m} \beta_j} \sum_{j=0}^{m} \left[\alpha_j \sum_{k=0}^{\infty} \frac{1}{k!} y^{(k)}(t_{n+1})(-jh)^k - \right.$$

$$\left. h\beta_j \sum_{k=0}^{\infty} \frac{1}{k!} \frac{d^k}{dt^k} [f(t_{n+1}, y(t_{n+1}))](-jh)^k \right]$$

$$= \frac{1}{h \sum_{j=0}^{m} \beta_j} \sum_{j=0}^{m} \left[\sum_{k=0}^{\infty} (-1)^k \frac{h^k}{k!} \alpha_j y^{(k)}(t_{n+1}) j^k + \right.$$

$$\left. \sum_{k=1}^{\infty} (-1)^k \frac{h^k}{(k-1)!} \beta_j y^{(k)}(t_{n+1}) j^{k-1} \right]$$

$$= \frac{1}{h \sum_{j=0}^{m} \beta_j} \left\{ y(t_{n+1}) \sum_{j=0}^{m} \alpha_j + \right.$$

$$\left. \sum_{k=1}^{\infty} (-h)^k y^{(k)}(t_{n+1}) \left[\frac{1}{k!} \sum_{j=1}^{m} j^k \alpha_j + \frac{1}{(k-1)!} \sum_{j=0}^{m} j^{k-1} \beta_j \right] \right\}$$

$$= \frac{1}{h \sum_{j=0}^{m} \beta_j} \left[y(t_{n+1}) C_0 + \sum_{k=1}^{\infty} (-h)^k y^{(k)}(t_{n+1}) C_k \right]$$

where

$$C_0 = \sum_{j=0}^{m} \alpha_j, \quad C_k = \frac{1}{k!} \sum_{j=1}^{m} j^k \alpha_j + \frac{1}{(k-1)!} \sum_{j=0}^{m} j^{k-1} \beta_j, \quad k = 1, 2, \ldots.$$

We find that $\tau_n(h) \to 0$ as $h \to 0$ if and only if $C_0 = C_1 = 0$. Furthermore, the method is of order p if and only if

$$C_0 = C_1 = C_2 = \cdots = C_p = 0, \quad C_{p+1} \neq 0. \tag{12.29}$$

Finally, we can conclude that the local truncation error for a method of order p is

$$\tau_n(h) = \frac{1}{\sum_{j=0}^{m} \beta_j} (-h)^p y^{(p+1)}(t_{n+1}) C_{p+1} + O(h^{p+1}).$$

Exploration 12.4.3 Use the conditions (12.29) to verify the order of accuracy of the four-step Adams-Bashforth Method (12.25). What is the local truncation error?

Further analysis is required to obtain the local truncation error of a predictor-corrector method that is obtained by combining two Adams Methods. The result of this analysis is the following theorem, which is proved in [Isaacson and Keller (1966)].

Theorem 12.4.3 Let the solution of the initial value problem

$$y' = f(t, y), \quad t_0 < t \leq T, \quad y(t_0) = y_0$$

be approximated by the m-step Adams-Moulton Predictor-Corrector Method with predictor

$$\tilde{y}_{n+1} = y_n + h \sum_{i=1}^{m} \tilde{\beta}_i f_{n+1-i}$$

and corrector

$$y_{n+1} = y_n + h \left[\beta_0 f(t_{n+1}, \tilde{y}_{n+1}) + \sum_{i=1}^{m} \beta_i f_{n+1-i} \right].$$

Then the local truncation error of the predictor-corrector method is

$$S_n(h) = \tilde{T}_n(h) + T_n(h) \beta_0 \frac{\partial f}{\partial y}(t_{n+1}, y(t_{n+1}) + \xi_{n+1})$$

where $T_n(h)$ and $\tilde{T}_n(h)$ are the local truncation errors of the predictor and corrector, respectively, and ξ_{n+1} is between 0 and $h T_n(h)$. Furthermore, there exist constants α and β such that

$$|y_n - y(t_n)| \leq \left[\max_{0 \leq i \leq m-1} |y_i - y(t_i)| + \beta S(h) \right] e^{\alpha(t_n - t_0)},$$

where $S(h) = \max_{m \leq n \leq (T-t_0)/h} |S_n(h)|$.

A single time step of a predictor-corrector method, as we have described it, can be viewed as an instance of Fixed-Point Iteration in which only one iteration is performed, with the initial guess being the prediction $\tilde{y}_{n+1}$ and the function $g(y)$ being the corrector. If desired, the iteration can be continued until convergence is achieved.

Exploration 12.4.4 Show that an m-step predictor-corrector method, in which the corrector is repeatedly applied until y_{n+1} converges, has local truncation error $O(h^{m+1})$.

12.4.2   Stability

We now specialize the definition of stability from Definition 12.4.1 to one-step and multistep methods, so that their stability (or lack thereof) can readily be determined.

12.4.2.1   *One-Step Methods*

We now show that a one-step method of the form (12.28) is **stable**, in the sense of Definition 12.4.1, if $\Phi(t, y, z, h)$ is Lipschitz continuous in y and z. That is, we require that $\Phi(t, y, z, h)$ satisfies, for $u, v, z, w \in \mathbb{R}$,

$$|\Phi(t, u, v, h) - \Phi(t, z, w, h)| \le L_\Phi[|u-z|+|v-w|], \quad t \in [t_0, T], \quad h \in [0, h_0], \quad (12.30)$$

for some constant L_Φ that is independent of h. Let $N = (T-t_0)/h$ and let $\{y_n\}_{n=0}^N$, $\{\tilde{y}_n\}_{n=0}^N$ be two computed solutions. We assume $h_0 = (1-\delta)/L_\Phi$ for some $\delta \in (0,1)$, so that $1 - hL_\Phi \ge \delta$. Then, from (12.30), we have

$$|y_{n+1} - \tilde{y}_n| \le \frac{1 + hL_\Phi}{1 - hL_\Phi} |y_n - \tilde{y}_n| \le \left(1 + \frac{2hL_\Phi}{\delta}\right) |y_n - \tilde{y}_n|,$$

which then yields

$$|y_n - \tilde{y}_n| \le e^{2hL_\Phi n/\delta} |y_0 - \tilde{y}_0| \le e^{2L_\Phi(T-t_0)/\delta} |y_0 - \tilde{y}_0|.$$

Therefore, the one-step method is stable in the sense of Definition 12.4.1, with the constant $K = e^{2L_\Phi(T-t_0)/\delta}$ that is independent of h.

Example 12.4.4 Consider the Midpoint Method (12.18). To investigate its stability, we check whether

$$\Phi(t, y, z, h) = f\left(t + \frac{h}{2}, y + \frac{h}{2} f(t, y)\right)$$

satisfies a Lipschitz condition in y. We assume that $f(t, y)$ satisfies a Lipschitz condition in y on $[t_0, T] \times (-\infty, \infty)$ with Lipschitz constant L. Then we have

$$|\Phi(t, y, z, h) - \Phi(t, \tilde{y}, \tilde{z}, h)| = \left| f\left(t + \frac{h}{2}, y + \frac{h}{2} f(t, y)\right) - f\left(t + \frac{h}{2}, \tilde{y} + \frac{h}{2} f(t, \tilde{y})\right) \right|$$

$$\le L \left| \left(y + \frac{h}{2} f(t, y)\right) - \left(\tilde{y} + \frac{h}{2} f(t, \tilde{y})\right) \right|$$

$$\le \left(L + \frac{1}{2} hL^2\right) |y - \tilde{y}|$$

$$\le \left(L + \frac{1}{2} hL^2\right) [|y - \tilde{y}| + |z - \tilde{z}|].$$

It follows that $\Phi(t, y, z, h)$ satisfies a Lipschitz condition on the domain $[t_0, T] \times (-\infty, \infty) \times [0, h_0]$ with Lipschitz constant $\tilde{L} = L + \frac{1}{2} h_0 L^2$. We conclude that it is stable. $\square$

Exploration 12.4.5 Prove that the Modified Euler Method (12.19) is stable.

12.4.2.2 *Multistep Methods*

We now examine the stability of a general m-step multistep method of the form

$$\sum_{i=0}^m \alpha_i y_{n+1-i} = h \sum_{i=0}^m \beta_i f(t_{n+1-i}, y_{n+1-i}).$$

If this method is applied to the initial value problem

$$y' = 0, \quad y(t_0) = y_0, \quad y_0 \neq 0, \tag{12.31}$$

for which the exact solution is $y(t) = y_0$, then for the method to be stable, the computed solution must remain bounded.

With this requirement in mind, we say that an m-step multistep method is stable, or **zero-stable**, if for h sufficiently small, there exists a constant K, independent of h, such that for any two sequences of values $\{y_n\}$ and $\{\tilde{y}_n\}$ produced by the method with step size h from different sets of starting values $\{y_0, y_1, \ldots, y_{m-1}\}$ and $\{\tilde{y}_0, \tilde{y}_1, \ldots, \tilde{y}_{m-1}\}$,

$$|y_n - \tilde{y}_n| \leq K \max_{0 \leq j \leq m-1} |y_j - \tilde{y}_j|, \quad m \leq n \leq (T - t_0)/h.$$

This is consistent with the definition of stability for a general time-stepping method given in Definition 12.4.1, provided that the starting values themselves are produced by a stable method.

The computed solution of (12.31) satisfies the *m-term recurrence relation*

$$\sum_{i=0}^{m} \alpha_i y_{n+1-i} = 0,$$

which has a solution of the form

$$y_n = \sum_{i=0}^{m} c_i n^{p_i} \lambda_i^n, \tag{12.32}$$

where the c_i are constants, and the λ_i are the roots of the **characteristic equation**

$$\alpha_0 \lambda^m + \alpha_1 \lambda^{m-1} + \cdots + \alpha_{m-1} \lambda + \alpha_m = 0. \tag{12.33}$$

For each root λ_i, the exponents p_i assume the values $0, 1, \ldots, m_i - 1$, where m_i is the multiplicity of λ_i. It follows that if λ_i is distinct from all other roots, then $p_i = 0$. Therefore, to ensure that the solution does not grow in magnitude as n increases, the multistep method must satisfy the **root condition:**

- All roots of (12.33) must satisfy $|\lambda_i| \leq 1$.
- If $|\lambda_i| = 1$ for any i, then it must be a **simple root**, meaning that its multiplicity is one.

From the form of y_n in (12.32), a multistep method is zero-stable if and only if it satisfies the root condition. Furthermore, $\lambda = 1$ is always a root, because in order to be consistent, a multistep method must have the property that $\sum_{i=0}^{m} \alpha_i = 0$. If this is the only root that has absolute value 1, then we say that the method is **strongly stable**, whereas if there are multiple roots that are distinct from one another, but have absolute value 1, then the method is said to be **weakly stable**. A multistep method is stable in the sense of Definition 12.4.1 if and only if it is zero-stable, whether weakly or strongly. This result is proved in [Isaacson and Keller (1966)], using the theory of inhomogeneous recurrence relations.

Because all Adams Methods have the property that $\alpha_0 = 1$, $\alpha_1 = -1$, and $\alpha_i = 0$ for $i = 2, 3, \ldots, m$, it follows that the roots of the characteristic equation are all zero, except for one root that is equal to 1. Therefore, all Adams Methods are strongly stable. The same is not true for BDFs; they are *zero-unstable* for $m > 6$ [Süli and Mayers (2003)].

Example 12.4.5 A multistep method that is neither an Adams Method, nor a backward differentiation formula, is an implicit 2-step method known as **Simpson's Method**:

$$y_{n+1} = y_{n-1} + \frac{h}{3}(f_{n+1} + 4f_n + f_{n-1}).$$

Although it is only a 2-step method, it is fourth-order accurate, due to the high degree of accuracy of Simpson's Rule.

This method is obtained from the relation satisfied by the exact solution,

$$y(t_{n+1}) = y(t_{n-1}) + \int_{t_{n-1}}^{t_{n+1}} f(t, y(t)) \, dt,$$

which is analogous to (12.21). Since the integral is over an interval of width $2h$, it follows that the coefficients β_i obtained by polynomial interpolation of f must satisfy the condition

$$\sum_{i=0}^{m} \beta_i = 2,$$

as opposed to summing to 1 for Adams Methods. For this method, we have $m = 2$, $\alpha_0 = 1$, $\alpha_1 = 0$ and $\alpha_2 = -1$, which yields the characteristic polynomial $\lambda^2 - 1$. This polynomial has two distinct roots, 1 and -1, that both have absolute value 1. It follows that Simpson's Method is only weakly stable. □

Exploration 12.4.6 Determine whether the 2-step BDF from Exploration 12.3.7 is strongly stable, weakly stable, or unstable.

12.4.3 Convergence

It can be shown that a consistent and stable one-step method of the form (12.28) is convergent. We assume $0 < h \leq h_0 = (1 - \delta)/L_\Phi$, where $0 < \delta < 1$ and L_Φ is the Lipschitz constant for Φ, as in (12.30). Using the same approach and notation as in the convergence proof of Euler's Method, and the fact that the method is stable, we obtain the following bound for the **global error** $e_n = y_n - y(t_n)$:

$$|e_n| \leq \left(\frac{e^{2L_\Phi(T-t_0)/\delta} - 1}{2L_\Phi} \right) \max_{0 \leq m \leq n-1} |\tau_m(h)|.$$

Because the method is consistent, we have

$$\lim_{h \to 0} \max_{0 \leq n \leq T/h} |\tau_n(h)| = 0.$$

It follows that as $h \to 0$ and $n \to \infty$ in such a way that $t_0 + nh = t$, we have

$$\lim_{n \to \infty} |e_n| = 0,$$

and therefore the method is convergent.

In the case of Euler's Method, we have

$$\Phi(t, y, z, h) = f(t, y), \quad \tau_n(h) = \frac{h}{2} y''(\tau), \quad \tau \in (t_0, T).$$

Therefore, there exists a constant K such that

$$|\tau_n(h)| \leq Kh, \quad 0 < h \leq h_0,$$

for some sufficiently small h_0. We say that Euler's Method is **first-order accurate**. More generally, we say that a one-step method has **order of accuracy** p if, for any sufficiently smooth solution $y(t)$, there exists constants K and h_0 such that

$$|\tau_n(h)| \leq Kh^p, \quad 0 < h \leq h_0.$$

Exploration 12.4.7 Prove that the Modified Euler Method (12.19) is convergent and second-order accurate.

As for multistep methods, we have the following theorem.

Theorem 12.4.6 (Dahlquist's Equivalence Theorem) A consistent multistep method with local truncation error $O(h^p)$ is convergent with global error $O(h^p)$ if and only if it is zero-stable.

This theorem shows that local error provides an indication of global error only for zero-stable methods. A proof can be found in [Gautschi (1997)]. Because Adams Methods are always strongly stable, it follows that all Adams Methods, as well as Adams-Moulton Predictor-Corrector Methods, are convergent.

12.4.4 Stiff Differential Equations

To this point, we have evaluated the accuracy of numerical methods for initial-value problems in terms of the rate at which the error approaches zero, when the step size h approaches zero. However, this characterization of accuracy is not always informative, because it neglects the fact that the local truncation error of any one-step or multistep method also depends on higher-order derivatives of the solution. In some cases, these derivatives can be quite large in magnitude, even when the solution itself is relatively small, which requires that h be chosen particularly small in order to achieve even reasonable accuracy.

This leads to the concept of a **stiff differential equation**. A differential equation of the form $y' = f(t, y)$ is said to be **stiff** if its exact solution $y(t)$ includes a term that decays to zero as t increases, but whose derivatives are much greater in magnitude than the term itself. An example of such a term is e^{-ct}, where c is a large, positive constant, because its kth derivative is $c^k e^{-ct}$. Because of the factor of c^k, this derivative decays to zero much more slowly than e^{-ct} as t increases. Because the error includes a term of this form, evaluated at a time less than t, the error can be quite large if h is not chosen sufficiently small to offset this large derivative. Furthermore, the larger c is, the smaller h must be to maintain accuracy.

Example 12.4.7 Consider the initial value problem

$$y' = -100y, \quad t > 0, \quad y(0) = 1.$$

The exact solution is $y(t) = e^{-100t}$, which rapidly decays to zero as t increases. If we solve this problem using Euler's Method, with step size $h = 0.1$, then we have

$$y_{n+1} = y_n - 100hy_n = -9y_n,$$

which yields the exponentially *growing* solution $y_n = (-9)^n$. On the other hand, if we choose $h = 10^{-3}$, we obtain the computed solution $y_n = (0.9)^n$, which is much more accurate, and correctly captures the qualitative behavior of the exact solution, in that it rapidly decays to zero. $\square$

The ODE in the preceding example is a special case of the **test equation**

$$y' = \lambda y, \quad y(0) = 1, \quad \text{Re}\,\lambda < 0.$$

The exact solution to this problem is $y(t) = e^{\lambda t}$. However, as λ increases in magnitude, the problem becomes increasingly stiff. By applying a numerical method to this problem, we can determine how small h must be, for a given value of λ, in order to obtain a qualitatively accurate solution.

When applying a one-step method to the test equation, the computed solution has the form

$$y_{n+1} = Q(h\lambda)y_n,$$

where $Q(h\lambda)$ is a polynomial in $h\lambda$ if the method is explicit, and a rational function if it is implicit. This function is meant to approximate $e^{h\lambda}$, since the exact solution satisfies $y(t_{n+1}) = e^{h\lambda}y(t_n)$. However, to obtain a qualitatively correct solution, that decays to zero as t increases, we must choose h so that $|Q(h\lambda)| < 1$.

Example 12.4.8 Consider the Modified Euler Method (12.19). Setting $f(t, y) = \lambda y$ yields the computed solution

$$y_{n+1} = y_n + \frac{h}{2}[\lambda y_n + \lambda(y_n + h\lambda y_n)] = \left(1 + h\lambda + \frac{1}{2}h^2\lambda^2\right)y_n,$$

so $Q(h\lambda) = 1 + h\lambda + \frac{1}{2}(h\lambda)^2$. If we assume λ is real, then in order to satisfy $|Q(h\lambda)| < 1$, we must have $-2 < h\lambda < 0$. It follows that the larger $|\lambda|$ is, the smaller h must be. $\square$

The test equation can also be used to determine how to choose h for a multistep method. The process is similar to the one used to determine whether a multistep method is zero-stable, except that we use $f(t, y) = \lambda y$, rather than $f(t, y) \equiv 0$. Given a general multistep method of the form

$$\sum_{i=0}^{m} \alpha_i y_{n+1-i} = h \sum_{i=0}^{m} \beta_i f_{n+1-i},$$

we substitute $f_n = \lambda y_n$ and obtain the recurrence relation

$$\sum_{i=0}^{m} (\alpha_i - h\lambda\beta_i) y_{n+1-i} = 0.$$

It follows that the computed solution has the form

$$y_n = \sum_{i=1}^{m} c_i n^{p_i} \mu_i^n,$$

where each μ_i is a root of the **stability polynomial**

$$Q(\mu, h\lambda) = (\alpha_0 - h\lambda\beta_0)\mu^m + (\alpha_1 - h\lambda\beta_1)\mu^{m-1} + \cdots + (\alpha_m - h\lambda\beta_m).$$

The exponents p_i range from 0 to the multiplicity of μ_i minus one, so if the roots are all distinct, all p_i are equal to zero. In order to ensure that the numerical solution y_n decays to zero as n increases, we must have $|\mu_i| < 1$ for $i = 1, 2, \ldots, m$. Otherwise, the solution will either converge to a nonzero value, or grow in magnitude.

Example 12.4.9 Consider the 3-step Adams-Bashforth Method (12.24). Applying this method to the test equation yields the stability polynomial

$$Q(\mu, h\lambda) = \mu^3 + \left(-1 - \frac{23}{12}h\lambda\right)\mu^2 + \frac{4}{3}h\lambda\mu - \frac{5}{12}h\lambda.$$

Let $\lambda = -100$. If we choose $h = 0.1$, so that $\lambda h = -10$, then $Q(\mu, h\lambda)$ has a root approximately equal to -18.884, so h is too large for this method. On the other hand, if we choose $h = 0.005$, so that $h\lambda = -1/2$, then the largest root of $Q(\mu, h\lambda)$ is approximately -0.924, so h is sufficiently small to produce a qualitatively correct solution.

Next, we consider the 2-step Adams-Moulton Method (12.27). In this case, we have

$$Q(\mu, h\lambda) = \left(1 - \frac{5}{12}h\lambda\right)\mu^2 + \left(-1 - \frac{2}{3}h\lambda\right)\mu + \frac{1}{12}h\lambda.$$

Setting $h = 0.05$, so that $h\lambda = -5$, the largest root of $Q(\mu, h\lambda)$ turns out to be approximately -0.906, so a larger step size can safely be chosen for this method. $\square$

In general, larger step sizes can be chosen for implicit methods than for explicit methods. However, the savings achieved from having to take fewer time steps can be offset by the expense of having to solve a nonlinear equation during every time step.

12.4.4.1 *Region of Absolute Stability*

The **region of absolute stability** of a one-step method or a multistep method is the region R of the complex plane such that if $h\lambda \in R$, then a solution of the test equation computed using h and λ will decay to zero, as desired. That is, for a one-step method, $|Q(h\lambda)| < 1$ for $h\lambda \in R$, and for a multistep method, the roots $\mu_1, \mu_2, \ldots, \mu_m$ of $Q(\mu, h\lambda)$ satisfy $|\mu_i| < 1$.

Because a larger region of absolute stability allows a larger step size h to be chosen for a given value of λ, it is preferable to use a method that has as large a region of absolute stability as possible. The ideal situation is when a method is **A-stable**, which means that its region of absolute stability contains the entire left half-plane, because then, the solution will decay to zero regardless of the choice of h.

An example of an A-stable one-step method is **Backward Euler's Method** (12.22)

$$y_{n+1} = y_n + h f(t_{n+1}, y_{n+1}),$$

an implicit method. For this method,

$$Q(h\lambda) = \frac{1}{1 - h\lambda},$$

and since $\operatorname{Re}\lambda < 0$, it follows that $|Q(h\lambda)| < 1$ regardless of the value of h. An A-stable multistep method is the **Implicit Trapezoidal Method** recalled from (12.23),

$$y_{n+1} = y_n + \frac{h}{2}[f_{n+1} + f_n].$$

The stability polynomial for this method is

$$Q(\mu, h\lambda) = \left(1 - \frac{h\lambda}{2}\right)\mu + \left(-1 - \frac{h\lambda}{2}\right),$$

which has the root

$$\mu = \frac{1 + \frac{h\lambda}{2}}{1 - \frac{h\lambda}{2}}.$$

The numerator and denominator have imaginary parts of the same magnitude, but because $\operatorname{Re}\lambda < 0$, the real part of the denominator has a larger magnitude than that of the numerator, so $|\mu| < 1$, regardless of h.

Implicit multistep methods, such as the Implicit Trapezoidal Method, are often used for stiff differential equations because of their larger regions of absolute stability. However, as the next explorations illustrate, it is important to properly estimate the largest possible value of λ for a given ODE in order to select an h such that $h\lambda$ actually lies within the region of absolute stability.

Exploration 12.4.8 Form the stability polynomial for the 2-step Adams-Moulton Method (12.27).

> **Exploration 12.4.9** Suppose the 2-step Adams-Moulton Method (12.27)
> is applied to the IVP
> $$y' = -2y, \quad y(0) = 1.$$
> How small must h be so that a bounded solution can be ensured?

> **Exploration 12.4.10** Now, suppose the same Adams-Moulton Method is
> applied to the IVP
> $$y' = -2y + e^{-100t}, \quad y(0) = 1.$$
> How does the addition of the source term e^{-100t} affect the choice of h?

> **Exploration 12.4.11** In general, for an ODE of the form $y' = f(t, y)$, how
> should the value of λ be determined for the purpose of choosing an h such
> that $h\lambda$ lies within the region of absolute stability?

12.4.4.2   *Dahlquist's Barrier Theorems*

We conclude our discussion of multistep methods with some important results, due
to Dahlquist, concerning the zero-stability of multistep methods. The first theorem
imposes a limit on the order of accuracy of zero-stable methods.

> **Theorem 12.4.10 (Dahlquist's Barrier Theorem)** The order of accu-
> racy of a zero-stable m-step method is at most $m + 1$ if m is odd, or $m + 2$
> if m is even.

For example, because of this theorem, it can be concluded that a 6th-order accurate
three-step method cannot be zero stable, whereas a 4th-order accurate, zero-stable
two-step method has the highest order of accuracy that can be achieved. A proof
can be found in [Gautschi (1997)].

Finally, we state a result, proved in [Dahlquist (1963)], concerning absolute
stability that highlights the trade-off between explicit and implicit methods.

> **Theorem 12.4.11 (Dahlquist's Second Barrier Theorem)** No ex-
> plicit multistep method is A-stable. Furthermore, no A-stable multistep
> method can have an order of accuracy greater than 2. The second-order
> accurate, A-stable multistep method with the smallest asymptotic error
> constant is the Implicit Trapezoidal Method.

While there are no A-stable methods with higher-order accuracy, that does not
mean that higher-order methods are not still useful. For example, Backward Differentiation formulas (BDF), described in Section 12.3.3, are efficient implicit methods
that are high-order accurate and have a region of absolute stability that includes a
large portion of the negative half-plane, including the entire negative real axis.

Exploration 12.4.12 Find a BDF of order greater than 1 that has a region of absolute stability that includes the entire negative real axis.

12.4.5 Concept Check

1. What does it mean for a time-stepping method to be consistent?
2. What does it mean for a time-stepping method to be stable? How does the general definition specialize to one-step and multistep methods?
3. Explain the relationship between consistency, stability and convergence.
4. What is the root condition for multistep methods, and how does it relate to stability?
5. What does it mean for a multistep method to be strongly or weakly stable?
6. What does it mean for a differential equation to be stiff?
7. What is the test equation, and what purpose does it serve?
8. What is the region of absolute stability of a one-step or multistep method?
9. What does it mean for a time-stepping method to be A-stable?
10. State the Dahlquist Equivalence Theorem and Dahlquist's Barrier Theorems.

12.5 Adaptive Methods*

So far, we have assumed that the time-stepping methods that we have been using for solving $y' = f(t, y)$, on the interval $t_0 < t \leq T$, compute the solution at times $t_1, t_2, \ldots$ that are equally spaced. That is, we define $t_{n+1} - t_n = h$ for some value of h that is *fixed* over the entire interval $(t_0, T]$ on which the problem is being solved. However, in practice, this is ill-advised because

- the chosen time step may be too large to resolve the solution with sufficient accuracy, especially if it is highly oscillatory, or
- the chosen time step may be too small when the solution is particularly smooth, thus wasting computational effort required for evaluations of f.

This is reminiscent of the problem of choosing appropriate subintervals when applying composite quadrature rules to approximate definite integrals. In that case, adaptive quadrature rules (see Section 9.6) were designed to get around this problem. These methods used estimates of the error to determine whether certain subintervals should be divided. In this section, we seek to develop an analogous strategy for time-stepping to solve initial value problems.

12.5.1 Error Estimation

Applying the approach behind adaptive quadrature to initial value problems would involve estimating the **global error**, measured by $\max_{0 \leq n \leq (T-t_0)/h} |y_n - y(t_n)|$, and then, if it is too large, repeating the time-stepping process with a smaller value of h. However, this is impractical, because it is difficult to obtain a sharp estimate of global error, and much of the work involved would be wasted due to overwriting of solution values, unlike with adaptive quadrature, where each subinterval can be handled independently. Instead, we propose to estimate the **local truncation error** at each time step, and use that estimate to determine whether h should be varied for the next time step. This approach minimizes the amount of extra work that is required to implement this kind of **adaptive time-stepping**, and it relies on an error estimate that is easy to compute.

We first consider error estimation for one-step methods. This error estimation is accomplished using a pair of one-step methods,

$$y_{n+1} = y_n + h\Phi_p(t_n, y_n, h), \tag{12.34}$$

$$\tilde{y}_{n+1} = \tilde{y}_n + h\Phi_{p+1}(t_n, \tilde{y}_n, h), \tag{12.35}$$

of orders p and $p+1$, respectively. Recall that their local truncation errors are

$$\tau_{n+1}(h) = \frac{1}{h}[y(t_{n+1}) - y(t_n)] - \Phi_p(t_n, y(t_n), h),$$

$$\tilde{\tau}_{n+1}(h) = \frac{1}{h}[y(t_{n+1}) - y(t_n)] - \Phi_{p+1}(t_n, y(t_n), h).$$

We make the assumption that both methods are exact at time t_n; that is, $y_n = \tilde{y}_n = y(t_n)$. It then follows from (12.34) and (12.35) that

$$\tau_{n+1}(h) = \frac{1}{h}[y(t_{n+1}) - y_{n+1}], \quad \tilde{\tau}_{n+1}(h) = \frac{1}{h}[y(t_{n+1}) - \tilde{y}_{n+1}].$$

Subtracting these equations yields

$$\tau_{n+1}(h) = \tilde{\tau}_{n+1}(h) + \frac{1}{h}[\tilde{y}_{n+1} - y_{n+1}].$$

Because $\tau_{n+1}(h)$ is $O(h^p)$ while $\tilde{\tau}_{n+1}(h)$ is $O(h^{p+1})$, we neglect $\tilde{\tau}_{n+1}(h)$ and obtain the simple error estimate

$$\tau_{n+1}(h) = \frac{1}{h}(\tilde{y}_{n+1} - y_{n+1}).$$

The approach for multistep methods is similar. We use a pair of Adams Methods, consisting of an m-step Adams-Bashforth (explicit) Method,

$$\sum_{i=0}^{m} \alpha_i y_{n+1-i} = h \sum_{i=1}^{m} \beta_i f_{n+1-i},$$

and an $(m-1)$-step Adams-Moulton (implicit) Method,

$$\sum_{i=0}^{m} \tilde{\alpha}_i \tilde{y}_{n+1-i} = h \sum_{i=0}^{m} \tilde{\beta}_i \tilde{f}_{n+1-i},$$

where $\tilde{f}_i = f(t_i, \tilde{y}_i)$, so that both are $O(h^m)$-accurate. We then have

$$\sum_{i=0}^{m} \alpha_i y(t_{n+1-i}) = h \sum_{i=1}^{m} \beta_i f(t_{n+1-i}, y(t_{n+1-i})) + h\tau_{n+1}(h),$$

$$\sum_{i=0}^{m} \tilde{\alpha}_i y(t_{n+1-i}) = h \sum_{i=0}^{m} \tilde{\beta}_i f(t_{n+1-i}, y(t_{n+1-i})) + h\tilde{\tau}_{n+1}(h),$$

where $\tau_{n+1}(h)$ and $\tilde{\tau}_{n+1}(h)$ are the local truncation errors of the explicit and implicit methods, respectively.

As before, we assume that $y_{n+1-m}, \ldots, y_n$ are exact, which yields

$$\tau_{n+1}(h) = \frac{1}{h}[y(t_{n+1}) - y_{n+1}], \quad \tilde{\tau}_{n+1}(h) = \frac{1}{h}[y(t_{n+1}) - \tilde{y}_{n+1}],$$

as in the case of one-step methods. It follows that

$$\tilde{y}_{n+1} - y_{n+1} = h[\tau_{n+1}(h) - \tilde{\tau}_{n+1}(h)].$$

The local truncation errors have the form

$$\tau_{n+1}(h) = Ch^m y^{(m+1)}(\xi_n), \quad \tilde{\tau}_{n+1}(h) = \tilde{C}h^m y^{(m+1)}(\tilde{\xi}_n),$$

where $\xi_n, \tilde{\xi}_n \in [t_{n+1-m}, t_{n+1}]$. We assume that these unknown values are equal, which yields

$$\tilde{\tau}_{n+1}(h) \approx \frac{\tilde{C}}{h(C - \tilde{C})}[\tilde{y}_{n+1} - y_{n+1}]. \tag{12.36}$$

Exploration 12.5.1 Formulate an error estimate of the form (12.36) for the case $m = 4$; that is, estimate the error in the 3-step Adams-Moulton Method (12.26) using the 4-step Adams-Bashforth Method (12.25). *Hint:* Use the result of Exploration 12.4.3.

12.5.2 Adaptive Time-Stepping

Now that we have a way to estimate the local truncation error in one-step and multistep methods, our goal is to determine how to modify h so that the local truncation error is approximately equal to a prescribed tolerance ε, and therefore is not too large nor too small.

When using two one-step methods as previously discussed, because $\tau_{n+1}(h)$ is the local truncation error of a method that is pth-order accurate, it follows that if we replace h by qh for some scaling factor q, the error is multiplied by q^p. Therefore, we relate the error obtained with step size qh to our tolerance, and obtain

$$|\tau_{n+1}(qh)| \approx \left| \frac{q^p}{h}(\tilde{y}_{n+1} - y_{n+1}) \right| \leq \varepsilon.$$

Solving for q yields

$$q \leq \left(\frac{\varepsilon h}{|\tilde{y}_{n+1} - y_{n+1}|} \right)^{1/p}.$$

In practice, though, the step size is kept bounded by chosen values $h_{\min}$ and $h_{\max}$ in order to avoid missing sensitive regions of the solution by using excessively large time steps, as well as expending too much computational effort on regions where $y(t)$ is oscillatory by using step sizes that are too small [Burden and Faires (2004)].

For one-step methods, if the error is small enough to accept y_{n+1}, but $\tilde{y}_{n+1}$ is obtained using a higher-order method, then it makes sense to instead use $\tilde{y}_{n+1}$ as input for the next time step, since it is ostensibly more accurate, even though the error estimate applies to y_{n+1}. Using $\tilde{y}_{n+1}$ instead is called **local extrapolation**.

The **Runge-Kutta-Fehlberg Method** [Fehlberg (1970)] is an example of an adaptive time-stepping method. It uses a four-stage, fourth-order Runge-Kutta Method and a five-stage, fifth-order Runge-Kutta Method. These two methods *share* some evaluations of $f(t, y)$, in order to reduce the number of evaluations of f per time step to six, rather than the nine that would normally be required from a pairing of fourth- and fifth-order methods. A pair of Runge-Kutta methods that can share stages in this way is called an **embedded pair**.

The **Bogacki-Shampine Method** [Bogacki and Shampine (1989)], which is used in the MATLAB function ode23, is an embedded pair consisting of a four-stage, second-order Runge-Kutta Method and a three-stage, third-order Runge-Kutta Method. As in the Runge-Kutta-Fehlberg Method, evaluations of $f(t, y)$ are shared, reducing the number of evaluations per time step from seven to four. However, unlike the Runge-Kutta-Fehlberg Method, its last stage, $k_4 = f(t_{n+1}, y_{n+1})$, is the same as the first stage of the next time step, $k_1 = f(t_n, y_n)$, if y_{n+1} is accepted, as local extrapolation is used. This reduces the number of *new* evaluations of f per time step from four to three. A Runge-Kutta Method that shares stages across time-steps in this manner is called a **FSAL** (First Same as Last) Method.

Exploration 12.5.2 Find the definitions of the two Runge-Kutta methods used in the Bogacki-Shampine Method (they can easily be found online). Use these definitions to write a MATLAB function [t,y]=rk23(f,tspan,y0,h) that implements the Bogacki-Shampine Method, using an initial step size specified in the input argument h. How does the performance of your method compare to that of ode23?

The MATLAB ODE solver ode45 uses the **Dormand-Prince Method** [Dormand and Prince (1980)], which consists of a 5-stage, 5th-order Runge-Kutta method and a 6-stage, 4th-order Runge-Kutta method. By sharing stages, the number of evaluations of $f(t, y)$ is reduced to 7. Like the Bogacki-Shampine Method, this method is FSAL, so in fact only six new evaluations per time step are required.

Exploration 12.5.3 Find the definitions of the two Runge-Kutta methods used in the Dormand-Prince method. Use these definitions to write a MATLAB function [t,y]=rk45(f,tspan,y0,h) that implements the Dormand-Prince method, using an initial step size specified in the input argument h. How does the performance of your method compare to that of ode45?

For multistep methods, we assume as before that an m-step predictor and $(m-1)$-step corrector are used. Recall that the error estimate $\tau_{n+1}(h)$ for the corrector is given in (12.36). As with one-step methods, we relate the error estimate $\tau_{n+1}(qh)$

to the error tolerance ε and solve for q, which yields

$$q \approx \left(\frac{\varepsilon}{\tau_{n+1}(h)} \right)^{1/m}.$$

Then, the time step can be adjusted to qh, but as with one-step methods, q is constrained to avoid drastic changes in the time step. Unlike one-step methods, a change in the time step is computationally expensive, as it requires the computation of new starting values at equally spaced times.

Exploration 12.5.4 Implement an adaptive multistep method based on the 4-step Adams-Bashforth Method (12.25) and 3-step Adams-Moulton Method (12.26). Use RK4 to obtain starting values.

12.5.3 Concept Check

1. What is the reason for designing adaptive time-stepping methods?
2. Briefly describe how adaptive methods work.
3. What is local extrapolation?
4. What is meant by an embedded pair?
5. What does it mean for a Runge-Kutta Method to be FSAL?

12.6 Higher-Order Equations and Systems of Equations

Numerical methods for solving a single, first-order ODE of the form $y' = f(t, y)$ can also be applied to more general ODEs, including systems of first-order equations, and equations with higher-order derivatives. We will now learn how to generalize these methods to such problems.

12.6.1 Systems of First-Order Equations

We consider a system of m first-order equations, that has the form

$$y_1' = f_1(t, y_1, y_2, \ldots, y_m),$$
$$y_2' = f_2(t, y_1, y_2, \ldots, y_m),$$
$$\vdots$$
$$y_m' = f_m(t, y_1, y_2, \ldots, y_m),$$

where $t_0 < t \leq T$, with initial conditions

$$y_1'(t_0) = y_{1,0}, \quad y_2'(t_0) = y_{2,0}, \quad \ldots, \quad y_m'(t_0) = y_{m,0}.$$

This problem can be written more conveniently in vector form

$$\mathbf{y}' = \mathbf{f}(y, \mathbf{y}), \quad \mathbf{y}'(t_0) = \mathbf{y}_0,$$

where $\mathbf{y}(t)$ is a vector-valued function with component functions

$$\mathbf{y}(t) = \begin{bmatrix} y_1(t) \ y_2(t) \ \cdots \ y_m(t) \end{bmatrix}^T,$$

$\mathbf{f}$ is a vector-valued function of t and $\mathbf{y}$, with component functions

$$\mathbf{f}(t,y) = \left[\, f_1(t,\mathbf{y}) \; f_2(t,\mathbf{y}) \cdots \; f_m(t,\mathbf{y}) \,\right]^T$$

and $\mathbf{y}_0$ is the vector of initial values,

$$\mathbf{y}_0 = \left[\, y_{1,0} \; y_{2,0} \cdots \; y_{m,0} \,\right]^T.$$

This initial-value problem has a unique solution $\mathbf{y}(t)$ on $[t_0, T]$ if $\mathbf{f}$ is continuous on the domain $D = [t_0, T] \times (-\infty, \infty)^m$, and satisfies a Lipschitz condition on D in each of the variables $y_1, y_2, \ldots, y_m$.

Applying a one-step method of the form

$$y_{n+1} = y_n + h\Phi(t_n, y_n, y_{n+1}, h)$$

to a system is straightforward. It just requires generalizing the function $\Phi(t_n, y_n, y_{n+1}, h)$ to a vector-valued function that evaluates $\mathbf{f}(t, \mathbf{y})$ in the same way as it evaluates $f(t, y)$ in the case of a single equation, with its arguments obtained from t_n and $\mathbf{y}_n$ in the same way as they are from t_n and y_n for a single equation.

Example 12.6.1 Consider the Modified Euler Method (12.19). To apply this method to a system of m equations of the form $\mathbf{y}' = \mathbf{f}(t, \mathbf{y})$, we compute

$$\mathbf{y}_{n+1} = \mathbf{y}_n + \frac{h}{2} \left[\mathbf{f}(t_n, \mathbf{y}_n) + \mathbf{f}\left(t_n + h, \mathbf{y}_n + \frac{h}{2}\mathbf{f}(t_n, \mathbf{y}_n) \right) \right],$$

where $\mathbf{y}_n$ is an approximation to $\mathbf{y}(t_n)$. The vector $\mathbf{y}_n$ has components

$$\mathbf{y}_n = \left[\, y_{1,n} \; y_{2,n} \cdots \; y_{m,n} \,\right]^T,$$

where, for $i = 1, 2, \ldots, m$, $y_{i,n}$ is an approximation to $y_i(t_n)$.

We illustrate this method on the system of two equations

$$y_1' = f_1(t, y_1, y_2) = -2y_1 + 3ty_2, \tag{12.37}$$
$$y_2' = f_2(t, y_1, y_2) = t^2 y_1 - e^{-t} y_2. \tag{12.38}$$

First, we rewrite the method in the more convenient form

$$k_1 = hf(t_n, y_n),$$
$$k_2 = hf(t_n + h, y_n + k_1),$$
$$y_{n+1} = y_n + \frac{1}{2}[k_1 + k_2].$$

Then, the Modified Euler Method, applied to this system of ODE, takes the form

$$k_{1,1} = hf_1(t_n, y_{1,n}, y_{2,n})$$
$$= h[-2y_{1,n} + 3t_n y_{2,n}],$$
$$k_{2,1} = hf_2(t_n, y_{1,n}, y_{2,n})$$
$$= h[t_n^2 y_{1,n} - e^{-t_n} y_{2,n}],$$
$$k_{1,2} = hf_1(t_n + h, y_{1,n} + k_{1,1}, y_{2,n} + k_{2,1})$$
$$= h[-2(y_{1,n} + k_{1,1}) + 3(t_n + h)(y_{2,n} + k_{2,2})],$$
$$k_{2,2} = hf_2(t_n + h, y_{1,n} + k_{1,1}, y_{2,n} + k_{2,1})$$
$$= h[(t_n + h)^2(y_{1,n} + k_{1,1}) - e^{-(t_n+h)}(y_{2,n} + k_{2,1})],$$
$$y_{1,n+1} = y_{1,n} + \frac{1}{2}[k_{1,1} + k_{1,2}],$$
$$y_{2,n+1} = y_{2,n} + \frac{1}{2}[k_{2,1} + k_{2,2}].$$

This can be written in vector form as follows:

$$\mathbf{k}_1 = h\mathbf{f}(t_n, \mathbf{y}_n),$$
$$\mathbf{k}_2 = h\mathbf{f}(t_n + h, \mathbf{y}_n + \mathbf{k}_1),$$
$$\mathbf{y}_{n+1} = \mathbf{y}_n + \frac{1}{2}[\mathbf{k}_1 + \mathbf{k}_2],$$

where $\mathbf{k}_i = \begin{bmatrix} k_{1,i} & k_{2,i} \end{bmatrix}^T$ for $i = 1, 2$. $\square$

> **Exploration 12.6.1** Try your `rk4` function from Exploration 12.2.6 on the system (12.37), (12.38) with initial conditions $y_1(0) = 1$, $y_2(0) = -1$. Write your time derivative function `yp=f(t,y)` for this system so that the input argument y and the value yp returned by f are both *column* vectors, and pass a column vector containing the initial values as the input argument y0. Do you even need to modify `rk4`?

Multistep methods generalize in a similar way. A general m-step multistep method for a system of first-order ODE $\mathbf{y}' = \mathbf{f}(t, \mathbf{y})$ has the form

$$\sum_{i=0}^{m} \alpha_i \mathbf{y}_{n+1-i} = h \sum_{i=0}^{m} \beta_i \mathbf{f}(t_{n+1-i}, \mathbf{y}_{n+1-i}),$$

where the constants α_i and β_i, for $i = 0, 1, \ldots, m$, are determined in the same way as in the case of a single equation.

Example 12.6.2 The explicit 3-step Adams-Bashforth Method applied to the system (12.37), (12.38) has the form

$$y_{1,n+1} = y_{1,n} + \frac{h}{12}[23f_{1,n} - 16f_{1,n-1} + 5f_{1,n-2}],$$
$$y_{2,n+1} = y_{2,n} + \frac{h}{12}[23f_{2,n} - 16f_{2,n-1} + 5f_{2,n-2}],$$

where

$$f_{1,i} = -2y_{1,i} + 3t_i y_{2,i}, \quad f_{2,i} = t_i^2 y_{1,i} - e^{-t_i} y_{2,i}, \quad i = 0, \ldots, n.$$

□

The order of accuracy for a one-step or multistep method, when applied to a system of equations, is the same as when it is applied to a single equation. For example, the Modified Euler's Method is second-order accurate for systems, and the 3-step Adams-Bashforth Method is third-order accurate. However, when using adaptive step size control for any of these methods, it is essential that the step size h is selected so that *all* components of the solution are sufficiently accurate, or it is likely that *none* of them will be.

12.6.2   Higher-Order Equations

The numerical methods we have learned are equally applicable to differential equations of higher order, that have the form

$$y^{(m)} = f(t, y, y', y'', \ldots, y^{(m-1)}), \quad t_0 < t \leq T,$$

because such equations are equivalent to systems of first-order equations, in which new variables are introduced that correspond to lower-order derivatives of y. Specifically, we define the variables

$$u_1 = y, \quad u_2 = y', \quad u_3 = y'', \quad \cdots \quad u_m = y^{(m-1)}.$$

Then, the above ODE of order m is equivalent to the system of first-order ODEs

$$u_1' = u_2,$$
$$u_2' = u_3,$$
$$\vdots$$
$$u_m' = f(t, u_1, u_2, \ldots, u_m).$$

The initial conditions of the original higher-order equation,

$$y(t_0) = y_0^{(0)}, \quad y'(t_0) = y_0^{(1)}, \quad \ldots, \quad y^{(m-1)}(t_0) = y_0^{(m-1)},$$

are equivalent to the following initial conditions of the first order system

$$u_1(t_0) = y_0^{(0)}, \quad u_2(t_0) = y_0^{(1)}, \quad \ldots, \quad u_m(t_0) = y_0^{(m-1)}.$$

We can then apply any one-step or multistep method to this first-order system.

Example 12.6.3 Consider the second-order equation

$$y'' + 3y' + 2y = \cos t, \quad y(0) = 2, \quad y'(0) = -1.$$

By defining $u_1 = y$ and $u_2 = y'$, we obtain the equivalent the first-order system

$$u_1' = u_2,$$
$$u_2' = -3u_2 - 2u_1 + \cos t,$$

with initial conditions

$$u_1(0) = 2, \quad u_2(0) = -1.$$

To apply RK4,

$$\mathbf{k}_1 = h\mathbf{f}(t_n, \mathbf{u}_n),$$

$$\mathbf{k}_2 = h\mathbf{f}\left(t_n + \frac{h}{2}, \mathbf{u}_n + \frac{1}{2}\mathbf{k}_1\right),$$

$$\mathbf{k}_3 = h\mathbf{f}\left(t_n + \frac{h}{2}, \mathbf{u}_n + \frac{1}{2}\mathbf{k}_2\right),$$

$$\mathbf{k}_4 = h\mathbf{f}(t_n + h, \mathbf{u}_n + \mathbf{k}_3),$$

$$\mathbf{u}_{n+1} = \mathbf{u}_n + \frac{1}{6}(\mathbf{k}_1 + 2\mathbf{k}_2 + 2\mathbf{k}_3 + \mathbf{k}_4),$$

to this system, we define

$$\mathbf{k}_i = \begin{bmatrix} k_{1,i} \\ k_{2,i} \end{bmatrix}, \quad i = 1, 2, \quad \mathbf{f}(t, \mathbf{u}) = \begin{bmatrix} f_1(t, u_1, u_2) \\ f_2(t, u_1, u_2) \end{bmatrix} = \begin{bmatrix} u_2 \\ -3u_2 - 2u_1 + \cos t \end{bmatrix}$$

and then compute

$$k_{1,1} = hf_1(t_n, u_{1,n}, u_{2,n}) = hu_{2,n},$$

$$k_{2,1} = hf_2(t_n, u_{1,n}, u_{2,n}) = h[-3u_{2,n} - 2u_{1,n} + \cos t_n],$$

$$k_{1,2} = hf_1\left(t_n + \frac{h}{2}, u_{1,n} + \frac{1}{2}k_{1,1}, u_{2,n} + \frac{1}{2}k_{2,1}\right) = h\left(u_{2,n} + \frac{1}{2}k_{2,1}\right),$$

$$k_{2,2} = hf_2\left(t_n + \frac{h}{2}, u_{1,n} + \frac{1}{2}k_{1,1}, u_{2,n} + \frac{1}{2}k_{2,1}\right)$$

$$= h\left[-3\left(u_{2,n} + \frac{1}{2}k_{2,1}\right) - 2\left(u_{1,n} + \frac{1}{2}k_{1,1}\right) + \cos\left(t_n + \frac{h}{2}\right)\right],$$

$$k_{1,3} = hf_1\left(t_n + \frac{h}{2}, u_{1,n} + \frac{1}{2}k_{1,2}, u_{2,n} + \frac{1}{2}k_{2,2}\right) = h\left(u_{2,n} + \frac{1}{2}k_{2,2}\right),$$

$$k_{2,3} = hf_2\left(t_n + \frac{h}{2}, u_{1,n} + \frac{1}{2}k_{1,2}, u_{2,n} + \frac{1}{2}k_{2,2}\right)$$

$$= h\left[-3\left(u_{2,n} + \frac{1}{2}k_{2,2}\right) - 2\left(u_{1,n} + \frac{1}{2}k_{1,2}\right) + \cos\left(t_n + \frac{h}{2}\right)\right],$$

$$k_{1,4} = hf_1(t_n + h, u_{1,n} + k_{1,3}, u_{2,n} + k_{2,3}) = h(u_{2,n} + k_{2,3}),$$

$$k_{2,4} = hf_2(t_n + h, u_{1,n} + k_{1,3}, u_{2,n} + k_{2,3})$$

$$= h[-3(u_{2,n} + k_{2,3}) - 2(u_{1,n} + k_{1,3}) + \cos(t_n + h)],$$

$$u_{1,n+1} = u_{1,n} + \frac{1}{6}(k_{1,1} + 2k_{1,2} + 2k_{1,3} + k_{1,4}),$$

$$u_{2,n+1} = u_{2,n} + \frac{1}{6}(k_{2,1} + 2k_{2,2} + 2k_{2,3} + k_{2,4}).$$

□

Exploration 12.6.2 Modify your `rk4` function from Exploration 12.2.6 so that it solves a single ODE of the form $y^{(m)} = f(t, y, y', y'', \ldots, y^{(m-1)})$ with initial conditions

$$y(t_0) = y_0, \quad y'(t_0) = y_0', \quad y''(t_0) = y_0'', \quad \ldots, \quad y^{(m-1)}(t_0) = y_0^{(m)}.$$

Assume that the input argument `ym = f(t, y)` treats the input argument `y` as a *row* vector consisting of the values of $y, y', y'', \ldots, y^{(m-1)}$ at time `t`, and that `f` returns a *scalar* value `ym` that represents the value of $y^{(m)}$. Your function should also assume that the argument `y0` containing the initial values is a row vector. The value of m indicating the order of the ODE can be automatically inferred from `length(y0)`.

Exploration 12.6.3 How would you use your function from Exploration 12.6.2 to solve a *system* of p ODEs, in which each ODE is of order m?

12.6.3   Concept Check

1. What is involved with time-stepping method for a system of ODEs?
2. How should the time-step be selected when using an adaptive method to solve a system of ODEs?
3. How can a one-step or multistep method be applied to a higher-order ODE?

12.7   Additional Resources

Textbooks on theory and analytical solution methods of ODEs that the authors have found to be particularly useful are [Betounes (2010); Boyce and DiPrima (2013); Braun (1993)]. There are several books devoted exclusively to numerical methods for differential equations, with varying emphases. The first author found the text [Lambert (1992)] very beneficial when first learning about the subject as a graduate student. A two-volume set, consisting of one volume [Hairer et al. (1993)] on non-stiff problems and a second volume [Hairer and Wanner (1996)] on stiff problems, provides a comprehensive reference for readers interested in either theory or algorithms. The text [Ascher and Petzold (1998)] focuses on practical software implementation. Essential contributions to the development of BDFs and methods for stiff problems are presented in [Gear (1971)]. Not covered in this text are implicit Runge-Kutta methods, which are motivated by the general unsuitability of explicit Runge-Kutta methods for stiff problems. An introduction to these methods can be found in [Iserles (2008); Süli and Mayers (2003)].

12.8   Exercises

1. Prove that Backward Euler (12.22) satisfies the same convergence result (12.13) as Forward Euler (12.8). Assume $hL < 1$, where L is the Lipschitz

constant for f.

2. Prove that the Implicit Trapezoidal Method (12.23) is second-order accurate.

3. Consider the IVP

$$x' = -y$$
$$y' = x$$

with initial conditions $x(0) = 1$, $y(0) = 0$.

(a) Differentiate either equation to obtain the exact solution $(x(t), y(t))$, which describes a parametric curve. What is this curve?

(b) Use (Forward) Euler's Method (12.8), Backward Euler (12.22) and the Implicit Trapezoidal Method (12.23) to solve this IVP with step size $h = 0.1$ and final time $T = 20$. For each method, plot $x(t)$ versus $y(t)$.

(c) For each of the three methods used in part 3b, compute

$$r_{n+1} = \sqrt{x(t_{n+1})^2 + y(t_{n+1})^2}$$

in terms of $r_n = \sqrt{x(t_n)^2 + y(t_n)^2}$. How do these results explain the behavior of the numerical solutions produced in part 3b?

4. Consider the IVP from Example 12.2.2,

$$y' = -2ty, \quad y(0) = 1.$$

(a) Solve this IVP with final time $T = 1$ and $h = 0.1, 0.01, 0.001$, using the Explicit Trapezoidal Method and RK4.

(b) For each method, assume the error at a given time T, for a given time step h, is of the form $e(h) = Ch^p$. Given the errors $e(h_1)$ and $e(h_2)$, derive a formula for the order of accuracy p.

(c) Use the errors in the solutions computed in part 4a, obtained by comparing these solutions to the exact solution at $T = 1$, to estimate the order of accuracy of the Explicit Trapezoidal Method and RK4, with the help of the formula for the order of accuracy derived in part 4b. Do the estimated orders of accuracy match theoretical expectations?

5. Prove that no explicit Runge-Kutta Method can be A-stable.

6. Consider the nonlinear IVP

$$y' = -y^3, \quad y(0) = 1.$$

(a) Suppose that we solve this IVP using Backward Euler (12.22). To solve the nonlinear equation required to compute y_{n+1} from y_n, we use Fixed-Point Iteration with initial guess $y_{n+1}^{(0)} = y_n$. Prove that if $y_n \leq 1$, then Backward Euler's Method can be written in the form

$$y_{n+1}^{(k+1)} = g(y_{n+1}^{(k)}),$$

where g maps $[0, y_n]$ into itself.

(b) Prove that if $h < 1/3$, then Fixed-Point Iteration with initial guess $y_{n+1}^{(0)} = y_n$, where $0 \le y_n \le 1$, is guaranteed to converge.

(c) Use Newton's Method to carry out each time step instead. Prove that if h is sufficiently small, then Newton's Method converges quadratically. Is quadratic convergence achieved if $h = 0.1$?

7. Consider the **Leap-Frog Method**

$$y_{n+1} = y_{n-1} + 2hf(t_n, y_n).$$

This is a multistep method, but is it an Adams Method, a BDF, or neither? Determine the order of accuracy and region of absolute stability of this method.

8. Consider the **Lotka-Volterra equations**

$$x' = \alpha x - \beta xy,$$
$$y' = \delta xy - \gamma y,$$

also known as the the **predator-prey equations**. The dependent variables x and y denote the populations of two species, which are the prey and predator, respectively. The parameters α, β, γ and δ are constants with the following meanings:

- α is the relative reproduction rate of the prey.
- βx is the relative rate of predation upon the prey.
- δx is the relative growth rate of the predator population.
- γ is the relative rate of decline of the predator population.

We will use the parameter values $\alpha = 2/3$, $\beta = 4/3$, and $\gamma = \delta = 1$. Use RK4 to solve the predator-prey equations with different initial conditions $(x(0), y(0))$, with chosen values of the time step h and final time T. For each set of initial conditions, plot the parametric curve $(x(t), y(t))$. What behavior do you observe? Is there any choice of initial conditions for which the predator and prey populations remain constant?

9. We consider the **two-body problem**

$$m_1 \mathbf{r}_1'' = \frac{Gm_1 m_2}{r_{12}^3}(\mathbf{r}_2 - \mathbf{r}_1), \qquad (12.39)$$

$$m_2 \mathbf{r}_2'' = \frac{Gm_1 m_2}{r_{12}^3}(\mathbf{r}_1 - \mathbf{r}_2), \qquad (12.40)$$

that describes the motion of two bodies, of mass m_1 and m_2, located at positions $\mathbf{r}_1$ and $\mathbf{r}_2$, due to the gravitational force they exert on each other. The gravitational constant is denoted by G.

The following change of variables provides a useful alternative perspective on the two-body problem.

> **Definition 12.8.1 (Jacobi Coordinates)** Let $\mathbf{r}_1$ and $\mathbf{r}_2$ be the solutions of the two-body system (12.39), (12.40). The variables $\mathbf{r}$ and $\mathbf{R}$, defined by
>
> $$\mathbf{R} = \frac{m_1}{M}\mathbf{r}_1 + \frac{m_2}{M}\mathbf{r}_2, \qquad (12.41)$$
>
> $$\mathbf{r} = \mathbf{r}_2 - \mathbf{r}_1, \qquad (12.42)$$
>
> are called the **Jacobi coordinates** of the system.

The variable $\mathbf{R}$ is the *center of mass* of the system; here, we assume it is a constant position vector. From (12.39), (12.40) and (12.42), we obtain the simpler system of ODEs

$$\mathbf{r}'' = -\frac{GM}{r^3}\mathbf{r}, \qquad (12.43)$$

where we use the shorthand $r = \|\mathbf{r}_2 - \mathbf{r}_1\|_2 = r_{12}$. Once we solve for $\mathbf{r}$, we can obtain $\mathbf{r}_1$ and $\mathbf{r}_2$ by inverting the transformation to Jacobi coordinates, which yields

$$\begin{bmatrix} \mathbf{r}_1 \\ \mathbf{r}_2 \end{bmatrix} = \begin{bmatrix} \mathbf{R} - \frac{m_2}{M}\mathbf{r} \\ \mathbf{R} + \frac{m_1}{M}\mathbf{r} \end{bmatrix}. \qquad (12.44)$$

(a) Write a MATLAB script that solves the two-body problem, after specifying m_1, m_2, and initial values for $\mathbf{r}_1, \mathbf{r}_2$, where $\mathbf{r}_i(t) = \begin{bmatrix} x_i(t) & y_i(t) \end{bmatrix}^T$ for $i = 1, 2$. For convenience, we assume the units are such that $G = 1$. Use your function rk4 from Exploration 12.2.6, that implements RK4 (12.20). For the case $m_1 = 1$, $m_2 = 10$, with initial conditions

$$x_1(0) = 1, \quad y_1(0) = 0, \quad x_1'(0) = 0, \quad y_1'(0) = 1,$$

$$x_2(0) = -1, \quad y_2(0) = 0, \quad x_2'(0) = 0, \quad y_2'(0) = 0,$$

how small must h be to obtain a physically reasonable solution?

(b) Use your function **predictcorrect** from Exploration 12.3.5, that implements a fourth-order Adams-Moulton Predictor-Corrector Method, in place of rk4. For the same parameter values as in part 9a, how small must h be to obtain a physically reasonable solution?

10. Consider the chemical reaction

$$\left. \begin{array}{l} A + B \underset{2}{\overset{1}{\rightleftharpoons}} X \\ X + B \overset{3}{\rightarrow} R + S \end{array} \right\} \Rightarrow A + 2B \rightarrow R + S$$

The following system of ODEs models this reaction:

$$a' = -k_1 ab + k_2 x,$$
$$b' = -k_1 ab + k_2 x - k_3 bx,$$
$$x' = k_1 ab - k_2 x - k_3 bx$$

where k_1, k_2 and k_3 are rate constants, and $a(t)$, $b(t)$ and $x(t)$ are are the concentrations of the species A, B and X at time t. Let $k_1 = k_2 = 1/10$ and $k_3 = 10$, and let the initial concentrations be $a(0) = b(0) = 1$ and $x(0) = 0$. Try solving this system of ODEs with final time $T = 10$, using the MATLAB ODE solvers ode45 and ode15s. Which solver is more efficient, in terms of execution time? Which solver uses fewer time steps, as determined by the number of rows in the output arguments?

Chapter 13

Two-Point Boundary Value Problems

Having learned about the numerical solution of initial value problems (IVPs) in Chapter 12, in this chapter we consider the **two-point boundary value problem** (BVP), which consists of a second-order ODE,

$$y'' = f(x, y, y'), \quad a < x < b, \tag{13.1}$$

and **boundary conditions**

$$y(a) = \alpha, \quad y(b) = \beta. \tag{13.2}$$

As proven in [Keller (1968)], this problem is guaranteed to have a unique solution if the following conditions hold:

- f, f_y, and $f_{y'}$ are continuous on the domain
  $$D = \{(x, y, y') \mid a \le x \le b, -\infty < y < \infty, -\infty < y' < \infty\},$$
- $f_y > 0$ on D, and
- $f_{y'}$ is bounded on D.

In this chapter, we will introduce several methods for solving this kind of problem. Section 13.1 presents the **Shooting Method**, which treats a BVP as an IVP to which methods from Chapter 12 can be applied. Section 13.2 shows how **finite difference formulas** for derivatives from Section 9.1 can be used to approximate BVPs by a system of equations. Such a system can be solved using methods from Chapter 3 or 5 in the linear case, or from Section 11.1 in the nonlinear case.

The remainder of the chapter considers methods in which the solution is represented as a linear combination of functions, thus reducing the solution of the BVP to that of a system of equations for the coefficients in the linear combination. Section 13.3 describes **collocation**, in which the solution is required to satisfy the BVP at selected points in (a, b). By contrast, the **Finite Element Method**, introduced in Section 13.4, requires the BVP to be satisfied in an "average" sense, meaning that its residual must be orthogonal to a collection of **test functions**. As will be seen in Chapter 14, most of the methods for solving BVPs presented in this chapter can be generalized to **partial differential equations** (PDEs) on higher-dimensional domains, or combined with time-stepping methods for IVPs from Chapter 12 to solve time-dependent PDEs.

13.1   The Shooting Method

The first method for solving BVPs that we will examine is called the **Shooting Method**. It treats the two-point boundary value problem as an **initial value problem** (IVP), in which x plays the role of the time variable, with a being the "initial time" and b being the "final time". Specifically, the Shooting Method solves the initial value problem

$$y'' = f(x, y, y'), \quad a < x < b,$$

with initial conditions

$$y(a) = \alpha, \quad y'(a) = t,$$

where t must be chosen so that the solution satisfies the remaining boundary condition, $y(b) = \beta$. Since t, being the first derivative of $y(x)$ at $x = a$, is the "initial slope" of the solution, this approach requires selecting the proper slope, or "trajectory", so that the solution will "hit the target" of $y(x) = \beta$ at $x = b$. This viewpoint indicates how the Shooting Method earned its name. Note that since the ODE associated with the IVP is second-order, it must be rewritten as a system of first-order equations, as in Section 12.6. Then, this system can be solved by standard time-stepping methods for IVPs of the kind seen in Chapter 12, such as Runge-Kutta methods or multistep methods.

13.1.1   Linear Problems

In the case where $y'' = f(x, y, y')$ is a *linear* ODE of the form

$$y'' = p(x)y' + q(x)y + r(x), \quad a < x < b, \tag{13.3}$$

selecting the slope t is relatively simple. Let $y_1(x)$ be the solution of the IVP

$$y'' = p(x)y' + q(x)y + r(x), \quad a < x \le b, \quad y(a) = \alpha, \quad y'(a) = 0, \tag{13.4}$$

and let $y_2(x)$ be the solution of the IVP

$$y'' = p(x)y' + q(x)y, \quad a < x \le b, \quad y(a) = 0, \quad y'(a) = 1. \tag{13.5}$$

Then, the solution of the original BVP has the form

$$y(x) = y_1(x) + ty_2(x), \tag{13.6}$$

where t is the correct slope, since any linear combination of solutions of (13.4) and (13.5) satisfies (13.3), and the initial values are linearly combined in the same manner as the solutions themselves.

Exploration 13.1.1 Assume $y_2(b) \neq 0$. Find the value of t in (13.6) such that the boundary conditions (13.2) are satisfied.

Exploration 13.1.2 Explain why the condition $y_2(b)$ is guaranteed to be satisfied, due to the previously stated assumptions about $f(x, y, y')$ that guarantee the existence and uniqueness of the solution.

Example 13.1.1 Consider the BVP

$$y'' = 2y' - y + xe^x - x, \quad 0 < x < 2, \quad y(0) = 0, \quad y(2) = -4. \qquad (13.7)$$

If we let $y_1(x)$ be the solution of the IVP

$$y'' = 2y' - y + xe^x - x, \quad 0 < x \le 2, \quad y(0) = 0, \quad y'(0) = 0,$$

and let $y_2(x)$ be the solution of the IVP

$$y'' = 2y' - y, \quad 0 < x \le 2, \quad y(0) = 0, \quad y'(0) = 1,$$

then the Method of Undetermined Coefficients yields

$$y_1(x) = \frac{1}{6}x^3 e^x - xe^x + 2e^x - x - 2, \quad y_2(x) = xe^x.$$

We then set $y(x) = y_1(x) + ty_2(x)$, where t satisfies

$$-4 = y(2) = y_1(2) + ty_2(2) = \frac{4}{3}e^2 - 4 + t(2e^2),$$

which yields $t = -2/3$. We conclude that the solution of the original BVP (13.7) is

$$y(x) = \frac{1}{6}x^3 e^x - \frac{5}{3}xe^x + 2e^x - x - 2.$$

□

Exploration 13.1.3 Write a MATLAB function

    Y=shootlinear(p,q,r,a,b,alpha,beta,n)

that solves the linear BVP (13.3), (13.2) using the Shooting Method. Use RK4 to solve the IVPs (13.4), (13.5). The input arguments p, q, and r are function handles for the coefficients $p(x)$, $q(x)$ and $r(x)$, respectively, of (13.3). The input arguments a, b, alpha and beta specify the boundary conditions (13.2), and n refers to the number of *interior* grid points; that is, a time step of $h = (b - a)/(n + 1)$ is to be used. The output Y is a vector consisting of $n + 2$ values, including both the boundary and interior values of the approximation of the solution $y(x)$ on $[a, b]$. Test your function on the BVP from Example 13.1.1. *Hint:* Consult Section 12.6 on solving second-order ODEs.

13.1.2 Nonlinear Problems

If the ODE is nonlinear, then t satisfies a nonlinear equation of the form

$$y(b; t) = \beta,$$

where $y(b; t)$ is the value of the solution, at $x = b$, of the IVP specified by the Shooting Method, with initial slope t. This nonlinear equation can be solved for t using an iterative method such as those covered in Chapter 10. The only difference is that each evaluation of the function $y(b; t)$, at a new value of t, is relatively expensive, since it requires the solution of an IVP over the interval $[a, b]$, for which $y'(a) = t$. The value of that solution at $x = b$ is taken to be the value of $y(b; t)$.

If **Newton's Method** (see Section 10.4) is used, then an additional complication arises, because it requires the derivative of $y(b, t)$, with respect to t, during each iteration. This can be computed using the fact that $z(x; t) = \partial y(x; t)/\partial t$ satisfies the ODE

$$z'' = f_y z + f_{y'} z', \quad a < x \le b, \quad z(a; t) = 0, \quad z'(a; t) = 1,$$

which can be obtained by differentiating the original BVP and its boundary conditions with respect to t. Therefore, each iteration of Newton's Method requires *two* IVPs to be solved, but this extra effort can be offset by the rapid convergence of Newton's Method.

Suppose that Euler's Method,

$$\mathbf{y}_{i+1} = \mathbf{y}_i + h\mathbf{f}(x, \mathbf{y}_i, h),$$

for the IVP $\mathbf{y}' = \mathbf{f}(x, \mathbf{y})$, $\mathbf{y}(x_0) = \mathbf{y}_0$, is to be used to solve any IVPs arising from the Shooting Method in conjunction with Newton's Method. Because each IVP, for $y(x, t)$ and $z(x, t)$, is of second order, we must rewrite each one as a first-order system, as in Section 12.6.2. We first define

$$y^1 = y, \quad y^2 = y', \quad z^1 = z, \quad z^2 = z'.$$

We then have the system of ODEs

$$\frac{\partial y^1}{\partial x} = y^2,$$

$$\frac{\partial y^2}{\partial x} = f(x, y^1, y^2),$$

$$\frac{\partial z^1}{\partial x} = z^2,$$

$$\frac{\partial z^2}{\partial x} = f_y(x, y^1, y^2)z^1 + f_{y'}(x, y^1, y^2)z^2,$$

with initial conditions

$$y^1(a) = \alpha, \quad y^2(a) = t, \quad z^1(a) = 0, \quad z^2(a) = 1.$$

The algorithm then proceeds as follows:

Algorithm 13.1.2 (Shooting Method with Newton's Method)
Given the function $f(x, y, y')$, an interval (a, b), boundary values α and β, and a number of interior points N, the following algorithm computes an approximate solution of the BVP (13.1), (13.2) using the Shooting Method in conjunction with Euler's Method (12.8).

Choose an initial guess $t^{(0)}$
$h = (b - a)/(N + 1)$
for $k = 0, 1, 2, \ldots$ until convergence **do**
$\quad y_0^1 = \alpha, \ y_0^2 = t^{(k)}, \ z_0^1 = 0, \ z_0^2 = 1$
$\quad$ for $i = 0, 1, 2, \ldots, N$ **do**
$\quad\quad x_i = a + ih$
$\quad\quad y_{i+1}^1 = y_i^1 + h y_i^2$
$\quad\quad y_{i+1}^2 = y_i^2 + h f(x_i, y_i^1, y_i^2)$
$\quad\quad z_{i+1}^1 = z_i^1 + h z_i^2$
$\quad\quad z_{i+1}^2 = z_i^2 + h[f_y(x_i, y_i^1, y_i^2)z_i^1 + f_{y'}(x_i, y_i^1, y_i^2)z_i^2]$
$\quad$ **end for**
$\quad t^{(k+1)} = t^{(k)} - (y_{N+1}^1 - \beta)/z_{N+1}^1$
end for

Example 13.1.3 We solve the BVP

$$y'' = y^3 - yy', \quad 1 < x < 2, \quad y(1) = \frac{1}{2}, \quad y(2) = \frac{1}{3}. \tag{13.8}$$

This BVP has the exact solution $y(x) = 1/(1+x)$. The following script sets up this BVP, calls `shootnewt` (see Exploration 13.1.5) to compute an approximate solution, and then visualizes the approximate solution and exact solution.

```
% set up BVP y'' = f(x,y,y')
f=@(x,y,yp)(y.^3-y.*yp);
fy=@(x,y,yp)(3*y.^2-yp);
fyp=@(x,y,yp)(-y);
% boundary conditions: y(a)=alpha, y(b)=beta
a=1;
b=2;
alpha=1/2;
beta=1/3;
% N: number of interior nodes
N=10;
% use Newton's Method
[x,y]=shootnewt(f,fy,fyp,a,b,alpha,beta,N,1e-8);
% compare to exact solution
yexact=1./(x+1);
plot(x,yexact,'b-o')
```

```
hold on
plot(x,y,'r--+')
hold off
xlabel('x')
ylabel('y(x)')
title('Shooting Method (Newton), n=10')
legend('Approximate','Exact')
```

Using an absolute error tolerance of 10^{-8}, Newton's Method converges in just three iterations, and does so quadratically. The resulting plot is shown in Figure 13.1. □

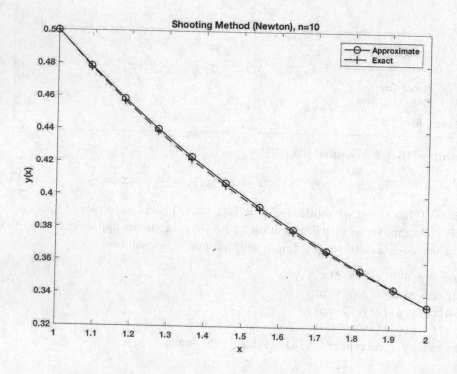

Fig. 13.1   Exact (solid curve with circles) and approximate (dashed curve with crosses) solutions of the BVP (13.8) from Example 13.1.3.

Exploration 13.1.4 What would be a logical choice of initial guess for the slope $t^{(0)}$, that would not require any information about the function $f(x, y, y')$?

Exploration 13.1.5 Write a MATLAB function

```
[x,y,niter]=shootnewt(f,fy,fyp,a,b,alpha,beta,N,tol)
```

that implements Algorithm 13.1.2 to solve the BVP (13.1), (13.2) using the Shooting Method in conjunction with Newton's Method and Euler's Method. The input argument `tol` is an error tolerance to check for the convergence of Newton's Method. The output argument `x` is a vector of $N + 2$ equally spaced points in $[a, b]$, including boundary points, and the output argument `y` consists of the values of the computed solution at these points. The third output argument `niter` is the number of iterations of Newton's Method needed for convergence. Use the code from Example 13.1.3 to test your function on the BVP (13.8).

Changing the implementation to use a different IVP solver, such as a Runge-Kutta Method or multistep method, in place of Euler's Method only changes the inner loop.

Exploration 13.1.6 Modify your code from Exploration 13.1.5 to use RK4 in place of Euler's Method. How does this affect the convergence of the Newton iteration?

Exploration 13.1.7 Modify your code from Exploration 13.1.5 to use the Secant Method instead of Newton's Method. How can the function $f(x, y, y')$ from the ODE be used to obtain a logical second initial guess $t^{(1)}$? *Hint:* Consider a solution that is a parabola. How is the efficiency of the iteration affected by the change to the Secant Method?

Exploration 13.1.8 Write a MATLAB function

```
[x,y,niter]=shootbvp(f,a,b,alpha,beta,N)
```

that solves the general nonlinear BVP (13.1), (13.2) using the Shooting Method in conjunction with the Secant Method. The input arguments `f` is a function handle for the function f. The output argument `x` is a vector of $N + 2$ equally spaced points in $[a, b]$, including boundary points, and the output argument `y` consists of the values of the computed solution at these points. Test your function on the BVP from Example 13.1.3. What happens to the performance as `N`, the number of interior points, increases?

13.1.3 Concept Check

1. What is the essential idea behind the Shooting Method?
2. Why does the Shooting Method only require iteration for nonlinear BVPs?
3. What is the drawback with using Newton's Method in conjunction with the Shooting Method, compared to other methods for solving nonlinear

equations?

13.2   Finite Difference Methods

The Shooting Method for a two-point BVP of the form (13.1), (13.2), while taking advantage of effective methods for IVPs, can not readily be generalized to boundary value problems in higher spatial dimensions. We therefore consider an alternative approach, in which the first and second derivatives of the solution $y(x)$ are approximated by **finite differences** presented in Section 9.1.

We **discretize** the problem by dividing the interval $[a, b]$ into $N+1$ subintervals of equal width $h = (b - a)/(N + 1)$. Each subinterval is of the form $[x_{i-1}, x_i]$, for $i = 1, 2, \ldots, N + 1$, where $x_i = a + ih$. We denote by y_i an approximation of the solution at x_i; that is, $y_i \approx y(x_i)$. Then, recalling (9.3) and (9.8), if we assume $y(x)$ is at least four times continuously differentiable, we can approximate y' and y'' at each x_i, $i = 1, 2, \ldots, N$, by the finite differences

$$y'(x_i) = \frac{y(x_{i+1}) - y(x_i)}{2h} - \frac{h^2}{6} y'''(\eta_i), \tag{13.9}$$

$$y''(x_i) = \frac{y(x_{i+1}) - 2y(x_i) + y(x_{i-1})}{h^2} - \frac{h^2}{12} y^{(4)}(\xi_i), \tag{13.10}$$

where η_i and ξ_i lie in (x_{i-1}, x_{i+1}). If we substitute these finite differences into the boundary value problem, and apply the boundary conditions to impose

$$y_0 = \alpha, \quad y_{N+1} = \beta, \tag{13.11}$$

then we obtain a system of equations

$$\mathcal{L}_h[\mathbf{y}]_i \equiv \frac{y_{i+1} - 2y_i + y_{i-1}}{h^2} - f\left(x_i, y_i, \frac{y_{i+1} - y_{i-1}}{2h}\right) = 0, \quad i = 1, 2, \ldots, N, \tag{13.12}$$

that we can solve for y_i, $i = 1, 2, \ldots, N$, which are the values of the approximate solution at the corresponding points x_i.

Convergence of finite difference methods for BVPs is analyzed in an analogous manner to that of time-stepping methods for IVPs. For convenience, we make use of the ∞-**norm** of an N-vector,

$$\|\mathbf{x}\|_\infty = \max_{1 \le i \le N} |x_i|,$$

defined in B.13.1.

Definition 13.2.1

- The **local truncation error** of the finite difference scheme (13.12) for the BVP (13.1), (13.2) is an N-vector $\tau(h)$ with components

$$\tau_i(h) = \mathcal{L}_h[y(\mathbf{x})]_i = \frac{y(x_{i+1}) - 2y(x_i) + y(x_{i-1})}{h^2} -$$
$$f\left(x_i, y(x_i), \frac{y(x_{i+1}) - y(x_{i-1})}{2h}\right),$$

for $i = 1, 2, \ldots, N$, where $y(x)$ is the exact solution of the BVP, and $y(\mathbf{x})$ is an N-vector consisting of the values of $y(x)$ at the interior grid points $x_1, x_2, \ldots, x_N$. We then say that the finite difference scheme is **consistent** if $\lim_{h \to 0} \|\tau(h)\|_\infty = 0$.

- We say the finite difference scheme (13.12) is **stable** if for any two vectors $\mathbf{y}, \tilde{\mathbf{y}} \in \mathbb{R}^n$, and for h sufficiently small, there exists a constant M, independent of h, such that

$$\|\mathbf{y} - \tilde{\mathbf{y}}\|_\infty \le M\left[\max\{|y_0 - \tilde{y}_0|, |y_N - \tilde{y}_N|\} + \|\mathcal{L}_h[\mathbf{y}] - \mathcal{L}_h[\tilde{\mathbf{y}}]\|_\infty\right].$$

- We say that the scheme (13.12) is **convergent** if the computed solution $\mathbf{y}$ satisfies

$$\lim_{N \to \infty} \max_{1 \le i \le N} |y(x_i) - y_i| = 0.$$

Because the finite difference formulas used are second-order accurate, it can be shown that if $y(x)$ is four times continuously differentiable on (a, b), then the scheme (13.11), (13.12) is consistent, with local truncation error that is $O(h^2)$.

Exploration 13.2.1 Find the local truncation error of (13.12).

Exploration 13.2.2 Use the Mean Value Theorem (Theorem A.5.3) to prove that the finite difference scheme (13.12) is stable with constant $M = \max\{1, 1/q\}$ if $h < 2/P$, where $|f_{y'}| \le P$ and $0 < q \le f_y$.

Exploration 13.2.3 Prove that if the finite difference scheme (13.12) is consistent and stable, then it is convergent.

13.2.1 Linear Problems

We first consider the case in which the boundary value problem includes a linear ODE of the form (13.3). Then, the system of equations (13.12) is also linear, and can therefore be expressed in matrix-vector form $A\mathbf{y} = \mathbf{r}$, where A is a tridiagonal matrix, since the approximations of y' and y'' at x_i only use y_{i-1}, y_i, and y_{i+1}, and $\mathbf{r}$ is a vector that includes the values of $r(x)$ at the grid points, as well as additional terms that account for the boundary conditions.

Specifically,

$$a_{ii} = 2 + h^2 q(x_i), \quad i = 1, 2, \ldots, N,$$
$$a_{i,i+1} = -1 + \frac{h}{2} p(x_i), \quad i = 1, 2, \ldots, N-1,$$
$$a_{i,i-1} = -1 - \frac{h}{2} p(x_i), \quad i = 2, 3, \ldots, N,$$
$$r_1 = -h^2 r(x_1) + \left(1 + \frac{h}{2} p(x_1)\right) \alpha,$$
$$r_i = -h^2 r(x_i), \quad i = 2, 3, \ldots, N-1,$$
$$r_N = -h^2 r(x_N) + \left(1 - \frac{h}{2} p(x_N)\right) \beta.$$

This system of equations is guaranteed to have a unique solution if A is strictly row diagonally dominant (see Section 5.1), which is the case if $q(x) \geq 0$ and $h < 2/P$, where P is an upper bound on $|p(x)|$.

Example 13.2.2 We revisit the BVP (13.7). The following script uses the function fdbvp (see Exploration 13.2.4) to solve this problem with $N = 10$ interior grid points, and then visualize the exact and approximate solutions, as well as the error.

```
% coefficients
p=@(x)(2*ones(size(x))); q=@(x)(-ones(size(x))); r=@(x)(x.*exp(x)-x);
% boundary conditions
a=0; b=2;
alpha=0; beta=-4;
% number of interior grid points
N=10;
% solve using finite differences
[x,y]=fdbvp(p,q,r,a,b,alpha,beta,N);
% exact solution: y = x^3 e^x/6 - 5xe^x/3 + 2e^x - x - 2
yexact=x.^3.*exp(x)/6-5*x.*exp(x)/3+2*exp(x)-x-2;
% plot exact and approximate solutions for comparison
subplot(121)
plot(x,y,'b-d')
hold on
plot(x,y,'r--+')
hold off
xlabel('x')
ylabel('y')
subplot(122)
plot(x,abs(yexact-y))
xlabel('x')
ylabel('error')
```

The plots are shown in Figure 13.2. We can see that the exact and computed solutions are visually indistinguishable from one another. □

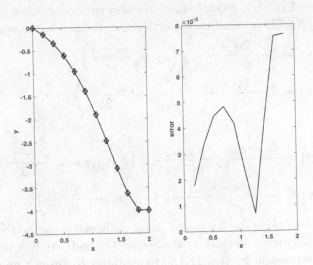

Fig. 13.2 Left plot: exact (solid curve with circles) and approximate (dashed curve with crosses) solutions of the BVP (13.7) computed using finite differences in Example 13.2.2. Right plot: error in the approximate solution.

Exploration 13.2.4 Write a MATLAB function

$$[x,y]=\texttt{fdbvp(p,q,r,a,b,alpha,beta,N)}$$

that solves the linear BVP (13.3), (13.2). The input arguments p, q and r are function handles that represent the functions $p(x)$, $q(x)$ and $r(x)$, respectively. The output argument x is a vector of $N+2$ equally spaced points in $[a, b]$, including boundary points, and the output argument y consists of the values of the computed solution at these points. Test your function on the BVP from Example 13.2.2 for different values of N. How does the error behave as N increases? Specifically, what is the order of accuracy? Does this value match the theoretical expectation? *Hint:* Use the diag function to set up the matrix A.

Exploration 13.2.5 After evaluating the coefficients $p(x)$, $q(x)$ and $r(x)$ from (13.3) at the grid points x_i, $i = 1, 2, \ldots, N$, how many floating-point operations are necessary to solve the system $Ay = r$? If the boundary conditions are changed but the ODE remains the same, how many additional floating-point operations are needed? *Hint:* Review the material in Section 3.3.1 on the solution of banded systems.

13.2.2 Nonlinear Problems

If the ODE is nonlinear, then we solve a system of nonlinear equations of the form $\mathbf{F}(\mathbf{y}) = \mathbf{0}$, where $\mathbf{F}(\mathbf{y})$ is a vector-valued function with coordinate functions $f_i(\mathbf{y})$, for $i = 1, 2, \ldots, N$. These coordinate functions are defined as follows:

$$F_1(\mathbf{y}) = \frac{y_2 - 2y_1 + \alpha}{h^2} - f\left(x_1, y_1, \frac{y_2 - \alpha}{2h}\right),$$

$$F_2(\mathbf{y}) = \frac{y_3 - 2y_2 + y_1}{h^2} - f\left(x_2, y_2, \frac{y_3 - y_1}{2h}\right),$$

$$\vdots$$

$$F_{N-1}(\mathbf{y}) = \frac{y_N - 2y_{N-1} + y_{N-2}}{h^2} - f\left(x_{N-1}, y_{N-1}, \frac{y_N - y_{N-2}}{2h}\right),$$

$$F_N(\mathbf{y}) = \frac{\beta - 2y_N + y_{N-1}}{h^2} - f\left(x_N, y_N, \frac{\beta - y_{N-1}}{2h}\right).$$

(13.13)

This system of equations can be solved using an iterative method such as Fixed-Point Iteration, Newton's Method, or the Secant Method. We first use Fixed-Point Iteration, as described in Section 11.1.1, to establish conditions under which this system can be solved.

While the Fixed-Point Iteration described in Exploration 13.2.7 converges, it does so slowly. Therefore, we consider Newton's Method, as described in Section 11.1.2, as an alternative. By the Chain Rule, the entries of the Jacobian matrix $J_{\mathbf{F}}(\mathbf{y})$, a tridiagonal matrix, are defined as follows:

$$J_{\mathbf{F}}(\mathbf{y})_{ii} = \frac{\partial f_i}{\partial y_i}(\mathbf{y})$$

$$= -\frac{2}{h^2} - f_y\left(x_i, y_i, \frac{y_{i+1} - y_{i-1}}{2h}\right), \quad i = 1, 2, \ldots, N,$$

$$J_{\mathbf{F}}(\mathbf{y})_{i,i+1} = \frac{\partial f_i}{\partial y_{i+1}}(\mathbf{y})$$

$$= \frac{1}{h^2} - \frac{1}{2h}f_{y'}\left(x_i, y_i, \frac{y_{i+1} - y_{i-1}}{2h}\right), \quad i = 1, 2, \ldots, N-1,\text{(13.14)}$$

$$J_{\mathbf{F}}(\mathbf{y})_{i,i-1} = \frac{\partial f_i}{\partial y_{i-1}}(\mathbf{y})$$

$$= \frac{1}{h^2} + \frac{1}{2h}f_{y'}\left(x_i, y_i, \frac{y_{i+1} - y_{i-1}}{2h}\right), \quad i = 2, 3, \ldots, N,$$

where, for convenience, we use $y_0 = \alpha$ and $y_{N+1} = \beta$. Then, during each iteration of Newton's Method, the system of equations

$$J_{\mathbf{F}}(\mathbf{y}^{(k)})\mathbf{s}_k = -\mathbf{F}(\mathbf{y}^{(k)})$$

is solved in order to obtain the next iterate

$$\mathbf{y}^{(k+1)} = \mathbf{y}^{(k)} + \mathbf{s}_k$$

from the previous iterate. An appropriate initial guess is the unique linear function that satisfies the boundary conditions,

$$\mathbf{y}^{(0)} = \alpha + \frac{\beta - \alpha}{b - a}(\mathbf{x} - a),$$

where $\mathbf{x}$ is the vector with coordinates $x_1, x_2, \ldots, x_N$.

Exploration 13.2.6 Derive the formulas (13.14) from (13.13).

Exploration 13.2.7 Suppose that we solve $\mathbf{F}(\mathbf{y}) = \mathbf{0}$, where $\mathbf{F}$ is defined by (13.13), using Fixed-Point Iteration to solve $\mathbf{y} = \mathbf{G}(\mathbf{y})$, where

$$\mathbf{G}(\mathbf{y}) = \frac{1}{1 + \mu}\left[(1 + \mu)\mathbf{y} + \frac{1}{2}\mathbf{F}(\mathbf{y})\right].$$

Show that Fixed-Point Iteration applied to this system of equations converges if

$$h < \frac{2}{P}, \quad \mu > \frac{h^2}{2}Q,$$

where $|f_{y'}| \le P$ and $0 < f_y \le Q$. *Hint:* Recall that the condition (11.5) is sufficient for convergence.

Example 13.2.3 We solve the BVP (13.8) from Example 13.1.3, which has the exact solution $y(x) = 1/(1 + x)$. To solve this problem using finite differences, we apply Newton's Method to solve the equation $\mathbf{F}(\mathbf{y}) = \mathbf{0}$, where

$$F_1(\mathbf{y}) = \frac{y_2 - 2y_1 + \frac{1}{2}}{h^2} - \left(y_1^3 - y_1 \frac{y_2 - 1/2}{2h}\right),$$

$$F_i(\mathbf{y}) = \frac{y_{i+1} - 2y_i + y_{i-1}}{h^2} - \left(y_i^3 - y_i \frac{y_{i+1} - y_{i-1}}{2h}\right), \quad i = 2, 3, \ldots, N - 1,$$

$$F_N(\mathbf{y}) = \frac{\frac{1}{3} - 2y_N + y_{N-1}}{h^2} - \left(y_N^3 - y_N \frac{1/3 - y_{N-1}}{2h}\right).$$

The following MATLAB function can be used to evaluate $\mathbf{F}(\mathbf{y})$ for a general BVP of the form (13.1). Its arguments are assumed to be vectors of x- and y-values, *including* boundary values, along with a function handle for the right-hand side $f(x, y, y')$ of the ODE (13.1) and the spacing h.

```
% newtF: evaluates F(y) for solving ODE
% y'' = f(x,y,y') with Newton's Method
function F=newtF(x,y,f,h)
% use only interior x-values
xi=x(2:end-1);
% y_i
yi=y(2:end-1);
% y_{i+1}
```

```
yip1=y(3:end);
% y_{i-1}
yim1=y(1:end-2);
% centered difference approximation of y':
% (y_{i+1} - y_{i-1})/(2h)
ypi=(yip1-yim1)/(2*h);
% evaluate F(y)
F=(yip1-2*yi+yim1)/h^2-f(xi,yi,ypi);
```

Using $f_y = 3y^2 - y'$ and $f_{y'} = -y$, we obtain

$$J_\mathbf{F}(\mathbf{y})_{ii} = -\frac{2}{h^2} - \left(3y_i^2 - \frac{y_{i+1} - y_{i-1}}{2h}\right), \quad i = 1, 2, \ldots, N,$$

$$J_\mathbf{F}(\mathbf{y})_{i,i+1} = \frac{1}{h^2} + \frac{1}{2h}y_i, \quad i = 1, 2, \ldots, N - 1,$$

$$J_\mathbf{F}(\mathbf{y})_{i,i-1} = \frac{1}{h^2} - \frac{1}{2h}y_i, \quad i = 2, 3, \ldots, N.$$

A MATLAB function similar to `newtF` can be written to construct $J_\mathbf{F}(\mathbf{y})$ for a general ODE of the form (13.1). This is left to Exploration 13.2.8.

The script from Example 13.1.3 can be modified to call `fdnlbvp` (see Exploration 13.2.9) instead of `shootnewt` to compute an approximate solution. As with the Shooting Method, using the same tolerance, Newton's Method converges in just three iterations, and does so quadratically. □

Exploration 13.2.8 Write a MATLAB function

$$\texttt{J=newtJ(x,y,fy,fyp,h)}$$

that uses (13.14) to construct the Jacobian matrix $J_\mathbf{F}(\mathbf{y})$ for use with Newton's Method. The input arguments `x` and `y` contain x- and y-values, respectively, including boundary values. The input arguments `fy` and `fyp` are assumed to be function handles that implement $f_y(x, y, y')$ and $f_{y'}(x, y, y')$, respectively. Use `newtF` from Example 13.2.3 as a model.

Exploration 13.2.9 Write a MATLAB function

```
[x,y,niter]=fdnlbvp(f,fy,fyp,a,b,alpha,beta,N)
```

that solves the general nonlinear BVP (13.1), (13.2) using finite differences in conjunction with Newton's Method. The input arguments `f`, `fy` and `fyp` are function handles for the functions f, f_y and $f_{y'}$, respectively. The output argument `x` is a vector of $N + 2$ equally spaced points in $[a, b]$, including boundary points, and the output argument `y` consists of the values of the computed solution at these points. The output argument `niter` is the number of iterations of Newton's Method required to achieve convergence. Use `newtF` from Example 13.2.3 and `newtJ` from Exploration 13.2.8 as helper functions. Test your function on the BVP from Example 13.2.3. What happens to the error as `N`, the number of interior grid points, increases? How does the accuracy compare to the methods implemented in Explorations 13.1.5 and 13.1.6? Explain your observations.

It is worth noting that for two-point boundary value problems that are discretized by finite differences, it is much more practical to use Newton's Method, as opposed to a quasi-Newton Method such as the Secant Method or Broyden's Method, than for a general system of nonlinear equations because the Jacobian matrix is tridiagonal. This reduces the expense of the computation of s_k from $O(N^3)$ operations in the general case to only $O(N)$ for two-point boundary value problems.

Exploration 13.2.10 Modify your function `fdnlbvp` from Exploration 13.2.9 to use Broyden's Method instead of Newton's Method. How does this affect the efficiency, when applied to the BVP from Example 13.2.3?

Exploration 13.2.11 Although Newton's Method is much more efficient for a BVP than for a general system of nonlinear equations, what is an advantage of using the Secant Method over Newton's Method or Broyden's Method?

It can be shown that regardless of the choice of iterative method used to solve the system of equations arising from discretization, the local truncation error of the finite difference method for nonlinear problems is $O(h^2)$, as in the linear case. The order of accuracy can be increased by applying Richardson Extrapolation (see Section 9.5).

13.2.3 Concept Check

1. How are finite difference approximations used to obtain an approximate solution of a two-point BVP?
2. What is the structure of the matrix A that arises from approximating a linear two-point BVP with a system of linear equations?

3. Describe the approach to solving a nonlinear two-point BVP using finite differences.

4. Using big-O notation, express the number of floating-point operations needed to solve a two-point BVP using finite differences. Assume N grid points are used.

5. Why is the solution of a nonlinear two-point BVP using finite differences with N grid points not as computationally expensive as the solution of a general system of N nonlinear equations?

6. What is the order of accuracy of finite difference methods for solving two-point BVPs, in terms of the grid spacing h?

13.3   Collocation*

While the finite-difference approach from Section 13.2 is generally effective for two-point boundary value problems, and is more flexible than the Shooting Method as it can be applied to higher-dimensional BVPs, it does have its drawbacks.

- First, the accuracy of finite difference approximations relies on the existence of the higher-order derivatives that appear in their error formulas. Unfortunately, the existence of these higher-order derivatives is not assured.
- Second, a matrix obtained from a finite-difference approximation can be ill-conditioned (see Section 3.4.1), and this conditioning worsens as the spacing h decreases.
- Third, it is best suited for problems in which the domain is relatively simple, such as a rectangular domain.

We now consider an alternative approach that, in higher dimensions, is more readily applied to problems on domains with complicated geometries.

First, we assume that the solution $y(x)$ is approximated by a function $y_N(x)$ that is a **linear combination** (see Section B.4) of chosen linearly independent functions $\phi_1(x), \phi_2(x), \ldots, \phi_N(x)$, called **basis functions** as they form a basis for an N-dimensional vector space. We then have

$$y_N(x) = \sum_{i=1}^{N} c_i \phi_i(x), \tag{13.15}$$

where the constants $c_1, c_2, \ldots, c_N$ are unknown. Substituting this form of the solution into (13.1), (13.2) yields the equations

$$\sum_{j=1}^{N} c_j \phi_j''(x) = f\left(x, \sum_{j=1}^{N} c_j \phi_j(x), \sum_{j=1}^{N} c_j \phi_j'(x)\right), \quad a < x < b, \tag{13.16}$$

$$\sum_{j=1}^{N} c_j \phi_j(a) = \alpha, \quad \sum_{j=1}^{N} c_j \phi_i(b) = \beta. \tag{13.17}$$

Already, the convenience of this assumption is apparent: instead of solving for a function $y(x)$ on the interval (a, b), we are instead having to solve for the N coefficients $c_1, c_2, \ldots, c_N$. However, it is not realistic to think that there is any choice of these coefficients that satisfies (13.16) on the *entire* interval (a, b), as well as the boundary conditions (13.17). Rather, we need to impose N conditions on these N unknowns, in the hope that the resulting system of N equations will have a unique solution that yields an accurate approximation of the exact solution $y(x)$.

To that end, we require that (13.16) is satisfied at $N-2$ points in (a, b), denoted by $x_1, x_2, \ldots, x_{N-2}$, and that the boundary conditions (13.17) are satisfied. The points $a = x_0, x_1, x_2, \ldots, x_{N-2}, x_{N-1} = b$ are called **collocation points**. This approach of approximating $y(x)$ by imposing (13.15) and solving the system of N equations given by

$$\sum_{j=1}^{N} c_j \phi_j''(x_i) = f\left(x_i, \sum_{j=1}^{N} c_j \phi_j(x_i), \sum_{j=1}^{N} c_j \phi_j'(x_i)\right), \quad i = 1, 2, \ldots, N-2, \quad (13.18)$$

and (13.17), is called **collocation**.

For simplicity, we assume that the BVP (13.1) is linear. We are then solving a problem of the form

$$y''(x) = p(x)y'(x) + q(x)y(x) + r(x), \quad a < x < b. \tag{13.19}$$

From (13.18), we obtain the system of linear equations

$$\sum_{j=1}^{N} c_j \phi_j''(x_i) = r(x_i) + \sum_{j=1}^{N} c_j q(x_i)\phi_j(x_i) + \sum_{j=1}^{N} c_j p(x_i)\phi_j'(x_i), \quad i = 1, 2, \ldots, N-2,$$

$$\tag{13.20}$$

along with (13.17). This system can be written in the form $A\mathbf{c} = \mathbf{b}$, where $\mathbf{c} = \begin{bmatrix} c_1 & \cdots & c_N \end{bmatrix}^T$. We can then solve this system using any of the methods from Chapter 3.

Exploration 13.3.1 Express the system of linear equations (13.20), (13.17) in the form $A\mathbf{c} = \mathbf{b}$, where $\mathbf{c}$ is defined as above. What are the entries a_{ij} and b_i of the matrix A and right-hand side vector $\mathbf{b}$, respectively?

Example 13.3.1 Consider the BVP

$$y'' = x^2, \quad 0 < x < 1, \tag{13.21}$$

$$y(0) = 0, \quad y(0) = 1. \tag{13.22}$$

We assume that our approximation of $y(x)$ has the form

$$y_4(x) = c_1 + c_2 x + c_3 x^2 + c_4 x^3.$$

That is, $N = 4$, since we are assuming that $y(x)$ is a linear combination of the four functions $1, x, x^2$ and x^3. Substituting this form into the BVP yields

$$2c_3 + 6c_4 x_i = x_i^2, \quad i = 1, 2,$$

$$c_1 = 0, \quad c_1 + c_2 + c_3 + c_4 = 1.$$

Writing this system of equations in matrix-vector form, we obtain

$$\begin{bmatrix} 1 & 0 & 0 & 0 \\ 0 & 0 & 2 & 6x_1 \\ 0 & 0 & 2 & 6x_2 \\ 1 & 1 & 1 & 1 \end{bmatrix} \begin{bmatrix} c_1 \\ c_2 \\ c_3 \\ c_4 \end{bmatrix} = \begin{bmatrix} 0 \\ x_1^2 \\ x_2^2 \\ 1 \end{bmatrix}. \tag{13.23}$$

For the system to be specified completely, we need to choose the two collocation points $x_1, x_2 \in (0, 1)$. As long as these points are chosen to be distinct, the matrix of the system will be nonsingular. For this example, we choose $x_1 = 1/3$ and $x_2 = 2/3$. We then have the system

$$\begin{bmatrix} 1 & 0 & 0 & 0 \\ 0 & 0 & 2 & 2 \\ 0 & 0 & 2 & 4 \\ 1 & 1 & 1 & 1 \end{bmatrix} \begin{bmatrix} c_1 \\ c_2 \\ c_3 \\ c_4 \end{bmatrix} = \begin{bmatrix} 0 \\ 1/9 \\ 4/9 \\ 1 \end{bmatrix}. \tag{13.24}$$

We can now solve this system in MATLAB:

```
>> x=[ 1/3 2/3 ];
>> A=[ 1 0 0 0;
       0 0 2 6*x(1);
       0 0 2 6*x(2);
       1 1 1 1 ];
>> b=[ 0; x(1)^2; x(2)^2; 1 ];
>> format rat
>> c=A\b
c =

     0
     17/18
    -1/9
     1/6
```

The `format rat` statement was used to obtain exact values of the entries of $\mathbf{c}$, since these entries are guaranteed to be rational numbers in this case. It follows that our approximate solution $y_N(x)$ is

$$y_4(x) = \frac{17}{18}x - \frac{1}{9}x^2 + \frac{1}{6}x^3.$$

The exact solution of the original BVP is easily obtained by integration:

$$y(x) = \frac{1}{12}x^4 + \frac{11}{12}x.$$

From these formulas, though, it is not easy to gauge how accurate $y_4(x)$ is. To visualize the error, we plot both solutions:

```
>> xp=0:0.01:1;
>> y4p=polyval(c(end:-1:1)',xp);
>> yp=polyval([ 1/12 0 0 11/12 0 ],xp);
>> plot(xp,yp,'b-d')
>> hold on
>> plot(xp,y4p,'r--+')
>> xlabel('x')
>> ylabel('y')
```

The result is shown in Figure 13.3. As we can see, even with only two collocation

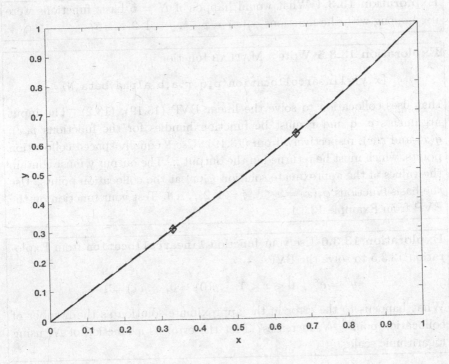

Fig. 13.3 Exact (solid curve with diamonds) and approximate (dashed curve with crosses) solutions of (13.21), (13.22) from Example 13.3.1. The collocation points $x_1 = 1/3$, $x_2 = 2/3$ are indicated by circles.

points, this approximate solution barely distinguishable from the exact solution on a plot. To get a numerical indication of the accuracy, we can measure the error at the points in xp that were used for plotting:

```
>> format short
>> norm(yp-y4p,'inf')
ans =
```

0.0023

☐

Exploration 13.3.2 Solve the BVP from Example 13.3.1 again, but with different collocation points $x_1, x_2 \in (0, 1)$. What happens to the error?

Exploration 13.3.3 Use MATLAB to compute the *relative* error in the ∞-norm and L^2-norm from the preceding example.

Exploration 13.3.4 What would happen if $N = 5$ basis functions were used, along with the functions $\phi_j(x) = x^{j-1}$, $j = 1, 2, \ldots, 5$?

Exploration 13.3.5 Write a MATLAB function

$$\texttt{[x,y]=linearcollocation(p,q,r,a,b,alpha,beta,N)}$$

that uses collocation to solve the linear BVP (13.19), (13.2). The input arguments `p`, `q` and `r` must be function handles for the functions $p(x)$, $q(x)$ and $r(x)$, respectively, from (13.19). Use `N` equally spaced collocation points, which must be returned in the output `x`. The output `y` must contain the values of the approximate solution $y_N(x)$ at the collocation points. Use the basis functions $\phi_j(x) = x^{j-1}$, $j = 1, 2, \ldots, N$. Test your function on the BVP from Example 13.3.1.

Exploration 13.3.6 Use your function `linearcollocation` from Exploration 13.3.5 to solve the BVP

$$y'' = e^x, \quad 0 < x < 1, \quad y(0) = 0, \quad y(1) = 1.$$

What happens to the error in the approximate solution as the number of collocation points, N, increases? Plot the error as a function of N, using logarithmic scales.

The choice of basis functions $\phi_j(x)$, $j = 1, 2, \ldots, N$, can significantly affect the process of solving the resulting system of equations. The choice used in Example 13.3.1, $\phi_j(x) = x^{j-1}$, while convenient, is not a good choice, especially when N is large. As shown in Section 8.2, these functions can be nearly linearly dependent on the interval $[a, b]$, which can lead to ill-conditioned systems.

Exploration 13.3.7 What happens to the condition number of the matrix used by your function `linearcollocation` from Exploration 13.3.5 as N increases?

Alternatives include orthogonal polynomials, such as Chebyshev polynomials, or trigonometric polynomials, as discussed in Section 8.4.

Exploration 13.3.8 Modify your function `linearcollocation` from Exploration 13.3.5 to use Chebyshev polynomials instead of the monomial basis, and the Chebyshev points as the collocation points instead of equally spaced points. What happens to the condition number of the matrix as N increases?

Collocation can be used for either linear or nonlinear BVPs. In the nonlinear case, choosing the functions $\phi_j(x)$, $j = 1, 2, \ldots, N$, and the collocation points x_i, $i = 0, 1, \ldots, N - 1$, yields a system of nonlinear equations for the unknowns $c_1, c_2, \ldots, c_N$. This system can then be solved using any of the techniques from Section 11.1, just as when using finite differences.

Exploration 13.3.9 Describe the system of nonlinear equations $\mathbf{F}(\mathbf{c}) = \mathbf{0}$ that must be solved at each iteration when using Newton's Method to solve a general nonlinear BVP of the form (13.1), (13.2). What is the Jacobian of $\mathbf{F}$, $J_{\mathbf{F}}(\mathbf{c})$?

Exploration 13.3.10 Write a MATLAB function

$$[\texttt{x,y}]=\texttt{nonlinearcollocation(f,a,b,alpha,beta,N)}$$

that solves a BVP of the form (13.1), (13.2) using Newton's Method. The input argument `f` is be a function handle for $f(x, y, y')$ from (13.1), and `N` is the number of collocation points. Use the Chebyshev points as the collocation points, and the Chebyshev polynomials as the basis functions. For the initial guess, use the coefficients corresponding to the unique linear function that satisfies the boundary conditions (13.2). Test your function on the BVP $\bar{y}'' = y^2$, $y(1) = 6$, $y(2) = 3/2$. What is the exact solution?

13.3.1 Concept Check

1. List three drawbacks of finite difference methods for two-point BVPs.
2. What does it mean to express a function as a linear combination of basis functions?
3. Describe how collocation is used to solve a BVP.
4. What are good or poor choices of basis functions for collocation?
5. Explain how collocation can be used to solve nonlinear two-point BVPs.

13.4 The Finite Element Method*

In collocation, the approximate solution $y_N(x)$ is defined to be an element of an N-dimensional function space, which is the *span* (see Section B.4) of the chosen basis functions $\phi_j(x)$, $j = 1, 2, \ldots, N$. In this section, we describe another method for solving a BVP in which the approximate solution is again restricted to an N-

dimensional function space, but instead of requiring the **residual**

$$R(x, y_N, y'_N, y''_N) \equiv y''_N - f(x, y_N, y'_N)$$

to vanish at selected points in (a, b), as in collocation, we require that the residual is *orthogonal* to a given function space, consisting of functions called **test functions**. That is, we require the residual to be zero in an "average" sense, rather than a pointwise sense. In fact, this approach is called the **weighted mean residual method**.

For concreteness and simplicity, we consider the linear boundary value problem

$$-u''(x) = f(x), \quad 0 < x < 1, \tag{13.25}$$

with boundary conditions

$$u(0) = 0, \quad u(1) = 0. \tag{13.26}$$

This equation can be used to model, for example, transverse vibration of a string due to an external force $f(x)$, or longitudinal displacement of a beam subject to a load $f(x)$. In either case, the boundary conditions prescribe that the endpoints of the object in question are fixed.

13.4.1   Derivation of the Weak Form

If we multiply both sides of (13.25) by a **test function** $w(x)$, and then integrate over the domain $[0, 1]$, we obtain

$$\int_0^1 -w(x)u''(x)\, dx = \int_0^1 w(x)f(x)\, dx.$$

Applying integration by parts, we obtain

$$\int_0^1 w(x)u''(x)\, dx = w(x)u'(x)\big|_0^1 - \int_0^1 w'(x)u'(x)\, dx.$$

As in Section B.13.3, let $H^1(0, 1)$ be the space of all functions that are absolutely continuous on $[0, 1]$ and have derivatives in $L^2(0, 1)$, and let $H_0^1(0, 1)$ be the space of all functions in $H^1(0, 1)$ that are equal to zero at the endpoints $x = 0$ and $x = 1$. If we require that our test function $w(x)$ belongs to $H_0^1(0, 1)$, then $w(0) = w(1) = 0$, and the boundary term in the above application of integration by parts vanishes. We then have

$$\int_0^1 w'(x)u'(x)\, dx = \int_0^1 w(x)f(x)\, dx.$$

This is called the **weak form** of the boundary value problem (13.25), (13.26), which is known as the **strong form** or **classical form**. The weak form only requires that the *first* derivative of $u(x)$ exist, as opposed to the original boundary value problem, or strong form, that requires the existence of the *second* derivative. The weak form also known as the **variational form**.

13.4.2 Spatial Discretization

To find an approximate solution of the weak form, we restrict ourselves to an N-dimensional *subspace* (see Section B.3) of $H_0^1(0,1)$, denoted by V_N, by requiring that the approximate solution $u_N(x)$ satisfies

$$u_N(x) = \sum_{j=1}^{N} c_j \phi_j(x), \qquad (13.27)$$

where the **trial functions** $\phi_1, \phi_2, \ldots, \phi_n$ form a *basis* for V_N. For now, we only assume that these trial functions belong to $H_0^1(0,1)$, and are linearly independent. Substituting this form into the weak form yields

$$\sum_{j=1}^{N} \left[\int_0^1 w'(x)\phi_j'(x)\, dx \right] c_j = \int_0^1 w(x)f(x)\, dx.$$

Since our trial functions and test functions come from the same space, this version of the weighted mean residual method is known as the **Galerkin Method**. As in collocation, we need N equations to uniquely determine the N unknowns $c_1, c_2, \ldots, c_N$. To that end, we use the basis functions $\phi_1, \phi_2, \ldots, \phi_N$ as test functions. This yields the system of equations

$$\sum_{j=1}^{N} \left[\int_0^1 \phi_i'(x)\phi_j'(x)\, dx \right] c_j = \int_0^1 \phi_i(x)f(x)\, dx, \quad i = 1, 2, \ldots, N. \qquad (13.28)$$

This system can be written in matrix-vector form

$$\mathbf{Ac} = \mathbf{f}$$

where $\mathbf{c}$ is a vector of the unknown coefficients $c_1, c_2, \ldots, c_N$ and

$$a_{ij} = \int_0^1 \phi_i'(x)\phi_j'(x)\, dx, \quad i, j = 1, 2, \ldots, N,$$

$$f_i = \int_0^1 \phi_i(x)f(x)\, dx, \quad i = 1, 2, \ldots, N.$$

By finding the coefficients $u_1, u_2, \ldots, u_N$ that satisfy (13.28), we ensure that the **residual** of the BVP, given by $R(x, u_N, u_N', u_N'') = u_N''(x) + f(x)$, satisfies

$$\langle w, R \rangle = -\int_0^1 w'(x)u_N'(x)\, dx + \int_0^1 w(x)f(x)\, dx = 0, \quad w \in V_N,$$

where $\langle \cdot, \cdot \rangle$ is the inner product of real-valued functions on $(0,1)$, as defined in Section B.15. That is, that the residual is orthogonal to V_N.

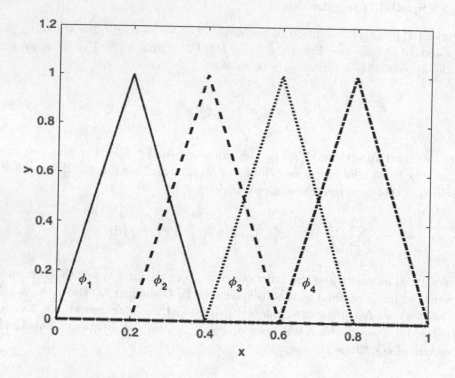

Fig. 13.4   Piecewise linear basis functions $\phi_j(x)$, as defined in (13.29), for $j = 1, 2, 3, 4$, with $N = 4$

13.4.3 Piecewise Linear Basis Functions

We must now choose trial (and test) functions $\phi_1, \phi_2, \ldots, \phi_N$. A simple choice is a set of piecewise linear "hat" functions, or "tent" functions, so named because of the shapes of their graphs, which are illustrated in Figure 13.4. We divide the interval $[0, 1]$ into $N + 1$ subintervals $[x_{i-1}, x_i]$, for $i = 1, 2, \ldots, N + 1$, with uniform spacing $h = 1/(N + 1)$, where $x_0 = 0$ and $x_{N+1} = 1$. Then we define

$$\phi_j(x) = \begin{cases} 0 & 0 \leq x \leq x_{j-1} \\ \frac{1}{h}(x - x_{j-1}) & x_{j-1} < x \leq x_j \\ \frac{1}{h}(x_{j+1} - x) & x_j < x \leq x_{j+1} \\ 0 & x_{j+1} < x \leq 1 \end{cases}, \quad j = 1, 2, \ldots, N. \tag{13.29}$$

These functions automatically satisfy the boundary conditions. Because they are only piecewise linear, their derivatives are piecewise constant. They are

$$\phi_j'(x) = \begin{cases} 0 & 0 \leq x \leq x_{j-1} \\ \frac{1}{h} & x_{j-1} < x \leq x_j \\ -\frac{1}{h} & x_j < x \leq x_{j+1} \\ 0 & x_{j+1} < x \leq 1 \end{cases}, \quad j = 1, 2, \ldots, N.$$

It follows from these definitions that $\phi_i(x)$ and $\phi_j(x)$ cannot simultaneously be nonzero at any point in $[0, 1]$ unless $|i - j| \leq 1$. This yields a symmetric tridiagonal matrix A, called the **stiffness matrix**, with entries

$$a_{ii} = \left(\frac{1}{h}\right)^2 \int_{x_{i-1}}^{x_i} 1 \, dx + \left(-\frac{1}{h}\right)^2 \int_{x_i}^{x_{i+1}} 1 \, dx = \frac{2}{h}, \quad i = 1, 2, \ldots, N,$$

$$a_{i,i+1} = -\frac{1}{h^2} \int_{x_i}^{x_{i+1}} 1 \, dx = -\frac{1}{h}, \quad i = 1, 2, \ldots, N - 1,$$

$$a_{i+1,i} = a_{i,i+1}, \quad i = 1, 2, \ldots, N - 1.$$

For the right-hand side vector $\mathbf{f}$, known as the **load vector**, we have

$$f_i = \frac{1}{h} \int_{x_{i-1}}^{x_i} (x - x_{i-1}) f(x) \, dx + \frac{1}{h} \int_{x_i}^{x_{i+1}} (x_{i+1} - x) f(x) \, dx, \quad i = 1, 2, \ldots, N.$$

$$(13.30)$$

When the Galerkin Method is used with basis functions such as these, that are only nonzero within a small portion of the spatial domain, the method is known as the **Finite Element Method**. In this context, the subintervals $[x_{i-1}, x_i]$ are called **elements**, and each x_i is called a **node**. As we have seen, an advantage of this choice of trial function is that the resulting matrix A is sparse.

13.4.4 Solution of the Linear System

It will be shown in Chapter 14 that the matrix A with entries defined from these approximate integrals is not only symmetric and tridiagonal, but also positive definite. It follows that the system $A\mathbf{c} = \mathbf{f}$ is stable with respect to roundoff error, and can be solved using methods such as the Conjugate Gradient Method (see Section 5.2.3) that are appropriate for sparse symmetric positive definite systems.

Example 13.4.1 We illustrate the Finite Element Method by solving (13.25), (13.26) with $f(x) = x$, with $N = 4$. The following MATLAB commands are used to help specify the problem.

```
>> % solve -u'' = f on (0,1), u(0)=u(1)=0, f polynomial
>> % represent f(x) = x as a polynomial
>> fx=[ 1 0 ];
>> % set number of interior nodes
>> N=4;
>> h=1/(N+1);
>> % compute vector containing all nodes, including boundary nodes
>> x=h*(0:N+1)';
```

This vector x of nodes will be convenient when constructing the load vector $\mathbf{f}$ and performing other tasks.

We need to solve the system $A\mathbf{c} = \mathbf{f}$, where

$$A = \frac{1}{h} \begin{bmatrix} 2 & -1 & 0 & 0 \\ -1 & 2 & -1 & 0 \\ 0 & -1 & 2 & -1 \\ 0 & 0 & -1 & 2 \end{bmatrix}, \quad \mathbf{c} = \begin{bmatrix} c_1 \\ c_2 \\ c_3 \\ c_4 \end{bmatrix},$$

with $h = 1/5$. The following MATLAB commands set up the stiffness matrix for a general value of N.

```
>> % construct stiffness matrix:
>> e=ones(N-1,1);
>> % use diag to place entries on subdiagonal and superdiagonal
>> A=1/h*(2*eye(N)-diag(e,1)-diag(e,-1));
```

The load vector $\mathbf{f}$ has elements

$$f_i = \frac{1}{h} \int_{x_{i-1}}^{x_i} (x - x_{i-1}) x \, dx + \frac{1}{h} \int_{x_i}^{x_{i+1}} (x_{i+1} - x) x \, dx, \quad i = 1, 2, \ldots, N.$$

The following statements compute these elements when f is a polynomial.

```
% construct load vector:
% pre-allocate column vector
f=zeros(N,1);
for i=1:N
    % note that in text, 0-based indexing is used
    % for x-values, while Matlab uses 1-based indexing
    % phi_{i-1}(x) = (x - x_{i-1})/h
    hat1=[ 1 -x(i) ]/h;
    % multiply hat function by f(x)
    integrand1=conv(fx,hat1);
    % anti-differentiate
    antideriv1=polyint(integrand1);
    % substitute limits of integration into antiderivative
    integral1=polyval(antideriv1,x(i+1))-polyval(antideriv1,x(i));
    % phi_i(x) = (x_{i+1} - x)/h
    hat2=[ -1 x(i+2) ]/h;
    % repeat integration process on [x_i,x_{i+1}]
    integrand2=conv(fx,hat2);
    % anti-differentiate
    antideriv2=polyint(integrand2);
    % substitute limits of integration into antiderivative
    integral2=polyval(antideriv2,x(i+2))-polyval(antideriv2,x(i+1));
    f(i)=integral1+integral2;
end
```

Now that the system $Ac = \mathbf{f}$ is set up, we can solve it in MATLAB using the command `c=A\f`.

For this BVP, we can obtain the exact solution analytically, to determine the accuracy of our computed solution. The following statements accomplish this for the BVP $-u'' = f$ on $(0,1)$, $u(x_0) = u(x_{N+1}) = 1$, when f is a polynomial.

```
% obtain exact solution by integration
u=-polyint(polyint(fx));
% to solve for constants of integration:
% u + d1 x + d2 = 0 at x = x_0, x_{N+1}
% in matrix-vector form:
% [ x_0     1 ] [ d1 ] = [ -u(x0)       ]
% [ x_{N+1} 1 ] [ d2 ]   [ -u(x_{N+1}) ]
V=[ x(1) 1; x(end) 1 ];
b=[ -polyval(u,x(1)); -polyval(u,x(end)) ];
d=V\b;
% modify two lowest-order coefficients of u
u(end-1:end)=d';
```

Now, we can visualize the exact solution $u(x)$ and approximate solution $u_N(x)$, which is a piecewise linear function due to the use of piecewise linear trial functions.

```
% make vector of x-values for plotting exact solution
xp=x(1):h/100:x(end);
plot(xp,polyval(u,xp),'b')
hold on
plot(x,[ 0; c; 0 ],'r--o')
hold off
xlabel('x')
ylabel('y')
```

Because each of the trial functions $\phi_j(x)$, $j = 1,2,3,4$, is equal to 1 at x_j and equal to 0 at x_i for $i \neq j$, each element c_j of $\mathbf{c}$ is equal to the value of $u_4(x)$ at the corresponding node x_j. The exact and approximate solutions are shown in Figure 13.5. It can be seen that there is very close agreement between the exact and approximate solutions at the nodes; in fact, in this example, they are exactly equal, though this does not occur in general. $\square$

In the preceding example, the integrals in (13.30) could be evaluated exactly. Generally, however, they must be approximated, using techniques such as those presented in Chapter 9.

> **Exploration 13.4.1** What is f_i if the Trapezoidal Rule (9.15) is used on each of the integrals in (13.30)? What if Simpson's Rule (9.16) is used?

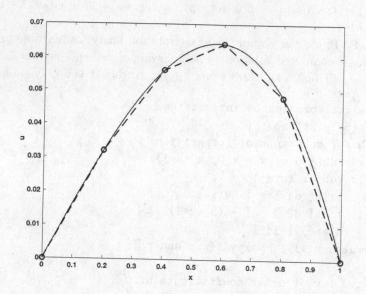

Fig. 13.5   Exact (solid curve) and approximate (dashed curve) solutions of (13.25), (13.26) from Example 13.4.1, with $f(x) = x$ and $N = 4$

Exploration 13.4.2 Write a MATLAB function `[x,u]=fembvp(f,N)` that solves the BVP (13.25), (13.26) with N *interior* nodes. The input argument `f` is a function handle. The outputs `x` and `u` are column vectors containing the nodes and values of the approximate solution at the nodes, respectively. Use the Trapezoidal rule to approximate the integrals (13.30). Test your function with $f(x) = e^x$. What happens to the error as N increases?

Exploration 13.4.3 Generalize your function `fembvp` from Exploration 13.4.2 so that it can be used to solve the BVP $-u'' + q(x)u = f(x)$ on $(0, 1)$, with boundary conditions $u(0) = u(1) = 0$, for a given function $q(x)$ that must be passed as an input argument that is a function handle. *Hint:* Re-derive the weak form of the BVP to determine how the matrix A must be modified. Use the Trapezoidal Rule to approximate any integrals.

Exploration 13.4.4 Modify `fembvp` from Exploration 13.4.3 so that it can be used to solve the BVP $-u'' + q(x)u = f(x)$ on $(0, 1)$, with boundary conditions $u(0) = u_0$, $u(1) = u_1$, where u_0 and u_1 must be passed as input arguments. *Hint:* Modify (13.27) to include additional basis functions $\phi_0(x)$ and $\phi_{N+1}(x)$, that are equal to 1 at $x = x_0$ and $x = x_{N+1}$, respectively, and equal to 0 at all other nodes. How must the load vector $\mathbf{f}$ be modified to account for these nonhomogeneous boundary conditions?

> **Exploration 13.4.5** Modify your function `fembvp` from Exploration 13.4.4 so that it can be used to solve the BVP $-(p(x)u')' + q(x)u = f(x)$ on $(0,1)$, with boundary conditions $u(0) = u_0$, $u(1) = u_1$, where the coefficient $p(x)$ must be passed as an input argument that is a function handle. *Hint:* Re-derive the weak form of the BVP to determine how the matrix A must be modified. Use the Trapezoidal Rule to approximate any integrals involving $p(x)$.

It can be shown that when using the Finite Element Method with piecewise linear trial functions, the error in the approximate solution is $O(h^2)$. Higher-order accuracy can be achieved by using higher-degree piecewise polynomials as basis functions, such as cubic B-splines [Süli and Mayers (2003)]. Such a choice also helps to ensure that the approximate solution is differentiable, unlike the solution computed using piecewise linear basis functions, which is continuous but not differentiable at the points x_i, $i = 1, 2, \ldots, N$.

With cubic B-splines, the error in the computed solution is $O(h^4)$ as opposed to $O(h^2)$ in the piecewise linear case, due to the two additional degrees of differentiability. However, the drawback is that the matrix arising form the use of higher-degree basis functions is no longer tridiagonal; the upper and lower bandwidth (see Section 3.3.1) are each equal to the degree of the piecewise polynomial that is used.

13.4.5 Concept Check

1. What is the residual of a two-point BVP? Use the residual to explain the difference between collocation and the weighted mean residual method.
2. What is the weak form of a boundary value problem?
3. What are test and trial functions, and what purpose do they serve?
4. What is the Galerkin Method?
5. What are "hat functions"?
6. What is the Finite Element Method? What distinguishes it from other Galerkin Methods?
7. Define the following terms, in the context of the Finite Element Method:
   (a) element
   (b) node
   (c) stiffness matrix
   (d) load vector

13.5 Additional Resources

An early treatment of numerical methods for two-point BVPs can be found in [Keller (1968)]. The text [Ascher and Petzold (1998)], previously cited as a resource for IVPs, serves the same purpose for BVPs, particularly for the Shooting Method and finite difference methods. A more comprehensive resource specifically about BVPs is [Ascher, et al. (1995)]. For more on the Shooting Method and finite difference methods, the reader is referred to [Bailey et al. (1968); Fox (1990); Pereyra (1984); Roberts and Shipman (1972)]. Additional references for the Finite Element Method are [Eriksson, et al. (1996); Hughes (1987); Strang and Fix (1973)].

13.6   Exercises

1. Consider the linear two-point BVP

$$-(e^x y')' + e^x y = x + (2 - x)e^x, \quad 0 < x < 1,$$

$$y(0) = 0, \quad y(1) = 0.$$

(a) Find the exact solution. *Hint:* The Method of Undetermined Coefficients [Boyce and DiPrima (2013)] is helpful.

(b) Solve this BVP using finite differences, with the help of your function fdbvp from Exploration 13.2.4, for different values of N. What is the observed order of accuracy?

(c) Solve this BVP using linearcollocation from Exploration 13.3.5, for different values of N. What is the observed order of accuracy?

(d) Solve this BVP using the Finite Element Method, with the help of your function fembvp from Exploration 13.4.5, for different values of N. What is the observed order of accuracy?

(e) Of the three methods used, which one is the most efficient in terms of execution time? As N increases, how does the running time of these methods increase? Is the rate of increase in line with what we know about the complexity of these methods?

2. In Exploration 13.1.4 we obtained a guess for the initial slope used in the Shooting Method that is independent of the function $f(x, y, y')$. Suppose a single step of Euler's Method is taken to solve the IVP associated with the Shooting Method, with "time" step $h = b - a$. Set the result of this single step equal to β to solve for the initial slope, and modify your implementation from Exploration 13.1.5 to use this initial guess. Does this yield any improvement in efficiency over the initial guess from Exploration 13.1.4?

3. Consider the **time-independent Schrödinger equation** in 1-D,

$$-\psi''(x) + V(x)\psi(x) = E\psi(x),$$

where the solution $\psi(x)$ is a *wave function* of a particle with energy E, subject to a *potential function* $V(x)$. Then $|\psi(x)|^2$ can be interpreted as the probability of finding the particle at position x.

If the particle is confined to the interval $[0, 1]$, the probability of finding the particle outside the interval is zero, which leads to the two-point BVP

$$-\psi''(x) + V(x)\psi(x) = E\psi(x), \quad 0 < x < 1,$$

$$\psi(0) = \psi(1) = 0.$$

However, if we try to solve this BVP using finite differences, then instead of obtaining a system of linear equations of the form $A\psi = \mathbf{f}$, we obtain an *eigenvalue problem* $A\psi = E\psi$, where E is an eigenvalue of A and ψ is a corresponding eigenvector (see Section B.12). The eigenvalues indicate the permitted energy levels.

Let $V(x)$ be a step function

$$V(x) = \begin{cases} 0 & 0 < x < 1/2, \\ 1 & 1/2 \leq x < 1 \end{cases}$$

Derive the matrix A for this BVP using finite differences. How do we know that the eigenvalues must be real? Solve the resulting eigenvalue problem using the MATLAB function `eig` for various values of N, the number of grid points. How do the energy levels behave as N increases?

4. Consider the eigenvalue problem from [Min and Gottlieb (2003)],

$$-u''(x) = \lambda \epsilon(x) u, \quad x \in (-\pi, \pi),$$

where

$$\epsilon(x) = \begin{cases} 1 & -\pi < x < 0 \\ \beta^2 & 0 \leq x < \pi \end{cases}$$

with $\beta \neq 1$. Let $\beta = 2$ and use finite differences to obtain a *generalized eigenvalue problem* of the form

$$A\mathbf{u} = \lambda B\mathbf{u},$$

where $\mathbf{u}$ is an N-vector, and A and B are $N \times N$ matrices. Use the MATLAB function `eig` to compute the eigenvalues and eigenvectors of the resulting matrix, after viewing its help page to learn how to use `eig` for solving generalized eigenvalue problems. How do the eigenvalues and eigenvectors behave as N, the number of grid points, increases? Fast numerical methods for closely related eigenvalue problems were developed by the first author of this book and his students in [Garon and Lambers (2018); Long et al. (2018)].

5. Show that the two-point BVP

$$y'' = -y, \quad 0 < x < 2\pi, \quad y(0) = \alpha, \quad y(2\pi) = \beta$$

does not have a solution unless $\alpha = \beta$. Furthermore, show that if $\alpha = \beta = 0$, then the solution is not unique. What happens if we try to solve this problem numerically, in the case where $\alpha \neq \beta$? How does the existence condition change if the interval changes to $(0, \pi)$?

6. Consider the two-point BVP from Example 13.2.2. Suppose that finite differences are used to solve this problem with $N = 8$, $N = 16$ and $N = 32$. Use Richardson Extrapolation to obtain a more accurate approximation of $y(1)$. How does the error decrease after the maximum number of extrapolations are performed? Recall that the finite difference formulas used in Example 13.2.2 are second-order accurate.

7. Consider the two-point BVP

$$-\frac{d}{dx}\left(\kappa(x)\frac{dy}{dx}\right) + q(x)y(x) = r(x), \quad a < x < b,$$

$$y(a) = \alpha, \quad y(b) = \beta,$$

where $\kappa(x) > 0$ on (a, b), and not a constant function. We also assume $q(x) > 0$ on (a, b).

(a) Assume that κ is differentiable on (a, b), and rewrite the ODE in the form 13.3. Show that the matrix A in the linear system $A\mathbf{y} = \mathbf{r}$ that results from replacing derivatives with centered finite differences, as in Section 13.2.1, is not symmetric.

(b) Consider an alternative approach, in which dy/dx is approximated by a backward difference, and the differentiation of $\kappa(x)dy/dx$ is approximated by a forward difference. Show that the resulting matrix A in the linear system $A\mathbf{y} = \mathbf{r}$ is symmetric positive definite. *Hint:* Try expressing the discretization of $d/dx(\kappa(x)dy/dx)$ as a product of three matrices.

(c) Suppose that we instead use a **staggered grid**, and the approximation

$$\frac{d}{dx}\left(\kappa(x)\frac{dy}{dx}\right)\Bigg|_{x=x_i} \approx \frac{\kappa_{i+1/2}(u_x)_{i+1/2} - \kappa_{i-1/2}(u_x)_{i-1/2}}{h},$$

with

$$\kappa_{i+1/2} = \kappa\left(x_i + \frac{h}{2}\right),$$

for $i = 1, 2, \ldots, N$, where $(u_x)_{i+1/2} \approx u'(x_i + h/2)$ is approximated by a centered difference using u_{i+1} and u_i. Show that the resulting matrix A, while not the same as the matrix A from part 7b, is also symmetric positive definite.

(d) What is the advantage of the matrix A being symmetric positive definite, as opposed to the unsymmetric matrix obtained in part 7a?

8. Consider an ODE of the form (13.3).

(a) Find a function $\kappa(x)$ such that the modified ODE

$$\kappa(x)y'' + \kappa(x)p(x)y' + \kappa(x)q(x)y = \kappa(x)r(x)$$

can be reduced to

$$(\kappa(x)y')' + \tilde{q}(x)y = \tilde{r}(x), \quad \tilde{q}(x) = \kappa(x)q(x), \quad \tilde{r}(x) = \kappa(x)r(x),$$

and therefore can be discretized by a linear system with a symmetric positive definite matrix.

(b) Suppose that the function $\kappa(x)$ from part 8a cannot be evaluated exactly at each grid point x_i. How can it be evaluated approximately in such a way that the local truncation error is still $O(h^2)$?

Chapter 14

Partial Differential Equations

Partial differential equations, or **PDEs**, are equations involving partial derivatives of an unknown function (the solution) of several variables. The most general form of a PDE with solution $u(x_1, x_2, \ldots, x_n)$ is

$$F(x_1, x_2, \ldots, x_n, u, u_{x_1}, u_{x_2}, \ldots, u_{x_n}, u_{x_1 x_1}, \ldots) = 0 \qquad (14.1)$$

for some function F. Recall from multivariable calculus that partial derivatives can be denoted using subscripts; that is, $u_{x_1} = \partial u / \partial x_1$. The solution is the *dependent* variable u; it depends on the *independent* variables $x_1, x_2, \ldots, x_n$.

In this chapter, we will use our knowledge of numerical methods for the solution of initial value problems (IVPs) from Chapter 12 and boundary value problems (BVPs) from Chapter 13 to develop methods for the solution of PDEs, including higher-dimensional BVPs and **initial-boundary value problems (IBVPs)**. We begin in Section 14.1 with an overview of certain PDEs of particular interest. Sections 14.2, 14.3, and 14.4 cover the design, analysis and implementation of **finite difference methods** for linear, second-order PDEs of elliptic, parabolic, and hyperbolic type, respectively. Section 14.5 generalizes the **Finite Element Method**, originally presented in Section 13.4 for 1-D BVPs, to higher spatial dimensions. Finally, **spectral methods**, which can deliver extraordinarily high accuracy for certain PDEs, are presented in Section 14.6.

14.1 Fundamentals of PDEs

Physical laws (and other laws) can be expressed mathematically in the form of PDEs. Derivatives represent quantities such as velocity, acceleration, flux, and so on. Some of the most fundamental equations we will study in this chapter are:

- The **heat equation**: If we let $F(a, b, c, d, e, f, g) = e - \alpha^2(f + g)$, where α is a constant, then the PDE $F(x, y, t, u, u_t, u_{xx}, u_{yy}) = 0$ is

$$u_t = \alpha^2(u_{xx} + u_{yy}), \qquad (14.2)$$

  which is the heat equation for a 2-D domain. The solution u describes the temperature, and the coefficient α^2 is the thermal conductivity of the medium.

- The **wave equation**: If we let $F(a, b, d, e, f, g, h) = f - c^2(g + h)$, where c is a constant, then the PDE $F(x, y, t, u, u_{tt}, u_{xx}, u_{yy}) = 0$ is

$$u_{tt} = c^2(u_{xx} + u_{yy}), \tag{14.3}$$

which is the wave equation for a 2-D domain. The solution u describes the displacement of a vibrating string or membrane (e.g., a drum) and the constant c describes the speed of propagation through the medium in which the membrane vibrates.

- **Laplace's equation**: If we let $F(a, b, c, d, e) = d + e$, then the PDE $F(x, y, u, u_{xx}, u_{yy}) = 0$ is

$$u_{xx} + u_{yy} = 0. \tag{14.4}$$

This is Laplace's equation for a 2-D domain; it generalizes in a natural way to other dimensions. Its solution can describe, among other things, a **steady-state** solution to the heat equation or wave equation, for which derivatives with respect to time are zero.

Analytical treatment of these PDEs, as well as other PDEs of importance, can be found in [Pinchover and Rubinstein (2005)].

The following terms are frequently used to classify PDEs:

- *Order:* the highest order of any partial derivative that appears in the PDE
- *Number of Independent Variables:* the number of independent variables on which the solution depends; this should be at least two for a PDE, as opposed to an ODE
- *Number of Dependent Variables:* One for a scalar equation, or greater than one for a system of equations
- *Linearity:* the PDE is linear if the function F in the general form of the PDE (14.1) is a linear function of u and its derivatives
- *Kinds of Coefficients:* constant or variable
- *Types of Linear Equations:* the general second-order linear equation

$$Au_{xx} + Bu_{xy} + Cu_{yy} + Du_x + Eu_y + Fu = G$$

is classified as **parabolic, hyperbolic**, or **elliptic** according to whether $B^2 - 4AC$ is 0, > 0, or < 0. These terms do not relate to the PDEs themselves, but rather the equations of the corresponding conic sections.

Exploration 14.1.1 Classify the heat equation, wave equation, and Laplace's equation as either elliptic, parabolic, or hyperbolic.

14.1.1 Concept Check

1. Name and describe three fundamental partial differential equations.
2. Give a physical interpretation of the solution of the heat equation.

3. Give a physical interpretation of the solution of the wave equation.

4. Describe five different ways to classify a PDE.

5. Explain what it means for a PDE to be parabolic, hyperbolic, or elliptic. To which class of PDEs do these labels apply?

14.2 Elliptic Equations

The **Laplacian** of u, defined in 3-D as

$$\Delta u = \nabla^2 u = u_{xx} + u_{yy} + u_{zz},$$

is one of the most important operators in mathematical physics, as it is featured in several essential PDEs. The Laplacian also has a very useful intuitive meaning. This meaning can easily be derived in the 1-D case, in which the Laplacian is simply the second derivative.

Exploration 14.2.1 Use a Taylor expansion to show that if u is twice continuously differentiable on the interval $(x_0 - h, x_0 + h)$, then

$$\text{Avg}_{[x_0 - h, x_0 + h]}\, u = u(x_0) + \frac{h^2}{6}\nabla^2 u(x_0) + O(h^3),$$

where $\text{Avg}_{[a,b]} f(x)$ is defined to be the average of $f(x)$ on $[a, b]$; that is,

$$\text{Avg}_{[a,b]} f(x) = \frac{1}{b - a}\int_a^b f(x)\, dx.$$

We can therefore conclude that

- When $\nabla^2 u > 0$, u is smaller than its average of neighboring values,
- When $\nabla^2 u < 0$, u is larger than its average of neighboring values, and
- When $\nabla^2 u = 0$, u is equal to its average of neighboring values.

A similar result can be obtained in higher dimensions by integrating over, say, the interior of a circle or a sphere.

We now consider the simplest PDE involving the Laplacian, which is the canonical form of an elliptic PDE, **Laplace's equation**

$$\Delta u = 0. \tag{14.5}$$

A solution of Laplace's equation is called a **harmonic function**. Laplace's equation is a special case of the more general **Poisson's equation**

$$\Delta u = f(x, y). \tag{14.6}$$

The solution of (14.6) can be interpreted as a steady state temperature distribution (that is, at equilibrium), and the nonhomogeneous term $f(x, y)$ represents the rate of heat production within the domain.

14.2.1   Spatial Discretization

To begin our exploration of the numerical solution of PDEs, we consider the example of solving Poisson's equation (14.6) on a square,

$$\Delta u = \frac{\partial^2 u}{\partial x^2} + \frac{\partial^2 u}{\partial y^2} = f(x, y), \quad 0 < x < 1, \quad 0 < y < 1, \tag{14.7}$$

with homogeneous **Dirichlet boundary conditions**

$$u(0, y) = u(1, y) = 0, \quad 0 < y < 1, \quad u(x, 0) = u(x, 1) = 0, \quad 0 < x < 1. \tag{14.8}$$

In this example, we let $f(x, y) = \sin(\pi x)\sin(2\pi y)$. To obtain a system of linear equations for this 2-D **boundary value problem (BVP)**, we proceed as we did in the 1-D case in Section 13.2.1 and use a **centered difference** approximation of the second derivative,

$$u''(x_i) = \frac{u(x_{i-1}) - 2u(x_i) + u(x_{i+1})}{h} - \frac{h^2}{12}u^{(4)}(\xi), \tag{14.9}$$

where the x_i are equally spaced points with spacing h and $\xi \in (x_{i-1}, x_{i+1})$. This finite difference approximation was derived using Taylor expansion, as presented in Section 9.1.1.

   This approach to constructing finite difference formulas can be generalized to higher space dimension, using a multivariable Taylor expansion (see Theorem A.6.6). For example, using centered difference approximations for both u_{xx} and u_{yy}, we obtain a **five-point stencil** for the Laplacian,

$$\nabla^2 u(x_i, y_j) = \frac{u(x_{i+1}, y_j) + u(x_i, y_{j+1}) - 4u(x_i, y_j) + u(x_{i-1}, y_j) + u(x_i, y_{j-1})}{h^2}$$

$$+O(h^2), \tag{14.10}$$

where h is the grid spacing in both x and y.

   When solving PDEs, because there is more than one independent variable, the design of finite difference schemes often includes a graphical depiction of a finite difference formula that entails plotting the points involved. Such a graphical depiction is called a **stencil**, which leads to the name of the finite difference formula (14.10), that requires values of $u(x, y)$ at five points to approximate the Laplacian. Stencils can be useful for implementation of finite difference methods, as well as understanding of the behavior of computed solutions. The five-point stencil is depicted in Figure 14.1.

Exploration 14.2.2 Show that the error in the centered difference approximation (14.10) of the 2-D Laplacian is

$$-\frac{h^2}{12}\left(u_{xxxx}(x_i, y_j) + u_{yyyy}(x_i, y_j)\right) + O(h^4).$$

As in Section 13.2.1, our goal is to spatially discretize the BVP, which yields a system of linear equations for our approximate solution. While these linear equations can be

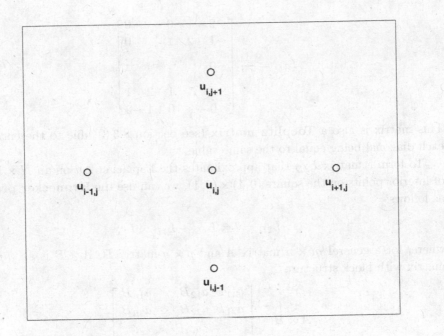

Fig. 14.1 Five-point stencil (14.10) for the Laplacian.

described using the five-point stencil (14.10) directly, in this case the matrix of the linear system for this 2-D BVP is obtained more easily by way of a **differentiation matrix** (see Section 9.1.5) for the 1-D case.

For Poisson's equation in 1-D, $u'' = f$, we use N equally spaced *interior* points in the interval $(0, 1)$, defined by

$$x_i = ih, \quad i = 1, 2, \ldots, N, \quad h = \frac{1}{N+1},$$

and impose the boundary conditions $u(0) = u(1) = 0$, by analogy with (14.8). Then $u''(x)$ can be approximated at each x_i using matrix-vector multiplication as follows:

$$\mathbf{f} = L_{1D}\mathbf{u}, \quad \mathbf{u} = \begin{bmatrix} u(x_1) \\ u(x_2) \\ \vdots \\ u(x_N) \end{bmatrix}, \quad \mathbf{f} \approx \begin{bmatrix} u''(x_1) \\ u''(x_2) \\ \vdots \\ u''(x_N) \end{bmatrix}$$

where L_{1D}, a discretization of the 1-D Laplacian, is the symmetric tridiagonal matrix

$$L_{1D} = \frac{1}{h^2} \begin{bmatrix} -2 & 1 & 0 & \cdots & & 0 \\ 1 & -2 & 1 & & & 0 \\ 0 & \ddots & \ddots & \ddots & & \vdots \\ \vdots & & & 1 & -2 & 1 \\ 0 & \cdots & & 0 & 1 & -2 \end{bmatrix}. \tag{14.11}$$

This matrix is also a **Toeplitz matrix** (see Section 8.3.3), due to the entries on each diagonal being equal to the same value.

To form a matrix L_{2D} that approximates the Laplacian Δu on an $N \times N$ grid of interior points on the square $(0, 1) \times (0, 1)$, we can use the **Kronecker product** as follows:

$$L_{2D} = I \otimes L_{1D} + L_{1D} \otimes I \tag{14.12}$$

where, for a general $m \times n$ matrix A and $p \times q$ matrix B, $A \otimes B$ is an $mp \times nq$ matrix with block structure

$$A \otimes B = \begin{bmatrix} a_{11}B & a_{12}B & \cdots & a_{1n}B \\ a_{21}B & a_{22}B & \cdots & a_{2n}B \\ \vdots & \ddots & \ddots & \vdots \\ a_{m1}B & \cdots & \cdots & a_{mn}B \end{bmatrix}.$$

That is, $A \otimes B$ creates an $m \times n$ "tiling" of B, in which the (i, j) "tile" is multiplied by a_{ij}. In MATLAB, this operation is performed using the function `kron`. If we order the points on the grid as follows:

$$(x_1, y_1), (x_1, y_2), \ldots, (x_1, y_N), (x_2, y_1), \ldots, (x_2, y_N), \ldots, (x_N, y_1), \ldots, (x_N, y_N),$$
$$\tag{14.13}$$

then $I \otimes L_{1D}$ represents $\partial^2 / \partial y^2$ and $L_{1D} \otimes I$ represents $\partial^2 / \partial x^2$. Generalization to 3-D is straightforward:

$$L_{3D} = L_{1D} \otimes I \otimes I + I \otimes L_{1D} \otimes I + I \otimes I \otimes L_{1D}. \tag{14.14}$$

Alternatively, the entries of the matrix can be obtained directly via multivariable Taylor expansions.

The right-hand side vector for our system of equations, denoted by $\mathbf{f}$, has the components

$$f_{(i-1)N+j} = f(x_i, y_j) = \sin(\pi ih) \sin(2\pi jh), \quad i, j = 1, 2, \ldots, N. \tag{14.15}$$

Thus the original BVP (14.7), (14.8) has been discretized by the system of linear equations

$$L_{2D}\mathbf{u} = \mathbf{f}, \tag{14.16}$$

where $\mathbf{u}$ represents the solution $u(x, y)$ by its values at the grid points, which are ordered in the same way as in $\mathbf{f}$. It is worth noting that L_{2D} is symmetric *negative* definite (see Exploration 14.2.11), meaning that $-L_{2D}$ is symmetric positive definite. Therefore, we can solve $-L_{2D}\mathbf{u} = -\mathbf{f}$ using the Cholesky factorization (see Section 3.3.3.2).

Exploration 14.2.3 Write a MATLAB function

$$u=\texttt{poissonsolve2d(f,N)}$$

that solves Poisson's equation (14.7), (14.8) on the square $(0,1)^2$ by solving the linear system $L_{2D}\mathbf{u} = \mathbf{f}$, where L_{2D} is defined in (14.12) and $\mathbf{f}$ is defined in (14.15). The input argument $\mathbf{f}$ is a function handle for the right-hand side $f(x,y)$, and the input argument $\mathbb{N}$ is the number of grid points per dimension. Test your function with $f(x,y) = \sin(\pi x)\sin(2\pi y)$, for which the exact solution is easily obtained, and verify that your approximate solution is second-order accurate by comparing the computed solution to the exact solution for various values of $\mathbb{N}$.

There are other stencils that can be used to approximate the Laplacian in 2-D, such as the **skewed five-point stencil**,

$$\nabla^2 u(x_i, y_j) = \frac{u_{i-1,j-1} + u_{i-1,j+1} - 4u_{ij} + u_{i+1,j-1} + u_{i+1,j+1}}{2h^2} + O(h^2), \quad (14.17)$$

where we use the shorthand $u_{ij} = u(x_i, y_j)$. This stencil uses values of $u(x,y)$ at the four "corners" of the square of side length $2h$ with center (x_i, y_j).

Exploration 14.2.4 Show that the error in the centered difference approximation (14.17) of the 2-D Laplacian is

$$\frac{h^2}{12}(u_{xxxx} + 6u_{xxyy} + u_{yyyy}) + O(h^4).$$

By linearly combining these stencils, we obtain the **nine-point stencil**

$$\nabla^2 u = \frac{1}{6h^2}\left[u_{i-1,j-1} + 4u_{i-1,j} + u_{i-1,j+1} + 4u_{i,j-1} - 20u_{i,j} + \right.$$
$$\left. 4u_{i,j+1} + u_{i+1,j-1} + 4u_{i+1,j} + u_{i+1,j+1}\right] + O(h^2). \quad (14.18)$$

An advantage of this stencil will be presented in the exercises.

Once a discretization of the BVP is chosen, there are several techniques that can be used for solving the resulting system of linear equations $L_{2D}\mathbf{u} = \mathbf{f}$. Various classes of solution techniques are:

- **direct methods**, covered in Sections 3.1 and 3.3, which use some form of elimination and substitution,
- **stationary iterative methods**, covered in Section 5.1, which use Fixed-Point Iteration (see Section 10.3). These are essential ingredients in **multigrid** methods, which are beyond the scope of this book.
- **non-stationary iterative methods**, better known as **Krylov subspace methods**, which are covered in Sections 5.2 and 5.3.

14.2.2   Implementing Boundary Conditions

For (14.7), (14.8), the homogeneous Dirichlet boundary conditions were trivial to implement, because the linear equations arising from discretization of (14.7) only included values of $u(x, y)$ at interior points. We now show how a finite difference stencil can be applied to obtain an approximate solution of Poisson's equation in 1-D with different boundary conditions. We consider the problem

$$u_{xx} = f, \quad 0 < x < 1, \tag{14.19}$$

with boundary conditions

$$u(0) = \alpha, \quad u_x(1) = \beta. \tag{14.20}$$

The condition at $x = 0$ is a *nonhomogeneous* Dirichlet boundary condition for $\alpha \neq 0$, while the condition at $x = 1$ is a **Neumann boundary condition**.

As before, we discretize this problem on a uniform grid with spacing $h = \frac{1}{N+1}$ and grid points $x_i = ih$, for $i = 0, 1, 2, \ldots, N+1$. We will also make use of the **ghost point** $x_{N+2} = 1 + h$, that is *outside* the domain of the BVP, due to the Neumann boundary condition. Our unknowns are $u_i \approx u(x_i)$, for $i = 1, 2, \ldots, N+1$, as we know that $u_0 = \alpha$ by the boundary condition at $x = 0$. We also let $f_i = f(x_i)$ for $i = 1, 2, \ldots, N+1$. At each *interior* point x_i, $i = 1, 2, \ldots, N$, as well as the right boundary point $x_{N+1} = 1$, we use a centered finite-difference approximation to obtain

$$u_{xx}(x_i) = \frac{u_{i-1} - 2u_i + u_{i+1}}{h^2} + O(h^2) = f_i, \quad i = 1, 2, \ldots, N+1. \tag{14.21}$$

From these equations, we seek a system of equations of the form $L\mathbf{u} = \mathbf{f}$, where L is a $(N+1) \times (N+1)$ matrix and $\mathbf{u}$ is a $(N+1)$-vector with entries $u_1, \ldots, u_{N+1}$. However, (14.21) includes values u_j that are not among these unknowns. Specifically, for $i = 1$ and $i = N+1$, we have

$$\frac{u_0 - 2u_1 + u_2}{h^2} = f_1, \tag{14.22}$$

$$\frac{u_N - 2u_{N+1} + u_{N+2}}{h^2} = f_{N+1}. \tag{14.23}$$

For (14.22), we have $u_0 = u(0) = \alpha$, and therefore we have

$$\frac{-2u_1 + u_2}{h^2} = f_1 - \frac{\alpha}{h^2}. \tag{14.24}$$

For (14.23), we eliminate u_{N+2} by approximating the boundary condition $u_x(1) = \beta$ with a centered difference (9.3) for the first derivative:

$$\beta = u_x(1) = \frac{u_{N+2} - u_N}{2h} + O(h^2). \tag{14.25}$$

Solving this equation for u_{N+2}, neglecting the $O(h^2)$ error, and substituting into (14.23) yields

$$\frac{u_N - u_{N+1}}{h^2} = \frac{1}{2}f_{N+1} - \frac{\beta}{h}. \tag{14.26}$$

Combining (14.21), (14.24), and (14.26) for $i = 1, 2, \ldots, N+1$ yields the system

$$\frac{1}{h^2} \begin{bmatrix} -2 & 1 & & & \\ 1 & -2 & 1 & & \\ & \ddots & \ddots & \ddots & \\ & & 1 & -2 & 1 \\ & & & 1 & -1 \end{bmatrix} \begin{bmatrix} u_1 \\ u_2 \\ \vdots \\ u_N \\ u_{N+1} \end{bmatrix} = \begin{bmatrix} f_1 - \frac{\alpha}{h^2} \\ f_2 \\ \vdots \\ f_N \\ \frac{1}{2} f_{N+1} - \frac{\beta}{h} \end{bmatrix}. \tag{14.27}$$

We see that the matrix is symmetric, which allows for a more efficient solution process, as was discussed in Section 3.3.2.

Exploration 14.2.5 Write a MATLAB function

$$u = \texttt{poissonsolve1d(f,alpha,beta,N)}$$

that solves (14.19), (14.20) by solving the linear system (14.27). The input arguments, in order, are a function handle for the right-hand side function $f(x)$ in (14.19), the values for the boundary conditions α and β in (14.20), and the number of interior grid points N. Test your function with several values of N and a choice of $f(x)$ for which the exact solution can be computed. Show that this approach is second-order accurate.

Exploration 14.2.6 Use Taylor expansion to prove that the solution obtained by solving (14.27) is second-order accurate. *Hint:* Taylor expansion must be performed on (14.25) as well.

14.2.3 Solution Using Iterative Methods

Now, suppose that we solve this system using stationary iterative methods, such as those covered in Section 5.1.

Exploration 14.2.7 Carry out 100 iterations of Jacobi, Gauss-Seidel, and SOR (with various choices of ω) to obtain an approximate solution of (14.7), (14.8) by solving the system of equations (14.16). Use $N = 100$ grid points per dimension. Describe the performance of the various methods, in terms of both accuracy and efficiency. What happens to the performance of these methods as N increases?

Exploration 14.2.8 As a continuation of Exploration 14.2.7, determine an approximate optimal value of ω for SOR by first trying m different values of ω, where $0 < \omega < 2$, recording the number of iterations needed to achieve $\|\mathbf{f} - L\mathbf{u}\|_\infty < 10^{-6}$ in each case, and then using polynomial interpolation of degree $m - 1$. Use MATLAB's polynomial functions from Section 1.2.18 to find the absolute minimum of the interpolating polynomial on $(0, 2)$. How is the optimal value of ω affected by the value of N?

We will now try to assess the performance of these methods by computing eigenvalues of their iteration matrices. To that end, it is very useful to be able to compute the eigenvalues of a symmetric, tridiagonal Toeplitz matrix such as in (14.11). Fortunately, these can be computed analytically.

The matrix A in (14.11) is of the form $\frac{1}{h^2}T(-2,1)$, where $T(a,b) = aI + bB$ and

$$
B = \begin{bmatrix}
0 & 1 & & & \\
1 & 0 & 1 & & \\
& \ddots & \ddots & \ddots & \\
& & 1 & 0 & 1 \\
& & & 1 & 0
\end{bmatrix}.
\tag{14.28}
$$

It is therefore sufficient to compute the eigenvalues of B, and then scale and shift them to obtain the eigenvalues of A.

Exploration 14.2.9 Let $\mathbf{v}$ be an eigenvector of the matrix B from (14.28), with corresponding eigenvalue λ, and without loss of generality let $v_1 = 1$. Show that the entries of $\mathbf{v}$ satisfy the three-term recurrence relation

$$
v_k = \lambda v_{k-1} - v_{k-2}, \quad k = 2, 3, \dots, N, \quad v_0 = 0
$$

with end condition $v_{N-1} = \lambda v_N$.

Exploration 14.2.10 Let $\mathbf{v}$ be defined as in Exploration 14.2.9. Assume that for $k = 1, 2, \dots, N$, $v_k = p_{k-1}(\lambda)$, where p_j is a polynomial of degree j. Show that $p_j(x) = U_j(x/2)$, where $U_j(x)$ is the **Chebyshev polynomial of the second kind** of degree j, defined by

$$
U_j(x) = \frac{\sin(j \cos^{-1} x)}{\sin(\cos^{-1} x)}, \quad -1 \leq x \leq 1.
$$

Use the end condition of the recurrence relation from Exploration 14.2.9 to conclude that the eigenvalues of the matrix L_{1D} from (14.11) are given by

$$
\lambda_j = \frac{1}{h^2}\left[2\cos(\pi j h) - 2\right], \quad j = 1, 2, \dots, N.
$$

Hint: What is the three-term recurrence relation for the Chebyshev polynomials of the second kind?

We see that L_{1D} is negative definite. We now generalize to higher dimensions.

Exploration 14.2.11 Use the result of Exploration 14.2.10 to compute the eigenvalues of the matrix L_{2D} from (14.12). Let $L_{1D} = V\Lambda V^T$ be the Schur Decomposition (6.3) of L_{1D}, and use a similarity transformation (see Section 6.1.2) of the form $M = Q^T L_{2D} Q$, where $Q = \text{diag}(V, V, \dots, V)$. Then use a second similarity transformation on M, involving a permutation matrix (see Section 3.2.5.3), to obtain an eigenvalue-revealing structure. *Hint:* Try applying permutations to M in MATLAB for a small value of N.

> **Exploration 14.2.12** Recall from (14.14) how the finite-difference approximation of the 3-D Laplacian can be obtained using Kronecker products. Assuming the centered difference is used, with homogeneous Dirichlet boundary conditions as in the 1-D case in (14.11), what are the eigenvalues of the resulting matrix?

In (5.6), the formula for the optimal value of ω for a symmetric tridiagonal matrix, such as the centered-difference discretization of the 1-D Laplacian, is given in terms of the spectral radius of the iteration matrix for the Jacobi method. In fact, this formula applies to any matrix A that has **Property A**, which means that there exists a permutation matrix P such that

$$P^T A P = \begin{bmatrix} D_1 & F \\ G & D_2 \end{bmatrix},$$

where D_1 and D_2 are diagonal. It can be shown that the matrix L_{2D} defined in (14.12) also has Property A.

> **Exploration 14.2.13** Use the result of Exploration 14.2.11 to prove the following:
>
> (a) The Jacobi method converges for the system (14.16). What happens to the convergence of this method as N increases?
> (b) SOR converges for this system for any ω in $(0, 2)$. What is the optimal ω as a function of N?

> **Exploration 14.2.14** Use SSOR (see Exercise 6 in Chapter 5) to solve Laplace's equation on an $(N + 1) \times (N + 1)$ grid. The problem is to be solved for $0 < x < 1$, $0 < y < 1$, with the following boundary conditions:
>
> $$u(0, y) = 0, \quad u(1, y) = y,$$
> $$u(x, 0) = 0, \quad u(x, 1) = x^2.$$
>
> Choose a random starting vector on $(0, 1)$, and iterate until the relative error is less than 10^{-5}. *Hint:* Use (14.13) to determine how to construct the right-hand side vector **b**.

14.2.4 Convergence Analysis

We will now learn how to analyze the accuracy of finite difference schemes for elliptic PDEs. For concreteness, we focus on the 1-D Poisson's equation with Dirichlet boundary conditions, but the analysis can be generalized to other linear PDEs.

The PDE $u''(x) = f(x)$ is discretized to obtain a system of linear equations $A_h \mathbf{u}_h = \mathbf{f}_h$, as described earlier in this section. The subscript h is used to emphasize the dependence of this system on the grid spacing h. We define the vector **u** to contain the values of the exact solution $u(x)$ at the interior grid points. That is,

$$\mathbf{u} = \begin{bmatrix} u(x_1) \\ u(x_2) \\ \vdots \\ u(x_N) \end{bmatrix}.$$

Then, by analogy with Definition 12.4.1 for IVPs, we define the **local truncation error** τ by

$$\tau(h) = A_h \mathbf{u} - \mathbf{f}_h.$$

That is, $\mathbf{u}$ is the exact solution of the "nearby" system $A_h \mathbf{u} = \mathbf{f}_h + \tau(h)$. Equivalently, $\tau(h)$ is obtained by substituting $\mathbf{u}$ into the finite difference scheme.

We assume that u is four times continuously differentiable on $(0, 1)$. If we use the centered difference, we have $A_h = L_{1D}$ from (14.11). From (14.9), we obtain, for $i = 1, 2, \ldots, N$,

$$\tau_i(h) = \frac{u(x_{i-1}) - 2u(x_i) + u(x_{i+1})}{h^2} - f(x_i)$$

$$= \frac{1}{12} u^{(4)}(\xi_i) h^2, \quad \xi_i \in (x_{i-1}, x_{i+1}). \tag{14.29}$$

We see that the local truncation error in the centered difference scheme is $O(h^2)$.

As in Section 12.4.3 for IVPs, the **global error** is defined to be the absolute error in the solution,

$$\mathbf{e}(h) = \mathbf{u}_h - \mathbf{u}.$$

From $A_h \mathbf{u}_h = \mathbf{f}_h$ and $A_h \mathbf{u} = \mathbf{f}_h + \tau(h)$, we obtain

$$A_h \mathbf{e}(h) = -\tau(h).$$

By the submultiplicative property of matrix norms (see Section B.13.2), we then have

$$\|\mathbf{e}(h)\| \leq \|A_h^{-1}\| \|\tau(h)\|.$$

However, as we have seen in Section 12.4 for IVPs, to ensure convergence, we must be certain that the numerical method does not amplify local truncation error by an arbitrarily large factor. With this is mind, we say that the finite difference method is **stable** if there exists a constant C and spacing h_0 such that A_h is invertible and $\|A_h^{-1}\| \leq C$ for $h < h_0$.

To discuss the behavior of the global error as $h \to 0$, we need the following definitions.

Definition 14.2.1 A finite difference scheme $A_h \mathbf{u}_h = \mathbf{f}_h$ is **consistent** if the local truncation error satisfies

$$\|\tau(h)\| \to 0 \quad \text{as} \quad h \to 0.$$

The scheme is **convergent** if the global error satisfies

$$\|\mathbf{e}(h)\| \to 0 \quad \text{as} \quad h \to 0.$$

If the finite difference scheme is consistent and stable, we then have $\|\mathbf{e}(h)\| \leq C\|\tau(h)\|$ due to stability, and since $\|\tau(h)\| \to 0$ as $h \to 0$ due to consistency, it follows that $\|\mathbf{e}(h)\| \to 0$ as well. That is, *a consistent and stable finite difference scheme is convergent.* Furthermore, if the local truncation error of a convergent scheme is $O(h^p)$, then the global error is also $O(h^p)$.

> **Exploration 14.2.15** Use the result of Exploration 14.2.10 to prove that the centered finite difference scheme $L_{1D}\mathbf{u}_h = \mathbf{f}_h$ is stable. Then, prove that it is convergent. What is the global order of accuracy?

14.2.5 Concept Check

1. What does the sign of the Laplacian tell us about a function?
2. State Laplace's equation and Poisson's equation.
3. Describe the matrix of the system of linear equations obtained through finite difference discretization of Poisson's equation in 1-D. How can this matrix be generalized to 2-D or 3-D problems?
4. What are three classes of methods that can be used to solve linear systems arising from elliptic PDEs?
5. How are Neumann boundary conditions handled when constructing a system of linear equations to solve an elliptic PDE?
6. Define the following terms, in the context of finite difference methods for elliptic PDEs:
   (a) local truncation error
   (b) global error
   (c) consistent
   (d) stable
   (e) convergent
7. How are consistency, stability, and convergence related to one another?

14.3 Parabolic Equations

Now that we have learned about how to solve *elliptic* PDEs with finite difference methods, we begin our exploration of numerical methods for *parabolic* PDEs with an overview of the **heat equation**. Suppose that $\mathbf{F}$ is a vector field that, at any point, represents the *flow rate* of heat energy, which is the rate of change, with respect to time, of the amount of heat energy flowing through that point. By *Fourier's Law*, $\mathbf{F} = -K\nabla T$, where K is a constant called *thermal conductivity*, and T is a function that indicates temperature.

Now, let E be a three-dimensional solid enclosed by a closed, positively oriented, surface S with *outward* unit normal vector $\mathbf{n}$. Then, by the law of conservation of energy, the *rate of change, with respect to time, of the amount of heat energy inside E is equal to the flow rate, or flux, or heat* **into** *E through S.* That is, if $\rho(x, y, z)$

is the density of heat energy, then

$$\frac{\partial}{\partial t} \iiint_E \rho \, dV = \iint_S \mathbf{F} \cdot (-\mathbf{n}) \, dS,$$

where we use $-\mathbf{n}$ because $\mathbf{n}$ is the *outward* unit normal vector, but we need to express the flux *into* E through S.

From the definition of $\mathbf{F}$, and the fact that $\rho = c\rho_0 T$, where c is the *specific heat* and ρ_0 is the *mass density*, which, for simplicity, we assume to be constant, we have

$$\frac{\partial}{\partial t} \iiint_E c\rho_0 T \, dV = \iint_S K \nabla T \cdot \mathbf{n} \, dS.$$

Next, we note that because c, ρ_0, and E do not depend on time, we can write

$$\iiint_E c\rho_0 \frac{\partial T}{\partial t} \, dV = \iint_S K \nabla T \cdot d\mathbf{S}.$$

Now, we apply the Divergence Theorem, and obtain

$$\iiint_E c\rho_0 \frac{\partial T}{\partial t} \, dV = \iiint_E K \operatorname{div} \nabla T \, dV = \iiint_E K \nabla^2 T \, dV.$$

That is,

$$\iiint_E \left(c\rho_0 \frac{\partial T}{\partial t} - K \nabla^2 T \right) dV = 0.$$

Since the solid E is arbitrary, it follows that

$$\frac{\partial T}{\partial t} = \alpha^2 \nabla^2 T, \quad \alpha^2 = \frac{K}{c\rho_0}.$$

Thus we have derived the heat equation.

A time-dependent PDE, combined with initial and boundary conditions, is called a **initial-boundary value problem (IBVP)**. We can see that the time-independent PDEs that we have studied are *steady-state* forms of time-dependent PDEs, meaning that we assume $u_t = 0$. For example, a steady-state solution of the heat equation satisfies either Laplace's equation or Poisson's equation, depending on whether there is a nonzero source term.

We can discretize a IBVP either in space first, or in time. In this book, we consider only the case of discretizing in space first, which is the **method of lines**. The solution $u(\mathbf{x}, t)$, where $\mathbf{x}$ is a vector of x-values taken from the spatial domain of the PDE, is represented by a vector-valued function of t, $\mathbf{u}(t)$. A linear differential operator L is represented a matrix A. That is, the PDE $u_t = Lu$ is semi-discretized by the system of ODEs

$$\mathbf{u}' = A\mathbf{u}.$$

The initial condition of the PDE is discretized to obtain a vector-valued initial condition for the system of ODEs, while boundary conditions are accounted for in the matrix A. If the boundary conditions are not homogeneous, then the system of ODEs will include a source term.

14.3.1 Solution via Finite Differences

We solve the IBVP
$$u_t = \alpha^2 u_{xx}, \quad 0 < x < 1, \quad t > 0, \tag{14.30}$$
with initial condition
$$u(x,0) = f(x), \quad 0 < x < 1, \tag{14.31}$$
and homogeneous Dirichlet boundary conditions
$$u(0,t) = 0, \quad u(1,t) = 0, \quad t > 0. \tag{14.32}$$
We begin with spatial discretization to obtain a system of ODEs of the form $\mathbf{u}' = A\mathbf{u}$, where

$$\mathbf{u}(t) \approx \begin{bmatrix} u(x_1,t) \\ u(x_2,t) \\ \vdots \\ u(x_N,t) \end{bmatrix}, \tag{14.33}$$

with $\Delta x = 1/(N+1)$ and $x_i = i\Delta x$. We then use a centered difference approximation for u_{xx},
$$u_{xx}(x_i,t_n) = \frac{u(x_{i+1},t_n) - 2u(x_i,t_n) + u(x_{i-1},t_n)}{\Delta x^2} - \frac{\Delta x^2}{12} u_{xxxx}(\xi,t_n),$$
for $i = 1,2,\ldots,N$, where $\xi \in (x_{i-1}, x_{i+1})$. This yields an *initial value problem*

$$\mathbf{u}' = A\mathbf{u}, \quad \mathbf{u}(0) = \mathbf{f}, \quad A = \alpha^2 L_{1D} = \frac{\alpha^2}{\Delta x^2} \begin{bmatrix} -2 & 1 & & & \\ 1 & -2 & 1 & & \\ & \ddots & \ddots & \ddots & \\ & & 1 & -2 & 1 \\ & & & 1 & -2 \end{bmatrix}. \tag{14.34}$$

To solve this problem, we use Forward Euler, adapted from (12.8):
$$\mathbf{u}^{n+1} = \mathbf{u}^n + \Delta t A \mathbf{u}^n, \quad n = 0,1,2,\ldots, \tag{14.35}$$
where $\mathbf{u}^n \approx \mathbf{u}(t_n)$ with $t_n = n\Delta t$, and $\mathbf{u}^0 = \mathbf{f}$. This yields the finite difference scheme
$$u_i^{n+1} = u_i^n + \frac{\Delta t \alpha^2}{\Delta x^2}(u_{i-1}^n - 2u_i^n + u_{i+1}^n), \quad i = 1,2,\ldots,N, \quad n = 0,1,2,\ldots, \tag{14.36}$$
where $u_i^n \approx u(x_i,t_n)$ for $i = 1,2,\ldots,N$ and $n = 0,1,2,\ldots$.

Exploration 14.3.1 Write a MATLAB function

$$\texttt{u=eulerheat(alpha,f,dt,N,T)}$$

to solve (14.30), (14.31), (14.32) with $\alpha = $ alpha using Forward Euler in time and centered differencing in space, as described in (14.36). The second input argument f is a function handle for the initial data $f(x)$. The third input argument dt is the time step Δt, the fourth input argument N is the number of *interior* grid points, which have equal spacing $\Delta x = 1/(N+1)$, and the fifth input argument T is the final time. Test your function with the initial data $f(x) = \sin \pi x$, for which the exact solution can readily be computed. How does the error behave as $\Delta t, \Delta x \to 0$?

To provide theoretical support for the convergence behavior observed in the preceding exploration, we seek the **local truncation error** for this finite difference scheme. To that end, we proceed as in Section 12.4.1 for time-stepping and Section 13.2 for spatial discretization, by substituting the exact solution $u(x,t)$ into (14.36) and using Taylor expansion around (x_j, t_n) to obtain the local truncation error

$$
\begin{aligned}
\tau_j^n(h) = \frac{1}{\Delta t} & \left[\left(u(x_j, t_n) + \Delta t u_t(x_j, t_n) + \frac{\Delta t^2}{2} u_{tt}(x_j, \eta) \right) - u(x_j, t_n) \right] - \\
& \frac{\alpha^2}{\Delta x^2} \left[\left(u(x_j, t_n) - \Delta x u_x(x_j, t_n) + \frac{\Delta x^2}{2} u_{xx}(x_j, t_n) - \right. \right. \\
& \frac{\Delta x^3}{6} u_{xxx}(x_j, t_n) + \frac{\Delta x^4}{24} u_{xxxx}(\xi_-, t_n) \bigg) - 2u(x_j, t_n) + \\
& \left(u(x_j, t_n) + \Delta x u_x(x_j, t_n) + \frac{\Delta x^2}{2} u_{xx}(x_j, t_n) + \right. \\
& \frac{\Delta x^3}{6} u_{xxx}(x_j, t_n) + \frac{\Delta x^4}{24} u_{xxxx}(\xi_+, t_n) \bigg) \bigg] \\
= & \frac{\Delta t}{2} u_{tt}(x_j, \eta) - \frac{\alpha^2 \Delta x^2}{12} u_{xxxx}(\xi, t_n) = O(\Delta t) + O(\Delta x^2), \quad (14.37)
\end{aligned}
$$

where $\eta \in (t_n, t_n + \Delta t)$, $\xi_- \in (x_j - \Delta x, x_j)$, $\xi_+ \in (x_j, x_j + \Delta x)$, and $\xi \in (x_j - \Delta x, x_j + \Delta x)$. We say that (14.36) is *first-order accurate in time*, and *second-order accurate in space*.

Exploration 14.3.2 Show that the scheme

$$
\mathbf{u}^{n+1} = \mathbf{u}^n + \alpha^2 \Delta t L_{1D} \mathbf{u}^n + \frac{\alpha^4 \Delta t^2}{2} L_{1D}^2 \mathbf{u}^n,
$$

applied to (14.30), (14.31), (14.32), that uses the **Explicit Trapezoidal Method** (12.19) in time and central differencing in space, has local truncation error that is $O(\Delta t^2) + O(\Delta x^2)$. Verify this order of accuracy by modifying your function `eulerheat` from Exploration 14.3.1 and testing it with initial data $f(x) = \sin \pi x$ again. How does the method behave as $\Delta t, \Delta x \to 0$?

As with IVPs and BVPs, the notion of stability is essential toward establishing that a finite difference scheme for a IBVP is convergent. For the IBVP (14.30), (14.31), (14.32), we must ensure that the computed solution possesses the same qualitative behavior as the exact solution. In particular, we have the following **energy estimate**

$$
\frac{1}{2} \| u(\cdot, T) \|_2^2 = \frac{1}{2} \| f \|_2^2 - \alpha^2 \int_0^T \| u_x(\cdot, t) \|^2 \, dt, \quad (14.38)
$$

where $\| \cdot \|_2$ is the L^2-norm on $(0,1)$ (see Section B.13.3). Therefore, at a minimum, the numerical scheme must produce a solution that does not grow in magnitude as n increases.

Exploration 14.3.3 Prove (14.38) by multiplying both sides of (14.30) by $u(x,t)$ and integrating over $(0,1) \times (0,T)$.

Exploration 14.3.4 Try your function `eulerheat` from Exploration 14.3.1 with increasing values of N but with the same time step $\Delta t = $ `dt`. What happens?

Recall from Section B.14 that a sequence of vectors $\{\mathbf{x}^{(k)}\}_{k=0}^{\infty}$, defined by $\mathbf{x}^{(k+1)} = A\mathbf{x}^{(k)}$ for some matrix A, converges to the zero vector if and only if $\rho(A) < 1$, where $\rho(A)$ is the **spectral radius** of A. Using the Schur Decomposition (6.3) it can be shown that if A is a **normal matrix**, then $\|\mathbf{x}^{(k+1)}\|_2 \leq \|\mathbf{x}^{(k)}\|_2$ if and only if $\rho(A) \leq 1$.

Exploration 14.3.5 Use the eigenvalues of the matrix $I + \Delta t A$ from (14.35) to show that the numerical solution computed using Forward Euler in time and centered differencing in space does not grow in magnitude over time if and only if

$$\frac{\alpha^2 \Delta t}{\Delta x^2} \leq \frac{1}{2}.$$

We say that Forward Euler is **conditionally stable**, meaning that the numerical solution of the heat equation can grow without bound over time unless the time step Δt is chosen sufficiently small relative to the grid spacing Δx.

14.3.2 Implicit Methods

The conditional stability of Forward Euler implies that the computational expense of this method increases substantially as the number of grid points, N, increases. To solve the heat equation more efficiently, we consider the use of **implicit methods**, as in Section 12.2.4 for IVPs, to solve the system of ODEs obtained through spatial discretization. We first consider **Backward Euler**, adapted from (12.22):

$$\mathbf{u}^{n+1} = \mathbf{u}^n + \Delta t A \mathbf{u}^{n+1}. \tag{14.39}$$

Rearranging, we see that each time step requires the solution of the system of linear equations

$$(I - \Delta t A)\mathbf{u}^{n+1} = \mathbf{u}^n. \tag{14.40}$$

Exploration 14.3.6 Explain the advantage of using either a direct or iterative method to solve the system (14.40). When using an iterative method, what should be used for an initial guess?

Exploration 14.3.7 Use the eigenvalues of the matrix A from (14.34) to show that Backward Euler (14.39) applied to (14.30), (14.31), (14.32) is **unconditionally stable**; that is, that the solution does not grow over time regardless of the choice of Δt.

Another implicit time-stepping method is the **Crank-Nicolson Method**, which is based on the Implicit Trapezoidal Method (12.23):

$$\mathbf{u}^{n+1} = \mathbf{u}^n + \frac{\Delta t}{2} A[\mathbf{u}^n + \mathbf{u}^{n+1}]. \tag{14.41}$$

In matrix-vector form, Crank-Nicolson can be written as

$$\left(I - \frac{\Delta t}{2} A\right) \mathbf{u}^{n+1} = \left(I + \frac{\Delta t}{2} A\right) \mathbf{u}^n.$$

Exploration 14.3.8 Show that Crank-Nicolson (14.41) applied to (14.30), (14.31), (14.32) is unconditionally stable.

Exploration 14.3.9 Modify your function `eulerheat` from Exploration 14.3.1 to obtain two new functions, `backeulerheat` and `crankheat`, that have the same input and output arguments, to solve (14.30), (14.31), (14.32) using centered differencing in space and Backward Euler in time from (14.39) and Crank-Nicolson from (14.41), respectively. For all three methods, experimentally determine the largest value of Δt that, for fixed N, yields relative error (in the ℓ_∞-norm sense) less than 10^{-5} at $T = 0.1$ with $\alpha = 1$ and initial data $u(x,0) = \sin \pi x$. How does this value of Δt change for the three methods as N increases?

Exploration 14.3.10 When using Crank-Nicolson to solve (14.30), (14.31), (14.32), how does the numerical solution behave for larger Δt? Use the eigenvalues of A from (14.34) to explain the observed behavior.

14.3.3 Convergence Analysis

In the previous section, we learned that a finite difference scheme for an elliptic PDE is convergent if it is consistent and stable. For a time-dependent PDE, the concepts of consistency, stability and convergence are defined in an analogous manner. By analogy with IVPs, we say a finite difference scheme of the form

$$\mathbf{u}^{n+1} = S(\Delta t)\mathbf{u}^n \tag{14.42}$$

for the linear PDE $u_t = Lu$, where $S(\Delta t)$ is an $N \times N$ matrix and L is a linear differential operator, is

- **consistent** if the **local truncation error** approaches zero as the grid spacing Δx and time step Δt approach zero.
- **stable** if the numerical solution depends continuously on the problem data, just as the exact solution of a well-posed problem does, by Hadamard's conditions (see Definition 12.1.6). That is, for any final time $T > 0$, there exists a constant K_T, independent of Δt and Δx, such that

$$\|\mathbf{u}^n\| \le K_T \|\mathbf{u}^0\|, \quad 0 \le n \le T/\Delta t.$$

- **convergent** if the **global error** approaches zero as Δx, Δt approach zero.

While it is straightforward to determine consistency using Taylor expansion, it is more difficult to determine stability using the above definition directly. We appeal to linear algebra for a more concrete criterion, and note that $\mathbf{u}^n = S(\Delta t)^n \mathbf{u}^0$. This leads to the following definition.

Definition 14.3.1 The finite difference scheme (14.42) for the IBVP $u_t = Lu$ is said to be **Lax-Richtmyer stable** if, for each $T > 0$, there exists a constant K_T such that

$$\|S(\Delta t)^n\| \leq K_T$$

for $0 \leq n \leq T/\Delta t$.

Using the result

$$\lim_{n \to \infty} \left(1 + \frac{1}{n}\right)^n = e,$$

it can be shown that if we let $K_T = e^{\alpha T}$ for some constant α, then we have Lax-Richtmyer stability if

$$\|S(\Delta t)\| \leq 1 + \alpha \Delta t.$$

If $S(\Delta t)$ is a normal matrix, then we can use the matrix ℓ_2-norm and examine the eigenvalues of $S(\Delta t)$ to determine stability, as we have in Explorations 14.3.5, 14.3.7, and 14.3.8.

In general, the eigenvalues of $S(\Delta t)$ might not be readily computable. However, in the case of linear PDEs with constant coefficients, we can use the fact that the function e^{ikx}, where k is a constant, is an eigenfunction of both differential and finite difference operators. This leads to **von Neumann stability analysis**, which involves examination of the behavior of the solution given initial data of the form e^{ikx}.

Example 14.3.2 Consider the heat equation $u_t = \alpha^2 u_{xx}$, where α is a constant. Given initial data $u(x, 0) = e^{ikx}$, the exact solution is $u(x, t) = e^{-\alpha^2 k^2 t} e^{ikx}$. That is, the initial data is amplified by the **growth factor** $G(k) = e^{-\alpha^2 k^2 t}$, also known as the **amplification factor**. The solution will not grow over time if $|G(k)| \leq 1$ for all k, which is required in view of the energy estimate (14.38).

If we use centered differencing in space and Forward Euler in time,

$$\frac{u_j^{n+1} - u_j^n}{\Delta t} = \alpha^2 \frac{u_{j-1}^n - 2u_j^n + u_{j+1}^n}{\Delta x^2},$$

or, equivalently,

$$u_j^{n+1} = u_j^n + r(u_{j-1}^n - 2u_j^n + u_{j+1}^n), \quad r = \frac{\alpha^2 \Delta t}{\Delta x^2},$$

then, with $u(x, t_n) = e^{ikx}$, we obtain

$$u_j^{n+1} = [1 + r(e^{ik\Delta x} - 2 + e^{-ik\Delta x})]e^{ikx}.$$

That is, the growth factor for the computed solution is

$$G(k) = 1 + r(e^{ik\Delta x} - 2 + e^{-ik\Delta x}) = 1 - 2r(1 - \cos(k\Delta x)).$$

To ensure that this growth factor satisfies $|G(k)| \leq 1$, we must have

$$r(1 - \cos(k\Delta x)) \leq 1$$

which yields the condition

$$r = \frac{\alpha^2 \Delta t}{\Delta x^2} \leq \frac{1}{2}.$$

That is, this scheme is **conditionally stable**. We note that this result is consistent with that of Exploration 14.3.5. $\square$

It is worth noting that the preceding result can be obtained more easily by noting that each term of the form u_m^ℓ is a multiple of e^{ikx}, so e^{ikx} can be factored out. We can then substitute $u_j^{n+1} = G(k)$, $u_j^n = 1$, $u_{j+1}^n = e^{i\theta}$, and $u_{j-1}^n = e^{-i\theta}$, where $\theta = k\Delta x$. Similarly, we can replace u_{j+1}^{n+1} with $G(k)e^{i\theta}$.

Example 14.3.3 We perform von Neumann stability analysis on the Backward Euler scheme (14.39),

$$\frac{u_j^{n+1} - u_j^n}{\Delta t} = \alpha^2 \frac{u_{j-1}^{n+1} - 2u_j^{n+1} + u_{j+1}^{n+1}}{\Delta x^2}.$$

We then have

$$\frac{G(k) - 1}{\Delta t} = \alpha^2 G(k) \frac{e^{-i\theta} - 2 + e^{i\theta}}{\Delta x^2}$$

or

$$G(k) = 1 - 2rG(k)(1 - \cos\theta)$$

which yields

$$G(k) = \frac{1}{1 + 2r(1 - \cos\theta)}.$$

It can be seen that regardless of θ, $|G(k)| \leq 1$, so the scheme is **unconditionally stable**. $\square$

Exploration 14.3.11 Use von Neumann stability analysis to show that the Crank-Nicolson scheme (14.41) is unconditionally stable.

Consider the heat equation in 2-D,

$$u_t = u_{xx} + u_{yy}, \quad 0 < x < 1, \quad 0 < y < 1, \tag{14.43}$$

with an appropriate initial condition and periodic boundary conditions. We use grid points (x_i, y_j), with spacings Δx and Δy. If we use Forward Euler in time and centered differencing in space, the resulting numerical scheme is

$$\frac{u_{i,j}^{n+1} - u_{i,j}^n}{\Delta t} = \frac{u_{i-1,j}^n - 2u_{i,j}^n + u_{i+1,j}^n}{\Delta x^2} + \frac{u_{i,j-1}^n - 2u_{i,j}^n + u_{i,j+1}^n}{\Delta y^2}. \tag{14.44}$$

To perform von Neumann stability analysis on this scheme, we assume $u(x, y, t_n) = e^{i(kx+\ell y)}$ so that the growth factor is defined by $u(x, y, t_{n+1}) = G(k, \ell)u(x, y, t_n)$.

Exploration 14.3.12 Use von Neumann stability analysis to show that the scheme (14.44) applied to (14.43) is stable if

$$\frac{\Delta t}{\Delta x^2} + \frac{\Delta t}{\Delta y^2} \leq \frac{1}{2}.$$

As with IVPs and BVPs, consistency and stability imply convergence. In fact, we have the following stronger theorem.

Theorem 14.3.4 (Lax Equivalence Theorem) A consistent finite difference scheme of the form (14.42) is convergent if and only if it is Lax-Richtmyer stable.

The sufficiency of consistency and stability for convergence is readily proved in a similar manner as for IVPs and BVPs. For a full proof of this theorem, including the necessity, the reader is referred to [Strikwerda (1989)].

14.3.4 Concept Check

1. What principle is used to derive the heat equation?
2. What is the method of lines?
3. What is the Crank-Nicolson method? What advantages does it have over other methods for solving the heat equation?
4. Give an example of a time-stepping method that is conditionally stable, and one that is unconditionally stable.
5. What is the purpose of von Neumann stability analysis? Briefly describe how it is carried out for a given PDE.

14.4 Hyperbolic Equations

Newton's Second Law applied to an arbitrary segment $[x, x + \Delta x]$ of a vibrating string yields

$$\Delta x \rho u_{tt} = T[u_x(x + \Delta x, t) - u_x(x, t)],$$

where $u(x, t)$ is the displacement of the string from equilibrium, ρ is the density and T is the tension. Other forces acting on the string, such as friction, have been neglected. Dividing by Δx and letting $\Delta x \to 0$ yields the *wave equation*

$$u_{tt} = c^2 u_{xx},$$

where $c^2 = T/\rho$. This equation describes the *transverse vibration* of the string. Other applications of the wave equation in one space dimension include 1) modeling the *longitudinal* vibration of a rod, in which case the constant c^2 is *Young's modulus*, a measure of the elasticity of the rod, and 2) modeling voltage and current along a wire, in which case the wave equation is a special case of the *transmission-line equation* [Farlow (1982)].

It should be noted that because of the second derivative with respect to time, the wave equation has *two* initial conditions, imposed on $u(x,0)$ and $u_t(x,0)$, which are the initial position and initial velocity, respectively. Adding boundary conditions, as with other types of PDEs, completes the specification of an initial-boundary value problem (IBVP).

14.4.1 Solution via Finite Differences

We now solve the IBVP

$$u_{tt} = c^2 u_{xx}, \quad 0 < x < 1, \quad t > 0, \tag{14.45}$$

with initial conditions

$$u(x,0) = f(x), \quad u_t(x,0) = g(x), \quad 0 < x < 1, \tag{14.46}$$

and homogeneous Dirichlet boundary conditions

$$u(0,t) = 0, \quad u(1,t) = 0, \quad t > 0. \tag{14.47}$$

The solution satisfies a different sort of **energy estimate** than that of the heat equation:

$$c^2 \|u_x(\cdot,t)\|_2^2 + \|u_t(\cdot,t)\|_2^2 = c^2 \|f_x\|_2^2 + \|g\|_2^2, \quad t > 0. \tag{14.48}$$

As such, it is again important that a numerical scheme not allow computed solutions to grow without bound.

As with the heat equation, we begin with spatial discretization to obtain a system of ODEs of the form $\mathbf{u}'' = A\mathbf{u}$, where $\mathbf{u}(t)$ is defined as in (14.33). As before, we use a uniform grid with spacing $\Delta x = 1/(N+1)$. We again use a centered difference approximation for u_{xx},

$$u_{xx}(x_i, t_n) = \frac{u(x_i+1, t_n) - 2u(x_i, t_n) + u(x_{i-1}, t_n)}{\Delta x^2} - \frac{\Delta x^2}{12} u_{xxxx}(\xi, t_n),$$

for $i = 1, 2, \ldots, N$, where $\xi \in (x_{i-1}, x_{i+1})$. This yields an initial value problem

$$\mathbf{u}'' = A\mathbf{u}, \quad \mathbf{u}(0) = \mathbf{f}, \quad \mathbf{u}'(0) = \mathbf{g}, \quad A = \frac{c^2}{\Delta x^2} \begin{bmatrix} -2 & 1 & & & \\ 1 & -2 & 1 & & \\ & \ddots & \ddots & \ddots & \\ & & 1 & -2 & 1 \\ & & & 1 & -2 \end{bmatrix}. \tag{14.49}$$

To solve this problem, we first use explicit time stepping, with a centered difference approximation of u_{tt}:

$$\mathbf{u}^{n+1} = 2\mathbf{u}^n - \mathbf{u}^{n-1} + \Delta t^2 A\mathbf{u}^n, \quad n = 1, 2, \ldots \tag{14.50}$$

where $\mathbf{u}^n \approx \mathbf{u}(t_n)$ with $t_n = n\Delta t$, and $\mathbf{u}^0 = \mathbf{f}$.

Unlike the heat equation, we have a second initial condition for which we must account, which affects the computation of $\mathbf{u}^1$. Using the Taylor expansion

$$u(x, \Delta t) = u(x, 0) + \Delta t u_t(x, 0) + \frac{\Delta t^2}{2} u_{tt}(x, 0) + O(\Delta t^3)$$

$$= f(x) + \Delta t g(x) + \frac{c^2 \Delta t^2}{2} f''(x) + O(\Delta t^3),$$

we obtain the second-order accurate approximation

$$\mathbf{u}^1 = \mathbf{u}^0 + \Delta t \mathbf{g} + \frac{\Delta t^2}{2} A \mathbf{u}^0, \tag{14.51}$$

where A is as defined in (14.49).

Exploration 14.4.1 Write a MATLAB function

$$\texttt{u=explicitwave(c,f,g,dt,N,T)}$$

to solve (14.45), (14.46), (14.47) with $c = \mathtt{c}$ using explicit time-stepping as described in (14.50), (14.51). The second input arguments $\mathtt{f}$ and $\mathtt{g}$ are function handles for the initial data $f(x)$ and $g(x)$, respectively. The fourth input argument $\mathtt{dt}$ is the time step Δt, the fifth input argument $\mathtt{N}$ is the number of *interior* grid points, which have equal spacing $\Delta x = 1/(\mathtt{N}+1)$, and the sixth input argument $\mathtt{T}$ is the final time.

Exploration 14.4.2 Show that explicit time-stepping with (14.50), (14.51) applied to (14.45), (14.46), (14.47) has local truncation error that is $O(\Delta t^2) + O(\Delta x^2)$. Verify this order of accuracy by trying your function $\texttt{explicitwave}$ from Exploration 14.4.1 with initial data $f(x) = \sin \pi x$, $g(x) = \sin 2\pi x$, for which the exact solution can readily be computed, and different values of Δt and N.

Exploration 14.4.3 Try your function $\texttt{explicitwave}$ from Exploration 14.4.1 with increasing values of N but with the same time step $\Delta t = \mathtt{dt}$. What happens?

We see that explicit time-stepping is **conditionally stable**, meaning that the numerical solution of the wave equation grows without bound unless the time step Δt is chosen sufficiently small relative to the grid spacing Δx.

To obtain a similar stability condition as the one from Exploration 14.3.5, more work is required due to the second-order time derivative. Let $A = V \Lambda V^T$ be the Schur decomposition of A. Then, if we let $\mathbf{y}^n = V^T \mathbf{u}^n$, (14.50) becomes

$$\mathbf{y}^{n+1} = 2\mathbf{y}^n - \mathbf{y}^{n-1} + \Delta t^2 \Lambda \mathbf{y}^n, \quad n = 1, 2, \ldots. \tag{14.52}$$

Because Λ is a diagonal matrix, each row of this vector equation is a three-term recurrence relation of the form

$$y_{n+1} = (2 + \Delta t^2 \lambda) y_n - y_{n-1},$$

where λ is an eigenvalue of A. To obtain an explicit formula for y_n, we seek the roots of the **characteristic polynomial** of the recurrence relation, which is

$$\rho(z) = z^2 - (2 + \Delta t^2 \lambda)z + 1. \tag{14.53}$$

This is analogous to the first characteristic polynomial for a multistep method (see Section 12.3). To ensure stability, we require the roots of $\rho(z)$ to satisfy the same conditions as for zero-stability for a multistep method for IVPs, for all $\lambda \in \lambda(A)$. That is, the roots z_1, z_2 must satisfy the **root condition** presented in Section 12.4.2.2: for $i = 1, 2$, $|z_i| \leq 1$, and if $|z_i| = 1$, then z_i must be a simple root.

Exploration 14.4.4 Use the eigenvalues of the matrix A from (14.49) to show that the roots of $\rho(z)$ from (14.53) satisfy the root condition for $\lambda \in \lambda(A)$ if

$$\frac{c\Delta t}{\Delta x} \leq 1.$$

The condition from Exploration 14.4.4 is a special case of what is known as the **Courant-Friedrichs-Lewy (CFL) Condition** for the convergence of a finite-difference method. While in Exploration 14.4.4 it appears to be a stability condition, as it ensures a bounded solution when using explicit time-stepping, it is actually about convergence, as a method may also fail to be *consistent* unless the CFL Condition is satisfied (consider, for example, the DuFort-Frankel scheme as described in [Gustafsson et al. (1995)]).

For the wave equation, the CFL Condition can be inferred from the concept of a **domain of dependence**. The domain of dependence for the wave equation at (x_0, t_0) is the triangle with vertices (x_0, t_0), $(x_0 - ct_0, 0)$, and $(x_0 + ct_0, 0)$. The domain of dependence for the explicit time-stepping scheme (14.50) at the point (x_j, t_n) is the triangle with vertices (x_j, t_n), $(x_j - n\Delta x, 0)$, and $(x_j + n\Delta x, 0)$. Both types of domains of dependence are illustrated in Figure 14.2, along with their relationship to the finite difference stencil used in (14.50). It can be seen from this figure that the CFL Condition is satisfied if and only if the domain of dependence for the numerical method contains the domain of dependence for the PDE.

Exploration 14.4.5 Use von Neumann stability analysis to confirm the result of Exploration 14.4.4. Then, show that in this case, the growth factor $G(k)$ satisfies $|G(k)| = 1$ for all k.

14.4.2 Implicit Time-Stepping

The conditional stability of explicit time-stepping implies that the computational expense of this method increases substantially as the number of grid points, N, increases. To solve the wave equation more efficiently, we consider the use of **implicit methods**, as in Section 14.3.2 for parabolic problems, to solve the system of ODEs obtained through spatial discretization. We first consider the scheme

$$\mathbf{u}^{n+1} = 2\mathbf{u}^n - \mathbf{u}^{n-1} + \Delta t^2 A \mathbf{u}^{n+1}. \tag{14.54}$$

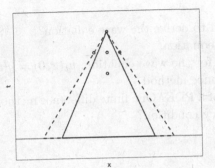 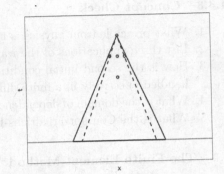

Fig. 14.2 Left plot: The dashed lines are the boundaries of a domain of dependence for the explicit time-stepping scheme (14.50) that satisfies the CFL Condition for the wave equation, for which the domain of dependence is depicted with solid lines. Right plot: Domain of dependence for the explicit scheme (14.50) that does not satisfy the CFL Condition. In both plots, the circles indicate the points on which the numerical solution at (x_j, t_n) directly depends, which constitute the stencil for (14.50).

Rearranging, we see that each time step requires the solution of the system of linear equations

$$\left(I - \Delta t^2 A\right) \mathbf{u}^{n+1} = 2\mathbf{u}^n - \mathbf{u}^{n-1}. \tag{14.55}$$

Exploration 14.4.6 Use the eigenvalues of the matrix A from (14.49) to show that implicit time-stepping with (14.54) applied to (14.45), (14.46), (14.47) is **unconditionally stable**; that is, the solution is bounded regardless of the choice of Δt.

Exploration 14.4.7 Suppose we express the wave equation as a first-order system:

$$u_t = v, \quad v_t = c^2 u_{xx}.$$

Show that the scheme obtained by applying Forward Euler (12.8) in time and centered differencing in space,

$$\mathbf{u}^{n+1} = \mathbf{u}^n + \Delta t \mathbf{v}^n,$$
$$\mathbf{v}^{n+1} = \mathbf{v}^n + \Delta t A \mathbf{u}^n,$$

where A is the matrix from (14.49), is "unconditionally unstable", meaning that it is not Lax-Richtmyer stable regardless of Δt.

A simple "fix" for the scheme in Exploration 14.4.7, that appears to be implicit but is actually not, will be investigated in the exercises.

14.4.3   Concept Check

1. What principle from physics is used to derive the wave equation?
2. List three applications of the wave equation.
3. How is the second initial condition for the wave equation, $u_t(x, 0) = g(x)$, handled effectively in a finite difference method?
4. What is the domain of dependence of a PDE? Of a finite difference method?
5. What is the Courant-Friedrichs-Lewy condition?

14.5   The Finite Element Method*

In Section 13.4, we introduced the **Finite Element Method** for solving a two-point boundary value problem. In this section, we illustrate how it can be applied to an elliptic PDE on a 2-D domain. Our model problem is

$$-\nabla \cdot (p(x, y)\nabla u(x, y)) + q(x, y)u(x, y) = r(x, y), \quad (x, y) \in \Omega \subset \mathbb{R}^2. \quad (14.56)$$

We assume the boundary of the domain, denoted by $\Gamma = \partial\Omega$, can be expressed as $\Gamma = \Gamma_1 \cup \Gamma_2$, and that the solution u satisfies the boundary conditions

$$u(x, y) = g(x, y), \quad (x, y) \in \Gamma_1, \quad \frac{\partial u}{\partial n}(x, y) = h(x, y), \quad (x, y) \in \Gamma_2. \quad (14.57)$$

That is, u satisfies a *Dirichlet* boundary condition on Γ_1, and a *Neumann* boundary condition on Γ_2. $\partial u/\partial n$ refers to the **normal derivative** of u, with the normal vector assumed to be pointing *outward*.

14.5.1   Derivation of the Weak Form

As with the two-point BVP, we multiply both sides of (14.56) by a **test function** $v \in H^1(\Omega)$ and integrate over Ω to obtain

$$\int_\Omega -v\nabla \cdot (p\nabla u) + qvu \, dA = \int_\Omega vr \, dA.$$

Applying one of *Green's identities* from vector calculus yields

$$\int_\Omega -\nabla \cdot (pv\nabla u) + p\nabla v \cdot \nabla u + qvu \, dA = \int_\Omega vr \, dA.$$

The *Divergence Theorem* then yields

$$-\int_\Gamma pv\frac{\partial u}{\partial n} \, ds + \int_\Omega p\nabla v \cdot \nabla u + qvu \, dA = \int_\Omega vr \, dA.$$

If we assume that $v = 0$ on Γ_1, we can apply the boundary conditions (14.57) to obtain

$$-\int_{\Gamma_2} pvh \, ds + \int_\Omega p\nabla v \cdot \nabla u + qvu \, dA = \int_\Omega vr \, dA.$$

Now, we express our approximate solution as a linear combination of **trial functions** $\varphi_1, \varphi_2, \ldots, \varphi_M$,

$$u_M(x, y) = \sum_{j=1}^{M} u_j \varphi_j(x, y),$$

where $\varphi_1, \ldots, \varphi_N$ vanish on Γ_1, and $\varphi_{N+1}, \ldots, \varphi_M$ are used to enforce the boundary condition $u = g$ on Γ_1. If we let $v = \varphi_i$, for $i = 1, 2, \ldots, N$, then we obtain a system of linear equations $K\mathbf{c} = \mathbf{f}$, where $\mathbf{c} = \begin{bmatrix} c_1 & \cdots & c_N \end{bmatrix}^T$ and

$$K_{ij} = \int_{\Omega} p\nabla\varphi_i \cdot \nabla\varphi_j + q\varphi_i\varphi_j \, dA, \tag{14.58}$$

$$f_i = \int_{\Omega} \varphi_i r \, dA + \int_{\Gamma_2} p\varphi_i h \, ds - \sum_{k=N+1}^{M} c_k \int_{\Omega} p\nabla\varphi_i \cdot \nabla\varphi_k + q\varphi_i\varphi_k \, dA. \tag{14.59}$$

As in Section 13.4, this is the **Galerkin Method**, in which test and trial functions come from the same function space.

14.5.2 Piecewise Linear Basis Functions

To choose our trial functions, we decompose our domain Ω into triangles. If Ω is not polygonal, then we settle for triangulating a polygonal approximation of Ω. The resulting triangles $T_1, T_2, \ldots, T_E$ are called **elements**. The vertices of the elements, of which there are M, are called **nodes**. Of these nodes, N are in the interior of the domain or on Γ_2, and the remaining $M - N$ lie on Γ_1. Then, we choose our trial functions $\varphi_1, \ldots, \varphi_M$ to be piecewise linear functions that are generalizations of the "hat" functions from Section 13.4, in that if $(x_1, y_1), (x_2, y_2), \ldots, (x_M, y_M)$ are the nodes, then the trial functions satisfy

$$\varphi_i(x_j, y_j) = \delta_{ij} = \begin{cases} 1 & i = j \\ 0 & i \neq j \end{cases}.$$

We say that $\varphi_1, \ldots, \varphi_M$ form a **nodal basis**, meaning that the values of the coefficients $c_1, \ldots, c_M$ are equal to the values of the approximate solution $u_M(x, y)$ at the nodes. It follows at once that for $k = N + 1, \ldots, M$, we have $c_k = g(x_k, y_k)$.

14.5.3 Element-wise Assembly

The easiest way to construct the matrix K, known as the **stiffness matrix**, and the vector $\mathbf{f}$, known as the **load vector**, is to proceed not by node, but rather by element. Only three basis functions $\varphi_{\ell_1}, \varphi_{\ell_2}, \varphi_{\ell_3}$ are nonzero on a given element T_ℓ, because T_ℓ is a triangle that has three nodes as vertices. On T_ℓ, we have

$$\varphi_{\ell_i}(x, y) = a_{\ell_i} + b_{\ell_i}x + c_{\ell_i}y,$$

where the constants $a_{\ell_1}, b_{\ell_1}, c_{\ell_1}$ satisfy the equations

$$a_{\ell_1} + b_{\ell_1}x_{\ell_1} + c_{\ell_1}y_{\ell_1} = 1,$$
$$a_{\ell_1} + b_{\ell_1}x_{\ell_2} + c_{\ell_1}y_{\ell_2} = 0,$$
$$a_{\ell_1} + b_{\ell_1}x_{\ell_3} + c_{\ell_1}y_{\ell_3} = 0,$$

with similar equations for the coefficients of φ_{ℓ_2} and φ_{ℓ_3}. By *Cramer's Rule*, the above system has the solutions

$$a_{\ell_1} = \frac{x_{\ell_2} y_{\ell_3} - y_{\ell_2} x_{\ell_3}}{D_\ell}, \quad b_{\ell_1} = \frac{y_{\ell_2} - y_{\ell_3}}{D_\ell}, \quad c_{\ell_1} = \frac{x_{\ell_3} - x_{\ell_2}}{D_\ell}$$

where

$$D_\ell = x_{\ell_2} y_{\ell_3} - y_{\ell_2} x_{\ell_3} + x_{\ell_3} y_{\ell_1} - x_{\ell_1} y_{\ell_3} + x_{\ell_1} y_{\ell_2} - y_{\ell_1} x_{\ell_2}$$

is the area of T_ℓ when its nodes are ordered counterclockwise.

We can therefore compute an **element stiffness matrix** K_ℓ, which in the case of triangular elements is a 3×3 matrix with entries

$$[K_\ell]_{ij} = \int_{T_\ell} p \nabla \varphi_{\ell_i} \cdot \nabla \varphi_{\ell_j} + q \varphi_{\ell_i} \varphi_{\ell_j} \, dA, \quad i, j = 1, 2, 3. \tag{14.60}$$

Using the above formulas for the coefficients $a_{\ell_i}, b_{\ell_i}, c_{\ell_i}$ for $i = 1, 2, 3$, we can readily evaluate these integrals. It is best to use quadrature rules, which are exact if the coefficients p and q are constant. After computing the element stiffness matrix for each element, their entries can be added to the (global) stiffness matrix K. Specifically, each entry $[K_\ell]_{ij}$ is added to K_{ℓ_i, ℓ_j}, if the nodes ℓ_i and ℓ_j are in the interior of Ω or lie on Γ_2.

Exploration 14.5.1 Write a function K=eltstiff(x,y,p,q) that computes the element stiffness matrix K for a triangular element T_ℓ with vertices specified by the 3-vectors x and y. The input arguments p and q are function handles for the coefficients $p(x, y)$ and $q(x, y)$ in (14.56). Use your function dblinttriangle from Exploration 9.8.6 to approximate the integrals in (14.61).

Similarly, the contributions to the load vector **f** due to T_ℓ are given by the element load vectors

$$[\mathbf{f}_\ell]_i = \int_{T_\ell} \varphi_{\ell_i} r \, dA + \int_{\Gamma_2 \cap T_\ell} p \varphi_{\ell_i} h \, ds -$$

$$\sum_{j=1}^{3} \sum_{k=N+1}^{m} \delta_{k\ell_j} c_{\ell_j} \int_{T_\ell} p \nabla \varphi_{\ell_i} \cdot \nabla \varphi_{\ell_j} + q \varphi_{\ell_i} \varphi_{\ell_j} \, dA, \quad i = 1, 2, 3. \tag{14.61}$$

It should be noted that the boundary terms will only appear in some of these element load vectors. Once the element load vectors are computed, each $[\mathbf{f}_\ell]_i$ is added to $\mathbf{f}_{\ell_i}$.

Exploration 14.5.2 Write a function

$$f=\text{eltload(x,y,b1,b2,p,q,r,g,h)}$$

that computes the element load vector f for a triangular element T_ℓ with vertices specified by the 3-vectors x and y. The input arguments b1 and b2 are *logical* vectors (see Section 1.2.11) indicating whether the vertices lie on the boundaries Γ_1 or Γ_2, respectively. The input arguments p, q, r, g and h are function handles for the coefficients $p(x, y)$, $q(x, y)$, $r(x, y)$, $g(x, y)$, and $h(x, y)$, respectively, in (14.56), (14.57). Use dblinttriangle from Exploration 9.8.6 to approximate the integrals in (14.61).

14.5.4 Putting it All Together

To efficiently perform the assembly of the stiffness matrix and load vector, it is necessary to have a data structure that lists, for each element, the indices of its vertices within the set of all nodes. This is the data structure returned by the MATLAB function `delaunay`, which produces a Delaunay triangulation [Delaunay (1934)] of a set of points in 2-D or 3-D space. The stiffness matrix K is sparse and symmetric, but not necessarily with the structure of a banded matrix. Generally, the system $K\mathbf{c} = \mathbf{f}$ is best solved using an iterative method such as those presented in Chapter 5.

Exploration 14.5.3 Write a function

$$[\texttt{K,f}]\texttt{=femsystem(x,y,b1,b2,p,q,r,g,h)}$$

that triangulates the polygonal domain with vertices specified by `x` and `y` and constructs the global stiffness matrix `K` and load vector `f`, using the functions `eltstiff` and `eltload` from Explorations 14.5.1 and 14.5.2, respectively. The input arguments `b1` and `b2` are logical vectors indicating whether the vertices lie on the boundaries Γ_1 or Γ_2, respectively. The input arguments `p`, `q`, `r`, `g` and `h` are function handles for the coefficients $p(x,y)$, $q(x,y)$, $r(x,y)$, $g(x,y)$ and $h(x,y)$, respectively, in (14.56), (14.57). *Note:* make sure that `K` and `f` only include entries corresponding to nodes in the interior of Ω or on Γ_2!

Exploration 14.5.4 Show that the stiffness matrix K defined in (14.58) is positive definite if, for each entry K_{ij}, the integral over Ω is approximated using a quadrature rule with positive weights.

Example 14.5.1 We now demonstrate the use of `femsystem` from Exploration 14.5.3 on the simple BVP

$$-u''(x,y) = 2, \quad x^2 + y^2 < 1, \tag{14.62}$$

$$u(x,y) = -x^2, \quad x^2 + y^2 = 1, \tag{14.63}$$

which has the exact solution $u(x,y) = -x^2$. Matching this PDE and boundary conditions with (14.56), (14.57), we have $p(x,y) = 1$, $q(x,y) = 0$, $r(x,y) = 2$, $g(x,y) = -x^2$, and $h(x,y)$ is not used since there are no Neumann boundary conditions.

We begin by creating a mesh on the unit disk, using equally-spaced polar coordinates (r_j, θ_i), where $\theta_i = 2\pi i/n_\theta$, $i = 0, 1, \ldots, n_\theta - 1$ and $r_j = j/n_r$, $j = 1, 2, \ldots, n_r$, with $n_\theta = 20$ and $n_r = 4$. We will also include the origin $(0,0)$.

```
nt=20; nr=4;
% theta-values
ts=linspace(0,2*pi,nt);
```

```
% r-values
rs=linspace(0,1,nr);
% remove 2pi due to periodicity
ts=ts(1:end-1);
% remove 0 due to singularity
rs=rs(2:end);
```

Next, to obtain x- and y-coordinates of our mesh points, we use the fact that the **outer product** of two vectors, as defined in Section B.7.3, computes all possible pairwise products of elements of the two vectors.

```
% get Cartesian coordinates, using outer product to get all
% possible pairwise products of, r-values with cos, sin values
x=cos(ts)'*rs;
y=sin(ts)'*rs;
% reshape into vectors
x=reshape(x,numel(x),1);
y=reshape(y,numel(y),1);
% add origin
x=[ 0; x ];
y=[ 0; y ];
```

Next, we specify that Dirichlet boundary conditions are imposed on the entire boundary, using `logical` vectors whose elements have the values `true` or `false`.

```
% record number of interior points
nint=(nr-2)*(nt-1)+1;
% make logical vectors to specify boundary conditions
% (in this case, all Dirichlet)
b1=false(numel(x),1);
% only set to true for boundary points
b1(nint+1:end)=true;
% no points have Neumann boundary conditions
b2=false(numel(x),1);
```

Then, we specify the functions from the BVP, as function handles:

```
% set coefficients, right-hand side, boundary functions
p=@(x,y)(ones(size(x)));
q=@(x,y)(zeros(size(x)));
r=@(x,y)(2*ones(size(x)));
g=@(x,y)(-x.^2);
h=@(x,y)(zeros(size(x)));
```

At last, we are ready to call `femsystem` to set up the stiffness matrix K and load vector f. Then we can solve the system K*u=f to obtain the values of our approximate solution at the interior mesh points.

```
% compute stiffness matrix and load vector
[K,f]=femsystem(x,y,b1,b2,p,q,r,g,h);
% compute solution values at grid points
u=K\f;
```

Finally, we can visualize both the approximate and exact solutions, and obtain the error. We wish to include the boundary points as well, which requires substituting the boundary mesh points into $g(x, y)$ and concatenating the resulting values onto the vector u.

```
% add boundary values for plot
u=[ u; g(x(nint+1:end),y(nint+1:end))] ;
plot3(x,y,u,'ko')
hold on
exact=g(x,y);
% plot exact solution too
plot3(x,y,exact,'k+')
hold off
xlabel('x')
ylabel('y')
zlabel('z')
legend('approximate','exact')
% show error
err=norm(u-exact)/norm(exact)
```

The solutions are shown in Figure 14.3. With the chosen number of mesh points, the error is 0.0117. □

Exploration 14.5.5 Implement Example 14.5.1, except with different numbers of mesh points. How does the error behave as n_r and n_θ increase? Try increasing either one of these individually, or both together. Can you explain your observations?

While finite difference methods, and spectral methods that are covered in Section 14.6, are best suited to problems on rectangular domains, the Finite Element Method is far more versatile. A general 2-D domain can be accurately approximated by a polygon, which can be decomposed into triangles. Similarly, 3-D domains can be approximated by polyhedrons. This facilitates numerical solution of PDEs defined on domains with complicated geometries, such as problems from structural analysis or fluid flow.

We note that in our presentation of the Finite Element Method, both in this section and in Section 13.4 for two-point boundary value problems, we have used only piecewise linear basis functions. Higher-degree piecewise polynomials can be

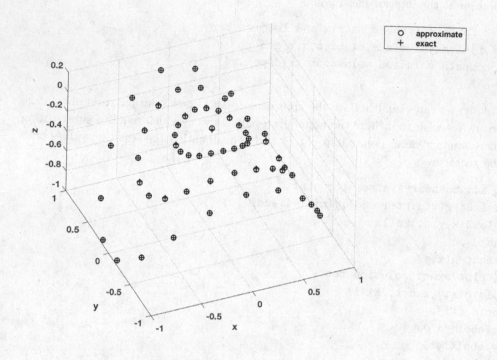

Fig. 14.3   Approximate and exact solutions of (14.62), (14.63) computed in Example 14.5.1

used instead, by prescribing more nodes per element. The use of higher-degree piecewise polynomials reduces the sparsity of the stiffness matrix, but allows for more accurate representation of smoothness in solutions.

14.5.5   Concept Check

1. Define the following terms related to the Finite Element Method:
   (a) element
   (b) node
   (c) stiffness matrix
   (d) element stiffness matrix
   (e) load vector
   (f) nodal basis

2. What is an advantage of the Finite Element Method over other numerical methods for PDEs?

3. Describe the process of deriving a system of linear equations from an elliptic PDE using the Finite Element Method.

4. What is the benefit of using "hat" functions as trial functions?

14.6 Spectral Methods*

We have seen that the error in finite difference methods is $O(h^p)$, where h is the grid spacing and p is some *fixed* positive integer. **Spectral methods**, on the other hand, have error that is $O(h^p)$ for *any* integer p. That is, as $h \to 0$, the error converges to 0 more rapidly than any power of h, provided that the solution is infinitely differentiable; that is, $u \in C^\infty$ (see Section B.13.3). In other words, the error decreases *exponentially*. If the solution is not infinitely smooth; that is, $u \in C^p$ but $u \notin C^{p+1}$, then the error is again $O(h^p)$.

Spectral methods are more limited in applicability than other classes of numerical methods. They are designed for linear PDEs, generally with constant coefficients, defined on domains with simple geometries, such as boxes or disks, with simple boundary conditions. However, for such problems, they tend to be the most effective methods. Ongoing work by the first author of this book and his collaborators (see, for example, [Garon and Lambers (2018); Long et al. (2018); Palchak, et al. (2015)]) is devoted to extending this effectiveness to other classes of PDEs.

Another key distinction between spectral methods and other classes of methods is that these other methods, such as finite difference or finite element methods, are *local* in character; that is, derivatives at a given point are approximated using values at nearby points. Spectral methods, on the other hand, are *global*; derivatives are approximated using values at *all* grid points.

14.6.1 Periodic Boundary Conditions

For concreteness, we illustrate how spectral methods work by deriving a numerical differentiation scheme for a 2π-periodic function $u(x)$ that uses all of its values at N equally spaced grid points $x_j = jh$, $j = 1, 2, \ldots, N$, with $h = 2\pi/N$. Recall from (8.24), (8.25) in Section 8.4.1 that the **Fourier series** of $u(x)$ is

$$u(x) = \frac{1}{2\pi} \sum_{\omega=-\infty}^{\infty} \hat{u}(\omega) e^{i\omega x},$$

where

$$\hat{u}(\omega) = \int_0^{2\pi} e^{-i\omega x} u(x) \, dx.$$

For convenience, we have moved the scaling factor of $1/\sqrt{2\pi}$ from (8.25) to (8.24).

Recall from Section 8.4.2 that by approximating this Fourier series with the values of $u(x)$ on the grid, we obtain the **Fourier interpolant**

$$u_N(x) = \frac{1}{2\pi} \sum_{\omega=-N/2+1}^{N/2} \tilde{u}(\omega) e^{i\omega x}, \tag{14.64}$$

where the coefficients $\tilde{u}(\omega)$ are given by the **Discrete Fourier Transform (DFT)**

$$\tilde{u}(\omega) = h \sum_{j=1}^{N} e^{-i\omega x_j} u(x_j), \quad \omega = -N/2+1, \ldots, N/2. \tag{14.65}$$

As presented in Section 8.4.3, these coefficients can be computed in $O(N \log N)$ floating-point operations using the **Fast Fourier Transform (FFT)**.

To obtain an approximation of $u'(x)$ that uses *all* of the grid points, we compute the exact derivative of the Fourier interpolant. For this purpose, we define the **discrete delta function** δ by

$$\delta(x_j) = \begin{cases} 1 & j = 0 \,(\text{mod } N) \\ 0 & j \neq 0 \,(\text{mod } N) \end{cases}$$

Then we can write

$$u(x_j) = \sum_{m=1}^{N} u(x_m)\delta(x_{j-m}), \quad j = 1, 2, \dots, N,$$

and therefore

$$u_N(x) = \sum_{m=1}^{N} u(x_m)p_N(x - x_m),$$

where $p_N(x)$ is the Fourier interpolant of $\delta(x)$. It follows that

$$u_N'(x_j) \approx \sum_{m=1}^{N} u(x_m)p_N'(x_j - x_m). \tag{14.66}$$

In Section 9.1.2, we computed the derivatives of Lagrange polynomials at x_0 to obtain the coefficients of a finite difference formula for $f'(x_0)$. Similarly, the coefficients of the $\{u(x_m)\}_{m=1}^{N}$ in our N-point formula are given by $p_N'(x_j - x_m)$. To compute these coefficients, we use (14.65) to obtain

$$\tilde{\delta}(\omega) = h \sum_{j=1}^{N} e^{-i\omega x_j}\delta(x_j) = h, \quad \omega = -N/2 + 1, \dots, N/2.$$

We then have

$$p_N(x) = \frac{1}{2\pi} \sum_{\omega=-N/2+1}^{N/2} \tilde{\delta}(\omega)e^{i\omega x} = \frac{h}{2\pi} \sum_{\omega=-N/2+1}^{N/2} e^{i\omega x}.$$

Exploration 14.6.1 Use Euler's identity $e^{i\theta} = \cos\theta + i\sin\theta$, and the trigonometric identities

$$\cos a \sin b = \frac{1}{2}[\sin(a+b) - \sin(a-b)], \quad \sin a + \sin b = 2\sin\frac{a+b}{2}\cos\frac{a-b}{2},$$

to show that

$$p_N(x) = \frac{h}{2\pi}\frac{\sin\frac{\pi x}{h}}{\tan\frac{x}{2}}, \tag{14.67}$$

and therefore

$$p_N'(x_j) = \begin{cases} \frac{1}{2}(-1)^j \cot\frac{jh}{2} & j \neq 0 \,(\text{mod } N) \\ 0 & j = 0 \,(\text{mod } N) \end{cases} \tag{14.68}$$

Hint: Use the fact that $p_N(x)$ is real.

From (14.66) and (14.68), we obtain the N-point formula

$$u'(x_j) \approx \frac{1}{2} \sum_{m=1, m\neq j}^{N} (-1)^{j-m} \cot \frac{(j-m)h}{2} u(x_m).$$

In matrix-vector form, we have $\mathbf{w} = D_N \mathbf{u}$, where $w_j \approx u'(x_j)$ and D_N is the **differentiation matrix**

$$D_N = \begin{bmatrix}
0 & \frac{1}{2}\cot\frac{h}{2} & -\frac{1}{2}\cot\frac{2h}{2} & \frac{1}{2}\cot\frac{3h}{2} & \ddots & & \ddots & \ddots \\
-\frac{1}{2}\cot\frac{h}{2} & 0 & \frac{1}{2}\cot\frac{h}{2} & -\frac{1}{2}\cot\frac{2h}{2} & \frac{1}{2}\cot\frac{3h}{2} & \ddots & & \ddots \\
\frac{1}{2}\cot\frac{2h}{2} & -\frac{1}{2}\cot\frac{h}{2} & 0 & \frac{1}{2}\cot\frac{h}{2} & -\frac{1}{2}\cot\frac{2h}{2} & \frac{1}{2}\cot\frac{3h}{2} & \ddots & \\
\ddots & \ddots & \ddots & \ddots & \ddots & \ddots & \ddots & \\
\ddots & -\frac{1}{2}\cot\frac{3h}{2} & \frac{1}{2}\cot\frac{2h}{2} & -\frac{1}{2}\cot\frac{h}{2} & 0 & \frac{1}{2}\cot\frac{h}{2} & -\frac{1}{2}\cot\frac{2h}{2} \\
& \ddots & -\frac{1}{2}\cot\frac{3h}{2} & \frac{1}{2}\cot\frac{2h}{2} & -\frac{1}{2}\cot\frac{h}{2} & 0 & \frac{1}{2}\cot\frac{h}{2} \\
& & \ddots & -\frac{1}{2}\cot\frac{3h}{2} & \frac{1}{2}\cot\frac{2h}{2} & -\frac{1}{2}\cot\frac{h}{2} & 0
\end{bmatrix} \tag{14.69}$$

Similarly, we can approximate the second derivative using the differentiation matrix

$$D_N^2 = \begin{bmatrix}
-\frac{\pi^2}{3h^2}-\frac{1}{6} & \frac{1}{2}\csc^2\frac{h}{2} & -\frac{1}{2}\csc^2\frac{2h}{2} & \frac{1}{2}\csc^2\frac{3h}{2} & \ddots & \ddots \\
\frac{1}{2}\csc^2\frac{h}{2} & -\frac{\pi^2}{3h^2}-\frac{1}{6} & \frac{1}{2}\csc^2\frac{h}{2} & -\frac{1}{2}\csc^2\frac{2h}{2} & \frac{1}{2}\csc^2\frac{3h}{2} & \ddots \\
-\frac{1}{2}\csc^2\frac{2h}{2} & \frac{1}{2}\csc^2\frac{h}{2} & -\frac{\pi^2}{3h^2}-\frac{1}{6} & \frac{1}{2}\csc^2\frac{h}{2} & -\frac{1}{2}\csc^2\frac{2h}{2} & \\
\ddots & \ddots & \ddots & \ddots & \ddots & \\
\ddots & & \frac{1}{2}\csc^2\frac{3h}{2} & -\frac{1}{2}\csc^2\frac{2h}{2} & \frac{1}{2}\csc^2\frac{h}{2} & -\frac{\pi^2}{3h^2}-\frac{1}{6} & \frac{1}{2}\csc^2\frac{h}{2} \\
\ddots & & & \frac{1}{2}\csc^2\frac{3h}{2} & -\frac{1}{2}\csc^2\frac{2h}{2} & \frac{1}{2}\csc^2\frac{h}{2} & -\frac{\pi^2}{3h^2}-\frac{1}{6}
\end{bmatrix} \tag{14.70}$$

Using these matrices, we can approximate derivatives with error $O(h^N) = O(h^{1/h})$, which converges to zero as $h \to 0$ more rapidly than any positive power of h.

Exploration 14.6.2 Use the differentiation matrix (14.69) and the differentiation matrix from Exploration 9.1.13(b), modified for the interval $[0, 2\pi]$, to compute the derivative of $\sin(5x)$ and $\sin(50x)$ for $N = 64$ and $N = 128$ grid points. What do you observe regarding the accuracy? Explain your observations.

For PDEs with constant coefficients, we can work with Fourier coefficients directly, instead of using a grid.

Example 14.6.1 Consider Poisson's equation in 1-D,

$$u_{xx} = f(x), \quad 0 < x < 2\pi,$$

with $u(x)$ required to be 2π-periodic. The Fourier series of $f(x)$ is

$$f(x) = \frac{1}{2\pi} \sum_{k=-\infty}^{\infty} \hat{f}_k e^{ikx},$$

with a similarly defined Fourier series for $u(x)$. It follows that

$$u_{xx}(x) = -\frac{1}{2\pi} \sum_{k=-\infty}^{\infty} \hat{u}_k k^2 e^{ikx}$$

and therefore

$$\hat{u}_k = -\frac{\hat{f}_k}{k^2}, \quad k \neq 0.$$

The coefficient $\hat{u}_0$ can be chosen arbitrarily, but for a solution to exist, it is necessary that $\hat{f}_0 = 0$. That is, f must have a zero mean on $(0, 2\pi)$. $\square$

In general, differentiation operators with constant coefficients on rectangular domains, with periodic boundary conditions, have eigenfunctions equal to the Fourier basis functions, thus making PDEs involving such operators very easy to solve. The following MATLAB code snippet solves the PDE from Example 14.6.1. We assume that the vector f contains the values of a function $f(x)$ at equally spaced points $x_j = jh$, $j = 1, 2, \ldots, N$.

```
>> Tf=fft(f);
>> w=[ 0:N/2 -N/2+1:-1 ]';
>> Tf=Tf./(-w.^2);
>> Tf(1)=0;
>> u=real(ifft(Tf));
```

The first line uses the built-in function fft to compute the DFT of f. The second line constructs a vector consisting of the values of ω that correspond to DFT coefficients in Tf. The third line computes the DFT coefficients of the second antiderivative of f. This causes a division by zero in the first element of Tf, so that element is set to zero on the fourth line, so that the solution will have a zero mean. We thus have the DFT of the solution. Finally, the fifth line uses the built-in function ifft to perform the inverse FFT, and thus obtain the solution in physical space. Because Tf is complex, roundoff error will cause the inverse DFT to be complex as well, even though it would be real in exact arithmetic. For this reason, we take the real part of the result of ifft before storing the solution in u.

> **Exploration 14.6.3** Use the above code snippet to solve the PDE from Example 14.6.1 with these choices of $f(x)$:
>
> (a)
> $$f(x) = -\sin(100x)$$
>
> (b)
> $$f(x) = \begin{cases} -1 & 0 < x \le \pi/2 \\ 1 & \pi/2 < x \le 3\pi/2 \\ -1 & 3\pi/2 < x \le 2\pi \end{cases}$$
>
> For both cases, use $N = 128$ and $N = 256$ grid points, and compare to the exact solution which can be found analytically (make sure that both your computed solution and the exact solution have zero mean). What do you observe about the accuracy in all cases? Explain your observations.

The FFT can also be used for differentiation. From (14.64), we have

$$\frac{\partial^k}{\partial x^k} u_N(x) = \frac{1}{2\pi} \sum_{\omega = -N/2+1}^{N/2} \tilde{u}(\omega)(i\omega)^k e^{i\omega x}. \qquad (14.71)$$

That is, the kth derivative of the Fourier interpolant of $u(x)$ can be computed exactly. This is accomplished by taking the FFT of the values of $u(x)$ at the grid points $x_j, j = 1, 2, \ldots, N$, then multiplying each coefficient $\tilde{u}(\omega)$ by $(i\omega)^k$, and then performing the inverse FFT. The entire process takes $O(N \log N)$ floating-point operations, assuming N is a number that factors completely into small prime numbers (that is, a *smooth number*). By contrast, using a differentiation matrix such as (14.69) requires $O(N^2)$ operations. It should be noted, however, that if k is odd, then $\tilde{u}(N/2)$ should be zeroed rather than multiplied by $(iN/2)^k$, so that the computed derivative will be real-valued.

> **Exploration 14.6.4** Use the FFT to compute the derivatives of the following functions on a grid with $N = 256$ equally spaced points in $(0, 2\pi]$.
> (a) $f(x) = e^{-(x-\pi)^2}$
> (b) $g(x) = \tan^{-1}(x - \pi)$.
> Compare the computed derivatives to the exact derivatives. Explain the results.

14.6.2 Non-Periodic Boundary Conditions

When a PDE has non-periodic boundary conditions, its solution cannot be represented accurately by a Fourier series, because its periodic extension is discontinuous. This leads to **Gibbs' phenomenon**, which was discussed in Section 8.4.4. Therefore, we use algebraic polynomials instead of trigonometric polynomials.

Unlike with periodic boundary conditions, polynomial interpolation with equally spaced points can lead to large errors, as demonstrated with **Runge's Example**

$f(x) = 1/(1 + x^2)$ on the interval $[-5, 5]$ in Section 7.4.1. Instead, we use **Chebyshev points**, which are equally spaced along the circle and then mapped to the x-axis. That is,

$$x_j = \cos \frac{\pi j}{N}, \quad j = 0, 1, \ldots, N.$$

Unlike in Section 7.4.2, where we used the *roots* of a Chebyshev polynomial $T_k(x)$ as the Chebyshev points, here we use the *extrema* of $T_N(x)$, which include the endpoints $x = \pm 1$.

Given function values $u_j = u(x_j)$, for $j = 0, 1, \ldots, N$, the polynomial $p_N(x)$ that interpolates u at the Chebyshev points is

$$p_N(x) = \sum_{j=0}^{N} u_j \mathcal{L}_{N,j}(x),$$

where

$$\mathcal{L}_{N,j}(x) = w_j \prod_{k=0, k \neq j}^{N} (x - x_k), \quad w_j = \prod_{k=0, k \neq j}^{N} \frac{1}{x_j - x_k}$$

is a **Lagrange polynomial**, as presented in Section 7.2.1. Recall that w_j is a **barycentric weight**, as defined in (7.5).

Proceeding as in Section 9.1.2, we differentiate these Lagrange polynomials to obtain

$$u'(x_i) \approx p_N'(x_i) = \sum_{i=0}^{N} [D_N]_{ij} u_j,$$

where, for $i, j = 0, 1, \ldots, N$,

$$[D_N]_{ij} = \mathcal{L}_{N,j}'(x_i) = \begin{cases} \dfrac{w_j}{w_i(x_i - x_j)} & i \neq j \\ \displaystyle\sum_{k=0, k \neq j}^{N} \dfrac{1}{x_j - x_k} & i = j \end{cases} \tag{14.72}$$

Exploration 14.6.5 Verify (14.72).

The matrix D_N serves as a differentiation matrix for *any* choice of distinct interpolation points $x_0, x_1, \ldots, x_N$. In the particular case of the Chebyshev points, it can be shown [Gottlieb et al. (1984)] that D_N has the form

$$D_N = \begin{bmatrix} \frac{2N^2+1}{6} & \cdots & 2\frac{(-1)^j}{1-x_j} & \cdots & \frac{1}{2}(-1)^N \\ \vdots & \ddots & & \frac{(-1)^{i+j}}{x_i-x_j} & \vdots \\ -\frac{1}{2}\frac{(-1)^i}{1-x_i} & & \frac{-x_j}{2(1-x_j^2)} & & \frac{1}{2}\frac{(-1)^{N+i}}{1+x_i} \\ \vdots & \frac{(-1)^{i+j}}{x_i-x_j} & & \ddots & \vdots \\ -\frac{1}{2}(-1)^N & \cdots & -2\frac{(-1)^{N+j}}{1+x_j} & \cdots & -\frac{2N^2+1}{6} \end{bmatrix}. \tag{14.73}$$

This matrix, and positive powers of it, can be used to compute derivatives of any order with spectral accuracy. One drawback, though, is that unlike differentiation matrices for finite difference schemes, Chebyshev differentiation matrices are not sparse. This means that matrix-vector multiplication requires $O(N^2)$ floating-point operations, while solution of linear systems requires $O(N^3)$. Fortunately, we can instead perform Chebyshev differentiation using the FFT. For details on this approach, the reader is referred to [Trefethen (2000)].

Exploration 14.6.6 Use the Chebyshev differentiation matrix D_N from (14.73) to solve the following BVP for Poisson's equation with homogeneous Dirichlet boundary conditions,

$$u_{xx} = e^{-2x}, \quad -1 < x < 1, \quad u(-1) = u(1) = 0.$$

Hint: Account for the Dirichlet boundary conditions by removing the first and last rows and columns of D_N. How does the accuracy improve as N increases?

Exploration 14.6.7 Repeat Exploration 14.6.6, but with boundary conditions $u(-1) = 1$, $u(1) = 2$.

Exploration 14.6.8 (a) Use **Fixed-Point Iteration** (see Section 10.3) to solve the nonlinear problem

$$u_{xx} = e^u, \quad -1 < x < 1, \quad u(-1) = u(1) = 0.$$

How is the convergence of Fixed-Point Iteration affected by the value of N?

(b) Explain why Fixed-Point Iteration is guaranteed to converge, in view of (11.5). *Hint:* Show that $u(x) < 0$ on $(-1, 1)$.

(c) What happens if you try the same approach on the equation $-u_{xx} = e^u$, with the same boundary conditions? Can you explain the outcome in terms of the sufficient conditions for a BVP to have a unique solution stated at the beginning of Chapter 13?

Exploration 14.6.9 Solve Poisson's equation on a square,

$$u_{xx} + u_{yy} = (x+1)\sin(\pi(1-y)), \quad -1 < x, y < 1,$$

with $u = 0$ on the boundary. *Hint:* Use (14.12) in conjunction with D_N^2 to set up the system of linear equations, and use the MATLAB function `meshgrid` to help construct the right-hand side vector.

Exploration 14.6.10 Consider the wave equation with a source term,

$$-v_{tt} + v_{xx} + v_{yy} = e^{ikt} f(x, y), \quad -1 < x < 1, \quad -1 < y < 1, \quad t > 0,$$

with homogeneous Dirichlet boundary conditions. Assume a solution of the form $v(x, y, t) = e^{ikt} u(x, y)$ and show that $u(x, y)$ solves the **Helmholtz equation**

$$u_{xx} + u_{yy} + k^2 u = f(x, y), \quad -1 < x, y < 1,$$

with $u(x, y)$ also satisfying homogeneous Dirichlet boundary conditions. Solve this problem with $f(x, y) = \cos(\pi x) \sin(\pi y)$ using a Chebyshev differentiation matrix. How is the behavior of the solution affected by the value of k?

14.6.3   Concept Check

1. What is an advantage of spectral methods compared to other classes of methods for solving PDEs? To which problems does this advantage apply?
2. Contrast spectral methods with finite difference methods in terms of how derivatives are approximated.
3. What is the order of accuracy of a spectral differentiation matrix?
4. What is the advantage of using the Fast Fourier Transform (FFT) to solve a PDE with periodic boundary conditions, instead of a differentiation matrix?
5. How can the benefits of spectral methods be realized for a problem that does not have periodic boundary conditions?

14.7   Additional Resources

There are a great many textbooks devoted to classes of numerical methods for solving PDEs. For finite differences, the reader is referred to [Leveque (2007)] and [Strikwerda (1989)]. Another text that focuses on time-dependent problems is [Gustafsson et al. (1995)]; see also [Gustafsson (2008)] which is devoted to high-order methods. An accessible text for using multigrid methods in MATLAB is [Briggs, et al. (2000)]; for more in-depth coverage, see [Trottenberg (2001)]. A comprehensive treatment of the Finite Element Method is given in [Hughes (1987)]. For more information on spectral methods, the reader is referred to [Boyd (2001); Canuto, et al. (2006); Hesthaven, et al. (2007); Shen, et al. (2011)]. A very accessible text for getting started with spectral methods in MATLAB is [Trefethen (2000)]. One class of methods not covered in this book is **finite volume methods**; see [Leveque (2002)] for a thorough treatment.

14.8   Exercises

1. In this exercise and the next, we examine the concept of **deferred correction**, in which the discretization of a PDE is modified so as to cancel the leading error term, thus yielding higher-order accuracy. Here, we apply this concept to the solution of Poisson's equation (14.7).

   (a) Show that the error in the nine-point stencil for the Laplacian defined in (14.18) can be expressed as

   $$\frac{h^2}{12}\nabla^2 f(x_i, y_j) + O(h^4),$$

   where h is the grid spacing in both x and y.

   (b) Modify your function `poissonsolve2d` from Exploration 14.2.3 so that it uses the nine-point stencil instead of the five-point stencil, and the right-hand side $\mathbf{f}$ of the resulting linear system $L_{2D}\mathbf{u} = \mathbf{f}$ is modified by subtracting the leading error term in the nine-point stencil, as it only depends on $f(x, y)$. Use Taylor expansion to show that this method is fourth-order accurate, and verify that your computed solution does indeed exhibit this order of accuracy as $h \to 0$.

2. Consider the IBVP for the heat equation in 1-D,

   $$u_t = u_{xx}, \quad 0 < x < 1, \quad t > 0,$$

   $$u(0, t) = u(1, t) = 0, \quad t > 0,$$

   $$u(x, 0) = \sin 2\pi x, \quad 0 < x < 1.$$

   (a) Compute the exact solution of this problem.
   (b) Solve this problem by using centered differencing in space and Backward Euler in time, for a fixed value of N, the number of grid points, and various values of Δt. Produce a convergence plot, in which the error is plotted against Δt using logarithmic scales on both axes (use the MATLAB function `loglog`). How does the error decrease as Δt decreases?
   (c) Compute the local truncation error. Is it consistent with the observed convergence behavior?
   (d) Improve the accuracy of your computed solution by modifying the numerical method to "cancel out" the leading terms in the local truncation error. This is another example of *deferred correction* as described in Exercise 1. How does the convergence behavior change? Does it match theoretical expectations?

3. Consider the following IBVP for the wave equation,

   $$u_{tt} = u_{xx}, \quad 0 < x < 1, \quad t > 0,$$

   $$u(0, t) = u(1, t), \quad t > 0,$$

   $$u(x, 0) = e^{-200(x-1/2)^2}, \quad u_t(x, 0) = 0, \quad 0 < x < 1.$$

(a) Rewrite the PDE as a system of two PDEs that are first-order in time by introducing a new dependent variable $v = u_t$.

(b) Solve the resulting system to $t = 1$ using centered differencing in space, Forward Euler in time for u, followed by Backward Euler in time for v. Show that the resulting method is actually explicit, even though an implicit method is used. Plot the computed solution u after each time step. How does u behave over time? Can you explain its behavior?

(c) Show that the method described in part 3b is conditionally stable, with the stability condition for this IBVP being $\Delta t \leq \Delta x$.

(d) Solve the same problem, in the same manner, except with homogeneous Dirichlet boundary conditions instead of periodic. How does the behavior of the computed solution u change? Can you explain this behavior?

4. Consider the BVP

$$-\epsilon u_{xx} + u_x = 1, \quad -1 < x < 1,$$

$$u(-1) = u(1) = 0.$$

Let $\epsilon = 10^{-4}$.

(a) Solve this BVP using centered differences for both u_{xx} and u_x, for several values of N, the number of interior grid points. Explain the behavior of your computed solution and how it behaves as N increases.

(b) Using techniques for solving second-order ODEs, solve the BVP analytically. What difficulty arises when trying to evaluate this analytical solution in MATLAB, and why?

(c) Replace $-\epsilon$ with ϵ. How does the behavior of the computed solution change, and why?

(d) Change the centered difference for u_x to a forward difference. How does this change the behavior of the computed solution as N increases? What about a backward difference? Can you explain the difference in behavior?

(e) Solve this BVP using a Chebyshev differentiation matrix, with several values of N. How does the performance compare to that of finite differences? Explain any discrepancy in performance.

5. Consider 2-D Poisson's equation on the unit square,

$$\Delta u = u_{xx} + u_{yy} = f(x, y), \quad 0 < x < 1, \quad 0 < y < 1,$$

with homogeneous Dirichlet boundary conditions.

(a) In MATLAB, create a matrix A that discretizes the Laplacian using centered differences for u_{xx} and u_{yy} on a uniform grid with N interior points. Use the **spy** command to visualize the pattern of nonzero entries in A.

(b) Choose N to be odd, and reorder the rows and columns of A so that odd-numbered and even-numbered rows and columns are grouped together. Visualize the structure of A again using **spy**. How is this structure advantageous for solving the system of equations $A\mathbf{u} = \mathbf{f}$? Describe an algorithm for solving this system efficiently. This ordering is called **red-black ordering**, as it has the effect of labeling the grid points in the domain like red and black spaces on a checkerboard, and then grouping the red and black spaces together.

(c) How can red-black ordering be implemented when N is even? Carry this out in MATLAB and use **spy** to verify that the same kind of structure is obtained as in the case where N is odd.

6. Consider 2-D Poisson's equation on the square $[0, 2\pi]^2$,

$$\Delta u = u_{xx} + u_{yy} = f(x, y), \quad 0 < x < 2\pi, \quad 0 < y < 2\pi,$$

with periodic boundary conditions

$$u(0, y) = u(2\pi, y), \quad u_x(0, y) = u_x(2\pi, y), \quad 0 < y < 2\pi,$$

$$u(x, 0) = u(x, 2\pi), \quad u_y(x, 0) = u_y(x, 2\pi), \quad 0 < x < 2\pi.$$

In this problem, we will let $f(x, y) = \sin 2x \cos 4y$.

(a) Use Fourier series representations of u and f to show that this problem only has a solution if f has zero mean on $[0, 2\pi]^2$. Furthermore, show that if a solution exists, it is not unique; rather, there are actually infinitely many solutions.

(b) Solve this problem using the standard five-point stencil. Verify that your computed solution is second-order accurate.

(c) Solve this problem using a 2-D FFT. Compare the accuracy of this solution to the one computed using finite differences. Explain any difference you observe.

(d) Use our knowledge of the eigenvalues and eigenvectors of the finite difference discretization of the Laplacian to compare, analytically, the solution computed via finite differences to the exact solution. Is the computed solution more accurate if it contains primarily low-frequency content, or high-frequency content? Confirm your conclusion by solving the problem numerically using $f(x, y) = \sin 20x \cos 20y$.

7. Consider the **inviscid Burgers' equation**

$$u_t + u u_x = 0, \quad 0 < x < 1, \quad t > 0$$

with initial and boundary conditions

$$u(x, 0) = 1 + \sin \pi x, \quad u(0) = u(1).$$

Solve this problem using (forward) Euler's Method (12.8) in time, and the following finite difference formulas in space, up to $t = 1$:

(a) forward differencing (9.1)
(b) backward differencing (9.2)
(c) centered differencing (9.3)

In all cases, use a uniform grid with $N = 100$ grid points and $\Delta t = 1/200$. How does the solution behave in each case? Can you explain the difference in behavior? For this problem, backward differencing is also **upwind differencing**. The difficulties encountered in solving Burgers' equation are caused by the exact solution developing a **shock**, which is a jump discontinuity in the solution.

8. Consider the **Allen-Cahn equation** in 2-D,

$$u_t = \alpha \Delta u + u - u^3, \quad 0 < x < 1, \quad 0 < y < 1, \quad t > 0,$$

$$u_x(0, y) = u_x(1, y) = 0, \quad 0 < y < 1,$$

$$u_y(x, 0) = u_y(x, 1) = 0, \quad 0 < x < 1,$$

$$u(x, y, 0) = 1 + \cos \pi x \cos 2\pi y,$$

with $\alpha = 0.2$.

(a) Solve this IBVP to $t = 0.2$ using central differencing in space and RK4 for time-stepping. Use several values of N, the number of grid points per dimension. For each value of N, determine the largest value of Δt for which the solution appears to have the correct qualitative behavior. How does this value of Δt change as N increases?

(b) For a fixed value of N, solve the problem to $t = 0.2$ with several values of Δt. Is fourth-order accuracy achieved? To measure error, since an exact solution is not available, compute an "exact" solution by first using an extremely small time step (note: this may take a while).

(c) Use the MATLAB ODE solvers ode45 and ode15s to compute solutions at $t = 0.2$. Compare the number of time steps required by the two solvers. How is the number of time steps affected by increasing N?

9. Consider the time-dependent **Schrödinger equation** in 1-D,

$$iu_t = -\Delta u + V(x)u, \quad 0 < x < 2\pi, \quad t > 0,$$

$$u(0, t) = u(2\pi, t), \quad t > 0.$$

We will use the *potential* $V(x) = (x - c)^4 + 0.1$.

(a) Solve this problem with $c = \pi/2$ and random initial data, using centered differencing in space and Backward Euler in time. Plot the real part of the solution versus its imaginary part. In separate figures, plot the real and imaginary parts versus x.

(b) Repeat with $c = \pi$. Use your plots to explain the difference in behavior.

10. In this exercise, we consider the variable-coefficient heat equation

$$\frac{\partial u}{\partial t} = \frac{1}{\alpha^2(x)} \frac{\partial}{\partial x} \left(\beta^2(x) \frac{\partial u}{\partial x} \right), \quad 0 < x < 1, \quad t > 0,$$

where $\alpha^2(x)$ is the specific heat and $\beta^2(x)$ is the diffusion coefficient. We impose homogeneous Dirichlet boundary conditions $u(0,t) = u(1,t) = 0$ and initial condition $u(x,0) = f(x)$. Let $\Delta x = 1/(N+1)$ for some positive integer N, and let $x_j = j\Delta x$ for $j = 1, 2, \ldots, N$. We will use finite differences in space to discretize the PDE and obtain a system of ODEs of the form

$$\mathbf{u}'(t) = L_N \mathbf{u}(t),$$

where L_N is an $N \times N$ matrix and

$$\mathbf{u}(t) = \begin{bmatrix} u_1(t) & u_2(t) & \cdots & u_N(t) \end{bmatrix},$$

with $u_j(t) \approx u(x_j, t)$. To construct the matrix L_N, we let D_N^+ and D_N^- denote the differentiation matrices for the forward and backward difference, respectively (see Section 9.1.5). Furthermore, we let A_N and B_N be diagonal matrices with the values of $\alpha^{-2}(x)$ and $\beta^2(x)$, respectively, at the grid points x_j, $j = 1, 2, \ldots, N$ on the main diagonal. Use the following approaches to obtain L_N:

(a) $L_N = A_N D_N^+ B_N D_N^-$
(b) $L_N = A_N D_N^- B_N D_N^+$
(c) Suppose we use the Product Rule to write the PDE in the form

$$u_t = \alpha^{-2}(x)\beta^2(x)u_{xx} + 2\alpha^{-2}(x)\beta(x)\beta'(x)u_x$$

and discretize the spatial differential operator on the right side with centered differences for both u_x and u_{xx}. What is the resulting matrix L_N in this case?

(d) Suppose we use a *staggered grid* $x_{j+1/2} = (j + 1/2)\Delta x$, $j = 0, 1, 2, \ldots, N$, consisting of the midpoints of the subintervals $[x_j, x_{j+1}]$ defined by the original grid. Then, suppose we use a centered difference to approximate $\beta^2(x)u_x$ on the staggered grid, using values of $\mathbf{u}(t)$ from the original grid. That is,

$$u_x(x_{j+1/2}, t) \approx \frac{u_{j+1}(t) - u_j(t)}{\Delta x}, \quad j = 0, 1, 2, \ldots, N.$$

If we then use these approximations in a centered difference to obtain approximations of $\alpha^{-2}(x_j)\left(\beta^2(x)u_x(x,t)\right)_x\big|_{x=x_j}$, $j = 1, 2, \ldots, N$, what is the corresponding matrix L_N?

Under what circumstances are each of these matrices symmetric? Why is symmetry desirable? Which matrix is preferable to the others? Explain. Then, modify your functions `eulerheat`, `backeulerheat` and `crankheat` from Explorations 14.3.1 and 14.3.9 to solve this IBVP. Add a new input parameter `beta` corresponding to $\beta(x)$. For both `alpha` and `beta`, use the function `isnumeric` (see Exploration 9.8.2) to determine whether to interpret these arguments as scalars or function handles. Do you observe any difference in the behavior of the computed solutions, depending on the choice of L_N? Use different values of Δx and Δt to determine the spatial and temporal order of accuracy. Do they match theoretical expectations? How should the condition from Exploration 14.3.5 be modified to ensure a bounded solution in the case of Forward Euler?

11. The **Perona-Malik equation** [Perona and Malik (1990)]

$$u_t = \nabla \cdot \left(\frac{1}{1 + k^2(u_x^2 + u_y^2)} \nabla u \right)$$

is a nonlinear diffusion equation used for denoising and deblurring images. It features both forward and backward diffusion, as the coefficients of u_{xx} and u_{yy} can be either positive or negative. Therefore, this equation is actually **ill-posed**, but has been found to be effective nonetheless when solved numerically. Choose a color image file, and then use the following commands to load it into MATLAB and add random noise to it.

```
>> A=imread(fname);
>> B=imnoise(A,'gaussian');
```

The variable `fname` is assumed to be a character array containing the filename. Then A is an $m \times n \times 3$ array, where the image size is $m \times n$ pixels. The third dimension accounts for the RGB channels of the image. Choose one of these channels (for example, `A(:,:,1)` for the red channel) and use it as initial data for the Perona-Malik equation. Use Backward Euler in time and centered differencing in space, and periodic boundary conditions. Find parameters, including k in the PDE itself, Δt, and a final time T, so that the quality of the noisy image is visibly improved. If you find that using the backslash operator to perform implicit time-stepping is impractical, then try using one of MATLAB's iterative solvers instead, such as `minres`. For spatial discretization, use an approach similar to that used in Exercise 10 to handle the spatially varying coefficient. To avoid having to solve a system of nonlinear equations, *linearize* the PDE by evaluating the diffusion coefficient at time t_n when computing $\mathbf{u}^{n+1}$. Explain why this simplification is reasonable when using first-order time-stepping, but not higher-order. To construct differentiation matrices for partial differentiation with respect to x and y, use the same approach as in (14.12) for the 2-D Laplacian.

12. Repeat the Finite Element discretization of (14.56), generalized to 3-D. Then, adapt the functions `eltstiff`, `eltload`, and `femsystem` from Explorations 14.5.1, 14.5.2 and 14.5.3, respectively, to the 3-D case. Use the MATLAB function `delaunay` to perform the 3-D analogue of triangulation of the spatial domain.

Appendices

Appendix A

Review of Calculus

Among the mathematical problems that can be solved using techniques from numerical analysis are the basic problems of differential and integral calculus:

- computing the instantaneous rate of change of one quantity with respect to another, which is a **derivative**, and
- computing the total change in a function from its rate of change, which is a **definite integral**.

Calculus also plays an essential role in the development and analysis of techniques used in numerical analysis, including those techniques that are applied to problems not arising directly from calculus. Therefore, it is appropriate to review some basic concepts from calculus to ensure that we have a sufficient foundation to begin our study of numerical analysis.

A.1 Limits and Continuity

A.1.1 Limits

The basic problems of differential and integral calculus described in the previous paragraph can be solved by computing a sequence of approximations to the desired quantity and then determining what value, if any, the sequence of approximations approaches. This value is called a **limit** of the sequence. As a sequence is a function, we begin by defining, precisely, the concept of the limit of a function.

Definition A.1.1 We write

$$\lim_{x \to a} f(x) = L$$

if, for any open interval I_1 containing L, there is some open interval I_2 containing a such that $f(x) \in I_1$ whenever $x \in I_2$, and $x \neq a$. We say that L is the **limit** of $f(x)$ as x **approaches** a.

Definition A.1.2 We write

$$\lim_{x \to a^-} f(x) = L$$

if, for any open interval I_1 containing L, there is an open interval I_2 of the form (c, a), such that $f(x) \in I_1$ whenever $x \in I_2$. We say that L is the **limit of $f(x)$ as x approaches a from the left**, or the **left-hand limit of $f(x)$ as x approaches a.**

Similarly, we write

$$\lim_{x \to a^+} f(x) = L$$

if, for any open interval I_1 containing L, there is an open interval I_2 of the form (a, c), such that $f(x) \in I_1$ whenever $x \in I_2$. We say that L is the **limit of $f(x)$ as x approaches a from the right**, or the **right-hand limit of $f(x)$ as x approaches a.**

We can make the definition of a limit a little more concrete by imposing sizes on the intervals I_1 and I_2, as long as the interval I_1 can still be of arbitrary size. It can be shown that the following definition is equivalent to the previous one.

Definition A.1.3 We write

$$\lim_{x \to a} f(x) = L$$

if, for any $\epsilon > 0$, there exists a number $\delta > 0$ such that $|f(x) - L| < \epsilon$ whenever $0 < |x - a| < \delta$.

Similar definitions can be given for the left-hand and right-hand limits. Note that in either definition, the point $x = a$ is specifically excluded from consideration when requiring that $f(x)$ be close to L whenever x is close to a. This is because the concept of a limit is only intended to describe the behavior of $f(x)$ *near $x = a$*, as opposed to its behavior *at $x = a$*. Later in this appendix, we discuss the case where the two distinct behaviors coincide.

A.1.2   Limits at Infinity

The concept of a limit defined above is useful for describing the behavior of a function $f(x)$ as x approaches a *finite* value a. However, suppose that the function f is a **sequence**, which is a function that maps $\mathbb{N}$, the set of natural numbers, to $\mathbb{R}$, the set of real numbers. We will denote such a sequence by $\{f_n\}_{n=0}^{\infty}$, or simply $\{f_n\}$. In numerical analysis, it is sometimes necessary to determine the value that the terms of a sequence $\{f_n\}$ approach as $n \to \infty$. Such a value, if it exists, is not a limit, as defined previously. However, it is natural to use the notation of limits to describe this behavior of a function. We therefore define what it means for a sequence $\{f_n\}$ to have a limit as n becomes infinite.

Definition A.1.4 (Limit at Infinity) Let $\{f_n\}$ be a sequence defined for all integers not less than some integer n_0. We say that the **limit of** $\{f_n\}$ **as** n **approaches** ∞ is equal to L, and write

$$\lim_{n\to\infty} f_n = L,$$

if for $\epsilon > 0$, there exists a number M such that $|f_n - L| < \epsilon$ whenever $x > M$.

Example A.1.5 Let the sequence $\{f_n\}_{n=1}^{\infty}$ be defined by $f_n = 1/n$ for every positive integer n. Then

$$\lim_{n\to\infty} f_n = 0,$$

since for any $\epsilon > 0$, no matter how small, we can find a positive integer n_0 such that $|f_n| < \epsilon$ for all $n \geq n_0$. In fact, for any given ϵ, we can choose $n_0 = \lceil 1/\epsilon \rceil$, where $\lceil x \rceil$, known as the **ceiling function**, denotes the smallest integer that is greater than or equal to x. $\square$

A.1.3   Continuity

In many cases, the limit of a function $f(x)$ as x approaches a can be obtained by simply computing $f(a)$. Intuitively, this indicates that f has to have a graph that is one continuous curve, because any "break" or "jump" in the graph at $x = a$ is caused by $f(x)$ approaching one value as x *approaches a*, only to actually assume a different value *at a*. This leads to the following precise definition of what it means for a function to be continuous at a given point.

Definition A.1.6 (Continuity) We say that a function f is **continuous at** a if

$$\lim_{x\to a} f(x) = f(a).$$

We also say that $f(x)$ has the **Direct Substitution Property** at $x = a$.

We say that a function f is **continuous from the right** at a if

$$\lim_{x\to a^+} f(x) = f(a).$$

Similarly, we say that f is **continuous from the left** at a if

$$\lim_{x\to a^-} f(x) = f(a).$$

The preceding definition describes continuity at a single point. In describing where a function is continuous, the concept of continuity over an interval is useful.

Definition A.1.7 (Continuity on an Interval) We say that a function f is **continuous on the interval** (a, b) if f is continuous at every point in (a, b). Similarly, we say that f is continuous on

1. $[a, b)$ if f is continuous on (a, b), and continuous from the right at a.
2. $(a, b]$ if f is continuous on (a, b), and continuous from the left at b.
3. $[a, b]$ if f is continuous on (a, b), continuous from the right at a, and continuous from the left at b.

In numerical analysis, it is often necessary to construct a continuous function, such as a polynomial, based on data obtained by measurements and problem-dependent constraints. In this book, we will learn some of the most basic techniques for constructing such continuous functions.

A.1.4   The Intermediate Value Theorem

Suppose that a function f is continuous on some closed interval $[a, b]$. The graph of such a function is a continuous curve connecting the points $(a, f(a))$ with $(b, f(b))$. If one were to draw such a graph, their pen would not leave the paper in the process, and therefore it would be impossible to "avoid" any y-value between $f(a)$ and $f(b)$. This leads to the following statement about such continuous functions.

Theorem A.1.8 (Intermediate Value Theorem) *Let f be continuous on $[a, b]$. Then, on (a, b), f assumes every value between $f(a)$ and $f(b)$; that is, for any value y between $f(a)$ and $f(b)$, $f(c) = y$ for some c in (a, b).*

The Intermediate Value Theorem has a very important application in the problem of finding solutions of a general equation of the form $f(x) = 0$, where f is a given continuous function. Often, methods for solving such an equation try to identify an interval $[a, b]$ where $f(a) > 0$ and $f(b) < 0$, or vice versa. In either case, the Intermediate Value Theorem states that f must assume every value between $f(a)$ and $f(b)$, and since 0 is one such value, it follows that the equation $f(x) = 0$ must have a solution somewhere in the interval (a, b).

We can find an approximation of this solution using a procedure called the **Bisection Method**, which repeatedly applies the Intermediate Value Theorem to smaller and smaller intervals that contain the solution. We will study the Bisection Method, and other methods for solving the equation $f(x) = 0$, in Chapter 10.

A.1.5   Limits of Functions of Several Variables

The notions of limit and continuity generalize to vector-valued functions and functions of several variables in a straightforward way.

Definition A.1.9 Given a function $f : D \subseteq \mathbb{R}^n \to \mathbb{R}$ and a point $\mathbf{x}_0 \in D$, we write

$$\lim_{\mathbf{x} \to \mathbf{x}_0} f(\mathbf{x}) = L$$

if, for any $\epsilon > 0$, there exists a $\delta > 0$ such that $|f(\mathbf{x}) - L| < \epsilon$ whenever $\mathbf{x} \in D$ and $0 < \|\mathbf{x} - \mathbf{x}_0\| < \delta$.

In this definition, we can use any appropriate vector norm $\| \cdot \|$, as discussed in Section B.13. For now, we assume the Euclidean norm $\|\mathbf{x}\| = \sqrt{x_1^2 + \cdots + x_n^2}$.

Definition A.1.10 We say that $f : D \subseteq \mathbb{R}^n \to \mathbb{R}$ is **continuous** at a point $\mathbf{x}_0 \in D$ if

$$\lim_{\mathbf{x} \to \mathbf{x}_0} f(\mathbf{x}) = f(\mathbf{x}_0).$$

It can be shown f is continuous at $\mathbf{x}_0$ if its partial derivatives are bounded near $\mathbf{x}_0$.

Having defined limits and continuity for scalar-valued functions of several variables, we can now define these concepts for vector-valued functions. Given a vector-valued function $\mathbf{F} : D \subseteq \mathbb{R}^n \to \mathbb{R}^n$, and $\mathbf{x} = (x_1, x_2, \ldots, x_n) \in D$, we write

$$\mathbf{F}(\mathbf{x}) = \begin{bmatrix} f_1(\mathbf{x}) \\ f_2(\mathbf{x}) \\ \vdots \\ f_n(\mathbf{x}) \end{bmatrix} = \begin{bmatrix} f_1(x_1, x_2, \ldots, x_n) \\ f_2(x_1, x_2, \ldots, x_n) \\ \vdots \\ f_n(x_1, x_2, \ldots, x_n) \end{bmatrix}$$

where $f_1, f_2, \ldots, f_n$ are the **component functions**, or **coordinate functions**, of $\mathbf{F}$.

Definition A.1.11 Given $\mathbf{F} : D \subseteq \mathbb{R}^n \to \mathbb{R}^n$, $\mathbf{x}_0 \in D$ and $\mathbf{L} \in \mathbb{R}^n$, we say that

$$\lim_{\mathbf{x} \to \mathbf{x}_0} \mathbf{F}(\mathbf{x}) = \mathbf{L}$$

if and only if

$$\lim_{\mathbf{x} \to \mathbf{x}_0} f_i(\mathbf{x}) = L_i, \quad i = 1, 2, \ldots, n.$$

Similarly, we say that $\mathbf{F}$ is **continuous** at $\mathbf{x}_0$ if and only if each coordinate function f_i is continuous at $\mathbf{x}_0$. Equivalently, $\mathbf{F}$ is continuous at $\mathbf{x}_0$ if

$$\lim_{\mathbf{x} \to \mathbf{x}_0} \mathbf{F}(\mathbf{x}) = \mathbf{F}(\mathbf{x}_0).$$

A.2 Derivatives

The basic problem of differential calculus is computing the instantaneous rate of change of one quantity y with respect to another quantity x. For example, y may represent the position of an object and x may represent time, in which case the instantaneous rate of change of y with respect to x is interpreted as the velocity of the object.

When the two quantities x and y are related by an equation of the form $y = f(x)$, it is certainly convenient to describe the rate of change of y with respect to x in terms of the function f. Because the instantaneous rate of change is so commonplace, it is practical to assign a concise name and notation to it, which we do now.

Definition A.2.1 (Derivative) The **derivative** of a function $f(x)$ at $x = a$, denoted by $f'(a)$, is

$$f'(a) = \lim_{h \to 0} \frac{f(a+h) - f(a)}{h},$$

provided that the above limit exists. When this limit exists, we say that f is **differentiable** at a.

Given a function $f(x)$ that is differentiable at $x = a$, the following numbers are all equal:

- the derivative of f at $x = a$, $f'(a)$,
- the slope of the tangent line of f at the point $(a, f(a))$, and
- the instantaneous rate of change of $y = f(x)$ with respect to x at $x = a$.

This can be seen from the fact that all three numbers are defined in the same way.

Many functions can be differentiated using differentiation rules learned in a calculus course. However, for some functions this is not the case. For example, we may need to compute the instantaneous rate of change of a quantity $y = f(x)$ with respect to another quantity x, where our only knowledge of the function f that relates x and y is a set of pairs of x-values and y-values that may be obtained using measurements. In Section 9.1 we will learn how to approximate the derivative of such a function using this limited information. The most common methods involve constructing a continuous function, such as a polynomial, based on the given data, using interpolation as in Chapter 7. The polynomial can then be differentiated using differentiation rules. Since the polynomial is an approximation of the function $f(x)$, its derivative is an approximation of $f'(x)$.

A.2.1   Differentiability and Continuity

Consider a tangent line of a function f at a point $(a, f(a))$. Because this tangent line is the limit of secant lines that can cross the graph of f at points on *either side* of a, it seems impossible that f can fail to be continuous at a. The following result confirms this: a function that is differentiable at a given point (and therefore has a tangent line at that point) *must* also be continuous at that point.

Theorem A.2.2 *If f is differentiable at a, then f is continuous at a.*

It is important to keep in mind, however, that the *converse* of the above statement, "if f is continuous, then f is differentiable", is not true. It is actually very easy to find examples of functions that are continuous at a point, but fail to be differentiable at that point.

As an extreme example, it is known that there is a function that is continuous everywhere, but is differentiable *nowhere*. Specifically, the **Weierstrass function** [Hardy (1917)], defined by

$$W(x) = \sum_{n=0}^{\infty} a^n \cos(b^n \pi x), \quad 0 < a < 1, \tag{A.1}$$

where b is a positive odd integer such that $ab > 1 + 3\pi/2$, has this property.

Example A.2.3 The functions $f(x) = |x|$ and $g(x) = x^{1/3}$ are examples of functions that are continuous for all x, but are not differentiable at $x = 0$. The graph of the *absolute value function* $|x|$ has a sharp corner at $x = 0$, since the one-sided limits

$$\lim_{h \to 0^-} \frac{f(h) - f(0)}{h} = -1, \quad \lim_{h \to 0^+} \frac{f(h) - f(0)}{h} = 1$$

do not agree, but in general these limits must agree in order for $f(x)$ to have a derivative at $x = 0$.

The cube root function $g(x) = x^{1/3}$ is not differentiable at $x = 0$ because the tangent line to the graph at the point $(0, 0)$ is vertical, so it has no finite slope. We can also see that the derivative does not exist at this point by noting that the function $g'(x) = (1/3)x^{-2/3}$ has a vertical asymptote at $x = 0$. □

A.3 Extreme Values

In many applications, it is necessary to determine where a given function attains its minimum or maximum value. For example, a business wishes to maximize profit, so it can construct a function that relates its profit to variables such as number of units produced, or wages for labor. We now consider the basic problem of finding a maximum or minimum value of a general function $f(x)$ that depends on a single independent variable x. First, we must precisely define what it means for a function to *have* a maximum or minimum value.

> **Definition A.3.1 (Absolute extrema)** A function f has a **absolute maximum** or **global maximum** at c if $f(c) \geq f(x)$ for all x in the domain of f. The number $f(c)$ is called the **maximum value** of f on its domain. Similarly, f has a **absolute minimum** or **global minimum** at c if $f(c) \leq f(x)$ for all x in the domain of f. The number $f(c)$ is then called the **minimum value** of f on its domain. The maximum and minimum values of f are called the **extreme values** of f, and the absolute maximum and minimum are each called an **extremum** of f.

Before computing the maximum or minimum value of a function, it is natural to ask whether it is possible to determine in advance whether a function even has a maximum or minimum, so that effort is not wasted in trying to solve a problem that has no solution. The following result is very helpful in answering this question.

> **Theorem A.3.2 (Extreme Value Theorem)** If f is continuous on $[a, b]$, then f has an absolute maximum and an absolute minimum on $[a, b]$.

Now that we can easily determine whether a function has a maximum or minimum on a closed interval $[a, b]$, we can develop a method for actually finding them. It turns out that it is easier to find points at which f attains a maximum or minimum value in a "local" sense, rather than a "global" sense. In other words, we can best find the absolute maximum or minimum of f by finding points at which f achieves a maximum or minimum with respect to "nearby" points, and then determine which of these points is the absolute maximum or minimum. The following definition makes this notion precise.

> **Definition A.3.3 (Local extrema)** A function f has a **local maximum** at c if $f(c) \geq f(x)$ for all x in an open interval containing c. Similarly, f has a **local minimum** at c if $f(c) \leq f(x)$ for all x in an open interval containing c. A local maximum or local minimum is also called a **local extremum**.

At each point at which f has a local maximum, the function either has a horizontal tangent line, or no tangent line due to not being differentiable. A similar statement applies to local minima. To state the formal result, we first introduce the following definition, which will also be useful when describing a method for finding local extrema.

> **Definition A.3.4 (Critical Point)** A number c in the domain of a function f is a **critical number** or **critical point** of f if $f'(c) = 0$ or $f'(c)$ does not exist.

The following result describes the relationship between critical points and local extrema.

> **Theorem A.3.5 (Fermat's Theorem)** If f has a local minimum or local maximum at c, then c is a critical point of f; that is, either $f'(c) = 0$ or $f'(c)$ does not exist.

This theorem suggests that the maximum or minimum value of a differentiable function $f(x)$ can be found by solving the equation $f'(x) = 0$. As mentioned previously, we will be learning techniques for solving such equations in Chapter 10. These techniques play an essential role in the solution of problems in which one must compute the maximum or minimum value of a function, subject to constraints on its variables. Such problems are called **optimization problems**, which are covered in Chapter 11.

A.4   Integrals

There are many cases in which some quantity is defined to be the product of two other quantities. For example, a rectangle of width w has uniform height h, and the area A of the rectangle is given by the formula $A = wh$. Unfortunately, in many applications, we cannot necessarily assume that certain quantities such as height are constant, and therefore formulas such as $A = wh$ cannot be used directly. However, they can be used indirectly to solve more general problems by employing the notation known as **integral calculus**.

Suppose we wish to compute the area A of a shape that is not a rectangle. To simplify the discussion, we assume that the shape is bounded by the vertical lines $x = a$ and $x = b$, the x-axis, and the curve defined by some continuous function $y = f(x)$, where $f(x) \geq 0$ for $a \leq x \leq b$. Then, we can approximate this shape by n rectangles that have width $\Delta x = (b-a)/n$ and height $f(x_i)$, where $x_i = a + i\Delta x$, for $i = 0, \dots, n$. We obtain the approximation

$$A \approx A_n = \sum_{i=1}^{n} f(x_i)\Delta x.$$

Intuitively, we can conclude that as $n \to \infty$, the approximate area A_n will converge to the exact area A of the given region. This can be seen by observing that as n increases, the n rectangles defined above constitute a more accurate approximation of the region.

More generally, suppose that for each $n = 1, 2, \dots$, we define the quantity R_n by choosing points $a = x_0 < x_1 < \cdots < x_n = b$, and computing the sum

$$R_n = \sum_{i=1}^{n} f(x_i^*)\Delta x_i, \quad \Delta x_i = x_i - x_{i-1}, \quad x_{i-1} \leq x_i^* \leq x_i.$$

The sum that defines R_n is known as a **Riemann sum**. Note that the interval $[a, b]$ need not be divided into subintervals of equal width, and that $f(x)$ can be evaluated at *arbitrary* points belonging to each subinterval.

If $f(x) \geq 0$ on $[a, b]$, then R_n converges to the area under the curve $y = f(x)$ as $n \to \infty$, provided that the widths of all of the subintervals $[x_{i-1}, x_i]$, for $i = 1, \dots, n$, approach zero. This behavior is ensured if we require that

$$\lim_{n \to \infty} \delta(n) = 0, \quad \text{where} \quad \delta(n) = \max_{1 \leq i \leq n} \Delta x_i.$$

This condition is necessary because if it does not hold, then, as $n \to \infty$, the region formed by the n rectangles will not converge to the region whose area we wish to compute. If f assumes negative values on $[a, b]$, then, under the same conditions on the widths of the subintervals, R_n converges to the *net* area between the graph of f and the x-axis, where area below the x-axis is counted negatively.

> **Definition A.4.1** *We define the **definite integral** of $f(x)$ from a to b by*
>
> $$\int_a^b f(x)\,dx = \lim_{n\to\infty} R_n,$$
>
> *where the sequence of Riemann sums $\{R_n\}_{n=1}^{\infty}$ is defined so that $\delta(n) \to 0$ as $n \to \infty$, as in the previous discussion. The function $f(x)$ is called the **integrand**, and the values a and b are the **lower** and **upper limits of integration**, respectively. The process of computing an integral is called **integration**.*

In Chapter 9, we will study the problem of computing an approximation of the definite integral of a given function $f(x)$ over an interval $[a, b]$. We will learn a number of techniques for computing such an approximation, and while all of these techniques can be viewed as akin to evaluating a Riemann sum, they yield far more accuracy relative to computational expense.

A.5 The Mean Value Theorem

While the derivative describes the behavior of a function at a point, we often need to understand how the derivative influences a function's behavior on an interval. This understanding is essential in numerical analysis, because it is often necessary to approximate a function $f(x)$ by a function $g(x)$ using knowledge of $f(x)$ and its derivatives at various points. It is therefore natural to ask how well $g(x)$ approximates $f(x)$ away from these points.

The following result, a consequence of Fermat's Theorem, gives limited insight into the relationship between the behavior of a function on an interval and the value of its derivative at a point.

> **Theorem A.5.1 (Rolle's Theorem)** If f is continuous on a closed interval $[a, b]$ and is differentiable on the open interval (a, b), and if $f(a) = f(b)$, then $f'(c) = 0$ for some number c in (a, b).

By applying Rolle's Theorem to a function f, then to its derivative f', its second derivative f'', and so on, we obtain the following more general result, which will be useful in analyzing the accuracy of methods for approximating functions by polynomials.

> **Theorem A.5.2 (Generalized Rolle's Theorem)** Let $x_0, x_1, x_2, \ldots, x_n$ be distinct points in an interval $[a, b]$. If f is n times differentiable on (a, b), and if $f(x_i) = 0$ for $i = 0, 1, 2, \ldots, n$, then $f^{(n)}(c) = 0$ for some number c in (a, b).

A more fundamental consequence of Rolle's Theorem is the Mean Value Theorem itself, which we now state.

Theorem A.5.3 (Mean Value Theorem) If f is continuous on a closed interval $[a, b]$ and is differentiable on the open interval (a, b), then

$$\frac{f(b) - f(a)}{b - a} = f'(c)$$

for some number c in (a, b).

The expression

$$\frac{f(b) - f(a)}{b - a}$$

is the slope of the secant line passing through the points $(a, f(a))$ and $(b, f(b))$, and it is also the average rate of change of f on $[a, b]$. The Mean Value Theorem therefore states that under the given assumptions, the slope of this secant line is equal to the slope of the tangent line of f at the point $(c, f(c))$, where $c \in (a, b)$.

The Mean Value Theorem has the following practical interpretation: the average rate of change of $y = f(x)$ with respect to x on an interval $[a, b]$ is equal to the instantaneous rate of change of y with respect to x at some point in (a, b).

A.5.1 The Mean Value Theorem for Integrals

Suppose that $f(x)$ is a continuous function on an interval $[a, b]$. Then, by the Fundamental Theorem of Calculus, $f(x)$ has an antiderivative $F(x)$ defined on $[a, b]$ such that $F'(x) = f(x)$. If we apply the Mean Value Theorem to $F(x)$, we obtain the following relationship between the integral of f over $[a, b]$ and the value of f at a point in (a, b).

Theorem A.5.4 (Mean Value Theorem for Integrals) If f is continuous on $[a, b]$, then

$$\int_a^b f(x)\, dx = f(c)(b - a)$$

for some c in (a, b).

In other words, f assumes its average value over $[a, b]$, defined by

$$f_{\text{ave}} = \frac{1}{b - a} \int_a^b f(x)\, dx,$$

at some point in $[a, b]$, just as the Mean Value Theorem states that the derivative of a function assumes its average value over an interval at some point in the interval.

The Mean Value Theorem for Integrals is also a special case of the following more general result.

Theorem A.5.5 (Weighted Mean Value Theorem for Integrals) If f is continuous on $[a, b]$, and g is a function that is integrable on $[a, b]$ and does not change sign on $[a, b]$, then

$$\int_a^b f(x)g(x)\,dx = f(c)\int_a^b g(x)\,dx$$

for some c in (a, b).

When $g(x)$ is a function that is easy to antidifferentiate and $f(x)$ is not, this theorem can be used to obtain an estimate of the integral of $f(x)g(x)$ over an interval.

Example A.5.6 Let $f(x)$ be continuous on the interval $[a, b]$. Then, for any $x \in [a, b]$, by the Weighted Mean Value Theorem for Integrals, we have

$$\int_a^x f(s)(s-a)\,ds = f(c)\int_a^x (s-a)\,ds = f(c)\frac{(s-a)^2}{2}\bigg|_a^x = f(c)\frac{(x-a)^2}{2},$$

where $c \in (a, x)$. We can apply the Weighted Mean Value Theorem because the function $g(x) = (x - a)$ does not change sign on $[a, b]$. □

A.6   Taylor's Theorem

In many cases, it is useful to approximate a given function $f(x)$ by a polynomial, because one can work much more easily with polynomials than with other types of functions. As such, it is necessary to have some insight into the accuracy of such an approximation. The following theorem, which is a consequence of the Weighted Mean Value Theorem for Integrals, provides this insight.

Theorem A.6.1 (Taylor's Theorem) Let f be n times continuously differentiable on $[a, b]$, and suppose $f^{(n+1)}$ exists on $[a, b]$. Let $x_0 \in [a, b]$. Then, for $x \in [a, b]$, $f(x)$ can be expressed as a **Taylor expansion** $f(x) = P_n(x) + R_n(x)$, where

$$P_n(x) = \sum_{j=0}^n \frac{f^{(j)}(x_0)}{j!}(x - x_0)^j$$

$$= f(x_0) + f'(x_0)(x - x_0) + \frac{1}{2}f''(x_0)(x - x_0)^2 + \cdots + \frac{f^{(n)}(x_0)}{n!}(x - x_0)^n$$

and

$$R_n(x) = \int_{x_0}^x \frac{f^{(n+1)}(s)}{n!}(x - s)^n\,ds = \frac{f^{(n+1)}(\xi(x))}{(n+1)!}(x - x_0)^{n+1},$$

where $\xi(x)$ is between x_0 and x.

The polynomial $P_n(x)$ is the nth **Taylor polynomial** of f with *center x_0*, and the expression $R_n(x)$ is called the **Taylor remainder** of $P_n(x)$. When the center x_0 is equal to zero, the nth Taylor polynomial is also known as the nth **Maclaurin polynomial**.

The final form of the remainder is obtained by applying the Mean Value Theorem for Integrals to the integral form. As $P_n(x)$ can be used to approximate $f(x)$, the remainder $R_n(x)$ is also referred to as the **truncation error** of $P_n(x)$. The accuracy of the approximation on an interval can be analyzed by using techniques for finding the extreme values of functions to bound the $(n+1)$-th derivative on the interval.

Because approximation of functions by polynomials is employed in the development and analysis of many techniques in numerical analysis, the usefulness of Taylor's Theorem cannot be overstated. In fact, it can be said that Taylor's Theorem is the "Fundamental Theorem of Numerical Analysis", just as the theorem describing inverse relationship between derivatives and integrals is called the Fundamental Theorem of Calculus.

The following examples illustrate how the nth-degree Taylor polynomial $P_n(x)$ and the remainder $R_n(x)$ can be computed for a given function $f(x)$.

Example A.6.2 If we set $n = 1$ in Taylor's Theorem, then we have

$$f(x) = P_1(x) + R_1(x)$$

where

$$P_1(x) = f(x_0) + f'(x_0)(x - x_0).$$

This polynomial is a linear function that describes the tangent line to the graph of f at the point $(x_0, f(x_0))$.

If we set $n = 0$ in the theorem, then we obtain

$$f(x) = P_0(x) + R_0(x),$$

where

$$P_0(x) = f(x_0)$$

and

$$R_0(x) = f'(\xi(x))(x - x_0),$$

where $\xi(x)$ lies between x_0 and x. If we use the integral form of the remainder,

$$R_n(x) = \int_{x_0}^{x} \frac{f^{(n+1)}(s)}{n!}(x - s)^n \, ds,$$

then we have

$$f(x) = f(x_0) + \int_{x_0}^{x} f'(s) \, ds,$$

which is equivalent to part of the Fundamental Theorem of Calculus. Using the Mean Value Theorem for integrals, we can see how the first form of the remainder $R_0(x)$ can be obtained from the integral form. $\square$

Example A.6.3 Let $f(x) = \sin x$. Then a Taylor expansion of $f(x)$ is
$$f(x) = P_3(x) + R_3(x),$$
where
$$P_3(x) = x - \frac{x^3}{3!} = x - \frac{x^3}{6},$$
and
$$R_3(x) = \frac{1}{4!}x^4 \sin \xi(x) = \frac{1}{24}x^4 \sin \xi(x),$$
where $\xi(x)$ is between 0 and x. The polynomial $P_3(x)$ is the 3rd Maclaurin polynomial of $\sin x$, or the 3rd Taylor polynomial with center $x_0 = 0$.

If $x \in [-1, 1]$, then
$$|R_n(x)| = \left| \frac{1}{24}x^4 \sin \xi(x) \right| = \left| \frac{1}{24} \right| |x^4| |\sin \xi(x)| \le \frac{1}{24},$$
since $|\sin x| \le 1$ for all x. This bound on $|R_n(x)|$ serves as an upper bound for the error in the approximation of $\sin x$ by $P_3(x)$ for $x \in [-1, 1]$. $\square$

Example A.6.4 Let $f(x) = e^x$. Then a Taylor expansion of $f(x)$ is
$$f(x) = P_2(x) + R_2(x),$$
where
$$P_2(x) = 1 + x + \frac{x^2}{2},$$
and
$$R_2(x) = \frac{x^3}{6}e^{\xi(x)},$$
where $\xi(x)$ is between 0 and x. The polynomial $P_2(x)$ is the 2nd Maclaurin polynomial of e^x, or the 2nd Taylor polynomial with center $x_0 = 0$. $\square$

Example A.6.5 Let $f(x) = x^2$. Then, for any real number x_0, a Taylor expansion of $f(x)$ is
$$f(x) = P_1(x) + R_1(x),$$
where
$$P_1(x) = x_0^2 + 2x_0(x - x_0) = 2x_0 x - x_0^2,$$
and
$$R_1(x) = (x - x_0)^2.$$
Note that the remainder does not include a "mystery point" $\xi(x)$ since the 2nd derivative of x^2 is only a constant. The linear function $P_1(x)$ describes the tangent line to the graph of $f(x)$ at the point $(x_0, f(x_0))$. If $x_0 = 1$, then we have
$$P_1(x) = 2x - 1,$$
and
$$R_1(x) = (x - 1)^2.$$
We can see that near $x = 1$, $P_1(x)$ is a reasonable approximation to x^2, since the error in this approximation, given by $R_1(x)$, would be small in this case. $\square$

Taylor's theorem can be generalized to functions of several variables, using partial derivatives. Here, we consider the case of two independent variables.

Theorem A.6.6 (Taylor's Theorem in Two Variables) Let $f(t, y)$ be $(n + 1)$ times continuously differentiable on a convex set D, and let $(t_0, y_0) \in D$. Then, for every $(t, y) \in D$, there exists ξ between t_0 and t, and μ between y_0 and y, such that

$$f(t, y) = P_n(t, y) + R_n(t, y),$$

where $P_n(t, y)$ is the nth **Taylor polynomial** of f about (t_0, y_0),

$$P_n(t, y) = f(t_0, y_0) + \left[(t - t_0)\frac{\partial f}{\partial t} + (y - y_0)\frac{\partial f}{\partial y} \right] +$$
$$\left[\frac{(t - t_0)^2}{2}\frac{\partial^2 f}{\partial t^2} + (t - t_0)(y - y_0)\frac{\partial^2 f}{\partial t \partial y} + \frac{(y - y_0)^2}{2}\frac{\partial^2 f}{\partial y^2} \right]$$
$$+ \cdots + \left[\frac{1}{n!} \sum_{j=0}^{n} \binom{n}{j} (t - t_0)^{n-j}(y - y_0)^j \frac{\partial^n f}{\partial t^{n-j} \partial y^j} \right],$$

where all partial derivatives are evaluated at (t_0, y_0), and $R_n(t, y)$ is the **remainder term** associated with $P_n(t, y)$,

$$R_n(t, y) = \frac{1}{(n + 1)!} \sum_{j=0}^{n+1} \binom{n+1}{j} (t - t_0)^{n+1-j}(y - y_0)^j \frac{\partial^{n+1} f}{\partial t^{n+1-j} \partial y^j}(\xi, \mu).$$

Appendix B

Review of Linear Algebra

B.1 Matrices

Writing a system of linear equations in the form

$$a_{11}x_1 + a_{12}x_2 + \cdots + a_{1n}x_n = b_1,$$
$$a_{21}x_1 + a_{22}x_2 + \cdots + a_{2n}x_n = b_2,$$
$$\vdots$$
$$a_{m1}x_1 + a_{m2}x_2 + \cdots + a_{mn}x_n = b_m$$

can be quite tedious, especially when the number of equations, m, or the number of unknowns, n, is large. Therefore, we instead represent a system of linear equations using a **matrix**, which is an array of elements, or entries. We say that a matrix A is $m \times n$ if it has m rows and n columns, and we denote the element in row i and column j by a_{ij}. We also denote the matrix A by $[a_{ij}]$. If the entries of A are real numbers, we write $A \in \mathbb{R}^{m \times n}$, and if the entries are complex numbers, we write $A \in \mathbb{C}^{m \times n}$.

With this notation, a general system of m equations with n unknowns can be represented using a matrix A that contains the coefficients of the equations, a vector $\mathbf{x}$ that contains the unknowns, and a vector $\mathbf{b}$ that contains the quantities on the right-hand sides of the equations. Specifically,

$$A = \begin{bmatrix} a_{11} & a_{12} & \cdots & a_{1n} \\ a_{21} & a_{22} & \cdots & a_{2n} \\ \vdots & & \vdots & \\ a_{m1} & a_{m2} & \cdots & a_{mn} \end{bmatrix}, \quad \mathbf{x} = \begin{bmatrix} x_1 \\ x_2 \\ \vdots \\ x_n \end{bmatrix}, \quad \mathbf{b} = \begin{bmatrix} b_1 \\ b_2 \\ \vdots \\ b_m \end{bmatrix}.$$

Note that the vectors $\mathbf{x}$ and $\mathbf{b}$ are represented by *column* vectors.

Example B.1.1 The coefficients in the linear system

$$3x_1 + 2x_2 = 4,$$
$$-x_1 + 5x_2 = -3$$

can be represented by the matrix

$$A = \begin{bmatrix} 3 & 2 \\ -1 & 5 \end{bmatrix}.$$

For example, the coefficient of x_2 in the first equation is represented by the entry in the first row and second column of A, which is $a_{12} = 2$. $\square$

B.2   Vector Spaces

Matrices are much more than notational conveniences for writing systems of linear equations. A matrix A can also be used to represent a linear function f whose domain and range are both sets of vectors called **vector spaces**. A vector space over a **field** (such as the field of real or complex numbers) is a set of vectors, together with two operations: addition of vectors, and multiplication of a vector by a scalar from the field.

Specifically, if $\mathbf{u}$ and $\mathbf{v}$ are vectors belonging to a vector space V over a field $\mathbb{F}$, then the *sum* of $\mathbf{u}$ and $\mathbf{v}$, denoted by $\mathbf{u} + \mathbf{v}$, is a vector in V, and the **scalar product** of $\mathbf{u}$ with a scalar α in $\mathbb{F}$, denoted by $\alpha\mathbf{u}$, is also a vector in V. Because the results of these operations on elements of V are also elements of V, we say that V is **closed** under addition and scalar multiplication. These operations have the following additional properties:

- **Commutativity**: For any vectors $\mathbf{u}$ and $\mathbf{v}$ in V,

$$\mathbf{u} + \mathbf{v} = \mathbf{v} + \mathbf{u}$$

- **Associativity**: For any vectors $\mathbf{u}$, $\mathbf{v}$ and $\mathbf{w}$ in V,

$$(\mathbf{u} + \mathbf{v}) + \mathbf{w} = \mathbf{u} + (\mathbf{v} + \mathbf{w})$$

- **Identity element for vector addition**: There is a vector $\mathbf{0}$, known as the *zero vector*, such that for any vector $\mathbf{u}$ in V,

$$\mathbf{u} + \mathbf{0} = \mathbf{0} + \mathbf{u} = \mathbf{u}$$

- **Additive inverse**: For any vector $\mathbf{u}$ in V, there is a unique vector $-\mathbf{u}$ in V such that

$$\mathbf{u} + (-\mathbf{u}) = -\mathbf{u} + \mathbf{u} = \mathbf{0}$$

- **Distributivity over vector addition**: For any vectors $\mathbf{u}$ and $\mathbf{v}$ in V, and scalar α in $\mathbb{F}$,

$$\alpha(\mathbf{u} + \mathbf{v}) = \alpha\mathbf{u} + \alpha\mathbf{v}$$

- **Distributivity over scalar multiplication**: For any vector $\mathbf{u}$ in V, and scalars α and β in $\mathbb{F}$,

$$(\alpha + \beta)\mathbf{u} = \alpha\mathbf{u} + \beta\mathbf{u}$$

- **Associativity of scalar multiplication**: For any vector **u** in V and any scalars α and β in $\mathbb{F}$,

$$\alpha(\beta\mathbf{u}) = (\alpha\beta)\mathbf{u}$$

- **Identity element for scalar multiplication**: For any vector **u** in V,

$$1\mathbf{u} = \mathbf{u}$$

Example B.2.1 Let V be the vector space $\mathbb{R}^3$. The vector addition operation on V consists of adding corresponding components of vectors, as in

$$\begin{bmatrix} 3 \\ 0 \\ 1 \end{bmatrix} + \begin{bmatrix} -2 \\ 4 \\ 5 \end{bmatrix} = \begin{bmatrix} 1 \\ 4 \\ 6 \end{bmatrix}.$$

The scalar multiplication operation consists of scaling each component of a vector by a scalar:

$$\frac{1}{2}\begin{bmatrix} 3 \\ 0 \\ 1 \end{bmatrix} = \begin{bmatrix} \frac{3}{2} \\ 0 \\ \frac{1}{2} \end{bmatrix}.$$

□

B.3 Subspaces

Before we can explain how matrices can be used to easily describe linear transformations, we must introduce some important concepts related to vector spaces. A **subspace** of a vector space V is a subset of V that is, itself, a vector space. In particular, a subset S of V is also a subspace if it is **closed** under the operations of vector addition and scalar multiplication. That is, if **u** and **v** are vectors in S, then the vectors $\mathbf{u} + \mathbf{v}$ and $\alpha\mathbf{u}$, where α is any scalar, must also be in S. In particular, S cannot be a subspace unless it includes the zero vector.

Example B.3.1 The set S of all vectors in $\mathbb{R}^3$ of the form

$$\mathbf{x} = \begin{bmatrix} x_1 \\ x_2 \\ 0 \end{bmatrix},$$

where $x_1, x_2 \in \mathbb{R}$, is a subspace of $\mathbb{R}^3$, as the sum of any two vectors in S, or a scalar multiple of any vector in S, must have a third component of zero, and therefore is also in S.

On the other hand, the set $\tilde{S}$ consisting of all vectors in $\mathbb{R}^3$ that have a third component of 1 is *not* a subspace of $\mathbb{R}^3$, as the sum of vectors in $\tilde{S}$ will not have a third component of 1, and therefore will not be in $\tilde{S}$. That is, $\tilde{S}$ is not *closed* under addition. It is not closed under scalar multiplication either. □

B.4   Linear Independence and Bases

Often a vector space or subspace can be characterized as the set of all vectors that can be obtained by adding and/or scaling members of a given set of specific vectors. For example, $\mathbb{R}^2$ can be described as the set of all vectors that can be obtained by adding and/or scaling the vectors

$$\mathbf{e}_1 = \begin{bmatrix} 1 \\ 0 \end{bmatrix}, \quad \mathbf{e}_2 = \begin{bmatrix} 0 \\ 1 \end{bmatrix}.$$

These vectors constitute what is known as the **standard basis** of $\mathbb{R}^2$.

More generally, given a set of vectors $\{\mathbf{v}_1, \mathbf{v}_2, \ldots, \mathbf{v}_k\}$ from a vector space V, a vector $\mathbf{v} \in V$ is called a **linear combination** of $\mathbf{v}_1$, $\mathbf{v}_2$, ..., $\mathbf{v}_k$ if there exist constants c_1, c_2, ..., c_k such that

$$\mathbf{v} = c_1\mathbf{v}_1 + c_2\mathbf{v}_2 + \cdots + c_k\mathbf{v}_k = \sum_{i=1}^{k} c_i\mathbf{v}_i.$$

We then define the **span** of $\{\mathbf{v}_1, \mathbf{v}_2, \ldots, \mathbf{v}_k\}$, denoted by $\mathrm{span}\,\{\mathbf{v}_1, \mathbf{v}_2, \ldots, \mathbf{v}_k\}$, to be the set of *all* linear combinations of $\mathbf{v}_1$, $\mathbf{v}_2$, ..., $\mathbf{v}_k$. From the definition of a linear combination, it follows that this set is a subspace of V.

Example B.4.1 Let

$$\mathbf{v}_1 = \begin{bmatrix} 1 \\ 0 \\ 1 \end{bmatrix}, \quad \mathbf{v}_2 = \begin{bmatrix} 3 \\ 4 \\ 0 \end{bmatrix}, \quad \mathbf{v}_3 = \begin{bmatrix} -1 \\ 2 \\ 1 \end{bmatrix}.$$

Then the vector

$$\mathbf{v} = \begin{bmatrix} 6 \\ 10 \\ 2 \end{bmatrix}$$

is a linear combination of $\mathbf{v}_1$, $\mathbf{v}_2$ and $\mathbf{v}_3$, as

$$\mathbf{v} = \mathbf{v}_1 + 2\mathbf{v}_2 + \mathbf{v}_3.$$

We can say that $\mathbf{v} \in \mathrm{span}\{\mathbf{v}_1, \mathbf{v}_2, \mathbf{v}_3\}$. $\square$

When a subspace is defined as the span of a set of vectors, it is helpful to know whether the set includes any vectors that are, in some sense, redundant, for if this is the case, the description of the subspace can be simplified. To that end, we say that a set of vectors $\{\mathbf{v}_1, \mathbf{v}_2, \ldots, \mathbf{v}_k\}$ is **linearly independent** if the equation

$$c_1\mathbf{v}_1 + c_2\mathbf{v}_2 + \cdots + c_k\mathbf{v}_k = \mathbf{0}$$

holds if and only if $c_1 = c_2 = \cdots = c_k = 0$. Otherwise, we say that the set is **linearly dependent**. If the set is linearly independent, then any vector $\mathbf{v}$ in the span of the set is a *unique* linear combination of members of the set; that is, there is only one way to choose the coefficients of a linear combination that is used to obtain $\mathbf{v}$.

Example B.4.2 The subspace S of $\mathbb{R}^3$ defined by

$$S = \left\{ \begin{bmatrix} x_1 \\ x_2 \\ 0 \end{bmatrix} \middle| \; x_1, x_2 \in \mathbb{R} \right\}$$

can be described as

$$S = \text{span} \left\{ \begin{bmatrix} 1 \\ 0 \\ 0 \end{bmatrix}, \begin{bmatrix} 0 \\ 1 \\ 0 \end{bmatrix} \right\}$$

or

$$S = \text{span} \left\{ \begin{bmatrix} 1 \\ 1 \\ 0 \end{bmatrix}, \begin{bmatrix} 1 \\ -1 \\ 0 \end{bmatrix} \right\},$$

but

$$S \neq \text{span} \left\{ \begin{bmatrix} 1 \\ 1 \\ 0 \end{bmatrix}, \begin{bmatrix} -1 \\ -1 \\ 0 \end{bmatrix} \right\},$$

as these vectors only span the subspace of vectors whose first two components are equal, and whose third component is zero, which does not account for every vector in S. It should be noted that the two vectors in the third set are linearly dependent, while the pairs of vectors in the previous two sets are linearly independent. $\square$

Example B.4.3 The vectors

$$\mathbf{v}_1 = \begin{bmatrix} 1 \\ 0 \\ 1 \end{bmatrix}, \quad \mathbf{v}_2 = \begin{bmatrix} 1 \\ 1 \\ 0 \end{bmatrix}$$

are linearly independent. It follows that the only way in which the vector

$$\mathbf{v} = \begin{bmatrix} 3 \\ 1 \\ 2 \end{bmatrix}$$

can be expressed as a linear combination of $\mathbf{v}_1$ and $\mathbf{v}_2$ is

$$\mathbf{v} = 2\mathbf{v}_1 + \mathbf{v}_2.$$

On the other hand, if

$$\mathbf{v}_1 = \begin{bmatrix} 1 \\ 1 \\ 0 \end{bmatrix}, \quad \mathbf{v}_2 = \begin{bmatrix} 2 \\ 2 \\ 0 \end{bmatrix}, \quad \mathbf{v} = \begin{bmatrix} 3 \\ 3 \\ 0 \end{bmatrix},$$

then, because $\mathbf{v}_1$ and $\mathbf{v}_2$ are linearly dependent, any linear combination of the form $c_1\mathbf{v}_1 + c_2\mathbf{v}_2$, such that $c_1 + 2c_2 = 3$, will equal $\mathbf{v}$. $\square$

Given a vector space V, if there exists a set of vectors $\{\mathbf{v}_1, \mathbf{v}_2, \ldots, \mathbf{v}_k\}$ such that V is the span of $\{\mathbf{v}_1, \mathbf{v}_2, \ldots, \mathbf{v}_k\}$, *and* $\{\mathbf{v}_1, \mathbf{v}_2, \ldots, \mathbf{v}_k\}$ is linearly independent, then we say that $\{\mathbf{v}_1, \mathbf{v}_2, \ldots, \mathbf{v}_k\}$ is a **basis** of V. Any basis of V must have the same number of elements, k. We call this number the **dimension** of V, which is denoted by $\dim(V)$.

Example B.4.4 The **standard basis** of $\mathbb{R}^3$ is

$$\mathbf{e}_1 = \begin{bmatrix} 1 \\ 0 \\ 0 \end{bmatrix}, \quad \mathbf{e}_2 = \begin{bmatrix} 0 \\ 1 \\ 0 \end{bmatrix}, \quad \mathbf{e}_3 = \begin{bmatrix} 0 \\ 0 \\ 1 \end{bmatrix}.$$

The set

$$\mathbf{v}_1 = \begin{bmatrix} 1 \\ 1 \\ 0 \end{bmatrix}, \quad \mathbf{v}_2 = \begin{bmatrix} 1 \\ -1 \\ 0 \end{bmatrix}, \quad \mathbf{v}_3 = \begin{bmatrix} 0 \\ 0 \\ 1 \end{bmatrix}$$

is also a basis for $\mathbb{R}^3$, as it consists of three linearly independent vectors, and the dimension of $\mathbb{R}^3$ is three. $\square$

In general, the jth vector in the standard basis of $\mathbb{R}^n$ or $\mathbb{C}^n$ is a vector whose components are all zero, except for the jth component, which is equal to one. These vectors are called the *standard basis vectors* of an n-dimensional space of real or complex vectors, and are denoted by $\mathbf{e}_j$, $j = 1, 2, \ldots, n$. From this point on, we will generally assume that V is $\mathbb{R}^n$, and that the field is $\mathbb{R}$, for simplicity.

B.5   Linear Transformations

A function $f : V \to W$, whose domain V and range W are vector spaces over a field $\mathbb{F}$, is a **linear transformation** if it has the properties

$$f(\mathbf{x} + \mathbf{y}) = f(\mathbf{x}) + f(\mathbf{y}), \quad f(\alpha \mathbf{x}) = \alpha f(\mathbf{x}),$$

where $\mathbf{x}$ and $\mathbf{y}$ are vectors in V and α is a scalar from $\mathbb{F}$. If V and W are the same vector space, then we say that f is a **linear operator** on V.

B.5.1   The Matrix of a Linear Transformation

If V is a vector space of dimension n over a field, such as $\mathbb{R}^n$ or $\mathbb{C}^n$, and W is a vector space of dimension m, then a linear transformation f with domain V and range W can be represented by an $m \times n$ matrix A whose entries belong to the field.

Suppose that the set of vectors $\{\mathbf{v}_1, \mathbf{v}_2, \ldots, \mathbf{v}_n\}$ is a *basis* for V, and the set $\{\mathbf{w}_1, \mathbf{w}_2, \ldots, \mathbf{w}_m\}$ is a basis for W. Then, a_{ij} is the scalar by which $\mathbf{w}_i$ is multiplied when applying the function f to the vector $\mathbf{v}_j$. That is,

$$f(\mathbf{v}_j) = a_{1j}\mathbf{w}_1 + a_{2j}\mathbf{w}_2 + \cdots + a_{mj}\mathbf{w}_m = \sum_{i=1}^{m} a_{ij}\mathbf{w}_i. \tag{B.1}$$

In other words, the jth column of A describes the **image** under f of the vector $\mathbf{v}_j$, in terms of the coefficients of $f(\mathbf{v}_j)$ in the basis $\{\mathbf{w}_1, \mathbf{w}_2, \ldots, \mathbf{w}_m\}$.

B.5.2 Matrix-Vector Multiplication

To describe the action of f on a general vector $\mathbf{x}$ from V, we can write

$$\mathbf{x} = x_1 \mathbf{v}_1 + x_2 \mathbf{v}_2 + \cdots + x_n \mathbf{v}_n = \sum_{j=1}^{n} x_j \mathbf{v}_j.$$

Then, because f is a linear function, (B.1) yields

$$\mathbf{y} = f(\mathbf{x}) = \sum_{j=1}^{n} x_j f(\mathbf{v}_j) = \sum_{j=1}^{n} x_j \sum_{i=1}^{m} a_{ij} \mathbf{w}_i = \sum_{i=1}^{m} \left(\sum_{j=1}^{n} a_{ij} x_j \right) \mathbf{w}_i.$$

The coefficients of $\mathbf{y}$ in the basis of W are obtained from the coefficients of $\mathbf{x}$ in the basis of V by the **matrix-vector product** of A and $\mathbf{x}$, which we denote by $A\mathbf{x}$. Each element of the vector $\mathbf{y} = A\mathbf{x}$, in the basis of W, is given by

$$y_i = [A\mathbf{x}]_i = a_{i1} x_1 + a_{i2} x_2 + \cdots + a_{in} x_n = \sum_{j=1}^{n} a_{ij} x_j.$$

The matrix of a linear transformation depends on the bases used. If V and W are spaces of real or complex vectors, then, by convention, the bases $\{\mathbf{v}_j\}_{j=1}^{n}$ and $\{\mathbf{w}_i\}_{i=1}^{m}$ are each chosen to be the **standard basis** for $\mathbb{R}^n$ and $\mathbb{R}^m$, respectively. In this case, the matrix A maps coordinates of vectors in V directly to coordinates of vectors in W.

Example B.5.1 Let

$$A = \begin{bmatrix} 3 & 0 & -1 \\ 1 & -4 & 2 \\ 5 & 1 & -3 \end{bmatrix}, \quad \mathbf{x} = \begin{bmatrix} 10 \\ 11 \\ 12 \end{bmatrix}.$$

Then

$$A\mathbf{x} = 10 \begin{bmatrix} 3 \\ 1 \\ 5 \end{bmatrix} + 11 \begin{bmatrix} 0 \\ -4 \\ 1 \end{bmatrix} + 12 \begin{bmatrix} -1 \\ 2 \\ -3 \end{bmatrix} = \begin{bmatrix} 18 \\ -10 \\ 25 \end{bmatrix}.$$

We see that $A\mathbf{x}$ is a linear combination of the columns of A, with the coefficients of the linear combination obtained from the components of $\mathbf{x}$. $\square$

B.5.3 Special Subspaces

Let A be an $m \times n$ matrix. Then the **range** of A, denoted by $\mathrm{range}(A)$, is the set of all vectors of the form $\mathbf{y} = A\mathbf{x}$, where $\mathbf{x} \in \mathbb{R}^n$. It follows that $\mathrm{ran}(A)$ is the span of the columns of A, which is also called the **column space** of A.

The dimension of $\mathrm{range}(A)$ is called the **column rank** of A. Similarly, the dimension of the **row space** of A is called the **row rank** of A. It can be shown that the row rank and column rank are equal; this common value is simply called the **rank** of A, and is denoted by $\mathrm{rank}(A)$. We say that A is **rank-deficient** if $\mathrm{rank}(A) < \min\{m, n\}$; otherwise, we say that A has **full rank**.

The **null space** of A, denoted by $\text{null}(A)$, is the set of all vectors $\mathbf{x} \in \mathbb{R}^n$ such that $A\mathbf{x} = \mathbf{0}$. Its dimension is called the **nullity** of A. It can be shown that for an $m \times n$ matrix A,

$$\dim(\text{null}(A)) + \text{rank}(A) = n.$$

B.6 Matrix-Matrix Multiplication

It follows from the definition of matrix-vector multiplication that a general system of m linear equations in n unknowns can be described in matrix-vector form by the equation

$$A\mathbf{x} = \mathbf{b},$$

where $A\mathbf{x}$ is a matrix-vector product of the $m \times n$ coefficient matrix A and the vector of unknowns $\mathbf{x}$, and $\mathbf{b}$ is the vector of right-hand side values.

Of course, if $m = n = 1$, the system of equations $A\mathbf{x} = \mathbf{b}$ reduces to the scalar linear equation $ax = b$, which has the solution $x = a^{-1}b$, provided that $a \neq 0$. As a^{-1} is the unique number such that $a^{-1}a = aa^{-1} = 1$, it is desirable to generalize the concepts of multiplication and identity element to square matrices, for which $m = n$.

The matrix-vector product can be used to define the *composition* of linear functions represented by matrices. Let A be an $m \times n$ matrix, and let B be an $n \times p$ matrix. Then, if $\mathbf{x}$ is a vector of length p, and $\mathbf{y} = B\mathbf{x}$, then we have

$$A\mathbf{y} = A(B\mathbf{x}) = (AB)\mathbf{x} = C\mathbf{x},$$

where C is an $m \times p$ matrix with entries

$$C_{ij} = \sum_{k=1}^{n} a_{ik}b_{kj}. \tag{B.2}$$

We define the **matrix product** of A and B to be the matrix $C = AB$ with entries defined in this manner. It should be noted that the product BA is not defined, unless $m = p$. Even if this is the case, in general, $AB \neq BA$. That is, matrix multiplication is not *commutative*. However, matrix multiplication is *associative*, meaning that if A is $m \times n$, B is $n \times p$, and C is $p \times k$, then $A(BC) = (AB)C$.

Example B.6.1 Consider the 2×2 matrices

$$A = \begin{bmatrix} 1 & -2 \\ -3 & 4 \end{bmatrix}, \quad B = \begin{bmatrix} -5 & 6 \\ 7 & -8 \end{bmatrix}.$$

Then

$$AB = \begin{bmatrix} 1 & -2 \\ -3 & 4 \end{bmatrix} \begin{bmatrix} -5 & 6 \\ 7 & -8 \end{bmatrix}$$

$$= \begin{bmatrix} 1(-5) - 2(7) & 1(6) - 2(-8) \\ -3(-5) + 4(7) & -3(6) + 4(-8) \end{bmatrix}$$

$$= \begin{bmatrix} -19 & 22 \\ 43 & -50 \end{bmatrix},$$

whereas

$$BA = \begin{bmatrix} -5 & 6 \\ 7 & -8 \end{bmatrix} \begin{bmatrix} 1 & -2 \\ -3 & 4 \end{bmatrix}$$

$$= \begin{bmatrix} -5(1) + 6(-3) & -5(-2) + 6(4) \\ 7(1) - 8(-3) & 7(-2) - 8(4) \end{bmatrix}$$

$$= \begin{bmatrix} -23 & 34 \\ 31 & -46 \end{bmatrix}.$$

We see that $AB \neq BA$. □

Example B.6.2 If

$$A = \begin{bmatrix} 3 & 1 \\ 1 & 0 \\ 2 & 4 \end{bmatrix}, \quad B = \begin{bmatrix} 1 & 5 \\ 4 & -1 \end{bmatrix},$$

then the matrix-matrix product of A and B is

$$C = AB = \begin{bmatrix} 3(1) + 1(4) & 3(5) + 1(-1) \\ 1(1) + 0(4) & 1(5) + 0(-1) \\ 2(1) + 4(4) & 2(5) + 4(-1) \end{bmatrix} = \begin{bmatrix} 7 & 14 \\ 1 & 5 \\ 18 & 6 \end{bmatrix}.$$

It does not make sense to compute BA, because the dimensions are incompatible.
□

B.7 Other Fundamental Matrix Operations

B.7.1 Vector Space Operations

The set of all matrices of size $m \times n$, for fixed m and n, is itself a vector space of dimension mn. The operations of vector addition and scalar multiplication for matrices are defined as follows: If A and B are $m \times n$ matrices, then the sum of A and B, denoted by $A + B$, is the $m \times n$ matrix C with entries

$$c_{ij} = a_{ij} + b_{ij}.$$

If α is a scalar, then the product of α and an $m \times n$ matrix A, denoted by αA, is the $m \times n$ matrix B with entries

$$b_{ij} = \alpha a_{ij}.$$

It is natural to identify $m \times n$ matrices with vectors of length mn, in the context of these operations.

Matrix addition and scalar multiplication have properties analogous to those of vector addition and scalar multiplication of vectors. In addition, matrix multiplication has the following properties related to these operations. We assume that A is an $m \times n$ matrix, B and D are $n \times k$ matrices, and α is a scalar.

- Distributivity: $A(B + D) = AB + AD$
- Commutativity of scalar multiplication: $\alpha(AB) = (\alpha A)B = A(\alpha B)$

B.7.2   The Transpose of a Matrix

An $n \times n$ matrix A is said to be **symmetric** if $a_{ij} = a_{ji}$ for $i, j = 1, 2, \ldots, n$. The $n \times n$ matrix B whose entries are defined by $b_{ij} = a_{ji}$ is called the **transpose** of A, which we denote by A^T. Therefore, A is symmetric if $A = A^T$. More generally, if A is an $m \times n$ matrix, then A^T is the $n \times m$ matrix B whose entries are defined by $b_{ij} = a_{ji}$. The transpose has the following properties:

- $(A^T)^T = A$
- $(A + B)^T = A^T + B^T$
- $(AB)^T = B^T A^T$

Example B.7.1 If

$$A = \begin{bmatrix} 3 & 1 \\ 1 & 0 \\ 2 & 4 \end{bmatrix},$$

then

$$A^T = \begin{bmatrix} 3 & 1 & 2 \\ 1 & 0 & 4 \end{bmatrix}.$$

□

Example B.7.2 Let A be the matrix from Example B.5.1,

$$A = \begin{bmatrix} 3 & 0 & -1 \\ 1 & -4 & 2 \\ 5 & 1 & -3 \end{bmatrix}.$$

Then

$$A^T = \begin{bmatrix} 3 & 1 & 5 \\ 0 & -4 & 1 \\ -1 & 2 & -3 \end{bmatrix}.$$

It follows that

$$A + A^T = \begin{bmatrix} 3+3 & 0+1 & -1+5 \\ 1+0 & -4-4 & 2+1 \\ 5-1 & 1+2 & -3-3 \end{bmatrix} = \begin{bmatrix} 6 & 1 & 4 \\ 1 & -8 & 3 \\ 4 & 3 & -6 \end{bmatrix}.$$

This matrix is symmetric. This can also be seen by the properties of the transpose, since

$$(A + A^T)^T = A^T + (A^T)^T = A^T + A = A + A^T.$$

□

Example B.7.3 The matrix

$$A = \begin{bmatrix} 3 & 1 & 5 \\ 1 & 2 & 0 \\ 5 & 0 & 4 \end{bmatrix}$$

is symmetric, while

$$B = \begin{bmatrix} 0 & 1 & 2 \\ -1 & 0 & -3 \\ -2 & 3 & 0 \end{bmatrix}$$

is **skew-symmetric**, meaning that $A^T = -A$. □

A related operation is the **Hermitian transpose** of A, denoted by A^H. It is defined by $A^H = \overline{A^T}$, where the bar indicates complex conjugation. Therefore, $A^H = A^T$ if A is real, and $A^H = -A^T$ is A is purely imaginary.

Example B.7.4 Let

$$A = \begin{bmatrix} 2 - 2i & -3 + 5i & i \\ 3 + 4i & -2 + 3i & 1 - 2i \\ -2 - 4i & -1 & -4 \end{bmatrix}.$$

Then

$$A^T = \begin{bmatrix} 2 - 2i & 3 + 4i & -2 - 4i \\ -3 + 5i & -2 + 3i & -1 \\ i & 1 - 2i & -4 \end{bmatrix}, \quad A^H = \begin{bmatrix} 2 + 2i & 3 - 4i & -2 + 4i \\ -3 - 5i & -2 - 3i & -1 \\ -i & 1 + 2i & -4 \end{bmatrix}.$$

□

B.7.3 Inner and Outer Products

We now define two operations on matrices and vectors that will be particularly useful in our study of numerical linear algebra. For simplicity, we work with real vectors and matrices.

Given two vectors $\mathbf{x}$ and $\mathbf{y}$ in $\mathbb{R}^n$, the **dot product**, or **inner product**, of $\mathbf{x}$ and $\mathbf{y}$ is the scalar

$$\mathbf{x}^T\mathbf{y} = x_1 y_1 + x_2 y_2 + \cdots + x_n y_n = \sum_{i=1}^{n} x_i y_i,$$

where

$$\mathbf{x} = \begin{bmatrix} x_1 \\ x_2 \\ \vdots \\ x_n \end{bmatrix}, \quad \mathbf{y} = \begin{bmatrix} y_1 \\ y_2 \\ \vdots \\ y_n \end{bmatrix}.$$

Note that $\mathbf{x}$ and $\mathbf{y}$ must both be defined to be *column* vectors, and they must have the same length. If $\mathbf{x}^T\mathbf{y} = 0$, then we say that $\mathbf{x}$ and $\mathbf{y}$ are **orthogonal**.

Let $\mathbf{x} \in \mathbb{R}^m$ and $\mathbf{y} \in \mathbb{R}^n$, where m and n are not necessarily equal. The term "inner product" suggests the existence of another operation called the **outer product**, which is defined by

$$
\mathbf{x}\mathbf{y}^T = \begin{bmatrix} x_1y_1 & x_1y_2 & \cdots & x_1y_n \\ x_2y_1 & x_2y_2 & \cdots & x_2y_n \\ \vdots & & \vdots & \\ x_my_1 & x_my_2 & \cdots & x_my_n \end{bmatrix}.
$$

Note that whereas the inner product is a scalar, the outer product is an $m \times n$ matrix.

Example B.7.5 Let

$$
\mathbf{x} = \begin{bmatrix} 1 \\ 0 \\ 2 \end{bmatrix}, \quad \mathbf{y} = \begin{bmatrix} 4 \\ -1 \\ 3 \end{bmatrix}.
$$

Then the inner (dot) product of $\mathbf{x}$ and $\mathbf{y}$ is

$$
\mathbf{x}^T\mathbf{y} = 1(4) + 0(-1) + 2(3) = 10,
$$

while the outer product of $\mathbf{x}$ and $\mathbf{y}$ is

$$
\mathbf{x}\mathbf{y}^T = \begin{bmatrix} 1(4) & 1(-1) & 1(3) \\ 0(4) & 0(-1) & 0(3) \\ 2(4) & 2(-1) & 2(3) \end{bmatrix} = \begin{bmatrix} 4 & -1 & 3 \\ 0 & 0 & 0 \\ 8 & -2 & 6 \end{bmatrix}.
$$

□

Example B.7.6 Let

$$
A = \begin{bmatrix} 1 & -3 & 7 \\ 2 & 5 & -8 \\ 4 & -6 & -9 \end{bmatrix}.
$$

To change a_{11} from 1 to 10, we can perform the **outer product update** $B = A + (10 - 1)\mathbf{e}_1\mathbf{e}_1^T$. Similarly, the outer product update $C = B + 5\mathbf{e}_2\mathbf{e}_1^T$ adds 5 to b_{21}, resulting in the matrix

$$
C = \begin{bmatrix} 10 & -3 & 7 \\ 7 & 5 & -8 \\ 4 & -6 & -9 \end{bmatrix}.
$$

Note that

$$
\mathbf{e}_2\mathbf{e}_1^T = \begin{bmatrix} 0 \\ 1 \\ 0 \end{bmatrix} \begin{bmatrix} 1 & 0 & 0 \end{bmatrix} = \begin{bmatrix} 0 \\ 1 \\ 0 \end{bmatrix} \begin{bmatrix} 0 \\ 1 \\ 0 \end{bmatrix} \begin{bmatrix} 0 \\ 0 \\ 0 \end{bmatrix} = \begin{bmatrix} 0 & 0 & 0 \\ 1 & 0 & 0 \\ 0 & 0 & 0 \end{bmatrix}.
$$

□

It is important to note that an outer product $\mathbf{u}\mathbf{v}^T$ has rank one, because each column is a scalar multiple of $\mathbf{u}$. In fact, every matrix of rank one can be expressed as an outer product. For this reason, an outer product update of the form $B = A + \mathbf{u}\mathbf{v}^T$ is also called a **rank-one update**.

It is important to note that these definitions of inner and outer product assume the vectors involved are real. To generalize to complex vectors, we use the Hermitian transpose introduced earlier in this section. That is, if $\mathbf{x}, \mathbf{y} \in \mathbb{C}^n$, then the inner product of $\mathbf{x}$ and $\mathbf{y}$ is $\mathbf{x}^H\mathbf{y}$. This definition is consistent with that of the absolute value, or *modulus*, of a complex number z that is defined by $|z| = (z\bar{z})^{1/2}$, in that we can define the magnitude of $\mathbf{x}$ by

$$\|\mathbf{x}\|_2 = (\mathbf{x}^H\mathbf{x})^{1/2} = \left(\sum_{i=1}^n |x_i|^2 \right)^{1/2}.$$

This is the ℓ_2-norm of $\mathbf{x}$, which will be discussed further in Section B.13.

Note that in the complex case, the inner product is *not* commutative, as $\mathbf{y}^H\mathbf{x} = \overline{\mathbf{x}^H\mathbf{y}}$. We say that $\mathbf{x}$ and $\mathbf{y}$ are orthogonal if $\mathbf{x}^H\mathbf{y} = 0$. Similarly, if $\mathbf{x} \in \mathbb{C}^m$ and $\mathbf{y} \in \mathbb{C}^n$, then the outer product of $\mathbf{x}$ and $\mathbf{y}$ is the $m \times n$ matrix $\mathbf{x}\mathbf{y}^H$.

B.7.4 Hadamard Product

If $\mathbf{x}, \mathbf{y} \in \mathbb{R}^n$, the **Hadamard product**, or *componentwise product*, of $\mathbf{x}$ and $\mathbf{y}$, denoted by $\mathbf{x} \circ \mathbf{y}$ or $\mathbf{x}.\ast\mathbf{y}$, is the vector $\mathbf{z}$ obtained by multiplying corresponding components of $\mathbf{x}$ and $\mathbf{y}$. That is, if $\mathbf{z} = \mathbf{x}.\ast\mathbf{y}$, then $z_i = x_i y_i$, for $i = 1, 2, \ldots, n$.

Example B.7.7 If $\mathbf{x} = \begin{bmatrix} 1 & -2 \end{bmatrix}^T$ and $\mathbf{y} = \begin{bmatrix} -3 & 4 \end{bmatrix}^T$, then

$$\mathbf{x}^T\mathbf{y} = 1(-3) + (-2)4 = -11, \quad \mathbf{x}\mathbf{y}^T = \begin{bmatrix} 1(-3) & 1(4) \\ -2(-3) & -2(4) \end{bmatrix} = \begin{bmatrix} -3 & 4 \\ 6 & -8 \end{bmatrix},$$

and

$$\mathbf{x}.\ast\mathbf{y} = \begin{bmatrix} 1(-3) \\ -2(4) \end{bmatrix} = \begin{bmatrix} -3 \\ -8 \end{bmatrix}.$$

□

B.7.5 Partitioning

It is useful to describe matrices as collections of row or column vectors. Specifically, a **row partition** of an $m \times n$ matrix A is a description of A as a "stack" of row vectors $\mathbf{r}_1^T, \mathbf{r}_2^T, \ldots, \mathbf{r}_m^T$. That is,

$$A = \begin{bmatrix} \mathbf{r}_1^T \\ \mathbf{r}_2^T \\ \vdots \\ \mathbf{r}_m^T \end{bmatrix}.$$

We can also view A as a "concatenation" of column vectors $\mathbf{c}_1, \mathbf{c}_2, \ldots, \mathbf{c}_n$:

$$A = \begin{bmatrix} \mathbf{c}_1 & \mathbf{c}_2 & \cdots & \mathbf{c}_n \end{bmatrix}.$$

This description of A is called a **column partition**.

Example B.7.8 If

$$A = \begin{bmatrix} 1 & 2 \\ 3 & 4 \end{bmatrix},$$

then a column partitioning of A is

$$A = \begin{bmatrix} \mathbf{c}_1 & \mathbf{c}_2 \end{bmatrix}, \quad \mathbf{c}_1 = \begin{bmatrix} 1 \\ 3 \end{bmatrix}, \quad \mathbf{c}_2 = \begin{bmatrix} 2 \\ 4 \end{bmatrix},$$

and a row partitioning of A is

$$A = \begin{bmatrix} \mathbf{r}_1^T \\ \mathbf{r}_2^T \end{bmatrix}, \quad \mathbf{r}_1 = \begin{bmatrix} 1 \\ 2 \end{bmatrix}, \quad \mathbf{r}_2 = \begin{bmatrix} 3 \\ 4 \end{bmatrix}.$$

□

B.7.6 Perspectives on Matrix Multiplication

The fundamental operation of matrix-matrix multiplication can be understood in three different ways, based on other operations that can be performed on matrices and vectors. Let A be an $m \times n$ matrix, and B be an $n \times p$ matrix, in which case $C = AB$ is an $m \times p$ matrix. We can then view the computation of C in the following ways:

- Inner product: each entry c_{ij} is the dot product of the ith row of A and the jth column of B.
- Matrix-vector multiplication: the jth column of C is a linear combination of the columns of A, where the coefficients are obtained from the jth column of B. That is, if

$$C = \begin{bmatrix} \mathbf{c}_1 & \mathbf{c}_2 & \cdots & \mathbf{c}_p \end{bmatrix}, \quad B = \begin{bmatrix} \mathbf{b}_1 & \mathbf{b}_2 & \cdots & \mathbf{b}_p \end{bmatrix}$$

  are column partitions of C and B, then $\mathbf{c}_j = A\mathbf{b}_j$, for $j = 1, 2, \ldots, p$.
- Outer product: given the partitions

$$A = \begin{bmatrix} \mathbf{a}_1 & \mathbf{a}_2 & \cdots & \mathbf{a}_n \end{bmatrix}, \quad B = \begin{bmatrix} \mathbf{b}_1^T \\ \mathbf{b}_2^T \\ \cdots \\ \mathbf{b}_n^T \end{bmatrix},$$

  we can write

$$C = \mathbf{a}_1\mathbf{b}_1^T + \mathbf{a}_2\mathbf{b}_2^T + \cdots + \mathbf{a}_n\mathbf{b}_n^T = \sum_{i=1}^{n} \mathbf{a}_i\mathbf{b}_i^T.$$

  That is, C is a sum of *outer product updates*.

B.8 The Identity Matrix

When $n = 1$, the identity element of 1×1 matrices, the number 1, is the unique number such that $a(1) = 1(a) = a$ for any number a. To determine the identity element for $n \times n$ matrices, we seek a matrix I such that $AI = IA = A$ for any $n \times n$ matrix A. That is, we must have

$$\sum_{k=1}^{n} a_{ik} I_{kj} = a_{ij}, \quad i, j = 1, \ldots, n.$$

This can only be the case for *any* matrix A if $I_{jj} = 1$ for $j = 1, 2, \ldots, n$, and $I_{ij} = 0$ when $i \neq j$. We call this matrix the **identity matrix**

$$I = \begin{bmatrix} 1 & 0 & \cdots & \cdots & 0 \\ 0 & 1 & 0 & & \vdots \\ & \ddots & \ddots & \ddots & \\ \vdots & & 0 & 1 & 0 \\ 0 & \cdots & \cdots & 0 & 1 \end{bmatrix}.$$

Note that the jth column of I is the standard basis vector $\mathbf{e}_j$.

B.9 The Inverse of a Matrix

Given an $n \times n$ matrix A, it is now natural to ask whether it is possible to find an $n \times n$ matrix B such that $AB = BA = I$. Such a matrix, if it exists, would then serve as the **inverse** of A, in the sense of matrix multiplication. We denote this matrix by A^{-1}, just as we denote the multiplicative inverse of a nonzero number a by a^{-1}. If the inverse of A exists, we say that A is **invertible** or **nonsingular**; otherwise, we say that A is **singular**. If A^{-1} exists, then we can use it to describe the solution of the system of linear equations $A\mathbf{x} = \mathbf{b}$, for

$$A^{-1}A\mathbf{x} = (A^{-1}A)\mathbf{x} = I\mathbf{x} = \mathbf{x} = A^{-1}\mathbf{b},$$

which generalizes the solution $x = a^{-1}b$ of a single linear equation in one unknown.

However, just as we can use the inverse to describe the solution to a system of linear equations, we can use systems of linear equations to characterize the inverse. Because A^{-1} satisfies $AA^{-1} = I$, it follows from multiplication of both sides of this equation by the jth standard basis vector $\mathbf{e}_j$ that

$$A\mathbf{b}_j = \mathbf{e}_j, \quad j = 1, 2, \ldots, n,$$

where $\mathbf{b}_j = A^{-1}\mathbf{e}_j$ is the jth column of $B = A^{-1}$. That is, we can compute A^{-1} by solving n systems of linear equations of the form $A\mathbf{b}_j = \mathbf{e}_j$, using a method such as Gaussian Elimination and back substitution (see Section 3.1). If Gaussian Elimination fails due to the inability to obtain a nonzero pivot element for each column, then A^{-1} does not exist, and we conclude that A is singular.

The inverse of a nonsingular matrix A has the following properties:

- A^{-1} is unique.
- A^{-1} is nonsingular, and $(A^{-1})^{-1} = A$.
- If B is also a nonsingular $n \times n$ matrix, then $(AB)^{-1} = B^{-1}A^{-1}$.
- $(A^{-1})^T = (A^T)^{-1}$. It is common practice to denote the transpose of A^{-1} by A^{-T}.

The set of all $n \times n$ matrices has an identity element, matrix multiplication is associative, and each nonsingular $n \times n$ matrix has a unique inverse with respect to matrix multiplication, that is also an $n \times n$ nonsingular matrix. Therefore, the set of all nonsingular $n \times n$ matrices over a field $\mathbb{F}$ (such as $\mathbb{R}$ or $\mathbb{C}$) forms a **group**, which is denoted by $GL(n, \mathbb{F})$, the **general linear group**.

If A^{-1} exists, then we know the following about the system of linear equations $A\mathbf{x} = \mathbf{b}$:

- The system $A\mathbf{x} = \mathbf{b}$ has a unique solution for any n-vector $\mathbf{b}$, which is given by $\mathbf{x} = A^{-1}\mathbf{b}$.
- The system $A\mathbf{x} = \mathbf{0}$ has only the **trivial solution $\mathbf{x} = \mathbf{0}$**.

The lack of an inverse is equally telling: if A is $n \times n$ and A^{-1} does *not* exist (that is, A is singular), then $A\mathbf{x} = \mathbf{0}$ has a nontrivial solution, and $A\mathbf{x} = \mathbf{b}$ has either no solution or infinitely many solutions, depending on whether $\mathbf{b}$ belongs to range(A).

B.10   Triangular and Diagonal Matrices

There are certain types of matrices for which the fundamental problems of numerical linear algebra, solving systems of linear equations or computing eigenvalues and eigenvectors, are relatively easy to solve. We now discuss a few such types.

Let A be an $m \times n$ matrix. We define the **main diagonal** of A to be the entries $a_{11}, a_{22}, \ldots, a_{pp}$, where $p = \min\{m, n\}$. That is, the main diagonal consists of all entries for which the row index and column index are equal. We then say that A is a **diagonal matrix** if the only nonzero entries of A lie on the main diagonal. That is, A is a diagonal matrix if $a_{ij} = 0$ whenever $i \neq j$.

We say that A is **upper triangular** if $a_{ij} = 0$ whenever $i > j$. That is, all nonzero entries of A are confined to the "upper triangle" of A, which consists of all entries on or "above" the main diagonal. Similarly, we say that A is **lower triangular** if $a_{ij} = 0$ whenever $i < j$. Such a matrix has all of its nonzero entries on or "below" the main diagonal.

We will see that a system of linear equations of the form $A\mathbf{x} = \mathbf{b}$ is easily solved if A is an upper or lower triangular matrix, and that the eigenvalues of a square matrix A are easy to obtain if A is triangular. As such, certain methods for solving both kinds of problems for a general matrix proceed by reducing A to triangular form.

Example B.10.1 The matrices

$$U = \begin{bmatrix} 1 & 2 & 3 \\ 0 & 4 & 5 \\ 0 & 0 & 6 \end{bmatrix}, \quad L = \begin{bmatrix} 1 & 0 & 0 \\ 2 & 3 & 0 \\ 4 & 5 & 6 \end{bmatrix}, \quad D = \begin{bmatrix} 1 & 0 & 0 \\ 0 & 2 & 0 \\ 0 & 0 & 3 \end{bmatrix}$$

are upper triangular, lower triangular, and diagonal, respectively. □

B.11 Determinants

A 1×1 matrix a is invertible if and only if its single entry, $a_{11} = a$, is nonzero. We now discuss the generalization of this determination of invertibility to general square matrices.

Let S_n be the **symmetric group** of order n, which consists of all possible **permutations** of the numbers $1, 2, \ldots, n$. For each permutation $\sigma \in S_n$, we define the **sign** of σ, denoted by $\text{sgn}(\sigma)$, to be $(-1)^p$, where p is the number of transpositions of the indices $1, 2, \ldots, n$ needed to obtain σ. Then, the **determinant** of an $n \times n$ matrix A, denoted by $\det(A)$, is defined by

$$\det(A) = \sum_{\sigma \in S_n} \text{sgn}(\sigma) \prod_{i=1}^{n} a_{i,\sigma(i)}. \tag{B.3}$$

That is, $\det(A)$ is the sum of all possible products of entries of A that can be obtained by selecting one entry of each row of A and choosing a different column index for each entry, times the sign of each resulting permutation.

Example B.11.1 The group S_2 contains only two permutations: the identity permutation $(1, 2)$, which has a sign of 1, and the transposition $(2, 1)$, that has a sign of -1. It follows that if A is a 2×2 matrix, then

$$\det(A) = a_{11}a_{22} + (-1)a_{12}a_{21} = a_{11}a_{22} - a_{12}a_{21}.$$

□

Because the order of S_n is $n!$, computing $\det(A)$ using the definition requires $O(n!)$ arithmetic operations (see Section 2.1.5 for the definition of big-O notation). This is a prohibitively high computational cost, but fortunately there are several properties of the determinant which can be exploited to obtain a much more efficient approach to computing $\det(A)$.

1. If a row of A is scaled by a factor c, then, because each term in (B.3) includes exactly one entry from each row, $\det(A)$ is scaled by c as well. An immediate consequence is that if A is $n \times n$, then $\det(cA) = c^n \det(A)$.
2. An immediate consequence of Property 1 is that if A has a row of zeros, then $\det(A) = 0$.
3. If two rows of A are interchanged, this has the effect of adding an additional transposition to each permutation in S_n, which changes its sign. It follows that the sign of $\det(A)$ is changed.

4. A consequence of Property 3 is that if A has two identical rows, then $\det(A) = 0$, because interchanging these rows changes the sign of the determinant, but the determinant cannot change because the matrix is still the same.

5. Suppose that A is modified by adding a multiple m of row j to row i, where $i \neq j$, to obtain a new matrix B. From (B.3), we have $\det(A) = \det(B) + m\det(C)$, where the matrix C is obtained by replacing row i with row j. By Property 4, $\det(C) = 0$, so $\det(B) = \det(A)$. That is, this row operation does not affect the determinant.

6. From (B.3), we have

$$\det(A^T) = \sum_{\sigma \in S_n} \text{sgn}(\sigma) \prod_{i=1}^{n} a_{\sigma(i),i} = \sum_{\sigma \in S_n} \text{sgn}(\sigma) \prod_{i=1}^{n} a_{i,\sigma^{-1}(i)}.$$

However, taking the inverse of all permutations in S_n yields S_n, because it is a group. Therefore $\det(A^T) = \det(A)$.

7. Suppose that A is an $n \times n$ upper triangular matrix, as defined in Section B.10. Then, the only permutations that can lead to a nonzero contribution to $\det(A)$ are those for which $\sigma(1) = 1$, because all other entries of the first column of A are equal to zero. It follows that only terms in (B.3) involving a_{11} need be included. Continuing this process, it can be shown that $\det(A) = a_{11}a_{22} \cdots a_{nn}$. That is, the determinant of an upper triangular matrix is simply the product of its diagonal entries. By Property 6, the same is true of a lower triangular matrix.

8. Let A, B be $n \times n$. It can be shown using the definition of an upper triangular matrix that the product of two upper triangular matrices is upper triangular, and that the diagonal entries of the product are the products of corresponding diagonal entries. Therefore, if A and B are upper triangular, we have $\det(AB) = \det(A)\det(B)$ by Property 7. For general $n \times n$ matrices A and B, we can obtain $\det(AB)$ by reducing A to upper triangular form using row operations, and B to upper triangular form using column operations. Applying Properties 3, 5 and 6, we see that $\det(AB) = \det(A)\det(B)$ in the general case as well.

9. An immediate consequence of Property 8 is that if A is a square invertible matrix, then $\det(A) \neq 0$ and $\det(A^{-1}) = (\det(A))^{-1}$.

It is shown in Section 3.2 that by applying these properties, it is possible to compute $\det(A)$ in $O(n^3)$ arithmetic operations, where A is $n \times n$.

The best-known application of the determinant is the fact that an $n \times n$ matrix A is invertible, or nonsingular, if and only if $\det(A) \neq 0$, but the determinant has other interesting applications. For example, the determinant of a 3×3 matrix is equal to the volume of a parallelepiped defined by the vectors that are the rows (or columns) of the matrix. A helpful relation between $\det(A)$ and *eigenvalues* of A is given in the next section.

Example B.11.2 Because the matrices

$$L = \begin{bmatrix} 1 & 0 & 0 \\ 2 & 4 & 0 \\ -3 & -5 & -6 \end{bmatrix}, \quad U = \begin{bmatrix} 1 & -1 & 0 & 5 \\ 0 & 2 & 3 & -6 \\ 0 & 0 & -4 & 7 \\ 0 & 0 & 0 & 8 \end{bmatrix}$$

are lower and upper triangular, respectively, their determinants are the products of their diagonal entries. That is,

$$\det(L) = 1(4)(-6) = -24, \quad \det(U) = 1(2)(-4)(8) = -64.$$

☐

B.12 Eigenvalues

Because matrix-vector multiplication is a complicated operation, it is helpful to identify circumstances under which this operation can be simply described. To that end, we say that a nonzero vector $\mathbf{x}$ is an **eigenvector** of an $n \times n$ matrix A if there exists a scalar λ such that

$$A\mathbf{x} = \lambda\mathbf{x}.$$

The scalar λ is called an **eigenvalue** of A corresponding to $\mathbf{x}$. Note that although $\mathbf{x}$ is required to be nonzero, it is possible that λ can be zero. It can also be complex, even if A is a real matrix.

If we rearrange the above equation, we have

$$(A - \lambda I)\mathbf{x} = \mathbf{0}.$$

That is, if λ is an eigenvalue of A, then $A - \lambda I$ is a singular matrix, and therefore $\det(A - \lambda I) = 0$. This equation is actually a polynomial in λ, which is called the **characteristic polynomial** of A. If A is an $n \times n$ matrix, then the characteristic polynomial is of degree n. This means that A has n eigenvalues, which may repeat.

The following properties of eigenvalues and eigenvectors are helpful to know:

- If λ is an eigenvalue of A, then there is at least one eigenvector of A corresponding to λ.
- If there exists an invertible matrix P such that $B = PAP^{-1}$, then A and B have the same eigenvalues. We say that A and B are **similar**, and the transformation PAP^{-1} is called a **similarity transformation**.
- If A is a *symmetric* matrix, then its eigenvalues are real.
- If A is a *skew-symmetric* matrix, meaning that $A^T = -A$, then its eigenvalues are either equal to zero, or are purely imaginary.
- If A is a real matrix, and $\lambda = u + iv$ is a complex eigenvalue of A, then $\bar{\lambda} = u - iv$ is also an eigenvalue of A.
- If A is a triangular matrix, then its diagonal entries are the eigenvalues of A.

- $\det(A)$ is equal to the product of the eigenvalues of A.
- The **trace** of A, denoted by $\mathrm{tr}(A)$, is defined to be the sum of the diagonal entries of A. It is also equal to the sum of the eigenvalues of A.

It follows that any method for computing the roots of a polynomial can be used to obtain the eigenvalues of a matrix A. However, in practice, eigenvalues are normally computed using iterative methods that employ similarity transformations to reduce A to upper triangular form, thus revealing the eigenvalues of A. In practice, such methods for computing eigenvalues are used to compute roots of polynomials, rather than using polynomial root-finding methods to compute eigenvalues, because they are much more robust with respect to roundoff error. Methods for computing eigenvalues are covered in Chapter 6.

B.13   Vector and Matrix Norms

Numerical analysis involves measuring error in quantities that are vectors or matrices. While the error in each entry of a vector or matrix is of interest, it is desirable to be able to quantify error in such quantities using a single number. For this reason, we need a notion of magnitude for vectors and matrices.

B.13.1   Vector Norms

Given vectors $\mathbf{x}$ and $\mathbf{y}$ of length one, which are simply scalars x and y, the most natural notion of distance between x and y is obtained from the absolute value; we define the distance to be $|x - y|$. We therefore define a distance function for vectors that has similar properties.

A function $\| \cdot \| : \mathbb{R}^n \to \mathbb{R}$ is called a **vector norm** if it has the following properties:

1. $\|\mathbf{x}\| \geq 0$ for any vector $\mathbf{x} \in \mathbb{R}^n$, and $\|\mathbf{x}\| = 0$ if and only if $\mathbf{x} = \mathbf{0}$
2. $\|\alpha\mathbf{x}\| = |\alpha|\|\mathbf{x}\|$ for any vector $\mathbf{x} \in \mathbb{R}^n$ and any scalar $\alpha \in \mathbb{R}$
3. $\|\mathbf{x} + \mathbf{y}\| \leq \|\mathbf{x}\| + \|\mathbf{y}\|$ for any vectors $\mathbf{x}, \mathbf{y} \in \mathbb{R}^n$.

The last property is called the **triangle inequality**. It should be noted that when $n = 1$, the absolute value function is a vector norm.

The most commonly used vector norms belong to the family of ℓ_p**-norms**, or p**-norms**, which are defined by

$$\|\mathbf{x}\|_p = \left(\sum_{i=1}^{n} |x_i|^p \right)^{1/p}.$$

It can be shown that for any $p > 0$, $\| \cdot \|_p$ defines a vector norm. The following p-norms are of particular interest:

- $p = 1$: The ℓ_1**-norm**

$$\|\mathbf{x}\|_1 = |x_1| + |x_2| + \cdots + |x_n|$$

- $p = 2$: The ℓ_2-**norm** or **Euclidean norm**

$$\|\mathbf{x}\|_2 = \sqrt{|x_1|^2 + |x_2|^2 + \cdots + |x_n|^2} = \sqrt{\mathbf{x}^H \mathbf{x}}$$

- $p = \infty$: The ℓ_∞-**norm**

$$\|\mathbf{x}\|_\infty = \max_{1 \le i \le n} |x_i|$$

Example B.13.1 Given the vector

$$\mathbf{x} = \begin{bmatrix} 1 \\ 2 \\ -3 \end{bmatrix},$$

we have

$$\|\mathbf{x}\|_1 = |1| + |2| + |-3| = 6,$$
$$\|\mathbf{x}\|_2 = \sqrt{1^2 + 2^2 + (-3)^2} = \sqrt{14},$$
$$\|\mathbf{x}\|_\infty = \max\{|1|, |2|, |-3|\} = 3.$$

□

It can be shown that the ℓ_2-norm satisfies the **Cauchy-Schwarz inequality**,

$$|\mathbf{x}^T \mathbf{y}| \le \|\mathbf{x}\|_2 \|\mathbf{y}\|_2 \tag{B.4}$$

for any vectors $\mathbf{x}, \mathbf{y} \in \mathbb{R}^n$. This inequality is useful for showing that the ℓ_2-norm satisfies the triangle inequality. It is a special case of the **Hölder inequality**

$$|\mathbf{x}^T \mathbf{y}| \le \|\mathbf{x}\|_p \|\mathbf{y}\|_q, \quad \frac{1}{p} + \frac{1}{q} = 1.$$

We will prove the Cauchy-Schwarz inequality for vectors in $\mathbb{R}^n$; the proof can be generalized to a complex vector space. For $\mathbf{x}, \mathbf{y} \in \mathbb{R}^n$ and $c \in \mathbb{R}$, with $\mathbf{y} \ne \mathbf{0}$, we have

$$(\mathbf{x} - c\mathbf{y})^T (\mathbf{x} - c\mathbf{y}) = \|\mathbf{x} - c\mathbf{y}\|_2^2 \ge 0.$$

It follows from the properties of the inner product that

$$0 \le (\mathbf{x} - c\mathbf{y})^T (\mathbf{x} - c\mathbf{y})$$
$$\le \mathbf{x}^T \mathbf{x} - \mathbf{x}^T (c\mathbf{y}) - (c\mathbf{y})^T \mathbf{x} + (c\mathbf{y})^T (c\mathbf{y})$$
$$\le \|\mathbf{x}\|_2^2 - 2c\mathbf{x}^T \mathbf{y} + c^2 \|\mathbf{y}\|_2^2.$$

We now try to find the value of c that minimizes this expression. Differentiating with respect to c and equating to zero yields the equation

$$-2\mathbf{x}^T \mathbf{y} + 2c\|\mathbf{y}\|_2^2 = 0,$$

and therefore the minimum occurs when $c = \mathbf{x}^T\mathbf{y}/\|\mathbf{y}\|_2^2$. It follows that

$$
\begin{aligned}
0 &\leq \|\mathbf{x}\|_2^2 - 2c\mathbf{x}^T\mathbf{y} + c^2\|\mathbf{y}\|^2 \\
&\leq \|\mathbf{x}\|_2^2 - 2\frac{\mathbf{x}^T\mathbf{y}}{\|\mathbf{y}\|_2^2}\mathbf{x}^T\mathbf{y} + \frac{(\mathbf{x}^T\mathbf{y})^2}{\|\mathbf{y}\|_2^4}\|\mathbf{y}\|_2^2 \\
&\leq \|\mathbf{x}\|_2^2 - 2\frac{(\mathbf{x}^T\mathbf{y})^2}{\|\mathbf{y}\|_2^2} + \frac{(\mathbf{x}^T\mathbf{y})^2}{\|g\|^2} \\
&\leq \|\mathbf{x}\|_2^2 - \frac{(\mathbf{x}^T\mathbf{y})^2}{\|\mathbf{y}\|_2^2}.
\end{aligned}
$$

It follows that

$$
(\mathbf{x}^T\mathbf{y})^2 \leq \|\mathbf{x}\|_2^2\|\mathbf{y}\|_2^2.
$$

Taking the square root of both sides yields the Cauchy-Schwarz inequality.

Now that we have defined various notions of the size, or magnitude, of a vector, we can discuss distance and convergence. Given a vector norm $\|\cdot\|$, and vectors $\mathbf{x}, \mathbf{y} \in \mathbb{R}^n$, we define the **distance** between $\mathbf{x}$ and $\mathbf{y}$, with respect to this norm, by $\|\mathbf{x} - \mathbf{y}\|$. Then, we say that a sequence of n-vectors $\{\mathbf{x}^{(k)}\}_{k=0}^{\infty}$ **converges** to a vector $\mathbf{x}$ if

$$
\lim_{k\to\infty} \|\mathbf{x}^{(k)} - \mathbf{x}\| = 0.
$$

That is, the distance between $\mathbf{x}^{(k)}$ and $\mathbf{x}$ must approach zero. It can be shown that regardless of the choice of norm, $\mathbf{x}^{(k)} \to \mathbf{x}$ if and only if

$$
x_i^{(k)} \to x_i, \quad i = 1, 2, \ldots, n.
$$

That is, each component of $\mathbf{x}^{(k)}$ must converge to the corresponding component of $\mathbf{x}$. This is due to the fact that for any vector norm $\|\cdot\|$, $\|\mathbf{x}\| = 0$ if and only if $\mathbf{x}$ is the zero vector.

Because we have defined convergence with respect to an arbitrary norm, it is important to know whether a sequence can converge to a limit with respect to one norm, while converging to a different limit in another norm, or perhaps not converging at all. Fortunately, for ℓ_p-norms, this is never the case. We say that two vector norms $\|\cdot\|_\alpha$ and $\|\cdot\|_\beta$ are **equivalent** if there exists constants C_1 and C_2, that are independent of $\mathbf{x}$, such that for any vector $\mathbf{x} \in \mathbb{R}^n$,

$$
C_1\|\mathbf{x}\|_\alpha \leq \|\mathbf{x}\|_\beta \leq C_2\|\mathbf{x}\|_\alpha.
$$

It follows that if two norms are equivalent, then a sequence of vectors that converges to a limit with respect to one norm will converge to the same limit in the other. It can be shown that all ℓ_p-norms are equivalent. In particular, if $\mathbf{x} \in \mathbb{R}^n$, then

$$
\|\mathbf{x}\|_2 \leq \|\mathbf{x}\|_1 \leq \sqrt{n}\|\mathbf{x}\|_2,
$$

$$
\|\mathbf{x}\|_\infty \leq \|\mathbf{x}\|_2 \leq \sqrt{n}\|\mathbf{x}\|_\infty,
$$

$$
\|\mathbf{x}\|_\infty \leq \|\mathbf{x}\|_1 \leq n\|\mathbf{x}\|_\infty.
$$

We will now prove the equivalence of $\| \cdot \|_1$ and $\| \cdot \|_2$. Let $\mathbf{x} \in \mathbb{R}^n$. First, we have

$$\|\mathbf{x}\|_2^2 = \sum_{i=1}^n |x_i|^2 \leq \sum_{i,j=1}^n |x_i||x_j| \leq \|\mathbf{x}\|_1^2,$$

and therefore $\|\mathbf{x}\|_2 \leq \|\mathbf{x}\|_1$. Then, we define the vector $\mathbf{y}$ by

$$y_i = \begin{cases} 1 & x_i \geq 0 \\ -1 & x_i < 0 \end{cases}.$$

It follows that $\|\mathbf{x}\|_1 = \mathbf{y}^T \mathbf{x}$. By the Cauchy-Schwarz inequality,

$$\|\mathbf{x}\|_1 = \mathbf{y}^T \mathbf{x} \leq \|\mathbf{y}\|_2 \|\mathbf{x}\|_2 \leq \sqrt{n} \|\mathbf{x}\|_2$$

and the equivalence of the norms has been established.

B.13.2 Matrix Norms

It is also very useful to be able to measure the magnitude of a matrix, or the distance between matrices. However, it is generally not sufficient to simply define the norm of an $m \times n$ matrix A as the norm of an mn-vector $\mathbf{x}$ whose components are the entries of A. We instead define a **matrix norm** to be a function $\| \cdot \| : \mathbb{R}^{m \times n} \to \mathbb{R}$ that has the following properties:

- $\|A\| \geq 0$ for any $A \in \mathbb{R}^{m \times n}$, and $\|A\| = 0$ if and only if $A = 0$
- $\|\alpha A\| = |\alpha| \|A\|$ for any $m \times n$ matrix A and scalar α
- $\|A + B\| \leq \|A\| + \|B\|$ for any $m \times n$ matrices A and B

Another property that is often, but not always, included in the definition of a matrix norm is the **submultiplicative property**: if A is $m \times n$ and B is $n \times p$, we require that

$$\|AB\| \leq \|A\| \|B\|.$$

This is particularly useful when A and B are square matrices.

Any vector norm **induces** a matrix norm. It can be shown that given a vector norm, defined appropriately for m-vectors and n-vectors, the function $\| \cdot \| : \mathbb{R}^{m \times n} \to \mathbb{R}$ defined by

$$\|A\| = \sup_{\mathbf{x} \neq 0} \frac{\|A\mathbf{x}\|}{\|\mathbf{x}\|} = \max_{\|\mathbf{x}\|=1} \|A\mathbf{x}\|$$

is a matrix norm. It is called the **natural**, or *induced*, matrix norm. Furthermore, if the vector norm is a ℓ_p-norm, then the induced matrix norm satisfies the submultiplicative property.

The following matrix norms are of particular interest:

- The ℓ_1-**norm**:

$$\|A\|_1 = \max_{\|\mathbf{x}\|_1=1} \|A\mathbf{x}\|_1 = \max_{1 \leq j \leq n} \sum_{i=1}^m |a_{ij}|.$$

That is, the ℓ_1-norm of a matrix is its maximum column sum of $|A|$. To see this, let $\mathbf{x} \in \mathbb{R}^n$ satisfy $\|\mathbf{x}\|_1 = 1$. Then

$$\|A\mathbf{x}\|_1 = \sum_{i=1}^{m} |(A\mathbf{x})_i|$$

$$= \sum_{i=1}^{m} \left| \sum_{j=1}^{n} a_{ij}x_j \right|$$

$$\leq \sum_{j=1}^{n} |x_j| \left(\sum_{i=1}^{m} |a_{ij}| \right)$$

$$\leq \sum_{j=1}^{n} |x_j| \max_{1 \leq j \leq n} \left(\sum_{i=1}^{m} |a_{ij}| \right)$$

$$\leq \max_{1 \leq j \leq n} \left(\sum_{i=1}^{m} |a_{ij}| \right).$$

Equality is achieved if $\mathbf{x} = \mathbf{e}_J$, where the index J satisfies

$$\max_{1 \leq j \leq n} \left(\sum_{i=1}^{m} |a_{ij}| \right) = \sum_{i=1}^{m} |a_{iJ}|.$$

It follows that the maximum column sum of $|A|$ is equal to the maximum of $\|A\mathbf{x}\|_1$ taken over all the set of all unit 1-norm vectors.

• The ℓ_∞-**norm**:

$$\|A\|_\infty = \max_{\|\mathbf{x}\|_\infty = 1} \|A\mathbf{x}\|_\infty = \max_{1 \leq i \leq m} \sum_{j=1}^{n} |a_{ij}|.$$

That is, the ℓ_∞-norm of a matrix is its maximum *row* sum. This formula can be obtained in a similar manner as the one for the matrix 1-norm.

• The ℓ_2-**norm**:

$$\|A\|_2 = \max_{\|\mathbf{x}\|_2 = 1} \|A\mathbf{x}\|_2.$$

To obtain a formula for this norm, we note that the function

$$g(\mathbf{x}) = \frac{\|A\mathbf{x}\|_2^2}{\|\mathbf{x}\|_2^2} = \frac{\mathbf{x}^T A^T A \mathbf{x}}{\mathbf{x}^T \mathbf{x}}$$

has a local maximum or minimum whenever $\mathbf{x}$ is a *unit* ℓ_2-norm vector (that is, $\|\mathbf{x}\|_2 = 1$) that satisfies

$$A^T A \mathbf{x} = \|A\mathbf{x}\|_2^2 \mathbf{x},$$

as can be shown by differentiation of $g(\mathbf{x})$. That is, $\mathbf{x}$ is an *eigenvector* of $A^T A$, with corresponding *eigenvalue* $\|A\mathbf{x}\|_2^2 = g(\mathbf{x})$. We conclude that

$$\|A\|_2 = \max_{1 \leq i \leq n} \sqrt{\lambda_i(A^T A)}. \tag{B.5}$$

That is, the ℓ_2-norm of a matrix is the square root of the largest eigenvalue of $A^T A$, which is guaranteed to be nonnegative, as can be shown using the vector 2-norm. We see that unlike the vector ℓ_2-norm, the matrix ℓ_2-norm is much more difficult to compute than the matrix ℓ_1-norm or ℓ_∞-norm.

- The **Frobenius norm**:

$$\|A\|_F = \left(\sum_{i=1}^{m} \sum_{j=1}^{n} a_{ij}^2 \right)^{1/2}. \tag{B.6}$$

It should be noted that the Frobenius norm is *not* induced by any vector ℓ_p-norm, but it is equivalent to the vector ℓ_2-norm in the sense that $\|A\|_F = \|\mathbf{x}\|_2$ where $\mathbf{x}$ is obtained by reshaping A into a vector.

Like vector norms, matrix norms are equivalent. For example, if A is an $m \times n$ matrix, we have

$$\|A\|_2 \leq \|A\|_F \leq \sqrt{n}\|A\|_2,$$

$$\frac{1}{\sqrt{n}}\|A\|_\infty \leq \|A\|_2 \leq \sqrt{m}\|A\|_\infty,$$

$$\frac{1}{\sqrt{m}}\|A\|_1 \leq \|A\|_2 \leq \sqrt{n}\|A\|_1.$$

Example B.13.2 Let

$$A = \begin{bmatrix} 1 & 2 & 3 \\ 0 & 1 & 0 \\ -1 & 0 & 4 \end{bmatrix}.$$

Then

$$\|A\|_1 = \max\{|1| + |0| + |-1|, |2| + |1| + |0|, |3| + |0| + |4|\} = 7,$$

and

$$\|A\|_\infty = \max\{|1| + |2| + |3|, |0| + |1| + |0|, |-1| + |0| + |4|\} = 6.$$

$\square$

B.13.3 Function Spaces and Norms

We now define norms on more general vector spaces. Let $\mathcal{V}$ be a vector space over the field of real numbers $\mathbb{R}$. A **norm** on $\mathcal{V}$ is a function $\|\cdot\| : \mathcal{V} \to \mathbb{R}$ that has the following properties:

1. $\|f\| \geq 0$ for all $f \in \mathcal{V}$, and $\|f\| = 0$ if and only if f is the zero vector of $\mathcal{V}$.
2. $\|cf\| = |c|\|f\|$ for any vector $f \in \mathcal{V}$ and any scalar $c \in \mathbb{R}$.
3. $\|f + g\| \leq \|f\| + \|g\|$ for all $f, g \in \mathcal{V}$.

As with norms of vectors in $\mathbb{R}^n$ and $m \times n$ matrices, the last property is known as the **triangle inequality**. A vector space $\mathcal{V}$, together with a norm $\|\cdot\|$, is called a **normed vector space** or **normed linear space**. In particular, we are interested in working with **function spaces**, which are vector spaces in which the vectors are functions.

Example B.13.3 The space $C[a, b]$ of functions that are continuous on the interval $[a, b]$ is a normed vector space with the norm

$$\|f\|_\infty = \max_{a \le x \le b} |f(x)|,$$

known as the ∞-**norm** or **maximum norm**. $\square$

Example B.13.4 We say that $f(x)$ is **square-integrable** on (a, b) if

$$\int_a^b |f(x)|^2 \, dx < \infty. \tag{B.7}$$

That is, the above integral must be finite; we also say that $f \in L^2(a, b)$. This function space is a normed vector space, equipped with the L^2-**norm** (B.7). $\square$

Example B.13.5 The space $C[a, b]$ can be equipped with a different norm, such as

$$\|f\|_2 = \left(\int_a^b |f(x)|^2 w(x) \, dx \right)^{1/2},$$

where the **weight function** $w(x)$ is positive and integrable on (a, b). It is allowed to be singular at the endpoints, as will be seen in certain examples. This norm is called the **weighted L^2-norm**. $\square$

The L^2-norm and L^∞-norm are related as follows:

$$\|f\|_2 \le W \|f\|_\infty, \quad W = \|1\|_2.$$

However, unlike the L^∞-norm and L^2-norm defined for the vector space $\mathbb{R}^n$, these norms are not **equivalent** in the sense that a function that has a small L^2-norm necessarily has a small L^∞-norm. In fact, given any $\epsilon > 0$, no matter how small, and any $M > 0$, no matter how large, there exists a function $f \in C[a, b]$ such that

$$\|f\|_2 < \epsilon, \quad \|f\|_\infty > M.$$

We say that a function f is **absolutely continuous** on $[a, b]$ if its derivative is finite almost everywhere in $[a, b]$ (meaning that it is not finite on at most a subset of $[a, b]$ that has *measure zero*), is integrable on $[a, b]$, and satisfies

$$\int_a^x f'(s) \, dx = f(x) - f(a), \quad a \le x \le b.$$

Any continuously differentiable function is absolutely continuous, but the converse is not necessarily true.

Example B.13.6 For example, $f(x) = |x|$ is absolutely continuous on any interval of the form $[-a, a]$, but it is not continuously differentiable on such an interval. $\square$

Next, we define the **Sobolev spaces** $H^k(a, b)$ as follows. The space $H^1(a, b)$ is the set of all absolutely continuous functions on $[a, b]$ whose derivatives belong to $L^2(a, b)$. Then, for $k > 1$, $H^k(a, b)$ is the subset of $H^{k-1}(a, b)$ consisting of functions whose $(k - 1)$st derivatives are absolutely continuous, and whose kth derivatives belong to $L^2(a, b)$. If we denote by $C^k[a, b]$ the set of all functions defined on $[a, b]$ that are k times continuously differentiable, then $C^k[a, b]$ is a proper subset of $H^k(a, b)$. For example, any piecewise linear belongs to $H^1(a, b)$, but does not generally belong to $C^1[a, b]$.

Example B.13.7 The function $f(x) = x^{3/4}$ belongs to $H^1(0, 1)$ because $f'(x) = \frac{3}{4}x^{-1/4}$ is integrable on $[0, 1]$, and also square-integrable on $[0, 1]$, since

$$\int_0^1 |f'(x)|^2\, dx = \int_0^1 \frac{9}{16}x^{-1/2} = \frac{9}{8}x^{1/2}\Big|_0^1 = \frac{9}{8}.$$

However, $f \notin C^1[a, b]$, because $f'(x)$ is singular at $x = 0$. $\square$

B.14 Convergence

We have learned in Section B.13 what it means for a sequence of vectors to converge to a limit. However, using the definition alone, it may still be difficult to determine, conclusively, whether a given sequence of vectors converges. For example, suppose a sequence of vectors is defined as follows: we choose the initial vector $\mathbf{x}^{(0)}$ arbitrarily, and then define the rest of the sequence by

$$\mathbf{x}^{(k+1)} = A\mathbf{x}^{(k)}, \quad k = 0, 1, 2, \ldots$$

for some matrix A. Such a sequence actually arises in the study of the convergence of various iterative methods for solving systems of linear equations, as shown in Chapter 5.

An important question is whether a sequence of this form converges to the zero vector. This will be the case if

$$\lim_{k \to \infty} \|\mathbf{x}^{(k)}\| = 0$$

in some vector norm. From the definition of $\mathbf{x}^{(k)}$, we must have

$$\lim_{k \to \infty} \|A^k\mathbf{x}^{(0)}\| = 0.$$

From the submultiplicative property of matrix norms,

$$\|A^k\mathbf{x}^{(0)}\| \leq \|A\|^k\|\mathbf{x}^{(0)}\|,$$

from which it follows that the sequence will converge to the zero vector if $\|A\| < 1$. However, this is only a *sufficient* condition; it is not *necessary*.

A necessary *and* sufficient condition can be described in terms of the *eigenvalues* of A, as described in Section B.12. It can be shown that if each eigenvalue λ of a matrix A satisfies $|\lambda| < 1$, then, for any vector $\mathbf{x}$,

$$\lim_{k \to \infty} A^k\mathbf{x} = \mathbf{0}.$$

Furthermore, the converse of this statement is also true: if there exists a vector $\mathbf{x}$ such that $A^k\mathbf{x}$ does not approach $\mathbf{0}$ as $k \to \infty$, then at least one eigenvalue λ of A must satisfy $|\lambda| \geq 1$.

Therefore, it is through the eigenvalues of A that we can describe a necessary and sufficient condition for a sequence of vectors of the form $\mathbf{x}^{(k)} = A^k\mathbf{x}^{(0)}$ to converge to the zero vector. Specifically, we need only check if the magnitude of the largest eigenvalue is less than 1. For convenience, we define the **spectral radius** of A, denoted by $\rho(A)$, to be $\max|\lambda|$, where λ is an eigenvalue of A. We can then conclude that the sequence $\mathbf{x}^{(k)} = A^k\mathbf{x}^{(0)}$ converges to the zero vector if and only if $\rho(A) < 1$.

Example B.14.1 Let

$$A = \begin{bmatrix} 2 & 3 & 1 \\ 0 & 4 & 5 \\ 0 & 0 & 1 \end{bmatrix}.$$

Because A is upper triangular, its eigenvalues are the diagonal entries, 2, 4 and 1. Because the largest eigenvalue in magnitude is 4, the spectral radius of A is $\rho(A) = 4$. □

The spectral radius is closely related to natural (induced) matrix norms. Let λ be the largest eigenvalue of A, with $\mathbf{x}$ being a corresponding eigenvector. Then, for any natural matrix norm $\| \cdot \|$, we have

$$\rho(A)\|\mathbf{x}\| = |\lambda|\|\mathbf{x}\| = \|\lambda\mathbf{x}\| = \|A\mathbf{x}\| \leq \|A\|\|\mathbf{x}\|.$$

Therefore, we have $\rho(A) \leq \|A\|$. When A is symmetric, we also have

$$\|A\|_2 = \rho(A).$$

For a general matrix A, we have

$$\|A\|_2 = [\rho(A^T A)]^{1/2},$$

which can be seen to reduce to $\rho(A)$ when $A^T = A$, since, in general, $\rho(A^k) = \rho(A)^k$. A proof of this is left to an exercise in Chapter 6.

Because the condition $\rho(A) < 1$ is necessary and sufficient to ensure that $\lim_{k \to \infty} A^k\mathbf{x} = \mathbf{0}$, it is possible that such convergence may occur even if $\|A\| \geq 1$ for some natural norm $\| \cdot \|$. However, if $\rho(A) < 1$, we can conclude that

$$\lim_{k \to \infty} \|A^k\| = 0,$$

even though $\lim_{k \to \infty} \|A\|^k$ may not even exist.

In view of the definition of a matrix norm, that $\|A\| = 0$ if and only if $A = 0$, we can conclude that if $\rho(A) < 1$, then A^k converges to the zero matrix as $k \to \infty$. In summary, the following statements are all equivalent:

1. $\rho(A) < 1$

2. $\lim\limits_{k \to \infty} \|A^k\| = 0$, for any natural norm $\|\cdot\|$

3. $\lim\limits_{k \to \infty} (A^k)_{ij} = 0,\ i, j = 1, 2, \ldots, n$

4. $\lim\limits_{k \to \infty} A^k \mathbf{x} = \mathbf{0}$

These results are very useful for analyzing the convergence behavior of various iterative methods for solving systems of linear equations, such as those covered in Chapter 5.

B.15 Inner Product Spaces

Recall that two real n-vectors $\mathbf{u} = \begin{bmatrix} u_1\ u_2\ \cdots\ u_n \end{bmatrix}^T$ and $\mathbf{v} = \begin{bmatrix} v_1\ v_2\ \cdots\ v_n \end{bmatrix}^T$ are *orthogonal* if

$$\mathbf{u}^T \mathbf{v} = \sum_{i=1}^{n} u_i v_i = 0,$$

where $\mathbf{u}^T \mathbf{v}$ is the *dot product*, or *inner product*, of $\mathbf{u}$ and $\mathbf{v}$.

Informally, by viewing functions defined on an interval $[a, b]$ as infinitely long vectors, we can generalize the inner product, and the concept of orthogonality, to functions. To that end, we define the **inner product** of two real-valued functions $f(x)$ and $g(x)$ defined on the interval $[a, b]$ by

$$\langle f, g \rangle = \int_a^b f(x) g(x)\, dx. \tag{B.8}$$

Then, we say f and g are **orthogonal** with respect to this inner product if $\langle f, g \rangle = 0$.

In general, an inner product on a vector space $\mathcal{V}$ over $\mathbb{R}$, be it continuous or discrete, has the following properties:

1. $\langle f + g, h \rangle = \langle f, h \rangle + \langle g, h \rangle$ for all $f, g, h \in \mathcal{V}$
2. $\langle cf, g \rangle = c \langle f, g \rangle$ for all $c \in \mathbb{R}$ and all $f \in \mathcal{V}$
3. $\langle f, g \rangle = \langle g, f \rangle$ for all $f, g \in \mathcal{V}$
4. $\langle f, f \rangle \geq 0$ for all $f \in \mathcal{V}$, and $\langle f, f \rangle = 0$ if and only if $f = \mathbf{0}$.

A vector space $\mathcal{V}$ equipped with such an inner product is called an **inner product space**.

An inner product naturally induces a norm. Recall that the ℓ_2-norm of a vector $\mathbf{v} \in \mathbb{R}^n$, denoted by $\|\mathbf{v}\|_2$, can be defined by

$$\|\mathbf{v}\| = (\mathbf{v}^T \mathbf{v})^{1/2}.$$

Along similar lines, the L^2-**norm** of a function $f(x)$ on $[a, b]$ can be defined by

$$\|f\|_2 = (\langle f, f \rangle)^{1/2} = \left(\int_a^b |f(x)|^2\, dx \right)^{1/2}.$$

One very important property that the inner product $\langle \cdot, \cdot \rangle$ has is that it satisfies the **Cauchy-Schwarz inequality**

$$|\langle f, g \rangle| \leq \|f\|_2 \|g\|_2, \quad f, g \in \mathcal{V}.$$

This can be proven in a similar manner as the Cauchy-Schwarz inequality (B.4) for vectors in $\mathbb{R}^n$.

Bibliography

Abramowitz, M. and Stegun, I. A., Eds.: *Handbook of Mathematical Functions with Formulas, Graphs, and Mathematical Tables*, 9th printing. New York: Dover (1972).

Ahlberg, J. H., Nilson, E. N. and Walsh, J. L.: *The Theory of Splines and Their Applications*, Academic Press, New York (1967).

Aitken, A. C.: "On interpolation by iteration of proportional parts, without the use of differences". *Proc. Edinburgh Math. Soc.* **3**(2) (1932), p. 56-76.

Alefeld, G. and Herzberger, J.: *Introduction to Interval Computations*, Academic Press, New York (1983).

Anderson, E., Bai, Z., Bischof, C., Blackford, S., Demmel, J., Dongarra, J., DuCroz, J., Greenbaum, A., Hammarling, S., McKenney, A. and Sorenson, D.: **LAPACK** *Users' Guide*, 3rd ed., SIAM, Philadelphia (1999).

Arfken, G. B., Weber, H. J. and Harris, F. E.: *Mathematical Methods for Physicists: A Comprehensive Guide*, 7th ed., Academic Press, New York (2012).

Ascher, U. M.: *Numerical Methods for Evolutionary Differential Equations*, SIAM, Philadelphia (2008).

Ascher, U. M. and Greif, C.: *A First Course in Numerical Methods*, SIAM, Philadelphia (2011).

Ascher, U. M., Mattheij, R. M. and Russell, R. D.: *Numerical Solution of Boundary Value Problems for Ordinary Differential Equations*, SIAM, Philadelphia (1995).

Ascher, U. M. and Petzold, L. R.: *Computer Methods for Ordinary Differential Equations and Differential-Algebraic Equations*, SIAM, Philadelphia (1998).

Atkinson, K..: *Introduction to Numerical Analysis*, 2nd ed., John Wiley & Sons, New York (1989).

Bailey, P. B., Shampine, L. F. and Waltman, P. E.: *Nonlinear Two Point Boundary Value Problems*, Academic Press, New York (1968).

Banach, S.: "Sur les opérations dans les ensembles abstraits et leur application aux équations intégrales." *Fund. Math.* **3** (1922), p. 133-181.

Bartels, R., Beatty, J. and Barksy, B.: *An Introduction to Splines for Use in Computer Graphics and Geometric Modeling*, Morgan Kaufmann, Los Altos, CA (1987).

Bashforth, F. and Adams, J. C.: *An Attempt to test the Theories of Capillary Action by comparing the theoretical and measured forms of drops of fluid. With an explanation of the method of integration employed in constructing the tables which give the theoretical forms of such drops*, Cambridge University Press (1883).

Bauer, F. L.: "Optimally scaled matrices", *Numer. Math.* **5** (1963), p. 73?87.

Bavely, C. and Stewart, G. W.: "An Algorithm for Computing Reducing Subspaces by Block Diagonalization", *SIAM J. Num. Anal.* **16** (1979), p. 359-367.

Berrut, J.-P. and Trefethen, L. N.: "Barycentric Lagrange Interpolation". *SIAM Review*

46(3) (2004), p. 501-517.

Berry, M. W. and Browne, M.: *Understanding Search Engines: Mathematical Modeling and Text Retrieval*, 2nd ed., SIAM, Philadelphia (2005).

Betounes, D.: *Differential Equations: Theory and Applications*, 2nd ed., Springer-Verlag, New York (2010).

Birkhoff, G. and De Boor, C.: "Error bounds for spline interpolation", *Journal of Mathematics and Mechanics* **13** (1964), p. 827-836.

Birkhoff, G. and Rota, G.: *Ordinary differential equations*, 4th ed., John Wiley & Sons, New York, 1989.

Björck, Å: *Numerical Methods for Least Squares Problems*, SIAM, Philadelphia (1996).

Björck, Å: "Solving Linear Least Squares Problems by Gram-Schmidt Orthogonalization", *BIT* **7** (1967), p. 1-21.

Blum, L., Cucker, F., Shub, M., and Smale, S.: *Complexity and Real Computation*, Springer-Verlag, Berlin (2001).

Bogacki, P. and Shampine, L. F.: "A 3(2) pair of Runge-Kutta formulas", *Applied Mathematics Letters* **2**(4) (1989), p. 321-325.

Boyce, W. E. and Di Prima, R. C.: *Elementary Differential Equations and Boundary Value Problems*, 10th ed., John Wiley & Sons, New York (1992).

Boyd, J. P.: *Chebyshev and Fourier Spectral Methods*, 2nd ed., Dover, New York (2001).

Braun, M.: *Differential Equations and Their Applications*, 4th ed., Springer-Verlag, New York (1993).

Brezinski, C. and Redivo Zaglia, M.: *Extrapolation Methods: Theory and Practice*, Elsevier, New York (1991).

Briggs, W. L., Henson, V. E. and McCormick, S. F.: *A Multigrid Tutorial*, 2nd ed., SIAM, Philadelphia (2000).

Brouwer, L. E. J.: "Über Abbildungen von Mannigfaltigkeiten", *Mathematische Annalen* **71** (1911), p. 97-115.

Burden, R. L. and Faires, J. D.: *Numerical Analysis*, 9th ed., Brooks/Cole, Pacific Grove, CA (2004).

Canuto, C., Hussaini, M. Y., Quarteroni, A. and Zang, Th. A.: *Spectral Methods: Fundamentals in Single Domains*, Springer-Verlag, New York (2006).

Cheney, W. and Kincaid, D.: *Numerical Mathematics and Computing*, 6th ed., Brooks/Cole, Pacific Grove, CA (2008).

Cipra, B. A.: "The Best of the 20th Century: Editors Name Top 10 Algorithms", *SIAM News* **33**(4) (2000).

Conte, S. D. and de Boor, C.: *Elementary Numerical Analysis: An Algorithmic Approach*, 3rd ed., McGraw-Hill, New York (1981).

Cools, R.: "Constructing cubature formulae: the science behind the art", *Acta Numerica* **6** (1997), p. 1-54.

Cormen, T. H., Leiserson, C. E. and Rivest, R. L.: *Introduction to Algorithms*, 3rd ed., MIT Press, Cambridge, MA (2009).

Dahlquist, G.: "A special stability problem for linear multistep methods", *BIT* **3** (1963), p. 27-43.

Dahlquist, G. and Björck, A.: *Numerical Methods*, Prentice-Hall, Englewood Cliffs, NJ (1974).

Daubechies, I: *Ten Lectures on Wavelets*, SIAM, Philadelphia (1992).

Davis, P. J.: *Interpolation and Approximation*, Dover, New York (1975).

Davis, P. J. and Rabinowitz, P.: *Methods of Numerical Integration*, 2nd ed., Academic Press, New York (1985).

Davis, T. A.: *Direct Methods for Sparse Linear Systems*, SIAM, Philadelphia (2006).

de Boor, C.: *A Practical Guide to Splines*, 2nd ed., Springer-Verlag, New York (1984).

Delaunay, B.: "Sur la sphère vide", *Bulletin de l'Académie des Sciences de l'URSS, Classe des sciences mathématiques et naturelles* 6 (1934), p. 793-800.

Demmel, J. W.: *Applied Numerical Linear Algebra*, SIAM (1997).

Demmel, J. W., Heath, M. T. and van der Vorst, H. H.: "Parallel numerical linear algebra", *Acta Numerica* 2 (1993), p. 111-197.

Dierckx, P.: *Curve and Surface Fitting with Splines*, Oxford University Press, New York (1993).

Dormand, J. R. and Prince, P. J.: "A family of embedded Runge-Kutta formulae", *Journal of Computational and Applied Mathematics* 6 (1) (1980), p. 19-26.

Dongarra, J. J., Bunch, J. R., Moler, C. B. and Stewart, G. W.: LINPACK *User's Guide*, 2nd ed., SIAM, Philadelphia (1979).

Dongarra, J. J., DuCroz, J., Duff, I. S. and Hammarling, S.: "A set of level-3 basic linear algebra subprograms", *ACM Trans. Math. Software* 16 (1990), p. 1-28.

Dongarra, J. J., DuCroz, J., Hammarling, S. and Hanson, R. J.: "An extended set of Fortran basic linear algebra subprograms", *ACM Trans. Math. Software* 14 (1988), p. 1-32.

Dongarra, J. J., Duff, I. S., Sorenson, D. C. and van der Vorst, H. H. *Numerical Linear Algebra for High-Performance Computers*, SIAM, Philadelphia (1998).

Dongarra, J. J., Gustavson, F. G. and Karp, A.: "Implementing linear algebra algorithms for dense matrices on a vector pipeline machine", *SIAM Review* 26 (1984), p. 91-112.

Dongarra, J. J. and Sorensen, D. C.: "A Fully Parallel Algorithm for the Symmetric Eigenvalue Problem", *SIAM J. Sci. and Stat. Comp.* 8 (1987), p. s139-s154.

Driscoll, T. A., Hale, N. and Trefethen, L. N., editors: *Chebfun Guide*, Pafnuty Publications, Oxford (2014).

Edelman, A. and Murakami, H.: "Polynomial roots from companion matrix eigenvalues", *Mathematics of Computation* 64(210) (1995), p. 763-776.

Elman, H. C., Silvester, D. J. and Wathen, A. J.: *Finite Elements and Fast Iterative Solvers: With Applications in Incompressible Fluid Dynamics*, Oxford University Press, New York (2005).

Engels, H.: *Numerical Quadrature and Cubature*, Academic Press, New York (1980).

Eriksson, K., Estep, D., Hansbo, P. and Johnson, C.: *Computational Differential Equations*, Cambridge University Press, New York (1996).

Evans, G.: *Practical Numerical Integration*, John Wiley & Sons, New York (1993).

Farebrother, R. W.: *Linear Least Squares Computations*, Marcel Dekker, New York (1988).

Farin, G. *Curves and Surfaces for Computer Aided Geometric Design*, 2nd ed., Academic Press, New York (1990).

Farlow, S. J.: *Partial Differential Equations for Scientists and Engineers*, John Wiley & Sons, New York (1982).

Fehlberg, E.: "Klassische Runge-Kutta Formeln vierter und niedrigerer Ordnung mit Schrittweiten-Kontrolle und ihre Anwendung auf Wrmeleitungsprobleme", *Computing* 6 (1970), p. 61-71.

Feng, X., Lewis, T. and Neilan, M.: "Discontinuous Galerkin finite element differential calculus and applications to numerical solutions of linear and nonlinear partial differential equations", *J. Comput. Appl. Math.* 299 (2016), p. 68-91.

Fletcher, R.: *Practical methods of optimization* (2nd ed.), New York: John Wiley & Sons (1987).

Foster, L. V.: "Gaussian elimination with partial pivoting can fail in practice", *SIAM J. Matrix Anal. Appl.* 15 (1994), p. 1354-1362.

Fox, L.: *The Numerical Solution of Two-Point Boundary Problems*, Dover, New York

(1990).

Fox, L. and Parker, I. B.: *Chebyshev Polynomials in Numerical Analysis*, Oxford University Press, New York (1968).

Francis, J. G. F.: "The QR Transformation: A Unitary Analogue to the LR Transformation, Parts I and II". *Comp. J.* **4** (1961), p. 265-272, 332-345.

Frazier, M. W.: *An Introduction to Wavelets through Linear Algebra*, Springer-Verlag, New York (1999).

Freund, R. W. and Nachtigal, N.: "QMR: A Quasi-Minimal Residual Method for Non-Hermitian Linear Systems", *Numer. Math.* **60** (1991), p. 315-339.

Frigo, M. and Johnson, S. G.: "The Design and Implementation of FFTW3", *Proceedings of the IEEE* **93**(2) (2005), p. 216?231.

Fritsch, F. N., Kahaner, D. K. and Lyness, J. N.: "Double integration using one-dimensional adaptive quadrature routines: A software interface problem", *ACM Trans. Math. Software* **7** (1981), p. 46-75.

Gander, W. and Gautschi, W.: "Adaptive quadrature–revisited", *BIT* **40** (2000), p. 84-101.

Garbow, B. S., Boyle, J. M., Dongarra, J. J. and Moler, C. B.: *Matrix Eigensystem Routines:* EISPACK *Guide Extension*, Springer-Verlag, New York (1972).

Garon, E. M. and Lambers, J. V.: "Modeling the Diffusion of Heat Energy within Composites of Homogeneous Materials using the Uncertainty Principle", *Comp. Appl. Math.* **37**(3) (2018), p. 2566-2587.

Gautschi, W. *Numerical Analysis: an Introduction*, Birkhäuser, Boston, MA (1997).

Gear, C. W.: *Numerical Initial Value Problems in Ordinary Differential Equations*, Prentice-Hall, Englewood Cliffs, NJ (1971)

Ghizzetti, A. and Ossicini, A.: *Quadrature Formulae*, Academic Press, New York (1970).

Gill, P. E., Murray, W. and Wright, M. H.: *Practical Optimization*, Academic Press, New York (1981).

Goldberg, D.: "What every computer scientist should know about floating-point arithmetic", *ACM Computing Surveys* **18**(1) (1991), p. 5-48.

Goldstine, H. H.: *A History of Numerical Analysis from the 16th century through the 19th century.* Springer-Verlag (1977).

Golub, G. H. and Kahan, W.: "Calculating the Singular Values and Pseudo-Inverse of a Matrix", *SIAM J. Num. Anal. Ser. B* **2** (1965), p. 205-224.

Golub, G. H. and Meurant, G.: *Matrices, Moments and Quadrature with Applications*, Princeton University Press, Princeton, NJ (2009).

Golub, G. H. and Ortega, J. M.: *Scientific Computing: An Introduction with Parallel Computing*, Academic Press, San Diego (1993).

Golub, G. H. and van Loan, C. F.: *Matrix Computations*, 4th ed., Johns Hopkins University Press (2012).

Golub, G. H. and Varah, J. M.: "On the characterization of the best L_2-scaling of a matrix", *SIAM J. Numer. Anal.* **11** (1974), p. 472-479.

Gottlieb, D., Hussaini, M. Y. and Orszag, S. A.: *Introduction: Theory and Applications of Spectral Methods, Spectral Methods for Partial Differential Equations*, R. G. Voigt, D. Gottlieb and M. Y. Hussaini, eds., SIAM, Philadelphia (1984).

Graham, R., Knuth, D. and Patashnik, O.: *Concrete Mathematics*, 2nd ed., Addison-Wesley, Reading, MA (1994).

Greenbaum, A.: *Iterative Methods for Solving Linear Systems*, SIAM, Philadelphia (1997).

Griewank, A: *Evaluating Derivatives: Principles and Techniques of Algorithmic Differentiation*, SIAM, Philadelphia (2000).

Gustafsson, B.: *High Order Difference Methods for Time Dependent PDE*, Springer-Verlag,

New York (2008).

Gustafsson, B., Kreiss, H.-O. and Oliger, J. E.: *Time-Dependent Problems and Difference Methods*, John WIley & Sons, New York (1995).

Haber, S.: "Numerical evaluation of multiple integrals", *SIAM Review* **12** (1970), p. 481-526.

Hairer, E., Nørsett, S. P., and Wanner, G.: *Solving Ordinary Differential Equations I: Nonstiff Problems*, 2nd ed., Springer-Verlag, Berlin (1993).

Hairer, E. and Wanner, G.: *Solving Ordinary Differential Equations II: Stiff and Differential-Algebraic Problems*, 2nd ed., Springer-Verlag, Berlin (1996).

Hansen, P. C.: *Rank-Deficient and Discrete Ill-Posed Problems: Numerical Aspects of Linear Inversion*, SIAM, Philadelphia (1998).

Hardy, G. H.: "Weierstrass's nondifferentiable function", *Trans. Amer. Math. Soc.* **17** (1916), p. 301-325.

Hastie, T., Tibshirani, R. and Friedman, J.: *The Elements of Statistical Learning*, Springer-Verlag, New York (2001).

Heath, M. T.: *Scientific Computing: An Introductory Survey*, 2nd ed., McGraw-Hill, New York (2002).

Hestenes, M. R. and Stiefel, E.: "Methods of Conjugate Gradients for Solving Linear Systems", *Journal of Research of the National Bureau of Standards* **49**(6) (1952), p. 409-436.

Hesthaven, J., Gottlieb, S. and Gottlieb, D.: *Spectral Methods for Time-Dependent Problems*, Cambridge University Press, New York (2007).

Higham, N. J.: "A survey of condition number estimation for triangular matrices", *SIAM Review* (1987), p. 575-596.

Higham, N. J.: *Accuracy and Stability of Numerical Algorithms*, 2nd ed., SIAM, Philadelphia, (2003).

Horn, R. A. and Johnson, C. R.: *Matrix Analysis*, Cambridge University Press, New York (1985).

Hotelling, H.: "Some new methods in matrix calculation", *Ann. Math. Stat.* **14** (1943), p. 1-34.

Hubbard, B. B.: *The World According to Wavelets*, 2nd ed., A K Peters, Natick, MA (1998).

Hughes, T. J. R.: *The Finite Element Method: Linear Static and Dynamic Finite Element Analysis*, Prentice-Hall, Englewood Cliffs, NJ (1987).

IEEE: "IEEE standard 754-1985 for binary floating-point arithmetic", *SIGPLAN Notices* **22**(2) (1987), p. 9-25.

Iserles, A.: *A First Course in the Numerical Analysis of Differential Equations*, 2nd Ed., Cambridge University Press, New York (2008).

Issacson, E. and Keller, H. B.: *Analysis of numerical methods*, John Wiley & Sons, New York (1966).

Jacobi, C. G. J.: "Uber ein Leichtes Verfahren Die in der Theorie der Sacularstroungen Vorkommendern Gleichungen Numerisch Aufzulosen", *Crelle's J.* **30** (1846), p. 51-94.

Jaulin, L., Keiffer, M., Didrit, O. and Walter, E.: *Applied Interval Analysis*, Springer-Verlag, New York (2001).

Johnson, C.: *Numerical Solution of Partial Differential Equations by the Finite Element Method*, Cambridge University Press, New York (1987).

Joyce, D. C.: "Survey of extrapolation processes in numerical analysis", *SIAM Review* **13** (1971), p. 435-488.

Kahan, W.: "Further remarks on reducing truncation errors", *Communications of the*

     ACM, **8**(1) (1965), p. 40.

Kahan, W.: "Numerical Linear Algebra", *Canadian Math. Bull.* **9** (1966), p. 757-801.

Kaniel, S.: "Estimates for Some Computational Techniques in Linear Algebra", *Math. Comp.* **20** (1966), p. 369-378.

Keller, H. B.: *Numerical Methods for Two-Point Boundary Value Problems*, Blaisdell, Waltham, MA (1968).

Keller, J. B.: "Probability of a shutout in racquetbell", *SIAM Review* **26**(2) (1984), p. 267-268.

Kelley, C. T.: *Iterative Methods for Optimization*, SIAM, Philadelphia (1999).

Kennedy, W. J. and Gentle, J. E.: *Statistical Computing*, Marcel Dekker, New York (1980).

Knott, G. D.: *Interpolating Cubic Splines*, Birkhäuser, Boston (2000).

Kogbetliantz, E. G.: "Solution of Linear Equations by Diagonalization of Coefficient Matrix", *Quart. Appl. Math.* **13** (1955), p. 123-132.

Koren, I.: *Computer Arithmetic Algorithms*, Prentice-Hall, Englewood Cliffs, NJ (1993).

Krommer, A. R. and Ueberhuber, C. W.: *Computational Integration*, SIAM, Philadelphia (1998).

Lambers, J. V. and Rice, J. R., "Numerical Quadrature for General Two-Dimensional Domains", *Computer Science Technical Reports*, Paper 906 (1991).

Lambert, J. D.: *Numerical Methods for Ordinary Differential Systems: The Initial Value Problem*, John Wiley & Sons, New York (1992).

Langville, A. N. and Meyer, C. D.: *Google?s PageRank and Beyond: The Science of Search Engine Rankings*, Princeton University Press, Princeton, NJ (2006).

Lapidus, L. and Pinder, G. F.: *Numerical Solution of Partial Differential Equations in Science and Engineering*, John Wiley & Sons, New York (1982).

Larkin, F. M.: "Root-finding by fitting rational functions", *Mathematics of Computation* **35** (1980), p. 803-816.

Laurie, D.: "Calculation of Gauss-Kronrod quadrature rules", *Mathematics of Computation of the American Mathematical Society* **66**(219) (1997), p. 1133-1145.

Lawson, C. L. and Hanson, R. J.: *Solving Least Squares Problems*, SIAM, Philadelphia (1995).

Lawson, C. L., Hanson, R. J., Kincaid, D. R. and Krogh, F. T.: "Basic linear algebra subprograms for Fortran usage", *ACM Trans. Math. Software* **5** (1979), p. 308-325.

Le Gendre, M.: "Recherches sur l'attraction des sphéroïdes homogènes", *Mémoires de Mathématiques et de Physique, présentés à l'Académie Royale des Sciences, par divers savans, et lus dans ses Assemblées*, Tome X (1785), p. 411-435.

Lehoucq, R. B., Sorensen, D. C. and Yang, C.: **ARPACK** *Users? Guide*, SIAM, Philadelphia (1998).

Leveque, R. J.: *Finite Difference Methods for Ordinary and Partial Differential Equations*, SIAM, Philadelphia (2007).

Leveque, R. J.: *Finite Volume Methods for Hyperbolic Problems*, Cambridge University Press, New York (2002).

Liesen, J. and Strakoš, Z.: *Krylov Subspace Methods: Principles and Analysis*, Oxford University Press, New York (2015).

Long, S. D., Sheikholeslami, S., Lambers, J. V. and Walker, C.: "Diagonalization of 1-D Differential Operators with Piecewise Constant Coefficients Using the Uncertainty Principle", *Math. Comput. Simulation* (2018), to appear.

Lyness, J. N.: "When not to use an automatic quadrature routine", *SIAM Review* **25** (1983), p. 63-88.

Lyness, J. N. and Cools, R.: "A survey of numerical cubature over triangles", *Proc. Symp. Appl. Math.* **48** (1993), p. 127-150.

Lyness, J. N. and Kaganove, J. J.: "Comments on the nature of automatic quadrature routines", *ACM Trans. Math. Software* **2** (1976), p. 65-81.

Mallat, S.: *A Wavelet Tour of Signal Processing: The Sparse Way*, 3rd ed., Academic Press, New York (2009).

MIn, M. S. and Gottlieb, D.: "On the convergence of the Fourier approximation for eigenvalues and eigenfunctions of discontinuous problems", *SIAM J. Numer. Anal.* **40**(6) (2003), p. 2254-2269.

Moler, C. B.: "Demonstration of a matrix laboratory". *Lecture notes in mathematics* (J. P. Hennart, ed.), Springer-Verlag, Berlin (1982), p. 84-98.

Moler, C. B.: "Iterative Refinement in Floating Point", *Journal of the ACM* **14**(2) (1967), p. 316-321.

Moler, C. B. and Stewart, G. W.: "An Algorithm for Generalized Matrix Eigenvalue Problems", *SIAM J. Num. Anal.* **10**, p. 241-256.

Moore, R. E., Kearfott, R. B., and Cloud, M. J.: *Introduction to Interval Analysis*, SIAM, Philadelphia (2009).

Moulton, F. R.: *New methods in exterior ballistics*, University of Chicago Press (1926).

Nash, J. C.: "A One-Sided Transformation Method for the Singular Value Decomposition and Algebraic Eigenproblem", *Comp. J.* **18** (1975), p. 74-76.

Nash, S. G., ed. *A History of Scientific Computing*. ACM Press (1990).

Neville, E. H.: "Iterative Interpolation", *J. Indian Math Soc.* **20** (1934), p. 87-120.

Nocedal, J. and Wright, S.: *Numerical Optimization*, 2nd ed., Springer-Verlag, New York (2006).

Omondi, A. R.: *Computer Arithmetic Systems*, Prentice-Hall, Englewood Cliffs, NJ (1994).

Ortega, J. M.: *Introduction to Parallel and Vector Solution of Linear Systems*, Plenum, New York (1988).

Ostrowski, A. M.: *Solution of Equations and Systems of Equations*, 2nd ed., Academic Press, New York (1966).

Overton, M. L.: *Numerical Computing with IEEE Floating Point Arithmetic*, SIAM, Philadelphia (2001).

Padé, H.: "Sur la répresentation approchée d'une fonction par des fractions rationelles", Thesis, *Ann. École Nor.* (3), **9** (1892), p. 1-93.

Paige, C. C.: "The Computation of Eigenvalues and Eigenvectors of Very Large Sparse Matrices", Ph.D. thesis, London University (1971).

Paige, C. C. and Saunders, M. A.: "LSQR: An Algorithm for Sparse Linear Equations and Sparse Least Squares", *ACM Trans. Math. Soft.* **8** (1982), p. 43-71.

Paige, C. C. and Saunders, M. A.: "Solution of Sparse Indefinite Systems of Linear Equations", *SIAM J. Num. Anal.* **12** (1975), p. 617-629.

Palchak, E. M., Cibotarica, A. and Lambers, J. V.: "Solution of Time-Dependent PDE Through Rapid Estimation of Block Gaussian Quadrature Nodes", *Linear Algebra and its Applications* **468** (2015), p. 233-259.

Parhami, B.: *Computer Arithmetic: Algorithms and Hardware Designs*, Oxford University Press, New York (1999).

Peressini, A. L., Sullivan, F. E. and Uhl, J. J.: *The Mathematics of Nonlinear Programming*, Springer (1988).

Pereyra, V.: "Finite difference solution of boundary value problems in ordinary differential equations", *Studies in Numerical Analysis*, G. H. Golub, ed. Math. Assoc. Amer., Washington, DC (1984), p. 243-269.

Perona, P. and Malik, J.: "Scale-space and edge detection using anisotropic diffusion", *IEEE Trans. Pattern Anal. Mach. Intell.* **12** (1990), p. 161-192.

Peters, G. and Wilkinson, J. H.: "Inverse Iteration, Ill-Conditioned Equations, and Newton's Method", *SIAM Review* **21** (1979), p. 339-360.

Piessens, R., de Doncker-Kapenga, E., Ueberhuber, C. W. and Kahaner, D.: QUADPACK: *A subroutine package for automatic integration*, Springer-Verlag, New York (1983).

Pinchover, Y. and Rubinstein, J.: *An Introduction to Partial Differential Equations*, Cambridge University Press (2005).

Powell, M. J. D.: *Approximation theory and methods*, Cambridge University Press, Cambridge (1981).

Press, W., Teukolsky, S., Vetterling, W. and Flannery, B.: *Numerical Recipes*, 3rd ed., Cambridge University Press, London (2007).

Quarteroni, A. and Saleri, F.: *Scientific Computing with MATLAB*, Texts in computational science and engineering **2**, Springer (2003), p. 66.

Ralston, A. and Rabinowitz, P.: *A first course in numerical analysis*, 2nd ed., McGraw-Hill, New York (1978).

Reid, J. K.: "On the Method of Conjugate Gradients for the Solution of Large Sparse Systems of Linear Equations", *Large Sparse Sets of Linear Equations*, J. K. Reid, ed., Academic Press, NY (1971), p. 231-254.

Reinsch, C. H.: "Smoothing by spline functions", *Numer. Math.* **10** (1967), p. 177-183.

Rice, J. R.: "A Metalgorithm for Adaptive Quadrature", *Journal of the ACM* **22**(1) (1975), p. 61-82.

Rice, J. R.: "A theory of condition", *SIAM Journal of Numerical Analysis* **3** (1966), p. 287-310.

Richardson, L. F.: "The approximate arithmetical solution by finite differences of physical problems including differential equations, with an application to the stresses in a masonry dam", *Philosophical Transactions of the Royal Society A.* **210**(459-470) (1911), p. 307-357.

Rivlin, T. J.: *Chebyshev Polynomials*, 2nd ed., John Wiley & Sons, New York (1990).

Roberts, S. and Shipman, J.: *Two-Point Boundary Value Problems: Shooting Methods*, Elsevier, New York (1972).

Romberg, W.: "Vereinfachte numerische Integration", *Det Kongelige Norske Videnskabers Selskab Forhandlinger*, Trondheim **28**(7) (1955), p. 30-36.

Runge, C.: "Über empirische Funktionen und die Interpolation zwischen quidistanten Ordinaten", *Zeitschrift für Mathematik und Physik* **46** (1901), p. 224-243.

Saad, Y.: *Iterative Methods for Sparse Linear Systems*, 2nd ed., SIAM, Philadelphia (2003).

Sauer, T.: *Numerical Analysis*, 2nd ed., Pearson, Boston (2012).

Schultz, M. H.: *Spline analysis*, Prentice-Hall, Englewood Cliffs, NJ (1973).

Schumaker, L. L.: *Spline Functions*, John Wiley & Sons, New York (1981).

Shampine, L. F. and Reichelt, M. W.: "The MATLAB ODE Suite", *SIAM J. Sci. Comput.* **18**(1) (1997), p. 1-22.

Shen, J., Tang, T. and Wang, L.-L.: *Spectral Methods: Algorithms, Analysis and Applications*, Springer-Verlag, New York (2011).

Shikin, E. V. and Plis, A. V.: *Handbook on Splines for the User*, CRC Press, Boca Raton, FL (1995).

Sloan, I. H. and Joe, S.: *Lattice Methods for Multiple Integration*, Oxford University Press, New York (1994).

Smith, B. T., Boyle, J. M., Ikebe, Y., Klema, V. C. and Moler, C. B.: *Matrix Eigensystem Routines: EISPACK Guide*, 2nd ed., Springer-Verlag, New York (1970).

Späth, H.: *One Dimensional Spline Interpolation Algorithms*, A K Peters, Ltd., Wellesley, MA (1995).

Stewart, G. W.: "Error and Perturbation Bounds for Subspaces Associated with Certain

Eigenvalue Problems", *SIAM Review* **15** (1973), p. 727-764.

Stewart, G. W.: *Matrix Algorithms. Vol II: Eigensystems*, SIAM, Philadelphia (2001).

Stigler, S. M.: "Gauss and the Invention of Least Squares", *Ann. Stat.* **9**(3) (1981), p. 465-474.

Strang, G.: *Linear Algebra and Its Applications*, 4th ed., Brooks/Cole, Pacific Grove, CA (2006).

Strang, G. and Fix, G.: *An Analysis of the Finite Element Method*, Prentice-Hall, Englewood Cliffs, NJ (1973).

Strassen, V.: "Gaussian elimination is not optimal", *Numer. Math.* **13** (1969), p. 354-356.

Strikwerda, J. C.: *Finite Difference Schemes and Partial Differential Equations*, Chapman and Hall, New York (1989).

Stroud, A. H.: *Approximate Calculation of Multiple Integrals*, Prentice-Hall, Englewood Cliffs, NJ (1972).

Stroud, A. H. and Secrest, D.: *Gaussian Quadrature Formulas*, Prentice-Hall, Englewood Cliffs, NJ (1966).

Süli, E. and Mayers, D.: *An Introduction to Numerical Analysis*, Cambridge University Press (2003).

Thisted, R. A.: *Elements of Statistical Computing*, Chapman and Hall, New York (1988).

Tikhonov, A. N., Arsenin, V. Y.: *Solution of ill posed problems*, John Wiley & Sons, New York (1977).

Traub, J. F.: *Iterative Methods for the Solution of Equations*, Prentice-Hall, Englewood Cliffs, NJ (1964).

Trefethen, L. N.: *Spectral Methods in MATLAB*, SIAM, Philadelphia (2000).

Trefethen, L. N.: "Three mysteries of Gaussian elimination", *SIGNUM Newsletter* **20**(4) (1985), p. 2-5.

Trefethen, L. N. and Bau, D.: *Numerical Linear Algebra*, SIAM, Philadelphia (1997).

Trefethen, L. N. and Schreiber, R. S.: "Average-case stability of Gaussian elimination", *SIAM J. Matrix Anal. Appl.* **11** (1990), p. 355-360.

Trottenberg, U., Oosterlee, C. and Schuller, A.: *Multigrid*, Academic Press, New York (2001).

Turing, A. M.: "Rounding-off errors in matrix processes", *Quarterly J. Mech. Appl. Math.* **1** (1948), p. 287-308.

van der Vorst, H.: *Iterative Krylov Methods for Large Linear Systems*, Cambridge University Press, London (2003).

Van Huffel, S. and Vandewalle, J.: *The Total Least Squares Problem*, SIAM, Philadelphia (1991).

Vavasis, S. A.: "Gaussian elimination with pivoting is P-complete", *SIAM J. Disc. Math.* **2** (1989), p. 413-423.

von Neumann, J. and Goldstine, H. H.: "Numerical inverting of matrices of high order", *Bull. Amer. Math. Soc.* **53** (1947), p. 1021-1099.

Watkins, D.: *Fundamentals of Matrix Computations*, 2nd ed., John Wiley & Sons, New York (2002).

Wilbraham, H.: "On a certain periodic function", *The Cambridge and Dublin Mathematical Journal* **3** (1848), p. 198-201.

Wilkinson, J. H.: *The Algebraic Eigenvalue Problem*, Clarendon Press, Oxford, England (1965).

Wilkinson, J. H.: "Error analysis of direct methods of matrix inversion", *J. ACM* **8** (1961), p. 281-330.

Wilkinson, J. H.: "Error analysis of floating-point computation", *Numerische Mathematik* **2**(1) (1960), p. 319-430.

Wilkinson, J. H.: "Global Convergence of Tridiagonal QR Algorithm With Origin Shifts", *Lin. Alg. and Its Applic.* **1** (1968), p. 409-420.

Wilkinson, J. H.: *Rounding Errors in Algebraic Processes*, Prentice-Hall, Englewood Cliffs, NJ (1963).

Wilkinson, J. H. and Reinsch, C., eds.: *Handbook for Automatic Computation, Linear Algebra*, volume 2, Springer-Verlag, New York (1971).

Wimp, J.: *Sequence Transformations and Their Applications*, Academic Press, New York (1981).

Wright, S. J.: "A collection of problems for which Gaussian elimination with partial pivoting is unstable", *SIAM J. Sci Comput.* **14** (1993), p. 231-238.

Yamaguchi, F. *Curves and Surfaces in Computer-Aided Geometric Design*, Springer-Verlag, New York (1988).

Young, D .M.: *Iterative methods for solving partial difference equations of elliptical type*, Ph.D. Thesis, Harvard University (1950).

Zwillinger, D., ed.: *Standard Mathematical Tables and Formulae*, 30th ed., CRC Press, Boca Raton, FL (1996).

Index

Printed in the United States
By Bookmasters